ADAPTIVE FILTERING
PREDICTION AND CONTROL

PRENTICE-HALL INFORMATION
AND SYSTEM SCIENCES SERIES
Thomas Kailath, *Editor*

ADAPTIVE FILTERING

PREDICTION AND CONTROL

GRAHAM C. GOODWIN AND **KWAI SANG SIN**

University of Newcastle
New South Wales, 2308
Australia

Computer Special Systems
Digital Equipment
New South Wales, 2064
Australia

PRENTICE-HALL, INC., Englewood Cliffs, New Jersey 07632

Library of Congress Cataloging in Publication Data

GOODWIN, GRAHAM C. (Graham Clifford), date
  Adaptive filtering prediction and control.

  Bibliography: p.
  Includes index.
  1. Discrete-time systems. 2. Filters (Mathematics)
3. Prediction theory. 4. Control theory. I. Sin,
Kwai Sang, date. II. Title.
QA402.G658 1984   003   83-23023
ISBN 0-13-004069-x

Editorial/production supervision and
    interior design: Virginia Huebner
Cover design: 20/20 Services, Inc./Mark Berghash
Manufacturing buyer: Anthony Caruso

Printed in the United States of America

10 9 8 7 6 5 4 3 2

ISBN 0-13-004069-X

Prentice-Hall International, Inc., *London*
Prentice-Hall of Australia Pty. Limited, *Sydney*
Editora Prentice-Hall do Brasil, Ltda., *Rio de Janeiro*
Prentice-Hall Canada Inc., *Toronto*
Prentice-Hall of India Private Limited, *New Delhi*
Prentice-Hall of Japan, Inc., *Tokyo*
Prentice-Hall of Southeast Asia Pte. Ltd., *Singapore*
Whitehall Books Limited, *Wellington, New Zealand*

Contents

3   PARAMETER ESTIMATION FOR DETERMINISTIC SYSTEMS   47

4   DETERMINISTIC ADAPTIVE PREDICTION   106

5   CONTROL OF LINEAR DETERMINISTIC SYSTEMS   118

6   ADAPTIVE CONTROL OF LINEAR DETERMINISTIC SYSTEMS   178

PART II: STOCHASTIC SYSTEMS

7   OPTIMAL FILTERING AND PREDICTION   245

8   PARAMETER ESTIMATION
          FOR STOCHASTIC DYNAMIC SYSTEMS   301

9   ADAPTIVE FILTERING AND PREDICTION   360

APPENDICES

Preface

The object of this book is to present in a unified fashion the theory of adaptive filtering, prediction, and control. The treatment is largely confined to linear discrete-time systems, although natural extensions to nonlinear systems are also explored. The emphasis on discrete-time systems reflects the growing importance of digital computers in practical applications of this theory.

Adaptive techniques in filtering, prediction, and control have been extensively studied for over a decade and numerous successful applications have been reported. The development of the theory over the years has led to a much better understanding of the performance of various adaptive algorithms. However, it was only recently that a rigorous and comprehensive theory of convergence of adaptive algorithms has emerged. This theory is very appealing from several points of view: it can be applied in a unified manner to both deterministic and stochastic systems, yet it is relatively simple and is easily understood with a minimum of background knowledge. Perhaps more important, there is a close link between the convergence theory and the performance of the algorithms in practice. This book summarizes the theoretical and practical aspects of a large class of adaptive algorithms for potential users.

The philosophy of the presentation is that the relatively new material on adaptive techniques is linked to standard results on design techniques applicable when the system parameters are known. The book is aimed at two major groups of readers: senior undergraduate students in engineering and applied mathematics, and graduate students and research workers.

The book is divided into two parts: Part I deals with deterministic systems and is suitable for both senior undergraduate and graduate students; Part II deals with stochastic systems and is more suitable for graduate students. The two parts are further subdivided into chapters covering different design techniques for adaptive filtering, prediction, and control. The book also contains appendices which summarize the relevant background material, thus making the book substantially self-contained. Throughout the book we have attempted to provide an adequate framework that readers can build on when pursuing their own individual goals. Some sections contain more difficult material or introduce ideas that are not heavily used in subsequent work. These sections have been marked by an asterisk and may be omitted on a first reading.

The book is an expanded version of lecture notes used for junior/graduate-level courses at the University of Newcastle and the University of Houston and as the basis of intensive courses on Adaptive Control at the University of Houston given in December 1980 and the University of California, Los Angeles, in June 1982 and June 1983. The book also includes the latest research results of the authors and others in the field and should be suitable as background reading and as reference for research workers.

We would like to pay tribute to a number of people who helped with the preparation of this book. Our first thanks go to the large number of people who motivated our interest in this topic by discussions, joint research, and correspondence. We would particularly like to mention Karl Åström, Peter Caines, Lennart Ljung, and Peter Ramadge. In writing the book, we were also helped by useful discussions with many others, including Brian Anderson, Howard Elliott, Tom Kailath, Bayliss McInnis, Tino Mingori, John Moore, and many others.

It is a pleasure to thank Tony Cantoni, for his expert advice on numerous practical questions, and Siew Wah Chan, who built the interface and organized several adaptive control experiments which are discussed in the book. Special thanks go to Lennart Ljung and Michel Gevers, who gave detailed and helpful comments on the first draft. Several people made special contributions to particular sections of the book. This includes Section 9.5.3, which was largely written by Michel Gevers and Vincent Wertz; Appendix C, which was largely written by David Hill; and Appendix E, which was initially prepared by Siew Wah Chan. The manuscript was superbly typed by Maureen Byrnes with assistance from Dianne Piefke and Betty Fewings, and the diagrams were expertly prepared by Phamie Sidell, Ann Pender, and Wanda Lis. Finally, the first author would like to thank his wife, Rosslyn Goodwin, and his children, Andrew and Sarah, for their generous support, understanding, and patience during the writing of the book.

GRAHAM C. GOODWIN

KWAI SANG SIN

New South Wales, Australia

ADAPTIVE FILTERING
PREDICTION AND CONTROL

1

Introduction to Adaptive Techniques

This book is concerned with the design of adaptive filters, predictors, and controllers. When the system model is completely specified, standard design techniques can be employed. Our emphasis here, however, is on design techniques that are applicable when the system model is only partially known. We shall use the term *adaptive* to describe this class of design techniques. *Ab initio*, these techniques will incorporate some form of on-line parameter adjustment scheme. We shall further distinguish two types of parameter adjustment algorithm: those applicable when the system parameters, although unknown, are time invariant; and those applicable when the system parameters are time varying.

A fact well appreciated by those involved in applications is that any realistic design problem is usually complex, involving many factors, such as engineering constraints, economic trade-offs, human considerations, and model uncertainty. However, just as classical control theory provides a useful way of thinking about control problems, we believe that the adaptive approach provides a useful set of tools and guidelines for design. The advantages of the approach have been borne out in numerous applications that have been reported in the literature.

We shall now outline briefly the types of problems intrinsic to filtering, prediction, and control.

1.1 FILTERING

Filtering is concerned with the extraction of signals from noise. Typical applications include:

- Reception and discrimination of radio signals
- Digital data transmission on telephone lines
- Beam-forming arrays
- Detection of radar signals
- Analysis of seismic data
- Processing of pictures sent from spacecraft
- Analysis of electrocardiogram (ECG) and electroencephalogram (EEG) signals

If models are available for the signal and noise, it is possible, at least in principle, to design a filter that optimally enhances the signal relative to the noise. A great deal is known about filter design procedures in simple cases, for example when the signal model is linear. However, some applications demand advanced nonlinear filtering algorithms, and less is known about these procedures.

When the signal and noise models are not completely specified, it seems plausible that appropriate models could be estimated by analyzing actual data. This is frequently done in practice, especially when the models are ill defined or are time varying. This leads to adaptive filters.

1.2 PREDICTION

Prediction is concerned with the problem of extrapolating a given time series into the future. As for filtering, if the underlying model of the time series is known, it is possible, in principle, to design an optimal predictor for future values. Typical applications include:

- Forecasting of product demand
- Predicting water consumption based on temperature and rainfall measurements
- Predicting population growth
- Encoding speech to conserve bandwidth
- Predicting future outputs as a guide to plant operators in industrial process control

When the model of the time series is not completely specified, it again seems plausible that a model could be estimated by analyzing past data from the time series. This leads to adaptive predictors.

1.3 CONTROL

Control is concerned with the manipulation of the inputs to a system so that the outputs achieve certain specified objectives. Typical applications include:

- Control of prosthetic devices and robots
- Control of the aileron and elevators on an aircraft

- Firing of retro rockets on spacecraft to control attitude
- Control of the flow of raw materials in an industrial plant to yield a desired product
- Control of interest and tariff rates to regulate the economy
- Control of anaesthetic dosage to produce a desired level of unconsciousness in a patient

Again there are a vast array of design techniques for generating control strategies when the model of the system is known. When the model is unknown, on-line parameter estimation could be combined with on-line control. This leads to adaptive, or self-learning controllers. The basic structure of an adaptive controller is shown in Fig. 1.3.1.

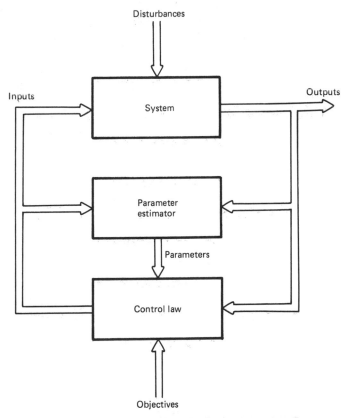

Figure 1.3.1 Basic structure of an adaptive controller.

One can distinguish the following types of control problem in increasing order of difficulty:

- Deterministic control (when there are no disturbances and the system model is known)

- Stochastic control (when there are disturbances and models are available for the system and disturbances)
- Adaptive control (when there may be disturbances and the models are not completely specified)

We shall emphasize the latter type of control problem in this book.

From the discussion above, it can be seen that underlying each of the problems of adaptive filtering, prediction, and control there is some form of parameter estimator. In fact, parameter estimation forms an integral part of any adaptive scheme. We shall discuss in detail various forms of parameter estimator.

It is useful to distinguish between situations in which the system parameters are constant and situations in which the parameters vary with time. Obviously, the former situation is easier to deal with than the latter. A feature of many of the parameter estimation algorithms used in the time-invariant case is that the gain of the algorithms ultimately decreases to zero. Heuristically, this means that when all the parameters have been estimated, the algorithm "turns off." On the other hand, if the parameters are time varying, it is necessary for the estimation algorithm to have the capability of tracking the time variations. Some of the algorithms aimed at the constant-parameter case will automatically have this capability. Other algorithms will need minor modifications to ensure that the turn-off phenomenon does not occur.

Part I

DETERMINISTIC SYSTEMS

The first part of the book is concerned primarily with adaptive algorithms applicable when noise and disturbances are of secondary importance relative to the system modeling errors. Thus, in much of the theoretical development, only disturbances that are perfectly predictable will be considered. This is not to say that the algorithms cannot be used in the presence of random noise. In fact, we will go to some lengths to establish the robustness properties of the algorithms when used in noisy environments. However, noise will be considered more of a nuisance than a factor of prime importance. The background required for the first part is relatively straightforward and the material should be readily understood by senior undergraduates having some previous exposure to systems theory.

In the second part of the book we shall turn to adaptive algorithms that incorporate explicit stochastic models for the noise and disturbances. This will require slightly deeper background since certain concepts from probability theory and stochastic processes will be called upon. Thus the second part of the book would be better suited to Master's-level students, although part of it would not be beyond an undergraduate audience.

2

Models for Deterministic Dynamical Systems

2.1 INTRODUCTION

In this chapter we give a brief account of various models for deterministic dynamical systems. By *deterministic* in this context, we mean that the model gives a complete description of the system response. Later we shall contrast this with *stochastic* models, when the response contains a random component defined on some probability space. The chapter gives a self-contained description of certain aspects of system modeling, with emphasis on those ideas that form the basis of our subsequent treatment of filtering, prediction, and control.

We will begin by looking at models for linear deterministic finite-dimensional systems. In particular, we discuss state-space models, difference operator representations, autoregressive moving-average models, and transfer functions. Models for certain classes of nonlinear systems will also be introduced. Our emphasis here is on the interelation between these different model formats. This is important since the choice of model is often the first step toward the prediction or control of a process. An appropriately chosen model structure can greatly simplify the parameter estimation procedure and facilitate the design of prediction and control algorithms for the process.

This chapter is intended to give a concise, but relatively complete picture of various models for linear systems. The chapter can be read in one of two ways: either in detail to gain a full appreciation of the results, or selectively to get the flavor of the results and to establish notation. If the latter approach is adopted, the reader's attention is directed in particular to Section 2.3.4, as the DARMA models covered in this section are used extensively in subsequent chapters.

Although many of the topics discussed here are covered in other books which deal specifically with systems theory, we believe that our treatment is interesting and distinctive in that it is tailored to the kind of application subsequently discussed in the book. We also present some novel insights and new material, especially concerning the modeling of deterministic disturbances and the interelationship between different types of models. A key point that we want to bring out is that certain model structures are more convenient for specific applications and not for others. It is important, therefore, to appreciate the interconnection between different model formats and their realm of applicability. For alternative viewpoints and further discussion on the topic of linear systems modeling, we refer the reader to standard books on the subject, including Brockett (1970), Desoer (1970), Franklin and Powell (1980), and Fortmann and Hitz (1977), and at a more advanced level Kailath (1980), Rosenbrock (1970), and Wolovich (1974).

2.2 STATE-SPACE MODELS

2.2.1 General

The internal and external behavior of a linear finite-dimensional system can be described by a state-space model. Some of the basic notions associated with state-space modeling are summarized in Section A.1 of Appendix A and those readers who would like to revise these ideas are encouraged to review this section before proceeding.

Here we shall use the following notation for a state-space model:

$$x(t + 1) = Ax(t) + Bu(t); \qquad x(t_0) = x_0 \qquad (2.2.1)$$

$$y(t) = Cx(t) \qquad (2.2.2)$$

where $\{u(t)\}$, $\{y(t)\}$, and $\{x(t)\}$ denote the $r \times 1$ input sequence, $m \times 1$ output sequence, and $n \times 1$ state sequence, respectively, and x_0 is the initial state.

We know from the canonical structure theorem (Section A.1) that a general state-space model can be decomposed into controllable (strictly speaking, reachable—see Appendix A) and observable subsystems. We shall study these subsystems in a little more detail in the sections to follow.

2.2.2 Controllable State-Space Models

In this section we turn our attention to the completely controllable part of a system. Given a finite-dimensional linear system, there exists a transformation that isolates the controllable and uncontrollable parts (see Lemma A.1.2 of Appendix A). Consequently, a general system can be partitioned as in Fig. 2.2.1.

It can be seen from Fig. 2.2.1 that the input signal influences only the completely controllable part of the system. In fact, we shall show later that the poles of the completely controllable part can be arbitrarily assigned by state-variable feedback (this

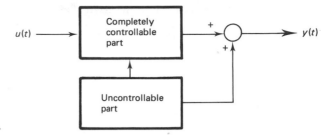

Figure 2.2.1 Decomposition into controllable and uncontrollable parts.

will be explored in Chapter 5). This will give further motivation for the term *completely controllable*.

For the remainder of this section we shall ignore the uncontrollable part of the system and assume complete controllability (strictly speaking, reachability—see Exercise 2.20). We shall explore some of the implications of this assumption. In particular, we shall describe two special structures that can be used to describe any completely controllable state-space model.

Consider the following completely controllable model (of order n, with r inputs and m outputs):

$$x(t+1) = Ax(t) + Bu(t); \quad x(0) = x_0 \tag{2.2.3}$$

$$y(t) = Cx(t) \tag{2.2.4}$$

An equivalent representation to (2.2.3)–(2.2.4) can be obtained by simply choosing a new basis for the state space, for example by using a transformation such that

$$\bar{x}(t) = P^{-1}x(t), \quad P \text{ nonsingular} \tag{2.2.5}$$

giving

$$\bar{x}(t+1) = \bar{A}\bar{x}(t) + \bar{B}u(t); \quad \bar{x}(0) = \bar{x}_0$$

$$y(t) = \bar{C}\bar{x}(t)$$

where $\bar{A} = P^{-1}AP$, $\bar{B} = P^{-1}B$, and $\bar{C} = CP$. An infinite number of choices exists for the transformation P, leading to an infinite number of equivalent completely controllable state-space models. Special forms for the state-space model can be achieved by forming P in particular ways, for example, by using any n linearly independent columns chosen from the controllability matrix. Note this can always be done, since for a completely controllable system the controllability matrix has rank n (see Lemma A.1.1). We shall find that certain choices of the transformation matrix result in the model having a specific structure which is convenient to use for particular applications. We discuss two particular structures: the *controllability form* and *controller form*.

Controllability Form

The single-input case. For ease of explanation, we first consider the *single-input* case and construct the transformation P from the columns of the controllability matrix as follows:

$$P = \mathcal{C} \triangleq [B \quad AB \quad \cdots \quad A^{n-1}B]; \quad \text{with dim } (B) = n \times 1$$

Since the rank of the controllability matrix is n, $A^n B$ will be a linear combination of the columns of P; that is, there exist scalars $\alpha_1, \ldots, \alpha_n$ such that

$$A^n B + \sum_{l=1}^{n} \alpha_l A^{n-l} B = 0$$

It can then be readily seen that the resulting state-space model for $\bar{x} = P^{-1} x$ has the following structure:

Controllability
Form

$$\bar{x}(t+1) = \begin{bmatrix} 0 & & & -\alpha_n \\ 1 & & & 0 \\ & \cdot & & \cdot \\ & & \cdot & \cdot \\ & & \cdot & \cdot \\ & & 1 & -\alpha_1 \end{bmatrix} \bar{x}(t) + \begin{bmatrix} 1 \\ 0 \\ \cdot \\ \cdot \\ \cdot \\ 0 \end{bmatrix} u(t)$$

$$y(t) = \bar{C}\bar{x}(t)$$

The structure above will, of course, be familiar to those readers who have taken an elementary course in linear systems theory.

The multi-input case. The ideas above can be extended to the multi-input case in a relatively straightforward way. (The remainder of this subsection may be omitted on a first reading.) For the multi-input case, we may construct the transformation matrix by simply searching the controllability matrix $\mathcal{C} = [B \quad AB \quad \cdots \quad A^{n-1}B]$, from left to right to isolate a set of linearly independent columns. The resulting columns are then rearranged into the following form:

$$P = [b_1 \quad \cdots \quad A^{k_1-1}b_1, b_2 \quad \cdots \quad A^{k_2-1}b_2, \quad \cdots \quad A^{k_r-1}b_r] \tag{2.2.6}$$

where $B = [b_1 \quad \cdots \quad b_r]$. The indices $k_1 \cdots k_r$ are called the *controllability indices* and $k_{max} \triangleq \max k_i$, where $i = 1, \ldots, r$, is called the *controllability index*.

We thus see that $A^{k_i}b_i$ is linearly dependent on preceding columns in $\mathcal{C}$ and we can write

$$A^{k_i}b_i + \sum_{j=1}^{r} \sum_{l=1}^{k_{ij}} \alpha_{jl}^{(i)}[A^{k_{ij}-l}b_j] = 0; \qquad i = 1, \ldots, r$$

where because of the order of the vectors in $\mathcal{C}$, the integers k_{ij} are given by

$$k_{ij} = \begin{cases} k_i & \text{for } i = j \\ \min(k_i + 1, k_j) & \text{for } i > j \\ \min(k_i, k_j) & \text{for } i < j \end{cases}$$

The total number of scalars, $\alpha_{jl}^{(i)}$, thus defined is n_s, where

$$n_s = \sum_{i=1}^{r} \sum_{j=1}^{r} k_{ij} \leq r \times n$$

The resulting state-space model has the following structure (illustrated for the three-input case in which no k_i is zero and $k_1 = k_3 > k_2 + 1$):

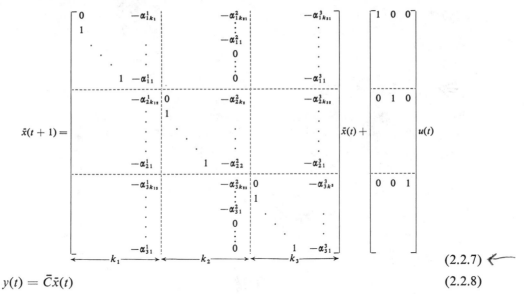

$$\bar{x}(t+1) = [\ldots] \bar{x}(t) + [\ldots] u(t) \tag{2.2.7}$$

$$y(t) = \bar{C}\bar{x}(t) \tag{2.2.8}$$

(The reader is asked to verify that the structure above results from the use of the transformation P. Note that in the structure above, $k_{12} = k_{32} = k_2$, $k_{21} = k_2 + 1$, $k_{23} = k_2$, and $k_{13} = k_{31} = k_1 = k_3$.)

An alternative in the multi-input case is to search the controllability matrix starting with b_1 and then to proceed to Ab_1, A^2b_1, ..., until the vector $A^{v_1}b_1$ can be expressed as a linear combination of previous vectors. Then one proceeds to b_2, Ab_2, ..., etc. Again, the resulting columns are then arranged in the form

$$P = [b_1 \cdots A^{v_1-1}b_1, b_2 \cdots A^{v_2-1}b_2, \cdots A^{v_s-1}b_s]; \quad s \leq r \tag{2.2.9}$$

The resulting state-space model then has the following structure (illustrated for $r = 4$ and $s = 3$):

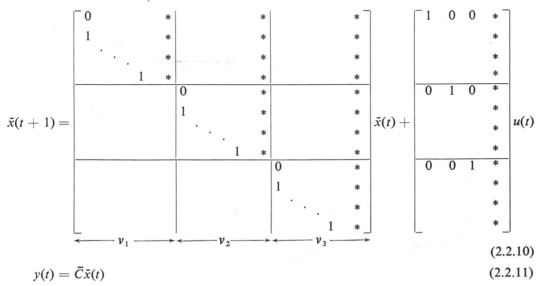

$$\bar{x}(t+1) = [\ldots] \bar{x}(t) + [\ldots] u(t) \tag{2.2.10}$$

$$y(t) = \bar{C}\bar{x}(t) \tag{2.2.11}$$

where $*$ denotes a possibly nonzero element.

Both of the forms above are called *controllability forms* [see Kailath (1980, p. 428)].

Controller Forms

The single-input case. Here we take the transformation found in the preceding section and postmultiply it by a nonsingular matrix, M. The matrix M is formed from the characteristic polynomial of A as follows.

Let

$$\det (zI - A) = z^n + \alpha_1 z^{n-1} + \cdots + \alpha_n$$

Then

$$M = \begin{bmatrix} 1 & \alpha_1 & \cdots & \alpha_{n-1} \\ 0 & 1 & \cdot & \cdot \\ & & \cdot & \cdot & \cdot \\ & & & \cdot & \alpha_1 \\ & & & & 1 \end{bmatrix}$$

The transformation is then given by $\bar{P} = \mathcal{C}M$, where $\mathcal{C}$ is the controllability matrix. The resulting state-space model for $\bar{x} = \bar{P}^{-1}x$ has the following structure:

Controller Form $\Rightarrow$

$$\bar{x}(t+1) = \begin{bmatrix} -\alpha_1 & \cdots & & -\alpha_n \\ 1 & & & \\ & \cdot & & \\ & & \cdot & \\ & & 1 & 0 \end{bmatrix} \bar{x}(t) + \begin{bmatrix} 1 \\ 0 \\ \cdot \\ \cdot \\ \cdot \\ 0 \end{bmatrix} u(t)$$

$$y(t) = \bar{C}\bar{x}(t)$$

The form above is commonly called a *controller form* (Kailath, 1980, p. 434) and is particularly convenient in determining state-variable feedback (See Chapter 5.) The structure of the state-space model is shown in Fig. 2.2.2.

*The multi-input case. The multi-input extension of the procedure above may be achieved as follows (Popov, 1972). Again, we shall form the transformation by further manipulation of P given in (2.2.6). We shall, for simplicity, assume that B has full column rank and thus $k_i > 0$ for $i = 1, \ldots, r$. In view of the arguments leading to (2.2.7) there exist scalars $\alpha_{ji}^{(i)}$ such that

$$A^{k_i}b_i + \sum_{j=1}^{r} \sum_{l=1}^{k_{ij}} \alpha_{ji}^{(i)}(A^{k_{ij}-1}b_j) = 0; \qquad i = 1, \ldots, r \qquad (2.2.12)$$

The new transformation $\bar{P}$ then has the following form (illustrated for $r = 3$):

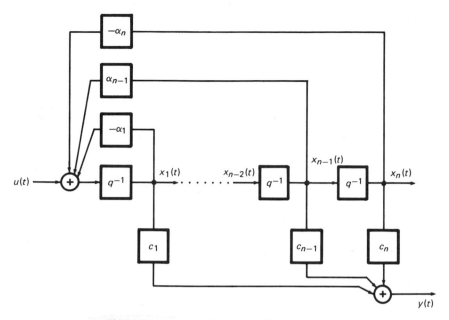

Figure 2.2.2 Structure of the controller state-space form.

$$\bar{P} = P \begin{bmatrix} 1 & \alpha^1_{11} & \alpha^1_{1(k_1-1)} & \cdot & \cdot & \alpha^2_{1(k_{21}-1)} & \cdot & \cdot & \alpha^3_{1(k_{31}-1)} \\ & \cdot & \cdot & & & \cdot & & & \cdot \\ & \cdot & \cdot & & & \cdot & & & \cdot \\ & \cdot & \alpha^1_{11} & & & \alpha^2_{11} & & & \alpha^3_{11} \\ & & 1 & & & 0 & & & 0 \\ \cdot & \cdot & \alpha^1_{2(k_{12}-1)} & 1 & & \alpha^2_{2(k_2-1)} & \cdot & \cdot & \alpha^3_{2(k_{32}-1)} \\ & & \cdot & & & \cdot & & & \cdot \\ & \cdot & \cdot & & & \cdot & & & \cdot \\ & \cdot & \alpha^1_{21} & & & \alpha^2_{21} & & & \alpha^3_{21} \\ & & 0 & & & 1 & & & 0 \\ \cdot & \cdot & \alpha^1_{3(k_{13}-1)} & \cdot & \cdot & \alpha^2_{3(k_{23}-1)} & 1 & & \alpha^3_{3(k_3-1)} \\ & & \cdot & & & \cdot & & & \cdot \\ & \cdot & \cdot & & & \cdot & & & \cdot \\ & \cdot & \alpha^1_{31} & & & \alpha^2_{31} & & & \alpha^3_{31} \\ & & 0 & & & 0 & & & 1 \end{bmatrix} \qquad (2.2.13)$$

$$P = \begin{bmatrix} b_1 & \cdots & A^{k_1-1}b_1 \,|\, b_2 & \cdots & A^{k_2-1}b_2 \,|\, b_3 & \cdots & A^{k_3-1}b_3 \end{bmatrix} \qquad (2.2.14)$$

Then applying the transformation as in (2.2.5) and using $\bar{P}$ in place of P, the resulting state-space model has the following structure (the controller form):

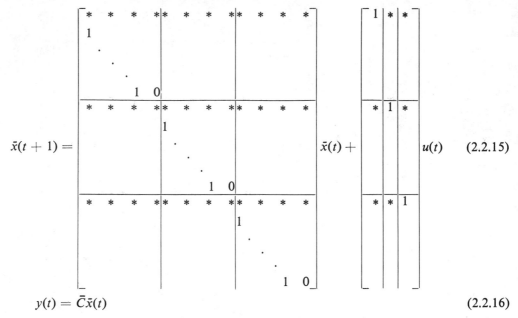

$$\bar{x}(t+1) = \begin{bmatrix} \ast & \ast & \ast & \ast|\ast & \ast & \ast & \ast|\ast & \ast & \ast & \ast \\ 1 & & & & & & & & & & & \\ & \ddots & & & & & & & & & & \\ & & 1 & 0 & & & & & & & & \\ \ast & \ast & \ast & \ast|\ast & \ast & \ast & \ast|\ast & \ast & \ast & \ast \\ & & & & 1 & & & & & & & \\ & & & & & \ddots & & & & & & \\ & & & & & & 1 & 0 & & & & \\ \ast & \ast & \ast & \ast|\ast & \ast & \ast & \ast|\ast & \ast & \ast & \ast \\ & & & & & & & & 1 & & & \\ & & & & & & & & & \ddots & & \\ & & & & & & & & & & 1 & 0 \end{bmatrix} \bar{x}(t) + \begin{bmatrix} 1 & \ast & \ast \\ & & \\ & & \\ & & \\ \ast & 1 & \ast \\ & & \\ & & \\ & & \\ \ast & \ast & 1 \\ & & \\ & & \\ & & \end{bmatrix} u(t) \qquad (2.2.15)$$

$$y(t) = \bar{C}\bar{x}(t) \qquad (2.2.16)$$

[Depending on the releative magnitude of the indices k_i, $i = 1, \ldots, r$, some of the elements denoted by $\ast$ in the transformed B matrix (i.e., in $\bar{B}$), will necessarily be zeros, giving an even simpler structure; see Antoulas (1981).] With the structure above (the controller form), the effect of state-variable feedback is particularly transparent (as we shall see in Chapter 5). In the next section we explore the dual of the results presented here as they apply to observable system.

2.2.3 Observable State-Space Models

Given a finite-dimensional linear system, there exists a transformation that isolates the observable and unobservable parts (see Lemma A.1.3). Consequently, a general system can be partitioned as in Fig. 2.2.3.

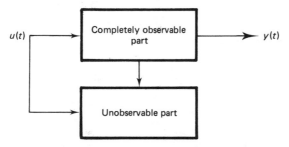

Figure 2.2.3 Decomposition into observable and unobservable parts.

The significance of the structure shown in Fig. 2.2.3 is that if one is confined to work with input–output data, it suffices to work only with the completely observable subsystem. This subsystem is input–output equivalent (has identical input output properties *for all initial states*) to the original system.

The unobservable subsystem may well be of importance in the overall operation of the system, but in so far as it is hidden from the output measurements, it is irrelevant to the input–output characteristics.

For the remainder of this subsection we shall assume a completely observable system (of order n, with r inputs and m outputs):

$$x(t+1) = Ax(t) + Bu(t); \qquad x(0) = x_0 \qquad (2.2.17)$$

$$y(t) = Cx(t) \qquad (2.2.18)$$

As we have seen in the case of controllable systems, there are an infinite number of representations of the form (2.2.17)–(2.2.18), all related by similarity transformations. If we follow the dual of the arguments presented in the preceding subsection we will arrive at *observability* and *observer* forms. The appropriate transformations can be obtained as follows. We construct a hypothetical system (A', B', C'), where

$$C' = B^T, \qquad B' = C^T, \qquad A' = A^T$$

We then construct the appropriate state-space transformation, P_c, for (A', B', C') as in Section 2.2.2 for *controllable* systems, giving

$$\bar{C}' = C'P_c$$
$$\bar{B}' = P_c^{-1}B'$$
$$\bar{A}' = P_c^{-1}A'P_c$$

Finally, we obtain the corresponding *observable* form for the original system by putting

$$\bar{C} = (\bar{B}')^T, \qquad \bar{B} = (\bar{C}')^T, \qquad \bar{A} = (\bar{A}')^T$$

The procedure above can be summarized as

$$\bar{C} = CP_0, \qquad \bar{B} = P_0^{-1}B, \qquad \bar{A} = P_0^{-1}AP_0$$

where $P_0 \triangleq P_c^{-T}$.

Alternatively, the controllability matrix for (A', B', C') is the transpose of the observability matrix for the original system (A, B, C). Thus P_0 can be directly determined from the original observability matrix by searching it by rows, constructing a matrix Q from the rows that are selected and finally putting $P_0 = Q^{-1}$. Note that $Q = P_c^T$ since Q is constructed by rows, whereas P_c is constructed by columns.

The procedure above can be used to construct *observability* and *observer* forms as described next.

Observability Forms

The single-output case. Following the reasoning above and by reference to Section 2.2.2, we see that the dual observability form is

Observability form ⇒

$$\bar{x}(t+1) = \begin{bmatrix} 0 & 1 & & & \\ & & \ddots & & \\ & & & \ddots & \\ & & & & 1 \\ -\alpha_n & \cdots\cdots & & -\alpha_1 \end{bmatrix} \bar{x}(t) + \bar{B}u(t)$$

$$y(t) = [1 \quad 0 \quad \cdots \quad 0]\bar{x}(t)$$

The multi-output case. In this case the observability form is obtained by searching the observability matrix by rows to isolate a set of linearly independent rows. The resulting rows are then rearranged into the form

$$Q = \begin{bmatrix} C_1 \\ \cdot \\ \cdot \\ \cdot \\ C_1 A^{\gamma_1 - 1} \\ C_2 \\ \cdot \\ \cdot \\ \cdot \\ C_m A^{\gamma_m - 1} \end{bmatrix} \qquad (2.2.19)$$

This gives a state-space model of the form (for $\gamma_1 = \gamma_3 > \gamma_2 + 1$)

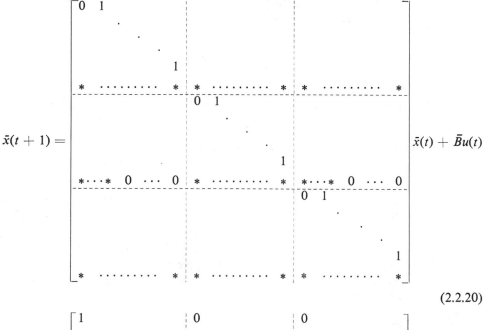

$$(2.2.20)$$

$$y(t) = \begin{bmatrix} 1 & & & | & 0 & & & | & 0 & & \\ 0 & & & | & 1 & & & | & 0 & & \\ 0 & & & | & 0 & & & | & 1 & & \end{bmatrix} \bar{x}(t)$$

$$\underset{\longleftarrow\gamma_1\longrightarrow}{}\underset{\longleftarrow\gamma_2\longrightarrow}{}\underset{\longleftarrow\gamma_3\longrightarrow}{}$$

$$(2.2.21)$$

Here γ_i, $i = 1, \ldots, m$, are known as the *observability indices*.

Observer Forms

The single-output case. Similarly, applying the dual of the construction of the controller forms, we obtain the observer form, which has the following structure:

Observer ⟹ Form

$$\bar{x}(t+1) = \begin{bmatrix} -\alpha_1 & 1 & & \\ & & \ddots & \\ & & & 1 \\ -\alpha_n & & & 0 \end{bmatrix} \bar{x}(t) + \bar{B}u(t)$$

$$y(t) = [1 \quad 0 \quad \cdots \quad 0]\bar{x}(t)$$

This is illustrated in Fig. 2.2.4.

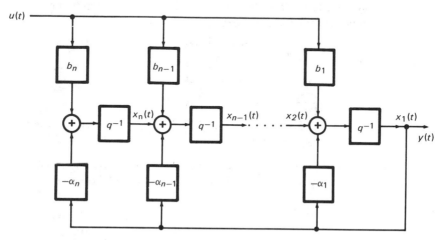

Figure 2.2.4 Structure of the observer state-space form.

The multi-output case. The multi-output *observer form* has the following structure [see (2.2.15)–(2.2.16)]:

$$\bar{x}(t+1) = \begin{bmatrix} * & 1 & & & & * & & & & * & & & \\ * & & \ddots & & & * & & & & * & & & \\ * & & & \ddots & & * & & & & * & & & \\ * & & & & 1 & * & & & & * & & & \\ * & 0 & \cdots & & 0 & * & & & & * & & & \\ * & & & & & * & 1 & & & * & & & \\ * & & & & & * & & \ddots & & * & & & \\ * & & & & & * & & & 1 & * & & & \\ * & & & & & * & 0 & \cdots & 0 & * & & & \\ * & & & & & * & & & & * & 1 & & \\ * & & & & & * & & & & * & & \ddots & \\ * & & & & & * & & & & * & & & 1 \\ * & & & & & * & & & & * & 0 & \cdots & 0 \end{bmatrix} \bar{x}(t) + \bar{B}u(t) \qquad (2.2.22)$$

$$y(t) = \begin{bmatrix} 1 & * & * \\ * & 1 & * \\ * & * & 1 \end{bmatrix} \bar{x}(t) \qquad (2.2.23)$$

Note that, as in the controller form, depending on the relative magnitude of the indices, some of the elements denoted by $*$ in $\bar{C}$ will necessarily be zero.

A Simple Application

Before leaving this subsection, we pause to show how observable models can be used to advantage to describe a system whose output is influenced by certain classes of persistent disturbances. For example, consider the situation depicted in Fig. 2.2.5.

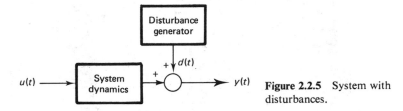

Figure 2.2.5 System with disturbances.

The disturbance, $d(t)$, is assumed to be described by a linear discrete-time model which is stable but not asymptotically stable. Typical examples would be:

1. A constant disturbance modeled by

$$\eta(t + 1) = \eta(t); \qquad \eta(0) = \eta_0$$
$$d(t) = \eta(t)$$

2. A sinusoidal disturbance modeled by

$$\eta_1(t + 1) = \eta_2(t); \qquad \eta_1(0) = \eta_1$$
$$\eta_2(t + 1) = [2 \cos \omega \Delta] \eta_2(t) - \eta_1(t); \qquad \eta_2(0) = \eta_2$$
$$d(t) = \eta_1(t)$$

It is obvious that the disturbance generator is uncontrollable; nonetheless, the output response $\{y(t)\}$, which includes the disturbance as a component, can be described by an observable model (see Exercises 2.10 and 2.22). The controllable part of the system can be manipulated to counter the effects of the (uncontrollable) disturbance on the output. We shall pursue this topic in more detail later in the book.

2.2.4 Minimal State-Space Models

In the preceding section we discussed observable models. These models were shown to be input–output equivalent to the original system for arbitrary initial states. In this section we further restrict attention to the completely controllable and completely observable part of the system.

It can be seen from the canonical structure theorem (Theorem A.1.1 of Appendix A) that for *zero initial state* the input–output properties of the system can be completely characterized by the completely controllable and observable subsystem, $R_1 = [\bar{A}_{11}, \bar{B}_1, \bar{C}_1]$:

$$x(t + 1) = \bar{A}_{11}x(t) + \bar{B}_1 u(t); \qquad x(0) = x_0, \quad t \geq 0 \qquad (2.2.24)$$

$$y(t) = \bar{C}_1 x(t) \qquad (2.2.25)$$

It turns out that no further reduction in the state dimension is possible while maintaining zero-state equivalence for all inputs. Thus the model (2.2.24)–(2.2.25) is said to be a *minimal realization* of the system. The model (2.2.24)–(2.2.25) is unique to within a similarity transformation (Kailath, 1980, p. 424).

Note that the minimal model (2.2.24)–(2.2.25) does not describe the uncontrollable part or unobservable part of the system. Of course, if these modes are asymptotically stable (see Appendix B), they will ultimately decay to zero and, in this case, little is lost by ignoring them from the start. The reader, however, should be aware that restriction to the controllable and observable part precludes certain important cases such as the situation previously described in Fig. 2.2.5, where the uncontrollable modes, associated with the disturbance, did not decay to zero. With this precaution in mind, we shall proceed to explore further zero-state equivalent models and their implications.

If we take z-transforms (see Section A.2) in (2.2.24)–(2.2.25), we obtain

$$zX(z) - zx_0 = \bar{A}_{11}X(z) + \bar{B}_1 U(z) \qquad (2.2.26)$$

$$Y(z) = \bar{C}_1 X(z) \qquad (2.2.27) \longleftarrow$$

or

$$Y(z) = \bar{C}_1 (zI - \bar{A}_{11})^{-1} \bar{B}_1 U(z) + \bar{C}_1 (zI - \bar{A}_{11})^{-1} z x_0 \qquad (2.2.28)$$

Thus setting the initial conditions to zero, it can be seen that the transform of the output, $Y(z)$, is related to the transform of the input, $U(z)$, by

$$Y(z) = H(z)U(z) \qquad (2.2.29)$$

where

$$H(z) = \bar{C}_1 (zI - \bar{A}_{11})^{-1} \bar{B}_1 \qquad (2.2.30)$$

We shall call the rational matrix, $H(z)$, the *system transfer function*. Alternatively, one can also obtain the transfer function from the original state-space model $[A, B, C]$ by taking z-transforms to give

$$H(z) = C(zI - A)^{-1}B \qquad (2.2.31)$$

The fact that (2.2.30) and (2.2.31) give the same transfer function can be easily seen by noting the special structure resulting from the canonical structure theorem. Consequently, only the controllable and observable part of the system is reflected in the system transfer function description.

In general, transfer functions derived from state-space models have the property that $\lim_{z \to \infty} H(z) = 0$; that is, the transfer function is *strictly proper*. If $\lim_{z \to \infty} H(z)$ is finite, the transfer function is said to be *proper*.

As a final remark, we note that equivalent systems related by a similarity transformation all have the same transfer function. (Verify it!)

2.3 DIFFERENCE OPERATOR REPRESENTATIONS

2.3.1 General

In Section 2.2 we discussed state-space models. These can be seen to be a set of first-order difference equations. An alternative description is to use a high-order difference equation. This leads us to the idea of a difference operator representation.

In order to describe these models in a succinct manner, we introduce the forward and backward shift operators q and q^{-1}. If $y(t)$ denotes the value of a sequence $\{y(t)\}$ at time t, where $t \in \{0, 1, \ldots\}$, then $qy(t)$ denotes the value of the sequence at time $(t + 1)$ and $q^{-1}y(t)$ denotes the value of the sequence at time $(t - 1)$. That is,

$$qy(t) \triangleq y(t + 1) \qquad \text{for } t \geq 0$$
$$q^{-1}y(t) \triangleq y(t - 1) \qquad \text{for } t \geq 1; \qquad q^{-1}y(0) \triangleq 0$$

and consequently,

$$q^i y(t) = y(t + i) \qquad \text{for } t \geq 0$$
$$q^{-i}y(t) = y(t - i) \qquad \text{for } t \geq i; \qquad q^{-i}y(t) \triangleq 0 \quad \text{for } 0 \leq t < i$$

Similarly, we can define the operation of q on a sequence $\{y(t)\} \triangleq y(0), y(1),$ ... as follows:

$$q\{y(t)\} = \{qy(t)\} = \{y(t + 1)\} \triangleq y(1), y(2), \ldots$$
$$q^{-1}\{y(t)\} = \{q^{-1}y(t)\} = \{y(t - 1)\} \triangleq 0, y(0), y(1), \ldots$$

As an example, consider a simple first-order difference equation

$$y(t + 1) + ay(t) = u(t); \qquad t \geq 0$$

with initial condition $y(0)$. This can be succinctly written in any one of the following forms (see also Exercises 2.11 and 2.12):

$$(q + a)\{y(t)\} = \{u(t)\}; \qquad y(0) \text{ given}$$
$$(q + a)y(t) = u(t); \qquad t \geq 0; \quad y(0) \text{ given}$$
$$(1 + aq^{-1})\{y(t)\} - \{y(0), 0, 0, \ldots\} = q^{-1}\{u(t)\}; \qquad y(0) \text{ given}$$
$$(1 + aq^{-1})y(t) = q^{-1}u(t); \qquad t \geq 1; \quad y(0) \text{ given}$$

Using the second alternative listed above, the general difference operator representation of a linear dynamical system can be written as

$$P(q)z(t) = Q(q)u(t); \qquad t \geq 0 \tag{2.3.1}$$
$$y(t) = R(q)z(t) \tag{2.3.2}$$

with appropriate initial conditions on $\{z(t)\}$, and where $P(q)$, $Q(q)$, and $R(q)$ are polynomial matrices in the forward shift operator q.

To ensure existence and uniqueness of the solution to (2.3.1)–(2.3.2), we require $P(q)$ to be square and nonsingular [det $P(q) \neq 0$ for almost all q]. In the model above, we call $z(t)$ the system *partial state*.

The difference operator representation includes the state-space model as a special case when

$$\begin{cases} P(q) = qI - A & \text{(2.3.3)} \\ Q(q) = B & \text{(2.3.4)} \\ R(q) = C & \text{(2.3.5)} \\ z(t) = x(t) & \text{(2.3.6)} \end{cases}$$

Here we shall be particularly interested in two special forms of the difference operator representation: right difference operator representations and left difference operator representations.

In a right difference operator representation, the model takes the following form:

$$\begin{cases} D_R(q)z(t) = u(t); & t \geq 0 & \text{(2.3.7)} \\ y(t) = N_R(q)z(t) & \text{(2.3.8)} \end{cases}$$

with appropriate initial conditions on $\{z(t)\}$. We shall see below that this is an equivalent description to a controllable state-space model.

In a left difference operator representation, the model takes the following form:

$$\begin{cases} D_L(q)z(t) = N_L(q)u(t); & t \geq 0 & \text{(2.3.9)} \\ y(t) = z(t) & \text{(2.3.10)} \end{cases}$$

or equivalently,

$$D_L(q)y(t) = N_L(q)u(t); \qquad t \geq 0 \qquad \text{(2.3.11)}$$

with appropriate initial conditions on $\{y(t)\}$. This form turns out to be equivalent to an observable state-space model.

For future reference, we define the *degree* of a square polynomial matrix as the degree of its determinant, and we define the degree of a difference operator representation as the degree of its *denominator matrix* [$P(q)$ in (2.3.1), $D_R(q)$ in (2.3.7), and $D_L(q)$ in (2.3.9)]. It turns out that the degree of a difference operator representation is the number of independent initial conditions that must be specified in order to solve the set of difference equations.

In the following subsections, we study the left and right difference operator representations in more detail. We first do the single-input single-output case, which is relatively straightforward. The multivariable case, on the other hand, depends on results from polynomial matrix theory. Some of these results are summarized in Section A.3.

2.3.2 Right Difference Operator Representations

The form of this model was given in (2.3.7)–(2.3.8). It can also be expressed in the following equivalent form:

$$\begin{bmatrix} D_R(q) \\ N_R(q) \end{bmatrix} z(t) = \begin{bmatrix} u(t) \\ y(t) \end{bmatrix}; \qquad t \geq 0 \qquad \text{(2.3.12)}$$

with appropriate initial conditions on $\{z(t)\}$. In (2.3.12), $D_R(q)$ is taken to be square and nonsingular.

The single-input case. In this case, $D_R(q)$ and $N_R(q)$ are scalar polynomials of the form

$$D_R(q) = a_0 q^n + a_1 q^{n-1} + \cdots + a_n$$
$$N_R(q) = b_1 q^{n-1} + \cdots + b_n$$

Naturally, a_0 can be taken to be nonzero by choice of the power n. (In fact, in the scalar case a_0 can be taken to be 1 by a simple scaling.) The degree of $N_R(q)$ is chosen as $(n-1)$, or less, to ensure that the corresponding transfer function, $H(z)$, is strictly proper, where

$$H(z) = \frac{b_1 z^{n-1} + \cdots + b_n}{a_0 z^n + a_1 z^{n-1} + \cdots + a_n}$$

Subject to the conditions above, a right difference operator representation is *controllable* in a sense that we now make precise by relating the model to a controllable state-space model.

We first note that the model (2.3.7) can be written in the form

$$[a_0 q^n + L(q)]z(t) = u(t); \qquad t \geq 0$$

where

$$L(q) = \bar{L}\psi(q)$$
$$\psi(q)^T = [q^{n-1} \quad q^{n-2} \quad \cdots \quad 1]$$
$$\bar{L} = [a_1 \quad \cdots \quad a_n]$$

Similarly, we can express $N(q)$ in (2.3.8) as

$$N(q) = \bar{N}\psi(q)$$
$$\bar{N} = [b_1 \quad \cdots \quad b_n]$$

Thus the system (2.3.7)–(2.3.8) can be described by

$$q^n z(t) = -\frac{1}{a_0}\bar{L}\psi(q)z(t) + \frac{1}{a_0}u(t); \qquad t \geq 0$$
$$y(t) = \bar{N}\psi(q)z(t)$$

Now defining the state vector as

$$x(t) = \psi(q)z(t) = [z(t+n-1) \quad \cdots \quad z(t)]^T$$

the model can be immediately expressed in state-space form as

$$x(t+1) = \begin{bmatrix} -\dfrac{a_1}{a_0} & \cdots & -\dfrac{a_n}{a_0} \\ 1 & & \\ & \cdot & \\ & & \cdot \\ & & \quad \cdot \\ & & 1 \quad 0 \end{bmatrix} x(t) + \begin{bmatrix} \dfrac{1}{a_0} \\ 0 \\ \cdot \\ \cdot \\ \cdot \\ 0 \end{bmatrix} u(t)$$

$$y(t) = \bar{N}x(t)$$

Note that the initial conditions for the right difference operator representation are $[z(n-1) \quad \cdots \quad z(0)]$ and these are in one-to-one correspondence with the initial condition $x(0)$ for the state equation. We also note that the equation above is in *controller form* and is completely controllable. In fact, it can be seen that the right difference operator representation is nothing more than an alternative (condensed) way of representing a controllable state-space model in controller form. The reverse procedure from controllable state space to right difference operator representation is straightforward by first transforming to controller form and then condensing to difference operator form by successive elimination of states in terms of $z(t)$.

We also note that there is a simple one-to-one relationship between the coefficients in the controller state-space form and the coefficients in the right difference operator representation and those in the system transfer function.

The corresponding transformation to the controllability form is explored in Exercise 2.25.

The multi-input case. In this case $D_R(q)$ and $N_R(q)$ are polynomial matrices. We shall make the following assumptions about the model (2.3.12):

Assumption 2.3.A

(i) $D_R(q)$ is column reduced.

(ii) $\partial c_i[N_R(q)] < \partial c_i[D_R(q)]$

where ∂c_i denotes the degree of the ith column. (For other notation and definitions, see Section A.3.)

We now give a brief interpretation of the assumptions above and show that they are justified in many cases of interest. We begin by noting that the set of equations (2.3.12) can be transformed to an equivalent representation without loss of information by using elementary column operations.

The notion of an elementary column operation is described in Section A.3. It can be seen from the appendix that elementary column operations can be described by right multiplication by a unimodular matrix.

Let $U(q)$ be any unimodular matrix and let $U'(q)$ be its inverse, also a polynomial matrix. Then (2.3.12) can be expressed as

$$\begin{bmatrix} D_R(q) \\ N_R(q) \end{bmatrix} U(q)z'(t) = \begin{bmatrix} u(t) \\ y(t) \end{bmatrix}; \qquad t \geq 0 \tag{2.3.13}$$

where

$$z'(t) = U'(q)z(t) \tag{2.3.14}$$

and multiplying by $U(q)$, we have

$$z(t) = U(q)z'(t) \tag{2.3.15}$$

It is clear from (2.3.14)–(2.3.15) that $z'(t)$ and $z(t)$ are equivalent in the sense that they can be obtained from each other by a finite number of difference operations. Also note that multiplication by a unimodular matrix does not change the degree of the difference operator representation (because the determinant of a unimodular matrix is a constant).

Now we can always find a unimodular matrix that transforms $D_R(q)$ to column-reduced form (see Lemma A.3.4). Thus Assumption 2.3.A(i) is without loss of generality. Assumption 2.3.A(ii) is related to the transfer function arising from (2.3.12) being strictly proper. Taking z-transforms (see Section A.2), the transfer function is given by

$$H(z) = N_R(z)D_R(z)^{-1} \tag{2.3.16}$$

The significance of Assumption 2.3.A(i) and (ii) in relation to $H(z)$ is summarized in the following lemma:

Lemma 2.3.1

(a) Assumption 2.3.A(ii) is a necessary condition for the corresponding transfer function to be strictly proper.
(b) Assumption 2.3.A(i) and (ii) are sufficient for this to hold.

Proof. (a) Postmultiplying (2.3.16) by $D_R(z)$, we have

$$H(z)D_R(z) = N_R(z)$$

and therefore for the jth column we can write

$$[N_R(z)]_{ij} = \sum_{k=1}^{r} [H(z)]_{ik}[D_R(z)]_{kj}$$

where $[\cdot]_{ij}$ denotes the ijth element.

Now, if every element of $H(z)$ is strictly proper, then all $[N_R(z)]_{ij}$ in the jth column must have degree less than the highest-degree polynomial in the jth column of $D_R(z)$. This establishes necessity.

(b) To establish sufficiency, we note that from Cramer's rule, we have

$$[H(z)]_{ij} = \frac{\det D_R^{ij}(z)}{\det D_R(z)}$$

where $D_R^{ij}(z)$ is the matrix obtained by replacing the jth row of $D_R(z)$ by the ith row of $N_R(z)$.

Now, as in (A.3.12) of Appendix A, we can write

$$D_R(z) = D_0 S(z) + L(z)$$

where

$$S(z) = \text{diag}\{z^{k_1} \quad \cdots \quad z^{k_r}\}$$
$$k_i = \text{degree of } i\text{th column of } D_R(z)$$
$$D_0 = \text{leading column coefficient matrix}$$
$$L(z) = \text{lower-order terms}$$

Similarly,

$$D_R^{ij}(z) = D_0^{ij} S(z) + L^{ij}(z)$$

where following Cramer's rule, D_0^{ij} is the same as D_0 save for the jth row, which is obtained from the ith row of $N_R(z)$ [the individual entries of which have degree less than the degree of the corresponding entry in the replaced row by Assumption 2.3.A(ii)]. Hence

$$\text{degree det } D_R(z) = \sum_{i=1}^{r} k_i$$

$$\text{degree det } D_R^{ij}(z) < \sum_{i=1}^{r} k_i$$

Thus $[H(z)]_{ij}$ is strictly proper as required.

▼▼▼

As in the scalar case, right difference operator representations of the form (2.3.7)–(2.3.8) satisfying Assumption 2.3.A(i) are *controllable*. To show this, we first note that, subject to Assumption 2.3.A(i), (2.3.7) can be written in the form

$$[D_0 S(q) + L(q)]z(t) = u(t); \qquad t \geq 0 \tag{2.3.17}$$

$S(q) = \text{diag}\{q^{k_1} \ \cdots \ q^{k_r}\}$ and D_0 is the leading column coefficient matrix [nonsingular by Assumption 2.3.A(i) and upper triangular without loss of generality by the dual of Lemma A.3.4].

Also, we can express $L(q)$ in the form

$$L(q) = \bar{L}\psi(q) \tag{2.3.18}$$

where

$$\psi(q)^T = \begin{bmatrix} q^{k_1-1} & \cdots & q & 1 & & & & & \\ & & & & q^{k_2-1} & \cdots & q & 1 & & \\ & & & & & & & \ddots & & \\ & & & & & & & & q^{k_r-1} & \cdots & q & 1 \end{bmatrix} \tag{2.3.19}$$

Similarly, in the light of Assumption 2.3.A(ii) we can write

$$N(q) = \bar{N}\psi(q) \tag{2.3.20}$$

Thus the system (2.3.7)–(2.3.8) can be described by

$$S(q)z(t) = -D_0^{-1}\bar{L}\psi(q)z(t) + D_0^{-1}u(t); \qquad t \geq 0 \tag{2.3.21}$$

$$y(t) = \bar{N}\psi(q)z(t) \tag{2.3.22}$$

We then define the state vector as

$$x(t) = \psi(q)z(t) = [z_1(t + k_1 - 1), \ldots, z_1(t), z_2(t + k_2 - 1), \ldots, z_r(t)]^T \tag{2.3.23}$$

Then (2.3.21)–(2.3.22) can immediately be expressed in state-space form for $t \geq 0$ as (illustrated for $r = 3$)

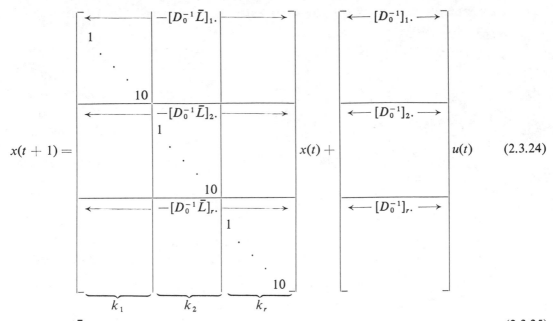

$$y(t) = \bar{N}x(t) \tag{2.3.25}$$

where $[\cdot]_i$ denotes the ith row. Note that the initial conditions for (2.3.21) are $[z_1(k_1 - 1), \ldots, z_1(0), z_2(k_2 - 1), \ldots, z_2(0), \ldots, z_r(0)]$ and that these are in one-to-one correspondence with the initial condition $x(0)$ for (2.3.24).

We observe that the model above is in *controller form* and consequently is completely controllable. The state dimension is $n = \sum_{i=1}^{r} k_i = $ degree $[D_R(q)]$. Again, the reverse procedure can be applied to construct a right difference operator representation via the controller state-space form.

2.3.3 Left Difference Operator Representations

The form of this model is as given in (2.3.9)–(2.3.10). We have seen that it can be expressed in the form

$$D_L(q)y(t) = N_L(q)u(t); \qquad t \geq 0 \tag{2.3.26}$$

with appropriate initial conditions on $y(t)$. In the equation above, $D_L(q)$ is square and nonsingular.

The single-output case. In this case $D_L(q)$ and $N_L(q)$ are scalar polynomials of the form

$$D_L(q) = a_0 q^n + a_1 q^{n-1} + \cdots + a_n$$
$$N_L(q) = b_1 q^{n-1} + \cdots + b_n$$

As in the case of right difference operator representations, a_0 can be taken to be nonzero by choice of the power n. The degree of $N_L(q)$ is chosen as $(n - 1)$, or less, to ensure strict properness for the transfer function.

Left difference operator representations correspond to *observable state-space models*. This is the dual of the fact established in the preceding section—that is, right difference operator representations correspond to controllable state-space models. We argue as follows. The model (2.3.26) can be written as

$$y(t) = -\frac{a_1}{a_0}y(t-1) \cdots -\frac{a_n}{a_0}y(t-n) + \frac{b_1}{a_0}u(t-1) \cdots + \frac{b_n}{a_0}u(t-n)$$

The model above can be expressed for $t \geq n$ as follows:

$$y(t) = -\frac{a_1}{a_0}y(t-1) + \frac{b_1}{a_0}u(t-1) + r_1(t-1)$$

$$r_1(t) = -\frac{a_2}{a_0}y(t-1) + \frac{b_2}{b_0}u(t-1) + r_2(t-1)$$

$$\vdots$$

$$r_{n-1}(t) = -\frac{a_n}{a_0}y(t-1) + \frac{b_n}{b_0}u(t-1)$$

We now define the state vector as

$$x(t)^T = [y(t) \quad r_1(t) \quad \cdots \quad r_{n-1}(t)]^T$$

leading to the following state-space model:

$$x(t+1) = \begin{bmatrix} -\dfrac{a_1}{a_0} & 1 & & \\ \cdot & & \cdot & \\ \cdot & & & \cdot \\ \cdot & & & 1 \\ -\dfrac{a_n}{a_0} & & & 0 \end{bmatrix} x(t) + \begin{bmatrix} \dfrac{b_1}{a_0} \\ \cdot \\ \cdot \\ \cdot \\ \dfrac{b_n}{a_0} \end{bmatrix} u(t)$$

$$y(t) = [1 \quad 0 \quad \cdots \quad 0]x(t)$$

We note that the equation above is in *observer form* and is completely observable. We thus see that a left difference operator representation is nothing more than an alternative (condensed) way of representing an observable state-space model in observer form. This is the dual of the corresponding result for right difference operator representations discussed earlier.

Again there is a one-to-one relationship between the coefficients in the observer state-space model and the coefficients in the left difference operator representation and those in the system transfer function. A succinct derivation of the observer state-space model from the left difference operator representation is explored in Exercise 2.26.

The multi-output case. Here we extended the analysis above to the multi-output case where $D_L(q)$ and $N_L(q)$ are polynomial matrices. We shall make the following assumptions about the model (2.3.26):

Assumption 2.3.B

(i) $D_L(q)$ is row reduced.
(ii) $\partial r_i[N_L(q)] < \partial r_i[D_L(q)]$

The assumptions above are natural and correspond to Assumption 2.3.A(i) and (ii). The rationale for the assumptions is basically as before: that if (i) is not satisfied, it can be achieved by transforming the model equations, and that (ii) is associated with the properness of the related transfer function. Note that elementary *row* operations on $D_L(q)$ and $N_L(q)$ can be carried out without changing the degree of $D_L(q)$.

As noted above for the scalar case, left difference operator representations correspond to *observable state-space models*. To explore this relationship further, we shall first express (2.3.26) in a slightly different form. To do this, we note that (2.3.26) can be written as

$$[S(q)D_0 + L(q)]y(t) = N_L(q)u(t); \qquad t \geq 0 \qquad (2.3.27)$$

where $S(q) = \text{diag}\,\{q^{k_1} \ \cdots \ q^{k_m}\}$, D_0 is the leading row coefficient matrix [nonsingular by Assumption 2.3.B(i) and lower triangular without loss of generality by Lemma A.3.4] and

$$L(q) = \psi(q)^T \bar{L} \qquad (2.3.28)$$
$$N_L(q) = \psi(q)^T \bar{N} \qquad (2.3.29)$$

where

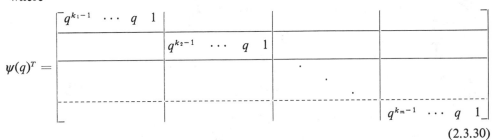

$$(2.3.30)$$

Thus the system (2.3.26) can be described by

$$S(q)y'(t) = -\psi(q)^T \bar{L} D_0^{-1} y'(t) + \psi(q)^T \bar{N} u(t); \qquad t \geq 0 \qquad (2.3.31)$$

where

$$y'(t) = D_0 y(t) \qquad (2.3.32)$$

and hence

$$y(t) = D_0^{-1} y'(t) \qquad (2.3.33)$$

The initial conditions required for (2.3.31) are the values of $y_i'(t)$; $0 \leq t \leq k_i - 1$; $i = 1, \ldots, m$.

We now introduce the shift operator matrix

$$S(q^{-1}) = \text{diag}\,\{q^{-k_1} \ \cdots \ q^{-k_m}\} \qquad (2.3.34)$$

where q^{-1} is a backward shift operator (see Section 2.3.1). Recall that

$$q^{-1}y(t) \triangleq y(t-1); \qquad t \geq 1$$
$$q^{-1}y(0) \triangleq 0$$

Operating on the left of (2.3.31) by $S(q^{-1})$ and noting that $S(q^{-1})S(q) \equiv I$, we have

$$y'(t) = -S(q^{-1})\psi(q)^T \bar{L} D_0^{-1} y'(t) + S(q^{-1})\psi(q)^T \bar{N} u(t) \qquad (2.3.35)$$

where the equation for $y'_i(t)$ is now defined for $t \geq k_i$ and the initial conditions are $y'_i(t)$, $0 \leq t \leq k_i - 1$, as before. Now the equation above can be written as

$$y'(t) = -\bar\psi(q^{-1})^T \bar{L} D_0^{-1} y'(t) + \bar\psi(q^{-1})^T \bar{N} u(t) \tag{2.3.36}$$

$$y(t) = D_0^{-1} y'(t) \tag{2.3.37}$$

where

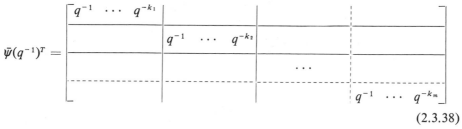

$$\tag{2.3.38}$$

The reader should note the dual relationship between (2.3.21)–(2.3.22) and (2.3.36)–(2.3.37).

The form (2.3.36)–(2.3.37) can be linked to an observable state-space model as shown in the following lemma:

Lemma 2.3.2

(a) The model (2.3.36)–(2.3.37) satisfying Assumption 2.3.B(i) is equivalent to a state-space model in observer form, having state dimension $n = \sum_{i=1}^m k_i =$ degree $D_L(q)$, and the structure (illustrated for $m = 3$)

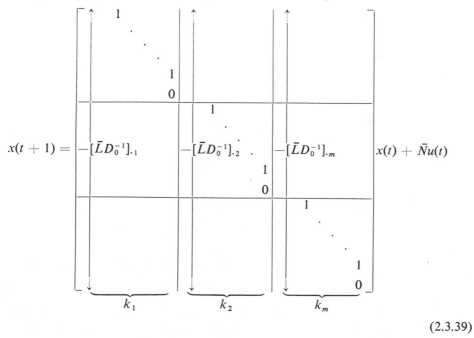

$$\tag{2.3.39}$$

$$y(t) = \begin{bmatrix} \uparrow \\ [D_0^{-1}]_{\cdot 1} \\ \downarrow \end{bmatrix} \quad \begin{matrix} \uparrow \\ [D_0^{-1}]_{\cdot 2} \\ \downarrow \end{matrix} \quad \begin{matrix} \uparrow \\ [D_0^{-1}]_{\cdot m} \\ \downarrow \end{matrix} \quad \end{bmatrix} x(t) \qquad (2.3.40)$$

where $[\cdot]_{\cdot i}$ denotes the ith column.

(b) The initial state $x(0)$ can be chosen to correspond to the initial conditions on (2.3.35).

Proof. (a) To simplify the notation, we let a_{ij} denote the ijth element of $-\bar{L}D_0^{-1}$. Then we note from (2.3.36) that $y_1'(t)$ for $t \geq k_1$ can be expressed as

$$y_1'(t) = a_{11}y_1'(t-1) + a_{12}y_2'(t-1) + \cdots + a_{1m}y_m'(t-1) + \bar{N}_{11}u_1(t-1) + \cdots$$
$$+ \bar{N}_{1r}u_r(t-1)$$
$$+ a_{21}y_1'(t-2) + a_{22}y_2'(t-2) + \cdots + a_{2m}y_m'(t-2) + \bar{N}_{21}u_1(t-2) + \cdots$$
$$+ \bar{N}_{2r}u_r(t-2)$$
$$\vdots$$
$$+ a_{k_1 1}y_1'(t-k_1) + a_{k_1 2}y_2'(t-k_1) + \cdots + ak_1 {}_m y_m'(t-k_1)$$
$$+ \bar{N}_{k_1 1}u_1(t-k_1) + \cdots + \bar{N}_{k_1 r}u_r(t-k_1)$$

$$(2.3.41)$$

Now define $r_{1i}(t-i)$ as the sum of the $(i+1)$th to k_1th rows in the equation above. We then note that

$$y_1'(t) = a_{11}y_1'(t-1) + a_{12}y_2'(t-1) + \cdots + a_{1m}y_m'(t-1) + \bar{N}_{11}u_1(t-1)$$
$$+ \cdots + \bar{N}_{1r}u_r(t-1) + r_{11}(t-1)$$
$$r_{11}(t) = a_{21}y_1'(t-1) + a_{22}y_2'(t-1) + \cdots + a_{2m}y_m'(t-1) + \bar{N}_{21}u_1(t-1)$$
$$+ \cdots + \bar{N}_{2r}u_r(t-1) + r_{12}(t-1)$$
$$\vdots$$
$$r_{1k_1-1}(t) = a_{k_1 1}y_1'(t-1) + a_{k_1 2}y_2'(t-1) + \cdots + a_{k_1 m}y_m'(t-1) + \bar{N}_{k_1 1}u_1(t-1)$$
$$+ \cdots + \bar{N}_{k_1 r}u_r(t-1)$$

Similarly, we can write for $t \geq k_2$,

$$y_2'(t) = a_{k_1+1,1}y_1'(t-1) + \cdots + a_{k_1+1,m}y_m'(t-1) + \bar{N}_{k_1+1,1}u_1(t-1) + \cdots$$
$$+ \bar{N}_{k_1+1,r}u_r(t-1) + r_{21}(t-1)$$

and so on.

The model (2.3.39)–(2.3.40) then follows immediately by defining

$$x(t)^T = [y_1'(t), r_{11}(t), \ldots, r_{1,k_1-1}(t), y_2'(t), r_{21}(t), \ldots, r_{m,k_m-1}(t)]$$

(b) We have seen from the development above that (2.3.39) and (2.3.40) are valid strictly only for $t \geq \max k_i$. If we insist that the equations hold for $t \geq 0$, we can solve for $y'(t)$ as follows:

Models for Deterministic Dynamical Systems   Chap. 2

$$
\begin{bmatrix} y'(0) \\ y'(1) \\ \cdot \\ \cdot \\ \cdot \\ y'(t) \end{bmatrix} = \begin{bmatrix} C \\ CA \\ \cdot \\ \cdot \\ \cdot \\ CA^t \end{bmatrix} x(0) + \begin{bmatrix} 0 \\ C\bar{N}u(0) \\ \cdot \\ \cdot \\ \cdot \\ \sum_{j=0}^{t-1} CA^{t-1-j}\bar{N}u(j) \end{bmatrix}
$$

where A and $\bar{N}$ are as in (2.3.39) and C has the structure (for $m = 2$)

$$
C = \begin{bmatrix} 1 & 0 & \cdots & 0 & \vdots & & & & \\ * & 0 & \cdots & 0 & 1 & 0 & \cdots & 0 \end{bmatrix}
$$

Now in view of the special structure of C and A, it can be shown that the matrix $(C^T, A^T C^T, \ldots, (A^t)^T C^T)^T$ has the following structure (illustrated for $m = 2$, $k_1 = 3$, $k_2 = 2$):

$$
\begin{bmatrix} C \\ \hline CA \\ \hline CA^2 \end{bmatrix} = \begin{bmatrix} 1 & 0 & 0 & 0 & 0 \\ * & 0 & 0 & 1 & 0 \\ \hline * & 1 & 0 & * & 0 \\ * & * & 0 & * & 1 \\ \hline * & * & 1 & * & * \\ * & * & * & * & * \end{bmatrix}
$$

Thus by choosing the appropriate rows corresponding to $(y'_1(0), \ldots, y'_1(k_1 - 1), y'_2(0), \ldots, y'_m(k_m - 1))^T$, we have (again for $m = 2$, $k_1 = 3$, $k_2 = 2$)

$$
\begin{bmatrix} y'_1(0) \\ y'_2(0) \\ \hline y'_1(1) \\ y'_2(1) \\ \hline y'_1(2) \end{bmatrix} = \begin{bmatrix} 1 & 0 & 0 & 0 & 0 \\ * & 0 & 0 & 1 & 0 \\ \hline * & 1 & 0 & * & 0 \\ * & * & 0 & * & 1 \\ \hline * & * & 1 & * & * \end{bmatrix} x(0) + \text{ terms in } \{u(t)\}
$$

It can be seen that the matrix on the right-hand side of the equation above is nonsingular (the corresponding general case follows a similar argument). Thus we can uniquely determine $x(0)$ corresponding to the given initial conditions in (2.3.35). This completes the proof of (b) and shows that the model can hold for all $t \geq 0$ *provided that the initial conditions are chosen appropriately*. Note that the main complicating factor in the proof above is the fact that we are required to express $x(0)$ in terms of *future* values of $\{y'(t)\}$. A succinct derivation using Tchirnhausen polynomials is explored in Exercises 2.26 and 2.27.

▼▼▼

The result above is the dual of the corresponding result for right difference operator representations described in (2.3.24) and (2.3.25). The reverse operation from observable state-space model to left difference operator form is straight forward by first converting the state-space model to observer form and then making the appropriate associations between (2.3.39)–(2.3.40) and (2.3.31) and (2.3.33). In fact, we once again see that a left difference operator representation satisfying Assumption 2.3.B(i) is nothing more than an alternative (condensed) way of representing an observable state-space model in observer form. (Further insights are given in Exercise 2.24.)

It is also possible to go directly from other canonical observable models (e.g., the observability form discussed in Section 2.2.3) to corresponding left difference operator representations. It has been shown by Guidorzi (1981) that simple algebraic relationships exist between the coefficients in the left difference form thus obtained and the observability state-space model. However, whereas the coefficients in the observer form appear directly in the corresponding left difference model, this is not true for the observability form.

2.3.4 Deterministic Autoregressive Moving-Average Models

In this section we introduce an alternative model format in which the current output vector is expressed as a linear combination of past outputs, $y(t)$, and past inputs, $u(t)$:

$$A_0 y(t) = -\sum_{j=1}^{n_1} A_j y(t-j) + \sum_{j=0}^{m_1} B_j u(t-j-d); \qquad t \geq 0 \qquad (2.3.42)$$

where A_0 is square and nonsingular, and d represents a time delay. As before, the dimensions of $y(t)$ and $u(t)$ are m and r, respectively.

We shall call the term in the past values of y the *autoregressive* component and the terms in u the *moving-average* component. The full model (2.3.42) will be termed the *deterministic autoregressive moving average* (DARMA) model. [We have introduced the qualifier "deterministic" since in the statistics literature, the term ARMA is used for a model of the form of (2.3.42) in which $\{u(t)\}$ is a white noise process.]

Using q^{-1} as the backward shift operator (as in Section 2.3.3), the model (2.3.42) can be expressed as

$$A(q^{-1})y(t) = B(q^{-1})u(t); \qquad t \geq 0 \qquad (2.3.43)$$

where

$$A(q^{-1}) = A_0 + A_1 q^{-1} + \cdots + A_{n_1} q^{-n_1}; \qquad A_0 \text{ nonsingular} \qquad (2.3.44)$$

$$B(q^{-1}) = (B_0 + B_1 q^{-1} + \cdots + B_{m_1} q^{-m_1})q^{-d} \qquad (2.3.45)$$

Comparing (2.3.43) with (2.3.36)–(2.3.37) we see that a DARMA model is simply a left difference operator representation [satisfying Assumption 2.3.B(i) and (ii)] expressed in terms of q^{-1}. Thus all the comments of Section 2.3.3 apply *mutatis mutandis* to DARMA models. In particular, a DARMA model is equivalent to an observable state-space model with arbitrary initial state. Also, it can describe the *input–output properties* of a general state-space model (which is not necessarily completely observable or completely controllable) having arbitrary initial state.

If the integers k_1 to k_m appearing in the observer state-space model are known, we note from (2.3.36) and (2.3.37) that $n_1 = m_1 = \max\{k_l; l = 1, \ldots, m\}$ and

$$\left.\begin{array}{r} [A_j]_{i.} = 0 \\ [B_j]_{i.} = 0 \end{array}\right\} \quad \text{for } k_i < j \leq n_1; \quad \text{for } i = 1, \ldots, m \qquad (2.3.46)$$

where $[A_j]_{i.}$ denotes the ith row of A_j.

Moreover, from the structure of the observer canonical form, it follows that A_0 will have ones on the main diagonal and depending on the relative magnitude of the $\{k_i\}$ indices will have scattered zeros elsewhere giving a simple structure (e.g., lower triangular). In the special case when the indices are equal, A_0 can be taken to be an identity matrix.

In general, (2.3.42) can be normalized so that $A_0 = I$ by multiplying both sides by A_0^{-1}, but then the structural constraints described in (2.3.46) no longer apply with a corresponding increase in the order and in the number of nonzero parameters.

With $A_0 = I$, the DARMA model can be expressed as

$$y(t) = \theta_0^T \phi(t - 1); \qquad t \geq 0 \tag{2.3.47}$$

where θ_0^T is an $m \times p$ matrix of parameters in $A(q^{-1})$ and $B(q^{-1})$ and $\phi(t)$ is a $p \times 1$ vector containing past values of the output and input vectors. We shall find models of the general form (2.3.47) particularly convenient in our subsequent development.

*2.3.5 Irreducible Difference Operator Representations

We have seen that right difference operator representations are equivalent to controllable state-space models and that left difference operator representations are equivalent to observable state-space models. The next question we wish to address is whether any simplification is possible if we only want to describe the system transfer function (see Section 2.2.4).

Consider the general difference operator representation given in (2.3.1)–(2.3.2); then by taking z-transforms and ignoring initial conditions, the corresponding transfer function is seen to be

$$H(z) = R(z)P(z)^{-1}Q(z) \tag{2.3.48}$$

We can then establish the following important existence result for left difference operator representations (a corresponding result also applies to right difference operator representations).

Lemma 2.3.3. Given a proper transfer function $H(z)$, one can always find a left difference operator representation

$$D_L(q)y(t) = N_L(q)u(t) \tag{2.3.49}$$

such that

$$H(z) = D_L(z)^{-1}N_L(z) \tag{2.3.50}$$

and satisfying

(i) $D_L(z)$ and $N_L(z)$ are relatively left prime.
(ii) $D_L(z)$ is row reduced.
(iii) $\partial r_i[N_L(z)] \leq \partial r_i[D_L(z)]$

where $\partial r_i[N_L(z)]$ denotes the degree of the ith row of $N_L(z)$.

Proof. (i) We can always construct a *right* difference operator representation $[D_R(q), N_R(q)]$ giving rise to the same transfer function by making $D_R(z)$ diagonal, with iith element being the least common denominator of the ith column of $H(z)$, and then to form $N_R(z)$ from the appropriate numerator polynomials. Note that $D_R(z)$ and $N_R(z)$ formed in this way will not, in general, be relatively right prime.

We can then use the dual of Lemma A.3.2 of Appendix A to find a unimodular matrix $U(z)$ such that

$$\begin{bmatrix} U_{11}(z) & U_{12}(z) \\ U_{21}(z) & U_{22}(z) \end{bmatrix} \begin{bmatrix} D_R(z) \\ N_R(z) \end{bmatrix} = \begin{bmatrix} R(z) \\ 0 \end{bmatrix} \tag{2.3.51}$$

and where $R(z)$ is a greatest common right divisor (gcrd) of $D_R(z)$ and $N_R(z)$. Hence

$$U_{21}(z)D_R(z) + U_{22}(z)N_R(z) = 0 \tag{2.3.52}$$

Now since $U(z)$ is unimodular, it follows that $U_{21}(z)$ and $U_{22}(z)$ are relatively left prime (see Exercise 2.9).

To show that $U_{22}(z)$ is nonsingular, we define $V(z)$ as the (polynomial) inverse of $U(z)$. Then applying $V(z)$ to (2.3.51), we obtain

$$\begin{bmatrix} D_R(z) \\ N_R(z) \end{bmatrix} = \begin{bmatrix} V_{11}(z) & V_{12}(z) \\ V_{21}(z) & V_{22}(z) \end{bmatrix} \begin{bmatrix} R(z) \\ 0 \end{bmatrix} \tag{2.3.53}$$

or $D_R(z) = V_{11}(z)R(z)$. Thus since $D_R(z)$ is nonsingular, it follows that $V_{11}(z)$ is nonsingular. Now from the inverse and determinant formulas for block matrices (see Exercise 3.5),

$$V_{11}(z) = [U_{11}(z) - U_{12}(z)U_{22}(z)^{-1}U_{21}(z)]^{-1} \tag{2.3.54}$$

and

$$\det U(z) = \frac{\det U_{22}(z)}{\det V_{11}(z)} \tag{2.3.55}$$

Thus

$$\det U_{22}(z) \neq 0 \tag{2.3.56}$$

Then, from (2.3.52),

$$N_R(z)D_R(z)^{-1} = U_{22}(z)^{-1}U_{21}(z) \tag{2.3.57}$$

where $U_{22}(z)$ and $U_{21}(z)$ are relatively left prime. This establishes (i).

(ii) If $U_{22}(z)$ is row reduced, we are done. If not, from Lemma A.3.4 there exists a unimodular matrix $U'(z)$ such that

$$U'_{22}(z) = U'(z)U_{22}(z) \tag{2.3.58}$$

and $U'_{22}(z)$ is row reduced.

Now define

$$U'_{21}(z) = U'(z)U_{21}(z) \tag{2.3.59}$$

Then

$$U'_{22}(z)^{-1}U'_{21}(z) = U_{22}(z)^{-1}U_{21}(z) \tag{2.3.60}$$

and hence $[U'_{22}(q), U'_{21}(q)]$ is a left difference operator representation satisfying (i) and (ii). (Note that multiplication by a unimodular matrix preserves left comprimeness, as is obvious from the Bezout identity; see Exercise 2.8.)

(iii) With $H(z) = U'_{22}(z)^{-1}U'_{21}(z)$ we have

$$U'_{22}(z)H(z) = U'_{21}(z)$$

and therefore for the ith row we can write

$$[U'_{21}(z)]_{ij} = \sum_{k=1}^{m} [U'_{22}(z)]_{ik}[H(z)]_{kj}$$

But every element of $H(z)$ is proper, so that all elements $[U'_{21}(z)]_{ij}$ must have degree less than or equal to the highest degree polynomial in the ith row of $U'_{22}(z)$.

▼▼▼

A left difference operator representation having property (i) of Lemma 2.3.3 is said to be *irreducible*. We now explore some of the implications of irreducibility.

Lemma 2.3.4. Suppose that $[D_1(q), N_1(q)]$ and $[D_2(q), N_2(q)]$ are two irreducible left difference operator representations of the one transfer function; then

(i) The representations are related by a unimodular matrix: that is,

$$[D_1(z) : N_1(z)] = U(z)[D_2(z) : N_2(z)] \qquad (2.3.61)$$

and as a consequence

(ii) The two representations have the same degree.

Proof. Since the two representations correspond to the same transfer function, we have

$$D_1(z)^{-1}N_1(z) = D_2(z)^{-1}N_2(z) \qquad (2.3.62)$$

From the Bezout identity (Exercise 2.8), there exist polynomial matrices $X(z)$ and $Y(z)$ such that

$$D_1(z)X(z) + N_1(z)Y(z) = I \qquad (2.3.63)$$

Using (2.3.62) to replace $N_1(z)$, we have

$$D_1(z)D_2(z)^{-1}D_2(z)X(z) + D_1(z)D_2(z)^{-1}N_2(z)Y(z) = I$$

or

$$D_1(z)D_2(z)^{-1}[D_2(z)X(z) + N_2(z)Y(z)] = I \qquad (2.3.64)$$

Similarly, there exist polynomial matrices $\bar{X}(z)$ and $\bar{Y}(z)$ such that

$$D_2(z)\bar{X}(z) + N_2(z)\bar{Y}(z) = I$$

Using (2.3.62) yields

$$D_2(z)D_1(z)^{-1}D_1(z)\bar{X}(z) + D_2(z)D_1(z)^{-1}N_1(z)\bar{Y}(z) = I$$

or

$$D_2(z)D_1(z)^{-1}[D_1(z)\bar{X}(z) + N_1(z)\bar{Y}(z)] = I \qquad (2.3.65)$$

We can now conclude from (2.3.64)–(2.3.65) that $D_1(z)D_2(z)^{-1}$ and its inverse are both polynomial matrices and hence are unimodular. Thus result (i) follows from (2.3.62) and (ii) is immediate.

▼▼▼

Finally, we establish the following result, which explains the use of the term "irreducible."

Lemma 2.3.5. An irreducible difference operator representation has the least degree among all representations for the same transfer function.

Proof. Let $[D(z), N(z)]$ be *any* left difference operator representation of a given transfer function. Then we can write

$$D(z) = R(z)D_1(z)$$
$$N(z) = R(z)N_1(z)$$

where $R(z)$ is a greatest common left divisor (gcld) and $D_1(z)$ and $N_1(z)$ are left coprime. Clearly,

$$\text{degree } D(z) \geq \text{degree } D_1(z)$$

with equality holding if and only if $R(z)$ is unimodular, implying the $D(z)$ and $N(z)$ are irreducible.

▼▼▼

The alert reader may have noticed that the unimodular matrix plays the same role for difference operator representations as does the similarity transformation for state-space models. Thus we see from Lemmas 2.3.4 and 2.3.5 that irreducible difference operator representations are duals of minimal state-space models.

In our earlier discussion of the relationship between state-space models and difference operator representations we have been careful to include in the model unobservable modes (in the case of right difference operators) and uncontrollable modes (in the case of left difference operators). The preservation of these modes may be significant to the problem under study. For example, we have seen earlier that deterministic disturbances (such as sine waves, offsets, etc.) can be modeled by an observable state-space model having uncontrollable modes on the unit circle. These modes would be lost if one attempted to use an irreducible difference operator representation (or equivalently, the transfer function or minimal state-space model).

*2.4 MODELS FOR BILINEAR SYSTEMS

Sor far in this chapter we have discussed only linear systems. However, in many cases of interest, the appropriate dynamical model may be nonlinear. In these circumstances, it is often desirable, if possible, to work with the true system description rather than a linearized approximation. We shall see the advantages of this approach later in the book when we discuss certain nonlinear control problems, including wastewater treatment and pH neutralization.

To be specific in our discussion of nonlinear models we shall consider a special class, discrete bilinear systems. In this section, for simplicity, we restrict ourselves to the single-input single-output case.

A single-input single-output discrete bilinear system can be expressed in state-space form as follows:

$$x(t + 1) = Ax(t) + u(t)Nx(t) + bu(t) \tag{2.4.1}$$

$$y(t) = cx(t) \tag{2.4.2}$$

where $x \in \mathbb{R}^n$; $u, y \in \mathbb{R}$, and $A, N, b,$ and c have appropriate dimensions.

We wish to investigate the question of the observability of the state of the system (2.4.1). This is more difficult than for the linear case, because of the direct interaction of u and x via the term uNx. Our treatment here will be brief. Further details may be found in Williamson (1977), Funahashi (1979), and Long, Goodwin, and Teoh (1982).

Definition 2.4.A.   The system (2.4.1)–(2.4.2) is said to be *uniformly observable* if any initial state $x(t_0)$ can be uniquely determined for all input sequences $\{u(t); t = t_0 + i, i = 0, 1, \ldots, n - 1\}$ given an input sequence and the corresponding output sequence $\{y(t); t = t_0 + i, i = 0, 1, \ldots, n - 1\}$.

▼▼▼

In the sequel we shall take $b = 0$ since it is easily established that observability is not affected by the term $bu(t)$.

A characterization of uniform observability is given in

Lemma 2.4.1.   The system (2.4.1)–(2.4.2) is uniformly observable if and only if the matrix $M(t)$ has rank n for all t and for all input sequences $\{u(t)\}$, where

$$M(t) = \begin{bmatrix} c \\ cL(t) \\ \cdot \\ \cdot \\ \cdot \\ cL(t + n - 2) & \cdots & L(t) \end{bmatrix} \tag{2.4.3}$$

$$L(t) = A + u(t)N \tag{2.4.4}$$

Proof. We note that, with $b = 0$,

$$[y(t) \quad \cdots \quad y(t + n - 1)]^T = M(t)x(t) \tag{2.4.5}$$

This equation is uniquely solvable for $x(t)$ if and only if rank $M(t) = n$.

▼▼▼

It can be shown that for a uniformly observable system, det $M(t)$ will be a constant for $n = 1, 2$ but that for $n \geq 3$ it may depend on $\{u(t)\}$ [for example, via a quadratic function which is guaranteed nonzero for all $\{u(t)\}$]. We therefore strengthen Definition 2.4.A to the following:

Definition 2.4.B.   The system (2.4.1)-(2.4.2) is *strongly uniformly observable* if det $M(t)$ is a (nonzero) constant independent of $\{u(t)\}$.

Returning to the notion of uniform observability, we have the following immediate results.

Lemma 2.4.2.   The property of uniform observability is invariant under any one-to-one transformation of the state, including similarity transformations.

Lemma 2.4.3.   If the system (2.4.1)–(2.4.2) is uniformly observable, there exists a linear transformation $\bar{x} = P^{-1}x$, P nonsingular, giving

$$\bar{x}(t+1) = P^{-1}AP\bar{x}(t) + u(t)P^{-1}NP\bar{x}(t) = \bar{A}\bar{x}(t) + u(t)\bar{N}\bar{x}(t) \qquad (2.4.6)$$

$$y(t) = cP\bar{x}(t) = \bar{c}\bar{x}(t) \qquad (2.4.7)$$

where $\bar{A}$ and $\bar{c}$ have the following structure:

$$\bar{A} = \begin{bmatrix} a_1 & 1 & & \\ \vdots & & \ddots & \\ \vdots & & & \ddots \\ & & & 1 \\ a_n & & & 0 \end{bmatrix}, \qquad \bar{c} = [1 \quad \cdots \quad 0 \quad 0] \qquad (2.4.8)$$

Proof. Note that if (2.4.1)–(2.4.2) is uniformly observable, it is observable for $u(t) = 0$, for all t. The form (2.4.8) then follows from the conversion to observer form in the linear case. (See Section 2.2.3.)

▼▼▼

Definition 2.4.C. We shall say that (2.4.6) to (2.4.8) are in the *first standard form*.

We now investigate the form of $\bar{N}$. We shall be particularly interested in cases when $\bar{N}$ is lower left triangular, since in many cases of interest this is without loss of generality. We therefore introduce the following definition:

Definition 2.4.D. We shall say that (2.4.6) to (2.4.8) are in the *second standard form* provided that $\bar{N}$ is lower left traingular.

We shall investigate the conditions under which a system can be modeled in the second standard form. An interesting fact is that this is always possible for uniformly observable continuous time bilinear systems [see Williamson (1977)]. However, Funahashi (1979) has shown that, in the discrete-time case, uniform observability ensures that the system may be put into the second standard form when the state dimension is 1 or 2, but this may not necessarily hold for state dimension greater than or equal to 3.

We shall call systems that can be described in the second standard form *strictly sequentially observable* since it turns out that with such systems the initial state can be determined one component at a time as successive values of $\{u(t)\}$ and $\{y(t)\}$ become available.

We have the following simple consequence of Definitions 2.4.B and 2.4.D.

Lemma 2.4.4. Strict sequential observability implies strong uniform observability. The converse is true for $n = 1$ and 2 but may not be true for $n \geq 3$.

Proof. It can be readily verified that $\det M = $ constant for a system in the second standard form. A counterexample to the converse for $n = 3$ has been given in Funahashi (1979).

▼▼▼

We now turn to the relationship between observable bilinear state-space models and input–output models. We have seen in Sections 2.3.3 and 2.3.4 that in the linear case, there is a one-to-one relationship between observable state-space models and left difference operator representations, (or equivalently, DARMA input–output models). The corresponding results for bilinear systems are summarized in the following theorem [we consider the general case of (2.4.1) where $b \neq 0$].

Theorem 2.4.1

(a) The external behavior of any uniformly observable bilinear system can be described by a nonlinear input–output model of the form

$$y(t+1) = \sum_{i=1}^{n} \beta_i y(t+1-i) + \beta_0 \qquad (2.4.9)$$

where

(i) Each β_i $(i = 0, \ldots, n)$ is a nonlinear function of $u(t), u(t-1), \ldots, u(t-n+1)$.

(ii) $y(t+1)$ is a linear function of $u(t)$.

(b) In the case of a strongly uniformly observable bilinear system, the input–output model (2.4.9) still applies, but in addition each β_i is a polynomial in $u(t) \cdots u(t-n+1)$.

Proof. (a) By successive substitution we can write

$$\begin{bmatrix} y(t) \\ y(t+1) \\ y(t+2) \\ \cdot \\ \cdot \\ \cdot \\ y(t+n-1) \end{bmatrix} = M(t)x(t) + \begin{bmatrix} 0 \\ cbu(t) \\ cL(t+1)bu(t) + cbu(t+1) \\ \cdot \\ \cdot \\ \cdot \\ \end{bmatrix} \qquad (2.4.10)$$

$$y(t+n) = cL(t+n-1) \cdots L(t)x(t) + \beta_0' \qquad (2.4.11)$$

where β_0' is a polynomial function of $u(t) \cdots u(t+n-1)$ which is linear in $u(t+n-1)$.

Uniform observability implies that $M(t)$ has rank n and thus $x(t)$ can be eliminated from (2.4.11) using (2.4.10). This gives an nth order input–output model in the form of (2.4.9).

Conditions (i) and (ii) follow from the construction.

(b) For a strongly uniformly observable system det $M(t) = k$ [independent of $u(t) \cdots$]. Hence

$$x(t) = \frac{\text{Adj}\,(M)}{k} \left\{ \begin{bmatrix} y(t) \\ y(t+1) \\ y(t+2) \\ \cdot \\ \cdot \\ \cdot \\ y(t+n-1) \end{bmatrix} - \begin{bmatrix} 0 \\ cbu(t) \\ cL(t+1)bu(t) + cbu(t+1) \\ \cdot \\ \cdot \\ \cdot \\ \end{bmatrix} \right\} \qquad (2.4.12)$$

It therefore follows that $x(t)$ is a polynomial function of $u(t) \cdots u(t+n-2)$. Substituting this into (2.4.11) yields the desired result.

▼▼▼

A key point about the form of (2.4.9) is that it can be written compactly as

$$y(t + 1) = \phi(t)^T \theta_0 \qquad (2.4.13)$$

where $\phi(t)$ is a vector whose components are functions of the input and output sequences $\{u(t)\}$ and $\{y(t)\}$. This property is also true of the linear DARMA models described in Section 2.3.4.

The simple structure of (2.4.13) will be used to advantage in Chapter 3 when we study parameter estimation. In fact, the key feature is that the model has been expressed as a linear function of a parameter vector, and this will allow us to treat parameter estimation for both linear and nonlinear dynamical systems in a unified fashion.

This completes our rather brief treatment of nonlinear models. One of the objectives in presenting these results has been to show that many of the ideas of linear systems theory (such as observability and controllability) carry over to the nonlinear case but that certain precautions are necessary. In practice it is often preferable to formulate these models with a specific application in mind and thus we will not attempt to present further general theory. Later in the book we shall consider a number of specific cases (see also Exercise 2.15).

EXERCISES

2.1. Consider a simple servomechanism under digital control as shown in Fig. 2.A. The continuous time transfer function of the servo is

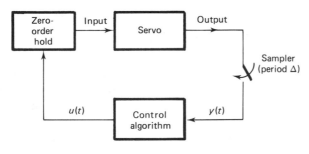

Figure 2.A Servomechanism under digital control.

$$G(s) = \frac{K}{s(Ts + 1)}$$

where s is the continuous time Laplace variable. Show that the sampled servo output $y(t)$ is related to the sampled input $u(t)$ by the following discrete-time transfer function:

$$H(z) = \frac{K_d(z - \beta)}{(z - 1)(z - \alpha)}$$

where

$$K_d = K(T\alpha - T + \Delta)$$

$$\beta = \frac{T\alpha + \alpha\Delta - T}{T\alpha - T + \Delta}$$

$$\alpha = e^{-\Delta/T}$$

2.2. Show that a linear system having a nilpotent A-matrix (i.e., $A^k = 0$ for some k) is always controllable (to the origin) whatever the rank of the controllability matrix.

2.3. Show that a necessary and sufficient condition for controllability (to the origin) of the system (2.2.1) is that $A^n x(0)$ is in the range space of the controllability matrix.

2.4. Prove Lemma A.1.1 of Appendix A. (*Hint:* Use the Cayley–Hamilton theorem.) See also Kailath (1980, p. 95).

2.5. Consider the two-sided z-transform

$$\bar{Z}\{u(i)\} \triangleq \bar{U}(z) = \frac{a}{a - z}$$

Show that

(a) If the region of convergence is $|z| > a$, then

$$u(i) = \begin{cases} -a^i, & i \geq 1 \\ 0, & \text{otherwise} \end{cases}$$

(b) If the region of convergence is $|z| < a$, then

$$u(i) = \begin{cases} a^i, & i \leq 0 \\ 0, & \text{otherwise} \end{cases}$$

2.6. Prove Lemma A.2.2.

2.7. Determine the two-sided z-transform of the following sequence and specify the region of convergence.

$$a_k = \begin{cases} (\frac{1}{8})^k & \text{for } k \geq 0 \\ 3^k & \text{for } k < 0. \end{cases}$$

2.8. (The Bezout Identity)   Show that the polynomial matrices $N(z)$ and $D(z)$ having the same number of columns will be right coprime if and only if there exist polynomial matrices $X(z)$ and $Y(z)$ such that

$$X(z)N(z) + Y(z)D(z) = I$$

inverse

2.9. Show that if $U(z)$ is unimodular, and is partitioned as $U(z) \; \overset{\vee}{V}(z) = I$
then use Bezout

$$\begin{bmatrix} U_{11}(z) & U_{12}(z) \\ U_{21}(z) & U_{22}(z) \end{bmatrix}$$

then $U_{21}(z)$ and $U_{22}(z)$ are left coprime. [*Hint:* Use the Bezout identity and the fact that $U(z)$ has a polynomial inverse.]

2.10. Consider a simple system with disturbances as in Fig. 2.2.5 with

$$\eta(t + 1) = \eta(t); \qquad t \geq 0 \qquad \text{(disturbance model)}$$
$$x(t + 1) = u(t); \qquad t \geq 0 \qquad \text{(system model)}$$
$$y(t) = x(t) + \eta(t) \qquad\qquad \text{(output)}$$

Express the system in state-space form and then use the dual of the transformation (2.2.13)–(2.2.14) to express the model in observer form. Now use the reverse of the procedure described in Lemma 2.3.2 to show that the corresponding DARMA model has the form

$$y(t) = y(t - 1) + u(t - 1) - u(t - 2); \qquad t \geq 0$$

with corresponding left difference operator representation

$$(q^2 - q)y(t) = (q - 1)u(t); \qquad t \geq 0$$

2.11. Show that the z-transform of the output, $\{y(t)\}$, satisfying the left difference operator given in Exercise 2.10 is

$$Y(z) = \frac{1}{z}U(z) + y(0) + \frac{y(1)}{z-1} - \frac{u(0)}{z-1}$$

Hence show that the solution can be written as

$$y(t) = u(t-1)S(t-1) + y(0)\delta(t) + y(1)S(t-1) - u(0)S(t-1); \qquad t \geq 0$$

where

$$\delta(j) = \begin{cases} 1 & j = 0 \\ 0 & j \neq 0 \end{cases}$$

$$S(j) = \begin{cases} 1 & j \geq 0 \\ 0 & j < 0 \end{cases}$$

(*Hint:* Everything that is needed for this question is contained in Section A.2.)

2.12. Multiplying the above left difference operator representation by q^{-2}, we obtain

$$y(t) - q^{-1}y(t) = q^{-1}u(t) - q^{-2}u(t); \qquad t \geq 2$$

Show that this equation can be written in terms of sequences as follows:

$$\{y(t)\} - \{y(0), y(1), 0, \ldots\} - q^{-1}\{y(t)\} + \{0, y(0), 0, \ldots\}$$
$$= q^{-1}\{u(t)\} - \{0, u(0), 0, \ldots\} - q^{-2}\{u(t)\}$$

Hence solve the equation using result (A.2.14) of Appendix and compare your answer with that obtained in Exercise 2.11.

2.13. Consider a second-order discrete-time bilinear system expressed in the first standard form as

$$\tilde{x}(t+1) = \begin{bmatrix} a_1 & 1 \\ a_2 & 0 \end{bmatrix} a(t) + u(t)\begin{bmatrix} n_{11} & n_{12} \\ n_{21} & n_{22} \end{bmatrix} \tilde{x}(t)$$

$$y(t) = [1 \qquad 0]\tilde{x}(t)$$

By using Lemma 2.4.1 show that the system is uniformly observable if and only if $n_{21} = 0$, that is, if and only if the system may be put into the second standard form.

2.14. Construct an example of a discrete-time bilinear system that is uniformly observable but which cannot be expressed in the second standard form [see Funahashi (1979)].

2.15. Consider the heat exchanger in Fig. 2.B. Assuming no heat loss from the walls and perfect mixing, show that the output $y(t)$ is modeled by

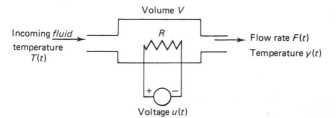

Figure 2.B Heat exchanger.

$$\frac{dy}{dt} = -\frac{1}{V}F(t)y(t) + \frac{1}{V}F(t)T(t) + \frac{1}{RCV}u(t)^2$$

where c is the specific heat of the fluid. Hence by using a first-order Euler expansion to the differential equation, construct a nonlinear DARMA model for the system. Show how a higher-order approximation to the solution (e.g., fourth-order Runge–Kutta) leads to a different nonlinear DARMA model.

2.16. Prove Lemma A.1.2. [*Hint:* Let $p_1, \ldots, p_k$ be a basis for the range space of $\mathcal{C}$; then define $P = [p_1 \quad \cdots \quad p_k, T]$, where $T = [P_{k+1} \cdots P_n]$ is arbitrary as long as P is nonsingular. Define $Q = P^{-1}$ and apply the transformation $\bar{x}(t) = Qx(t)$.]

2.17. Prove Lemma A.1.3. (*Hint:* Let $p_1, \ldots, p_n$ be a basis for the null space of Θ.)

2.18. Give a discrete-time second-order system which is completely controllable but which is not completely reachable.

2.19. Consider an nth-order linear continuous-time system of the form

$$\frac{d}{dt} x(t) = Fx(t) + Gu(t)$$

$$y(t) = Hx(t)$$

Assume that the input is of zero-order sample–hold type, that is,

$$u(\tau) = u(k\Delta) \qquad \text{for } k\Delta \leq \tau < (k + 1)\Delta$$

Write down an explicit equation for the corresponding discrete state-space model.

2.20. Using the results of Exercise 2.19, show that the A-matrix in a discrete state-space model derived from a continuous-time model is always nonsingular. What does this imply about reachability and controllability?

2.21. Consider the servo system described in Exercise 2.1. Write down the DARMA model relating $\{y(k\Delta)\}$ to $\{u(k\Delta)\}$, where Δ is the sampling period and the input is of sample–hold type. (*Hint:* The DARMA model can be obtained directly from the transfer function found in Exercise 2.1.)

2.22. Consider the following discrete-time system:

$$y(t) = bu(t - 1) + G \sin(pt + \phi); \qquad t = 0, 1, 2, \ldots$$

Convince yourself that an appropriate state-space model is

$$\begin{bmatrix} x_1(t + 1) \\ x_2(t + 1) \\ x_3(t + 1) \end{bmatrix} = \begin{bmatrix} 2\cos p & -1 & 0 \\ 1 & 0 & 0 \\ 0 & 0 & 0 \end{bmatrix} \begin{bmatrix} x_1(t) \\ x_2(t) \\ x_3(t) \end{bmatrix} + \begin{bmatrix} 0 \\ 0 \\ 1 \end{bmatrix} u(t)$$

$$y(t) = \begin{bmatrix} 1 & 0 & 1 \end{bmatrix} \begin{bmatrix} x_1(t) \\ x_2(t) \\ x_3(t) \end{bmatrix}$$

(a) Is this system completely controllable? Why?

(b) Is this system completely observable? Why?

(c) Transform the system to observability form.

(d) Transform the system to observer form.

(e) Determine a DARMA model for the system (find explicit expressions for a_1, a_2, b_0, b_1, and b_2).

(f) What is the transfer function of the system?

(g) Determine a minimal state-space model for the system.

(h) Under what circumstances does the minimal model give the same output as the original system?

2.23. Explain why a sinusoidal disturbance of the type discussed in Exercise 2.22 always gives rise to a DARMA model $A(q^{-1})y(t) = B(q^{-1})u(t)$ in which $A(q^{-1})$, $B(q^{-1})$ have common roots on the unit circle.

2.24. Investigate further the relationship between observer form state-space models [see (2.2.22)] and left difference operator representations by carrying out the following steps:

(a) Show that the state-space model can immediately be written as a general difference operator representation as follows:

$$D(q)x(t) = Nu(t)$$
$$y(t) = Rx(t)$$

where $D(q)$ and R have the following structure (illustrated for $m = 2$):

$$D(q) = \begin{bmatrix} q + a_1^{11} & -1 & & & \Big| & a_1^{12} & & & \\ \cdot & q & \cdot & & \Big| & \cdot & & & \\ \cdot & & \cdot & \cdot & \Big| & \cdot & & & \\ \cdot & & & \cdot & -1 & \Big| & \cdot & & \\ a_{k_1}^{11} & & & q & \Big| & a_{k_1}^{12} & & & \\ \hline a_1^{21} & & & & \Big| & q + a_1^{22} & -1 & & \\ \cdot & & & & \Big| & \cdot & q & \cdot & \\ \cdot & & & & \Big| & \cdot & & \cdot & \cdot \\ \cdot & & & & \Big| & \cdot & & & -1 \\ a_{k_2}^{21} & & & & \Big| & a_{k_2}^{22} & & & q \end{bmatrix}$$

$$R = \begin{bmatrix} r_{11} & & & & \Big| & r_{12} & & & \\ r_{21} & & & & \Big| & r_{22} & & & \end{bmatrix}$$

(b) Now consider the following unimodular matrix:

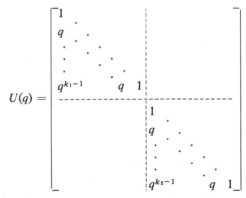

$$U(q) = \begin{bmatrix} 1 & & & & & \Big| & & & & \\ q & \cdot & & & & \Big| & & & & \\ \cdot & \cdot & & & & \Big| & & & & \\ \cdot & & \cdot & & & \Big| & & & & \\ \cdot & & & \cdot & & \Big| & & & & \\ q^{k_1-1} & & & q & 1 & \Big| & & & & \\ \hline & & & & & \Big| & 1 & \cdot & & \\ & & & & & \Big| & q & \cdot & & \\ & & & & & \Big| & \cdot & \cdot & & \\ & & & & & \Big| & \cdot & & \cdot & \\ & & & & & \Big| & q^{k_2-1} & & q & 1 \end{bmatrix}$$

By multiplication on the left by the unimodular matrix above, show that the difference operator representation found in part (a) is transformed into an equivalent form where

$$D'(q) = U(q)D(q) = \begin{bmatrix} p_{k_1-1}^{11}(q) & -1 & & & \vdots & p_{k_1-1}^{12}(q) & & & \\ \cdot & & \cdot & & \vdots & \cdot & & & \\ \cdot & & & \cdot & \vdots & \cdot & & & \\ \cdot & & & & \vdots & \cdot & & & \\ p_1^{11}(q) & & & -1 & \vdots & \cdot & & & \\ p_0^{11}(q) & & & & \vdots & p_0^{12}(q) & & & \\ \hdashline p_{k_2-1}^{21}(q) & & & & \vdots & p_{k_2-1}^{22}(q) & -1 & & \\ \cdot & & & & \vdots & \cdot & & \cdot & \\ \cdot & & & & \vdots & \cdot & & & \cdot \\ \cdot & & & & \vdots & p_1^{22}(q) & & & -1 \\ p_0^{21}(q) & & & & \vdots & p_0^{22}(q) & & & \end{bmatrix}$$

Give explicit expressions for all polynomials in $D'(q)$ in terms of the entries in $D(q)$. Also show the form of the matrix multiplying $u(t)$ and give explicit expressions for its entries in terms of the entries in N.

(c) By reading off the k_1th and $(k_1 + k_2)$th rows of the model obtained in part (b), show that one obtains the corresponding left difference operator representation for $y(t)$ having the form

$$\begin{bmatrix} p^{11}(q) & p^{12}(q) \\ p^{21}(q) & p^{22}(q) \end{bmatrix} D_0 y(t) = \begin{bmatrix} b^{11}(q) & b^{12}(q) \\ b^{21}(q) & b^{22}(q) \end{bmatrix} u(t)$$

where

$$D_0 = \begin{bmatrix} r_{11} & r_{12} \\ r_{21} & r_{22} \end{bmatrix}^{-1}$$

(d) Use the development above to discuss the statement that a left difference operator representation is a compact way of writing an observer state-space model.

2.25. Consider the right difference operator representation (2.3.7), (2.3.8) where

$$D_R(q) = a_0 q^n + a_1 q^{n-1} + \cdots + a_n; \qquad a_0 = 1$$
$$N_R(q) = b_1 q^{n-1} + \cdots + b_n$$

Define a state vector $x(t)$ as

$$x(t) = \begin{bmatrix} p_D^{(1)}(q) \\ p_D^{(n)}(q) \end{bmatrix} z(t)$$

where $p_D^{(1)}(q), \ldots, p_D^{(n)}(q)$ are the Tchirnhausen polynomials for $D_R(q)$ defined as follows:

$$p_D^{(i)}(q) = \sum_{k=0}^{n-i} a_k q^{n-k-i}$$

Show that the state space model for $x(t)$ is in the Controllability form.

2.26. Consider the left difference operator representation (2.3.26) where

$$D_L(q) = a_0 q^n + a_1 q^{n-1} + \cdots + a_n; \qquad a_0 = 1$$
$$N_L(q) = b_1 q^{n-1} + \cdots + b_n$$

Define a state vector $x(t)$ as

$$x(t) = \begin{bmatrix} p_D^{(n)}(q) \\ \cdot \\ \cdot \\ \cdot \\ p_D^{(1)}(q) \end{bmatrix} y(t) - \begin{bmatrix} p_N^{(n)}(q) \\ \\ \\ p_N^{(1)}(q) \end{bmatrix} u(t)$$

where $p_D^{(i)}(q)$, $p_N^{(i)}(q)$ are the Tchirnhausen polynomials for $D_L(q)$ and $N_L(q)$ respectively (see Exercise 2.25).

Show that the state space model for $x(t)$ is in the Observer form.

2.27. Extend Exercise 2.26 to the multi-input multi-output case.

Models for Deterministic Dynamical Systems   Chap. 2

3

Parameter Estimation for Deterministic Systems

3.1 INTRODUCTION

In Chapter 2 we discussed various types of model structure that could be used to describe the dynamical behavior of deterministic systems. Although the main emphasis was on linear systems, mention was made of the nonlinear case as well. It has been seen that the precise nature of the model response is determined by the values of certain parameters. In some cases these parameter values may be obtained from the laws of physics, chemistry, and so on. In other cases, however, it might not be feasible to determine the precise values of the system parameters in this way. In these instances it may be possible to deduce the values of the parameters by observing the nature of the system's response under appropriate experimental conditions. We shall call this procedure *parameter estimation*. In this chapter we explore some of the ramifications of parameter estimation for deterministic dynamic systems.

As was mentioned in Chapter 1, the principle of parameter estimation is central to the processes of adaptive filtering, prediction, and control, and as such, forms a key concept within the theme of this book. The essential ingredients of a parameter estimation problem are:

1. Class of model
2. Criteria of best fit
3. Experimental conditions
4. Estimation algorithms
5. Use of a priori knowledge

We briefly discuss each of these below.

Class of model.   The problem here is to select an appropriate class of model structures. Often this entails choosing between two or more competing structures on the basis of experimental evidence. It is frequently the case that complete description of the physical system is either not feasible, or is impractical, and thus a compromise has to be made between the model complexity and its adequacy for a given application.

Criteria of best fit.   Having chosen an appropriate class of model, we must now determine the criteria by which the relative performance of different models within the class will be judged. The obvious criterion to use is one that can measure how well the final objective is achieved. For example, in a control problem, different models could be judged on the basis of how well the controller developed from each model meets the design specifications.

The main difficulty associated with this approach is that it may be necessary to implement fully each and every design (developed from its own model) before a decision can be made on what is the best model. Consequently, it is generally more practicable to decide between models on the basis of simpler indirect criteria, such as minimization of mean square one-step-ahead prediction error, minimization of maximum prediction error, and so on. These criteria can be used to form a decision hierarchy involving the choice of model structure (e.g., linear, bilinear, nonlinear, polynomial, etc.), the dimension of the chosen model structure (e.g., the number of parameters in a linear model), and the estimation of the unknown parameters.

Experimental conditions.   Clearly, the relative performance of different models will depend, to some extent, on the conditions under which they operate. Thus the choice of experimental conditions for parameter estimation is of considerable importance. It is clear that the best experimental conditions are those that appertain to the final application of the model. This may occur naturally in some cases; for example, in adaptive control the model is adjusted under normal operating conditions. In other cases it may be necessary to perform separate experiments which are divorced from the final application environment. Either way, there could be advantages in contriving an artificial experiment which subjects the system to a rich and informative set of different conditions in the shortest possible time. Further discussion on the topic of experimental design may be found in Goodwin and Payne (1977) and Ljung and Söderström (1982).

Estimation algorithms.   We may distinguish two main classes of algorithms: on-line and off-line. In the *off-line case*, it is presumed that all the data are available prior to analysis. Consequently, the data may be treated as a complete block of information, with no strict time limit on the process of analysis.

In contrast to the off-line case, the *on-line case* deals with sequential data, which requires that the parameter estimates be recursively updated within the time limit imposed by the sampling period. Thus, in many applications, it is necessary to use a relatively simple algorithm to meet the imposed time constraint.

The performance of different estimation algorithms may be compared under various criteria. For example, if we assume that the chosen model structure corresponds to the true system structure, we can ask such questions as: Do the outputs of the model converge to the outputs of the system? Do the estimated parameters converge to the "true" system parameters? How fast does the algorithm converge (if at all)? And: How robust is the algorithm to various sources of errors, for example, noise, unmodeled dynamics, and numerical precision?

Use of a priori knowledge. The final factor to be considered is that of making use of one's prior knowledge of the system. It is usually advantageous to incorporate as much prior knowledge as possible into the estimation algorithm. Typically, this information might include structural constraints, parameter values, feasible ranges of parameter values, and so on.

In this chapter we are concerned primarily with on-line algorithms for the estimation of parameters within a given model structure. In particular, we emphasize prediction error methods because of their direct applicability to adaptive filtering, prediction, and control. However, the reader should bear in mind the general comments made above so as to view these algorithms within their proper perspective.

In the next section we introduce a particular class of on-line parameter estimation algorithms which are particularly attractive because of their simplicity.

3.2 ON-LINE ESTIMATION SCHEMES

As we mentioned in the preceding section, on-line estimation schemes produce an updated parameter estimate within the time span between successive samples. Thus it is highly desirable that the algorithm be simple and easy to implement.

One particular class of on-line algorithms that is attractive in practice is where the current parameter estimate, $\hat{\theta}(t)$, is computed in terms of the previous estimate, $\hat{\theta}(t-1)$, as follows:

$$\hat{\theta}(t) = f(\hat{\theta}(t-1), D(t), t) \tag{3.2.1}$$

where $D(t)$ denotes data available at time t, and $f(\cdot, \cdot, \cdot)$ denotes an algebraic function, the form of which determines the specific algorithm. In the case of dynamical systems, the data, $D(t)$, normally take the form of present and past observations of the system outputs and inputs, which we denote by

$$\mathcal{Y}(t) \triangleq \{y(t), y(t-1), \ldots\} \quad \text{and} \quad \mathcal{U}(t) \triangleq \{u(t), u(t-1), \ldots\}$$

respectively.

A special form of (3.2.1) that is widely used in practice is as follows:

$$\hat{\theta}(t) = \hat{\theta}(t-1) + M(t-1)\bar{\phi}(t-d)\bar{e}(t) \tag{3.2.2}$$

where $\hat{\theta}(t)$ denotes the parameter estimate at time t
    $M(t-1)$ denotes an algorithm gain (possibly a matrix)
    $\bar{\phi}(t-d)$ denotes a regression vector of some kind composed of selected elements from $\mathcal{Y}(t-d)$, $\mathcal{U}(t-d)$; d is an integer
    $\bar{e}(t)$ denotes a modeling error of some kind [e.g., the model prediction error arising from the use of $\hat{\theta}(t-1)$]

Although (3.2.2) appears very simple, the algorithm can in fact be motivated by many different objective functions. Also, depending on the precise meanings of $\hat{\theta}(t)$,

$M(t)$, $\bar{\phi}(t)$, and $\bar{e}(t)$, the algorithm can take many different forms. We explore this in detail below.

In much of the discussion in this chapter we restrict our attention to single-input single-output systems for pedagogical reasons. However, the extension to multiple-output systems is straightforward, as well shall see in Section 3.8.

3.3 EQUATION ERROR METHODS FOR DETERMINISTIC SYSTEMS

We have seen in Chapter 2 that the input–output characteristics of a wide class of linear and nonlinear deterministic dynamical systems can be described by a model that may be expressed succinctly in the following simple form:

$$y(t) = \phi(t-1)^T \theta_0 \tag{3.3.1}$$

where $y(t)$ denotes the (scalar) system output at time t

$\phi(t-1)$ denotes a vector that is a linear or nonlinear function of

$$\mathcal{Y}(t-1) = \{y(t-1), y(t-2), \ldots\}$$
$$\mathcal{U}(t-1) = \{u(t-1), u(t-2), \ldots\}$$

θ_0 denotes a parameter vector (unknown)

A simple example is given below.

Example 3.3.1 (First-Order DARMA Model)

Consider the situation in which $y(t)$ is a linear function of $y(t-1)$ and a known input $u(t-1)$. Then the corresponding first-order DARMA model (see Section 2.3.4) is

$$y(t) = -ay(t-1) + bu(t-1) \tag{3.3.2}$$

Equation (3.3.2) can be expressed in the form of (3.3.1), where

$$\phi(t-1)^T = [-y(t-1), u(t-1)] \tag{3.3.3}$$
$$\theta_0^T = [a, b] \tag{3.3.4}$$

▼▼▼

Based on the model (3.3.1), which is linear in the parameters, we can now introduce our first on-line parameter estimation scheme and study it in some detail. We shall call this algorithm the projection algorithm:

Projection Algorithm

$$\hat{\theta}(t) = \hat{\theta}(t-1) + \frac{\phi(t-1)}{\phi(t-1)^T\phi(t-1)}[y(t) - \phi(t-1)^T\hat{\theta}(t-1)] \tag{3.3.5}$$

with the initial estimate, $\hat{\theta}(0)$, given. This is of the form of (3.2.2), with

$$M(t-1) = \frac{1}{\phi(t-1)^T\phi(t-1)}; \quad \bar{\phi}(t) = \phi(t); \quad d = 1 \tag{3.3.6}$$
$$\bar{e}(t) = y(t) - \phi(t-1)^T\hat{\theta}(t-1) \tag{3.3.7}$$

The algorithm above can be motivated geometrically as follows. Given $y(t)$ and $\phi(t-1)$, all possible values of θ_0 satisfying the model (3.3.1) lie on the hypersurface

$$H = \{\theta : y(t) = \phi(t-1)^T\theta\} \qquad (3.3.8)$$

Among all these candidate values for θ_0, we choose the one denoted by $\hat{\theta}(t)$, which is closest to $\hat{\theta}(t-1)$. Hence the criterion, J, given below, is minimized:

$$J = \tfrac{1}{2}\|\hat{\theta}(t) - \hat{\theta}(t-1)\|^2 \qquad (3.3.9)$$

This gives the algorithm (3.3.5), as shown below.

Lemma 3.3.1. The projection algorithm results from the following optimization problem. Given $\hat{\theta}(t-1)$ and $y(t)$, determine $\hat{\theta}(t)$ so that

$$J = \tfrac{1}{2}\|\hat{\theta}(t) - \hat{\theta}(t-1)\|^2 \qquad (3.3.10)$$

is minimized subject to

$$y(t) = \phi(t-1)^T\hat{\theta}(t) \qquad (3.3.11)$$

Proof. Introducing a Lagrange multiplier for the constraint (3.3.11), we have

$$J_c = \tfrac{1}{2}\|\hat{\theta}(t) - \hat{\theta}(t-1)\|^2 + \lambda[y(t) - \phi(t-1)^T\hat{\theta}(t)] \qquad (3.3.12)$$

Hence the necessary conditions for a minimum are

$$\frac{\partial J_c}{\partial \hat{\theta}(t)} = 0 \qquad (3.3.13)$$

$$\frac{\partial J_c}{\partial \lambda} = 0 \qquad (3.3.14)$$

These equations become

$$\hat{\theta}(t) - \hat{\theta}(t-1) - \lambda\phi(t-1) = 0 \qquad (3.3.15)$$

$$y(t) - \phi(t-1)^T\hat{\theta}(t) = 0 \qquad (3.3.16)$$

Substituting (3.3.15) into (3.3.16) gives

$$y(t) - \phi(t-1)^T[\hat{\theta}(t-1) + \lambda\phi(t-1)] = 0 \qquad (3.3.17)$$

or

$$\lambda = \frac{y(t) - \phi(t-1)^T\hat{\theta}(t-1)}{\phi(t-1)^T\phi(t-1)} \qquad (3.3.18)$$

Substituting (3.3.18) into (3.3.15) gives (3.3.5).

▼▼▼

The foregoing geometric interpretation of the algorithm is illustrated for a second-order case in Fig. 3.3.1.

A potential problem with the basic algorithm (3.3.5) is that there is the (remote) possibility of division by zero. This can be easily overcome by adding a 1 to the regression vector (see Exercise 3.21). This has the added advantage that the model now applies to systems of the form

$$A(q^{-1})y(t) = B(q^{-1})u(t) + k$$

where k is a constant. Thus arbitrary offsets between input and output can be accommodated.

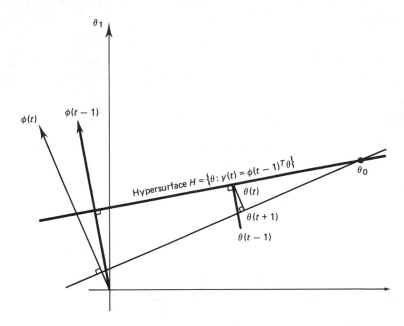

Figure 3.3.1 Geometric interpretation of the projection algorithm.

An alternative scheme for avoiding division by zero is to add a small constant, c, to the denominator of the algorithm. This leads to the following slightly modified form of the algorithm:

Projection Algorithm

$$\hat{\theta}(t) = \hat{\theta}(t-1) + \frac{a\phi(t-1)}{c + \phi(t-1)^T\phi(t-1)}[y(t) - \phi(t-1)^T\hat{\theta}(t-1)] \qquad (3.3.19)$$

with $\hat{\theta}(0)$ given and $c > 0$; $0 < a < 2$.

This algorithm is also known as the normalized least-mean-squares (NLMS) algorithm in some of the filtering literature (where the choice of a is usually such that $0 < a \ll 1$).

We introduce the following notation:

$$\tilde{\theta}(t) \triangleq \hat{\theta}(t) - \theta_0 \qquad (3.3.20)$$

$$e(t) \triangleq y(t) - \phi(t-1)^T\hat{\theta}(t-1)$$
$$= -\phi(t-1)^T\tilde{\theta}(t-1) \qquad (3.3.21)$$

Elementary properties of the projection algorithm are summarized in

Lemma 3.3.2. For the algorithm (3.3.19) and subject to (3.3.1), it follows that

(i) $\|\hat{\theta}(t) - \theta_0\| \le \|\hat{\theta}(t-1) - \theta_0\| \le \|\hat{\theta}(0) - \theta_0\|$; $\quad t \ge 1$ $\qquad (3.3.22)$

(ii) $\lim\limits_{N \to \infty} \sum\limits_{t=1}^{N} \dfrac{e(t)^2}{c + \phi(t-1)^T\phi(t-1)} < \infty$ $\qquad (3.2.23)$

and this implies

(a) $\lim_{t \to \infty} \dfrac{e(t)}{[c + \phi(t - 1)^T \phi(t - 1)]^{1/2}} = 0$ (3.3.24)

(b) $\lim_{N \to \infty} \sum_{t=1}^{N} \dfrac{\phi(t - 1)^T \phi(t - 1) e(t)^2}{[c + \phi(t - 1)^T \phi(t - 1)]^2} < \infty$ (3.3.25)

(c) $\lim_{N \to \infty} \sum_{t=1}^{N} \| \hat{\theta}(t) - \hat{\theta}(t - 1) \|^2 < \infty$ (3.3.26)

(d) $\lim_{N \to \infty} \sum_{t=k}^{N} \| \hat{\theta}(t) - \hat{\theta}(t - k) \|^2 < \infty$ (3.3.27)

(e) $\lim_{t \to \infty} \| \hat{\theta}(t) - \hat{\theta}(t - k) \| = 0$ (3.3.28)

for any finite k.

Proof. (i) Subtracting θ_0 from both sides of (3.3.19) and using (3.3.1) and (3.3.20), we obtain

$$\tilde{\theta}(t) = \tilde{\theta}(t - 1) - \frac{a\phi(t - 1)}{c + \phi(t - 1)^T \phi(t - 1)} \phi(t - 1)^T \tilde{\theta}(t - 1)$$

Hence, using (3.3.21),

$$\| \tilde{\theta}(t) \|^2 - \| \tilde{\theta}(t - 1) \|^2 = a\left[-2 + \frac{a\phi(t - 1)^T \phi(t - 1)}{c + \phi(t - 1)^T \phi(t - 1)} \right] \frac{e(t)^2}{c + \phi(t - 1)^T \phi(t - 1)}$$

(3.3.29)

Now since $0 < a < 2$, $c > 0$, we have

$$a\left[-2 + \frac{a\phi(t - 1)^T \phi(t - 1)}{c + \phi(t - 1)^T \phi(t - 1)} \right] < 0$$ (3.3.30)

and then (3.3.22) follows from (3.3.29).

(ii) We observe that $\| \tilde{\theta}(t) \|^2$ is a bounded nonincreasing function, and by summing (3.3.29), we have

$$\| \tilde{\theta}(t) \|^2 = \| \tilde{\theta}(0) \|^2 + \sum_{j=1}^{t} a\left[-2 + \frac{a\phi(j - 1)^T \phi(j - 1)}{c + \phi(j - 1)^T \phi(j - 1)} \right] \frac{e(j)^2}{c + \phi(j - 1)^T \phi(j - 1)}$$

Since $\| \tilde{\theta}(t) \|^2$ is nonnegative, and since (3.3.30) holds, we can conclude (3.3.23).

(a) Equation (3.3.24) follows immediately from (3.3.23).

(b) Noting that

$$\frac{e(t)^2}{c + \phi(t - 1)^T \phi(t - 1)} = \frac{[c + \phi(t - 1)^T \phi(t - 1)] e(t)^2}{[c + \phi(t - 1)^T \phi(t - 1)]^2}$$

we establish (3.3.25) using (3.3.23).

(c) Equation (3.3.25) immediately implies (3.3.26) by noting the form of the algorithm (3.3.19).

(d) It is clear that

$$\| \hat{\theta}(t) - \hat{\theta}(t - k) \|^2$$
$$= \| \hat{\theta}(t) - \hat{\theta}(t - 1) + \hat{\theta}(t - 1) - \hat{\theta}(t - 2) \cdots \hat{\theta}(t - k + 1) - \hat{\theta}(t - k) \|^2$$

Then, using the Schwarz inequality,

$$\|\hat{\theta}(t) - \hat{\theta}(t-k)\|^2$$
$$\leq k(\|\hat{\theta}(t) - \hat{\theta}(t-1)\|^2 + \cdots + \|\hat{\theta}(t-k+1) - \hat{\theta}(t-k)\|^2)$$

Then the result follows immediately from (3.3.26) since k is finite.

(e) Equation (3.3.28) follows immediately from (3.3.27).

▼▼▼

Lemma 3.3.2 is a slightly modified version of a result that was first given in Goodwin, Ramadge, and Gaines (1978a) in the context of discrete-time adaptive control.

We pause to discuss briefly the implications of the properties described in the lemma. Note that we have not said anything about $\hat{\theta}(t)$ necessarily converging to θ_0. In fact, we have not said that $\hat{\theta}(t)$ converges to anything. However, the properties as given above are of great importance since they have been derived under extremely weak assumptions. For example, no restriction has been imposed on the nature of the data [in particular, $\phi(t)$ need not necessarily be bounded].

Property (i) ensures that $\hat{\theta}(t)$ is never further from θ_0 than $\hat{\theta}(0)$ is.

Property (ii) implies that the modeling error, $e(t)$, when appropriately normalized is square summable. This turns out to be a sufficient condition to establish global convergence of an important class of adaptive control algorithms.

Property (e) shows that the parameter estimates get closer together as $t \longrightarrow \infty$. We shall make considerable use of this property in our subsequent development. Note, however, that this property does *not* mean that $\hat{\theta}(t)$ is a Cauchy sequence (Rudin, 1960) and hence does not necessarily converge (see Exercise 3.22).

Referring back to Fig. 3.3.1, it can be seen (in the two-dimensional case illustrated) that $\hat{\theta}(t+1)$ would actually coincide with θ_0 provided that $\phi(t)$ was orthogonal to $\phi(t-1)$. This suggests that an improved algorithm could be obtained by projecting in a direction orthogonal to previous $\phi(\cdot)$ vectors. This notion leads to

Orthogonalized Projection Algorithm

$$\hat{\theta}(t) = \hat{\theta}(t-1) + \frac{P(t-2)\phi(t-1)}{\phi(t-1)^T P(t-2)\phi(t-1)}[y(t) - \phi(t-1)^T\hat{\theta}(t-1)] \qquad (3.3.31)$$

where

$$P(t-1) = P(t-2) - \frac{P(t-2)\phi(t-1)\phi(t-1)^T P(t-2)}{\phi(t-1)^T P(t-2)\phi(t-1)} \qquad (3.3.32)$$

with the initial estimate, $\hat{\theta}(1)$, given and

$$P(0) = I \qquad (3.3.33)$$

[When $\phi(t-1)^T P(t-2)\phi(t-1) = 0$, we put $\hat{\theta}(t) = \hat{\theta}(t-1)$ and $P(t-1) = P(t-2)$.]

As we shall prove formally below, the vector $P(t-2)\phi(t-1)$ in the algorithm above is the component of $\phi(t-1)$ which is orthogonal to all previous $\phi(\cdot)$ vectors. The matrix $P(t-1)$ is a projection operator that ensures the property above. In effect, the algorithm above is nothing more than a way of sequentially solving a set of linear equations for the unknown vector θ_0. We shall now summarize some interesting properties of the algorithm. These properties have simple geometric interpretations.

Lemma 3.3.3. Consider the algorithm (3.3.31), (3.3.30), and (3.3.33) subject to the model (3.3.1), that is, $y(t) = \phi(t-1)^T \theta_0$; then

(i) $P(t)$ is idempotent, that is,
$$P(t)^2 = P(t) \tag{3.3.34}$$

(ii) $P(0) \cdots P(t) = P(t)$ (3.3.35)

(iii) $P(t-1)\phi(t) \in R[\phi(1) \quad \phi(2) \quad \cdots \quad \phi(t)]$ (3.3.36)
that is, $P(t-1)\phi(t)$ is a linear combination of $\phi(1) \cdots \phi(t)$.

(iv) $N[P(t)] = R[\phi(1) \quad \phi(2) \quad \cdots \quad \phi(t)]$ (3.3.37)
that is, if and only if $P(t)x = 0$, then x is a linear combination of $\phi(1), \phi(2),$ $\ldots, \phi(t)$.

(v) $P(t-1)\phi(t)$ is orthogonal to $\phi(1) \cdots \phi(t-1)$. (3.3.38)

(vi) $\tilde{\theta}(t)$ is orthogonal to $\phi(1) \cdots \phi(t-1)$ where
$$\tilde{\theta}(t) = \hat{\theta}(t) - \theta_0 \tag{3.3.39}$$

(vii) $\eta(t) \triangleq y(t) - \phi(t-1)^T \hat{\theta}(t) = 0$

Proof. (The techniques used in this proof do not appear elsewhere in the book, and thus the proof may be omitted without loss of continuity.)

(i) We use induction. We note that the result is true for $t = 0$. We assume that it is true for $(t-1)$ and prove that it is true for t. If $\phi(t)^T P(t-1)\phi(t) = 0$, then $P(t) = P(t-1)$ and the result follows from the induction hypothesis. If $\phi(t)^T P(t-1)\phi(t) \neq 0$, then

$$P(t)^2 = P(t-1)^2 - \frac{P(t-1)^2\phi(t)\phi(t)^T P(t-1)}{\phi(t)^T P(t-1)\phi(t)} - \frac{P(t-1)\phi(t)\phi(t)^T P(t-1)^2}{\phi(t)^T P(t-1)\phi(t)}$$
$$+ \frac{P(t-1)\phi(t)\phi(t)^T P(t-1)^2\phi(t)\phi(t)^T P(t-1)}{[\phi(t)^T P(t-1)\phi(t)]^2}$$

$$= P(t-1) - \frac{P(t-1)\phi(t)\phi(t)^T P(t-1)}{\phi(t)^T P(t-1)\phi(t)} \quad \text{using the induction hypothesis}$$

$$= P(t) \quad \text{by (3.3.32)}$$

(ii) We again use induction. The result is true for $t = 1$. We assume that it is true for $(t-1)$ and prove that it is true for t. When $\phi(t)^T P(t-1)\phi(t) \neq 0$,

$$P(0) \cdots P(t-1)P(t) = P(t-1)P(t) \quad \text{by the induction hypotheses}$$

$$= P(t-1)\left[P(t-1) - \frac{P(t-1)\phi(t)\phi(t)^T P(t-1)}{\phi(t)^T P(t-1)\phi(t)} \right]$$

$$= P(t-1)^2 - \frac{P(t-1)^2\phi(t)\phi(t)^T P(t-1)}{\phi(t)^T P(t-1)\phi(t)}$$

$$= P(t) \quad \text{by (i) and (3.3.32)}$$

When $\phi(t)^T P(t-1)\phi(t) = 0$, then $P(t) = P(t-1)$:

$$P(0) \cdots P(t) = P(t-1)P(t) \quad \text{by the induction hypotheses}$$

$$= P(t-1)^2$$

$$= P(t-1) \quad \text{by (i)}$$

$$= P(t)$$

(iii) Again use induction. Assume the result for $P(t-2)\phi(t-1)$. Then from (3.3.32),

$$P(t-1) = P(0) - \sum_{j=2}^{t} \frac{P(t-j)\phi(t-j+1)\phi(t-j+1)^T P(t-j)}{\phi(t-j+1)^T P(t-j)\phi(t-j+1)} a(j); \qquad t \geq 2$$

$$(3.3.40)$$

where

$$a(j) = \begin{cases} 1 & \text{if } \phi(t-j+1)^T P(t-j)\phi(t-j+1) \neq 0 \\ 0 & \text{otherwise} \end{cases}$$

$P(t-1)\phi(t)$

$$= P(0)\phi(t) - \sum_{j=2}^{t} P(t-j)\phi(t-j+1)\frac{\phi(t-j+1)^T P(t-j)\phi(t)a(j)}{\phi(t-j+1)^T P(t-j)\phi(t-j+1)}$$

Since $P(0) = I$, the result follows from the induction hypothesis.

(iv) Let x be any vector such that

$$P(t)x = 0$$

Then from (3.3.40),

$$0 = P(t)x = x - \sum_{j=1}^{t} P(t-j)\phi(t-j+1)\left[\frac{\phi(t-j+1)^T P(t-j)x}{\phi(t-j+1)^T P(t-j)\phi(t-j+1)} a(j)\right]$$

Then from part (iii), $x \in R[\phi(1) \quad \phi(2) \quad \cdots \quad \phi(t)]$. To show that if $x \in R[\phi(1)$ $\phi(2) \quad \cdots \quad \phi(t)]$, then $P(t)x = 0$, we first note that when $\phi(t)^T P(t-1)\phi(t) \neq 0$, we have

$$P(t)\phi(t) = \left[P(t-1) - \frac{P(t-1)\phi(t)\phi(t)^T P(t-1)}{\phi(t)^T P(t-1)\phi(t)}\right]\phi(t)$$

$$(3.3.41)$$

$$= 0$$

When $\phi(t)^T P(t-1)\phi(t) = 0$, then by (i) $\phi(t)^T P(t-1)^2\phi(t) = 0$ and this implies that $P(t-1)\phi(t) = 0$. Also, when $\phi(t)^T P(t-1)\phi(t) = 0$ we set $P(t) = P(t-1)$ and hence $P(t)\phi(t) = 0$. Thus

$$P(t)\phi(t) = 0 \quad \text{always} \tag{3.3.42}$$

Finally, for $i = 1, \ldots, t-1$, we have (3.3.35) that $P(t) = P(0) \cdots P(t-i)$ $\cdots P(t) = P(t-i) \cdots P(t)$. Hence

$$P(t)\phi(t-i) = P(t) \cdots P(t-i)\phi(t-i) \qquad \text{from (ii)}$$

$$= 0 \qquad \text{from (3.3.42)}$$

(v) $\phi(t-i)^T P(t-1)\phi(t) = \phi(t-i)^T P(t-i) \cdots P(t-1)\phi(t)$

$$\text{for } i = 1, \ldots, t-1$$

$$= 0 \qquad \text{from (3.3.42)}$$

(vi) We again proceed by induction. First, for $\tilde{\theta}(2)$, if $\phi(1)^T P(0)\phi(1) = 0$, then since $P(0) = I$, $\phi(1) = 0$, and $\tilde{\theta}(2)^T\phi(1) = 0$. If $\phi(1)^T P(0)\phi(1) \neq 0$, then from (3.3.31) and (3.3.1),

$$\tilde{\theta}(2) = \tilde{\theta}(1) - \frac{P(0)\phi(1)}{\phi(1)^T P(0)\phi(1)}\phi(1)^T\tilde{\theta}(1)$$

Thus

$$\phi(1)^T\tilde{\theta}(2) = 0$$

Now assume that the result is true for $\tilde{\theta}(t-1)$ and prove for $\tilde{\theta}(t)$. If

$$\phi(t-1)^T P(t-2)\phi(t-1) \neq 0$$

then

$$\tilde{\theta}(t) = \tilde{\theta}(t-1) - \frac{P(t-2)\phi(t-1)}{\phi(t-1)^T P(t-2)\phi(t-1)}\phi(t-1)^T\tilde{\theta}(t-1)$$

Thus

$$\phi(t-1)^T\tilde{\theta}(t) = 0 \qquad \text{by calculation}$$

and for $i = 2, \ldots, t-1$,

$$\phi(t-i)^T\tilde{\theta}(t) = \phi(t-i)^T\tilde{\theta}(t-1) - \frac{\phi(t-i)^T P(t-2)\phi(t-1)\phi(t-1)^T\tilde{\theta}(t-1)}{\phi(t-1)^T P(t-2)\phi(t-1)}$$

$$= 0 \qquad \text{by the induction hypothesis and part (iv)}$$

If

$$\phi(t-1)^T P(t-2)\phi(t-1) = 0$$

Then

$$P(t-2)\phi(t-1) = 0 \qquad \text{and} \qquad \tilde{\theta}(t) = \tilde{\theta}(t-1)$$

But $P(t-2)\phi(t-1) = 0$ implies that $\phi(t-1)$ is a linear combination of $\phi(1) \cdots \phi(t-2)$ from (iv).

The result follows from the induction hypotheses.

(vii) Follows from (vi) and (3.3.1).

▼▼▼

An immediate consequence of the result above is the following convergence theorem:

Theorem 3.3.1. The algorithm (3.3.31), (3.3.32), and (3.3.33) and subject to the model (3.3.1) will converge to θ_0 in m steps provided that

$$\text{rank } [\phi(1) \quad \cdots \quad \phi(m)] = n = \text{dimension of } \theta_0 \tag{3.3.43}$$

Proof. Immediate from Lemma 3.3.3, part (vi).

▼▼▼

We will show in Section 3.4 how the rank condition above can be achieved by manipulation of the input $\{u(t)\}$ only.

The algorithm above is compared with the projection algorithm [equation (3.3.5)] for a simple two-dimensional case in Fig. 3.3.2.

For the orthogonalized projection algorithm, the necessity of checking $\phi(t-1)^T P(t-2)\phi(t-1)$ for zero at each step can be avoided by introducing a constant $c > 0$ in the denominator of (3.3.31)–(3.3.32) to give

$$\hat{\theta}(t) = \hat{\theta}(t-1) + \frac{P(t-2)\phi(t-1)}{c + \phi(t-1)^T P(t-2)\phi(t-1)}[y(t) - \phi(t-1)^T\hat{\theta}(t-1)]$$

$$\tag{3.3.44}$$

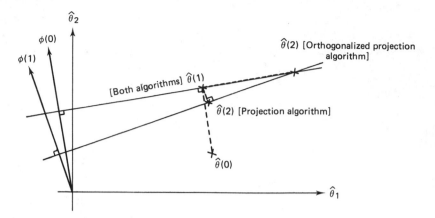

Figure 3.3.2 Comparison of the projection algorithm and the orthogonalized projection algorithm.

$$P(t-1) = P(t-2) - \frac{P(t-2)\phi(t-1)\phi(t-1)^T P(t-2)}{c + \phi(t-1)^T P(t-2)\phi(t-1)} \qquad (3.3.45)$$

When c takes the value 1, the algorithm above becomes the following very well known algorithm, which allegedly dates from the time of Gauss. This algorithm is the least-squares algorithm and is described as follows:

Least-Squares Algorithm

$$\hat{\theta}(t) = \hat{\theta}(t-1) + \frac{P(t-2)\phi(t-1)}{1 + \phi(t-1)^T P(t-2)\phi(t-1)}[y(t) - \phi(t-1)^T\hat{\theta}(t-1)];$$

$$t \geq 1 \qquad (3.3.46)$$

$$P(t-1) = P(t-2) - \frac{P(t-2)\phi(t-1)\phi(t-1)^T P(t-2)}{1 + \phi(t-1)^T P(t-2)\phi(t-1)} \qquad (3.3.47)$$

with $\hat{\theta}(0)$ given and $P(-1)$ is any positive definite matrix P_0.

The algorithm (3.3.46) obviously has the form of (3.2.2). We shall show presently that the algorithm results from minimization of the following quadratic cost function:

$$J_N(\theta) = \tfrac{1}{2}\sum_{t=1}^{N}(y(t) - \phi(t-1)^T\theta)^2 + \tfrac{1}{2}(\theta - \hat{\theta}(0))^T P_0^{-1}(\theta - \hat{\theta}(0)) \qquad (3.3.48)$$

Basically, the cost in (3.3.48) represents the sum of squares of the errors $e(t) = y(t) - \phi(t-1)^T\theta$ [which is the difference between the actual observation $y(t)$ and the value predicted by the model with parameter vector θ]. The second term on the right-hand side of (3.3.48) has been included to account for the initial conditions. [P_0 can be seen to be a measure of confidence in the initial estimate $\hat{\theta}(0)$; see Chapter 9 for further discussion.] Before we prove this connection, we need the following well-known result. (A more general form is given in Exercise 3.2.)

Lemma 3.3.4 (Matrix Inversion Lemma). If

$$P(t-1)^{-1} = P(t-2)^{-1} + \phi(t-1)\phi(t-1)^T a(t-1) \qquad (3.3.49)$$

where the scalar $a(t-1) > 0$, then $P(t-1)$ is related to $P(t-2)$ via

$$P(t-1) = P(t-2) - \frac{P(t-2)\phi(t-1)\phi(t-1)^T P(t-2)a(t-1)}{1 + \phi(t-1)^T P(t-2)\phi(t-1)a(t-1)} \quad (3.3.50)$$

Also,

$$P(t-1)\phi(t-1) = \frac{P(t-2)\phi(t-1)}{1 + \phi(t-1)^T P(t-2)\phi(t-1)a(t-1)} \quad (3.3.51)$$

$$P(t-2)\phi(t-1) = \frac{P(t-1)\phi(t-1)}{1 - \phi(t-1)^T P(t-1)\phi(t-1)a(t-1)} \quad (3.3.52)$$

$$\phi(t-1)^T P(t-1)\phi(t-1) = \frac{\phi(t-1)^T P(t-2)\phi(t-1)}{1 + \phi(t-1)^T P(t-2)\phi(t-1)a(t-1)}$$

$$\phi(t-1)^T P(t-2)\phi(t-1) = \frac{\phi(t-1)^T P(t-1)\phi(t-1)}{1 - \phi(t-1)^T P(t-1)\phi(t-1)a(t-1)} \quad (3.3.53)$$

Proof. By direct verfication (see also Exercise 3.6, where other similar results are also given).

▼▼▼

The significance of result (3.3.50) is that the inverse of a matrix, modified by addition of a diad (outer product of two vectors), can be computed in terms of the inverse of the original matrix with only a scalar division. Equation (3.3.50) gives a nice iterative form of the inverse of the relationship given in (3.3.49).

We can now verify the statement made before (3.3.48):

Lemma 3.3.5. The algorithm (3.3.46)–(3.3.47) minimizes the cost function (3.3.48).

Proof. Let $Y_N^T \triangleq [y(1) \quad y(2) \quad \cdots \quad y(N)]$

$$\Phi_{N-1} \triangleq [\phi(0) \quad \cdots \quad \phi(N-1)]^T \quad (3.3.54)$$

Then (3.3.48) can be written as

$$J_N(\theta) = \tfrac{1}{2}[Y_N - \Phi_{N-1}\theta]^T[Y_N - \Phi_{N-1}\theta] + \tfrac{1}{2}[\theta - \hat{\theta}(0)]^T P_0^{-1}[\theta - \hat{\theta}(0)]$$

Differentiating with respect to θ and setting the result to zero gives

$$-\Phi_{N-1}^T[Y_N - \Phi_{N-1}\theta] + P_0^{-1}[\theta - \hat{\theta}(0)] = 0$$

or

$$[\Phi_{N-1}^T\Phi_{N-1} + P_0^{-1}]\theta = P_0^{-1}\hat{\theta}(0) + \Phi_{N-1}^T Y_N$$

We shall denote the value of θ satisfying this equation as $\hat{\theta}(N)$. Then

$$\hat{\theta}(N) = [\Phi_{N-1}^T\Phi_{N-1} + P_0^{-1}]^{-1}[P_0^{-1}\hat{\theta}(0) + \Phi_{N-1}^T Y_N]$$
$$= P(N-1)[P_0^{-1}\hat{\theta}(0) + \Phi_{N-1}^T Y_N] \quad (3.3.55)$$

where $P(N-1)^{-1} = \Phi_{N-1}^T\Phi_{N-1} + P_0^{-1}$. Then using the form of (3.3.54),

$$P(N-1)^{-1} = P(N-2)^{-1} + \phi(N-1)\phi(N-1)^T \quad (3.3.56)$$

From (3.3.55),

$$\hat{\theta}(N) = P(N-1)[P_0^{-1}\hat{\theta}(0) + \Phi_{N-2}^T Y_{N-1} + \phi(N-1)y(N)]$$
$$= P(N-1)[P(N-2)^{-1}\hat{\theta}(N-1) + \phi(N-1)y(N)] \quad \text{using (3.3.55)}$$
$$= P(N-1)[P(N-1)^{-1} - \phi(N-1)\phi(N-1)^T]\hat{\theta}(N-1)$$
$$+ P(N-1)\phi(N-1)y(N) \quad \text{using (3.3.56)}$$
$$= \hat{\theta}(N-1) + P(N-1)\phi(N-1)[y(N) - \phi(N-1)^T\hat{\theta}(N-1)]$$

This establishes (3.3.46) using (3.3.51). Equation (3.3.47) follows from (3.3.56) by use of Lemma 3.3.4.

▼▼▼

As before, we define
$$e(t) = y(t) - \phi(t-1)^T\hat{\theta}(t-1)$$
$$= -\phi(t-1)^T\tilde{\theta}(t-1)$$
$$\tilde{\theta}(t-1) = \hat{\theta}(t-1) - \theta_0$$

The basic convergence properties of the least-squares algorithm are summarized in the following lemma.

Lemma 3.3.6. For the algorithm (3.3.46) and (3.3.47) and subject to (3.3.1) it follows that

(i) $\|\hat{\theta}(t) - \theta_0\|^2 \leq \kappa_1 \|\hat{\theta}(0) - \theta_0\|^2; \quad t \geq 1$ (3.3.57)
where

$$\kappa_1 = \text{condition number of } [P(-1)^{-1}] \triangleq \frac{\lambda_{\max}P(-1)^{-1}}{\lambda_{\min}P(-1)^{-1}}$$

(ii) $\displaystyle\lim_{N\to\infty} \sum_{t=1}^{N} \frac{e(t)^2}{1 + \phi(t-1)^T P(t-2)\phi(t-1)} < \infty$ (3.3.58)
and this implies

(a) $\displaystyle\lim_{t\to\infty} \frac{e(t)}{[1 + \kappa_2\phi(t-1)^T\phi(t-1)]^{1/2}} = 0$ (3.3.59)
where $\kappa_2 = \lambda_{\max}P(-1)$ ($\lambda_{\max}$ denotes the maximum eigenvalue).

(b) $\displaystyle\lim_{N\to\infty} \sum_{t=1}^{N} \frac{\phi(t-1)^T P(t-2)\phi(t-1)e(t)^2}{[1 + \phi(t-1)^T P(t-2)\phi(t-1)]^2} < \infty$ (3.3.60)

(c) $\displaystyle\lim_{N\to\infty} \sum_{t=1}^{N} \|\hat{\theta}(t) - \hat{\theta}(t-1)\|^2 < \infty$ (3.3.61)

(d) $\displaystyle\lim_{N\to\infty} \sum_{t=k}^{N} \|\hat{\theta}(t) - \hat{\theta}(t-k)\|^2 < \infty$ (3.3.62)

(e) $\displaystyle\lim_{t\to\infty} \|\hat{\theta}(t) - \hat{\theta}(t-k)\| = 0$ for any finite k (3.3.63)

Proof. (i) Subtracting θ_0 from both sides of (3.3.46) and using (3.3.1), we obtain
$$\tilde{\theta}(t) = \tilde{\theta}(t-1) - \frac{P(t-2)\phi(t-1)\phi(t-1)^T\tilde{\theta}(t-1)}{1 + \phi(t-1)^T P(t-2)\phi(t-1)}$$ (3.3.64)
where $\tilde{\theta}(t) = \hat{\theta}(t) - \theta_0$.

Then, using (3.3.47) and the matrix inversion lemma (Lemma 3.3.4), we have
$$\tilde{\theta}(t) = P(t-1)P(t-2)^{-1}\tilde{\theta}(t-1)$$ (3.3.65)

Hence introducing $V(t) = \tilde{\theta}(t)^T P(t-1)^{-1}\tilde{\theta}(t)$, we have

$$V(t) - V(t-1) = [\tilde{\theta}(t) - \tilde{\theta}(t-1)]^T P(t-2)^{-1}\tilde{\theta}(t-1) \qquad (3.3.66)$$

$$= -\frac{\tilde{\theta}(t-1)^T \phi(t-1)\phi(t-1)^T\tilde{\theta}(t-1)}{1 + \phi(t-1)^T P(t-2)\phi(t-1)}$$

Thus $V(t)$ is a nonnegative, nonincreasing function and hence

$$\tilde{\theta}(t)^T P(t-1)^{-1}\tilde{\theta}(t) \le \tilde{\theta}(0)^T P(-1)\tilde{\theta}(0) \qquad (3.3.67)$$

Now from the matrix inversion lemma (Lemma 3.3.4),

$$P(t)^{-1} = P(t-1)^{-1} + \phi(t)\phi(t)^T$$

It follows that

$$\lambda_{\min}[P(t)^{-1}] \ge \lambda_{\min}[P(t-1)^{-1}]$$
$$\ge \lambda_{\min}[P(-1)^{-1}] \qquad (3.3.68)$$

Equation (3.3.68) implies

$$\lambda_{\min}[P(-1)^{-1}] \,\|\tilde{\theta}(t)\|^2 \le \lambda_{\min}[P(t-1)^{-1}] \,\|\tilde{\theta}(t)\|^2$$
$$\le \tilde{\theta}(t)^T P(t-1)^{-1}\tilde{\theta}(t)$$
$$\le \tilde{\theta}(0)^T P(-1)^{-1}\tilde{\theta}(0) \qquad \text{by (3.3.67)} \qquad (3.3.69)$$
$$\le \lambda_{\max}[P(-1)^{-1}] \,\|\tilde{\theta}(0)\|^2$$

This establishes part (i).

(ii) Returning to (3.3.66) and summing from 1 to N gives

$$V(N) = V(0) - \sum_{t=1}^{N} \frac{\tilde{\theta}(t-1)^T \phi(t-1)\phi(t-1)^T\tilde{\theta}(t-1)}{1 + \phi(t-1)^T P(t-2)\phi(t-1)}$$

Since $V(N)$ is nonnegative, we immediately have (3.3.58).

(a) Equation (3.3.59) follows immediately from (3.3.58) and (3.3.68), which implies that

$$\lambda_{\max}[P(t)] \le \lambda_{\max}[P(t-1)] \le \lambda_{\max}[P(-1)] \qquad (3.3.70)$$

(b) Noting that

$$\frac{e(t)^2}{[1 + \phi(t-1)^T P(t-2)\phi(t-1)]} = \frac{[1 + \phi(t-1)^T P(t-2)\phi(t-1)]e(t)^2}{[1 + \phi(t-1)^T P(t-2)\phi(t-1)]^2}$$

we establish (3.3.60) using (3.3.58).

(c) From the algorithm (3.3.46),

$$\|\hat{\theta}(t) - \hat{\theta}(t-1)\|^2 = \frac{\phi(t-1)^T P(t-2)^2\phi(t-1)e(t)^2}{[1 + \phi(t-1)^T P(t-2)\phi(t-1)]^2}$$
$$\le \frac{\phi(t-1)^T P(t-2)\phi(t-1)e(t)^2}{[1 + \phi(t-1)^T P(t-2)\phi(t-1)]^2} \lambda_{\max}[P(-1)]$$

$$\text{using (3.3.70)}$$

Then (3.3.61) follows immediately from (3.3.60).

(d) This then follows by use of the Schwarz inequality, as in the proof of part (d) of Lemma 3.3.2.

(e) Immediate from (d).

▼▼▼

Lemma 3.3.6 is a slightly modified version of a result that was first given in Goodwin, Ramadge, and Caines (1978a) in the context of discrete-time adaptive control.

The least-squares algorithm generally has much faster convergence than the projection algorithm. Again, the orthogonalized projection algorithm has even faster convergence than the least-squares algorithm. However, this is true only for the purely deterministic case and the orthogonalized projection algorithm will be extremely sensitive to the presence of noise. By way of contrast, we shall see later in the book that the least-squares algorithm can be used essentially unaltered with noisy signals!

We now discuss a number of variants of the least-squares algorithm. We distinguish four broad cases. In the first case, the data are selectively weighted based on their perceived information content. In the second case, past data are gradually discarded on the assumption that more recent data are more informative. In the third case, we reset the covariance matrix, P, at various times. Finally, in case 4, we add a matrix to the right-hand side of the covariance update equation. An important effect of the modifications of the latter three cases is that they render the least-squares algorithm applicable to time-varying systems.

Case 1 : Least Squares with Selective Data Weighting

Here we consider a slight modification of the cost function (3.3.48) to the form

$$\bar{J}_N(\theta) = \tfrac{1}{2} \sum_{t=1}^{N} a(t-1)[y(t) - \phi(t-1)^T\theta]^2 + \tfrac{1}{2}(\theta - \hat{\theta}(0))^T P_0^{-1}(\theta - \hat{\theta}(0)) \quad (3.3.71)$$

where $\{a(t-1)\}$ is a nonnegative sequence of weighting coefficients.

One possible way of thinking about the sequence $\{a(t-1)\}$ is when there are measurement errors in $y(t)$. Then $a(t-1)$ might be chosen as the inverse of the expected mean-square error. This has the effect of attaching smaller weight to those terms in which $y(t)$ is expected to have larger errors. Later in the book, when we introduce stochastic considerations, we shall see that this is the optimal (in a certain sense) thing to do.

It can be readily seen (Exercise 3.7) that the following sequential algorithm minimizes $\bar{J}_N(\theta)$.

Weighted Least-Squares Algorithm

$$\hat{\theta}(t) = \hat{\theta}(t-1) + \frac{a(t-1)P(t-2)\phi(t-1)}{1 + a(t-1)\phi(t-1)^T P(t-2)\phi(t-1)}$$

$$[y(t) - \phi(t-1)^T\hat{\theta}(t-1)] \quad (3.3.72)$$

$$P(t-1) = P(t-2) - \frac{a(t-1)P(t-2)\phi(t-1)\phi(t-1)^T P(t-2)}{1 + a(t-1)\phi(t-1)^T P(t-2)\phi(t-1)} \quad (3.3.73)$$

with $\hat{\theta}(0)$ given and $P(-1) = P_0 > 0$.

We can establish the following result (analogous to Lemma 3.3.6).

Lemma 3.3.7. For the algorithm (3.3.72)–(3.3.73) and subject to (3.3.1), it follows that

(i) $\|\hat{\theta}(t) - \theta_0\| \le \kappa_1 \|\hat{\theta}(0) - \theta_0\|; \quad t \ge 1$ (3.3.74)

    $\kappa_1 = $ condition number of $[P(-1)^{-1}]$

(ii) $\displaystyle \lim_{N\to\infty} \sum_{t=1}^{N} \frac{a(t-1)e(t)^2}{1 + a(t-1)\phi(t-1)^T P(t-2)\phi(t-1)} < \infty$ (3.3.75)

Proof. See Exercise 3.8.

                                               ▼▼▼

In applying Lemma 3.3.7, one has to be careful of the range of $\{a(t-1)\}$. For example, if $a(t-1)$ is arbitrarily set to zero, the corresponding data point will not be included in the parameter update, and this might lead to difficulties.

We now look at various special cases of the algorithm.

1. An interesting observation is that if we let $a(t-1)$ become very large (corresponding to diminishing measurement errors), then in the limiting case the algorithm above reduces to the orthogonalized projection algorithm [equations (3.3.31) and (3.3.32)]. In fact, the orthogonalized projection algorithm corresponds to the following selection of $\{a(t-1)\}$:

$$a(t-1) = \begin{cases} \infty & \text{for } \phi(t-1)^T P(t-2)\phi(t-1) \ne 0 \\ 0 & \text{for } \phi(t-1)^T P(t-2)\phi(t-1) = 0 \end{cases} \qquad (3.3.76)$$

2. The choice above for $\{a(t-1)\}$ is an extreme case. However, it is clearly not robust since infinite numerical accuracy is required. The key thing is that the data are divided into two sets, one being accepted and the other being rejected. A more robust form of the algorithms can be obtained by modifying the selection criterion as follows:

$$a(t-1) = \begin{cases} \kappa_1 & \text{for } \phi(t-1)^T P(t-2)\phi(t-1) \ge \epsilon \\ \kappa_2 & \text{for } \phi(t-1)^T P(t-2)\phi(t-1) < \epsilon \end{cases} \qquad (3.3.77)$$

with $\kappa_1 \gg \kappa_2 > 0$.

We note that the test on $\phi(t-1)^T P(t-2)\phi(t-1)$ is now numerically robust. Moreover, the algorithm (3.3.77) now incorporates all data points, albeit with different weights, and this allows us to establish convergence properties similar to those given in Lemma 3.3.6; see Exercise 3.9.

3. A further choice [see e.g., Fuchs (1980b)] for the weighting sequence is

$$a(t-1) = \begin{cases} \dfrac{\phi(t-1)^T P(t-2)\phi(t-1)}{\phi(t-1)^T \phi(t-1)}; & \phi(t-1)^T \phi(t-1) \ne 0 \\ \text{arbitrary otherwise} \end{cases} \qquad (3.3.78)$$

[Note that $a(t-1)$ can be interpreted as a measure of the information content of $\phi(t-1)$ in terms of its relative orthogonality to previous data.]

For this algorithm, part (ii) of Lemma 3.3.7 becomes (see Exercise 3.10)

$$\lim_{N\to\infty} \sum_{t=1}^{N} \frac{e(t)^2}{\text{trace } P(-1)^{-1} + \sum_{j=1}^{t} \phi(j-1)^T P(j-2)\phi(j-1)} > \infty \qquad (3.3.79)$$

The convergence property (3.3.79) is related to the property (3.3.58) and can be useful in certain cases (Fuchs, 1980b).

The reader is invited to contemplate the relationship between (3.3.76), (3.3.77), and (3.3.78), especially in regard to the use of $\phi(t-1)^T P(t-2)\phi(t-1)$ as an indicator of the new information contained in $\phi(t-1)$.

4. In certain applications where an upper bound is known for the errors due to noise, inaccurate modeling, computer round-off error, and so on, then a sensible choice for $\{a(t-1)\}$ is

$$a(t-1) = \begin{cases} 1 & \text{if } \dfrac{|y(t) - \phi(t-1)^T\hat{\theta}(t-1)|^2}{1 + \phi(t-1)^T P(t-2)\phi(t-1)} > \Delta^2 > 0 \\ 0 & \text{otherwise} \end{cases} \qquad (3.3.80)$$

The interpretation for the choice of $a(t-1)$ above is that only those data points where the prediction error exceeds a prespecified bound are used in refining the parameter estimate. We shall consider this class of algorithm in more detail in Section 3.6.

Case 2: Least Squares with Exponential Data Weighting

In case 1, discussed above, greater weighting was selectively applied to those data points that were deemed to be more informative based on some criteria. Here we consider the case where the most recent data are assumed to be more informative than past data and hence we exponentially discard old data, leading to the following exponentially weighted cost function:

$$S_N(\theta) = \alpha(N-1)S_{N-1}(\theta) + [y(N) - \phi(N-1)^T\theta]^2 \qquad (3.3.81)$$

where $0 < \alpha(\cdot) < 1$. Note that $\alpha = 1$ gives the standard least-squares cost function.

It can be readily shown (see Exercise 3.11) that minimization of the cost function $S_N(\theta)$ leads to the following sequential algorithm:

$$\hat{\theta}(t) = \hat{\theta}(t-1) + \frac{P(t-2)\phi(t-1)}{\alpha(t-1) + \phi(t-1)^T P(t-2)\phi(t-1)} \qquad (3.3.82)$$
$$[y(t) - \phi(t-1)^T\hat{\theta}(t-1)]$$

$$P(t-1) = \frac{1}{\alpha(t-1)}\left[P(t-2) - \frac{P(t-2)\phi(t-1)\phi(t-1)^T P(t-2)}{\alpha(t-1) + \phi(t-1)^T P(t-2)\phi(t-1)}\right] \qquad (3.3.83)$$

with $\hat{\theta}(0)$, $P(-1) > 0$ given.

The algorithm above is frequently used in the case of time-varying systems because greater weighting is attached to more recent data. (We shall show later that care must be exercised when using the algorithm if the data are not rich in information.) Another use of the sequence $\{\alpha(t)\}$ is to discard initial data in nonlinear estimation problems. These initial data may deteriorate the performance of the algorithm unless it is discarded once the algorithm is under way. It has been shown in Söderström, Ljung, and Gustavsson (1978) that a good choice for $\alpha(t)$ in such cases is

$$\alpha(t) = \alpha_0\alpha(t-1) + (1 - \alpha_0) \qquad (3.3.84)$$

with typical values $\alpha(0) = 0.95$, $\alpha_0 = 0.99$. The effect of this is to impose exponential data weighting for a transient period during algorithm startup.

Case 3: Least Squares with Covariance Resetting

In practice the ordinary least-squares algorithm has very fast initial convergence rate, but the algorithm gain reduces dramatically when the P matrix gets small after a few interations (typically 10 or 20). This motivates a related shceme in which the covariance P is reset at various times. This will revitalize the algorithm and is helpful in maintaining an overall fast convergence rate. This is particularly useful when tracking time-varying parameters. In this case, the obvious time to reset P is when one suspects that a significant parameter change has occurred.

The least-squares algorithm with covariance resetting is described as follows:

$$\hat{\theta}(t) = \hat{\theta}(t-1) + \frac{P(t-2)\phi(t-1)}{1 + \phi(t-1)^T P(t-2)\phi(t-1)}[y(t) - \phi(t-1)^T \theta(t-1)]$$

(3.3.85)

$$P(-1) = k_0 I; \qquad k_0 > 0$$

(3.3.86)

Let $\{Z_s\} = \{t_1 \quad t_2 \quad t_3 \quad \ldots\}$ be the times at which resetting occurs; then for $t \notin \{Z_s\}$ an ordinary sequential least-squares update is used: that is,

$$P(t-1) = P(t-2) - \frac{P(t-2)\phi(t-1)\phi(t-1)^T P(t-2)}{1 + \phi(t-1)^T P(t-2)\phi(t-1)}$$

(3.3.87)

Otherwise, for $t = t_i \in \{Z_s\}$, $P(t_i - 1)$ is reset as follows:

$$P(t_i - 1) = k_i I \qquad \text{where } 0 < k_{\min} \le k_i \le k_{\max} < \infty$$

(3.3.88)

Lemma 3.3.8. For the algorithm (3.3.85) to (3.3.88) and subject to the system (3.3.1), it follows that

(i) $\|\hat{\theta}(t) - \theta_0\| \le \|\hat{\theta}(t-1) - \theta_0\| \le \|\hat{\theta}(0) - \theta_0\|; \quad t \ge 1$ (3.3.89)

(ii) $\displaystyle \lim_{N \to \infty} k_{\min} \sum_{t=1}^{N} \frac{e(t)^2}{1 + k_{\max}\phi(t-1)^T \phi(t-1)} < \infty$ (3.3.90)

where

$$e(t) = y(t) - \phi(t-1)^T \hat{\theta}(t-1)$$

(3.3.91)

and this implies

(a) $\displaystyle \lim_{t \to \infty} \frac{e(t)}{[1 + k_{\max}\phi(t-1)^T \phi(t-1)]^{1/2}} = 0$ (3.3.92)

(b) $\displaystyle \lim_{t \to \infty} \|\hat{\theta}(t) - \hat{\theta}(t-k)\| = 0 \qquad$ for any finite k (3.3.93)

Proof. [The result can be more readily appreciated when one realizes that if P is reset at every time instant, the algorithm simply reduces to the projection algorithm. Thus the scheme (3.3.85) to (3.3.88) allows one to take advantage of the merits of both the least-squares and projection algorithms.]

For $t = 0, \ldots, t_1$ we argue as for the least-squares analysis using the following Lyapunov function (see the proof of Lemma 3.3.6):

$$V(t) = \tilde{\theta}(t)^T \bar{P}(t-1)^{-1}\tilde{\theta}(t)$$

(3.3.94)

where $\bar{P}(t-1)$ is equal to $P(t-1)$ as given in (3.3.87) for $0 < t \le t_1$ (i.e., prior to resetting), and

$$\tilde{\theta}(t) = \hat{\theta}(t) - \theta_0 \qquad (3.3.95)$$

Arguing as in the proof of Lemma 3.3.6, it is readily seen that an elementary property of the algorithm (3.3.85)–(3.3.87) is that

$$V(t) = V(0) - \sum_{j=1}^{t} \frac{e(j)^2}{1 + \phi(j-1)^T P(j-2)\phi(j-1)}; \qquad t = 1, \ldots, t_1 \qquad (3.3.96)$$

where $e(j)$ is as in (3.3.91).

Now in view of (3.3.86),

$$V(0) = k_0^{-1} \tilde{\theta}(0)^T \tilde{\theta}(0) \qquad (3.3.97)$$

Also,

$$V(t) = \tilde{\theta}(t)^T \bar{P}(t-1)^{-1} \tilde{\theta}(t); \qquad t = 1, \ldots, t_1$$

$$\geq [\lambda_{min} \bar{P}(t-1)^{-1}] \tilde{\theta}(t)^T \tilde{\theta}(t); \qquad t = 1, \ldots, t_1$$

From (3.3.87) it follows that $\lambda_{max} P(t-1) \leq \lambda_{max} P(t-2)$ and hence

$$V(t) \geq [\lambda_{min} P(-1)^{-1}] \tilde{\theta}(t)^T \tilde{\theta}(t)$$
$$= k_0^{-1} \tilde{\theta}(t)^T \tilde{\theta}(t); \qquad t = 1, \ldots, t_1 \qquad (3.3.98)$$

Substituting (3.3.97) and (3.3.98) into (3.3.96) gives

$$\bar{V}(t) \leq \bar{V}(0) - k_0 \sum_{j=0}^{t} \frac{e(j)^2}{1 + \phi(j-1)^T P(j-2)\phi(j-1)} \qquad \text{for } t = 1, \ldots, t_1 \qquad (3.3.99)$$

where

$$\bar{V}(t) = \tilde{\theta}(t)^T \tilde{\theta}(t) \qquad (3.3.100)$$

Again using (3.3.87), we have

$$\phi(j-1)^T P(j-2)\phi(j-1) \leq [\lambda_{max} P(j-2)]\phi(j-1)^T \phi(j-1)$$
$$\leq [\lambda_{max} P(-1)]\phi(j-1)^T \phi(j-1) \qquad (3.3.101)$$
$$= k_0 \phi(j-1)^T \phi(j-1)$$

Substituting (3.3.101) into (3.3.99) gives

$$\bar{V}(t) \leq \bar{V}(0) - k_0 \sum_{j=1}^{t} \frac{e(j)^2}{1 + k_0\phi(j-1)^T \phi(j-1)}; \qquad t = 1, \ldots, t_1 \qquad (3.3.102)$$

By the same argument, we can readily show that

$$\bar{V}(t) \leq \bar{V}(t_1) - k_1 \sum_{j=t_1+1}^{t} \frac{e(j)^2}{1 + k_1\phi(j-1)^T \phi(j-1)}; \qquad t = t_1 + 1, \ldots, t_2 \qquad (3.3.103)$$

and so on. Hence for all t it is clear that

$$\bar{V}(t) \leq \bar{V}(0) - k_{min} \sum_{j=1}^{t} \frac{e(j)^2}{1 + k_{max}\phi(j-1)^T \phi(j-1)} \qquad (3.3.104)$$

Thus $\bar{V}(t)$ is a nonincreasing function bounded below by zero and hence converges. Thus we may conclude that

$$k_{min} \sum_{j=1}^{t} \frac{e(j)^2}{1 + k_{max}\phi(j-1)^T \phi(j-1)} < \infty \qquad (3.3.105)$$

The remainder of the proof follows immediately from (3.3.105), as in the proof of Lemma 3.3.2.

▼▼▼

The algorithm above has been primarily designed for use with time-varying parameters. However, Lemma 3.3.8 shows that the algorithm has useful convergence properties when used in the time-invariant case. This is clearly a desirable "compatibility" property for the algorithm. Similar "compatibility" arguments have been advanced by Cordero and Mayne (1981) for an interesting adaptive control algorithm with variable foregetting factor due to Fortescue, Kershenbaum, and Ydstie (1981). We shall discuss the latter algorithm in more detail in Section 6.7. A similar result has also been established by Lozano (1981) for an exponentially weighted least-squares algorithm in which the weighting is adjusted depending on the covariance matrix.

Case 4: Least Squares with Covariance Modification

There are many other possible algorithms for the time-varying problem. For example, Vogel and Edgar (1982) suggest simply using ordinary least squares but to add an additional term into the covariance matrix when parameter changes are detected. This actually achieves a similar effect as the previous algorithm since P is kept from converging to zero.

The algorithm with covariance modification is

$$\hat{\theta}(t) = \hat{\theta}(t-1) + \frac{P(t-2)\phi(t-1)}{1 + \phi(t-1)^T P(t-2)\phi(t-1)}[y(t) - \phi(t-1)^T\hat{\theta}(t-1)]$$

$$(3.3.106)$$

$$\bar{P}(t-1) = P(t-2) - \frac{P(t-2)\phi(t-1)\phi(t-1)^T P(t-2)}{1 + \phi(t-1)^T P(t-2)\phi(t-1)} \qquad (3.3.107)$$

$$P(t-1) = \bar{P}(t-1) + Q(t-1) \qquad (3.3.108)$$

where $0 \leq Q(t-1) < \infty$.

A minor precaution is necessary to ensure that $\lambda_{max}P(t-1)$ is kept finite. This can be achieved by monitoring the trace of $P(t-1)$ and by putting $P(t-1) = \bar{P}(t-1)$ whenever the trace of $P(t-1)$ exceeds a preset upper bound (k_{max}, say).

[Those readers who are familiar with the Kalman filter will appreciate that the algorithm above can be motivated by assuming that the parameters undergo a random walk, that is,

$$\theta(t+1) = \theta(t) + w(t)$$

where $\{w(t)\}$ is a "white sequence" having covariance $Q(t)$].

The algorithm above can be shown to have the same convergence properties as in Lemma 3.3.8 using the Lyapunov function given in (3.3.94). The key observation is that

$$\tilde{\theta}(t)^T \bar{P}(t-1)^{-1}\tilde{\theta}(t)$$

$$= \tilde{\theta}(t-1)^T P(t-2)^{-1}\tilde{\theta}(t-1) - \frac{e(t)^2}{1 + \phi(t-1)^T P(t-2)\phi(t-1)} \qquad (3.3.109)$$

and also that (3.3.108) implies

$$P(t-1)^{-1} \leq \bar{P}(t-1)^{-1} \qquad (3.3.110)$$

Hence we can conclude that

$$V(t) \leq V(t-1) - \frac{e(t)^2}{1 + k_{\max}\phi(t-1)^T\phi(t-1)} \qquad (3.3.111)$$

Convergence of the algorithm then follows as in Lemma 3.3.8.

▼▼▼

In developing the convergence properties of the algorithms discussed so far, we have deliberately imposed weak assumptions in order to widen the applicability of the result. In particular, we shall find that the results are immediately useful in adaptive filtering, prediction, and control. However, the results fall short of establishing convergence of the parameter estimates to the true values. In the next section we strengthen the assumptions so as to establish parameter convergence.

3.4 PARAMETER CONVERGENCE

In Section 3.3 we introduced two basic parameter estimation algorithms (projection and least squares) and established some of their elementary properties under weak assumptions. Our motivation for keeping the assumptions relatively weak was to ensure that the results could be applied to a wide range of circumstances. For example, the results apply to adaptive control *where the signals are not assumed to be bounded a priori*. However, under such weak assumptions it is not possible to establish convergence of the parameter estimates to the true values of the parameters. In many cases, this is not a significant limitation and the performance of the algorithm in a particular application can be inferred from the properties established in Section 3.3. However, in other circumstances, it is important to know that the parameter estimates converge to the true parameters. For example, we shall find that this property is useful in certain adaptive filtering algorithms.

To establish convergence, we shall need to strengthen the assumptions. In particular, we shall need to impose conditions on the nature of the input signals. We consider the various algorithms next.

3.4.1 The Orthogonalized Projection Algorithm

Conditions for convergence of this algorithm were given in Theorem 3.3.1. The result is intuitively reasonable since basically what one is doing is solving a set of m equations for n unknowns ($m \geq n$). This it is simply necessary to ensure that n of the equations are linearly independent. This is guaranteed by the condition

$$\text{rank } [\phi(1) \cdots \phi(m)] = n = \text{dimension of } \theta_0 \qquad (3.4.1)$$

The physical interpretation of the condition above can be seen in Fig. 3.3.2. The basic idea is that the regression vectors $\{\phi(t)\}$ span the full dimension of the parameter space.

We shall show in Section 3.4.4 how the foregoing rank condition on $\{\phi(t)\}$ can be related to the nature of the system input $\{u(t)\}$.

3.4.2 The Least-Squares Algorithm

It was shown in the proof of Lemma 3.3.6 that a basic property of the least-squares algorithm is that the error function $V(t)$ is nonincreasing and hence converges where

$$V(t) = \tilde{\theta}(t)^T P(t-1)^{-1} \tilde{\theta}(t)$$
$$\tilde{\theta}(t) = \hat{\theta}(t) - \theta_0$$

Also, from the matrix inversion lemma we have that

$$P(t-1)^{-1} = P(-1)^{-1} + \sum_{j=0}^{t-1} \phi(j)\phi(j)^T$$

It is clear from the equations above that $\tilde{\theta}(t)$ will converge to zero provided that

$$\lim_{t\to\infty} \lambda_{\min} P(t-1)^{-1} = \infty$$

and this is guaranteed provided that

$$\lim_{t\to\infty} \lambda_{\min} \sum_{j=0}^{t-1} \phi(j)\phi(j)^T = \infty \tag{3.4.2}$$

It can be seen that condition (3.4.2) is related to condition (3.4.1) although (3.4.2) is more demanding. The reason for the different conditions is basically to render the initial term $\frac{1}{2}(\theta - \hat{\theta}(0))^T P_0^{-1}(\theta - \hat{\theta}(0))$ in the cost function (3.3.48) asymptotically negligible.

Again, we shall see in Section 3.4.4 how the foregoing condition on $\{\phi(t)\}$ can be related to the input $\{u(t)\}$.

3.4.3 The Projection Algorithm

We first prove the following result regarding the propagation of the parameter estimation error, $\tilde{\theta}(t)$, for the projection algorithm (3.3.5).

Lemma 3.4.1. Using the projection algorithm, the estimation error $\tilde{\theta}(t) \triangleq \hat{\theta}(t) - \theta_0$ satisfies

(i) $\tilde{\theta}(t) = F(t-1)\tilde{\theta}(t-1)$ (3.4.3)

where $F(t-1)$ is the following idempotent matrix (a projection operator):

$$F(t-1) = I - \frac{\phi(t-1)\phi(t-1)^T}{\phi(t-1)^T\phi(t-1)} \tag{3.4.4}$$

(ii) $\tilde{\theta}(t+j) = \psi(t+j, t)\tilde{\theta}(t)$ (3.4.5)

where

$$\psi(t+j, t) = \prod_{k=0}^{j-1} F(t+k); \qquad j \geq 1$$
$$= F(t+j-1) \cdots F(t) \tag{3.4.6}$$

Proof. (i) Immediate from (3.3.5) by subtracting θ_0 from both sides
(ii) Follows from (3.4.3) by successive substitution

▼▼▼

The basic idea in establishing convergence is to show that the parameter error $\tilde{\theta}(t)$ is appropriately reflected in the prediction error, $y(t) - \phi(t-1)^T\hat{\theta}(t)$, via an observable condition and hence to show that convergence of the prediction error to zero ensures that $\tilde{\theta}(t)$ converges to zero.

We first relate the prediction error to the parameter error as follows:

Lemma 3.4.2

$$\sum_{i=0}^{l-1} \epsilon(t+i)^2 = \tilde{\theta}(t)^T G_l(t)\tilde{\theta}(t) \tag{3.4.7}$$

(i) where $\epsilon(t)$ is the normalized prediction error given by

$$\epsilon(t) = \frac{1}{\|\phi(t)\|}[y(t+1) - \phi(t)^T\hat{\theta}(t)]$$

$$= \phi_n(t)^T\tilde{\theta}(t) \tag{3.4.8}$$

$$\phi_n(t) \triangleq \frac{\phi(t)}{\|\phi(t)\|}$$

(ii) $G_l(t)$ is the "observability grammian" for the state-space system (3.4.3) and (3.4.8), that is,

$$G_l(t) \triangleq \sum_{i=0}^{l-1} \frac{\psi(t+i, t)^T\phi(t+i)\phi(t+i)^T\psi(t+i, t)}{\phi(t+i)^T\phi(t+i)} \tag{3.4.9}$$

Proof. Immediate from (3.4.3), (3.4.6), and (3.4.8).

▼▼▼

We now have the following convergence result.

Lemma 3.4.3. The projection algirithm is globally exponentially convergent to θ_0 provided that the following "observability condition" is satisfied.

$$G_l(t) \geq cI, \quad c > 0 \quad \text{for all } t \text{ and some fixed } l > 0 \tag{3.4.10}$$

Proof. Define

$$V(t) = \tilde{\theta}(t)^T\tilde{\theta}(t)$$

Now from (3.4.3),

$$V(t+1) = \tilde{\theta}(t)^T F(t)^T F(t)\tilde{\theta}(t)$$

$$= V(t) - \frac{\tilde{\theta}(t)^T\phi(t)\phi(t)^T\tilde{\theta}(t)}{\phi(t)^T\phi(t)}$$

Summing l terms yields

$$V(t+l) = V(t) - \sum_{i=0}^{l-1} \frac{\tilde{\theta}(t+i)^T\phi(t+i)\phi(t+i)^T\tilde{\theta}(t+i)}{\phi(t+i)^T\phi(t+i)}$$

$$= V(t) - \sum_{i=0}^{l-1} \tilde{\theta}(t)^T\frac{\psi(t+i, t)^T\phi(t+i)\phi(t+i)^T\psi(t+i, t)\tilde{\theta}(t)}{\phi(t+i)^T\phi(t+i)}$$

$$\text{using (3.4.5)} \tag{3.4.11}$$

$$= V(t) - \tilde{\theta}(t)^T G_l(t)\tilde{\theta}(t)$$

$$\leq (1 - c)V(t), \quad c > 0 \quad \text{using (3.4.10)}$$

Clearly, $c < 1$ [since $V(t + l) > 0$]. Thus $(1 - c) < e^{-c}$ and hence

$$V(t + l) \leq e^{-c}V(t) \tag{3.4.12}$$

Hence

$$V(lN) \leq e^{-cN}V(0) \tag{3.4.13}$$

and thus from the definition of $V(\cdot)$ we see that $\tilde{\theta}(t)$ converges to zero exponentially fast. (Note that the equation above is expressed in terms of time steps of length l.)

▼▼▼

In contrast to the previous conditions (3.4.1) and (3.4.2), the observability condition (3.4.10) is difficult to check. We therefore further investigate the observability grammian $G_l(t)$. We can see from (3.4.9) that the condition (3.4.10) is closely related to the condition (3.4.2) save for the additional complication of the dependence of (3.4.10) on the state transition matrix $\psi(t)$. We now note two important facts:

1. Observability is invariant under bounded output feedback [see Kailath (1980, p. 217)].
2. There exists a simple feedback law in (3.4.3) which gives I as the state transition matrix of the resulting system.

We therefore consider the following fictitious system obtained from (3.4.3) by adding a feedback term:

$$\tilde{\theta}'(t + 1) \triangleq F(t)\tilde{\theta}'(t) + K(t)\epsilon'(t) \tag{3.4.14}$$

where $F(t)$ is as in (3.4.4), $K(t)$ is a feedback gain given by

$$K(t) \triangleq \frac{\phi(t)}{\|\phi(t)\|} = \phi_n(t) \tag{3.4.15}$$

and $\epsilon'(t)$ is the output as in (3.4.8) but with $\tilde{\theta}(t)$ replaced by $\tilde{\theta}'(t)$, that is,

$$\epsilon'(t) \triangleq \phi_n(t)^T \tilde{\theta}'(t) \tag{3.4.16}$$

Substituting (3.4.15) and (3.4.16) into (3.4.14) gives

$$\tilde{\theta}'(t + 1) = \tilde{\theta}'(t) \tag{3.4.17}$$

and hence the observability grammian for (3.4.14) and (3.4.16) is

$$G_l'(t) = \sum_{i=0}^{l-1} \frac{\phi(t + i)\phi(t + i)^T}{\phi(t + i)^T\phi(t + i)} \tag{3.4.18}$$

The reasoning above is summarized in the following result:

Lemma 3.4.4. The projection algorithm is globally exponentially convergent to θ_0 provided that

$$\sum_{i=0}^{l-1} \frac{\phi(t + i)\phi(t + i)^T}{\phi(t + i)^T\phi(t + i)} \geq cI; \quad c > 0 \tag{3.4.19}$$

for all t and some fixed $l > 0$.

Proof. The proof formalizes the statement made above that bounded output feedback does not affect observability; that is, we relate the observability grammian

$G'_i(t)$ for the feedback system (3.4.14) and (3.4.16) to the observability grammian $G_i(t)$ for the original system (3.4.3) and (3.4.8).

We may write the two observability grammians as

$$G_i(t) = LL^T \tag{3.4.20}$$

$$G'_i(t) = LMM^TL^T \tag{3.4.21}$$

where

$$L = [\phi_n(t) \quad F(t)^T\phi_n(t+1) \quad F(t)^TF(t+1)^T\phi_n(t+2) \quad \cdots \quad] \tag{3.4.22}$$

$$M = \begin{bmatrix} I & K(t)^T\phi_n(t+1) & K(t)^T[F(t+1) + K(t+1)\phi_n(t+1)^T]^T\phi_n(t+2) & \cdots \\ & I & K(t+1)^T\phi_n(t+2) & \cdots \\ & & I & \\ & & & \ddots \\ & & & \end{bmatrix}$$

$$\tag{3.4.23}$$

M is upper triangular, nonsingular, and has bounded entries. Similarly, M^{-1} is upper triangular, nonsingular, and has bounded entries, that is,

$$\alpha I \le MM^T \le \beta I \tag{3.4.24}$$

for some $\alpha, \beta > 0$. The result then follows from (3.4.20)–(3.4.21) and Lemma 3.4.3.

▼▼▼

The result above establishes exponential convergence of the projection algorithm. This is a desirable property and tends to improve the robustness of the algorithm [see Anderson and Johnson (1980)]. However, the reader is cautioned that the exponent in the exponential decay may be very small. In practice the least-squares algorithm and its variants are much preferable to the projection algorithm.

3.4.4 Persistent Excitation

In summary, the conditions for parameter convergence derived in Sections 3.4.1, 3.4.2, and 3.4.3 are as follows:

Orthogonalized projection algorithm

$$\text{rank } [\phi(1) \cdots \phi(m)] = n = \dim \theta_0 \tag{3.4.25}$$

Least squares

$$\lim_{N\to\infty} \lambda_{\min} \sum_{t=1}^{N} \phi(t)\phi(t)^T = \infty \tag{3.4.26}$$

Projection algorithm

$$\sum_{i=0}^{l-1} \frac{\phi(t+i)\phi(t+i)^T}{\phi(t+i)^T\phi(t+i)} \ge cI, \quad c > 0 \qquad \text{for all } t \text{ and some fixed } l > 0 \tag{3.4.27}$$

Obviously, the conditions are closely related although they differ in detail for the different algorithms. We shall next relate these conditions to corresponding conditions on the *input* signal. For simplicity, we shall restrict attention to the

DARMA models described in Section 2.3.4. We begin with a special case in which the AR terms are absent.

Deterministic Moving-Average Models

For this model class, the output $\{y(t)\}$ is described in terms of the input by

$$y(t) = \sum_{j=1}^{n} b_j u(t - j) \qquad (3.4.28)$$
$$= \phi(t - 1)^T \theta_0$$

where

$$\theta_0^T = [b_1 \quad \cdots \quad b_n]$$
$$\phi(t - 1)^T = [u(t - 1) \quad \cdots \quad u(t - n)]$$

In this case, conditions (3.4.25) to (3.4.27) are explicitly expressible in terms of the input $\{u(t)\}$. For example, condition (3.4.27) is implied by the following "strong persistent excitation" condition:

Definition 3.4.A. A scalar input signal $\{u(t)\}$ is said to be *strongly persistently exciting* of order n if for all t, there exists an integer l such that

$$\rho_1 I > \sum_{k=t}^{t+l} \begin{bmatrix} u(k + n) \\ u(k + n - 1) \\ \cdot \\ \cdot \\ \cdot \\ u(k + 1) \end{bmatrix} [u(k + n) \quad \cdots \quad u(k + 1)] > \rho_2 I \qquad (3.4.29)$$

where $\rho_1, \rho_2 > 0$.

▼▼▼

The weaker condition (3.4.26) is implied by the following "weak persistent excitation" condition.

Definition 3.4.B. A scalar input signal $\{u(t)\}$ is said to be *weakly persistently exciting* of order n if

$$\rho_1 I \geq \lim_{N \to \infty} \frac{1}{N} \sum_{t=1}^{N} \begin{bmatrix} u(t + n) \\ \cdot \\ \cdot \\ \cdot \\ u(t + 1) \end{bmatrix} [u(t + n) \quad \cdots \quad u(t + 1)] \geq \rho_2 I \qquad (3.4.30)$$

where $\rho_1, \rho_2 > 0$.

▼▼▼

The condition above has an interesting interpretation in the frequency domain.

Lemma 3.4.5. A stationary input $\{u(t)\}$ is weakly persistently exciting of order n if its two-sided spectrum is nonzero at n points (or more).

Proof. From the Herglotz theorem (Burill, 1973) the time-domain limit in (3.4.30) can be expressed in the frequency domain as

$$M = \frac{1}{2\pi} \int_{-\pi}^{\pi} \begin{bmatrix} 1 \\ e^{-jw} \\ \cdot \\ \cdot \\ \cdot \\ e^{-j(n-1)w} \end{bmatrix} dF_u(w)[1 \quad e^{jw} \quad \cdots \quad e^{j(n-1)w}] \qquad (3.4.31)$$

where $F_u(w)$ is the spectral distribution function for $\{u(t)\}$.

We will assume the contrary to the result and show how this leads to a contradiction. Thus assume that $F_u(w)$ is nonzero at n points but that there exists a nonzero λ such that

$$\lambda^T M \lambda = 0 \qquad (3.4.32)$$

Equation (3.4.32) implies that

$$\frac{1}{2\pi} \int_{-\pi}^{\pi} |\lambda^T v(e^{jw})|^2 \, dF_u(w) = 0 \qquad (3.4.33)$$

where

$$v(e^{jw}) = [1 \quad e^{-jw} \quad \cdots \quad e^{-j(n-1)w}]^T \qquad (3.4.34)$$

Equation (3.4.33) implies that $\lambda^T v(e^{jw})$ must be zero at the points of support of $F_u(w)$, that is,

$$\lambda_1 + \lambda_2 e^{-jw} + \cdots + \lambda_n e^{-j(n-1)w} = 0 \qquad \text{at } n \text{ or more points} \qquad (3.4.35)$$

However, the polynomial equation $\lambda_1 + \lambda_2 z + \cdots + \lambda_n z^{n-1} = 0$ has at most $(n-1)$ roots. This contradicts (3.4.35), thus establishing the lemma.

▼▼▼

In particular, the lemma above shows that a combination of $n/2$ sinusoids (having frequency $\neq 0$ or π and random phase) will be weakly persistently exciting of order n since each sine wave has two points of support in the two-sided spectrum.

The above discussion is summarized in

Lemma 3.4.6. For deterministic moving-average models:

(i) The projection algorithm is exponentially convergent provided that the system input is strongly persistently exciting of order n.

(ii) The least-squares algorithm converges (at rate $1/t$) provided that the system input is weakly persistently exciting of order n; in particular, it suffices to use a stationary input whose spectral distribution is nonzero at n points or more.

Proof. Immediate from the discussion above and Lemma 3.4.5.

▼▼▼

We next consider the general case of DARMA models.

Deterministic Autoregressive Moving-Average Models

Consider the single-input single-output time-invariant system in the DARMA form [equation (2.3.42)]:

$$y(t) = -\sum_{j=1}^{n} a_j y(t-j) + \sum_{j=1}^{n} b_j u(t-j) \qquad (3.4.36)$$

When the model (3.4.36) is expressed in the usual form as

$$y(t) = \phi(t-1)^T \theta_0 \qquad (3.4.37)$$

the vector $\phi(t-1)$ has the form

$$\phi(t-1) = [y(t-1), y(t-2), \ldots, y(t-n), u(t-1), \ldots, u(t-n)]^T \qquad (3.4.38)$$

A key difficulty now is to relate the conditions on $\{\phi(t)\}$ to conditions on $\{u(t)\}$ alone. The essential feature of the conditions (3.4.25), (3.4.26), and (3.4.27) is that $\{\phi(t)\}$ should span the whole space in some sense. Since $\{y(t)\}$ and hence $\{\phi(t)\}$ are related to $\{u(t)\}$ by a dynamical system, this naturally raises the question of the reachability of the model generating $\{\phi(t)\}$. This question is addressed in the following lemma.

Lemma 3.4.7 (Key Reachability Lemma)

(i) The vector $\phi(t)$ is generated by the following state-space model:

$$\phi(t) = \left[\begin{array}{ccc|ccc}
-a_1 & \cdots & -a_n & b_1 & \cdots & b_n \\
1 & & & & & \\
& \ddots & & & & \\
& & 1 \quad 0 & & & \\
0 & \cdots & 0 & 0 & \cdots & 0 \\
& & 1 & & & \\
& & & \ddots & & \\
& & & 1 & 0 &
\end{array}\right] \phi(t-1) + \left[\begin{array}{c}
0 \\ \cdot \\ \cdot \\ \cdot \\ 0 \\ \hline 1 \\ 0 \\ \cdot \\ \cdot \\ 0
\end{array}\right] u(t) \qquad (3.4.39)$$

$$\triangleq F\phi(t-1) + Gu(t) \qquad (3.4.40)$$

(ii) The $2n$-dimensional state-space model above is completely reachable if and only if $A(q^{-1})$ and $B(q^{-1})$ are relatively prime.

Proof. (i) Immediate from the definition of $\phi(t)$.

(ii) A standard result (Kailath, 1980) shows that a pair $\{F, G\}$ is completely reachable if and only if

$$\text{rank } [\lambda I - F \quad G] = \dim F \qquad \text{for all } \lambda$$

Now from (3.4.40) we have

$$[\lambda I - F \quad G] = \left[\begin{array}{cccc|cccc|c}
(\lambda + a_1) & a_2 & \cdots & a_n & -b_1 & \cdots & -b_n & 0 \\
-1 & \lambda & & & & & & \cdot \\
& \ddots & \ddots & & & & & \cdot \\
& & -1 & \lambda & & & & 0 \\
\hline
& & & 0 & \lambda & & & 1 \\
& & & & -1 & \ddots & & 0 \\
& & & & & \ddots & \ddots & \cdot \\
& & & & & -1 & \lambda & 0
\end{array}\right] \qquad (3.4.41)$$

The dependence of the matrix above on the polynomials $A(\lambda)$ and $B(\lambda)$ is somewhat implicit. However, by a sequence of elementary column operations (which do not alter the rank) we make this dependence more transparent. The basic idea is to multiply the first column by λ and add the result to the second column, thus eliminating λ from the second column. This procedure can be continued to remove all the λ's on the "diagonal." The set of transformations required is conveniently summarized in the following unimodular matrix:

$$U(\lambda) = U_1(\lambda)U_2(\lambda)U_3(\lambda) \tag{3.4.42}$$

where

$$U_1(\lambda) = \begin{bmatrix} 1 & \lambda & \cdots\cdots & \lambda^{n-1} & & \\ 0 & 1 & \lambda & \cdots & \lambda^{n-2} & & \\ & & \cdot & & \cdot & & 0 \\ & & & \cdot & \cdot & & \\ & & & & \cdot & & \\ & & & & 1 & & \\ \hline & & 0 & & & I_{(n+1)\times(n+1)} \end{bmatrix} \tag{3.4.43}$$

$$U_2(\lambda) = \begin{bmatrix} I_{n\times n} & 0 \\ \hline & 1 \\ & \hline \\ 0 & 0 \\ & -\lambda & I_{n\times n} \end{bmatrix} \tag{3.4.44}$$

$$U_3(\lambda) = \begin{bmatrix} I_{n\times n} & 0 & & 0 \\ \hline & 1 & \lambda & \cdots\cdots & \lambda^{n-1} & 0 \\ & 0 & 1 & \lambda & \cdots & \lambda^{n-2} & 0 \\ 0 & & & \cdot & & & \cdot \\ & & & & \cdot & & \cdot \\ & & & & & \cdot & \cdot \\ & & & & & 1 & 1 \end{bmatrix} \tag{3.4.45}$$

This yields

$$[\lambda I - F \quad G]U(\lambda) = \begin{bmatrix} A^{n-1}(\lambda) & \cdots & A^1(\lambda) & A^0(\lambda) & -B^{n-1}(\lambda) & \cdots & -B^0(\lambda) & 0 \\ -1 & & & & & & & 0 \\ & \cdot & & & & 0 & & \cdot \\ & & \cdot & & & & & \cdot \\ & & & -1 & 0 & & & 0 \\ \hline & & & & & 0 & 0 & 1 \\ & & & & -1 & & 0 & 0 \\ & & & & & \cdot & & \cdot \\ & & & & & \cdot & & \cdot \\ & & & & & & -1 & 0 & 0 \end{bmatrix}$$

where

$$A^0(\lambda) \triangleq \lambda^n + a_1\lambda^{n-1} + \cdots + a_n \tag{3.4.46}$$

$$B^0(\lambda) \triangleq b_1\lambda^{n-1} + \cdots + b_n \tag{3.4.47}$$

and $A^1(\lambda)$, $A^2(\lambda)$, and so on, are the Tschirnhausen polynomials defined as follows:

$$A^i(\lambda) \triangleq \lambda^{n-i} + a_1\lambda^{n-i-1} \cdots a_{n-i} \tag{3.4.48}$$

and similarly for $B^i(\lambda)$.

It is obvious by inspection of (3.4.45) that rank $[\lambda I - F \quad G]$ is always $(2n - 1)$ due to the location of the 1's and is $2n$ if and only if $A^0(\lambda)$ and $B^0(\lambda)$ do not vanish simultaneously, that is, if and only if $A(q^{-1})$ and $B(q^{-1})$ are relatively prime.

▼▼▼

The lemma above allows us to relate conditions on $\{\phi(t)\}$ to corresponding conditions on $\{u(t)\}$. We illustrate this for condition (3.4.26) applicable to the least-squares algorithm. We treat first the case of stable systems.

Lemma 3.4.8. For the DARMA model (3.4.36), the least-squares algorithm converges to θ_0 provided that

(i) The system is stable.
(ii) The input $\{u(t)\}$ has a spectral distribution function which is nonzero at $2n$ points (or more).
(iii) $A(q^{-1})$ and $B(q^{-1})$ are relatively prime.

Proof. From Lemma 3.4.7 and the Herglotz theorem (Burill, 1973), the time-domain limit in (3.4.30) can be expressed in the frequency domain as

$$M = \frac{1}{2\pi} \int_{-\pi}^{\pi} H(e^{jw})H(e^{-jw})^T \, dF_u(w) \tag{3.4.49}$$

where $F_u(w)$ is the input spectral distribution function and $H(e^{jw})$ is the transfer function of the model generating $\{\phi(t)\}$; that is, from (3.4.40),

$$H(e^{jw}) \triangleq (zI - F)^{-1}G; \qquad z = e^{jw} \tag{3.4.50}$$

Now, in view of Lemma 3.4.7 and part (iii) of the theorem statement, the pair (F, G) is completely reachable. Thus from Section 2.2.2 these exists a similarity transformation P which converts (F, G) to controller form; that is,

$$\phi(t) = P\phi'(t) \tag{3.4.51}$$

where $\phi'(t)$ satisfies a model of the form

$$\phi'(t + 1) = F'\phi'(t) + G'(t)u(t) \tag{3.4.52}$$

and

$$F' = \begin{bmatrix} * & * & * & * & * \\ 1 & & & & \\ & 1 & & & \\ & & 1 & & \\ & & & 1 & 0 \end{bmatrix}; \qquad * \text{ a nonzero element} \tag{3.4.53}$$

$$G' = \begin{bmatrix} 1 \\ 0 \\ 0 \\ 0 \\ 0 \end{bmatrix} \tag{3.4.54}$$

Hence M can be expressed as

$$M = P\left[\frac{1}{2\pi}\int_{-\pi}^{\pi} H'(e^{jw})H'(e^{-jw})^T \, dF_u(w)\right]P^T \qquad (3.4.55)$$

and

$$H'(e^{jw}) = (zI - F')^{-1}G'; \qquad z = e^{jw} \qquad (3.4.56)$$

In view of the structure of (3.4.53)–(3.4.54), it can be seen that (Exercise 3.26)

$$H'(z) = \frac{1}{a(z)}\begin{bmatrix} z^{2n-1} \\ z^{2n-2} \\ \cdot \\ \cdot \\ \cdot \\ 1 \end{bmatrix} \qquad (3.4.57)$$

where

$$a(z) \triangleq \det\,[zI - F'] \qquad (3.4.58)$$

Now, as in the proof of Lemma 3.4.5, we assume the contrary to the result; that is, we assume that $F_u(w)$ is nonzero at $2n$ points but that there exists a nonzero λ such that

$$\lambda^T M \lambda = 0 \qquad (3.4.59)$$

This implies, using (3.4.55) and (3.4.57), that

$$\frac{1}{2\pi}\int_{-\pi}^{\pi} |\gamma^T v(e^{jw})|^2 \frac{1}{|a(e^{jw})|^2} \, dF_u(w) = 0 \qquad (3.4.60)$$

where $\gamma = P^T \lambda \neq 0$ since P is nonsingular and $\lambda \neq 0$ and

$$v(e^{jw}) = [1 \quad e^{-jw} \quad \cdots \quad e^{-j(2n-1)w}]^T \qquad (3.4.61)$$

The remainder of the proof is exactly as for the proof of Lemma 3.4.5 from (3.4.34) onward.

▼▼▼

The result above assumes a stable system. In the case of an unstable system, one needs to establish that the initial conditions will not invalidate the result. In the following lemma we generalize Lemma 3.4.8 to the case where the system may be unstable. We consider the input to be a combination of sinewaves. The result is general since a set of N samples has a discrete Fourier expansion with N terms!

Lemma 3.4.9. (Persistency of Excitation in the presence of possibly unbounded signals):
Consider the system (3.4.36) then, provided

(i) $A(q^{-1})$, $B(q^{-1})$ are relatively prime.
(ii) The input is generated by a feedback control law of the following general form:

$$u(t) = -K\phi(t-1) + v(t) \qquad (3.4.62)$$

such that K is constant over the interval $I(t_0) = [t_0, t_0 + N - 1]$

(iii) the external input, $v(t)$, is of the form:

$$v(t) = \sum_{k=1}^{s} \Gamma_k \sin(\omega_k t + \sigma_k) \tag{3.4.63}$$

where $\omega_k \in (0, \pi)$; $\Gamma_k \neq 0$ and $\omega_j \neq \omega_k$; $k = 1, \ldots, s$; $j = 1, \ldots, s$

(iv) the length, N, of the interval and number, s, of non-zero coefficients satisfy

(a) $N \geq 10n$ (3.4.64)

(b) $s \geq 4n$ (n is the system order)   (3.4.65)

then we can conclude that

$$\lambda_{\min}[X(t_0 + 1)X(t_0 + 1)^T] \geq \epsilon_1 > 0 \tag{3.4.66}$$

where ϵ_1 is *independent* of t_0 and the initial conditions, $\phi(t_0)$ and where

$$X(t_0 + 1) = [\phi(t_0 + 1), \phi(t_0 + 2), \ldots, \phi(t_0 + N)] \tag{3.4.67}$$

Proof. Substituting (3.4.62) into (3.4.40) gives

$$\phi(t) = \tilde{F}\phi(t - 1) + Gv(t) \tag{3.4.68}$$

where

$$\tilde{F} = F - GK \tag{3.4.69}$$

It follows from the Key Reachability Lemma (Lemma 3.4.7) and assumption (i) that $\{F, G\}$ is completely reachable. Since reachability is not altered by state variable feedback (Kailath, 1980), then $(\tilde{F}, G)$ is also completely reachable.

Since $\tilde{F}$ is real $(2n \times 2n)$, it has at most n eigenvalues of the form $\lambda = e^{j\omega_k}$; $\omega_k \in (0, \pi)$. Thus in view of assumption (iv) (b) and (3.4.63) we can be sure that, at least, $3n$ of the $\{\omega_k\}$ in (3.4.63) satisfy

$$\rho(e^{j\omega_k}) \triangleq \det(e^{j\omega_k}I - \tilde{F}) \neq 0 \tag{3.4.70}$$

Thus, without loss of generality, we can assume (3.4.70) holds for $k = 1, \ldots, s'$ (where $s' \geq s - n$). Hence, by building a state space model for the remaining sinusoids, we can represent the system as follows:

$$\begin{bmatrix} \phi(t) \\ \eta(t) \end{bmatrix} = \begin{bmatrix} \tilde{F} & \tilde{F}_1 \\ 0 & \tilde{F}_2 \end{bmatrix} \begin{bmatrix} \phi(t - 1) \\ \eta(t - 1) \end{bmatrix} + \begin{bmatrix} G \\ 0 \end{bmatrix} v'(t) \tag{3.4.71}$$

where $\{\eta(t)\}$ has maximum dimension $2n$ and $\tilde{F}_2$ has roots on the unit circle at $\{e^{\pm j\omega_k}: \det(e^{j\omega_k}I - \tilde{F}) = 0\}$ and

$$v'(t) = \sum_{k=1}^{s'} \Gamma_k \sin(\omega_k t + \sigma_k) \tag{3.4.72}$$

The solution of (3.4.71) can now be written as (for some g_0)

$$\phi(t) = [I \quad 0](\tilde{F})^\tau g_0 + \frac{1}{2} \sum_{\substack{k=-s' \\ k \neq 0}}^{s'} (e^{j\omega_k}I - \tilde{F})^{-1} G\Gamma_k e^{j(\omega_k \tau + \sigma_k)} \tag{3.4.73}$$

where

$$\omega_{-k} = -\omega_k; \; \sigma_{-k} = -\sigma_k; \; \bar{F} = \begin{bmatrix} \tilde{F} & \tilde{F}_1 \\ 0 & \tilde{F}_2 \end{bmatrix}; \; \tau = t - t_0 \tag{3.4.74}$$

so that $\bar{F}$ has dimension $l \leq 4n$.

Now since $(\tilde{F}, G)$ is completely reachable, there exists a nonsingular similarity transformation, Q, that converts $(\tilde{F}, G)$ into controller canonical form (see Chapter 2). Thus, we can write (3.4.73) as

$$\phi(t) = [I \quad 0](\bar{F})^\tau g_0 + \frac{1}{2} \sum_{\substack{k=-s' \\ k \neq 0}}^{s'} Q(e^{j\omega_k}I - \tilde{F}_c)^{-1}G_c\Gamma_k e^{j(\omega_k \tau + \sigma_k)} \qquad (3.4.75)$$

where $(\tilde{F}_c, G_c)$ is the controller form of $(\tilde{F}, G)$.

Now using the result in Exercise (2.26) on controller forms as in Lemma 3.4.8, we have

$$(e^{j\omega_k}I - \tilde{F}_c)^{-1}G_c = \frac{1}{\rho(e^{j\omega_k})}\begin{bmatrix} 1 \\ e^{j\omega_k} \\ (e^{j\omega_k})^{2n-1} \end{bmatrix} \qquad (3.4.76)$$

and $\rho(e^{j\omega_k})$ is as defined in (3.4.70) and is nonzero for $k = 1, \ldots, s'$.

Now to establish (3.4.66) it suffices to show that

$$r^T[X(t_0 + 1)X(t_0 + 1)^T]r \geq \epsilon_2 \text{ for all } r \neq 0 \qquad (3.4.77)$$

where $\epsilon_2 > 0$ independent of t_0 and $\phi(t_0)$.

Now we see that

$$[\lambda_{\max} MM^T]h^T h > h^T MM^T h \text{ for any matrix } M \qquad (3.4.78)$$

Hence (3.4.77) will follow with $\epsilon_2 = \dfrac{\epsilon_3}{\lambda_{\max}[MM^T]}$, if we can show

$$r^T[X(t_0 + 1)MM^T X(t_0 + 1)^T]r \geq \epsilon_3 \text{ for all } r \neq 0, \text{ for some}$$

$M \neq 0$, and for some

$$\epsilon_3 > 0 \text{ (independent of } t_0 \text{ and } \phi(t_0)). \qquad (3.4.79)$$

A key difficulty is to eliminate the initial condition terms in (3.4.75). We do this by introducing a particular matrix, M, into (3.4.78) of dimension ($N \times q$ where $q = N - \ell$) defined as follows:

$$M = \begin{bmatrix} 1 & 0 \cdot \cdot \cdot 0 \\ & & \\ \alpha_1 & & 0 \\ & & \\ \alpha_\ell & \cdot & 1 \\ & & \\ 0 & & \alpha_1 \\ & & \\ 0 \cdot \cdot 0 & & \alpha_\ell \end{bmatrix} \qquad (3.4.80)$$

where $1 + \alpha_1 z + \cdots + \alpha_\ell z^\ell$ is the characteristic polynomial of $\bar{F}$ as defined in (3.4.74).

Now from (3.4.75), (3.4.76) we have

$$r^T X(t_0 + 1) = W_1^T[(\bar{F})^1 g_0, \ldots, (\bar{F})^N g_0]$$

$$+ \frac{1}{2} \sum_{\substack{k=-s' \\ k \neq 0}}^{s'} \frac{\Gamma_k e^{j\sigma_k}}{\rho(e^{j\omega_k})} W_2^T \begin{bmatrix} e^{j\omega_k} & \cdots & e^{j\omega_k N} \\ \cdot & & \cdot \\ \cdot & & \cdot \\ \cdot & & \cdot \\ e^{j\omega_k(2n)} & \cdots & e^{j\omega_k(N+2n-1)} \end{bmatrix} \quad (3.4.81)$$

where

$$W_1^T \triangleq r^T[I, 0]; \quad W_2^T \triangleq r^T Q = [W_{2,1} \quad \cdots \quad W_{2,2n}] \quad (3.4.82)$$

We can now write (3.4.81) as

$$r^T X(t_0 + 1) = W_1^T[(\bar{F})^1 g_0 \quad \cdots \quad (\bar{F})^N g_0]$$

$$+ \frac{1}{2} \sum_{\substack{k=-s' \\ k \neq 0}}^{s'} \frac{\Gamma_k e^{j\sigma_k}}{\rho(e^{j\omega_k})} p_1(e^{j\omega_k})(e^{j\omega_k} \quad \cdots \quad e^{j\omega_k N}) \quad (3.4.83)$$

where

$$p_1(z) = W_{2,1} + W_{2,2} z + \cdots W_{2,2n} z^{2n-1} \quad (3.4.84)$$

Now multiplying (3.4.83) on the right by M as in (3.4.80) gives

$$r^T X(t_0 + 1)M = W_1^T[(\bar{F})^1 g_0 \quad \cdots \quad (\bar{F})^N g_0]M$$

$$+ \frac{1}{2} \sum_{\substack{k=-s' \\ k \neq 0}}^{s'} \frac{\Gamma_k e^{j\sigma_k}}{\rho(e^{j\omega_k})} p_1(e^{j\omega_k})[e^{j\omega_k} \quad \cdots \quad e^{j\omega_k N}]M \quad (3.4.85)$$

Now, from the Cayley Hamilton theorem, we know every matrix satisfies its own characteristic equation and hence, from the definition of M, it can readily be seen that the first term on the right hand side of (3.4.85) is identically zero for all W_1 and g_0! Finally using the structure of M as in (3.4.80) we have

$$r^T X(t_0 + 1)M = \sum_{\substack{k=-s' \\ k \neq 0}}^{s'} \frac{\Gamma_k e^{j\sigma_k}}{2\rho(e^{j\omega_k})} p_1(e^{j\omega_k}) p_2(e^{j\omega_k})[e^{j\omega_k} \quad \cdots \quad e^{j\omega_k q}] \quad (3.4.86)$$

where $p_2(z) = 1 + \alpha_1 z + \cdots \alpha_\ell z^\ell$ is the characteristic polynomial of $\bar{F}$, and $q = N - \ell$ as in (3.4.80).

Now the summation in (3.4.86) can be written as an inner product as follows:

$$r^T X(t_0 + 1)M = [\delta_{-s'}, \ldots, \delta_{-1}, \delta_1, \ldots, \delta_{s'}] \begin{bmatrix} 1 & (z_{-s'})^1 & \cdots & (z_{-s'})^{q-1} \\ \cdot & \cdot & & \cdot \\ \cdot & \cdot & & \cdot \\ \cdot & \cdot & & \cdot \\ 1 & (z_{s'})^1 & & (z_{s'})^{q-1} \end{bmatrix} \quad (3.4.87)$$

where

$$\delta_k = \frac{\Gamma_k e^{j\sigma_k}}{2\rho(z_k)} p_1(z_k) p_2(z_k) z_k; \quad z_k \triangleq e^{j\omega_k}; \quad k = -s', \ldots, s'; \quad k \neq 0 \quad (3.4.88)$$

Now $p_1(z)p_2(z)$ is a $(2n - 1 + \ell) \leq (6n - 1)$ order polynomial and can have at most $(6n - 1)$ roots. However in evaluating $[\delta_{-s'}, \ldots, \delta_{s'}]$ we use $2s'$ values of z. Hence all terms cannot be zero so long as $2s' \geq 6n$. Also, the matrix on the far right hand side of (3.4.87) is a Vandermonde matrix and since $z_j \neq z_k$ for $j \neq k$, it follows this matrix has rank $2s'$ provided $q \geq 2s'$. Thus in summary, $r^T X(t_0 + 1)M$ is a nonzero vector provided

$$q \geq 2s' \geq 6n \tag{3.4.89}$$

However, (3.4.89) is automatically satisfied, provided assumption (iv) (a) and (b) hold, since $s' \geq s - n$ and $q = N - \ell$.

Finally, the scalar on the left hand side of (3.4.79) is the sum of the squares of the elements of the vector $r^T X(t_0 + 1)M$ and hence (3.4.79) holds subject to the assumptions of the theorem. The result then follows using the argument given in (3.4.77) and sequel.

▼▼▼

The result above was inspired by the joint work of Elliott, Christi and Das (1982).

The lemma immediately implies that (3.4.26) is satisfied and hence parameter convergence follows for the least squares algorithm applied to unstable systems. This is an important result in that it lends further credibility to the application of adaptive algorithms to *not necessarily stable systems* having unknown parameters. Clearly, if it is possible to estimate the parameters for systems that are initially unstable, it will not be difficult to ultimately achieve stability and tracking by using a suitable control system design strategy employing the estimated parameters. This will be explored in Chapter 6.

In the next section we shall return to the weaker assumptions of Section 3.3 (i.e., without persistent excitation.) However, we shall now consider a slightly different class of algorithm. These algorithms are called *output error methods*. We shall analyze these algorithms in a deterministic framework. However, our main purpose in studying these algorithms is to pave the way for a deeper study of their properties in a stochastic setting in Chapter 9. (In fact, we shall find that they give unbiased estimation in the presence of white output noise.)

3.5 OUTPUT ERROR METHODS

Consider again the general parameter estimation algorithm given in (3.2.2), that is,

$$\hat{\theta}(t) = \hat{\theta}(t - 1) + M(t - 1)\bar{\phi}(t - 1)\bar{e}(t) \tag{3.5.1}$$

Here we shall be concerned with systems described by a DARMA model:

$$A(q^{-1})y(t) = B(q^{-1})u(t) \tag{3.5.2}$$

where $A(q^{-1})$ and $B(q^{-1})$ are polynomials in the delay operator q^{-1}. We shall therefore think of $\hat{\theta}(t)$ as providing the estimated coefficients in an autoregressive polynomial $\hat{A}(t, q^{-1})$ and moving-average polymomial $\hat{B}(t, q^{-1})$.

The parameter estimation methods of Section 3.3 are often called *equation error* methods. The reason for this name is that $\bar{e}(t)$ in (3.5.1) is taken to be the error in the estimated model equation, that is,

$$
\begin{aligned}
\bar{e}(t) &= \hat{A}(t - 1, q^{-1})y(t) - \hat{B}(t - 1, q^{-1})u(t) \\
&= y(t) + \hat{a}_1(t - 1)y(t - 1) + \cdots + \hat{a}_n(t - 1)y(t - n) \\
&\quad - \hat{b}_1(t - 1)u(t - 1) - \cdots - \hat{b}_m(t - 1)u(t - m) \\
&= y(t) - \phi(t - 1)^T\hat{\theta}(t - 1)
\end{aligned} \tag{3.5.3}
$$

where

$$\phi(t-1)^T = [-y(t-1), \ldots, -y(t-n), u(t-1), \ldots, u(t-m)] \qquad (3.5.4)$$

$$\hat{\theta}(t-1)^T = [\hat{a}_1(t-1), \ldots, \hat{a}_n(t-1), \hat{b}_1(t-1), \ldots, \hat{b}_m(t-1)] \qquad (3.5.5)$$

The generation of $\bar{e}(t)$ is depicted in Fig. 3.5.1, which motivates the term "equation error."

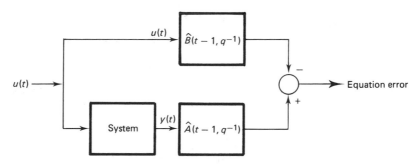

Figure 3.5.1 Diagrammatic representation of equation error.

In this section we will use a different driving error term, $\bar{e}(t)$, in the algorithm. We first define a model output $\bar{y}(t)$ at time t in terms of $\hat{A}(t, q^{-1})$, $\hat{B}(t, q^{-1})$ and past values of $\bar{y}(t-1)$, $\bar{y}(t-2)$, $\ldots$ (which are assumed given) as

$$\hat{A}(t, q^{-1})\bar{y}(t) = \hat{B}(t, q^{-1})u(t) \qquad (3.5.6)$$

Note that $\bar{y}(t)$ is a function of $\bar{y}(t-1)$, $\ldots$, and *not* of $y(t-1)$, $\ldots$. This is in contrast to the equation error method, where the model output can be thought of as $\hat{y}(t) \triangleq \phi(t-1)^T \hat{\theta}(t-1)$ with $\phi(t-1)$ as in (3.5.4). The model (3.5.6) is depicted diagrammatically in Fig. 3.5.2, which motivates the term "*output error.*"

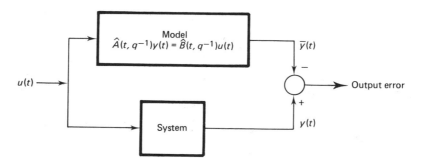

Figure 3.5.2 Diagrammatic representation of output error.

Since $\bar{y}(t)$ depends on $\hat{\theta}(t)$ rather than $\hat{\theta}(t-1)$, we shall call $\bar{y}(t)$ the *a posteriori model output*. We can then define an *a posteriori model output error* as

$$\eta(t) = y(t) - \bar{y}(t) \qquad (3.5.7)$$

We now wish to investigate parameter estimation algorithms (*output error methods*), which, in some sense, use $\eta(t)$ as the basis of parameter update. We first write (3.5.6) in our standard form as

$$\bar{y}(t) = \bar{\phi}(t-1)^T \hat{\theta}(t) \tag{3.5.8}$$

where

$$\bar{\phi}(t-1)^T = [-\bar{y}(t-1), \ldots, -\bar{y}(t-n), u(t-1), \ldots, u(t-m)] \tag{3.5.9}$$

$$\hat{\theta}(t) = [\hat{a}_1(t), \ldots, \hat{a}_n(t), \hat{b}_1(t), \ldots, \hat{b}_m(t)] \tag{3.5.10}$$

Note that $\bar{y}(t), \bar{y}(t-1), \ldots$ are functions of $\hat{\theta}(t), \hat{\theta}(t-1), \ldots$, respectively.
We define the following:

A priori model output

$$\hat{y}(t) \triangleq \bar{\phi}(t-1)^T \hat{\theta}(t-1) \tag{3.5.11}$$

A priori model output error

$$e(t) \triangleq y(t) - \hat{y}(t) \tag{3.5.12}$$

Generalized a posteriori output error

$$\bar{\eta}(t) \triangleq D(q^{-1})\eta(t) \tag{3.5.13}$$

where $D(q^{-1}) \triangleq 1 + d_1 q^{-1} + \cdots + d_\ell q^{-\ell}$ is a fixed moving-average filter. Note that $\bar{\eta}(t)$ depends on $\bar{y}(t)$ and thus on $\hat{\theta}(t)$.

Generalized a priori output error

$$\bar{v}(t) \triangleq e(t) + [D(q^{-1}) - 1]\eta(t) \tag{3.5.14}$$

Note that $\bar{v}(t)$ depends on $\hat{y}(t)$ and $\bar{y}(t-1), \ldots$ and thus on $\hat{\theta}(t-1)$.

We are now in a position to define the output error algorithm. For simplicity, we illustrate the notions involved through the projection algorithm.

Output Error Estimation Method

$$\hat{\theta}(t) = \hat{\theta}(t-1) + \frac{\bar{\phi}(t+1)}{1 + \bar{\phi}(t-1)^T \bar{\phi}(t-1)} \bar{v}(t) \tag{3.5.15}$$

where $\bar{\phi}(t-1)$ is as in (3.5.9) and $\bar{v}(t)$ is the generalized a priori output error defined in (3.5.14).

The properties of the algorithm are discussed in the following theorem, which uses the notion of passivity (see Appendix C).

Theorem 3.5.1. Consider the algorithm (3.5.15) applied to the DARMA model (3.5.2); then, provided that the system $H(q^{-1}) = D(q^{-1})/A(q^{-1})$ is very strictly passive (VSP),

(i) $\displaystyle\lim_{N\to\infty} \sum_{t=1}^{N} \eta(t)^2 < \infty$ \hfill (3.5.16)

this implies

$$\lim_{N\to\infty} \sum_{t=1}^{N} \bar{\eta}(t)^2 < \infty \tag{3.5.17}$$

(ii) $\displaystyle\lim_{N\to\infty} \sum_{t=1}^{N} \bar{\phi}(t-1)^T \bar{\phi}(t-1)\bar{\eta}(t)^2 < \infty$ \hfill (3.5.18) this implies

$$\lim_{N \to \infty} \sum_{t=1}^{N} \| \hat{\theta}(t) - \hat{\theta}(t - k) \|^2 < \infty \qquad (3.5.19)$$

(iii) If $\{u(t)\}$ is bounded, then

$$\lim_{t \to \infty} \bar{v}(t) = 0 \qquad (3.5.20)$$

$$\lim_{t \to \infty} |y(t) - \hat{y}(t)| = 0 \qquad (3.5.21)$$

Proof

Step 1: Define

$$b(t) \triangleq -\bar{\phi}(t - 1)^T \tilde{\theta}(t) \qquad (3.5.22)$$

where

$$\tilde{\theta}(t) \triangleq \hat{\theta}(t) - \theta_0 \qquad (3.5.23)$$

We now find a relationship between $b(t)$ and $\bar{\eta}(t)$. From (3.5.2),

$$A(q^{-1})y(t) = B(q^{-1})u(t) \qquad (3.5.24)$$

and from (3.5.6),

$$\hat{A}(t, q^{-1})\bar{y}(t) - \hat{B}(t, q^{-1})u(t) = 0 \qquad (3.5.25)$$

Combining (3.5.24) and (3.5.25) and introducing $A(q^{-1})\bar{y}(t)$, we have

$$A(q^{-1})[y(t) - \bar{y}(t)] = B(q^{-1})u(t) - A(q^{-1})\bar{y}(t) + \hat{A}(t, q^{-1})\bar{y}(t) - \hat{B}(t, q^{-1})u(t)$$
$$= -\bar{\phi}(t - 1)^T \tilde{\theta}(t)$$

or

$$A(q^{-1})\eta(t) = b(t) \qquad (3.5.26)$$

Thus, using (3.5.13), $\bar{\eta}(t)$ is related to $b(t)$ by

$$A(q^{-1})\bar{\eta}(t) = D(q^{-1})b(t) \qquad (3.5.27)$$

which is very strictly passive by assumption.

Step 2: Multiplying (3.5.15) by $\bar{\phi}(t - 1)^T$ and then subtracting from $y(t)$ gives

$$\eta(t) = e(t) - \frac{\bar{\phi}(t - 1)^T \bar{\phi}(t - 1)}{1 + \bar{\phi}(t - 1)^T \bar{\phi}(t - 1)} \bar{v}(t) \qquad (3.5.28)$$

But from (3.5.13) and (3.5.14),

$$\eta(t) = \bar{\eta}(t) - \bar{v}(t) + e(t)$$

Substituting for $\eta(t)$ into (3.5.28) and simplifying gives

$$\bar{\eta}(t) = \frac{\bar{v}(t)}{1 + \bar{\phi}(t - 1)^T \bar{\phi}(t - 1)} \qquad (3.5.29)$$

Step 3: Substituting (3.5.29) into (3.5.15) gives

$$\hat{\theta}(t) = \hat{\theta}(t - 1) + \bar{\phi}(t - 1)\bar{\eta}(t) \qquad (3.5.30)$$

[*Note:* This is a more convenient form of the algorithm for analysis than the implementable form (3.5.15).] Or, subtracting θ_0 from both sides and rearranging yields

$$\tilde{\theta}(t) - \bar{\phi}(t - 1)\bar{\eta}(t) = \tilde{\theta}(t - 1) \qquad (3.5.31)$$

Now let
$$V(t) = \tilde{\theta}(t)^T \tilde{\theta}(t) \tag{3.5.32}$$
Then from (3.5.31)
$$V(t) - 2\tilde{\phi}(t-1)^T\tilde{\theta}(t)\bar{\eta}(t) + \tilde{\phi}(t-1)^T\tilde{\phi}(t-1)\bar{\eta}(t)^2 = V(t-1) \tag{3.5.33}$$
or
$$\begin{aligned} V(t) &= V(t-1) - 2b(t)\bar{\eta}(t) - \tilde{\phi}(t-1)^T\tilde{\phi}(t-1)\bar{\eta}(t)^2 \\ &= V(t-1) - 2b(t)[\bar{\eta}(t) - pb(t)] - 2pb(t)^2 - \tilde{\phi}(t-1)^T\tilde{\phi}(t-1)\bar{\eta}(t)^2 \end{aligned} \tag{3.5.34}$$
for some $p > 0$. Let $g(t) \triangleq \bar{\eta}(t) - pb(t)$ and note from (3.5.27) that $g(t)$ is given by
$$A(q^{-1})g(t) = [D(q^{-1}) - pA(q^{-1})]b(t) \tag{3.5.35}$$
Substituting into (3.5.34) gives us
$$V(t) = V(t-1) - 2b(t)g(t) - 2pb(t)^2 - \tilde{\phi}(t-1)^T\tilde{\phi}(t-1)\bar{\eta}(t)^2 \tag{3.5.36}$$
Now define
$$S(t) \triangleq 2\sum_{j=1}^{t} b(t)g(t) + K \tag{3.5.37}$$
and note that $p > 0$, $K < \infty$ can be chosen so that $S(t) \geq 0$ [this is a consequence of the very strict passivity of the system (3.5.35) (see Appendix C)].
Let $G(t) \triangleq V(t) + S(t) \geq 0$; then from (3.5.36) and (3.5.37),
$$G(t) = G(t-1) - 2pb(t)^2 - \tilde{\phi}(t-1)^T\tilde{\phi}(t-1)\bar{\eta}(t)^2 \tag{3.5.38}$$
Thus $G(t)$ is a nonnegative, nonincreasing function of time and therefore converges. Hence from (3.5.38),
$$\lim_{N\to\infty} \sum_{t=1}^{N} b(t)^2 < \infty \tag{3.5.39}$$
$$\lim_{N\to\infty} \sum_{t=1}^{N} \tilde{\phi}(t-1)^T\tilde{\phi}(t-1)\bar{\eta}(t)^2 < \infty \tag{3.5.40}$$
Step 4: Now very strict passivity of (3.5.35) implies that $A(z^{-1})$ is stable (Lemma C.2.1 of Appendix C). Then from (3.5.26) and (3.5.27) we have
$$\lim_{N\to\infty} \sum_{t=1}^{N} \eta(t)^2 < \infty \tag{3.5.41}$$
This establishes part (i). Equation (3.5.17) follows from (3.5.13) and (3.5.41). Part (ii) follows from (3.5.40).
From (3.5.30)
$$\tilde{\phi}(t-1)^T\tilde{\phi}(t-1)\bar{\eta}(t)^2 = \|\hat{\theta}(t) - \hat{\theta}(t-1)\|^2$$
Thus (3.5.19) follows from (ii) using the Schwarz inequality. Now from (3.5.29),
$$\bar{v}(t) = [1 + \tilde{\phi}(t-1)^T\tilde{\phi}(t-1)]\bar{\eta}(t) \tag{3.5.42}$$
Thus (3.5.20) follows from (3.5.17), the stability of $A(z^{-1})$, and the assumed boundedness of $\{u(t)\}$.
Finally, from (3.5.14),
$$\begin{aligned} e(t) &= \bar{v}(t) - [D(q^{-1}) - 1]\eta(t) \\ &= \bar{v}(t) - \bar{\eta}(t) + \eta(t) \end{aligned} \tag{3.5.43}$$
Thus (3.5.21) follows from part (i), (3.5.17), and (3.5.20).

▼▼▼

Theorem 3.5.1 parallels Lemma 3.3.2 for the equation error method and has a similar interpretation.

The result was first proved by Landau (1976, 1978a) via *hyperstability theory* (Popov, 1973) and later applied to infinite impulse response filtering by Johnson (1979, 1980). [The equivalence between hyperstability and the approach used above is discussed in Narendra and Valavani (1980).] Note that nothing has been said about the convergence of $\hat{\theta}(t)$ to θ_0 in the theorem above.

It is straightforward to show using the same proof techniques that similar results hold for the least-squares algorithm, save that the very strict passivity condition has to be strengthened. This is discussed in the following corollary:

Corollary 3.5.1. The least-squares variant of the algorithm (3.5.15), that is,

$$\hat{\theta}(t) = \hat{\theta}(t-1) + \frac{P(t-2)\bar{\phi}(t-1)}{1 + \bar{\phi}(t-1)P(t-2)\bar{\phi}(t-1)} \bar{v}(t) \tag{3.5.44}$$

$$P(t-1) = P(t-2) - \frac{P(t-2)\bar{\phi}(t-1)\bar{\phi}(t-1)^T P(t-2)}{1 + \bar{\phi}(t-1)^T P(t-2)\bar{\phi}(t-1)}; \qquad P(0) > 0 \tag{3.5.45}$$

when applied to the DARMA model (3.5.2), and provided that the transfer function $[D(z^{-1})/A(z^{-1}) - \frac{1}{2}]$ is very strictly passive, gives

(i) $\lim_{N\to\infty} \sum_{t=1}^{N} \eta(t)^2 < \infty$ $\qquad\qquad$ (3.5.46)

this implies

$$\lim_{N\to\infty} \sum_{t=1}^{N} \bar{\eta}(t)^2 < \infty \tag{3.5.47}$$

(ii) $\lim_{N\to\infty} \sum_{t=1}^{N} \bar{\phi}(t-1)^T P(t-2)\bar{\phi}(t-1)\bar{\eta}(t)^2 < \infty$ $\qquad$ (3.5.48)

this implies

$$\lim_{N\to\infty} \sum_{t=1}^{N} ||\hat{\theta}(t) - \hat{\theta}(t-k)||^2 < \infty \qquad \text{for any finite } k. \tag{3.5.49}$$

(iii) If $\{u(t)\}$ is bounded, then

$$\lim_{t\to\infty} \bar{v}(t) = 0 \tag{3.5.50}$$

$$\lim_{t\to\infty} |y(t) - \hat{y}(t)| = 0 \tag{3.5.51}$$

Proof. Essentially as for Theorem 3.5.1 (see Exercise 3.14).

▼▼▼

A key notion in the output error algorithms is the use of a posteriori model outputs in the regression vector $\bar{\phi}(t-1)$. We shall see later that this idea is also useful for parameter estimation in the stochastic case. In the stochastic ARMA case, the notion was first introduced by Young (1974), who called the resulting algorithm AML (approximate maximum likelihood). The algorithm has also been studied by Moore and Ledwich (1977) and Solo (1979).

A brief comment is in order regarding the very strict passivity condition in the theorem statement. We shall find this kind of condition occurs very often when the regression vector depends on estimated states such as $\{\bar{y}(t)\}$ in the output error algorithm. Since $A(q^{-1})$ is unknown, it may be difficult to choose $D(q^{-1})$ a priori to satisfy the very strict passivity condition. A unit transfer function $(H(z) = 1)$ is always very strictly passive and thus $D(q^{-1})$ is generally taken to be an estimate of $A(q^{-1})$. Landau (1978b) has argued that the very strictly passivity condition can be eliminated by setting $D(q^{-1})$ equal to $\hat{A}(q^{-1})$, the on-line estimate of $A(q^{-1})$. This is a very interesting idea and leads to a locally convergent algorithm (see Exercise 3.15).

Discussion. The use of output error methods is often motivated by a model of the following form:

$$y(t) = z(t) + n(t) \tag{3.5.52}$$

where $z(t)$ is a noise-free output related to $u(t)$ by

$$A(q^{-1})z(t) = B(q^{-1})u(t) \tag{3.5.53}$$

and $n(t)$ is a white noise sequence. If (3.5.52) and (3.5.53) are true, we obtain the following stochastic model for $y(t)$:

$$A(q^{-1})y(t) = B(q^{-1})u(t) + A(q^{-1})n(t) \tag{3.5.54}$$

It is known [see Goodwin and Payne (1977)] that equation error techniques of the type discussed in Section 3.3 will lead to biased parameter estimates since the equivalent noise in the equation error is $A(q^{-1})n(t)$, which is no longer white noise. We shall see later that the output error method can overcome this problem in the stochastic case (see Chapter 9).

Of course, in the deterministic case there seems little advantage, if any, in using output error method. In fact, in this case, the need for the very strictly passivity condition can be eliminated by applying the equation error methods of Section 3.3. However, we shall find the ideas of a priori error, a posteriori error, and passivity very useful in the analysis of stochastic algorithms presented later.

3.6 PARAMETER ESTIMATION WITH BOUNDED NOISE

In this section we investigate the performance of our two prototype estimation schemes (projection and least squares) in the presence of bounded noise. We shall find that, in the presence of bounded noise, the robustness of the algorithms can be enhanced by introducing a dead zone in the parameter update equation. This simple idea also leads us to remind the reader of the importance of always being on the lookout for problem-specific modifications to the basic algorithms to improve their performance and robustness in a given situation.

Consider the following system model:

$$y(t) = \phi(t - 1)^T\theta_0 + w(t) \tag{3.6.1}$$

where $w(t)$ denotes a bounded "noise" term that can account for measurement noise, inaccurate modeling, computer round-off error, and so on. The first term of the right-hand side of (3.6.1) is our usual model format, which, as we have seen previously, can describe both linear and nonlinear systems depending on the choice of $\phi(\cdot)$.

We introduce the following projection algorithm to handle the system (3.6.1).

Projection Algorithm with Dead Zone

$$\hat{\theta}(t) = \hat{\theta}(t-1) + \frac{a(t-1)\phi(t-1)}{c + \phi(t-1)^T\phi(t-1)}[y(t) - \phi(t-1)^T\hat{\theta}(t-1)] \qquad (3.6.2)$$

where $\hat{\theta}(0)$ is given, $c > 0$, and

$$a(t-1) = \begin{cases} 1 & \text{if } |y(t) - \phi(t-1)^T\hat{\theta}(t-1)| > 2\Delta \\ 0 & \text{otherwise} \end{cases} \qquad (3.6.3)$$

The motivation for the choice of $\{a(t)\}$ above is to turn off the algorithm when the prediction error is small compared with the size of the noise.

We can now establish the following extension of Lemma 3.3.2:

Lemma 3.6.1. Consider the system model given in (3.6.1), where $\{w(t)\}$ is a bounded sequence such that

$$\sup |w(t)| \leq \Delta \qquad (3.6.4)$$

Then the algorithm (3.6.2)–(3.6.3) has the following properties:

(i) $\|\hat{\theta}(t) - \theta_0\| \leq \|\hat{\theta}(t-1) - \theta_0\| \leq \|\hat{\theta}(0) - \theta_0\|; \quad t > 1$ $\qquad (3.6.5)$

(ii) $\displaystyle \lim_{N\to\infty} \sum_{t=1}^{N} \frac{a(t-1)[e(t)^2 - 4\Delta^2]}{c + \phi(t-1)^T\phi(t-1)} < \infty$ $\qquad (3.6.6)$

where $e(t)$ is the modeling error given by

$$e(t) = y(t) - \phi(t-1)^T\hat{\theta}(t-1) = -\phi(t-1)^T\tilde{\theta}(t-1) + w(t) \qquad (3.6.7)$$

and this implies

(a) $\displaystyle \lim_{t\to\infty} \frac{a(t-1)[e(t)^2 - 4\Delta^2]}{c + \phi(t-1)^T\phi(t-1)} = 0$ $\qquad (3.6.8)$

(b) $\displaystyle \limsup_{t\to\infty} \|\hat{\theta}(t) - \hat{\theta}(t-1)\| \leq \frac{2\Delta}{\sqrt{c}}$ $\qquad (3.6.9)$

Also, provided that $\{\phi(t)\}$ is a bounded sequence, we may conclude that

(c) $\limsup e(t) \leq 2\Delta$ $\qquad (3.6.10)$

Proof. Subtracting θ_0 from both sides of (3.6.2) and using (3.6.1) gives

$$\tilde{\theta}(t) = \tilde{\theta}(t-1) - \frac{a(t-1)\phi(t-1)}{c + \phi(t-1)^T\phi(t-1)}[\phi(t-1)^T\tilde{\theta}(t-1) - w(t)] \qquad (3.6.11)$$

Then, noting that $a(t-1) = 0$ or 1,

$$\|\tilde{\theta}(t)\|^2 = \|\tilde{\theta}(t-1)\|^2 - \frac{2a(t-1)[e(t) - w(t)]e(t)}{c + \phi(t-1)^T\phi(t-1)}$$

$$+ \frac{a(t-1)^2\phi(t-1)^T\phi(t-1)e(t)^2}{[c + \phi(t-1)^T\phi(t-1)]^2}$$

$$\leq \|\tilde{\theta}(t-1)\|^2 + \frac{a(t-1)}{c + \phi(t-1)^T\phi(t-1)}[2e(t)w(t)]$$

$$- \frac{a(t-1)e(t)^2}{c + \phi(t-1)^T\phi(t-1)}$$

$$\leq ||\tilde{\theta}(t-1)||^2 + \frac{a(t-1)}{c + \phi(t-1)^T \phi(t-1)} \left[\frac{e(t)^2}{2} + 2w(t)^2 \right] \qquad (3.6.12)$$

$$- \frac{a(t-1)e(t)^2}{c + \phi(t-1)^T \phi(t-1)} \qquad \text{since } 2ab \leq ka^2 + b^2/k \text{ for any } k \text{ (see Exercise 3.17)}$$

$$\leq ||\tilde{\theta}(t-1)||^2 - \frac{1}{2} \frac{a(t-1)e(t)^2}{c + \phi(t-1)^T \phi(t-1)}$$

$$+ \frac{2a(t-1)\Delta^2}{c + \phi(t-1)^T \phi(t-1)}$$

In view of (3.6.3), $\{||\tilde{\theta}(t)||^2\}$ is a nonincreasing sequence bounded below by zero. This establishes (i).

Summing both sides of (3.6.12), we obtain (ii). Then (a) immediately follows since $e(t)^2 - 4\Delta^2 \geq 0$. To establish (b), we note from (3.6.2) and (3.6.7) that

$$\hat{\theta}(t) - \hat{\theta}(t-1) = \frac{-a(t-1)\phi(t-1)}{c + \phi(t-1)^T \phi(t-1)} e(t)$$

Hence

$$||\hat{\theta}(t) - \hat{\theta}(t-1)||^2 = \frac{a(t-1)\phi(t-1)^T \phi(t-1)}{[c + \phi(t-1)^T \phi(t-1)]^2} e(t)^2 \qquad (3.6.13)$$

From (3.6.8)

$$\limsup_{t \to \infty} \frac{a(t-1)e(t)^2}{c + \phi(t-1)^T \phi(t-1)} \leq \frac{4\Delta^2}{C}$$

or

$$\limsup_{t \to \infty} \frac{a(t-1)(c + \phi(t-1)^T \phi(t-1))e(t)^2}{[c + \phi(t-1)^T \phi(t-1)]^2} \leq \frac{4\Delta^2}{C}$$

or

$$\limsup_{t \to \infty} \frac{a(t-1)\phi(t-1)^T \phi(t-1)e(t)^2}{[c + \phi(t-1)^T \phi(t-1)]^2} \leq \frac{4\Delta^2}{C}$$

Then noting (3.6.13), we see that (b) follows. Result (c) is immediate from (3.6.8); see Exercise 3.18.

▼▼▼

We shall make use of the result above in the context of adaptive control later in the book.

The algorithm above has been employed by Goodwin, Long, and McInnis (1980) for the adaptive control of bilinear systems with bounded errors. A similar algorithm has been employed by Narendra and Peterson (1980) for the adaptive control of continuous-time systems.

It is relatively straightforward to extend the discussion above to the case of the least-squares algorithm:

Least Squares with Dead Zone

$$\hat{\theta}(t) = \hat{\theta}(t-1) + \frac{a(t-1)P(t-2)\phi(t-1)}{1 + a(t-1)\phi(t-1)^T P(t-2)\phi(t-1)}$$

$$[y(t) - \phi(t-1)^T \hat{\theta}(t-1)] \qquad (3.6.14)$$

$$P(t-1) = P(t-2) - \frac{a(t-1)P(t-2)\phi(t-1)\phi(t-1)^T P(t-2)}{1 + a(t-1)\phi(t-1)^T P(t-2)\phi(t-1)} \qquad (3.6.15)$$

with $\hat{\theta}(0)$ given and $P(-1) = P_0 > 0$.

$$a(t-1) = \begin{cases} 1 & \text{if } \dfrac{|y(t) - \phi(t-1)^T\hat{\theta}(t-1)|^2}{1 + \phi(t-1)^T P(t-2)\phi(t-1)} > \Delta^2 > 0 \\ 0 & \text{otherwise} \end{cases} \qquad (3.6.16)$$

The reader may recall that the algorithm above was briefly introduced in Section 3.3 [equation (3.3.80)] in the context of weighted least-squares algorithms.

The analysis of the algorithm above follows similar lines to the proof of Lemma 3.6.1 and thus we will not go into the details. We invite the reader to show (Exercise 3.19) that

$$\limsup_{t \to \infty} \frac{e(t)^2}{1 + \phi(t-1)^T P(t-2)\phi(t-1)} \leq \Delta^2 \qquad (3.6.17)$$

and that, if $\{\phi(t)\}$ is bounded, then $\limsup e(t)^2$ is bounded.

3.7 CONSTRAINED PARAMETER ESTIMATION

It is often the case that one wishes to constrain the parameter estimates to lie within a certain region. For example, one may have priori knowledge about the allowable range of certain parameters or it may be known that the model possess a certain property (e.g., stability).

We shall consider the constrained region to be a closed convex region in parameter space denoted by $\mathcal{C}$. For example, if the first element of the parameter $\theta = [\theta_1 \quad \cdots \quad \theta_n]^T$ is known to be positive, then

$$\mathcal{C} = \{\theta : \theta_1 \geq b > 0\} \qquad (3.7.1)$$

We now consider the projection and least-squares algorithms separately.

The Projection Algorithm

If the algorithm leads to a $\hat{\theta}(t)$ outside $\mathcal{C}$, one simply projects $\hat{\theta}(t)$ orthogonally onto the surface of $\mathcal{C}$ before continuing. This is illustrated in Fig. 3.7.1.

The key point about the algorithm is that the projected estimate, $\hat{\theta}'(t)$, is closer (in the sense of the Euclidean norm) to θ_0 than is $\hat{\theta}(t)$ for all $\theta_0 \in \mathcal{C}$. This can be readily proved and is shown diagrammatically in Fig. 3.7.1. Thus the Lyapunov function, $V(t) = \tilde{\theta}(t)^T \tilde{\theta}(t)$, used in the convergence analysis, maintains its nonincreasing property and hence the convergence results are retained precisely as in the nonconstrained case. Of course, the qualitative properties of the algorithms will be improved by constraining $\hat{\theta}(t)$ to remain within a region in which it is known that θ_0 lies.

For example, if $\mathcal{C}$ is as in (3.7.1), then the constrained algorithm simply becomes:

1. If $\hat{\theta}_1(t) \geq b$ go to 3.
2. If $\hat{\theta}_1(t) < b$, put $\hat{\theta}_1(t) = b$ and go to 3.
3. Continue with the algorithm.

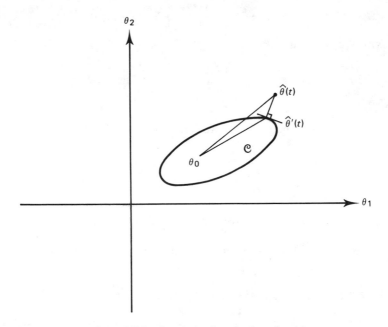

Figure 3.7.1 Constrained projection algorithm.

The Least-Squares Algorithm

A difficulty in this case is that the Lyapunov function used in the convergence analysis is

$$V(t) = \tilde{\theta}(t)^T P(t-1)^{-1} \tilde{\theta}(t) \qquad (3.7.2)$$

Thus one must devise a scheme for projecting $\hat{\theta}(t)$ onto $\mathcal{C}$ while ensuring that $V(t)$ is nonincreasing. This can be achieved using the following simple scheme.

Constrained Least-Squares Algorithm

$$\hat{\theta}(t) = \hat{\theta}(t-1) + \frac{P(t-2)\phi(t-1)}{1 + \phi(t-1)^T P(t-2)\phi(t-1)}[y(t) - \phi(t-1)^T \hat{\theta}(t-1)]$$
$$(3.7.3)$$

$$P(t-1) = P(t-2) - \frac{P(t-2)\phi(t-1)\phi(t-1)^T P(t-2)}{1 + \phi(t-1)^T P(t-2)\phi(t-1)} \qquad (3.7.4)$$

If $\hat{\theta}(t) \in \mathcal{C}$, then continue; else:

1. Transform the coordinate basis for the parameter space by defining

$$\rho = P(t-1)^{-1/2}\theta \qquad (3.7.5)$$

   where

$$P(t-1)^{-1} \triangleq P(t-1)^{-T/2} P(t-1)^{-1/2}$$

   and denote by $\bar{\mathcal{C}}$ the image of $\mathcal{C}$ under the linear transformation $P(t-1)^{-1/2}$.

2. Orthogonally project the image, $\hat{\rho}(t)$, of $\hat{\theta}(t)$ under $P(t-1)^{-1/2}$ onto the

Parameter Estimation for Deterministic Systems   Chap. 3

boundary of $\bar{\mathcal{C}}$ to yield $\hat{\rho}'(t)$ where

$$\hat{\rho}(t) = P(t-1)^{-1/2}\hat{\theta}(t) \qquad (3.7.6)$$

3. Put

$$\hat{\theta}(t) = \hat{\theta}'(t) \triangleq P(t-1)^{1/2}\hat{\rho}'(t) \qquad (3.7.7)$$

and continue.

This is illustrated in Fig. 3.7.2.

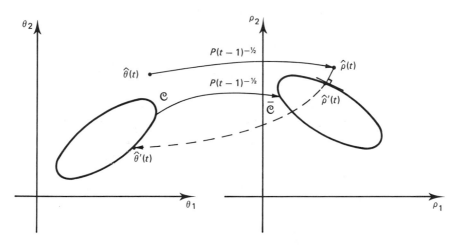

Figure 3.7.2 Constrained least-squares algorithm.

Remark 3.7.1. The transformation $\rho = P(t-1)^{-1/2}\theta$ has the effect of yielding

$$
\begin{aligned}
V(t) &= [\hat{\theta}(t) - \theta_0]^T P(t-1)^{-1}[\hat{\theta}(t) - \theta_0] \\
&= [\hat{\rho}(t) - \rho_0]^T [\hat{\rho}(t) - \rho_0]
\end{aligned} \qquad (3.7.8)
$$

where

$$\rho_0 = P(t-1)^{-1/2}\theta_0 \in \bar{\mathcal{C}} \qquad (3.7.9)$$

Thus, since $\hat{\rho}'(t)$ is an orthogonal projection of $\hat{\rho}(t)$ on $\bar{\mathcal{C}}$ and since $\rho_0 \in \bar{\mathcal{C}}$, then as for the projection algorithm

$$\| \hat{\rho}'(t) - \rho_0 \|^2 \leq \| \hat{\rho}(t) - \rho_0 \|^2 \qquad (3.7.10)$$

and hence

$$(\hat{\theta}'(t) - \theta_0)^T P(t-1)^{-1}(\hat{\theta}'(t-1) - \theta_0) \leq (\hat{\theta}(t) - \theta_0)^T P(t-1)^{-1}(\hat{\theta}(t) - \theta_0) \qquad (3.7.11)$$

This enables the usual Lyapunov argument to be used as before and all properties of the least-squares algorithm are retained as in Lemma 3.3.6, and so on.

Of course, when the constrained region is simple and regular (e.g., bounded by hyperplanes), then the additional complexity in implementing the algorithm above will only be marginal. In practice, very simply regions are all that we need to consider. For example, upper and lower bounds for certain parameter values can be implemented by making $\mathcal{C}$ a hypercube. We illustrate by a simple example.

Example 3.7.1

Consider again the constrained region given in (3.7.1) and assume $\hat{\theta}_1(t)$ is negative. The constraint boundary is clearly given by

$$e_1^T \theta = b \tag{3.7.12}$$

where

$$e_1^T = [1 \quad 0 \quad \cdots \quad 0] \tag{3.7.13}$$

In the transformed space, the constraint boundary is

$$e_1^T P(t-1)^{1/2} \rho = b \tag{3.7.14}$$

The projection onto this boundary is simply achieved by using the projection algorithm of Section 3.3, that is,

$$\hat{\rho}'(t) = \hat{\rho}(t) + \frac{P(t-1)^{T/2} e_1}{e_1^T P(t-1) e_1} [b - e_1^T P(t-1)^{1/2} \hat{\rho}(t)] \tag{3.7.15}$$

Hence using (3.7.7), the corresponding constrained estimate in the original space is given by

$$\hat{\theta}'(t) = P(t-1)^{1/2} \hat{\rho}'(t) \tag{3.7.16}$$

The equations above can be combined with (3.7.6) to give

$$\hat{\theta}'(t) = \hat{\theta}(t) + \frac{P_{1.}(t-1)}{P_{11}(t-1)} [b - \hat{\theta}_1(t)] \tag{3.7.17}$$

where

$$P_{1.}(t-1) = \text{first column of } P(t-1)$$

$$P_{11}(t-1) = (1,1)\text{th element of } P(t-1)$$

▼▼▼

3.8 PARAMETER ESTIMATION FOR MULTI-OUTPUT SYSTEMS

In the previous sections of this chapter we have taken the system output to be a scalar to simplify the presentation. In this section we show how the parameter estimation algorithms can be extended to the multi-output case. This extension is very straightforward and the reader should have no difficulty in formulating the appropriate multi-output versions of the algorithms described in Sections 3.2 to 3.6.

As the basis of our discussion we assume that the system is described by a model of the following form:

$$y_i(t) = \phi_i(t-1)^T \theta_0; \qquad i = 1, \ldots, m \tag{3.8.1}$$

The equations above can be written in matrix form as follows:

$$y(t) = \Phi(t-1)^T \theta_0 \tag{3.8.2}$$

where

$$y(t) = m \times 1 \text{ output vector}$$

$$\Phi(t-1)^T = m \times p \text{ matrix of past values of } \{y(t)\} \text{ and } \{u(t)\}$$

$$\theta_0 = p \times 1 \text{ parameter vector}$$

The parameters can be estimated by minimizing the following cost function:

$$J_N(\theta) \triangleq \tfrac{1}{2}(\theta - \theta_0)^T P_0^{-1}(\theta - \theta_0) + \tfrac{1}{2} \sum_{t=1}^{N} [y(t) - \Phi(t-1)^T \theta]^T R^{-1} [y(t) - \Phi(t-1)^T \theta] \tag{3.8.3}$$

It is readily seen as in Lemma 3.3.5 that the value of θ minimizing (3.8.3) can be computed sequentially using the following algorithm.

Multivariable Least-Squares Algorithm

$$\hat{\theta}(t) = \hat{\theta}(t-1) + P(t-2)\Phi(t-1) \times$$
$$[\Phi(t-1)^T P(t-2)\Phi(t-1) + R]^{-1}[y(t) - \Phi(t-1)^T\hat{\theta}(t-1)] \quad (3.8.4)$$

$$P(t-1) = P(t-2) - P(t-2)\Phi(t-1) \times$$
$$[\Phi(t-1)^T P(t-2)\Phi(t-1) + R]^{-1}\Phi(t-1)^T P(t-2) \quad (3.8.5)$$

$$P(-1) = P_0 \quad (3.8.6)$$

Note that the algorithm above is essentially as for the scalar case save that we now need to invert the $m \times m$ matrix $[\Phi(t-1)^T P(t-2)\Phi(t-1) + R]$.

The algorithm can be simplified in a number of important special cases. These are briefly discussed below:

1. When θ_0 can be partitioned into disjoint sets of parameters each associated with one output

An example of where this occurs is when one uses the DARMA model of Section 2.3.4. The key observation is that $\Phi(t-1)^T$ now has the following special form:

$$\Phi(t-1)^T = \begin{bmatrix} \phi^{(1)}(t-1)^T & 0 & & 0 \\ 0 & \phi^{(2)}(t-1)^T & & \\ & & \ddots & \\ & & & \phi^{(m)}(t-1)^T \end{bmatrix} \quad (3.8.7)$$

We now make the following simplifying assumptions regarding the form of P_0 and R.

$$P_0 \triangleq \begin{bmatrix} P_0^{(1)} & & \\ & \ddots & \\ & & P_0^{(m)} \end{bmatrix} \quad (3.8.8)$$

$$R \triangleq \begin{bmatrix} r_{11} & 0 & \cdots & & 0 \\ 0 & \ddots & & & \\ \vdots & & \ddots & & \vdots \\ & & & \ddots & 0 \\ 0 & & \cdots & 0 & r_{mm} \end{bmatrix} \quad (3.8.9)$$

Substituting (3.8.7) to (3.8.9) into (3.8.4) to (3.8.6) gives the following m decoupled algorithms:

$$\hat{\theta}^{(i)}(t) = \hat{\theta}^{(i)}(t-1) + \frac{P^{(i)}(t-2)\phi^{(i)}(t-1)}{r_{ii} + \phi^{(i)}(t-1)^T P^{(i)}(t-2)\phi^{(i)}(t-1)} \quad (3.8.10)$$
$$\times [y_i(t) - \phi^{(i)}(t-1)^T\hat{\theta}^{(i)}(t-1)]; \quad i = 1, \ldots, m$$

$$P^{(i)}(t-1) = P^{(i)}(t-2) - \frac{P^{(i)}(t-2)\phi^{(i)}(t-1)\phi^{(i)}(t-1)^T P^{(i)}(t-2)}{r_{ii} + \phi^{(i)}(t-1)^T P^{(i)}(t-2)\phi^{(i)}(t-1)} \qquad (3.8.11)$$

$$P^{(i)}(-1) = P_0^{(i)} \qquad (3.8.12)$$

A key feature of the algorithm above is that one need only use m separate scalar output algorithms.

A final simplification is achieved as follows:

2. When θ_0 can be partitioned into disjoint sets of parameters each associated with one output and the regression vectors $\phi^{(i)}(t) \cdots \phi^{(m)}(t)$ are identical

We now make the following additional assumption:

$$r_{ii} = r_{jj}; \qquad i,j = 1, \ldots, m \qquad (3.8.13)$$

$$P_0^{(i)} = P_0^{(j)}; \qquad i,j = 1, \ldots, m \qquad (3.8.14)$$

Equation (3.8.11) now becomes the same for each of the m algorithms and hence we can replace $P^{(i)}(t-1)$, $i = 1, \ldots, m$, by a common $P(t-1)$. We write

$$y(t) = \theta_0^T \phi(t-1) \qquad (3.8.15)$$

where

$$\theta_0^T = \begin{bmatrix} \theta_0^{(1)T} \\ \cdot \\ \cdot \\ \cdot \\ \theta_0^{(m)T} \end{bmatrix} \qquad (3.8.16)$$

The algorithm then simplifies to

Simplified Multivariable Least-Squares Algorithm

$$\hat{\theta}(t) = \hat{\theta}(t-1) + \frac{P(t-2)\phi(t-1)}{1 + \phi(t-1)^T P(t-2)\phi(t-1)}[y(t)^T - \phi(t-1)^T\hat{\theta}(t-1)]$$

$$(3.8.17)$$

$$P(t-1) = P(t-2) - \frac{P(t-2)\phi(t-1)\phi(t-1)^T P(t-2)}{1 + \phi(t-1)^T P(t-2)\phi(t-1)} \qquad (3.8.18)$$

with $\hat{\theta}(0)$ and $P(-1) > 0$ given.

The algorithm above is applicable, for example, to the DARMA model of Section 2.3.4 after multiplication by A_0^{-1} to give $A_0 = I$. [Note that if the structural constraints given in (2.3.46) are to be retained, it is necessary to use the form of the algorithm given in (3.8.10) to (3.8.12) unless $K_i = K_j$ $(i,j = 1, \ldots, m)$.] It is always possible to use the simple form (3.8.17)–(3.8.18) by introducing dummy zero parameters. But introduction of these extra parameters will lead to slower algorithm convergence.

It is also possible to develop a projection version of the algorithms above. For example, the projection form of (3.8.17)–(3.8.18) is

Projection Algorithm (Multi-output Case)

$$\hat{\theta}(t) = \hat{\theta}(t-1) + \frac{a\phi(t-1)[y(t)^T - \phi(t-1)^T\hat{\theta}(t-1)]}{c + \phi(t-1)^T\phi(t-1)} \qquad (3.8.19)$$

with $\hat{\theta}(0)$ given and $c > 0$; $0 < a < 2$.

Comparison of (3.8.19) with (3.3.19) indicates that the multi-output case is almost identical with the scalar case save that in the multi-output case it must be remembered that $\hat{\theta}(t)$ represents an $n_1 \times m$ matrix.

The previous convergence results can be readily extended to the multi-output case. We illustrate by repsenting the multivariable version of Lemma 3.3.2 for the algorithm (3.8.19). As before, we introduce the notation

$$\tilde{\theta}(t) = \hat{\theta}(t) - \theta_0 \qquad (3.8.20)$$

$$e(t) = y(t) - \hat{\theta}(t-1)^T\phi(t-1) \qquad (3.8.21)$$

We then have

Lemma 3.8.1 (Multi-output Version of Lemma 3.3.2). For the algorithm (3.8.19) and subject to (3.8.15) it follows that

(i) $\|\hat{\theta}(t) - \theta_0\|^2 \leq \|\hat{\theta}(t-1) - \theta_0\|^2 \leq \|\hat{\theta}(0) - \theta_0\|^2 \qquad (3.8.22)$
where the symbol $\|\cdot\|$ is defined as

$$\|M\|^2 \triangleq \text{trace}\,(M^TM) = \sum_i \sum_j M_{ij}^2 \qquad (3.8.23)$$

(ii) $\displaystyle\lim_{N\to\infty} \sum_{t=1}^{N} \frac{e(t)^Te(t)}{[c + \phi(t-1)^T\phi(t-1)]} < \infty \qquad (3.8.24)$
and this implies

(a) $\displaystyle\lim_{t\to\infty} \frac{e_i(t)}{(c + \phi(t+1)^T\phi(t-1))^{1/2}} = 0; \quad i = 1, \ldots, m \qquad (3.8.25)$

(b) $\displaystyle\lim_{N\to\infty} \sum_{t=1}^{N} \frac{\phi(t-1)^T\phi(t-1)e(t)^Te(t)}{[c + \phi(t-1)^T\phi(t-1)]^2} < \infty \qquad (3.8.26)$

(c) $\displaystyle\lim_{N\to\infty} \sum_{t=1}^{N} \|\hat{\theta}(t) - \hat{\theta}(t-1)\|^2 < \infty \qquad (3.8.27)$

(d) $\displaystyle\lim_{N\to\infty} \sum_{t=k}^{N} \|\hat{\theta}(t) - \hat{\theta}(t-k)\|^2 < \infty \qquad (3.8.28)$

(e) $\displaystyle\lim_{t\to\infty} \|\hat{\theta}(t) - \hat{\theta}(t-k)\|^2 = 0 \qquad$ for any finite k. $\qquad (3.8.29)$

Proof. We follow the proof of Lemma 3.3.2.
(i) Subtracting θ_0 from both sides of (3.8.19) and using (3.8.15), we have

$$\tilde{\theta}(t) = \tilde{\theta}(t-1) - \frac{a\phi(t-1)\phi(t-1)^T\tilde{\theta}(t-1)}{c + \phi(t-1)^T\phi(t-1)} \qquad (3.8.30)$$

Hence using (3.8.21) and (3.8.23) yields

$$\|\tilde{\theta}(t)\|^2 - \|\tilde{\theta}(t-1)\|^2 = a\left[-2 + \frac{a\phi(t-1)^T\phi(t-1)}{c + \phi(t-1)^T\phi(t-1)}\right]\frac{e(t)^Te(t)}{c + \phi(t-1)^T\phi(t-1)} \qquad (3.8.31)$$

Now since $0 < a < 2$, $c > 0$, we have

$$a\left[-2 + \frac{a\phi(t-1)^T\phi(t-1)}{c + \phi(t-1)^T\phi(t-1)}\right] < 0 \qquad (3.8.32)$$

and then (3.8.22) follows from (3.8.31).

(ii) We observe that $\|\tilde{\theta}(t)\|^2$ is a bounded nonincreasing function, and by using (3.8.31) we conclude that since (3.8.32) holds, then (3.8.24) is true.

(a) Follows immediately from (3.8.24) by noting that

$$e(t)^T e(t) = \sum_{i=1}^{m} e_i(t)^2$$

(b) Immediate from (3.8.24) since $\phi(t-1)^T\phi(t-1) < c + \phi(t-1)^T\phi(t-1)$.

(c) Equation (3.8.26) immediately implies (3.8.27) by noting the form of the algorithm (3.8.19).

(d) Follows from the Schwarz inequality since k is finite.

(e) Immediate from (3.8.28).

▼▼▼

Corresponding results can also be established for the least-squares algorithms. For example, Lemma 3.3.6 can also be extended to the multi-output case. We will not go into the details since the proof is essentially identical to the proof of Lemma 3.3.6 (see Exercise 3.20).

3.9 CONCLUDING REMARKS

In the previous sections we have described a number of alternative algorithms. The alert reader probably suspects that there is a common basis for all these schemes. This is indeed the case as we shall demonstrate below. An appreciation of this underlying notion has the advantage of unifying the different schemes and of showing how new schemes can be developed which have desirable convergence properties built into them.

We have motivated the various algorithms by geometric considerations. Here we adopt an alternative point of view and look at the algorithms on the basis of the convergence theory and particularly the stability of the parameter error equation.

1. Returning to the projection algorithm [equation (3.3.19)], we have seen that (3.3.19) for the parameter update can be written in the following error form:

$$\tilde{\theta}(t) = \tilde{\theta}(t-1) - \frac{\phi(t-1)}{c + \phi(t-1)^T\phi(t-1)}\phi(t-1)^T\tilde{\theta}(t-1) \qquad (3.9.1)$$

This equation can be simplified for the purpose of convergence analysis. If we multiply by $\phi(t-1)^T$, we obtain

$$\phi(t-1)^T\tilde{\theta}(t) = \frac{c\phi(t-1)^T\tilde{\theta}(t-1)}{c + \phi(t-1)^T\phi(t-1)} \qquad (3.9.2)$$

Substituting (3.9.2) into (3.9.1), we obtain the following simplified error equation:

$$\tilde{\theta}(t) = \tilde{\theta}(t-1) - \frac{1}{c}\phi(t-1)\phi(t-1)^T\tilde{\theta}(t) \qquad (3.9.3)$$

with $1/c$ positive. Equation (3.9.3) can also be written as

$$\left[I + \frac{1}{c}\phi(t-1)\phi(t-1)^T\right]\tilde{\theta}(t) = \tilde{\theta}(t-1) \qquad (3.9.4)$$

The error equation above is the basic structure underlying all the algorithms that we have considered and has the desirable convergence properties described in Lemma 3.3.2. For example, in the case of least squares it can be seen that the error equation becomes

$$\tilde{\theta}(t) = \tilde{\theta}(t-1) - P(t-2)\phi(t-1)\phi(t-1)^T\tilde{\theta}(t) \qquad (3.9.5)$$

with $P(t-2)$ a positive semidefinite matrix.

In the output error schemes of Section 3.5, the basic structure above is generalized to [see (3.5.22), (3.5.27), and (3.5.30)]

$$\tilde{\theta}(t) = \tilde{\theta}(t-1) - \phi(t-1)G(q)\{\phi(t-1)^T\tilde{\theta}(t)\} \qquad (3.9.6)$$

where $G(q)$ is a very strictly passive transfer function (i.e., is positive in a transfer function sense; see Section C.3 of Appendix C), $\phi(t-1)$ is actually $\bar{\phi}(t-1)$, and $G(q)$ is actually $D(q^{-1})/A(q^{-1})$. Comparing (3.9.6) with (3.9.1) we see that (3.9.6) requires the additional assumption that $G(q)$ be "positive" (i.e., very strictly passive). Of course, if $G(q^{-1}) = 1/c$, the two forms become equivalent.

Given any algorithm, provided that the associated error equation can be expressed in one of the forms described above, the convergence properties will follow straightforwardly. An example of how an algorithm can be massaged into the right form is given below.

2. In some cases [see, e.g., Ionescu and Monopoli (1977) and Narendra and Lin (1980)] it can occur that we end up with an error equation of the form

$$\tilde{\theta}(t) = \tilde{\theta}(t-1) - \frac{\{H(q)\phi(t)\}}{c + \{H(q)\phi(t)\}^T\{H(q)\phi(t)\}}H(q)\{\phi(t)^T\tilde{\theta}(t)\} \qquad (3.9.7)$$

where $H(q)$ is a known filter transfer function. We shall find a simple example of this in Chapter 4 when we discuss d-step-ahead prediction [where $H(q) = q^{-d}$].

Equation (3.9.7) is not in the form of (3.9.1) or (3.9.6). However, a clever technique due to Monopoli (1974) can be used to adjust the error equation to the required format by addition of compensating terms. We define an *augmented error*, $\bar{e}(t)$, as follows:

$$\bar{e}(t) \triangleq e(t) + \check{e}(t) \qquad (3.9.8)$$

where

$$e(t) \triangleq H(q)\{\phi(t)^T\tilde{\theta}(t)\} \qquad \text{as before} \qquad (3.9.9)$$

$$\check{e}(t) \triangleq H(q)\{\phi(t)^T\hat{\theta}(t)\} - \{H(q)\phi(t)\}^T\hat{\theta}(t-1) \qquad (3.9.10)$$

The motivation for the definition above is that

$$
\begin{aligned}
\bar{e}(t) &= H(q)[\phi(t)^T\theta_0 - \phi(t)^T\hat{\theta}(t)] + H(q)\{\phi(t)^T\hat{\theta}(t)\} - [H(q)\phi(t)]^T\hat{\theta}(t-1) \\
&= H(q)[\phi(t)^T\theta_0] - [H(q)\phi(t)]^T\hat{\theta}(t-1) \\
&= [H(q)\phi(t)]^T\theta_0 - [H(q)\phi(t)]^T\hat{\theta}(t-1) \qquad \text{since } \theta_0 \text{ is time invariant} \\
&= -[H(q)\phi(t)]^T\tilde{\theta}(t-1)
\end{aligned}
\qquad (3.9.11)
$$

Thus the use of the augmented error $\bar{e}(t)$ in place of $e(t)$ leads to

$$\tilde{\theta}(t) = \tilde{\theta}(t-1) - \frac{\{H(q)\phi(t)\}}{c + \{H(q)\phi(t)\}^T\{H(q)\phi(t)\}}\{H(q)\phi(t)\}^T\tilde{\theta}(t-1) \qquad (3.9.12)$$

or equivalently,

$$\tilde{\theta}(t) = \tilde{\theta}(t-1) - \frac{1}{c}\bar{\phi}(t)\bar{\phi}(t)^T\tilde{\theta}(t) \qquad (3.9.13)$$

where

$$\bar{\phi}(t) \triangleq \{H(q)\phi(t)\} \qquad (3.9.14)$$

We thus see that (3.9.13) is now in the form of (3.9.3) and the convergence analysis proceeds as before.

Regarding the convergence properties of the various algorithms studied in this chapter, we have gone to some trouble to avoid assuming that the data are bounded a priori. In particular, the reader's attention is drawn to Lemma 3.3.2, Lemma 3.3.6 and Lemma 3.4.9. Because of the weak assumptions used, these results have wide applicability in adaptive filtering, prediction and control.

3. This completes our discussion of parameter estimation algorithms for deterministic systems. In Chapters 4 and 6 we shall apply these methods to problems in adaptive prediction and adaptive control. Later in the book we shall turn to stochastic systems. We shall see that many of the ideas introduced in this chapter generalize to the stochastic case in a very straightforward and natural way.

EXERCISES

3.1. Prove Lemma 3.3.4 (the matrix inversion lemma).

3.2. Prove the following more general form of the matrix inversion lemma:

$$(A + BC)^{-1} = A^{-1} - A^{-1}B(I + CA^{-1}B)^{-1}CA^{-1}$$

3.3. Show that the projection algorithm will have slow convergence if successive $\phi(t)$ vectors are almost collinear.

3.4. Show that with $P(0) = kI$, $k \gg I$, the least-squares algorithm has the following two properties:

1. The initial step is the same as that of projection algorithm with $c = 0$ ($\phi(0) \neq 0$).
2. It reaches θ_0 in n, steps where $n = $ dimension of θ_0 provided that rank $[\phi(0) \cdots \phi(n-1)] = n$.

3.5. If A is partitioned as

$$A = \begin{bmatrix} A_{11} & A_{12} \\ A_{21} & A_{22} \end{bmatrix}$$

show that

(a) $A^{-1} = \begin{bmatrix} (A_{11} - A_{12}A_{22}^{-1}A_{21})^{-1} & -(A_{11} - A_{12}A_{22}^{-1}A_{21})^{-1}A_{12}A_{22}^{-1} \\ -(A_{22} - A_{21}A_{11}^{-1}A_{12})^{-1}A_{21}A_{11}^{-1} & (A_{22} - A_{21}A_{11}^{-1}A_{12}) \end{bmatrix}$

(b) $\det A = \det A_{11} \det(A_{22} - A_{21}A_{11}^{-1}A_{12})$

3.6. Consider the algorithm (3.3.49). Show that

(a) $P(t-1)\phi(t-1) = \dfrac{P(t-2)\phi(t-1)}{1 + \phi(t-1)^T P(t-2)\phi(t-1)}$

(b) $P(t-2)\phi(t-1) = \dfrac{P(t-1)\phi(t-1)}{1 - \phi(t-1)^T P(t-1)\phi(t-1)}$

(c) $\phi(t-1)^T P(t-1)\phi(t-1) = \dfrac{\phi(t-1)^T P(t-2)\phi(t-1)}{1 + \phi(t-1)^T P(t-2)\phi(t-1)}$

(d) $\phi(t-1)^T P(t-2)\phi(t-1) = \dfrac{\phi(t-1)^T P(t-1)\phi(t-1)}{1 - \phi(t-1)^T P(t-1)\phi(t-1)}$

(e) $\phi(t-1)^T P(t-1)^2\phi(t-1) = \dfrac{\phi(t-1)^T P(t-1)P(t-2)\phi(t-1)}{1 + \phi(t-1)^T P(t-2)\phi(t-1)}$

$\qquad\qquad\qquad\qquad\quad \le \phi(t-1)^T P(t-1)P(t-2)\phi(t-1)$

(f) $\phi(t-1)^T P(t-2)^2\phi(t-1) = \dfrac{\phi(t-1)^T P(t-1)P(t-2)\phi(t-1)}{1 - \phi(t-1)^T P(t-1)\phi(t-1)}$

(g) $\phi(t-1)^T P(t-2)P(t-1)\phi(t-1) = \dfrac{\phi(t-1)^T P(t-1)^2\phi(t-1)}{1 - \phi(t-1)^T P(t-1)\phi(t-1)}$

3.7. Establish the algorithm given in (3.3.72) and (3.3.73).

3.8. Prove Lemma 3.3.7.

3.9. Show that Lemma 3.3.6 applies essentially unaltered to the algorithm (3.3.72)–(3.3.73) with $\{a(t-1)\}$ chosen as in (3.3.78).

3.10. Establish (3.3.79) using Lemma 3.3.7.
[*Hint:*

$$\frac{1}{a(t-1)} = \frac{\phi(t-1)^T\phi(t-1)}{\phi(t-1)^T P(t-2)\phi(t-1)}$$

$$\le \frac{1}{\lambda_{\min} P(t-2)}$$

$$\le \operatorname{trace} P(t-2)^{-1}.]$$

3.11. Establish (3.3.82)–(3.3.83). [*Hint:* See Goodwin and Payne (1977, p. 180).]

3.12. Show that a sinusoidal input

$$u(t) = A \cos w_0 t; \qquad 0 < w_0 < \pi$$

is weakly persistently exciting of order 2.

3.13. Repeat Exercise 3.12 for a linear combination of m sinusoids of different frequencies.

3.14. Prove Corollary 3.5.1. [*Hint:* See Landau (1976, 1978a).]

3.15. Consider the following algorithm due to Landau (1978b) aimed at eliminating the positive real condition in output error methods. The model is

$$A(q^{-1})y(t) = q^{-1}B(q^{-1})u(t); \qquad A, B \text{ of order } n, m, \text{ respectively}$$

or

$$y(t) \triangleq \psi(t-1)^T\theta_0$$
$$\psi(t-1)^T = [-y(t-1), \ldots, u(t-1), \ldots]$$
$$\theta_0 = [a_1 \ldots, a_n, b_0, \ldots, b_m]$$

The a priori predicted output and error are

$$\hat{y}(t) \triangleq \phi(t-1)^T\hat{\theta}(t-1); \qquad \epsilon(t) \triangleq y(t) - \hat{y}(t)$$
$$\phi(t-1)^T \triangleq [-\hat{y}(t-1), \ldots, u(t-1), \ldots]$$

The a posteriori predicted output and error are

$$\bar{y}(t) \triangleq \phi(t-1)^T \hat{\theta}(t); \qquad \eta(t) \triangleq y(t) - \bar{y}(t)$$

Define an estimated filter

$$\hat{C}(t, q^{-1}) \triangleq 1 + \hat{c}_1(t)q^{-1} + \cdots + \hat{c}_n(t)q^{-n}$$

The filtered error is

$$\bar{\eta}(t) \triangleq \hat{C}(t, q^{-1})\eta(t)$$

The a priori filtered error is

$$v(t) \triangleq \epsilon(t) + [\hat{C}(t, q^{-1}) - 1]\eta(t)$$

The extended regression vector is

$$\bar{\phi}(t-1)^T = [-\bar{y}(t-1), \ldots, u(t-1) \ldots, -\eta(t-1), \ldots]$$

The extended parameter vector is

$$\beta_0 = [a_1, \ldots, a_n, b_0, \ldots, b_m, a_1, \ldots, a_n]$$
$$\hat{\beta}(t) = [\hat{a}_1(t), \ldots, \hat{a}_n(t), \hat{b}_0(t), \ldots, \hat{b}_m(t), \hat{c}_1(t), \ldots, \hat{c}_n(t)]$$

The algorithm is

$$\hat{\beta}(t) = \hat{\beta}(t-1) + \frac{P(t-2)\bar{\phi}(t-1)}{1 + \bar{\phi}(t-1)^T P(t-2)\bar{\phi}(t-1)}v(t)$$

$$P(t-1) = P(t-2) - \frac{P(t-2)\bar{\phi}(t-1)\bar{\phi}(t-1)^T P(t-2)}{1 + \bar{\phi}(t-1)^T P(t-2)\bar{\phi}(t-1)}$$

Show that

(a) $\eta(t) = -\phi(t-1)^T \tilde{\theta}(t) + [1 - A(q^{-1})]\eta(t); \quad \tilde{\theta}(t) = \hat{\theta}(t) - \theta_0$

(b) $\bar{\eta}(t) = -\bar{\phi}(t-1)^T \tilde{\beta}(t); \quad \tilde{\beta}(t) = \hat{\beta}(t) - \beta_0$

(c) $\epsilon(t) = -\phi(t-1)^T \tilde{\theta}(t-1) + [1 - A(q^{-1})]\eta(t)$

(d) $v(t) = -\bar{\phi}(t-1)^T \tilde{\beta}(t-1)$

(e) $\bar{\eta}(t) = \dfrac{v(t)}{1 + \bar{\phi}(t-1)^T P(t-2)\bar{\phi}(t-1)}$

Hence, by analyzing the algorithm, show that

(f) $\sum\limits_{t=1}^{\infty} \bar{\eta}(t)^2 < \infty$

(g) $\sum\limits_{t=1}^{\infty} \bar{\phi}(t-1)^T P(t-2)\bar{\phi}(t-1)\bar{\eta}(t)^2 < \infty$

(h) Finally, show that the algorithm is globally convergent provided that a projection facility is included as in Section 3.7 to ensure that $\hat{C}(t, q^{-1})$ is stable for all t and as a result

$$\lim_{t=\infty} \epsilon(t) = 0$$

[For further discussion of the need for $\hat{C}(t, q^{-1})$ to be stable, see Goodwin, Saluja, and Sin (1979) and Johnson and Taylor (1979).]

3.16. In the algorithm (3.3.106)–(3.3.108) (least squares with covariance modification), what happens if (3.3.106) is replaced by

$$\hat{\theta}(t) = \hat{\theta}(t-1) + P(t-1)\phi(t-1)[y(t) - \phi(t-1)^T \hat{\theta}(t-1)]?$$

3.17. Show that $2ab \leq ka^2 + b^2/k$ for arbitrary k. (*Hint:* First show that $2xy \leq x^2 + y^2$. Then put $a = x/\sqrt{k}$, $b = y\sqrt{k}$.)

3.18. Verify that $\limsup |e(t)| < 2\Delta$ for the algorithm (3.6.2)–(3.6.3) when $\{\phi(t)\}$ is bounded.

3.19. Establish property (3.6.17) for the least squares algorithm with dead zone.

3.20. Extend Lemma 3.3.6 to the multi-output algorithm (3.8.17)–(3.8.18). [*Hint:* Consider the following nonnegative function (as in Lemma 3.3.6):

$$V(t) = \text{trace } [\tilde{\theta}(t)^T P(t-1)\tilde{\theta}(t)].]$$

3.21. Consider the projection algorithm (3.3.5) applied to the system

$$y(t) = \phi(t-1)^T \theta_0$$

Show that devision by zero in the algorithm can be avoided by augmenting $\phi(t-1)$ and θ_0 as follows:

$$\phi'(t-1) = [\phi(t-1)^T \quad 1]^T$$
$$\theta_0' = [\theta_0^T \quad 0]^T$$

Discuss the advantages of this approach as compared with adding a constant to the denominator as in (3.3.19).

3.22. Give an example of a scalar sequence $\{\theta(t)\}$ satisfying (3.3.28) which is not a Cauchy sequence and which does not converge.

3.23. Consider the following algorithm:

Projection Algorithm with Filtered Errors

$$\hat{\theta}(t) = \hat{\theta}(t-1) + \frac{\bar{\phi}(t-d)}{c + \bar{\phi}(t-d)^T\bar{\phi}(t-d)}\bar{e}(t); \quad c > 0$$

where

$$\bar{\phi}(t-d) = G(q)\phi(t)$$
$$\bar{e}(t) = e(t) + \tilde{e}(t)$$
$$\tilde{e}(t) = G(q)[\phi(t)^T\hat{\theta}(t)] - [G(q)\phi(t)]^T\hat{\theta}(t-1)$$
$$G(q) = \frac{q^{-d}F(q^{-1})}{E(q^{-1})}, \text{ a given stable filter such that}$$
$$F(q^{-1}) = f_0 + f_1 q^{-1} + \cdots + f_l q^{-l}$$
$$E(q^{-1}) = e_0 + e_1 q^{-1} + \cdots + e_l q^{-l}; \quad e_0 = 1$$

and $e(t)$ is a measured error quantity that can be expressed as

$$e(t) = G(q)[\phi(t)^T\theta_0 - \phi(t)^T\hat{\theta}(t)] = -G(q)\phi(t)^T\tilde{\theta}(t)$$

Establish the following relationships:
(a) $\bar{e}(t) = -\bar{\phi}(t-d)^T\tilde{\theta}(t-1)$
(b) $\bar{\eta}(t) \triangleq \dfrac{\bar{e}(t)}{c + \bar{\phi}(t-d)^T\bar{\phi}(t-d)} = -\dfrac{1}{c}\bar{\phi}(t-d)^T\tilde{\theta}(t)$
(c) $\tilde{\theta}(t) = \tilde{\theta}(t-1) + \bar{\phi}(t-d)\bar{\eta}(t)$
where $\tilde{\theta}(t) = \hat{\theta}(t) - \theta_0$.

3.24. For the algorithm above, and subject to stability and inverse stability of $G(q)$, show that
(a) $\|\tilde{\theta}(t)\| \le \|\tilde{\theta}(t-1)\| \le \|\tilde{\theta}(0)\|; \quad t > 1$
(b) $\displaystyle\lim_{N\to\infty} \sum_{t=1}^{N} \frac{\bar{e}(t)^2}{c + \bar{\phi}(t-d)^T\bar{\phi}(t-d)} < \infty$
which implies that

$$\lim_{t\to\infty} \frac{\bar{e}(t)}{[c + \bar{\phi}(t-d)^T\bar{\phi}(t-d)]^{1/2}} = 0$$

or equivalently,

$$\bar{e}(t) = 0 \|\bar{\phi}(t-d)\|$$

(for notation see below) and thus

$$\lim_{N \to \infty} \sum_{t=1}^{N} \| \hat{\theta}(t) - \hat{\theta}(t-1) \|^2 < \infty$$

(c) $e'(t) = \phi(t)^T \tilde{\theta}(t) = 0 \ [\sup_{t \geq J} \| \bar{\phi}(j) \|]$

   [*Hint:* $E(q^{-1})\bar{\phi}(t-d) = F(q^{-1})\phi(t-d)$

$$\bar{\phi}(t) \leq k_1 [\sup_{t \geq J} \| \phi(j) \|] + K_2 \lambda^t; \qquad \phi(t) \leq K_3 [\sup_{t \geq J} \bar{\phi}(j) \|] + K_4 \lambda^t$$

$$0 \leq \lambda \leq 1 \qquad\qquad\qquad 0 < K_i < \infty$$

   and

$$\tilde{\theta}(t)^T E(q^{-1}) \bar{\phi}(t-d) = \tilde{\theta}(t)^T F(q^{-1}) \phi(t-d)$$

   or

$$\sum_{j=0}^{l_2} e_j \tilde{\theta}(t)^T \bar{\phi}(t-j-d) = \sum_{j=0}^{l_1} f_j \tilde{\theta}(t)^T \phi(t-d-j).]$$

(d) $\tilde{e}(t) = 0 \ [\sup_{t-d \geq J} \| \bar{\phi}(j) \|]$

(e) $e(t) = 0 \ [\sup_{t-d \geq J} \| \bar{\phi}(j) \|]$

   where the notation means $\bar{e}(t) = \alpha(t) \| \bar{\phi}(t-d) \| + \beta(t)$, where $\lim_{t \to \infty} \alpha(t) = 0$ and $\lim_{t \to \infty} \beta(t) = 0$.

3.25. Consider the model (3.4.39) for generating $\{\phi(t)\}$. Define $x(t) = \phi(t)$ and add the output equation $y(t) = Hx(t)$, where $H = [1 \ \ 0 \ \ \cdots \ \ 0]$. This lead to the following $2n$-dimensional model for the system:

$$x(t+1) = Fx(t) + Gu(t+1)$$

$$y(t) = Hx(t)$$

Lemma 3.4.7 establishes that the model above is completely reachable if $A(q^{-1})$ and $B(q^{-1})$ are relatively prime. Show that the model is not observable by evaluating the observability matrix. Why does this result make sense?

3.26. Verify that $(zI - F)^{-1}G$ has the form shown in (3.4.57) for a state-space model in controller form.

3.27. Use the results of Exercise 3.26 to show that $\{A, b\}$ is completely reachable if and only if the matrix P defined by

$$(zI - A)^{-1}b = \frac{P}{a(z)} \begin{bmatrix} z^{n-1} \\ \cdot \\ \cdot \\ \cdot \\ z \\ 1 \end{bmatrix}$$

is nonsingular [see Kailath (1980, p. 115)].

3.28. Develop a least-squares algorithm for estimating the parameters in the multivariable DARMA model obtained directly from the observer state-space form (i.e., having observability indices not necessarily equal and A_0 lower triangular with 1's on the main diagonal).

3.29. Consider the following scalar DARMA model:

$$A(q^{-1})y(t) = B(q^{-1})u(t)$$

where $A(q^{-1})$ is second order, that is,

$$A(q^{-1}) = 1 + a_1 q^{-1} + a_2 q^{-2}$$

(a) Show that $A(q^{-1})$ is guaranteed stable provided that a_1 and a_2 are constrained to lie in the closed convex region, $\mathcal{C}$, shown in Fig. 3.A.

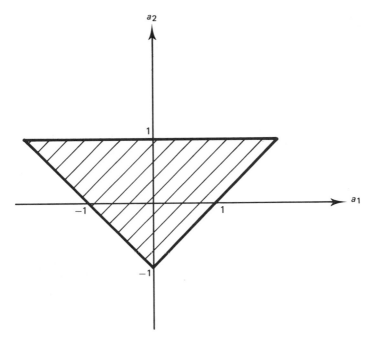

Figure 3.A

(b) Hence using the results of Section 3.7, develop a least-squares algorithm having guaranteed convergence properties which ensures that the estimated $A(q^{-1})$ polynomial remains asymptotically stable.

3.30. Repeat Exercise 3.29 for a third-order system.

3.31. Consider the least-squares algorithm with exponential data weighting [algorithm (3.3.82)–(3.3.83)], that is,

$$\hat{\theta}(t) = \hat{\theta}(t-1) + \frac{P(t-2)\phi(t-1)}{\lambda + \phi(t-1)^T P(t-2)\phi(t-1)}[y(t) - \phi(t-1)^T\hat{\theta}(t-1)]$$

$$P(t-1) = \frac{1}{\lambda}\left[P(t-2) - \frac{P(t-2)\phi(t-1)\phi(t-1)^T P(t-2)}{\lambda + \phi(t-1)^T P(t-2)\phi(t-1)}\right]$$

where $0 < \lambda < 1$. Consider the case where $\{\phi(t)\}$ lies in a hyperplane of dimension lower than θ. Show that P diverges. Discuss the implications of this observation and the advisability of using the algorithm in the absence of persistent excitation.

4

Deterministic Adaptive
Prediction

4.1 INTRODUCTION

In this chapter we introduce briefly the idea of adaptive prediction. The results presented here are of interest in their own right. However, they also provide a framework for the discussion of prediction for stochastic systems which we will address later in the book.

In the deterministic case, the problem of adaptive prediction can be viewed as a special type of parameter estimation problem where the estimated model is regarded as the predictor. Thus the ideas introduced in this chapter should reinforce the concepts introduced in the preceding chapter.

In our general discussion of parameter estimation in Chapter 3, the underlying objective was to estimate the parameters of a model so that the model output was close to the output of a given system. We shall see below that the problem of adaptive prediction can be thought of in the same way, except that the measure of "closeness" of the system and model outputs is tailored to the specific problem under study.

In Section 4.2 we consider prediction when the underlying model for a time series is known. In Section 4.3 we show how the parameter estimation techniques of Chapter 3 can be combined with the predictors developed in Section 4.2 to yield adaptive predictors.

4.2 PREDICTOR STRUCTURES

Prediction is concerned with extrapolating a given time series into the future. As such it is inherent to many problems that arise in practice. For example, the decision to expand the manufacturing facilities of a company could be based on the prediction of future product demand. The structure of the predictor can be obtained in two ways. First, the precictor can be obtained by massaging the known model of the system into an appropriate form. Alternatively, the predictor structure can be simply assumed without regard to the true nature of the underlying system. We discuss each of these viewpoints briefly below.

4.2.1 Prediction with Known Models

If a model for the evolution of the time series of interest is known, it is possible to construct suitable predictors from the model. In fact, in the case that the model is linear and finite-dimensional, suitable predictors can be obtained by simple algebraic manipulations of the model. We show how this is achieved below.

Linear Systems

A convenient way of describing a time series is to regard it as the output, $y(t)$, of a dynamical system driven by some known input, $u(t)$ [later when we discuss the stochastic case we will replace $u(t)$ by an unmeasured white noise term]. We begin our discussion by considering the linear case. We recall from Chapter 2 that the input–output behavior of a finite-dimensional linear dynamical system can be described by a DARMA model. The following lemma shows how a DARMA model can be expressed in an alternative *predictor form*.

Lemma 4.2.1. Consider a multi-input multi-output linear discrete-time system (with m outputs and r inputs) described by the following DARMA model:

$$A(q^{-1})y(t) = B(q^{-1})u(t) \tag{4.2.1}$$

where

$$A(q^{-1}) = I + A_1 q^{-1} + A_2 q^{-2} + \cdots + A_n q^{-n} \tag{4.2.2}$$

$$B(q^{-1}) = q^{-d}(B_0 + B_1 q^{-1} + \cdots + B_l q^{-l})$$
$$= q^{-d}B'(q^{-1}) \tag{4.2.3}$$

Then the output of the system at time $t + d$ can be expressed in the following predictor form:

$$y(t + d) = \alpha(q^{-1})y(t) + \beta(q^{-1})u(t) \tag{4.2.4}$$

where

$$\alpha(q^{-1}) = G(q^{-1}), \quad \beta(q^{-1}) = F(q^{-1})B'(q^{-1}) \tag{4.2.5}$$

and $F(q^{-1})$ and $G(q^{-1})$ are the unique polynomial matrices satisfying

$$I = F(q^{-1})A(q^{-1}) + q^{-d}G(q^{-1}) \tag{4.2.6}$$

$$F(q^{-1}) = I + F_1 q^{-1} + \cdots + F_{d-1}q^{-d+1}$$
$$G(q^{-1}) = G_0 + G_1 q^{-1} + \cdots + G_{n-1}q^{-n+1} \tag{4.2.7}$$

Further, the coefficients in $F(q^{-1})$ and $G(q^{-1})$ can be computed as follows:

$$F_0 = I \tag{4.2.8}$$

$$F_i = -\sum_{j=0}^{i-1} F_j A_{i-j}; \qquad i = 1, \ldots, d-1 \qquad (4.2.9)$$

$$G_i = -\sum_{j=0}^{d-1} F_j A_{i+d-j}; \qquad i = 0, \ldots, n-1 \qquad (4.2.10)$$

(*Note:* In the above, whenever $A_{n+1}, A_{n+2}, \ldots$ appear in the formulas, we simply put them to zero.)

Proof. Multiplying (4.2.1) by $F(q^{-1})$ gives

$$F(q^{-1})A(q^{-1})y(t) = F(q^{-1})B(q^{-1})u(t) \qquad (4.2.11)$$

or using (4.2.6),

$$y(t + d) = \alpha(q^{-1})y(t) + \beta(q^{-1})u(t) \qquad (4.2.12)$$

where

$$\alpha(q^{-1}) = G(q^{-1}) \qquad (4.2.13)$$

$$\beta(q^{-1}) = F(q^{-1})B'(q^{-1}) \qquad (4.2.14)$$

On expanding (4.2.6), we obtain (4.2.8) to (4.2.10). This establishes the lemma.

▼▼▼

The predictor form (4.2.4) is simply an alternative way of writing the system model. It can be expressed in our usual format as [see (2.3.47)]

$$y(t + d) = \theta_0^T \phi(t) \qquad (4.2.15)$$

where

$$\phi(t)^T = (y(t)^T, \ldots, y(t - n + 1)^T, u(t)^T, \ldots, u(t - l - d + 1)^T) \qquad (4.2.16)$$

and θ_0^T is a matrix of parameters consisting of coefficients in the matrices $\alpha(q^{-1})$ and $\beta(q^{-1})$.

Equation (4.2.15) immediately suggests the following structure for a d-step-ahead predictor:

$$\hat{y}(t + d, \theta) = \theta^T \phi(t) \qquad (4.2.17)$$

where θ is a parameter vector. Note that when $\theta = \theta_0$, then $\hat{y}(t + d, \theta) = y(t + d)$ for all t. Later we shall study algorithms for choosing values of θ such that $\hat{y}(t + d, \theta)$ is "close" to $y(t + d)$ when θ_0 is not known.

A feature of the predictor above is that the predicted output at time $t + d$ is a function of both the system outputs and the system inputs up to and including time t. This form of predictor will be well suited to many applications. However, in other applications it may be desirable to have a predictor that is driven by the input alone. An example of this is when one desires to have a long-term prediction based on a known or assumed input sequence, such as the prediction of a city's water consumption over a period of one or two years given typical rainfall and temperature profiles (Vizwanathan, 1981). This contrasts with short-term prediction, where the water consumption a few days ahead might be predicted on the basis of current consumption as well as rainfall and temperature.

If a model of the form (4.2.1) is assumed to represent the dynamics of the system of interest, then, provided that $A(q^{-1})$ is asymptotically stable, a possible long-term predictor driven by $\{u(t)\}$ alone could be formed by simply utilizing the model in the following way:

Open-Loop Predictor

$$A(\theta, q^{-1})\hat{y}(t, \theta) = B(\theta, q^{-1})u(t) \tag{4.2.18}$$

where $A(\theta, q^{-1})$, $B(\theta, q^{-1})$ denote polynomial matrices having the same structure as $A(q^{-1})$, $B(q^{-1})$ in (4.2.1) and having entries parameterized by some parameter matrix θ. It is clear that if θ is chosen so that

$$A(\theta, q^{-1}) = A(q^{-1}) \tag{4.2.19}$$

$$B(\theta, q^{-1}) = B(q^{-1}) \tag{4.2.20}$$

Then the prediction error, $e(t) = y(t) - \hat{y}(t, \theta)$, satisfies

$$A(q^{-1})e(t) = 0 \tag{4.2.21}$$

and if $A(q^{-1})$ is asymptotically stable, then

$$\lim_{t \to \infty} e(t) = 0 \tag{4.2.22}$$

Thus the predictor (4.2.18) will asymptotically yield a good prediction of the system output.

It is instructive to relate the open-loop predictor (4.2.18) to the predictor (4.2.4). Heuristically, one would expect that the open-loop predictor would be a limiting case of (4.2.4) when the "look-ahead" interval becomes large.

For simplicity, we consider the scalar case, in which case the open predictor (4.2.18) is basically as follows:

$$\hat{y}(t, \theta) = \frac{B(\theta, q^{-1})}{A(\theta, q^{-1})}u(t) \tag{4.2.23}$$

Now from (4.2.6) we see that for arbitrary d' we have

$$\frac{1}{A(q^{-1})} = F(q^{-1}) + q^{-d'}\frac{G(q^{-1})}{A(q^{-1})} \tag{4.2.24}$$

Thus $F(q^{-1})$ is the first d' terms in the expansion of $1/A(q^{-1})$. Now provided that $A(q^{-1})$ is stable, the series $F(q^{-1})$ will converge as $d \to \infty$ and the remainder $G(q^{-1})$ will approach zero. Thus, for large d',

$$\frac{1}{A(q^{-1})} \simeq F(q^{-1}) \tag{4.2.25}$$

Now arguing as in (4.2.11)–(4.2.12), the d'-step-ahead predictor is given by

$$y(t) = G(q^{-1})y(t - d') + F(q^{-1})B(q^{-1})u(t) \tag{4.2.26}$$

For large d', the predictor above is approximately given by

$$y(t) \simeq F(q^{-1})B(q^{-1})u(t)$$

$$\simeq \frac{B(q^{-1})}{A(q^{-1})}u(t) \quad \text{using (4.2.25)}$$

Thus, as expected, the open-loop predictor for a stable system is a limiting case of the d-step-ahead predictor when the "look-ahead" interval gets large.

Nonlinear Systems

The input–output behavior of many systems is inherently nonlinear (see the discussion in Section 2.4). In these cases it is more appropriate to use a nonlinear predictor than a linear predictor. A general nonlinear predictor has the form

$$\hat{y}(t + d) = f(t, \theta, D(t)) \tag{4.2.27}$$

where $f(\cdot, \cdot, \cdot)$ denotes a specified linear or nonlinear function such that $\hat{y}(t)$ equals $y(t)$ (for all t) for some θ, say θ_0, and $D(t)$ denotes the data from which $\hat{y}(t, \theta)$ is computed.

If (4.2.27) is *linear in* θ, the predictor can be expressed as

$$\hat{y}(t + d) = \theta^T \phi(t) \tag{4.2.28}$$

where $\phi(t)$ is a (possibly nonlinear) function of $D(t)$.

4.2.2 Restricted Complexity Predictors

In the discussion above we have motivated the predictors by hypothesizing that the structure of the model describing the underlying system dynamics was known. An alternative viewpoint is simply to nominate a form for the predictor without regard to the nature of the system model. Thus, in general, we might assume (arbitrarily) a form for the predicted output as follows:

$$\hat{y}(t, \theta) = g(t, \theta, D(t)) \tag{4.2.29}$$

where θ denotes a set of parameters
    $g(\cdot, \cdot, \cdot)$ denotes an arbitrarily specified linear or nonlinear function
    $D(t)$ denotes the data from which $\hat{y}(t, \theta)$ is computed

We shall term a predictor of the form (4.2.29) a *restricted complexity predictor* on the grounds that its structure is prespecified and does not necessarily correspond to the true system structure. A consequence of the restricted nature of (4.2.29) is that there may not exist any value of θ for which $\hat{y}(t, \theta)$ is precisely equal to $y(t)$.

In the restricted complexity case, the data $D(t)$ can be quite arbitrary. It can include various time series which are believed to be informative in the prediction of $y(t)$ without there necessarily existing a strict functional relationship as in (4.2.1). For example, in the prediction of water consumption mentioned earlier, it is likely that rainfall and temperature are helpful indicators of consumption, but it is most unlikely that a strict relationship of the form (4.2.1) exists in this case.

In the next section we shall show how the parameter estimation techniques of Chapter 3 can be used to determine suitable values for θ such that $\hat{y}(t, \theta)$ gives a "good" prediction of $y(t)$.

4.3 ADAPTIVE PREDICTION

The basic idea of adaptive prediction is that given the data up to the present time, one adjusts the parameters in the predictor so that past predictions closely match the observed data. Then these parameters are used to generate future predictions. A number of obvious questions arise:

1. How should the parameters be adjusted?
2. What range of data should be used in adjusting the coefficients—more recent data being more relevant if the system is time varying?
3. Should the parameters in the predictor be estimated directly, or should one first estimate a model for the system and then derive the corresponding predictor by algebraic manipulation?

One approach is to estimate the parameters of the optimal predictor directly. The key advantage of this approach is that no additional calculations are required to determine the predictor from the estimated model.

The alternative approach is first to estimate the parameters in an arbitrary model for the system and then to convert this model into the required predictor format. This gives an *indirect* algorithm. The advantages of this approach are: (1) it is likely to involve fewer parameters, and (2) it would be possible to generate a range of predictors, for example 1, 2, . . . steps ahead based on the one model.

We discuss *direct* and *indirect* approaches to adaptive prediction in more detail below.

4.3.1 Direct Adaptive Prediction

We consider a simple case when the predicted output is a linear function of the parameters θ and we describe the output of the predictor as

$$\hat{y}(t + d, \theta) = \phi(t)^T \theta \tag{4.3.1}$$

where $\phi(t)$ is a linear or nonlinear function of the data,

$$D(t) \triangleq \{y(t), y(t - 1), \ldots, u(t), u(t - 1), \ldots\}$$

We will consider the use of the predictor structure of (4.3.1) in two situations. The first situation is when there exists a value of θ (say θ_0) such that the predictor output, $\hat{y}$, is equal to the true system output, y (the nonrestricted case). The second situation is when there does not necessarily exist a value of θ such that $\hat{y} = y$ (the restricted complexity case).

Case 1: The Nonrestricted Case

Here we assume that there exists a value of θ (say θ_0) such that for all t,

$$\hat{y}(t + d, \theta_0) = \phi(t)^T \theta_0 = y(t + d) \tag{4.3.2}$$

In view of (4.3.2), we can simply view the predictor as a special model for the system. Thus the parameter estimation methods of Chapter 3 are immediately applicable to the problem of sequentially estimating θ. For example, the projection algorithm [see (3.3.19)] suggests the following adaptive predictor:

Adaptive d-step-Ahead Predictor Based on Projection

$$\hat{\theta}(t) = \hat{\theta}(t - 1) + \frac{\phi(t - d)}{c + \phi(t - d)^T \phi(t - d)}[y(t) - \phi(t - d)^T \hat{\theta}(t - 1)]. \qquad c > 0$$

$$\tag{4.3.3}$$

The d-step-ahead prediction of $y(t)$ is then given by

$$\hat{y}(t) = \phi(t - d)^T \hat{\theta}(t - d) \qquad (4.3.4)$$

The predictor above has the appropriate causality properties since $\hat{\theta}(t - d)$ is a function of $y(t - d)$, $y(t - d - 1)$, $\ldots$, as is evident from (4.3.3). Properties of the predictor above are discussed in the following lemma.

Lemma 4.3.1.   If the sequence $\{y(t)\}$ satisfies a "model" of the form (4.3.2), and provided that

(i) the dimension of $\phi(\cdot)$ is chosen commensurately with the system order, and
(ii) $\{y(t)\}$, and $\{u(t)\}$ are bounded, then the adaptive predictor (4.3.3)–(4.3.4) has the convergence property

$$\lim_{N \to \infty} \sum_{t=1}^{N} (y(t) - \hat{y}(t))^2 < \infty \qquad (4.3.5)$$

and this implies

$$\lim_{t \to \infty} (y(t) - \hat{y}(t)) = 0 \qquad (4.3.6)$$

Proof. We can immediately apply Lemma 3.3.2 to show that the algorithm (4.3.3)–(4.3.4) has the following properties:

$$\lim_{N \to \infty} \sum_{t=1}^{N} \frac{[\phi(t - d)^T \tilde{\theta}(t - 1)]^2}{1 + \phi(t - d)^T \phi(t - d)} < \infty \qquad (4.3.7)$$

and

$$\lim_{N \to \infty} \sum_{t=k}^{N+k} \| \hat{\theta}(t) - \hat{\theta}(t - k) \|^2 < \infty \qquad (4.3.8)$$

for finite k.

Now we note that the prediction error is given by

$$e(t) \triangleq y(t) - \hat{y}(t)$$
$$= -\phi(t - d)^T \tilde{\theta}(t - d)$$
$$= -\phi(t - d)^T \tilde{\theta}(t - 1) - \phi(t - d)^T [\hat{\theta}(t - d) - \hat{\theta}(t - 1)]$$

Thus from the Schwarz inequality and the triangle inequality,

$$e(t)^2 \le 2[\phi(t - d)^T \tilde{\theta}(t - 1)]^2 + 2\phi(t - d)^T \phi(t - d) \| \hat{\theta}(t - 1) - \hat{\theta}(t - d) \|^2$$

The result (4.3.5) then follows from (4.3.7)–(4.3.8) and the boundedness of $\{\phi(t)\}$, which is implied by assumption (ii).

Equation (4.3.6) follows from (4.3.5).

▼▼▼

Note that Lemma 4.3.1, and in particular (4.3.6) shows that the adaptive prediction $\hat{y}(t)$ converges to the true system output $y(t)$! The result has been established under very weak assumptions. For example, arbitrary feedback between $\{y(t)\}$ and $\{u(t)\}$ is allowed and no persistent excitation condition is needed.

Analogous to the algorithm above, we also have the following least-squares-based algorithm.

Adaptive *d*-step-Ahead Predictor Based on Least Squares

$$\hat{\theta}(t) = \hat{\theta}(t-1) + P(t-d)\phi(t-d)]y(t) - \phi(t-d)^T\hat{\theta}(t-1)] \qquad (4.3.9)$$

$$P(t-d) = P(t-d-1) - \frac{P(t-d-1)\phi(t-d)\phi(t-d)^T P(t-d-1)}{1 + \phi(t-d)^T P(t-d-1)\phi(t-d)}$$

$$(4.3.10)$$

(All of the modifications to the least-squares algorithm discussed in Chapter 3 are also applicable here.)

The *d-step-ahead* prediction of $y(t)$ is given as before by

$$y(t) = \phi(t-d)^T\hat{\theta}(t-d) \qquad (4.3.11)$$

It is straightforward to establish properties similar to those in Lemma 4.3.1 for the algorithm above (see Exercise 4.1).

Case 2: The Restricted Complexity Case

Here we do not necessarily assume that there exists a value of θ such that $\hat{y} = y$. Instead, we seek a value of θ minimizing some function of the past prediction errors.

In the case when the predictor is linear in θ and we use a quadratic criterion, θ at time t is determined by minimizing

$$J_t(\theta) = \sum_{j=1}^{t} [y(j) - \phi(j-d)^T\theta]^2 + (\theta - \hat{\theta}(0))^T P_0^{-1}(\theta - \hat{\theta}(0)) \qquad (4.3.12)$$

where the second term has been included to account for initial information about θ.

We have seen in Chapter 3 (Lemma 3.3.5) that the cost function (4.3.12) is sequentially minimized by the algorithm (4.3.9)–(4.3.10). (Note that this occurs even in the restricted complexity case.)

A difficulty in the restricted complexity case is that the value of θ minimizing (4.3.12) may be highly data dependent. This is because there does not necessarily exist a single value of θ giving good predictions under all conditions. Thus, in the restricted complexity case, it is desirable to discount old data in some way, for example, by resetting P periodically (see Section 3.3).

4.3.2 Indirect Adaptive Prediction

As described previously, the basic idea here is to first fit a simple model to the data and then to manipulate the estimated model into the form of a *d*-step-ahead predictor. We shall discuss the algorithm in the nonrestricted and restricted complexity cases.

Case 1: The Nonrestricted Case

We assume that the system is described by a DARMA model of the form

$$A(q^{-1})y(t) = q^{-d}B'(q^{-1})u(t) \qquad (4.3.13)$$

We then estimate the coefficients in $A(q^{-1})$, $B(q^{-1})$ and apply the procedure described in Section 4.2 to the estimated model. This leads to the following algorithm (we use the projection algorithm for illustration):

$$\hat{\theta}(t) = \hat{\theta}(t-1) + \frac{\phi(t-1)}{c + \phi(t-1)^T \phi(t-1)} [y(t) - \phi(t-1)^T \hat{\theta}(t-1)] \qquad (4.3.14)$$

where

$$\phi(t-1)^T = [-y(t-1), \ldots, -y(t-n), u(t-d), \ldots, u(t-d-m)] \qquad (4.3.15)$$

Given $\hat{\theta}(t)$, we form $\hat{A}(t, q^{-1})$ and $\hat{B}'(t, q^{-1})$ as follows:

$$\hat{A}(t, q^{-1}) = 1 + \hat{a}_1(t)q^{-1} + \cdots + \hat{a}_n(t)q^{-n} \qquad (4.3.16)$$

$$\hat{B}'(t, q^{-1}) = \hat{b}_0(t) + \cdots + \hat{b}_m(t)q^{-m} \qquad (4.3.17)$$

Next, we determine $\hat{F}(t, q^{-1})$ and $\hat{G}(t, q^{-1})$ by solving

$$\hat{F}(t, q^{-1})\hat{A}(t, q^{-1}) + q^{-d}\hat{G}(t, q^{-1}) = 1 \qquad (4.3.18)$$

Finally, we form $\alpha(t, q^{-1})$ and $\beta(t, q^{-1})$ as

$$\alpha(t, q^{-1}) = \hat{G}(t, q^{-1}) \qquad (4.3.19)$$

$$\beta(t, q^{-1}) = \hat{F}(t, q^{-1})\hat{B}'(t, q^{-1}) \qquad (4.3.20)$$

and the adaptive d-step-ahead prediction is given by

$$\hat{y}(t + d) = \alpha(t, q^{-1})y(t) + \beta(t, q^{-1})u(t) \qquad (4.3.21)$$

Convergence properties of the indirect algorithm above are given in the following lemma.

Lemma 4.3.2. If the sequence $y(t)$ satisfies the model (4.3.13) and provided that $y(t)$ and $u(t)$ are bounded, then

$$\lim_{N \to \infty} \sum_{t=1}^{N} (y(t) - \hat{y}(t))^2 < \infty \qquad (4.3.22)$$

and this implies that

$$\lim_{t \to \infty} [y(t) - \hat{y}(t)] = 0 \qquad (4.3.23)$$

Proof. In the following proof we will need to manipulate time-varying operators. This is facilitated by the following notation. Given time-varying polynomials $\hat{B}(t, q^{-1})$, $(\hat{A}t, q^{-1})$, define

$$\hat{A}\hat{B} \triangleq \sum_i \sum_j \hat{a}_i(t)\hat{b}_j(t)q^{-i-j} = \hat{B}\hat{A} \qquad (4.3.24)$$

$$\hat{A} \cdot \hat{B} \triangleq \sum_i \sum_j \hat{a}_i(t)\hat{b}_j(t-i)q^{-i-j} \neq \hat{B} \cdot \hat{A} \qquad (4.3.25)$$

We also define

$$\check{\hat{B}} = \hat{B}(t-1, q^{-1}) \qquad (4.3.26)$$

The key equations for future reference are (we suppress q^{-1})

$$Ay(t) = Bu(t) \qquad (4.3.27)$$

$$\hat{y}(t+d) = \hat{G}y(t) + \hat{F}\hat{B}u(t) \qquad (4.3.28)$$

$$\hat{F}\hat{A} + q^{-d}\hat{G} = 1 \qquad (4.3.29)$$

$$e(t) \triangleq y(t) - \phi(t-1)^T\hat{\theta}(t-1)$$
$$= \hat{A}y(t) - \hat{B}u(t) \qquad (4.3.30)$$

$$v(t+d) \triangleq y(t+d) - \hat{y}(t+d) \qquad (4.3.31)$$

Now operating on (4.3.30) by $\hat{F}$ gives

$$\hat{F}e(t + d) = \hat{F} \cdot \hat{A}y(t + d) - \hat{F} \cdot \hat{B}u(t + d) \tag{4.3.32}$$

$$= \hat{F}\hat{A}y(t + d) - \hat{F}\hat{B}u(t + d) + (\hat{F} \cdot \hat{A} - \hat{F}\hat{A})y(t + d)$$
$$- (\hat{F} \cdot \hat{B} - \hat{F}\hat{B})u(t + d) \tag{4.3.33}$$

Using (4.3.28), (4.3.29), and (4.4.31) gives us

$$\hat{F}e(t + d) = v(t + d) + (\hat{F} \cdot \hat{A} - \hat{F}\hat{A})y(t + d) - (\hat{F} \cdot \hat{B} - \hat{F}\hat{B})u(t + d) \tag{4.3.34}$$

Now the algorithm ensures (see Lemma 3.3.2) that $\hat{A}$ is bounded and that

$$\lim_{N \to \infty} \sum_{t=1}^{N} \frac{e(t)^2}{c + \phi(t - 1)^T \phi(t - 1)} < \infty \tag{4.3.35}$$

$$\lim_{N \to 8} \sum_{t=k}^{N} \| \hat{\theta}(t) - \hat{\theta}(t - k) \|^2 < \infty, \qquad k \text{ finite} \tag{4.3.36}$$

Since $\{y(t)\}$ and $\{u(t)\}$ are bounded, it immediately follows that $\{\phi(t)\}$ is bounded and hence from (4.3.35),

$$\lim_{N \to \infty} \sum_{t=1}^{N} e(t)^2 < \infty \tag{4.3.37}$$

Also, (4.3.29) is always solvable and gives bounded $\hat{F}, \hat{G}$ since $\hat{A}$ is bounded.

Finally, squaring both sides of (4.3.34) and using the Schwarz inequality, the triangle inequality, and (4.3.36)–(4.3.37) we can conclude that

$$\sum_{t=1}^{\infty} v(t)^2 < \infty \tag{4.3.38}$$

This establishes (4.3.22). Equation (4.3.23) follows immediately.

▼▼▼

Case 2: The Restricted Complexity Case

In this case the algorithm is exactly as above. However, we can now not say anything definite about the convergence properties of the prediction error. In the first place, the one-step-ahead modeling error, $e(t)$, cannot be brought to zero due to the model mismatch. Second, the validity of the derivation of the d-step prediction from the one-step-ahead model relies on the accuracy of the given model. Thus all one can say is that if the modeling error is small, the prediction error will be correspondingly small.

4.4 CONCLUDING REMARKS

We see from the discussion above that there is a very close connection between parameter estimation and adaptive prediction. Indeed, the algorithms of this chapter are identical with those discussed in Chapter 3. However, there is an important philosophical distinction. In the case of parameter estimation our prime focus is on how close the estimated parameters are to the true values in the parameter space; whereas in prediction, the real measure of performance is the size of the prediction error, irrespective of the parameter estimates. We shall find that this distinction carries over to other applications, such as filtering and control.

EXERCISES

4.1. Establish the result corresponding to Lemma 4.3.1 for the algorithm (4.3.9)–(4.3.11).

4.2. Why can't the adaptive prediction algorithms of Section 4.3.1 be applied to the problem of finding a good predictor of the form

$$\hat{y}(t + d) = \frac{B(q^{-1})}{A(q^{-1})} u(t)$$

where

$$B(q^{-1}) = \theta_1 + \theta_2 q^{-1} + \cdots + \theta_n q^{-n}$$
$$A(q^{-1}) = 1 + \theta_{n+1} q^{-1} + \cdots + \theta_{n+m} q^{-m}?$$

4.3. Massage the following predictor into the form where the algorithms of Section 4.3.1 can be used (i.e., redefine θ so that the model is linear in the parameters).

$$y(t + d) = \theta_1[y(t)^3 + \theta_2 y(t)u(t) + e^{\theta_3}u(t)]$$

(What is θ and what is ϕ in the resulting algorithm?)

4.4. Consider a nonlinear system

$$y(t) = b_0 u(t)^2$$

Suppose that we use a restricted complexity linear predictor of the form

$$\hat{y}(t) = \phi(t)\theta \qquad \text{where } \phi(t) = u(t)$$

Show that optimal value of θ depends on the data.

4.5. Consider again the predictor given in Exercise 4.2.

(a) Write down an output error prediction error algorithm using the parameter estimator of Section 3.5 (using a posteriori predictions in the regression vector).

(b) Show that the algorithm ensures that

$$\lim_{t \to \infty} [y(t) - \hat{y}(t)] = 0$$

(c) Predictors of the form given in Exercise 4.2 are often said to be useful when long-term predictions are required since the predicted output is a function of $u(t)$ only [i.e., $\hat{y}(t)$ does not depend upon past observations of $y(t)$]. Why is this *not* true of the adaptive predictor constructed in part (a)? [*Hint:* What is $\hat{\theta}(t)$ a function of?]

4.6. Consider a periodic waveform of period p.

(a) Show that the signal can be modeled by

$$(1 - q^{-p})y(t) = 0$$

(b) By evaluating the roots of $(z^p - 1)$ show that the model above incorporates all Fourier components of period p.

(c) Consider a p-step-ahead predictor of the form

$$(1 - \alpha q^{-p})\hat{y}(t) = (1 - \alpha)y(t - p)$$

Show that
(i) The predictor is stable for all $\alpha: |\alpha| < 1$.
(ii) The predictor error converges to zero exponentially fast for all $|\alpha| < 1$ and for all $\hat{y}(0)$.

(d) Show that $\hat{y}(t)$ is given by

$$\hat{y}(t) = (1 - \alpha)[y(t - p) + \alpha y(t - 2p) + \alpha^2 y(t - 3\phi) \cdots]$$

Hence show that the algorithm incorporates exponential weighted smoothing of past data.

(e) Using the results of part (d), why is it desirable to use $\alpha = 0$ when the system is time varying and there is no noise, and α near 1 when the system is time invariant but there is noise? Comment on the trade-off between averaging over past data to minimize the effects of noise and having the ability to respond to lack of phase coherency.

4.7. Convince yourself that a sine wave of frequency p rad/sec can be modeled as

$$(1 + a_1 q^{-1} + a_2 q^{-2})y(t) = 0$$

(a) What are the values of a_1 and a_2?
(b) Use Lemma 4.2.1 to write down a two-step-ahead predictor for $y(t)$.
(c) Write down an adaptive predictor for $y(t)$ when the frequency of the sine wave, p, is unknown. (Use two steps ahead.)
(d) What would be the form of a 65-step-ahead predictor for $y(t)$?
(e) Write down the form of an adaptive 65-step-ahead predictor.
(f) Theoretically, what are the convergence properties of the algorithm in part (e)?
(g) Do you anticipate any practical difficulties with the algorithm in part (e)? (*Hint:* Consider $p \simeq \pi/50$ and $p \simeq \pi/3$.)
(h) Can you think of a way to overcome the practical difficulties for at least one of the values of p given in part (g)?

5

Control of Linear
Deterministic Systems

5.1 INTRODUCTION

In this chapter we review some aspects of the control of linear deterministic systems to provide a basis for our subsequent discussion of adaptive control.

In any control system design problem one can distinguish five important considerations: stability, transient response, tracking performance, constraints, and robustness.

1. *Stability.* This is concerned with stability of the system, including boundedness of inputs, outputs, and states. This is clearly a prime consideration in any control system.

2. *Transient response.* Roughly this is concerned with how fast the system responds. For linear systems, it can be specified in the time domain in terms of rise time, settling time, percent overshoot, and so on, and in the frequency domain in terms of bandwidth, damping, resonance, and so on.

3. *Tracking performance.* This is concerned with the ability of the system to reproduce desired output values. A special case is when the desired outputs are constant, in which case we often use the term *output regulation*, in lieu of tracking.

4. *Constraints.* It turns out that in the linear case if the system model is known exactly and there is no noise, then arbitrarily good performance can be achieved by use of a sufficiently complex controller (see Section 5.4). However, one must bear in mind that there are usually physical constraints that must be adhered to, such as limits in the magnitude of the allowable control effort, limits in the rate of change of control signals, limits on internal variables such as temperatures

and pressures, and limits on controller complexity. These factors ultimately place an upper limit on the achievable performance.

5. *Robustness.* This is concerned with the degradation of the performance of the system depending on certain contingencies, such as unmodeled dynamics, including disturbances, parameter variations, component failure, and so on.

We will primarily treat sampled data control systems so that our inputs and outputs will be sequences. Typically, the actual process will be continuous in nature. Suitable sampled data models can then be obtained by the techniques discussed in Section A.1 of Appendix A (see also Exercise 2.1). It must be borne in mind that a design based on discrete-time models refers only to the sampled values of the input and output. Usually, the corresponding continuous response will also be satisfactory, but in certain cases it may be important to check that the response between samples is acceptable. This aspect will be discussed in some detail later.

We shall begin our discussion by emphasizing tracking performance and introduce a very simple form of control law which we call a one-step-ahead controller. The basic principle behind this controller is that the control input is determined at each point in time so as to bring the system output at a future time instant to a desired value. This is a very simple idea but, nonetheless, has considerable intuitive appeal. We shall show that the scheme works well not only for linear systems but can also be applied in a straightforward fashion to a large class of nonlinear systems. We shall subsequently develop an adaptive control algorithm based on the one-step-ahead principle. However, as with all design techniques, one must be aware of its limitations as well as its applications. Potential difficulties with this approach are that an excessively large effort may be called for to bring the output to the desired value in one step, and resonances may be excited in lightly damped systems.

We are therefore motivated to look at alternative control schemes which allow us to limit the control effort needed to achieve the objective. This will then lead us to consider weighted one-step-ahead controllers wherein a penalty is placed on excessive control effort. We shall subsequently show that the one-step-ahead controller is a special case of the weighted one-step-ahead controller. Again, we shall point to its limitations and show that it may still lead to difficulty in some cases. To some extent these difficulties are overcome by introducing a reference model to generate the desired output sequence, thereby smoothing the overall response. This leads to the idea of model reference control. Finally, we turn to more complicated algorithms, which at least for the linear case, resolve many of the difficulties encountered by the simpler algorithms discussed above. In particular, we shall study algorithms based on closed-loop pole assignment. We shall briefly discuss the relationship between pole locations and the nature of the system response and we shall show that model reference control is a special case of pole assignment. Where appropriate, our discussion will include such practical considerations as response time, steady-state error, and robustness.

Our main purpose here is to set the scene for our subsequent development of adaptive control. For further discussion of linear deterministic control theory, the reader is referred to the following books (among others): Anderson and Moore (1971), Brockett (1970), Chen (1970), Desoer (1970), Fortmann and Hitz (1977),

Franklin and Powell (1980), Kailath (1980), Kuo (1980), Rosenbrock (1970), and Wolovich (1974).

5.2 MINIMUM PREDICTION ERROR CONTROLLERS

We have seen in Chapter 4 that we can think of a model as providing a way of predicting the future outputs of a system based on past outputs and on past and present inputs. It therefore seems reasonable to turn this around and ask what control action at the present instant of time would bring the future output to a desired value. This will be especially simple if the future outputs are a linear function of the *present* control, since then the determination of the control effort involves the solution of a set of linear equations.

We consider first the case of linear dynamical system. We have seen in Chapter 2 that the input–output properties of these systems (for arbitrary initial states) can be described by one of the following equivalent model formats: an observable state-space model, a left difference operator representation, or a DARMA model. We shall use the last of these as the basis of our discussion. The general DARMA model can be expressed as

$$A(q^{-1})y(t) = B(q^{-1})u(t) \tag{5.2.1}$$

where

$$A(q^{-1}) = I + A_1 q^{-1} + \cdots + A_n q^{-n} \tag{5.2.2}$$

$$B(q^{-1}) = q^{-d}(B_0 + B_1 q^{-1} + \cdots + B_{n_1} q^{-n_1}) \tag{5.2.3}$$

$$= q^{-d} B'(q^{-1}) \tag{5.2.4}$$

We have seen in Lemma 4.2.1 that the DARMA model above can be expressed in d-step-ahead predictor form as

$$y(t + d) = \alpha(q^{-1})y(t) + \beta(q^{-1})u(t) \tag{5.2.5}$$

where

$$\alpha(q^{-1}) = \alpha_0 + \alpha_1 q^{-1} + \cdots + \alpha_{n-1} q^{-(n-1)} = G(q^{-1}) \tag{5.2.6}$$

$$\beta(q^{-1}) = \beta_0 + \beta_1 q^{-1} + \cdots + \beta_{n_1+d-1} q^{-(n_1+d-1)} = F(q^{-1})B'(q^{-1}) \tag{5.2.7}$$

To simplify our presentation, we shall first treat the single-input single-output case. We shall see later that the results extend in a natural way to the multi-input multi-output case.

5.2.1 One-Step-Ahead Control (The SISO Case)

In the single-input single-output case, we choose the time delay d so that the leading coefficient of $B'(q^{-1})$ in the model (5.2.4) is nonzero. This, in turn, implies that β_0 in the corresponding predictor (5.2.7) is nonzero.

Here we will consider one-step-ahead control, which brings $y(t + d)$ to a desired value $y^*(t + d)$ in one step. We have the following result regarding the stability, transient performance, and tracking performance of the one-step-ahead control law.

Theorem 5.2.1 (One-Step-Ahead Control). Consider the system described by the DARMA model (5.2.1),

(a) The feedback control law which brings the output at time $t + d$, $y(t + d)$, to some desired bounded value $y^*(t + d)$ has the form

$$\beta(q^{-1})u(t) = y^*(t + d) - \alpha(q^{-1})y(t); \quad t \geq 0 \quad (5.2.8)$$

(b) The resulting closed-loop system is described by

$$y(t) = y^*(t); \quad t \geq d \quad (5.2.9)$$

$$B(q^{-1})u(t) = A(q^{-1})y^*(t); \quad t \geq d + n \quad (5.2.10)$$

(c) The resulting closed-loop system has bounded inputs and outputs provided that

(i) All modes of the "inverse" model (5.2.10) [i.e., the zeros of the polynomial $z^d B(z^{-1})$] lie inside or on the closed unit disk.

(ii) All controllable modes of the "inverse" model (5.2.10) [i.e., the zeros of the transfer function $z^d B(z^{-1})/A(z^{-1})$] lie strictly inside the unit circle.

(iii) Any modes of the "inverse" model (5.2.10) on the unit circle have a Jordan block size of 1.

In particular, if there are no roots on the unit circle, $z^{d+n_1}B(z^{-1})$ should have all its roots strictly inside the unit disk.

Proof. (a) Immediate from the predictor (5.2.5) by setting the predicted output $y(t + d)$ to $y^*(t + d)$ and solving for $u(t)$ in terms of $u(t - 1), \ldots, y(t), y(t - 1)$, ... Note that this is possible since $\beta_0 \neq 0$ by construction.

(b) Equation (5.2.9) follows from (a) by the choice of the control input. Equation (5.2.10) then follows from (5.2.1) by substituting (5.2.9).

(c) We first note that the inverse DARMA model (5.2.10) [relating $y^*(t)$ to $u(t)$] is equivalent to an observable state-space model. The result then follows from Lemma B.3.3 of Appendix B and the fact that $\{y^*(t)\}$ is bounded.

▼▼▼

Example 5.2.1

We illustrate the discussion above by reference to the system introduced in Exercise 2.10. The system is a simple system with disturbance described by the following state equations:

$$\eta(t + 1) = \eta(t); \quad t \geq 0 \quad \text{(disturbance model)}$$

$$x(t + 1) = u(t); \quad t \geq 0 \quad \text{(system model)}$$

$$y(t) = x(t) + \eta(t) \quad \text{(output)}$$

As shown in Exercise 2.10, the corresponding DARMA model is

$$y(t) = y(t - 1) + u(t - 1) - u(t - 2)$$

The one-step-ahead control law (5.2.8) is

$$u(t) = y^*(t + 1) - y(t) + u(t - 1); \quad t \geq 0$$

and this gives a closed-loop system characterized by

$$y(t) = y^*(t); \qquad t \geq 1$$
$$u(t) - u(t-1) = y^*(t+1) - y^*(t)$$

Note that conditions (i) to (iii) of Theorem 5.2.1 are satisfied and thus $\{u(t)\}$ and $\{y(t)\}$ will be bounded.

▼▼▼

We note in passing that the one-step-ahead feedback control law (5.2.8) minimizes the following cost function comprising the squared prediction error:

$$J_1(t+d) = \tfrac{1}{2}[y(t+d) - y^*(t+d)]^2 \qquad (5.2.11)$$

As noted in Section 5.1, excessive control effort may be called for to bring $y(t+d)$ to $y^*(t+d)$ in one step. We therefore consider a slight generalization of the cost function (5.2.11) to the form (5.2.12) given below, which aims to achieve a compromise between bringing $y(t+d)$ to $y^*(t+d)$ and the amount of effort expended.

$$J_2(t+d) = \left\{ \frac{1}{2}(y(t+d) - y^*(t+d))^2 + \frac{\lambda}{2}u(t)^2 \right\} \qquad (5.2.12)$$

We shall call the control law minimizing (5.2.12) at each time instant a *weighted one-step-ahead controller*. The term "one-step-ahead" is still used because we aim to minimize (5.2.12) on a one-step basis rather than in an average sense over an extended time horizon. We now have the following extension of Theorem 5.2.1:

Theorem 5.2.2 (Weighted One-Step-Ahead Control)

(a) The control law minimizing $J_2(t+d)$ has the form

$$u(t) = \frac{\beta_0\{y^*(t+d) - \alpha(q^{-1})y(t) - \beta'(q^{-1})u(t-1)\}}{\beta_0^2 + \lambda} \qquad (5.2.13)$$

where

$$\beta'(q^{-1}) = q[\beta(q^{-1}) - \beta_0]$$
$$= \beta_1 + \beta_2 q^{-1} + \cdots + \beta_{n_1+d-1}q^{-(n_1+d-2)}$$

(b) The closed-loop system is given by

$$\left[B'(q^{-1}) + \frac{\lambda}{\beta_0} A(q^{-1}) \right] y(t+d) = B'(q^{-1})y^*(t+d) \qquad (5.2.14)$$

$$\left[B'(q^{-1}) + \frac{\lambda}{\beta_0} A(q^{-1}) \right] u(t) = A(q^{-1})y^*(t+d) \qquad (5.2.15)$$

where $B'(q^{-1})$ is as in (5.2.4).

(c) The resulting closed-loop system has bounded inputs and outputs provided that:

  (i) All modes of the "inverse" models (5.2.14) and (5.2.15) [i.e., the zeros of the polynomial $[B'(z^{-1}) + (\lambda/\beta_0)A(z^{-1})]$] lie inside or on the closed unit disk.

  (ii) All controllable modes of the "inverse" models (5.2.14) and (5.2.15) (i.e., the zeros of the transfer functions

$$\frac{1}{B'(z^{-1})}\left[B'(z^{-1}) + \frac{\lambda}{\beta_0} A(z^{-1}) \right]$$

and

$$\frac{1}{A(z^{-1})}\left[B'(z^{-1}) + \frac{\lambda}{\beta_0}A(z^{-1})\right]$$

lie strictly inside the unit circle.

(iii) Any modes of the "inverse" models (5.2.14) and (5.2.15) on the unit circle have a Jordan block size of 1.

Proof. (a) Substituting (5.2.5) into (5.2.12) gives

$$J_2(t + d) = \left\{\frac{1}{2}(\alpha(q^{-1})y(t) + \beta(q^{-1})u(t) - y^*(t + d))^2 + \frac{\lambda}{2}u(t)^2\right\} \qquad (5.2.16)$$

Differentiating with respect to $u(t)$ and setting the result to zero gives

$$\beta_0(\alpha(q^{-1})y(t) + \beta(q^{-1})u(t) - y^*(t + d)) + \lambda u(t) = 0 \qquad (5.2.17)$$

Equation (5.2.13) immediately follows from (5.2.17) by rearrangement.

(b) The control law is (5.2.17). This equation can also be written using the predictor representation (5.2.5) as

$$\beta_0\{y(t + d) - y^*(t + d)\} + \lambda u(t) = 0 \qquad (5.2.18)$$

Multiplying by $A(q^{-1})$ gives

$$\beta_0\left\{A(q^{-1})y(t + d) - A(q^{-1})y^*(t + d) + \frac{\lambda}{\beta_0}A(q^{-1})u(t)\right\} = 0$$

or using (5.2.1) and (5.2.4),

$$\beta_0\left\{B'(q^{-1})u(t) - A(q^{-1})y^*(t + d) + \frac{\lambda}{\beta_0}A(q^{-1})u(t)\right\} = 0$$

Equation (5.2.15) then follows immediately. Similarly, multiplying (5.2.18) by $B'(q^{-1})$ gives (5.2.14).

(c) As for Theorem 5.2.1, part (c).

▼▼▼

It is important to note that the closed-loop system is stable if $B'(q^{-1}) + (\lambda/\beta_0)A(q^{-1})$ is stable. This represents a slight relaxation of the requirement for the one-step-ahead controller of Theorem 5.2.1. We now have an additional degree of freedom. If $\lambda = 0$, the result reduces to the one-step-ahead result. A simple root-locus argument shows that $B'(q^{-1}) + (\lambda/\beta_0)A(q^{-1})$ can be stabilized by suitable choice of λ for all systems having a stable inverse [i.e., $B'(q^{-1})$ having roots inside or on the unit circle], all stable systems and *some* systems that are neither stable nor stably invertible.

Example 5.2.2

Consider a system in the form of (5.2.1) with $d = 1$ and

$$A(q^{-1}) = 1 - 2q^{-1}$$
$$B(q^{-1}) = 1 + 3q^{-1}$$

This system does not satisfy the conditions of Theorem 5.2.1 and thus $\{u(t)\}$ will not be bounded for a one-step-ahead control law. The weighted one-step-ahead controller gives a closed-loop system whose modes are the zeros of

$$(1 + 3z^{-1}) + \frac{\lambda}{\beta_0}(1 - 2z^{-1})$$

Thus the closed-loop modes satisfy

$$\frac{\lambda}{\beta_0}\frac{z-2}{z+3} = -1$$

A simple root-locus argument (see Exercise 5.2) shows that the closed loop system is asymptotically stable for $\frac{2}{3} < \lambda/\beta_0 < 4$.

▼▼▼

The example above raises the question as to whether or not all systems can be stabilized by a weighted one-step-ahead controller with λ chosen appropriately. The answer is unfortunately no. A concrete example of this failure is given in Exercise 5.3. The cause of the trouble is that the design may not have enough degrees of freedom. We therefore introduce the following cost function, J_3, which aims to maintain the essential simplicity of the one-step-ahead principle but incorporate sufficient degrees of freedom in the design to ensure stability for *all* systems. Thus consider

$$J_3(t+d) = \left\{\frac{1}{2}(y(t+d) - y^*(t+d))^2 + \frac{\lambda}{2}\bar{u}(t)^2\right\} \tag{5.2.19}$$

where $\{\bar{u}(t)\}$ is related to $\{u(t)\}$ by a linear transfer function:

$$P(q^{-1})\bar{u}(t) = R(q^{-1})u(t)$$
$$P(q^{-1}) = 1 + p_1 q^{-1} + \cdots + p_l q^{-l} \tag{5.2.20}$$
$$R(q^{-1}) = 1 + r_1 q^{-1} + \cdots + r_l q^{-l}$$

We then have the following result, which is a generalization of Theorem 5.2.2.

Theorem 5.2.3

(a) The control law minimizing (5.2.19) is given by

$$u(t)$$
$$= \frac{\beta_0\{y^*(t+d) - \alpha(q^{-1})y(t) - \beta'(q^{-1})u(t-1)\} + \lambda P'(q^{-1})\bar{u}(t-1) - \lambda R'(q^{-1})u(t-1)}{\beta_0^2 + \lambda} \tag{5.2.21}$$

where

$$\beta'(q^{-1}) = q[\beta(q^{-1}) - \beta_0]$$
$$P'(q^{-1}) = q[P(q^{-1}) - 1]$$
$$R'(q^{-1}) = q[R(q^{-1}) - 1]$$

and $\bar{u}(t)$ is given by (5.2.20).

(b) The closed-loop system is described by

$$\left[P(q^{-1})B'(q^{-1}) + \frac{\lambda}{\beta_0}R(q^{-1})A(q^{-1})\right]u(t) = P(q^{-1})A(q^{-1})y^*(t+d) \tag{5.2.22}$$

$$\left[P(q^{-1})B'(q^{-1}) + \frac{\lambda}{\beta_0}R(q^{-1})A(q^{-1})\right]y(t+d) = P(q^{-1})B'(q^{-1})y^*(t+d) \tag{5.2.23}$$

(c) The closed-loop system has bounded inputs and bounded outputs with conditions on the models (5.2.22) and (5.2.23) similar to those in Theorems 5.2.1 and 5.2.2.

Control of Linear Deterministic Systems   Chap. 5

Note the modes of the system are now the zeros of

$$\left[P(z^{-1})B'(z^{-1}) + \frac{\lambda}{\beta_0} R(z^{-1})A(z^{-1}) \right]$$

Proof. As for Theorem 5.2.2. (See Exercise 5.4.)

▼▼▼

Remark 5.2.1. A natural question that arises in conjunction with the weighted one-step-ahead controller is how does one choose the polynomials $P(q^{-1})$ and $R(q^{-1})$. The following points are helpful guidelines:

1. As we shall see later, provided only that $A(q^{-1})$ and $B(q^{-1})$ have no unstable common modes, there always exists a choice of λ, $P(q^{-1})$, and $R(q^{-1})$ such that the closed-loop system is guaranteed stable. If $A(q^{-1})$ and $B(q^{-1})$ are roughly known, suitable values for $\lambda, P(q^{-1})$, and $R(q^{-1})$ can often be obtained by drawing root-locus diagrams for the equation

$$P(z^{-1})B'(z^{-1}) + \frac{\lambda}{\beta_0} R(z^{-1})A(z^{-1}) = 0$$

2. It is important to note that the weighted one-step-ahead criterion $J_2(t + d)$ given in (5.2.12) will in general lead to a steady-state tracking error unless the system itself includes a pure integrator. This is because a compromise is made between bringing $y(t + d)$ to $y^*(t + d)$ and keeping $u(t)$ small. Zero steady-state tracking error for a *constant* $y^*(t)$ sequence can be achieved by weighting the change in control from instant to instant rather than the control itself. This can be achieved with the cost function $J_3(t + d)$ given in (5.2.19) provided that $(1 - q^{-1})$ is a factor of $R(q^{-1})$. (See Exercise 5.5.) This introduces integral action. The reader is cautioned about the effect of *integral wind-up*, especially in the presence of saturation in the input or output (see Section 5.3.6).

We illustrate the idea of one-step-ahead control for the servo-kit example discussed in Exercise 2.1.

▼▼▼

Example 5.2.3

As shown in Exercise 2.1, the transfer function of the servo is

$$G(s) = \frac{K}{s(Ts + 1)}; \qquad K = 30, \quad T = 0.1 \text{ sec}$$

with corresponding discrete transfer function

$$H(z) = \frac{K_d(z - \beta)}{(z - 1)(z - \alpha)}$$

At a sampling rate of 100 msec, the transfer function is approximately

$$H(z) = \frac{0.2(z + 0.885)}{(z - 1)(z - 0.67)}$$

The corresponding DARMA form follows immediately and is given by

$$y(t) = 1.67y(t - 1) - 0.67y(t - 2) + 0.2u(t - 1) + 0.18u(t - 2) \qquad (5.2.24)$$

Now due to the time required to evaluate an input signal, it is necessary to use the technique inherent in the development of the predictor equation (5.2.5) to replace $y(t-1)$ on the right-hand side of (5.2.24) by $y(t-2)$, ... This can be done by simply substituting for $y(t-1)$ on the right-hand side of (5.2.24).

The resulting model now takes the following equivalent form:

$$y(t) = 2.1y(t-2) - 1.11y(y-3) + 0.2u(t-1) + 0.51u(t-2) + 0.3u(t-3)$$

We then use the weighted one-step-ahead cost function

$$J(t) = \frac{1}{2}[y(t+1) - y^*(t+1)]^2 + \frac{\lambda}{2}[u(t) - u(t-1)]^2 \qquad (5.2.25)$$

The control minimizing (5.2.25) is

$$u(t) = \frac{0.2[y^*(t+1) - 2.1y(t-1) + 1.11y(t-2) - 0.51u(t-1) - 0.3u(t-2)] + \lambda u(t-1)}{(0.2)^2 + \lambda}$$

The control law above was implemented on a servo kit using an on-line computer with sampling rate 100 msec. The *sampled* response for $y^*(t)$, a square wave of period 10 sec, was then obtained for different values of λ. (See Figs. 5.2.1 to 5.2.3.) Note that the performance of the controller depends rather critically on having the correct values for the system parameters; see the discussion later in Section 5.2.4. The precise nature of the responses shown in Figs. 5.2.1 to 5.2.3 can be explained using root-locus argu-

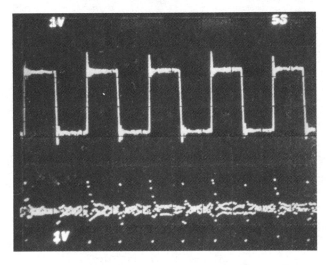

Figure 5.2.1 Response for $\lambda = 0.1$.
Upper trace, output; lower trace, input.

ments to roughly find the location of the closed-loop poles for different values of λ; see also the example is Section 5.4.

▼▼▼

Remark 5.2.2. The idea of bringing a predicted system output to a desired value is not restricted to linear models. The idea can be used with nonlinear models as well. This is particularly straightforward in the case of bilinear systems described in Section 2.4 because in this case the output is a *linear* function of the latest control input. For example, in the first-order case, the bilinear model has the simple form

$$y(t+1) = ay(t) + bu(t) + nu(t)y(t) \qquad (5.2.26)$$

         Control of Linear Deterministic Systems   Chap. 5

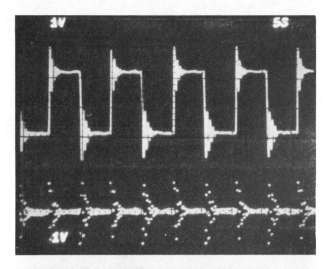

Figure 5.2.2 Response for $\lambda = 0.5$.
Upper trace, output; lower trace, input.

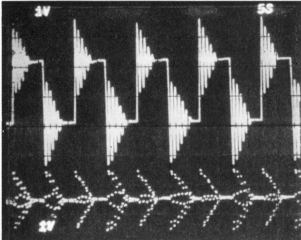

Figure 5.2.3 Response for $\lambda = 1.0$.
Upper trace, output; lower trace, input.

The corresponding one-step-ahead bilinear control law is

$$u(t) = \frac{y^*(t+1) - ay(t)}{b + ny(t)} \qquad (5.2.27)$$

Of course, it is necessary to avoid outputs such that $b + ny(t) = 0$ since this will result in loss of controllability.

The one-step-ahead control principle is very simple and may not be as general as other control laws, such as those based on pole assignment or linear optimal control. However, as we have seen above, the one-step-ahead control principle does have the advantage of being able to utilize the true nonlinear dynamical model of the process. Thus it may be better able to capture the intrinsic features of the control problem than a control law based on a linear approximation. A successful application of this idea to the control of a variable speed induction motor (a highly nonlinear system) has been reported by Webster (1981).

The reader may better appreciate the principles involved by reference to an example.

Example 5.2.4

Consider the heat exchanger discussed in Exercise 2.15. Using a first-order Euler approximation to the solution of the model equations leads to the following discrete-time model:

$$y(t+1) = y(t) - \frac{\Delta}{V}F(t)y(t) + \frac{\Delta}{V}F(t)T(t) + \frac{\Delta}{RCV}u(t)^2$$

where $y(t) =$ output temperature
$\quad F(t) =$ flow rate
$\quad T(t) =$ input temperature
$\quad u(t) =$ voltage applied to the heating element (the input)
$\quad \Delta =$ sampling period
$\quad V, R, C$ are constants

The one-step-ahead control law for the model above is

$$u(t) = \left[\frac{y^*(t+1) - y(t) + (\Delta/V)F(t)y(t) - (\Delta/V)F(t)T(t)}{\Delta/RCV} \right]^{1/2}$$

A possible advantage associated with the control law above is that changes in the input temperature and flow rate are automatically compensated for in the appropriate manner via the product terms $F(t)y(t)$ and $F(t)T(t)$. The controller automatically has feedforward action and a variable gain depending on the flow rate.

▼▼▼

Remark 5.2.3. A minor variant of the one-step-ahead design method arises when one wishes to bring y to y^* and, in addition, to simultaneously ensure that the rate of change of y (i.e., dy/dt) is zero on reaching y^*. As we shall see later, this may lead to improved performance especially for lightly damped systems. Of course, to achieve these multiple objectives we will, in general, need more degrees of freedom than is provided by a single input. One possibility is to construct the input $\{u(t)\}$ in blocks to give the requisite number of degrees of freedom. For example, in the case of $y(t)$ and $\dot{y}(t)$, one needs to use two successive controls to bring both $y(t)$ and $\dot{y}(t)$ to their desired values. This is illustrated below.

Let $y(t)$ and $\dot{y}(t)$ be described by the following predictor:

$$\begin{bmatrix} y(t) \\ \dot{y}(t) \end{bmatrix} = \alpha(q^{-1}) \begin{bmatrix} y(t-1) \\ \dot{y}(t-1) \end{bmatrix} + \beta(q^{-1})u(t-1)$$

where $\alpha(q^{-1})$ and $\beta(q^{-1})$ are a 2×2 and 2×1 polynomial matrices, respectively, of the form

$$\alpha(q^{-1}) = \alpha_1 + \alpha_1 q^{-1} + \cdots + \alpha_{n_1} q^{-n_1}; \qquad \beta(q^{-1}) = \beta_0 + \beta_1 q^{-1} + \cdots + \beta_{n_2} q^{-n_2}$$

Substituting the predictor above into itself once gives

$$\begin{bmatrix} y(t+2) \\ \dot{y}(t+2) \end{bmatrix} = \alpha(q^{-1})^2 \begin{bmatrix} y(t) \\ \dot{y}(t) \end{bmatrix} + \beta(q^{-1})u(t+1) + \alpha(q^{-1})\beta(q^{-1})u(t)$$

$$\triangleq M \begin{bmatrix} u(t+1) \\ u(t) \end{bmatrix} + \begin{bmatrix} r_1(t) \\ r_2(t) \end{bmatrix}$$

where $M = [\beta_0 \mid \beta_1 + \alpha_0\beta_0]$, and $r_1(t)$ and $r_2(t)$ are functions of $y(t)$, $y(t-1)$, ...,
$\dot{y}(t)$, $\dot{y}(t-1)$, ..., $u(t-1)$, $u(t-2)$.

The one-step-ahead design principle then gives the following feedback control law:

$$\begin{bmatrix} u(t+1) \\ u(t) \end{bmatrix} = M^{-1}\begin{bmatrix} y^*(t+2) - r_1(t) \\ \dot{y}^*(t+2) - r_2(t) \end{bmatrix}; \qquad t = 0, 2, 4$$

Typically, we will have $y^* = $ constant, $\dot{y}^* = 0$. We shall then call this a *velocity compensated dead-beat controller*.

We shall discuss the foregoing idea further under the more general framework of multiple-input multiple-output systems in Section 5.2.3. An example is given in Section 5.4.

▼▼▼

In the next section we consider a slightly different form of minimum prediction error controller in which the desired output $\{y^*(t)\}$ is generated by a reference model.

5.2.2 Model Reference Control (The SISO Case)

As before, the system is described as in (5.2.1), and we require that the output of the system tracks a desired output sequence $\{y^*(t)\}$. Note that, up to this point, this is the same problem formulation as previously. Here, however, we further specify that the desired output $\{y^*(t)\}$ will be the output of a linear dynamical system driven by a reference input $\{r(t)\}$ and having *known* transfer function $G(z)$. Specifically, we have:

Reference Model Assumption

1. The desired output $y^*(t)$ satisfies the following reference model:

$$E(q^{-1})y^*(t) = q^{-d'}gH(q^{-1})r(t) \tag{5.2.28}$$

with associated transfer function $G(z) = z^{-d'}H(z^{-1})g/E(z^{-1})$, where g is a constant gain and

$$H(z^{-1}) = h_0 + h_1z^{-1} + \cdots + h_lz^{-l}; \qquad h_0 = 1 \tag{5.2.29}$$

$$E(z^{-1}) = e_0 + e_1z^{-1} + \cdots + e_lz^{-l}; \qquad e_0 = 1 \tag{5.2.30}$$

We also impose the following additional constraints.

2. $E(z^{-1})$ is *stable* (having no zeros for $|z| \geq 1$).
3. The delay in the model, d', should be greater than or equal to the delay, d, in the system. For simplicity, we will take $d' = d$ in the sequel.

We wish to apply feedback to the system so that the output of the system, when driven by $\{r(t)\}$, is equal to the output of the model $G(z)$ driven by the same reference input. This is illustrated in Fig. 5.2.4. In the discrete-time case, the setup of Fig. 5.2.4 is restrictive, since as we have seen in the preceding section, a controller can actually be designed to track an arbitrary desired output sequence $y^*(t)$ as long as it is bounded. To make the link with the previous approach more transparent, we note that, in Fig. 5.2.4, the signal denoted $z(t) = y^*(t+d)$ is available at time t. Thus this signal can be used instead of $\{r(t)\}$ to drive the system. This gives the previous setup as in Fig. 5.2.5. Thus we see that the model reference set up serves as a particular way of specifying the desired output $y^*(t + d)$.

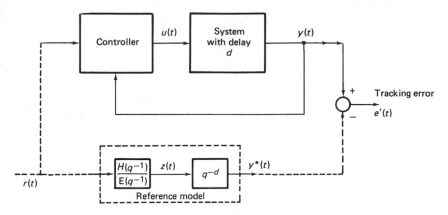

Figure 5.2.4 Model reference control system.

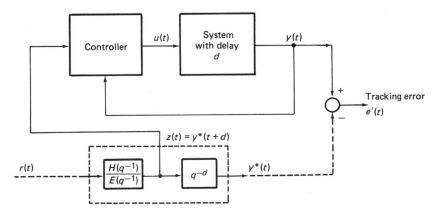

Figure 5.2.5 Reformulation of the model reference control system of Fig. 5.2.4.

Now, given that the desired output $y^*(t)$ is specified by a reference model, and if the one-step-ahead approach is adopted, we attempt to apply feedback to the system so that it faithfully reproduces $y^*(t)$ as the output when $y^*(t + d)$ is applied to the input. The overall characteristics of the control system [relating $r(t)$ to $y(t)$] then match those of the reference model since the reference model generating $y^*(t)$ appears in series with the system. An alternative to this is to apply feedback to the system so that the closed-loop poles are assigned to those of the reference model and only the zeros of the reference model appear in series. The latter approach has the advantage (depending on the location of the poles of the reference model) that the feedback controller gains can be less than in the previous case, and this may result in a more robust controller in the presence of unmodeled dynamics, and so on. Of course, the two approaches are identical when the reference model has all its poles at the origin (a delay line).

In line with our previous derivation of the one-step-ahead control law, we shall derive the model reference controller based on forming a prediction of the output. Previously, we predicted $y(t + d)$ and then choose the control so as to set it equal to $y^*(t + d)$, [since the objective was to achieve $y(t + d) = y^*(t + d)$]. Here we predict $E(q^{-1})y(t)$ and choose the control so as to set it equal to $q^{-d}H(q^{-1})gr(t)$ [since the objective is to achieve $E(q^{-1})y(t) = q^{-d}H(q^{-1})gr(t)$].

We develop an appropriate predictor for $E(q^{-1})y(t + d)$ in the following lemma:

Lemma 5.2.1. The system (5.2.1) can be expressed in predictor form as

$$E(q^{-1})y(t + d) = \alpha(q^{-1})y(t) + \beta(q^{-1})u(t) \tag{5.2.31}$$

where

$$\alpha(q^{-1}) = G(q^{-1})$$

$$\beta(q^{-1}) = F(q^{-1})B'(q^{-1})$$

and $F(q^{-1})$ and $G(q^{-1})$ are the unique polynomials of order $d - 1, n - 1$, respectively, satisfying

$$E(q^{-1}) = F(q^{-1})A(q^{-1}) + q^{-d}G(q^{-1}) \tag{5.2.32}$$

Proof. Multiplying (5.2.1) by $F(q^{-1})$ and using (5.2.32) leads immediately to the result.

▼▼▼

Based on the result above [in particular, (5.2.31)] it is clear that the design objective can be achieved by setting $u(t)$ according to

$$\alpha(q^{-1})y(t) + \beta(q^{-1})u(t) = gH(q^{-1})r(t) \tag{5.2.33}$$

thereby giving the closed-loop system:

$$E(q^{-1})y(t + d) = gH(q^{-1})r(t) \tag{5.2.34}$$

[Note that the control law (5.2.33) allows us to handle systems in which $A(q^{-1})$ and $B'(q^{-1})$ have common roots on the unit circle as in the preceding section. This permits us to treat uncontrollable disturbances in the model. This is a relatively straightforward but important extension of the usual approach in model reference control in which $A(q^{-1})$ and $B'(q^{-1})$ are assumed relatively prime (or at most to have common roots that are asymptotically stable). This is explored further in Exercise 5.1.]

Theorem 5.2.1 applies *mutatis mutandis* to the control law (5.2.33) except that the resulting closed-loop system is described by

$$E(q^{-1})y(t + d) = gH(q^{-1})r(t) = E(q^{-1})y^*(t + d) \tag{5.2.35}$$

and

$$E(q^{-1})B(q^{-1})u(t) = E(q^{-1})A(q^{-1})y^*(t) \tag{5.2.36}$$

The discussion above gives a concise treatment of model reference control which is a slight generalization of the one-step-ahead control principle.

5.2.3 One-Step-Ahead Design
for Multi-input Multi-output Systems

In this section we examine the extension of the one-step-ahead controller design technique to multi-input multi-output linear systems. The extension is relatively straightforward, although we shall find that the multi-input multi-output case has some special features.

We shall assume, without loss of generality, that the system is described by a DARMA model of the form (see Section 2.3.4)

$$A(q^{-1})y(t) = B(q^{-1})u(t) \tag{5.2.37}$$

with $A(q^{-1})$ and $B(q^{-1})$ as in (5.2.2) to (5.2.4) and where $\{y(t)\}$ and $\{u(t)\}$ denote the $m \times 1$ output vector and $r \times 1$ input vector, respectively. $A(q^{-1})$ and $B(q^{-1})$ are $m \times m$ and $m \times r$ polynomial matrices.

As we have seen in Chapter 2, the model (5.2.37) is equivalent to a left difference operator representation of the form

$$D_L(q)y(t) = N_L(q)u(t) \tag{5.2.37a}$$

We shall denote by $T(z)$ the input–output transfer function corresponding to the model (5.2.37). Further, we shall assume that the transfer function is strictly proper; that is, there exists a delay of at least one unit between every input and every output.

In the sequel we shall be interested in causing the output $\{y(t)\}$ to track a given desired output sequence $\{y^*(t)\}$. If this is possible, we shall say that the system is *output function controllable*.

We shall need the following preliminary result.

Lemma 5.2.2.   The system (5.2.37) is output function controllable if and only if the rank of the transfer function, $T(z)$, is equal to the number of outputs, m, for almost all z.

Proof. The z-transform of the output, $Y(z)$, can be expressed in terms of the z-transform of the input, $U(z)$, as

$$Y(z) = T(z)U(z) + I_c(z) \tag{5.2.38}$$

where $I_c(z)$ denotes the transform of the initial condition response.

If rank $T(z) = m$, there exists m linear independent columns of $T(z)$. Denote these columns by $T \cdot_{i_1}(z), \ldots, T \cdot_{i_m}(z)$ and define the following submatrix of $T(z)$:

$$\bar{T}(z) = [T \cdot_{i_1}(z) \cdots T \cdot_{i_m}(z)] \tag{5.2.39}$$

and the corresponding subvector of $U(z)$ as

$$\bar{U}(z) = [U_{i_1}(z) \cdots U_{i_m}(z)]^T \tag{5.2.40}$$

where $U_{i_j}(z)$ denotes the i_jth row of $U(z)$. Note that $\bar{T}(z)$ is nonsingular. Hence if we put

$$\bar{U}(z) = [\bar{T}(z)]^{-1}[Y^*(z) - I_c(z)] \tag{5.2.41}$$

and all other components of $U(z)$ equal to 0, it is clear that

$$Y(z) = Y^*(z) \tag{5.2.42}$$

where $Y^*(z)$ is the z-transform of the desired output sequence $\{y^*(t)\}$.

Conversely, if rank $T(z) < m$, then $T(z)$ will have a null space and it is clearly impossible to find a $U(z)$ to give $Y(z) = Y^*(z)$ if $Y^*(z) - I_c(z)$ is in the null space of $T(z)$.

▼▼▼

Initially, we will assume, for simplicity, that the number of inputs is equal to the number of outputs (i.e., $m = r$). If this is not the case, then, from the proof of Lemma 5.2.2, one need only select those inputs for which the corresponding columns of $T(z)$ are linearly independent. Later we shall consider the cases $m > r$ and $m < r$.

In the light of the discussion above, we introduce the following assumptions.

Assumption 5.2.A.   The number of inputs, r, is equal to the number of outputs, m, and the system transfer function, $T(z)$, satisfies

$$\det T(z) \neq 0 \qquad \text{almost all } z$$

▼▼▼

Assumption 5.2.B.   The transfer function is strictly proper.

▼▼▼

Note that Assumption 5.2.A ensures output function controllability and Assumption 5.2.B ensures that there is a delay of at least one unit between each input and each output.

The reader will recall that in the case of single-input single-output systems the delay structure was very transparent. In fact, one simply chose the delay, d, so that the leading coefficient of $B'(q^{-1})$ was nonzero. This can be stated slightly more formally by saying that in the single-input single-output case, there exists a scalar function $\xi(q)$ of the form $\xi(q) = q^d$ such that

$$\lim_{z \to \infty} \xi(z)A(z^{-1})^{-1}B(z^{-1}) = k \qquad (5.2.43)$$

where k is a nonzero scalar.

In the multivariable case it turns out that the delay structure of the transfer function matrix can also be specified in terms of a polynomial matrix $\xi(q)$. The following result applies to the multivariable case.

Lemma 5.2.3.   Given any transfer function satisfying Assumptions 5.2.A and 5.2.B, there exists a polynomial matrix, $\xi(q)$, known as the *interactor matrix*, satisfying:

(i) $\det \xi(q) = q^{\bar{m}}$,   where $\bar{m}$ is an integer   (5.2.44)
(ii) $\lim_{z \to \infty} \xi(z)T(z) = K$,   where K is a nonsingular matrix   (5.2.45)

In general, $\xi(z)$ can be taken to have the following structure:

$$\xi(z) = H(z)D(z) \qquad (5.2.46)$$

where

$$D(z) = \text{diag } [z^{f_1} \quad \cdots \quad z^{f_m}]$$

$f_i \geq d_i \triangleq \min_{1 \leq j \leq m} d_{ij}$ and d_{ij} is the delay between the jth input and ith output. $H(z)$ is a unimodular matrix (see Section A.3) of the form

$$H(z) = \begin{bmatrix} 1 & & & \\ h_{21}(z) & & \ddots & 0 \\ \vdots & & & \ddots \\ \vdots & & & & \ddots \\ h_{m_1}(z) & h_{m_2}(z) & \cdots & 1 \end{bmatrix} \qquad (5.2.47)$$

and $h_{ij}(z)$ is divisible by z (or is zero).

▼▼▼

We shall prove this result below but first we give a brief interpretation of the operator $\xi(q)$.

In many cases of interest, $\xi(q)$ can be taken to have one of the following two simple forms:

Form 1: $\xi(q) = q^d I$, where $d = \min\limits_{i,j} d_{ij}$ \hfill (5.2.48)

where d is a single delay associated with every output.

Form 2: $\xi(q) = \text{diag}\,[d_1 \;\; \cdots \;\; d_m]$ \hfill (5.2.49)

where $d_{ij} =$ delay between the jth input and the ith output,

$$d_i = \min_j d_{ij}$$

Here d_i represents the delay to the ith output.

It is readily seen that the choices above satisfy condition (5.2.44) and gives

$$\lim_{z \to \infty} \xi(z)T(z) = K \text{ nonzero} \qquad (5.2.50)$$

In elementary discussions of multivariable systems, the delay structures discussed above are usually adequate. However, for complete generality and to guarantee that K is *nonsingular* in all cases, we occasionally need to use the more general form for $\xi(q)$, as in (5.2.47). With the remarks above in mind, we now turn to the proof of Lemma 5.2.3.

Proof of Lemma 5.2.3. We shall establish the existence of $\xi(z)$ as in Wolovich and Falb (1976). There exist unique integers d_i, $i = 1, \ldots, m$, such that

$$\lim_{z \to \infty} z^{d_i} T_i \cdot (z) = \tau_i \qquad (5.2.51)$$

where $T_i \cdot (z)$ denotes the ith row of $T(z)$ and τ_i is not identically zero in view of the definition of d_i.

We define the first row, $\xi(z)_1$, of $\xi(z)$ by

$$\xi(z)_1 = [z^{d_1} \;\; 0 \;\; \cdots \;\; 0] \qquad (5.2.52)$$

so that

$$\lim_{z \to \infty} \xi(z)_1 T(z) = \xi_1 = \tau_1 \qquad (5.2.53)$$

If τ_2 is linearly independent of ξ_1, then set

$$\xi(z)_2 = (0 \;\; z^{d_2} \;\; 0 \;\; \cdots \;\; 0] \qquad (5.2.54)$$

so that

$$\lim_{z \to \infty} \xi(z)_2 T(z) = \xi_2 = \tau_2 \qquad (5.2.55)$$

On the other hand, if τ_2 and ξ_1 are linearly dependent so that $\tau_2 = \alpha_1^1 \xi_1$ with $\alpha_1^1 \neq 0$, then let

$$\tilde{\xi}^1(z)_2 = z^{d_2^1}[(0 \quad z^{d_2} \quad 0 \quad \cdots \quad 0) - \alpha_1^1 \xi(z)_1] \tag{5.2.56}$$

where d_2^1 is the unique integer for which $\lim_{z \to \infty} \tilde{\xi}^1(z)_2 T(z) = \tilde{\xi}_2^1$ is both finite and nonzero. If $\tilde{\xi}_2^1$ is linearly independent of ξ_1, we set

$$\xi(z)_2 = \tilde{\xi}^1(z)_2 \tag{5.2.57}$$

and note that

$$\lim_{z \to \infty} \xi(z)_2 T(z) = \tilde{\xi}_2^1 \tag{5.2.58}$$

is linearly independent of ξ_1. If not, then $\tilde{\xi}_2^1 = \alpha_1^2 \xi_1$ and we let

$$\tilde{\xi}^2(z)_2 = z^{d_2^2}[\tilde{\xi}^1(z)_2 - \alpha_1^2 \xi(z)_1]$$

where d_2^2 is the unique integer for which $\lim_{z \to \infty} \tilde{\xi}^2(z)_2 T(z) = \tilde{\xi}_2^2$ is both finite and nonzero. If $\tilde{\xi}_2^2$ and ξ_1 are linearly independent, we set $\xi(z)_2 = \tilde{\xi}^2(z)_2$, and if not, we repeat the procedure until linear independence is obtained. The procedure must terminate since $\det T(z) \neq 0$ almost all z (a.e. z) and since $d_i \geq 1, i = 1, \ldots, m$. The remaining rows of $\xi(z)$ are obtained in an analogous fashion. Finally, we see that (5.2.45) is satisfied since the ξ_i are linearly independent.

Uniqueness of $\xi(z)$ will not be used in our subsequent development but can be established as in Wolovich and Falb (1976).

▼▼▼

The lemma above is illustrated in the following examples.

Example 5.2.5

Consider the following transfer function:

$$T(z) = \begin{bmatrix} \dfrac{z^{-2}}{1 + z^{-1}} & \dfrac{2z^{-2}}{1 + 3z^{-1}} \\ \dfrac{z^{-3}}{1 + z^{-1}} & \dfrac{3z^{-3}}{1 + 4z^{-1}} \end{bmatrix}$$

Clearly, for this example, $\xi(q)$ can be taken to be as in (5.2.49); that is,

$$\xi(q) = \begin{bmatrix} z^2 & 0 \\ 0 & z^3 \end{bmatrix}$$

giving

$$K = \begin{bmatrix} 1 & 2 \\ 1 & 3 \end{bmatrix} \quad \text{(nonsingular)}$$

*Example 5.2.6

Consider the following transfer function:

$$T(z) = \begin{bmatrix} \dfrac{z^{-1}}{1 + z^{-1}} & \dfrac{z^{-1}}{1 + 2z^{-1}} \\ \dfrac{z^{-1}}{1 + 3z^{-1}} & \dfrac{z^{-1}}{1 + 4z^{-1}} \end{bmatrix} \tag{5.2.59}$$

Clearly, $\det T(z) \neq 0$ for a.e. z and thus we see that Assumption 5.2.A is satisfied. Also, we note that Assumption 5.2.B is satisfied. We shall find that for this example, $\xi(q)$ should have the more complex structure given in Lemma 5.2.2.

Using the procedure described in Lemma 5.2.3, we obtain

$$\xi(z)_1 = [z \quad 0]$$
$$\tau_1 = [1 \quad 1] = \xi_1$$

If we put

$$\xi(z)_2 = [0 \quad z]$$

then

$$\tau_2 = [1 \quad 1]$$

Note that ξ_1 and τ_2 are linearly dependent

$$\tau_2 = \alpha_1^1 \xi_1 \qquad \text{with } \alpha_1^1 = 1$$

Let

$$\tilde{\xi}^1(z)_2 = z[(0 \quad z) - \alpha_1^1 \xi(z)_1]$$
$$= [-z^2 \quad z^2]$$

and $\lim_{z \to \infty} \tilde{\xi}^1(z)_2 T(z) = \tilde{\xi}_2^1 = (-2, -2) = \alpha_1^2 \xi_1$ with $\alpha_1^2 = -2$. Since $\tilde{\xi}_2^1$ depends linearly on ξ_1, we continue by defining

$$\tilde{\xi}^2(z)_2 = z[(-z^2 \quad z^2) - \alpha_1^2 \xi(z)_1]$$
$$= [-z^3 + 2z^2 \quad z^3]$$

The $\lim_{z \to \infty} \tilde{\xi}^2(z)_2 T(z) = \tilde{\xi}_2^2 = (6, 8)$, which is linearly independent of ξ_1. Hence

$$\xi(z) = \begin{bmatrix} z & 0 \\ -z^3 + 2z^2 & z^3 \end{bmatrix}$$

$$T'(z) = \xi(z)T(z) = \begin{bmatrix} \dfrac{1}{1 + z^{-1}} & \dfrac{1}{1 + 2z^{-1}} \\ \dfrac{6}{(1 + z^{-1})(1 + 3z^{-1})} & \dfrac{8}{(1 + 2z^{-1})(1 + 4z^{-1})} \end{bmatrix}$$

$$\lim_{z \to \infty} \xi(z)T(z) = \begin{bmatrix} 1 & 1 \\ 6 & 8 \end{bmatrix} \qquad \text{(nonsingular)}$$

▼▼▼

We shall see later how the interactor matrix can be incorporated in the design of one-step-ahead controllers. We first investigate some properties of the interactor matrix in the general case:

Lemma 5.2.4

(i) In the discrete-time case, $\xi(z)^{-1}$ is a stable operator.
(ii) For a strictly proper transfer matrix, $T(z)$, f_1 to f_m in (5.2.46) and the sequel are nonzero:

$$f_i > 0; \qquad i = 1, \ldots, m \tag{5.2.60}$$

Proof. (i) $\det \xi(z) = z^f$ 　　　　　　　　　　　　　　　　　　(5.2.61)

where

$$f = \sum_{i=1}^{m} f_i \tag{5.2.62}$$

Hence

$$\det \xi(z) \neq 0 \qquad \text{for } |z| \geq 1$$

(ii) Immediate from the definition of f_i.

▼▼▼

We have the following important result, which leads directly to the multi-input multi-output one-step-ahead control law:

Theorem 5.2.4. If we define a variable $\bar{y}(t)$ by the difference equation

$$\bar{y}(t) = \xi(q)y(t) \qquad (5.2.63)$$

then, subject to Assumptions 5.2.A and 5.2.B, $\bar{y}(t)$ is related to $\{u(t)\}$ and $\{y(t)\}$ by a model of the following predictor form:

$$\bar{y}(t) = \alpha(q^{-1})y(t) + \beta(q^{-1})u(t) \qquad (5.2.64)$$

where

(i) $\alpha(q^{-1}) = \alpha_0 + \alpha_1 q^{-1} + \cdots + \alpha_{n'} q^{-n'}$ (5.2.65)

(ii) $\beta(q^{-1}) = \beta_0 + \beta_1 q^{-1} + \cdots + \beta_{n'} q^{-n'}$ (5.2.66)

(iii) β_0 is nonsingular.   (5.2.67)

(iv) $\bar{y}(t - d)$ is a known function of $y(t), y(t - 1), \ldots$.

Proof. Consider the system as in (5.2.37). Since we are constructing a predictor for the quantity $\xi(q)y(t)$, we use the standard predictor equality,

$$\xi(q) = F(q)A(q^{-1}) + G(q^{-1}) \qquad (5.2.68)$$

where

$$F(q) = F_0 q^{d'} + \cdots + F_{d'-1} q \qquad (5.2.69)$$

$$G(q^{-1}) = G_0 + G_1 q^{-1} + \cdots + G_{n'} q^{-n'} \qquad (5.2.70)$$

$$d' = \text{maximum advance in } \xi(q)$$

Matrices satisfying (5.2.68) can be readily found by equating coefficients in (5.2.68) (see Exercise 5.22).

Multiplying (5.2.37) by $F(q)$ gives

$$F(q)A(q^{-1})y(t) = F(q)B(q^{-1})u(t) \qquad (5.2.71)$$

Using (5.2.68), we have

$$\xi(q)y(t) = G(q^{-1})y(t) + F(q)B(q^{-1})u(t) \qquad (5.2.72)$$

or

$$\xi(q)y(t) = \alpha(q^{-1})y(t) + \beta(q^{-1})u(t) \qquad (5.2.73)$$

where $\alpha(q^{-1}) = G(q^{-1})$ has the form shown in (5.2.65):

$$\beta(q^{-1}) = \beta_{[-d'+1]} q^{d'-1} + \cdots + \beta_{[-1]} q + \beta_0 + \beta_1 q^{-1} + \cdots + \beta_{n'} q^{-n'} \qquad (5.2.74)$$

Now

$$\lim_{z \to \infty} \xi(z)T(z) = K \qquad \text{finite and nonsingular} \qquad (5.2.75)$$

$$T(z) = [\xi(z) - \alpha(z^{-1})]^{-1} \beta(z^{-1}) \qquad (5.2.76)$$

Using (5.2.75) in (5.2.76), we see that

$$\beta_{[-d'+1]} = \cdots = \beta_{[-1]} = 0$$

and $\beta_0 = K$ nonsingular.

Equations (5.2.66) and (5.2.67) follow immediately.

▼▼▼

Note that in the theorem above we have been careful to include all observable modes, whether or not they are controllable. (See further discussion in Chapter 2.)

To use Theorem 5.2.4 to develop a one-step-ahead control law, we first define a filtered desired output, $\bar{y}^*(t)$, by

$$\bar{y}^*(t) = \xi(q)y^*(t) \tag{5.2.77}$$

We then choose $u(t)$ to bring $\bar{y}(t)$ to $\bar{y}^*(t)$. Properties of the resulting feedback control system are described in the following theorem:

Theorem 5.2.5

(a) The feedback control law that brings $\bar{y}(t)$ to $\bar{y}^*(t)$ has the following form:

$$\beta(q^{-1})u(t) = \bar{y}^*(t) - \alpha(q^{-1})y(t) \tag{5.2.78}$$

[Solvability of the equation above for $u(t)$ is guaranteed by part (iii) of Theorem 5.2.4.]

(b) The resulting closed-loop system is described by

$$\bar{y}(t) = \bar{y}^*(t); \qquad t \geq d \tag{5.2.79}$$

The closed-loop system is characterized by

$$\begin{bmatrix} \xi(q) & 0 \\ A(q^{-1}) & -B(q^{-1}) \end{bmatrix} \begin{bmatrix} y(t) \\ u(t) \end{bmatrix} = \begin{bmatrix} \xi(q)y^*(t) \\ 0 \end{bmatrix} \tag{5.2.80}$$

(c) $\lim_{t \to \infty} [y(t) - y^*(t)] = 0 \tag{5.2.81}$

(d) The closed-loop system is stable provided that $\det \xi(z) \det B(z^{-1})$ is nonzero for all $|z| > 1$.

(Actually, a more general condition covering uncontrollable roots on the unit circle follows as in Theorem 5.2.1.)

Proof. Equation (5.2.81) follows from (5.2.79) since $\xi(q)$ has all its roots at the origin (i.e., is stable).

Part (d) follows by using the formula for a partitioned matrix determined in (5.2.80).

▼▼▼

We note that the one-step-ahead control law does not now lead to perfect tracking of the desired output $y^*(t)$. Instead, the filtered outputs $\bar{y}(t)$ track the filtered desired outputs $\bar{y}^*(t)$. We see from (5.2.81) that $\{y(t)\}$ will converge to $y^*(t)$ after a possibly nonzero transient produced by the dynamics of $\xi(q)$.

It is also straightforward to extend the result above to model reference control, where the desired output $y^*(t)$ is generated by

$$E(q^{-1})\bar{y}*(t) = H(q^{-1})r(t) \tag{5.2.82}$$

with

$$E(q^{-1}) = E_0 + E_1 q^{-1} + \cdots + E_n q^{-n}; \quad E_0 \text{ nonsingular} \tag{5.2.83}$$

$$\bar{y}*(t) = \xi(q)y*(t)$$

We now simply have to build a predictor for $E(q^{-1})\xi(q)y(t)$, which can be done by a straightforward extension of Theorem 5.2.4. (See Exercise 5.23.)

We conclude our discussion of one-step-ahead control by completing Example 5.2.6.

Example 5.2.6 (Continued)

We define $\bar{y}(t)$ by

$$\bar{y}(t) = \xi(q)y(t)$$

where

$$\xi(q) = \begin{bmatrix} q & 0 \\ -q^3 + 2q^2 & q^3 \end{bmatrix}$$

The DARMA model for the system can be written as

$$A(q^{-1})y(t) = B(q^{-1})u(t)$$

$$A(q^{-1}) = \begin{bmatrix} (1 + q^{-1})(1 + 2q^{-1}) & 0 \\ 0 & (1 + 3q^{-1})(1 + 4q^{-1}) \end{bmatrix} \tag{5.2.84}$$

$$B(q^{-1}) = \begin{bmatrix} q^{-1} + 2q^{-2} & q^{-1} + q^{-2} \\ q^{-1} + 4q^{-2} & q^{-1} + 3q^{-2} \end{bmatrix}$$

The prediction equality gives

$$\xi(q) = F(q)A(q^{-1}) + G(q^{-1})$$

where by equating coefficients

$$F_0 = \begin{bmatrix} 0 & 0 \\ -1 & 1 \end{bmatrix}$$

$$F_1 = \begin{bmatrix} 0 & 0 \\ 5 & -7 \end{bmatrix}$$

$$F_2 = \begin{bmatrix} 1 & 0 \\ -13 & 37 \end{bmatrix}$$

$$G_0 = \begin{bmatrix} -3 & 0 \\ 29 & -175 \end{bmatrix}$$

$$G_1 = \begin{bmatrix} -2 & 0 \\ 26 & -444 \end{bmatrix}$$

giving

$$\xi(q)y(t) = \alpha(q^{-1})y(t) + \beta(q^{-1})u(t)$$

$$\alpha(q^{-1}) = G_0 + G_1 q^{-1}$$

$$\beta(q^{-1}) = [F_0 q^3 + F_1 q^2 + F_2 q][B(q^{-1})]$$

$$= \beta_0 + \beta_1 q^{-1}$$

$$\beta_0 = \begin{bmatrix} 1 & 1 \\ 6 & 8 \end{bmatrix} = K \text{ nonsingular}$$

$$\beta_1 = \begin{bmatrix} 2 & 1 \\ 122 & 98 \end{bmatrix}$$

Finally, the one-step-ahead control law is

$$\begin{bmatrix} u_1(t) \\ u_2(t) \end{bmatrix} = \begin{bmatrix} 1 & 1 \\ 6 & 8 \end{bmatrix}^{-1} \left\{ \begin{pmatrix} y_1^*(t+1) \\ -y_1^*(t+3) + 2y_1^*(t+2) + y_2^*(t+3) \end{pmatrix} \right.$$
$$\left. - \begin{bmatrix} -3 & 0 \\ 29 & -175 \end{bmatrix} \begin{bmatrix} y_1(t) \\ y_2(t) \end{bmatrix} - \begin{bmatrix} -2 & 0 \\ 26 & -444 \end{bmatrix} \begin{bmatrix} y_1(t-1) \\ y_2(t-1) \end{bmatrix} - \begin{bmatrix} 2 & 1 \\ 122 & 98 \end{bmatrix} \begin{bmatrix} u_1(t-1) \\ u_2(t-1) \end{bmatrix} \right\}$$

▼▼▼

In the discussion above, we have treated only the basic one-step-ahead control law. However, the extension to the weighted one-step-ahead control law is straightforward. We leave the details to the reader (see Exercise 5.7).

We will now briefly discuss the situation when the number of inputs is not equal to the number of outputs.

Nonsquare systems. We shall assume that the transfer function $T(z)$ is strictly proper and satisfies rank $T(z) = \min(r, m)$. In this case, there exists an interactor matrix $\xi(z)$ having the following property [see Wolovich and Falb (1976)]:

$$T'(z) \triangleq \xi(z)T(z)$$

then

$$\lim_{z \to \infty} T'(z) = K'; \qquad \text{rank } K' = \min(r, m) \qquad (5.2.85)$$

(The result above is a simple extension of Lemma 5.2.3.)

Our control objective can now be stated as causing the system output $\{y(t)\}$ to "track" (in some sense) a function $\{y^*(t)\}$ which is the output of a stable reference model. The reference model is assumed to have transfer function $T^*(z)$ and reference input $\{r(t)\}$. To ensure causality of the resulting controller, we shall require that $T^*(z)$ have delay structure related to $T(z)$. In particular we require that $\xi(z)T^*(z)$ be proper, that is,

$$\lim_{z \to \infty} \xi(z)T^*(z) < \infty \qquad (5.2.86)$$

In view of (5.2.86), $\{\xi(q)y^*(t)\}$ can be related to $\{r(t)\}$ by a row-reduced proper left difference operator realization, that is,

$$E(q^{-1})[\xi(q)y^*(t)] = H(q^{-1})r(t) \qquad (5.2.87)$$

where

$$\xi(z)T^*(z) = E(z^{-1})^{-1}H(z^{-1})$$

and

$$E(q^{-1}) = E_0 + E_1 q^{-1} + \cdots + E_l q^{-l}; \qquad \det E_0 \neq 0$$
$$H(q^{-1}) = H_0 + H_1 q^{-1} + \cdots + H_l q^{-l}$$

Control of Linear Deterministic Systems   Chap. 5

As before, the minimum prediction error control laws can be evaluated by forming a predictor for the quantity $E(q^{-1})\xi(q)y(t)$. A simple extension of Theorem 5.2.4 shows that the predictor can be expressed as

$$[E(q^{-1})\xi(q)y(t)] = \alpha(q^{-1})y(t) + \beta(q^{-1})u(t) \qquad (5.2.88)$$

where

$$\alpha(q^{-1}) = \alpha_0 + \alpha_1 q^{-1} + \cdots + \alpha_{l_1} q^{-l_1}$$
$$\beta(q^{-1}) = \beta_0 + \beta_1 q^{-1} + \cdots + \beta_{l_2} q^{-l_2}; \qquad \beta_0 \text{ having rank min } (r, m)$$

Case 1: The Number of Outputs Exceeds the Number of Inputs

A possible problem falling into this category is the development of an automatic control system for an artificial heart (McInnis, Wang, and Akutsu, 1981; McInnis et al., 1981). This system has three outputs (aortic pressure, pulmonary arterial pressure, and cardiac output) and two inputs (left drive pressure and right drive pressure). It is clearly impossible to control the system so that the three outputs are driven to arbitrary desired levels in steady state. However, the control objective can be stated as that of keeping two of the outputs near their nominal desired values while maximizing the remaining variable (the cardiac output in this case). The compromises arising from this objective can be conveniently expressed in terms of a quadratic criterion:

$$J_1 = \| L(q^{-1})[y(t) - y^*(t)] \|_Q^2 + \| S(q^{-1})u(t) \|_R^2 \qquad (5.2.89)$$

where $L(q^{-1}) \triangleq E(q^{-1})\xi(q)$
$\qquad Q =$ positive definite matrix
$\qquad R =$ positive *semi*definite matrix (possibly zero)
$\qquad \xi(q) =$ system interactor matrix
$\qquad S(q^{-1}) =$ filter of the form

$$S(q^{-1}) = S_0 + S_1 q^{-1} + \cdots + S_l q^{-l}; \qquad \det S_0 \neq 0 \qquad (5.2.90)$$

A possible choice for $S(q^{-1})$ is

$$S(q^{-1}) = I - q^{-1}I \qquad (5.2.91)$$

which ensures that zero weighting is attached to $\{u(t)\}$ in steady state (i.e., zero steady-state error is achieved in the case $r = m$).

Lemma 5.2.5

(i) The one-step-ahead feedback control law minimizing (5.2.89) is given by

$$[\beta_0^T Q\beta(q^{-1}) + S_0^T RS(q^{-1})]u(t) = -\beta_0^T Q\alpha(q^{-1})y(t) + \beta_0^T QH(q^{-1})r(t) \qquad (5.2.92)$$

(ii) The closed-loop system resulting from the use of (5.2.92) is

$$\begin{bmatrix} A(q^{-1}) & -B(q^{-1}) \\ \beta_0^T QE(q^{-1})\xi(q) & S_0^T RS(q^{-1}) \end{bmatrix}\begin{bmatrix} y(t) \\ u(t) \end{bmatrix} = \begin{bmatrix} 0 \\ \beta_0^T QH(q^{-1}) \end{bmatrix}r(t) \qquad (5.2.93)$$

Proof. (i) Immediate from (5.2.89) by differentiating and using (5.2.87) and (5.2.88). Note that the feedback law is well defined since the coefficient of $u(t)$ is $[\beta_0^T Q\beta_0 + S_0^T RS_0]$, which is nonsingular in view of the rank of β_0.

(ii) Follows from the system model (5.2.88) and (5.2.92).

▼▼▼

Note that the result above specializes to all previous cases (e.g., when $r = m$, etc.).

Remark 5.2.4. From the discussion above we can see that, in general, it is impossible to achieve perfect tracking of more outputs than there are inputs available. Thus a compromise design objective has been used. However, in special circumstances it is possible to achieve steady-state tracking of more outputs than there are inputs provided that the outputs are compatible. An example of this possibility is in output regulation when y is brought to y^* (a constant) and simultaneously $\dot{y}$ is zero. In this case it is possible to achieve perfect tracking of more outputs than inputs. A possible design method was given in Remark 5.2.3.

▼▼▼

Case 2: The Number of Outputs Is Less Than the Number of Inputs

For cases where the number of inputs exceeds the number of outputs, output function controllability can be retained if some of the inputs are discarded. However, this approach does not take account of the relative cost of applying each input. Thus one may ask: What control brings $E(q^{-1})\xi(q)y(t)$ to $E(q^{-1})\xi(q)y^*(t)$ while using minimum energy? Here "energy" is measured by the following quadratic criterion:

$$J_2 = \| u(t) \|_R^2; \qquad R \text{ positive definite} \tag{5.2.94}$$

The input minimizing (5.2.94) and bringing $E(q^{-1})\xi(q)y(t)$ to $E(q^{-1})\xi(q)y^*(t)$ is described in the following lemma.

Lemma 5.2.6

(i) The one-step-ahead control law for the problem described above is given by

$$u(t) = -R^{-1}\beta_0^T(\beta_0 R^{-1}\beta_0^T)^{-1}[\alpha(q^{-1})y(t) + q[\beta(q^{-1}) - \beta_0]u(t-1) - H(q^{-1})r(t)] \tag{5.2.95}$$

(ii) The closed-loop system resulting from the use of (5.2.95) is

$$\begin{bmatrix} A(q^{-1}) & -B(q^{-1}) \\ \beta_0^T(\beta_0 R^{-1}\beta_0^T)^{-1}E(q^{-1})\xi(q) & R - \beta_0^T(\beta_0 R^{-1}\beta_0^T)^{-1}\beta_0 \end{bmatrix}\begin{bmatrix} y(t) \\ u(t) \end{bmatrix}$$

$$= \begin{bmatrix} 0 \\ \beta_0^T(\beta_0 R^{-1}\beta_0^T)^{-1}H(q^{-1}) \end{bmatrix}r(t) \tag{5.2.96}$$

Proof. (i) Equation (5.2.95) can be derived by using Lagrange multipliers to minimize the criterion J_2 subject to the constraint that $E(q^{-1})\xi(q)y(t) = H(q^{-1})r(t)$. Note that the feedback law (5.2.95) is well defined since $(\beta_0 R^{-1}\beta_0^T)$ is nonsingular in view of the rank conditions on β_0.

(ii) Follows from the system model, (5.2.88) and (5.2.95).

▼▼▼

Control of Linear Deterministic Systems   Chap. 5

In the next section we discuss some practical considerations associated with one-step-ahead controller design for both single-input single-output and multi-input multi-output systems.

5.2.4 Robustness Considerations

Model Structure

In the preceding section we found it necessary to introduce the interactor matrix to achieve one-step-ahead control for general multi-input multi-output systems. One might wonder if this is really of theoretical importance only or whether the result has genuine practical significance.

In fact, if we look at the derivation of the interactor matrix in Lemma 5.2.3, we find that the interactor matrix is nondiagonal if and only if certain linear dependencies arise in the extraction of successive delays from the transfer function matrix. In a real plant, it would be necessary to determine the interactor matrix by a preliminary experiment and thus *exact* linear dependencies would never be found even if they were present in the true system description. This leads us to ask the following robustness question: If during the determination of the interactor matrix a nearly linear dependency is found, is it better to assume that the linear dependency is exact (leading to a nondiagonal interactor matrix) or to ignore the near linear dependency (leading to a diagonal interactor matrix)?

The interactor matrix describes the delay structure for multivariable systems and reduces to the system delay in the single-input single-output case. Thus before we attempt to answer the question posed above, we pause to consider the corresponding robustness question for single-input single-output systems: If, in determining the delay of a single-input single-output system, there is a very small response at delay $d_1 - 1$ and a larger response at delay d_1 following a step change in the input, is it better to take the delay as $d_1 - 1$ or d_1?

A partial answer to this question is provided in

Lemma 5.2.7. If the leading coefficient of the numerator polynomial in a system description is smaller than the last significant coefficient, a one-step-ahead control will lead to an unbounded input.

Proof. The product of the zeros of the system is equal to the last significant coefficient divided by the leading coefficient. Thus if this ratio is greater than 1, it is impossible for all the zeros to be inside the closed unit disk. Thus the conditions of Theorem 5.2.1, part (c), will not be satisfied.

▼▼▼

The lemma above essentially says that if you have a small leading coefficient in the numerator, one-step-ahead control will probably fail. This suggests that a possible remedy could be to ignore a small leading coefficient. However, we have yet to show that this does not lead to other difficulties. This is verified in the following lemma.

Lemma 5.2.8. Consider a single-input single-output system

$$A(q^{-1})y(t) = [\epsilon q^{-d+1} + q^{-d}B(q^{-1})]u(t); \quad |\epsilon| \ll 1 \quad (5.2.97)$$

$$A(q^{-1}) = 1 + a_1 q^{-1} + \cdots + a_n q^{-n} \quad (5.2.98)$$

$$B(q^{-1}) = b_0 + b_1 q^{-1} + \cdots + b_m q^{-m}; \quad b_0 > 0 \quad (5.2.99)$$

Then if a one-step-ahead controller is designed by ignoring ϵ, and the resulting control law is implemented on the system (5.2.97) in which ϵ is nonzero, the resulting closed-loop system is described by

$$[B(q^{-1}) + \epsilon G(q^{-1})q^{-d+1}]y(t) = [B(q^{-1}) + q\epsilon]y^*(t) \quad (5.2.100)$$

$$[B(q^{-1}) + \epsilon G(q^{-1})q^{-d+1}]u(t) = A(q^{-1})y^*(t+d) \quad (5.2.101)$$

where $G(q^{-1})$ in the unique polynomial satisfying

$$1 = q^{-d}G(q^{-1}) + F(q^{-1})A(q^{-1}) \quad (5.2.102)$$

$$F(q^{-1}) = 1 + f_1 q^{-1} + \cdots + f_{d-1}q^{-d+1} \quad (5.2.103)$$

Proof. If we ignore ϵ in the design of the controller, the one-step-ahead control law is

$$F(q^{-1})B(q^{-1})u(t) = y^*(t+d) - G(q^{-1})y(t) \quad (5.2.104)$$

Substituting (5.2.104) into (5.2.97) and eliminating first $\{u(t)\}$ and then $\{y(t)\}$ gives (5.2.100)–(5.2.101). (See Exercise 5.8.)

▼▼▼

From the lemma above, it can be seen that the necessary condition for closed-loop stability is that the zeros of $[B(z^{-1}) + \epsilon G(z^{-1})z^{-d+1}]$ lie strictly inside the closed unit disk. This can be guaranteed provided that $B(z^{-1})$ is asymptotically stable and provided that ϵ is small compared with the stability margin of $B(z^{-1})$. These conditions are stronger than in Theorem 5.2.1, part (c), but the strengthened conditions give *robustness* of the one-step-ahead control law in the presence of the "unmodeled" term ϵq^{-d+1}. In view of (5.2.100)–(5.2.101), the resulting closed-loop system obtained by ignoring ϵ in the design of the controller will behave essentially the same as if ϵ were actually zero.

Returning to the multivariable case, we see that a similar argument applies as in the single-input single-output case discussed above. The multivariable analog of Lemma 5.2.8 is

Lemma 5.2.9. If the determinant of the leading coefficient matrix of the numerator polynomial matrix in the left difference operator representation (5.2.37a) is smaller than the coefficient of the last significant term in det $N_L(z)$, the system has zeros outside the unit circle.

Proof. As for Lemma 5.2.8.

▼▼▼

By inspection of the development of the interactor matrix in Lemma 5.2.3, we see that the leading coefficient matrix of the numerator polynomial will be near singular (i.e., have a small determinant) if a nearly linear dependence is ecountered during successive delay extraction. Thus we see from Lemma 5.2.9 that a nearly linear dependence (if taken as independent) will result in zeros outside the unit circle and the one-step-ahead controller will fail. This suggests that in forming the interactor matrix, one should extract sufficient delay from successive rows of the transfer function to ensure that nearly linear dependence is avoided. This means that if a nearly linear dependence is encountered, it should be taken as an exact dependence leading to a nondiagonal interactor matrix. Of course, this is likely to lead to modeling errors, but the multivariable counterpart of Lemma 5.2.8 shows that these will not lead to difficulties provided that the inverse system has adequate stability margin.

Parameter Sensitivity

Inherent in the philosophy of the one-step-ahead design is the dependence of the tracking performance on the knowledge of the system parameters. Consequently, one might expect that the one-step-ahead design would be sensitive to plant parameter variations. We show that this is indeed the case.

Let the true system description be

$$A(q^{-1})y(t) = q^{-d}B'(q^{-1})u(t) \tag{5.2.105}$$

and let the control system be designed on the basis of a *nominal model* with $A(q^{-1})$ and $B'(q^{-1})$ replaced by $A_n(q^{-1})$ and $B'_n(q^{-1})$, respectively.

Then the resulting one-step-ahead control law (in the single-input single-output case) is (from Theorem 5.2.1)

$$F_n(q^{-1})B'_n(q^{-1})u(t) = y^*(t + d) - G_n(q^{-1})y(t) \tag{5.2.106}$$

where

$$A_n(q^{-1})F_n(q^{-1}) + q^{-d}G_n(q^{-1}) = 1 \tag{5.2.107}$$

If we substitute (5.2.106) into (5.2.105), the resulting closed-loop system has transfer function

$$G_c(z) = \frac{B'(z^{-1})}{F_n(z^{-1})B'_n(z^{-1})A(z^{-1}) + z^{-d}B'(z^{-1})G_n(z^{-1})} \tag{5.2.108}$$

In the case of exact modeling, when $A_n(z^{-1}) = A(z^{-1})$ and $B'_n(z^{-1}) = B'(z^{-1})$, we have (as expected)

$$G_c^0(z) = \frac{B'_n(z^{-1})}{B'_n(z^{-1})[F_n(z^{-1})A_n(z^{-1}) + z^{-d}G_n(z^{-1})]} = \frac{B'(z^{-1})}{B'(z^{-1})} = 1 \tag{5.2.109}$$

where the superscript 0 indicates that $G_c(z)$ is evaluated at the true values for $A_n(z^{-1})$ and $B'_n(z^{-1})$.

However, it is instructive to see how $G_c(z)$ varies with changes in the system parameters. We therefore evaluate the following two sensitivity functions at $A_n = A$ and $B_n = B$:

$$\frac{\partial G_c(z)}{\partial b_i} = \frac{z^{-i}F_n(z^{-1})A(z^{-1})}{B'(z^{-1})} \tag{5.2.110}$$

$$\frac{\partial G_c(z)}{\partial a_i} = -F_n(z^{-1})z^{-i} \tag{5.2.111}$$

In general, there is no way of ensuring that these sensitivities are small since they depend directly on the system parameters.

A very important point is that this sensitivity problem is mitigated using the *adaptive* form of the one-step-ahead controller to be discussed later. The reason is that the model parameters are continuously adjusted so that the model output coincides with the system output. This greatly strengthens the applicability of the one-step-ahead principle. These observations were, in fact, substantiated in Example 5.2.3 presented earlier in this section. The one-step-ahead controller was originally implemented with the nominal parameter values given in (5.2.24). This did not perform very well. However, after a few iterations of the adaptive version of the algorithm, the parameters converged to new values, which resulted in the excellent performance reported earlier in Figs. 5.2.1 to 5.2.3 (see also Fig. 6.3.1).

In the next section we present a second design technique as an alternative to the one-step-ahead principle.

5.3 CLOSED-LOOP POLE ASSIGNMENT

5.3.1 Introduction

We begin by interpreting the one-step-ahead controller in the framework of pole assignment. In the preceding section, we motivated the control law by aiming to achieve $y(t) = y^*(t)$. In its simplest form, the resulting control law had the following structure [see (5.2.8)] (for clarity we consider the single-input single-output case):

$$L(q^{-1})u(t) = -P(q^{-1})y(t) + M(q^{-1})y^*(t + d) \qquad (5.3.1)$$

where in (5.2.8), $L(q^{-1})$, $P(q^{-1})$, and $M(q^{-1})$ have the following special values:

$$L(q^{-1}) = \beta(q^{-1}) \triangleq F(q^{-1})B'(q^{-1}); \qquad P(q^{-1}) = \alpha(q^{-1}) \triangleq G(q^{-1}); \qquad M(q^{-1}) = 1 \qquad (5.3.2)$$

We shall assume that the system is described by a DARMA model of the form of (5.2.1) to (5.2.4): that is,

$$A(q^{-1})y(t) = B(q^{-1})u(t); \qquad B(q^{-1}) \triangleq q^{-d}B'(q^{-1}) \qquad (5.3.3)$$

Multiplying (5.3.3) by $L(q^{-1})$ and using (5.3.1) gives the following closed-loop system:

$$[L(q^{-1})A(q^{-1}) + q^{-d}B'(q^{-1})P(q^{-1})]y(t) = B'(q^{-1})M(q^{-1})y^*(t) \qquad (5.3.4)$$

Now as before, we aim to achieve $y(t) = y^*(t)$, and in the light of (5.3.4), it seems sensible to put $M(q^{-1}) = 1$ and then attempt to set the denominator of the closed-loop system equal to $B'(q^{-1})$;

$$L(q^{-1})A(q^{-1}) + q^{-d}B'(q^{-1})P(q^{-1}) = B'(q^{-1}) \qquad (5.3.5)$$

The question now is: Can $L(q^{-1})$, $P(q^{-1})$ be found such that (5.3.5) is satisfied?

Since $B'(q^{-1})$ is a factor of the right-hand side of (5.3.5) and of the second term on the left-hand side, it seems reasonable to assume that $L(q^{-1})$ has $B'(q^{-1})$ as a factor. Thus we write $L(q^{-1})$ as

$$L(q^{-1}) = F(q^{-1})B'(q^{-1}) \qquad (5.3.6)$$

Substituting (5.3.6) into (5.3.5) and canceling $B'(q^{-1})$ from both sides of the resulting equation gives

$$F(q^{-1})A(q^{-1}) + q^{-d}P(q^{-1}) = 1 \qquad (5.3.7)$$

If $F(q^{-1})$ has order $(d-1)$, the equation above has a unique solution exactly as in the proof of Lemma 4.2.1, with $G(q^{-1})$ replaced by $P(q^{-1})$ [see (4.2.6)].

Substituting into (5.3.1) gives precisely the same control law as in Theorem 5.2.1. We thus see that the one-step-ahead control law can be thought of as a technique for setting the denominator of the closed-loop system equal to the open-loop numerator polynomial. Thus the closed-loop poles have been *assigned* to the open-loop zeros. The resulting closed-loop system involves pole–zero cancellation and gives a pure delay from reference input to output. It will be stable only if the conditions of Theorem 5.2.1, part (c), are satisfied [roughly that $B'(q^{-1})^{-1}$ is a stable operator].

5.3.2 The Pole Assignment Algorithm (Difference Operator Formulation)

The discussion above leads us to ask: Is it possible to assign the closed-loop poles to positions other than the open-loop zeros? This would require us to choose $L(q^{-1})$ and $P(q^{-1})$ in the control law (5.3.1) so that (5.3.5) was satisfied, with $B'(q^{-1})$ on the right-hand side replaced by a desired closed-loop polynomial, say, $A^*(q^{-1})$; that is, we need to solve

$$L(q^{-1})A(q^{-1}) + q^{-d}B'(q^{-1})P(q^{-1}) = A^*(q^{-1}) \tag{5.3.8}$$

If this is possible, it is clear that the closed-loop poles can be arbitrarily assigned. Now if $A(q^{-1})$ and $B'(q^{-1})$ have common factors, it is clear from (5.3.8) that these factors must also appear in $A^*(q^{-1})$. Thus for the moment we shall assume that $A(q^{-1})$ and $B'(q^{-1})$ are relatively prime. As we have seen in Chapter 2, deterministic disturbances will give rise to common roots in $A(q^{-1})$ and $B'(q^{-1})$. We shall study this case later. For $A(q^{-1})$ and $B'(q^{-1})$ relatively prime, we have the following result.

Theorem 5.3.1. If $A(q^{-1})$ and $B(q^{-1})$ are relatively prime and $n = \max$ degree $(A(q^{-1}), B(q^{-1}))$, any arbitrary polynomial $A^*(q^{-1})$ of degree $(2n-1)$ can be obtained as the sum

$$A(q^{-1})L(q^{-1}) + B(q^{-1})P(q^{-1}) = A^*(q^{-1}) \tag{5.3.9}$$

for unique polynomials $L(q^{-1})$ and $P(q^{-1})$ of degree $(n-1)$.

Proof. For given $L(q^{-1})$ and $P(q^{-1})$, define $A_c(q^{-1})$ by

$$A(q^{-1})L(q^{-1}) + B(q^{-1})P(q^{-1}) = A_c(q^{-1}) \tag{5.3.10}$$

where

$$A_c(q^{-1}) = a_0^1 + a_1^1 q^{-1} + \cdots + a_{2n-1}^1 q^{-(2n-1)} \tag{5.3.11}$$

Equating coefficients on either side of (5.3.10) gives

$$M_e \begin{bmatrix} l_0 \\ \cdot \\ \cdot \\ \cdot \\ l_{n-1} \\ p_0 \\ \cdot \\ \cdot \\ p_{n-1} \end{bmatrix} = \begin{bmatrix} a_0^1 \\ \cdot \\ \cdot \\ \cdot \\ \cdot \\ \cdot \\ \cdot \\ a_{2n-1}^1 \end{bmatrix} \tag{5.3.12}$$

where

$$
M_e = \begin{bmatrix}
a_0 & & & & & b_0 & & & \\
a_1 & a_0 & & & & b_1 & b_0 & & \\
\cdot & a_1 & \cdot & & & \cdot & b_1 & \cdot & \\
\cdot & \cdot & \cdot & \cdot & & & \cdot & \cdot & \\
\cdot & \cdot & \cdot & a_0 & & b_0 & & \cdot & b_0 \\
a_n & \cdot & \cdot & a_1 & b_n & \cdot & & \cdot & b_1 \\
& a_n & & \cdot & & b_n & & \cdot & \cdot \\
& & \cdot & \cdot & & & \cdot & & \cdot \\
& & & \cdot & & & & \cdot & \cdot \\
& & & a_n & & & & b_n &
\end{bmatrix} 2n
\tag{5.3.13}
$$

$$\longleftarrow \quad n \quad \longrightarrow | \longleftarrow \quad n \quad \longrightarrow$$

Now if $A(q^{-1})$ and $B(q^{-1})$ are relatively prime, we have from Theorem A.4.1 of Appendix A that $\det M_e \neq 0$. Thus from (5.3.12) it is clear that, given $A^*(q^{-1})$, we can determine $L(q^{-1})$ and $P(q^{-1})$ as follows to satisfy (5.3.9):

$$
\begin{bmatrix}
l_0 \\
\cdot \\
\cdot \\
\cdot \\
l_{n-1} \\
p_0 \\
\cdot \\
\cdot \\
\cdot \\
p_{n-1}
\end{bmatrix} = M_e^{-1}
\begin{bmatrix}
a_0^* \\
a_1^* \\
\cdot \\
\cdot \\
\cdot \\
\cdot \\
\cdot \\
\cdot \\
\cdot \\
a_{n-1}^*
\end{bmatrix}
$$

▼▼▼

The implication of the theorem above is that provided that $A(q^{-1})$ and $B(q^{-1})$ are relatively prime, $L(q^{-1})$ and $P(q^{-1})$ can be designed by an algebraic procedure to assign arbitrarily the closed-loop poles. Any closed-loop pole locations are possible. In the special case where we choose $A^*(q^{-1}) = 1$ (corresponding to all closed-loop poles at the origin), we have a form of controller commonly called *dead-beat control*, in which the system response settles in at most n steps. Other special cases will be discussed in Section 5.3.4.

The pole assignment control law has the general form shown in (5.3.1). This is illustrated in Fig. 5.3.1.

It is usually desirable, although not essential, to put $M(q^{-1}) = P(q^{-1})$, in which case the control system shown in Fig. 5.3.1 reduces to the usual output-error-driven form, as shown in Fig. 5.3.2 with improved sensitivity properties (Exercise 5.32).

5.3.3 Rapprochement with State-Variable Feedback

In this section we give an alternative interpretation of the pole assignment algorithm. In particular, we show that the control law can be thought of as an implementation of a state estimator plus state-variable feedback. We can see this as follows.

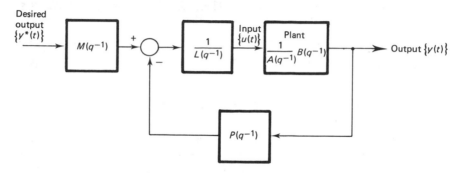

Figure 5.3.1 General feedback structure.

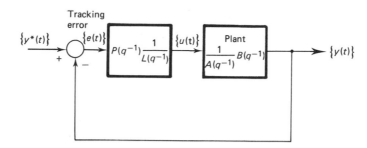

Figure 5.3.2 Error-driven control structure.

1. We first give an alternative interpretation of the difference operator formulation of the pole assignment algorithm in terms of partial state feedback. To do this, we express $A^*(q^{-1})$ as the product of two polynomials $Q(q^{-1})\bar{A}(q^{-1})$ of order $(n-1)$ and n, respectively. We further express $\bar{A}(q^{-1})$ as $A(q^{-1}) + K(q^{-1})$, where $K(q^{-1}) = k_1 q^{-1} + \cdots + k_n q^{-n}$. (The motivation for this factorization will become evident presently.) Thus

$$A^*(q^{-1}) = Q(q^{-1})[A(q^{-1}) + K(q^{-1})] \tag{5.3.14}$$

We now define a filter $R(q^{-1})$ in terms of $L(q^{-1})$ in the control law (5.3.1) and $Q(q^{-1})$ in (5.3.14) as follows:

$$R(q^{-1}) \triangleq L(q^{-1}) - Q(q^{-1}) \tag{5.3.15}$$

where we assume without loss of generality that $L(q^{-1})$ and $Q(q^{-1})$ are both monic and hence $R(q^{-1})$ has zero leading coefficient. We then rewrite the control law (5.3.1) as

$$[R(q^{-1}) + Q(q^{-1})]u(t) = -P(q^{-1})y(t) + M(q^{-1})y^*(t+d) \tag{5.3.16}$$

or

$$Q(q^{-1})u(t) = M(q^{-1})y^*(t+d) - (R(q^{-1})u(t) + P(q^{-1})y(t)) \tag{5.3.17}$$

The motivation for rewriting the control law as in (5.3.17) is that the second term on the right-hand side can be expressed as a linear function of a "state" vector. To see this we express the system (5.3.3) in right difference operator form (expressed in terms of q^{-1} rather than q):

$$A(q^{-1})z(t) = u(t) \tag{5.3.18}$$

$$y(t) = B(q^{-1})z(t) \tag{5.3.19}$$

where $z(t)$ is a partial state vector. Then (5.3.17) can be expressed as state-variable feedback in terms of $z(t)$ as follows:

$$Q(q^{-1})u(t) = M(q^{-1})y^*(t + d) - [R(q^{-1})A(q^{-1}) + P(q^{-1})B(q^{-1})]z(t) \qquad (5.3.20)$$

where we have made use of the following relationship, which is a consequence of (5.3.18) and (5.3.19):

$$[R(q^{-1})A(q^{-1}) + P(q^{-1})B(q^{-1})]z(t) = R(q^{-1})u(t) + P(q^{-1})y(t) \qquad (5.3.21)$$

It is important to note that while $z(t)$ is not directly measured, the filtered value as given in (5.3.21) is readily available via $u(t)$ and $y(t)$. The interpretation of (5.3.20) as state-variable feedback becomes particularly transparent on noting from the pole assignment equation (5.3.9) and using (5.3.14) that

$$L(q^{-1})A(q^{-1}) + P(q^{-1})B(q^{-1}) = A^*(q^{-1}) = Q(q^{-1})\bar{A}(q^{-1}) = Q(q^{-1})[A(q^{-1}) + K(q^{-1})]$$

Hence using (5.3.15), we have

$$R(q^{-1})A(q^{-1}) + P(q^{-1})B(q^{-1}) = Q(q^{-1})K(q^{-1}) \qquad (5.3.22)$$

where

$$Q(q^{-1}) = 1 + Q_1 q^{-1} + \cdots + Q_{n-1} q^{-n+1}, \qquad K(q^{-1}) = k_1 q^{-1} + \cdots + k_n q^{-n}$$

Then (5.3.20) becomes

$$Q(q^{-1})u(t) = M(q^{-1})y^*(t + d) - Q(q^{-1})K(q^{-1})z(t) \qquad (5.3.23)$$

Provided that $Q(q^{-1})$ is stable, this is asymptotically equivalent to

$$u(t) = \left[\frac{M(q^{-1})}{Q(q^{-1})}\right]y^*(t + d) - K(q^{-1})z(t) \qquad (5.3.24)$$

The additional dynamics $Q(q^{-1})$ allow us to estimate the state vector as in (5.3.21). We shall call $Q(q^{-1})$ the *observer dynamics* and we shall denote $[1/Q(q^{-1})]$ $[Q(q^{-1})K(q^{-1})z(t)]$ as $K(q^{-1})\hat{z}(t)$.

The closed-loop system resulting from the control law (5.3.20) can be found by substituting (5.3.24) into (5.3.18) to yield

$$Q(q^{-1})[A(q^{-1}) + K(q^{-1})]z(t) = M(q^{-1})y^*(t + d) \qquad (5.3.25)$$

$$y(t) = B(q^{-1})z(t) \qquad (5.3.26)$$

The closed-loop denominator is thus $Q(q^{-1})[A(q^{-1}) + K(q^{-1})]$ [i.e., $A^*(q^{-1})$] and by appropriate choice of $Q(q^{-1})$ and $K(q^{-1})$ we can arbitrarily assign the closed-loop poles. Indeed, this is precisely what the pole assignment algorithm does. It is interesting to note that if $M(q^{-1})$ in the feedback law (5.3.1) is chosen as

$$M(q^{-1}) = Q(q^{-1}) \qquad (5.3.27)$$

then (5.3.24) asymptotically becomes

$$u(t) = y^*(t + d) - K(q^{-1})z(t) \qquad (5.3.28)$$

that is, in this case the observer dynamics $Q(q^{-1})$ are canceled in forming the closed-loop transfer function. However, it is more usual in practice to choose $M(q^{-1}) = P(q^{-1})$ to give the error feedback form shown in Fig. 5.3.2.

It should now be clear that the control law can be implemented either as in (5.3.1) or in the equivalent state-variable feedback form as in (5.3.20)–(5.3.21). The state-variable feedback interpretation is illustrated in Fig. 5.3.3.

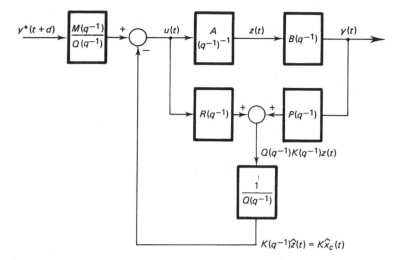

Figure 5.3.3 State-variable feedback interpretation of pole assignment controller.

2. We now give a state-space formulation of the pole assignment algorithm.

The preceding development can also be described in state-space terms. The key observation is that (5.3.18) and (5.3.19) are simply a compact way of writing a state-space model in controller form (see Section 2.3.2) as follows:

$$x_c(t+1) = \begin{bmatrix} -a_1 & \cdots & -a_n \\ 1 & & \\ & \ddots & \\ & & 1 & 0 \end{bmatrix} x_c(t) + \begin{bmatrix} 1 \\ 0 \\ \vdots \\ 0 \end{bmatrix} u(t) \qquad (5.3.29)$$

$$y(t) = [\; b_1 \quad \cdots \quad b_n \;] x_c(t) \qquad (5.3.30)$$

where

$$x_c(t) = [z(t-1) \quad z(t-2) \quad \cdots \quad z(t-n)] \qquad (5.3.31)$$

Then if $z(t)$ is available, the feedback law (5.3.28) becomes

$$u(t) = -Kx_c(t) + y^*(t+d) \qquad (5.3.32)$$

where $K = [k_1 \quad k_2 \quad \cdots \quad k_n]$.

Substituting (5.3.32) into (5.3.29) gives

$$x_c(t+1) = \begin{bmatrix} (-a_1 - k_1) & \cdots & (-a_n - k_n) \\ 1 & & \\ & \ddots & \\ & & 1 & 0 \end{bmatrix} x_c(t) + \begin{bmatrix} 1 \\ 0 \\ \vdots \\ 0 \end{bmatrix} y^*(t+d) \qquad (5.3.33)$$

$$\triangleq \bar{A}x_c(t) + \bar{b}y^*(t+d)$$

Now we observe that the closed-loop characteristic polynomial is given by

$$p_c(z) = \det(zI - \bar{A}) = z^n + (a_1 + k_1)z^{n-1} + \cdots + (a_n + k_n) \quad (5.3.34)$$

and thus it follows that the closed-loop poles can be arbitrarily assigned by appropriate choice of $k_1, \ldots, k_n$. This gives further insight into the significance of $K(q^{-1})$ in (5.3.24).

If the state $x_c(t)$ is not directly measured, we proceed as follows. We first assume that the state-space model (5.3.29)–(5.3.30) is observable and thus can be transformed into observer form [see (2.3.39)]:

$$x_0(t+1) = \begin{bmatrix} -a_1 & 1 & & \\ \cdot & & \cdot & \\ \cdot & & & \cdot \\ \cdot & & & 1 \\ -a_n & & & 0 \end{bmatrix} x_0(t) + \begin{bmatrix} b_1 \\ \cdot \\ \cdot \\ \cdot \\ b_n \end{bmatrix} u(t) \quad (5.3.35)$$

$$y(t) = \begin{bmatrix} 1 & 0 & \cdots & 0 \end{bmatrix} x_0(t) \quad (5.3.36)$$

We now form an estimate of the state $\hat{x}_0(t)$ by using the estimator

$$\hat{x}_0(t+1) = \begin{bmatrix} -a_1 & 1 & & \\ \cdot & & \cdot & \\ \cdot & & & \cdot \\ \cdot & & & 1 \\ -a_n & & & 0 \end{bmatrix} \hat{x}_0(t) + \begin{bmatrix} b_1 \\ \cdot \\ \cdot \\ \cdot \\ b_n \end{bmatrix} u(t) + \begin{bmatrix} l_1 \\ \cdot \\ \cdot \\ \cdot \\ l_n \end{bmatrix} (y(t) - \begin{bmatrix} 1 & 0 & \cdots & 0 \end{bmatrix}\hat{x}_0(t))$$

$$(5.3.37)$$

Now by subtracting (5.3.35) from (5.3.37) and introducing the state estimation error $\tilde{x}_0(t) = \hat{x}_0(t) - x_0(t)$, we have

$$\tilde{x}_0(t+1) = \begin{bmatrix} -a_1 - l_1 & 1 & & \\ \cdot & & \cdot & \\ \cdot & & & \cdot \\ \cdot & & & 1 \\ -a_n - l_n & & & 0 \end{bmatrix} \tilde{x}_0(t) \quad (5.3.38)$$

Thus we can see that the state estimation error can be made to decay as quickly as desired by choosing $l_1 \cdots l_n$ appropriately.

Now since the models (5.3.29) and (5.3.35) are both controllable and observable (minimal), the states are related by a similarity transformation, say $x_0(t) = Px_c(t)$. Then the estimator (5.3.37) for the observer state, $x_0(t)$, can be transformed into an estimator for the controller state, $x_c(t)$, as follows:

$$\hat{x}_c(t+1) = A'x_c(t) + B'u(t) + L'y(t) \quad (5.3.39)$$

where

$$A' = P^{-1} \begin{bmatrix} -a_1 - l_1 & 1 & & \\ \vdots & & \ddots & \\ \vdots & & & 1 \\ -a_n - l_n & & & 0 \end{bmatrix} P \tag{5.3.40}$$

$$B' = P^{-1} \begin{bmatrix} b_1 \\ \vdots \\ \vdots \\ b_n \end{bmatrix} \tag{5.3.41}$$

$$L' = P^{-1} \begin{bmatrix} l_1 \\ \vdots \\ \vdots \\ l_n \end{bmatrix} \tag{5.3.42}$$

3. Finally, we display the connections between the difference operator and state-space approaches to pole assignment. For state-variable feedback we are actually interested only in a linear combination of the controller state estimate, $\hat{x}_c(t)$, that is,

$$f(t) = K\hat{x}_c(t) \tag{5.3.43}$$

We then note that the *scalar* quantity $f(t)$ described by (5.3.39) and (5.3.43) can be compactly expressed by an equivalent left difference operator representation of the form

$$Q(q^{-1})f(t) = R(q^{-1})u(t) + P(q^{-1})y(t) \tag{5.3.44}$$

or

$$Q(q^{-1})[K\hat{x}_c(t)] = R(q^{-1})u(t) + P(q^{-1})y(t) \tag{5.3.45}$$

Thus again, the feedback control law can be represented as in Fig. 5.3.3 or equivalently as in Fig. 5.3.4. The analogy between the state-space and difference operator formulations of the pole assignment algorithm is thus evident (Exercise 5.33).

▼▼▼

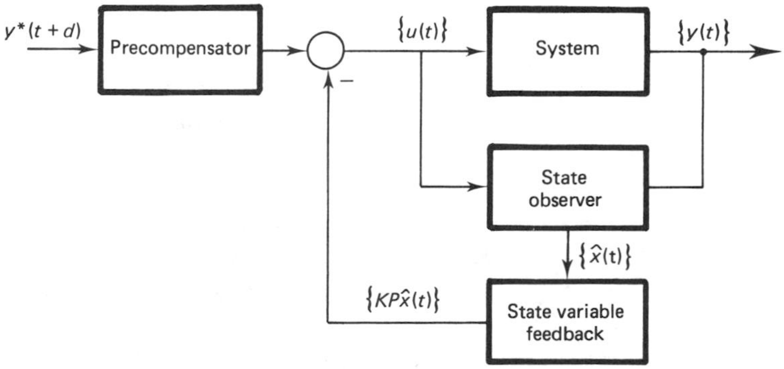

Figure 5.3.4 Combined state observer controller.

Having established that we can arbitrarily assign the closed-loop poles, a natural question is: Where should we aim to put the poles? One obvious requirement is that the closed-loop poles should be inside the unit circle to ensure stability. Of course, there are other considerations, such as transient response, tracking error, and so on. We shall discuss these briefly in Section 5.3.6.

5.3.4 Rapprochement with Minimum Prediction Error Control

In Section 5.2 we have derived the minimum predictor error control law from the d-step-ahead predictor. Here we shall interpret the resulting control law as a special case of closed-loop pole assignment.

Recall the pole assignment equation (5.3.9):

$$A(q^{-1})L(q^{-1}) + B(q^{-1})P(q^{-1}) = A^*(q^{-1}) \qquad (5.3.46)$$

Now consider the special case where we choose the closed-loop polynomial, $A^*(q^{-1})$, as

$$A^*(q^{-1}) = Q(q^{-1})\bar{A}(q^{-1}) \qquad (5.3.47)$$

where

$$\bar{A}(q^{-1}) = B'(q^{-1})$$
$$Q(q^{-1}) = 1$$

Substituting into the pole assignment equation (5.3.46) gives

$$A(q^{-1})L(q^{-1}) + q^{-d}B'(q^{-1})P(q^{-1}) = B'(q^{-1}) \qquad (5.3.48)$$

Since $B'(q^{-1})$ is a factor of the second term on the left-hand side and of the right-hand side, then $B'(q^{-1})$ must be a factor of the first term on the left; that is, we can write

$$L(q^{-1}) = F(q^{-1})B'(q^{-1}) \qquad (5.3.49)$$

Substituting (5.3.49) into (5.3.48) and canceling $B'(q^{-1})$ from both sides gives

$$F(q^{-1})A(q^{-1}) + q^{-d}P(q^{-1}) = 1 \qquad (5.3.50)$$

This is clearly identical to the d-step-ahead prediction equation for the system (see Lemma 4.2.1).

Thus we see that the one-step-ahead controller is a special case of pole assignment in which the observer poles are chosen at the origin [i.e., $Q(q^{-1}) = 1$] (this is often called a dead-beat observer) and the closed-loop poles resulting from feedback are assigned to the open-loop zeros.

An interesting feature of (5.3.50) is that this equation is always solvable, whereas solution of (5.3.46) requires, in general, that $A(q^{-1})$, $B(q^{-1})$ be relatively prime. The limitation of using (5.3.50) is that the resulting controller can only be used when $B'(q^{-1})$ is stable.

The same comments apply to model reference control except that we choose

$$\bar{A}(q^{-1}) = B'(q^{-1})$$
$$Q(q^{-1}) = E(q^{-1})$$

and implement the control law as

$$L(q^{-1})u(t) = -P(q^{-1})y(t) + E(q^{-1})y^*(t + d)$$

We see that the model reference controller is a special case of pole assignment in which an observer is used having poles equal to those of the reference model and the closed-loop poles resulting from feedback are assigned to the open-loop zeros.

5.3.5 The Internal Model Principle

So far in our discussion of pole assignment we have stressed closed-loop pole location in the absence of deterministic disturbances. We now wish to show how disturbances can be incorporated and how perfect tracking can be achieved for certain classes of desired output signals.

We shall consider an error-driven control system as in Fig. 5.3.2. Adding an output disturbance $\{d(t)\}$, we are led to the structure shown in Fig. 5.3.5.

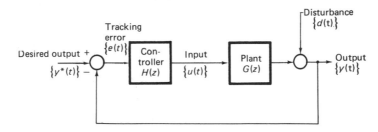

Figure 5.3.5 Control system with disturbance.

Using z-transforms, the output $Y(z)$ can be expressed in terms of the transform of the set-point sequence, $\{Y^*(z)\}$, and of the disturbance, $D(z)$, as

$$Y(z) = \frac{G(z)H(z)}{1 + G(z)H(z)} Y^*(z) + \frac{1}{1 + G(z)H(z)} D(z)$$

The corresponding frequency response is obtained by putting $z = e^{jw}$, that is,

$$Y(e^{jw}) = \frac{G(e^{jw})H(e^{jw})}{1 + G(e^{jw})H(e^{jw})} Y^*(e^{jw}) + \frac{1}{1 + G(e^{jw})H(e^{jw})} D(e^{jw})$$

Thus we see that the output will faithfully reproduce the desired output, provided that the loop gain, $[G(e^{jw})H(e^{jw})]$, is large over the range of frequencies dominant in both $Y^*(e^{jw})$ and $D(e^{jw})$.

In fact, the loop gain can be made infinite at a set of frequencies (and very large and neighboring frequencies) by placing the corresponding modes in the denominator polynomial of the controller, $H(z)$. This idea is sometimes called the *internal model principle* (Francis and Wonham, 1976). In the case of a constant disturbance or constant desired output, the frequency of interest is zero, and thus one should include $(1 - q^{-1})$ in the controller. This clearly gives integral action as commonly employed in practical control system design. The extension to more general models for disturbances and desired outputs is discussed below.

We have seen in Chapter 2 that in the presence of deterministic disturbances, the system model can be written in the form

$$A(q^{-1})y(t) = B(q^{-1})u(t) \qquad (5.3.51)$$

where

$$A(q^{-1}) \triangleq \tilde{A}(q^{-1})D(q^{-1}) \qquad (5.3.52)$$

$$B(q^{-1}) \triangleq \tilde{B}(q^{-1})D(q^{-1}) \qquad (5.3.53)$$

where $D(q^{-1})$ has roots on the unit circle corresponding to the deterministic distur-bances. We shall further assume that $\tilde{A}(q^{-1})$ and $\tilde{B}(q^{-1})$ are relatively prime.

We define the class of desired output sequences, $\{y^*(t)\}$, as those that can be generated by a linear finite-dimensional dynamical system.

$$S(q^{-1})y^*(t) = 0 \qquad (5.3.54)$$

It is further assumed that $S(q^{-1})D(q^{-1})$ and $\tilde{B}(q^{-1})$ are relatively prime.

We consider the feedback control law (5.3.8), that is,

$$L(q^{-1})u(t) = -P(q^{-1})y(t) + M(q^{-1})y^*(t) \qquad (5.3.55)$$

and choose $M(q^{-1})$ and $L(q^{-1})$ in accordance with the internal model principle as follows:

$$M(q^{-1}) = P(q^{-1}); \qquad L(q^{-1}) = L'(q^{-1})S(q^{-1})D(q^{-1}) \qquad (5.3.56)$$

When then determine $L'(q^{-1})$ and $P(q^{-1})$ by solving the usual closed-loop pole assign-ment equation,

$$L'(q^{-1})S(q^{-1})D(q^{-1})\tilde{A}(q^{-1}) + P(q^{-1})\tilde{B}(q^{-1}) = A^*(q^{-1}) \qquad (5.3.57)$$

Note that the control law denominator, $L(q^{-1})$, includes $S(q^{-1})$ (the reference model poles) and $D(q^{-1})$ (the disturbance poles), as suggested by the internal model principle.

Equation (5.3.57) is always solvable for $L'(q^{-1})$ and $P(q^{-1})$ since $S(q^{-1})D(q^{-1})\tilde{A}(q^{-1})$ and $\tilde{B}(q^{-1})$ are relatively prime. Properties of the resulting closed-loop system are described in the following:

Lemma 5.3.1. The closed-loop system resulting from the feedback law (5.3.55)–(5.3.56) has the following properties:

(i) $A^*(q^{-1})y(t) = \tilde{B}(q^{-1})P(q^{-1})y^*(t)$ (5.3.58)

(ii) $A^*(q^{-1})D(q^{-1})u(t) = \tilde{A}(q^{-1})D(q^{-1})P(q^{-1})y^*(t)$ (5.3.59)

(iii) $A^*(q^{-1})[y(t) - y^*(t)] = 0$ (5.3.60)

(iv) $\lim_{t \to \infty} [y(t) - y^*(t)] = 0$ provided that $A^*(q)$ is stable.   (5.3.61)

(v) $\{u(t)\}$ and $\{y(t)\}$ are bounded provided that $D(q^{-1})$ has no repeated roots.

Proof. (i) Multiplying (5.3.51) by $L'(q^{-1})S(q^{-1})$ gives

$$L'(q^{-1})S(q^{-1})A(q^{-1})y(t) = L'(q^{-1})S(q^{-1})B(q^{-1})u(t)$$

or

$$L'(q^{-1})S(q^{-1})D(q^{-1})\tilde{A}(q^{-1})y(t) = L(q^{-1})\tilde{B}(q^{-1})u(t) = \tilde{B}(q^{-1})[P(q^{-1})y^*(t) - P(q^{-1})y(t)]$$

or

$$[L'(q^{-1})S(q^{-1})D(q^{-1})\tilde{A}(q^{-1}) + \tilde{B}(q^{-1})P(q^{-1})]y(t) = \tilde{B}(q^{-1})P(q^{-1})y^*(t)$$

This automatically gives (5.3.58) using (5.3.57).

(ii) Equation (5.3.59) follows a similar argument.

(iii) From (5.3.58),

$$A^*(q^{-1})[y(t) - y^*(t) + y^*(t)] = \tilde{B}(q^{-1})P(q^{-1})y^*(t)$$

or

$$A^*(q^{-1})[y(t) - y^*(t)] = -[A^*(q^{-1}) - \tilde{B}(q^{-1})P(q^{-1})]y^*(t)$$
$$= -L'(q^{-1})S(q^{-1})D(q^{-1})\tilde{A}(q^{-1})y^*(t)$$
$$= 0 \quad \text{using (5.3.54)}$$

(iv) Immediate from (iii).

(v) Follows from (i) and (ii).

▼▼▼

Note that the analysis above has been carried out under ideal assumptions. If the input or output goes into saturation, one must be careful of integral wind-up; see Section 5.3.6.

An interesting interpretation of the result above is that the uncontrollable modes corresponding to disturbances are effectively rendered unobservable by the feedback action.

5.3.6 Some Design Considerations

We discuss some of the practical features of the pole assignment algorithm:

1. *Stability.* An obvious feature of the pole assignment design procedure is that stability can be guaranteed for arbitrary linear systems [provided only that $A(q^{-1})$ and $B'(q^{-1})$ have no common unstable roots]. This is in contrast to the one-step-ahead design method which (in its simplest form) requires inverse stability of the system.

2. *Transient response.* The transient performance of the system depends on the locations to which the closed-loop poles are assigned as we explain in detail below.

In the continuous-time case, the transient response is often described in terms of the damping ratio, ξ, and natural frequency ω_0, of the dominant poles (those closest to the origin). Figure 5.3.6 shows a continuous-time complex-conjugate root pair and explains the significance of ξ and ω_0 in terms of root location.

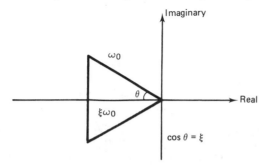

Figure 5.3.6 Roots of second-order continuous-time system showing damping ratio, ξ, and natural frequency, ω_0.

The following rules of thumb apply to the *continuous-time case* [see Fortmann and Hitz (1977) and Franklin and Powell (1980)].

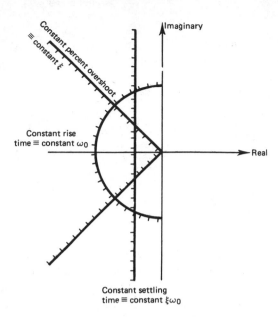

Figure 5.3.7 Loci of constant percent overshoot, rise time, and settling time for continuous systems. The shaded side gives the acceptable region.

a. Percentage overshoot to a step input

$$\simeq \left(1 - \frac{\xi}{0.6}\right)100\% \tag{5.3.62}$$

b. Rise time (time to go from 10% to 90% of the final value)

$$\simeq \frac{2.5}{\omega_0} \tag{5.3.63}$$

c. Settling time (time to settle to within 1% of the final value following a step)

$$\simeq \frac{4.6}{\xi\omega_0} \tag{5.3.64}$$

Figure 5.3.7 shows the loci of constant percent overshoot, rise time, and settling time for a *continuous-time* system.

We see in Section A.1 of Appendix A that with a zero-order hold input and a sampling period of Δ seconds, the discrete pole $\bar{z}$ corresponding to a continuous-time pole $\bar{s}$ is given by

$$\bar{z} = e^{\bar{s}\Delta} \tag{5.3.65}$$

We can use the relationship above to construct the loci of constant percent overshoot, rise time, and settling time in the *discrete-time case*. These are shown in Fig. 5.3.8.

Now given a specification of desired transient performance we can translate this into a feasible set of closed-loop pole locations. We can then use the design procedure described in Theorem 5.3.1 to ensure that the closed-loop poles are in the appropriate locations. In terms of pole location, any specification can be satisfied. Of course, this pertains to an ideal situation only. In practice, one must be careful not to violate the constraints on acceptable system operation such as the allowable size of the input signal. Such constraints must be taken into account in the overall design.

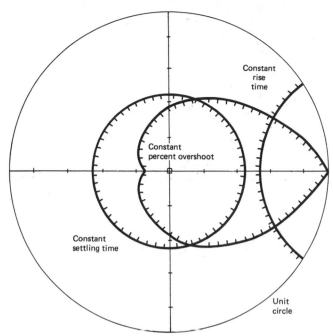

Constant rise time

Constant percent overshoot

Constant settling time

Unit circle

Figure 5.3.8 Loci of constant percent overshoot, rise time, and settling time for discrete systems. The shaded side gives the acceptable region.

3. *Tracking performance.* Referring to Fig. 5.3.2, the closed-loop transfer function is

$$G_c(z) = \frac{G(z)H(z)}{1 + G(z)H(z)} \tag{5.3.66}$$

where $G(z)$ and $H(z)$ are the plant and controller transfer functions, respectively, and are given by

$$G(z) = \frac{1}{A(z^{-1})}B(z^{-1}) \tag{5.3.67}$$

$$H(z) = P(z^{-1})\frac{1}{L(z^{-1})} \tag{5.3.68}$$

The corresponding closed-loop frequency response is obtained by putting $z = e^{j\omega\Delta}$ in (5.3.66), giving

$$G_c(e^{j\omega\Delta}) = \frac{G(e^{j\omega\Delta})H(e^{j\omega\Delta})}{1 + G(e^{j\omega\Delta})H(e^{j\omega\Delta})} \tag{5.3.69}$$

To achieve ideal tracking we would like $G_c(e^{j\omega\Delta})$ to be 1 for all ω. This cannot be achieved in general. However, it is clear from (5.3.69) that good tracking can be achieved over a wide bandwidth provided that loop gain $|G(e^{j\omega\Delta})H(e^{j\omega\Delta})|$ is kept large over a wide bandwidth.

Perfect tracking for constant desired outputs can be simply achieved by introducing a pole into the control at $z = 1$. This effectively gives $G(e^{j\omega\Delta})H(e^{j\omega\Delta}) = \infty$ at $\omega = 0$. We call this *integral action* since a discrete pole at $z = 1$ corresponds to an integrator and gives infinite gain for constant inputs.

The result above can be generalized to other types of desired output, as pointed out in the preceding section. Basically, the idea is to put the poles of $H(z)$ at the roots of $Y^*(z)$. The result extends to disturbance rejection, as we shall see below.

4. *Robustness*. We shall investigate the response of the closed-loop system of Fig. 5.3.2 in the presence of unmodeled output disturbances, parameter variations, and unmodeled dynamics.

We consider the presence of disturbances as modeled previously in Fig. 5.3.5. The z-transform of the error sequence $\{e(t)\}$ is given by

$$E(z) = \frac{1}{1 + G(z)H(z)}[Y^*(z) - D(z)] \qquad (5.3.70)$$

Thus it is desirable to have large loop gain for disturbance rejection, in the same way that it is desirable to have high loop gain for output tracking. In fact, the general comments made earlier about tracking performance carry over to disturbance rejection.

As a simple example, consider the case when $y^*(t) = y^*$, a constant, and $d(t) = d$, a constant; then, subject to stability, the steady-state value of $e(t)$ is given by Lemma A.2.2, part (5) as

$$\lim_{t \to \infty} e(t) = \lim_{z \to 1} (z - 1)E(z)$$

$$= \lim_{z \to 1} (z - 1)\frac{1}{1 + G(z)H(z)}\frac{(y^* - d)z}{z - 1} \qquad (5.3.71)$$

$$= \lim_{z \to 1} \frac{z}{1 + G(z)H(z)}(y^* - d)$$

Thus provided that $G(z)$ or $H(z)$ has a pole at $z = 1$, which is not canceled by a zero in $G(z)H(z)$, the steady-state error will be zero.

The generalization of this result is that to achieve zero steady-state tracking error, $H(z)$ should have a pole at the roots of the denominator of $Y^*(z)$ and $D(z)$ as discussed in Section 5.3.5. Note that this result is independent of variations in the system model.

Note that the roots of the denominator of $Y^*(z)$ and $D(z)$ will, in general, lie on the unit circle. Thus introducing these poles into the controller gives a generalized form of integral action and results in the controller itself being input–output unstable. If the input or output saturates, the integrator output can build up to a very large value, leading to poor control system performance or even instability. This effect is generally known as *integral wind-up*, *reset wind-up*, or *integrator saturation*. The effect often goes unnoticed in analog control systems due to the natural physical limitations on the output of the integrator. In digital control systems the problem can be overcome by either limiting the output of the integrator or by turning the integrator off when the tracking error is larger than some threshold.

On the question of parameter variations, let us assume that the system transfer function, $G(z)$, can change to $G'(z)$ due to parameter variations. Then the corresponding change in the output closed-loop transfer function is

$$\Delta G_c(z) = \frac{G'(z)}{1 + G'(z)H(z)} - \frac{G(z)}{1 + G(z)H(z)}$$

$$= \frac{G'(z) - G(z)}{[1 + G(z)H(z)][1 + G'(z)H(z)]} \qquad (5.3.72)$$

Thus it can be seen that we will have reduced sensitivity with respect to parameter variations compared to the open loop in the frequency range in which $[1 + G(z)H(z)][1 + G'(z)H(z)]$ (with $z = e^{jw}$) is large. This is essentially the same condition as was found for tracking and disturbance rejection.

Finally, we ask what will be the effect of unmodeled dynamics. Usually, these dynamics will be at the high-frequency end of the spectrum. One consequence of unmodeled dynamics is that the phase of the system response will become indeterminate at high frequencies. This is potentially bad for closed-loop stability and the best solution is to ensure that the controller has low gain at these frequencies. Thus we see that a compromise has to be reached between having a wide bandwidth for low-frequency disturbance rejection, tracking, and fast response time, and limiting the bandwidth to avoid problems due to unmodeled dynamics and high frequency noise at higher frequencies.

▼▼▼

In conclusion, design by closed-loop pole assignment is simple, and as we have seen above, the resulting system has many desirable features. We end this section by presenting two simple examples to illustrate the procedure.

Example 5.3.1

(a) Design a closed-loop control system for the following plant so that the closed-loop poles are at the origin

$$G(z) = \frac{z^{-1}(1 - 2z^{-1})}{1 - 3z^{-1}}$$

(b) Determine the steady-state error if $y^* = 1$.

(c) Modify the controller so that all closed-loop poles are at the origin and the steady-state error is zero.

Solution (a) Let the control law be as in Fig. 5.3.2, that is,

$$L(q^{-1})u(t) = P(q^{-1})[y^*(t) - y(t)]$$

Then the closed-loop system is described by [see (5.3.4)]

$$[(1 - 3q^{-1})L(q^{-1}) + (q^{-1} - 2q^{-2})P(q^{-1})]y(t) = (q^{-1} - 2q^{-2})P(q^{-1})y^*(t)$$

For all poles to be the origin, we require that

$$A^*(q^{-1}) = 1$$

Then (5.3.12) becomes

$$\begin{bmatrix} 1 & 0 & & \\ -3 & 1 & 1 & 0 \\ 0 & -3 & -2 & 1 \\ & 0 & & -2 \end{bmatrix} \begin{bmatrix} l_0 \\ l_1 \\ p_0 \\ p_1 \end{bmatrix} = \begin{bmatrix} 1 \\ 0 \\ 0 \\ 0 \end{bmatrix}$$

from which we obtain

$$l_0 = 1, \quad l_1 = -6, \quad p_0 = 9, \quad p_1 = 0$$

giving the control law

$$u(t) = 6u(t - 1) + 9[y^*(t) - y(t)]$$

(b) From part (a) the closed-loop system is

$$y(t) = 9y^*(t - 1) - 18y^*(t - 2)$$

Hence if $y^* = 1$, the steady-state value of $y(t)$ is -9. (Note that this represents a 1000% steady-state error!)

(c) To eliminate the steady-state error, we simply make $(1 - q^{-1})$ a factor of $L(q^{-1})$. Then the closed-loop characteristic polynomial becomes

$$L(q^{-1})(1 - 3q^{-1}) + P(q^{-1})(q^{-1} - 2q^{-2})$$

where

$$L(q^{-1}) = (1 - q^{-1})(e_0 + e_1 q^{-1})$$
$$P(q^{-1}) = f_0 + f_1 q^{-1}$$

We shall lump the factor $(1 - q^{-1})$ together with the $A(q^{-1})$ polynomial to yield the effective denominator polynomial for the system as $(1 - q^{-1})A(q^{-1}) = (1 - q^{-1})$ $(1 - 3q^{-1})$, which has order 2. This implies that $L(q^{-1})$ and $P(q^{-1})$ again have order 1. [In general, the order of $L(q^{-1})$ and $P(q^{-1})$ will have to be increased by 1.] Resolving the pole assignment equation gives $e_0 = 1$, $e_1 = -10$, $f_0 = 14$, and $f_1 = -15$. This the final control law is

$$u(t) = 11u(t - 1) - 10u(t - 2) + 14[y^*(t) - y(t)] - 15[y^*(t - 1) - y(t - 1)]$$

▼▼▼

Finally, we present a very well known example of control system design by pole assignment.

Example 5.3.2 (Inverted Pendulum)

A practical realization of an inverted pendulum control system is shown in Fig. 5.3.9. The x and y directions are approximately decoupled for small angular displacements. It can then readily be shown [see Fortmann and Hitz (1977)] that the linearized continuous-time state-space model for motion in the y direction is

$$\frac{d}{dt}\begin{bmatrix} x_1 \\ x_2 \\ x_3 \\ x_4 \end{bmatrix} = \begin{bmatrix} 0 & 1 & 0 & 0 \\ 0 & 0 & \dfrac{-mg}{M} & 0 \\ 0 & 0 & 0 & 1 \\ 0 & 0 & \dfrac{(M+m)g}{Ml} & 0 \end{bmatrix}\begin{bmatrix} x_1 \\ x_2 \\ x_3 \\ x_4 \end{bmatrix} + \begin{bmatrix} 0 \\ \dfrac{1}{M} \\ 0 \\ -\dfrac{1}{Ml} \end{bmatrix} u \qquad (5.3.73)$$

$$y = [\, 1 \quad 0 \quad 0 \quad 0 \,]\begin{bmatrix} x_1 \\ x_2 \\ x_3 \\ x_4 \end{bmatrix} \qquad (5.3.74)$$

where x_1, x_2, x_3, and x_4 denote the displacement in the y direction, the velocity in the y direction, the angle in the y plane, and the angular velocity in the y plane, respectively; M, m, l, and g denote the mass of carriage, mass at the end of the pendulum, length of the pendulum, and gravitational constant, respectively; y denotes the y displacement; and u is the force supplied in the y direction. The control objective is to balance the pendulum at $y = 0$.

From (5.3.73)–(5.3.74), the continuous-time transfer function y to u is

$$\frac{Y(s)}{U(s)} = \frac{K(s - b)(s + b)}{s^2(s - a)(s + a)} \qquad (5.3.75)$$

where

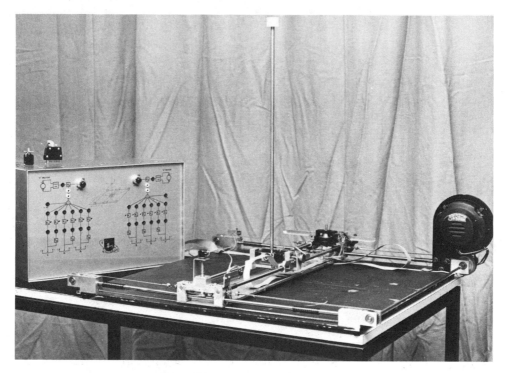

Figure 5.3.9 Inverter pendulum system.

$$K = \frac{1}{M} \tag{5.3.76}$$

$$b = +\sqrt{\frac{g}{l}} \tag{5.3.77}$$

$$a = +\sqrt{\frac{(M + m)g}{Ml}} \tag{5.3.78}$$

It is clear from (5.3.75) that the system is both unstable and nonstably invertible (has poles and zeros in the right half-plane). One consequence of this is that positive feedback from y is required to stabilize the system.

It is easy to measure x_1 and x_3 directly. The other states x_2 and x_4 need to be estimated, but this can be done by using a bandlimited differentiation of x_1 and x_3, respectively (a simple observer).

Then state-variable feedback can be used to stabilize the system. A working system is shown in Fig. 5.3.9.

5.4 AN ILLUSTRATIVE EXAMPLE

In this section we present a simple example to illustrate some features of the different design methods. It turns out that the performance of the various methods on heavily damped systems is roughly comparable.

For lightly damped systems some marked differences appear. To make the discussion simple and clear, we have chosen a system having zero damping. This renders some of the design methods not strictly applicable, but the introduction of a small amount of damping results in practical performance which can readily be inferred from the theoretical zero-damping case. Therefore, consider the following continuous-time system:

$$\frac{d^2}{dt^2}y = ku \tag{5.4.1}$$

With a sampling interval of Δ and a zero-order hold input, the corresponding discrete-time model is

$$(1 - 2q^{-1} + q^{-2})y(t) = \frac{k\Delta^2}{2}q^{-1}(1 + q^{-1})u(t) \tag{5.4.2}$$

Clearly, the system above is not stably invertible. [However, a small amount of damping in (5.4.1) will shift the zero in (5.4.2) from -1 to just inside the unit circle. The only effect of this is to slightly dampen input signals that otherwise might persist.]

Our design objective is to track a unit step input and to settle in approximately 1 sec. We also take $k = 2$ by choice of units and assume that the system is initially at rest.

Method 1: One-Step-Ahead Control

Choosing a sampling interval of Δ, the one-step-ahead control is immediately given by [see (5.4.2)]

$$u(t) = \frac{1}{\Delta^2}[y^* - 2y(t) + y(t - \Delta)] - u(t - \Delta) \tag{5.4.3}$$

Taking $\Delta = 1$ second gives the continuous-time response to the one-step-ahead controller depicted in Fig. 5.4.1. Note that at the sampling times, the response coincides with the desired value of 1 as required. However, we note that there is 50% overshoot between samples.

The observation above also applies to practical systems having light damping. In the latter case the oscillations gradually die out and the overshoot is slightly less than 50%. However, the general nature of the response is retained. It is therefore clear that one-step-ahead control is, in practice, limited to systems that are dominated by a first-order-type response or have heavy damping.

To illustrate the effect of different sampling rates, Fig. 5.4.2 shows the corresponding results for $\Delta = \frac{1}{2}$ sec. This results in larger control effort and faster rise time. Note, however, that the continuous-time overshoot is identical.

Method 2: Dead-Beat Control

We use the pole assignment algorithm of Section 5.3 and place all the closed-loop poles at the origin. This results in the following feedback control law (with $\Delta = 1$):

$$(1 + \tfrac{3}{4}q^{-1})u(t) = (\tfrac{5}{4} - \tfrac{3}{4}q^{-1})[y^* - y(t)] \tag{5.4.4}$$

The resulting continuous-time response is shown in Fig. 5.4.3.

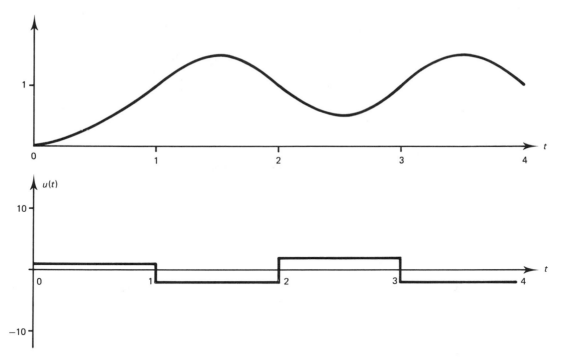

Figure 5.4.1 One-step-ahead control with $\Delta = 1$.

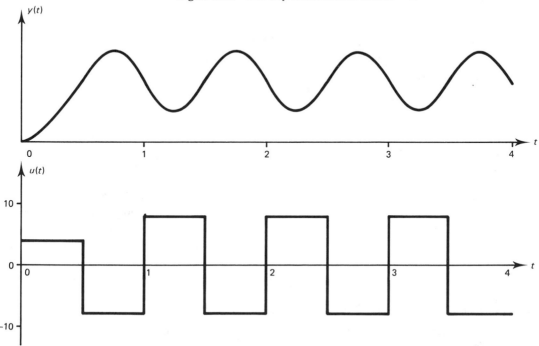

Figure 5.4.2 One-step-ahead control with $\Delta = \frac{1}{2}$.

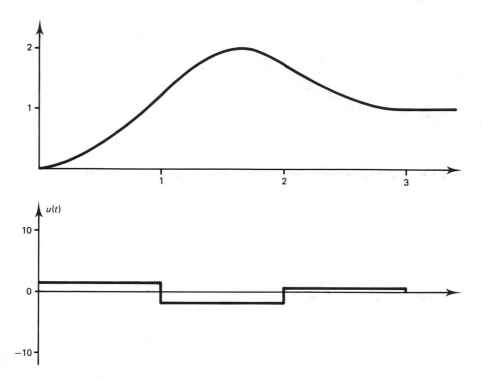

Figure 5.4.3 Dead-beat control with $\Delta = 1$.

Method 3: Weighted One Step Ahead

Here the sampling interval was set at $\frac{1}{10}$ sec and the following one-step-ahead cost function was used:

$$J = \frac{1}{2}[y(t + 1) - y^*]^2 + \frac{\lambda}{2}u(t)^2 \tag{5.4.5}$$

where $\lambda = 0.001$. (This value was determined by a simple root-locus argument.)

The resulting feedback control law is

$$u(t) = 9.1[y^* - 2y(t) + y(t - 1) - 0.01u(t - 1)] \tag{5.4.6}$$

Note that no zero steady-state error occurs in this case due to the fact that the plant itself has a pure integrator. The resulting continuous-time response is shown in Fig. 5.4.4a. The overshoot is 70%. Figure 5.4.4b shows the corresponding result when λ is chosen as 0.0002. The overshoot is now about 50%. The corresponding control law is

$$u(t) = 33[y^* - 2y(t) + y(t - 1) - 0.01u(t - 1)]$$

Method 4: Model Reference Control

Here the sampling rate was chosen as $\frac{1}{10}$ sec and the following reference model was used:

$$(1 - 0.9q^{-1})y^*(t) = 0.1r; \qquad r = 1 \tag{5.4.7}$$

This results in the following feedback control law:

$$u(t) = -110y(t) + 100y(t-1) - u(t-1) + 10 \qquad (5.4.8)$$

The continuous-time response is shown in Fig. 5.4.5. The output response appears acceptable but the input is highly oscillatory.

Method 5: Pole Assignment with Fast Sampling

We can think of this design in two equivalent ways. We can do the design entirely in continuous time and then make an approximate discrete-time implementation or we can map the continuous-time desired pole locations into the discrete domain and then do a discrete-time pole assignment. Both of these methods give roughly the same answer provided that the sampling rate is high relative to the closed-loop pole locations. The motivation for these continuous-time considerations is to better account for the continuous-time response which has not been especially considered in the earlier designs. In the continuous-time design we shall place the closed-loop poles at $s = -0.47 \pm j0.6$ (with damping 0.6).

This gives a continuous time control law (of the proportional plus derivative type) as follows:

$$u = (0.6 + 0.94s)\{y^* - y\} \qquad (5.4.9)$$

An approximate discrete-time implementation of this law at sampling interval, $\Delta = \frac{1}{10}$ sec, is

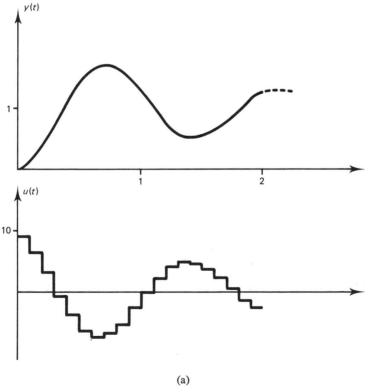

(a)

Figure 5.4.4 Weighted one step ahead with $\Delta = \frac{1}{10}$. (a) $\lambda = 0.001$;

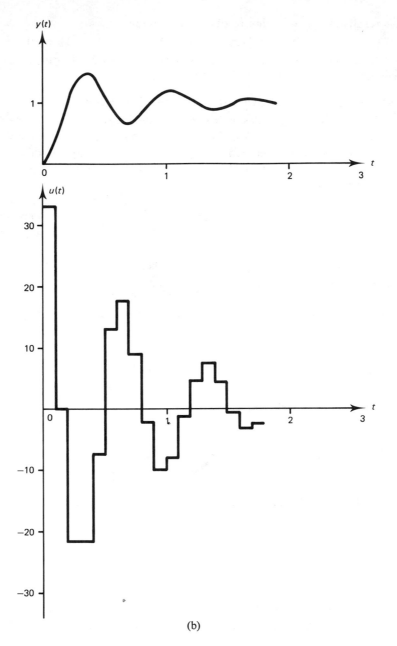

(b)

Figure 5.4.4 (cont'd.) Weighted one step ahead with $\Delta = \frac{1}{10}$. (b) $\lambda = 0.0002$.

$$u(t) = \left[0.6 + \frac{0.94}{\Delta}(1 - q^{-1})\right]\{y^* - y(t)\}; \qquad \Delta = \frac{1}{10}$$
$$= (10 - 9.4q^{-1})\{y^* - y(t)\}$$

(5.4.10)

If one applies the discrete control above to the appropriate discrete-time model, one finds that the corresponding discrete-time pole locations are approximately at

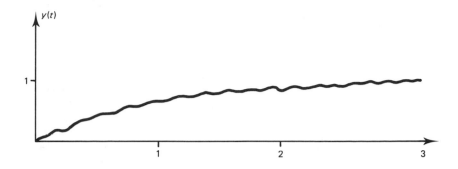

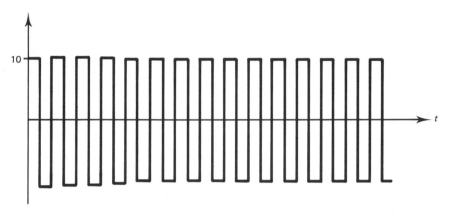

Figure 5.4.5 Model reference control with $\Delta = \frac{1}{10}$.

0.12, 0.94 $\pm$ j0.06. The dominant poles at 0.94 $\pm$ j0.06 are at the image of $s = -0.47$ $+ j$0.06 under the transformation $z = e^{s\Delta}$, indicating that the design can be equally well done in the discrete domain. (Recall that the principal requirement is that the sampling rate be fast relative to the desired continuous-time pole locations or equivalently that the discrete-time poles have magnitude of the order of 0.9 or 0.95.)

The continuous-time response is shown in Fig. 5.4.6.

Method 6: Velocity Compensated Dead-Beat Controller

This design method was briefly described in Remark 5.2.3. The motivation is to try to retain the simplicity of dead-beat design without incurring some of the disadvantages inherent in using this method on lightly damped systems. The idea is to bring, in two steps, y to y^* and $\dot{y}$ to zero.

For the system (5.4.1) we have the following two-step-ahead model for y and $\dot{y}$:

$$y(t + 2) = 3y(t) - 2y(t - 1) + \frac{k\Delta^2}{2}u(t + 1) + \frac{3k\Delta^2}{2}u(t) + \frac{k\Delta^2}{2}u(t - 1) \quad (5.4.11)$$

$$\dot{y}(t + 2) = \dot{y}(t) + k\,\Delta u(t + 1) + k\,\Delta u(t) \quad (5.4.12)$$

Thus the velocity compensated dead-beat control law is

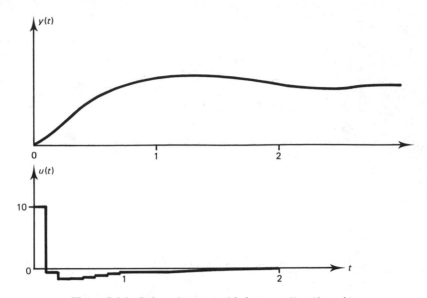

Figure 5.4.6 Pole assignment with fast sampling ($\Delta = \frac{1}{10}$).

$$\begin{bmatrix} u(t+1) \\ u(t) \end{bmatrix} = \begin{bmatrix} \dfrac{k\Delta^2}{2} & \dfrac{3k\Delta^2}{2} \\ k\Delta & k\Delta \end{bmatrix}^{-1} \left\{ \begin{pmatrix} y^* \\ 0 \end{pmatrix} - \begin{pmatrix} 3y(t) - 2y(t-1) + \dfrac{k\Delta^2}{2}u(t-1) \\ \dot{y}(t) \end{pmatrix} \right\}$$
(5.4.13)

Implementation of this control law clearly requires a tachogenerator (or equivalent) to be provided to measure $\dot{y}$. Alternatively, faster samples could be used to estimate $\dot{y}$. The continuous-time response for this example is shown in Fig. 5.4.7 with $\Delta = 0.5$ sec.

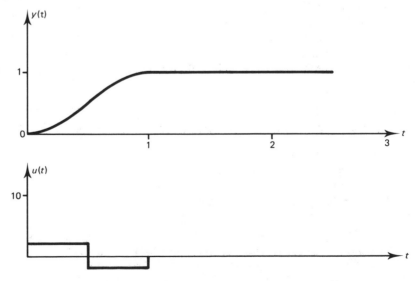

Figure 5.4.7 Velocity-compensated dead-beat controller.

Control of Linear Deterministic Systems   Chap. 5

Discussion

1. The reader should keep in mind that the example above is based on a specific system and thus care is needed in extrapolating the results to other systems with different characteristics.

2. It is important for a designer to be aware of the range of methods available since each has its strength and weakness depending on the problem under study.

3. Some physical insight is highly useful is picking an appropriate control law.

4. In broad terms, if one asks a system to respond quickly, larger control signals will be called for. For example, in one-step-ahead control, the choice of Δ directly affects the response time and hence the size of the input and the bandwidth of the controller.

5. For this example (a lightly damped system) the one-step-ahead controller leads to poor continuous-time response even though the sampled response coincides with the desired output. The crux of the problem is that the one-step-ahead controller results in a closed loop bandwidth which is excessively high relative to the sampling rate.

6. For this example, the dead-beat controller gives the largest overshoot (100%) but settles within three samples.

7. In the weighted one-step-ahead controller, the choice of λ markedly affects the nature of the response and the input signal. The response, as shown in Fig. 5.4.4a and b, is poor but can be improved by choosing appropriate filters in the cost function. However, a more direct and preferable way of achieving precisely the same end result is to use the pole assignment algorithm.

8. In model reference control the effect of the reference model is to smooth the desired trajectory and hence to place lower demands on the control system. The output response now appears acceptable, but the control signal is highly oscillatory due to the fact that the system numerator is effectively canceled in forming the closed-loop response.

9. The pole assignment algorithm with fast sampling gives what appears to be excellent performance. An interesting fact is that for a second-order system, the method gives a proportional and derivative controller with well-known robustness characteristics (integral control is automatically included since the plant is a double integrator).

10. The velocity-compensated controller is a hybrid controller in that it combines continuous-time and discrete-time characteristics. The response appears excellent for this example.

11. The reader is asked to check the sensitivity of the designs to parameter variations in the control law. The most robust design is the pole assignment algorithm with fast sampling.

12. The reader should note that the sampling periods have been chosen differently for some of the methods. The basic principle followed has been to ensure that the rise time is roughly the same for all the methods.

EXERCISES

5.1. Consider the following SISO system:

$$A(q^{-1})y(t) = q^{-d}B'(q^{-1})u(t)$$

where $A(q^{-1})$ and $B'(q^{-1})$ (of orders n and l) can be expressed in the form

$$A(q^{-1}) = \tilde{A}(q^{-1})C(q^{-1}), \qquad B'(q^{-1}) = \tilde{B}'(q^{-1})C(q^{-1})$$

where $\tilde{A}(q^{-1})$ and $\tilde{B}'(q^{-1})$ are relative prime, $\tilde{B}(q^{-1})$ has roots inside the unit circle, and $C(q^{-1})$ (of order s) has roots on the unit circle (corresponding to sinusoidal disturbances).

(a) Use the method of Section 5.2.2 to determine expressions for $L(q^{-1})$, $P(q^{-1})$, and and $Q(q^{-1})$ in the following feedback control law so that the closed-loop system behaves the same as the reference model: $E(q^{-1})y^*(t) = q^{-d}H(q^{-1})r(t)$.

$$L(q^{-1})u(t) = -P(q^{-1})y(t) + Q(q^{-1})r(t)$$

(b) Would the resulting closed-loop system have bounded inputs and bounded outputs? Why? Does the resulting system asymptotically yield $y(t) = y^*(t)$? Why?

(c) An alternative design method for model reference control is:

  (i) Solve the following equation for $\hat{L}$ and $\breve{P}$:

$$\tilde{A}(q^{-1})\hat{L}(q^{-1}) + q^{-d}\tilde{B}'(q^{-1})\breve{P}(q^{-1}) = E(q^{-1})\tilde{B}'(q^{-1})$$

  (ii) Implement the control law as

$$\hat{L}(q^{-1})u(t) = -\breve{P}(q^{-1})y(t) + \tilde{B}'(q^{-1})H(q^{-1})r(t)$$

Show that this leads to the same feedback control law as in part (a) when $C(q^{-1}) = 1$.

(d) When $C(q^{-1}) \neq 1$, show that the controller in part (c) has lower dimension than in (a).

(e) Show that the controller in part (c) asymptotically leads to the following closed-loop system:

$$C(q^{-1})E(q^{-1})y(t) = q^{-d}C(q^{-1})H(q^{-1})r(t)$$

(f) Show that, in general, for the controller derived in part (c), $y(t)$ does not track $y^*(t)$ asymptotically.

(g) Compute the two solutions as in parts (a) and (c) for the following problem:

$$y(t) = u(t-1) + d$$
$$y^*(t) = r(t-1)$$

Show that method (a) gives

$$u(t) = u(t-1) - y(t) + r(t)$$

with the resulting closed-loop system being

$$y(t) = y^*(t); \qquad u(t) = y^*(t+1) - d$$

Show that method (b) gives

$$u(t) = r(t)$$

with the resulting closed-loop system being

$$y(t) = y^*(t) + d; \qquad u(t) = y^*(t+1)$$

5.2. Complete the root-locus argument given in Example 5.2.2.

5.3. Consider the closed-loop characteristic polynomial resulting from weighted one-step-ahead control, that is,

$$A_c(q^{-1}) = B'(q^{-1}) + \frac{\lambda}{\beta_0} A(q^{-1})$$

(a) Show by a root-locus argument that a given system cannot be stabilized by weighted one-step-ahead control if either of the following conditions hold:
1. $A(z^{-1})$ has a zero lying between two zeros of $B'(z^{-1})$ on the real axis outside the unit circle ($|z| > 1$).
2. $B'(z^{-1})$ has a zero lying between two zeros of $A(z^{-1})$ on the real axis outside the unit circle ($|z| > 1$).

(b) Show that the control law required to stabilize the systems in condition 1 must have an unstable pole in the controller. What is the corresponding result for the system in condition 2?

5.4. Complete the proof of Theorem 5.2.3.

5.5. Consider the weighted one-step-ahead controller of Theorem 5.2.3. Show that if $(1 - q^{-1})$ is a factor of $R(q^{-1})$, there is zero steady-state error to a constant desired output.

5.6. Develop the one-step-ahead controller for Example 5.2.5.

5.7. Develop a weighted one-step-ahead controller for multi-input multi-output systems. What is the condition on $A(q^{-1})$, $B(q^{-1})$, and λ for closed-loop stability?

5.8. Complete the proof of Lemma 5.2.8.

5.9. Show by use of the final value theorem [Lemma A.2.2(5)] that zero-state error is achieved provided that the closed-loop system is stable and that the controller has poles on the unit circle at the same locations as there are poles of $Y^*(z)$ and $D(z)$ on the unit circle [where $Y^*(z)$ and $D(z)$ denote the z-transform of the desired output and disturbance, respectively].

5.10. Can the following system be stabilized by state-variable feedback?

$$x(t + 1) = \begin{bmatrix} 1.5 & 0.5 \\ -0.5 & 0.5 \end{bmatrix} x(t) + \begin{bmatrix} 0.5 \\ 0.5 \end{bmatrix} u(t)$$

$$y(t) = [\ 2 \quad 0\ ] x(t)$$

5.11. Consider the following system:

$$x(t + 1) = \begin{bmatrix} 1.5 & 0.5 \\ -0.5 & 0.5 \end{bmatrix} x(t) + \begin{bmatrix} 0.5 \\ -0.5 \end{bmatrix} u(t)$$

$$y(t) = [\ 2 \quad 0\ ] x(t)$$

(a) Can the closed-loop poles be arbitrarily assigned using state-variable feedback?
(b) Can a feedback system be designed to give bounded outputs and bounded inputs?
(c) Will a dead-beat controller work for the system to give perfect tracking of a given bounded desired output?

5.12. Design a model reference controller for the following system:

$$z(t) = 3z(t - 1) + u(t - 1)$$
$$y(t) = z(t) + A \sin (\omega t + \phi); \qquad 0 \le \omega < \pi$$

so that the system behaves in the same way as the following reference model:

$$y^*(t) = \tfrac{1}{2} y^*(t - 1) + r(t - 1)$$

5.13. Would you advise the use of a dead-beat controller on the system

$$A(q^{-1})y(t) = q^{-d} B'(q^{-1})y(t)$$

where

$$A(q^{-1}) = 1 + 3q^{-1} + 7q^{-2} + 8q^{-3}$$
$$B'(q^{-1}) = 2q^{-1} + q^{-2} + 4q^{-3} + 3q^{-4}?$$

Why?

5.14. The equation describing the neutralization of a strong acid by a strong base is

$$V\frac{dy}{dt} = F(t)[a - y(t)] - u(t)[b + y(t)]$$

where y = distance from neutrality = $10^{-p} - 10^{p-14}$
p = pH
F = acid flow rate
a = acid concentration (a constant)
u = base flow rate
b = base concentration (a constant)
V = volume

By using a first-order Euler discrete-time model, develop a dead-beat controller for bringing $y(t)$ to a pH of 7 [Assume that pH and $F(t)$ are measured, a and b are known, and $u(t)$ is the control.]

5.15. Design a feedback control system to place the closed-loop poles of the following system at the origin.

$$[1 - 4q^{-1} + 3q^{-2}]y(t) = [q^{-1} - 2q^{-2}]u(t)$$

5.16. Consider a third-order system of the form

$$A(q^{-1})y(t) = B(q^{-1})u(t)$$

where

$$A(q^{-1}) = 1 + a_1 q^{-1} + a_2 q^{-2} + a_3 q^{-3}$$
$$B(q^{-1}) = q^{-1}(b_0 + b_1 q^{-1} + b_2 q^{-2}); \qquad b_0 \neq 0$$

(a) Show that the corresponding observer form realization is

$$\begin{bmatrix} x_1(t+1) \\ x_2(t+1) \\ x_3(t+1) \end{bmatrix} = \begin{bmatrix} -a_1 & 1 & 0 \\ -a_2 & 0 & 1 \\ -a_3 & 0 & 0 \end{bmatrix} \begin{bmatrix} x_1(t) \\ x_2(t) \\ x_3(t) \end{bmatrix} + \begin{bmatrix} b_0 \\ b_1 \\ b_2 \end{bmatrix} u(t)$$

$$y(t) = [\ 1 \quad 0 \quad 0\ ]x(t) = x_1(t)$$

(b) Using the result of part (a), show that the dead-beat controller can be implemented in the form of state-variable feedback as

$$u(t) = Kx(t) + gy^*(t+1)$$

where

$$K = \begin{bmatrix} \frac{a_1}{b_0} & -\frac{1}{b_0} & 0 \end{bmatrix}; \qquad g = \frac{1}{b_0}$$

(c) Show that the state-variable feedback control law used above gives the following closed-loop system:

$$\begin{bmatrix} x_1(t+1) \\ x_2(t+1) \\ x_3(t+1) \end{bmatrix} = \begin{bmatrix} 0 & 0 & 0 \\ -a_2 + \frac{b_1 a_1}{b_0} & \frac{-b_1}{b_0} & 1 \\ -a_3 + \frac{b_2 a_1}{b_0} & \frac{-b_2}{b_0} & 0 \end{bmatrix} \begin{bmatrix} x_1(t) \\ x_2(t) \\ x_3(t) \end{bmatrix} + \begin{bmatrix} y^*(t+1) \\ 0 \\ 0 \end{bmatrix}$$

$$y(t) = x_1(t)$$

(d) Show that the closed-loop system found in part (c) has eigenvalues at the roots of

$$z[b_0 z^2 + b_1 z + b_2] = 0$$

Why is this obviously the correct solution?

(e) Generalize the discussion above to an nth-order ARMA model.

5.17. Consider the system shown in Fig. 5.A. Assume that the desired output $y^*(t)$ is a constant y^* and that the disturbance $d(t)$ is a sine wave of known frequency.

$$d(t) = A \sin(\omega_0 t + \theta); \qquad t = 1, 2, \ldots$$

What form should $H(z)$ take to ensure that the steady-state error is zero?

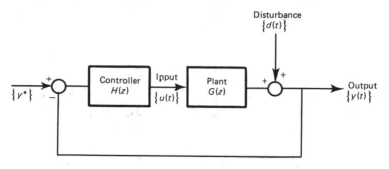

Figure 5.A

5.18. Consider the velocity-compensated dead-beat controller described in Remark 5.2.3. Why might it be a good idea to bring y to y^* and simultaneously achieve $\dot{y} = 0, \ddot{y} = 0$ for a third-order system?

5.19. Compare the feedback gains obtained by evaluating the model reference control for the example of Section 5.4 by the two methods illustrated in Figs. 5.2.4 and 5.2.5. Comment on the relative merits of each approach.

5.20. Consider the system of Example 5.2.6 having transfer function

$$T(z) = \begin{bmatrix} \dfrac{z^{-1}}{1+z^{-1}} & \dfrac{z^{-1}}{1+2z^{-1}} \\[2mm] \dfrac{z^{-1}}{1+3z^{-1}} & \dfrac{z^{-1}}{1+4z^{-1}} \end{bmatrix}$$

Show that the first-order predictor for the model above is

$$\begin{bmatrix} y_1(t+1) \\ y_2(t+1) \end{bmatrix} = -\begin{bmatrix} 3+2q^{-1} & 0 \\ 0 & 7+12q^{-1} \end{bmatrix}\begin{bmatrix} y_1(t) \\ y_2(t) \end{bmatrix} + \begin{bmatrix} 1+2q^{-1} & 1+q^{-1} \\ 1+4q^{-1} & 1+3q^{-1} \end{bmatrix}\begin{bmatrix} u_1(t) \\ u_2(t) \end{bmatrix}$$

Show that $y(t)$ can be brought to $y^*(t)$ in d steps by choosing the input $u(t) \cdots u(t+d-1)$ so as to minimize

$$J = \frac{1}{2}\sum_{j=0}^{d-1} \|u(t+j)\|^2$$

where d is the maximum advance in the interactor matrix (3 in this case).

5.21. Generalize Exercise 5.20 to arbitrary systems having a full rank-square transfer function. Hence show that it suffices to know a lower bound on the maximum advance in the interactor matrix (and not the interactor itself) if one is prepared to take d steps to achieve the desired objective.

5.22. Consider the prediction equality given in (5.2.68). A slightly more general form of this equality used in model reference control is

$$E(q^{-1})\xi(q) = F(q)A(q^{-1}) + G(q^{-1})$$

where

$$E(q^{-1}) = E_0 + E_1 q^{-1} + \cdots$$
$$\xi(q) = \Lambda_0 q^d + \cdots + \Lambda_{d-1} q$$
$$F(q) = F_0 q^d + \cdots + F_{d-1} q$$
$$G(q^{-1}) = G_0 + G_1 q^{-1} + \cdots$$
$$A(q^{-1}) = A_0 + A_1 q^{-1} + \cdots; \qquad A_0 \text{ nonsingular}$$

Show by equating coefficients that

$$F_0 = E_0 \Lambda_0 A_0^{-1}$$

$$F_i = \left(\sum_{j=0}^{i} E_j \Lambda_{i-j} - \sum_{j=0}^{i-1} F_j A_{i-j} \right) A_0^{-1}; \qquad i = 1, \ldots, d-1$$

and $G(q^{-1})$ is determined as the remainder

$$G(q^{-1}) = E(q^{-1})\xi(q) - F(q)A(q^{-1})$$

5.23. Use the result above to build a predictor for $E(q^{-1})\xi(q)y(t)$, where $y(t)$ satisfies

$$A(q^{-1})y(t) = B(q^{-1})u(t)$$

5.24. (a) What are the conditions for closed-loop stability of the control law described in Lemma 5.2.5?

(b) Show that the results reduce to the earlier results when $r = m = 1$, $L(q^{-1}) = 1$, and $R = 0$.

5.25. Repeat Exercise 5.24 for the control law described in Lemma 5.2.6.

5.26. Extend the discussion in Section 5.3.4 to cover the case when $A(q^{-1})$ and $B(q^{-1})$ have common roots on the unit circle (i.e., in the presence of deterministic disturbances).

5.27. Consider a periodic disturbance of period p sampling intervals.

(a) Show that the disturbance can be described by the following model:

$$(1 - q^{-p}) \, d(t) = 0$$

(b) Using the result of part (a), show that zero tracking error to a constant desired output can be achieved by putting $(1 - q^{-1})$ in the denominator of the controller (see Section 5.3.5).

(c) Give an interpretation of the result above as p interlaced ordinary integral controllers working in parallel.

(d) Use a root-locus argument to show why it may be helpful to include $(1 - \alpha q^{-p})$, where $|\alpha| < 1$ in the numerator of the controller, if $1 - q^{-p}$ is included in the denominator.

5.28. (a) Verify that the overshoot for design method 1 in Section 5.4 is 50% independent of Δ.

(b) Repeat the calculations in Section 5.4, design method 1, for a lightly damped system.

5.29. (a) Consider the model reference control method in Section 5.4. Explain why the control signal is oscillatory by reference to the system zeros.

(b) What happens if the system is lightly damped?

5.30. Consider the pole assignment design with fast sampling method in Section 5.4. Show that arbitrary continuous-time pole assignment is possible using a PD (proportional + derivative) controller.

5.31. Show that the velocity compensated dead-beat controller can be implemented on the system of method 2 in Section 5.4 by using $u(t)$ only [i.e., we need not calculate $u(t + 1)$ but can simply reevaluate $u(t)$ at each step]. Does this result extend to either (1) general second-order systems or (2) nth-order systems? Why?

5.32. For a constant y^*, compare: (i) the design equation, (ii) the minimum response time and (iii) the sensitivity, for one step control as in Section 5.2.1 and the corresponding pole assignment solution as in Lemma 5.3.1 with $S(q^{-1}) = (1 - q^{-1})$, $A^*(q^{-1}) = B'(q^{-1})$ and $L'(q^{-1}) = F(q^{-1})B'(q^{-1})$.

5.33. Modify the polynomial pole assignment solution of Section 5.3.2 by restricting $u(t)$ to be a function of $y(t - 1)$ but not $y(t)$. Hence, reconcile the orders of the control laws of Sections 5.3.2 and 5.3.3.

6

Adaptive Control of

Linear Deterministic Systems

6.1 INTRODUCTION

In this chapter we turn our attention to the control of linear systems whose parameters are unknown. Essentially, the approach we adopt is to combine the parameter estimation techniques of Chapter 3 with the control system design techniques of Chapter 5. This will lead to self-tuning or adaptive controllers. These controllers offer certain advantages over conventional controllers. For example, they are capable of tuning themselves to optimal settings and they are capable of retuning themselves should the process dynamics subsequently change. We shall show that the techniques have a sound theoretical design basis and work well in practice. We hope to convince the reader that adaptive control techniques provide a very useful approach to control system design. Of course, as is the case with any design technique, the theory must be augmented with practical considerations relevant to the system under study.

The idea of adaptive control has been around for at least two decades. In the 1950s the topic was enthusiastically persued by many people, especially in relation to autopilot design (Gregory, 1959). However, the supporting theory was essentially nonexistent; there were also difficulties in implementation since computer technology was still in its infancy. Thus early attempts at adaptive control design were largely unsuccessful.

During the 1960s there were major developments in system identification and control theory. These led to an improved understanding of the general problem of adaptive control and spurred renewed interest in this topic.

Following vigorous development of the theory in the 1970s by many research workers, a much better understanding of the underlying principles of the design and operation of adaptive control systems has gradually emerged. There is now an established theoretical basis for the design of adaptive controllers, and indeed several successful applications have been reported in the literature. It is important to have a comprehensive theory, not only to provide guidelines for design, but also to point to pitfalls and limitations of the algorithms. Both of these are essential to successful practical applications.

Many apparently different approaches to adaptive control have been proposed in the literature. Two schemes in particular have attracted much interest: model reference adaptive control (MRAC) and self-tuning regulators (STR). These two approaches actually turn out to be special cases of a more general design philosophy.

In model reference adaptive control the basic idea is to cause the system to behave like a given reference model. The idea was originally proposed by Whitaker, Yamron, and Kezer (1958) and was further developed by Parks (1966), Monopoli (1974), and Landau (1974). The book by Landau (1979) gives a comprehensive study of work up to 1977.

A key question in model reference adaptive control concerns the stability of the resulting system (i.e., the boundedness of the system's inputs and outputs). This remained an open question for many years but was finally resolved in the late 1970s by the composite work of Egardt (1978, 1980a, 1980b), Feuer and Morse (1978), Fuchs 1980a), Goodwin, Ramadge, and Caines (1978a), Jeanneau and de Larminat (1975), Morse (1980), Narendra and Valavani (1978), and Narendra and Lin (1980).

An alternative approach that has attracted much interest is the self-tuning regulator. The basic idea of the self-tuning regulator is inherent in the early work of Kalman (1958) and Young (1965a, 1965b, 1969). The idea was further developed by Chang and Rissanen (1968), who proposed the use of a least-squares parameter estimator together with a minimum variance controller and showed that this technique was one-step-ahead optimal. A very exciting step forward was made by Åström and Wittenmark (1973), who showed that if the parameter estimates converged to some value (not necessarily the true value), then a self-tuning minimum variance regulator would be optimal for a general ARMAX model even though the noise parameters were not explicitly estimated. A key observation of Åström and Wittenmark was that, under certain conditions, it is not necessary to identify the parameters of the process, but instead it is sufficient to directly adjust the control law.

A very readable account of the self-tuning regulator philosophy is contained in Åström et al. (1977). The self-turning regulator is generally viewed in a stochastic framework, and this adds an extra level of difficulty to the convergence and stability question. Important background work on the convergence of recursive stochastic algorithms was done by Tsypkin (1971), Kushner (1977, 1978a, 1978b), Ljung (1974, 1975, 1977a, 1977b), Solo (1978), and Sternby (1977). The question of overall stability and convergence of a stochastic self-tuning regulator was resolved by Goodwin, Ramadge, and Caines (1978b). We shall discuss this in more detail later when we come to stochastic adaptive control. However, it turns out, in hindsight, that a very thorough understanding of the deterministic adaptive control problem leads very naturally to a resolution of the stochastic case without much additional effort. In fact, the underlying principles are common to both the deterministic and stochastic cases.

The foregoing introduction to adaptive control reviews some of the approaches that we will subsequently take up in more detail. For alternative viewpoints and further discussion, see Everleigh (1967), Asher, Andrisani, and Dorato (1976), Lainiotis (1967a, 1976b), Martin-Sanchez (1976), Kutz, Isermann, and Schumann (1980), and the recent excellent survey by Åström (1981).

In the remainder of this chapter we develop deterministic adaptive control and highlight some of the practical and theoretical aspects of the algorithms. We shall concentrate on the discrete-time case since this is in line with the application requirements using modern computer technology, including microprocessors.

The design of an adaptive control system is conceptually simple. A very natural approach is to combine a particular parameter estimation technique with any control law. This approach of using the estimates as if they were the true parameters for the purpose of design is called *certainty equivalence adaptive control*. With this approach we can conceive of generating a wide spectrum of algorithms, depending on which parameter estimation scheme is chosen and which control law is used. Of course, some combinations will work better than others and some may not work at all. We will concentrate on those schemes that have proven convergence and stability properties.

The general block diagram of an adaptive control system is shown in Fig. 6.1.1. We can distinguish two broad classes of algorithms depending on the complexity of

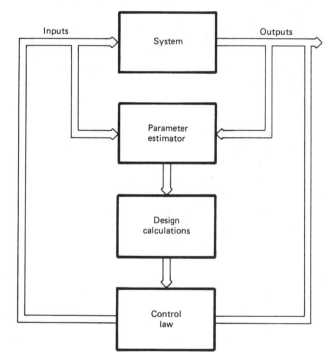

Figure 6.1.1 Block diagram of adaptive control system.

the design calculation block. The simplest conceptual scheme is when the system is parameterized in a natural way and the design calculations are carried out based on the estimated system model. For example, one might adaptively solve the closed-loop pole assignment problem by estimating the system parameters on-line and calculating the corresponding feedback law as described in Section 5.3. This class of algorithms is commonly called *indirect* since the evaluation of the control law is indirectly achieved via the system model. These schemes are also sometimes called *explicit* since the design is based on an explicit process model.

In some cases it is possible to parameterize the system directly in terms of the control law parameters. If this is done, the design calculations necessary to determine the control law become essentially trivial. An example is the one-step-ahead design, in which the control law parameters are simply the parameters in a one-step-ahead predictor. This class of algorithms is commonly called *direct* since the control law is directly estimated. These schemes are also called *implicit* since the design is based on estimation of an implicit process model. In terms of this classification, the model reference adaptive control scheme and the self-tuning control scheme described above are both direct algorithms.

The reader may well find all of these terms confusing. We have introduced them mainly because they are terms that have been used in the literature on adaptive control. However, the origin of the terms is largely of historical significance. We believe that much of the confusion associated with the various terms can be avoided provided that the basic block diagram of Fig. 6.1.1 is kept in mind.

6.2 THE KEY TECHNICAL LEMMA

Our analysis of discrete-time adaptive control algorithms will be based on the following fundamental result. This result will be called upon repeatedly in the convergence analysis of various adaptive control schemes throughout the book.

Lemma 6.2.1 (The Key Technical Lemma). If the following conditions are satisfied for some given sequences $\{s(t)\}$, $\{\sigma(t)\}$, $\{b_1(t)\}$, and $\{b_2(t)\}$:

(1)
$$\lim_{t \to \infty} \frac{s(t)^2}{b_1(t) + b_2(t)\sigma(t)^T\sigma(t)} = 0 \tag{6.2.1}$$

where $\{b_1(t)\}$, $\{b_2(t)\}$, and $\{s(t)\}$ are real scalar sequences and $\{\sigma(t)\}$ is a real $(p \times 1)$ vector sequence.

(2) Uniform boundedness condition
$$0 < b_1(t) < K < \infty \text{ and } 0 \le b_2(t) < K < \infty \tag{6.2.2}$$

for all $t \ge 1$.

(3) Linear boundedness condition
$$\|\sigma(t)\| \le C_1 + C_2 \max_{0 \le \tau \le t} |s(\tau)| \tag{6.2.3}$$

where $0 < C_1 < \infty$ and $0 < C_2 < \infty$, it follows that

(i) $\lim_{t \to \infty} s(t) = 0$

(ii) $\{\|\sigma(t)\|\}$ is bounded.

Proof. If $\{s(t)\}$ is a bounded sequence, then by (6.2.3) $\{\|\sigma(t)\|\}$ is a bounded sequence. Then by (6.2.1) and (6.2.2) it follows that

$$\lim_{t \to \infty} s(t) = 0$$

Now assume that $\{s(t)\}$ is unbounded.

It follows that there exists a subsequence $\{t_n\}$ such that

$$\lim_{t_n \to \infty} |s(t_n)| = \infty$$

and

$$|s(t)| \leq |s(t_n)| \qquad \text{for } t \leq t_n$$

Now along the subsequence $\{t_n\}$

$$\left| \frac{s(t_n)}{[b_1(t_n) + b_2(t_n)\sigma(t_n)^T\sigma(t_n)]^{1/2}} \right| \geq \frac{|s(t_n)|}{[K + K\|\sigma(t_n)\|^2]^{1/2}} \qquad \text{using (6.2.2)}$$

$$\geq \frac{|s(t_n)|}{K^{1/2} + K^{1/2}\|\sigma(t_n)\|}$$

$$\geq \frac{|s(t_n)|}{K^{1/2} + K^{1/2}[C_1 + C_2|s(t_n)|]} \qquad \text{using (6.2.3)}$$

Hence

$$\lim_{t_n \to \infty} \left| \frac{s(t_n)}{[b_1(t_n) + b_2(t_n)\sigma(t_n)^T\sigma(t_n)]^{1/2}} \right| \geq \frac{1}{K^{1/2}C_2} > 0$$

but this contradicts (6.2.1) and hence the assumption that $\{s(t)\}$ is unbounded is false and the result follows.

▼▼▼

6.3 MINIMUM PREDICTION ERROR ADAPTIVE CONTROLLERS (DIRECT APPROACH)

We begin our discussion of adaptive control by developing adaptive version of the minimum prediction error algorithms of Chapter 5. As explained in the introduction, one can take a direct or indirect approach to adaptive control. In the direct approach, one rearranges the model so that it is expressed directly in terms of the control law parameters. Thus, in effect, one directly estimates the control law parameters. This will turn out to be particularly simple for the minimum prediction error algorithms. In the indirect approach one first estimates the parameters in a given model for the system and these are subsequently used to generate the feedback control law via an intermediate calculation.

In this section we discuss the direct approach to adaptive control. In Section 6.4 we look at the indirect approach.

6.3.1 One-Step-Ahead Adaptive Control (The SISO Case)

We shall consider the direct adaptive forms of the one-step-ahead controllers discussed in Section 5.2. As before we shall assume that the system is described by a DARMA model of the form

$$\bar{A}(q^{-1})y(t) = q^{-d}\bar{B}'(q^{-1})u(t) \tag{6.3.1}$$

where $\bar{A}(q^{-1})$ and $\bar{B}'(q^{-1})$ are given by

$$\bar{A}(q^{-1}) = 1 + \bar{a}_1 q^{-1} + \cdots + \bar{a}_{\bar{n}} q^{-\bar{n}}$$
$$\bar{B}'(q^{-1}) = \bar{b}_0 + \bar{b}_1 q^{-1} + \cdots + \bar{b}_{\bar{m}} q^{-\bar{m}}; \qquad \bar{b}_0 \neq 0$$

and d represents a pure time delay. [Note that the model (6.3.1) can be used to describe the input–output properties of a general dynamic system with a state-space description; see Chapter 2 for further discussion.]

In the development of the adaptive minimum prediction error control algorithms, we will require knowledge of only an upper bound on the orders of $\bar{A}(q^{-1})$ and $\bar{B}'(q^{-1})$ in (6.3.1) rather than the exact order. Thus it is possible to replace (6.3.1) by the following overparameterized model:

$$A(q^{-1})y(t) = q^{-d} B'(q^{-1})u(t) \tag{6.3.2}$$

where

$$A(q^{-1}) = 1 + a_1 q^{-1} + \cdots + a_n q^{-n}; \qquad n \geq \bar{n}$$
$$B'(q^{-1}) = b_0 + b_1 q^{-1} + \cdots + b_m q^{-m}; \qquad m \geq \bar{m}$$

Since the minimum prediction error control laws depend on the parameters in the d-step-ahead predictor, we shall first convert the model to d-step-ahead predictor form to facilitate direct control. As shown in Lemma 4.2.1, the d-step-ahead predictor for the model (6.3.2) is

$$y(t + d) = \alpha(q^{-1})y(t) + \beta(q^{-1})u(t) \tag{6.3.3}$$

where

$$\alpha(q^{-1}) = \alpha_0 + \alpha_1 q^{-1} + \cdots + \alpha_{n-1} q^{-(n-1)}$$
$$\beta(q^{-1}) = \beta_0 + \beta_1 q^{-1} + \cdots + \beta_{m+d-1} q^{-(m+d-1)}$$
$$\beta_0 = b_0 \neq 0$$

With these preliminary notions in mind, we turn to the adaptive control problem. We are given a desired output sequence $\{y^*(t)\}$ and our objective is to design an adaptive control law to achieve closed-loop stability and to asymptotically achieve zero tracking error; that is, we desire

$$\lim_{t \to \infty} [y(t) - y^*(t)] = 0$$

Referring to (6.3.3) we denote by θ_0 the vector of parameters in the predictor, that is,

$$\theta_0^T = (\alpha_0, \ldots, \alpha_{n-1}, \beta_0, \ldots, \beta_{m+d-1}) \tag{6.3.4}$$

Then (6.3.3) can be written in the following regression form:

$$y(t + d) = \phi(t)^T \theta_0 \tag{6.3.5}$$

where

$$\phi(t)^T = (y(t), \ldots, y(t - n + 1), u(t), \ldots, u(t - m - d + 1)) \tag{6.3.6}$$

Now define the output tracking error as

$$\xi(t + d) \triangleq y(t + d) - y^*(t + d)$$
$$= \phi(t)^T \theta_0 - y^*(t + d) \tag{6.3.7}$$

If we could choose $\{u(t)\}$ to satisfy

$$\phi(t)^T \theta_0 = y^*(t + d) \tag{6.3.8}$$

it is evident that the tracking error would be identically zero (see Theorem 5.2.1). Here we assume that θ_0 is unknown and we replace (6.3.8) by an adaptive algorithm that uses the parameter estimators of Chapter 3 to generate a sequence of parameter estimates for use in the control law; that is, using the certainty equivalence principle, we replace (6.3.8) by

$$\phi(t)^T \hat{\theta}(t) = y^*(t + d)$$

where $\hat{\theta}(t)$ denotes an estimate of θ_0 at time t.

Prototype One-Step-Ahead Adaptive Control

Using the simple gradient or projection algorithm for parameter estimation, we are led to the following adaptive control law.

One-Step-Ahead Adaptive Controller

$$\hat{\theta}(t) = \hat{\theta}(t - 1) + a(t)\phi(t - d)[c + \phi(t - d)^T\phi(t - d)]^{-1}[y(t) - \phi(t - d)^T\hat{\theta}(t - 1)];$$

$$0 < a(t) < 2; \quad c > 0 \tag{6.3.9}$$

The control law is given by

$$\phi(t)^T \hat{\theta}(t) = y^*(t + d) \tag{6.3.10}$$

In (6.3.9), $\hat{\theta}(t)$ is a p-vector of reals depending on an initial estimate $\hat{\theta}(0)$ and on $y(\tau), 0 \leq \tau \leq t, u(\tau), 0 \leq \tau \leq t - d$.

Remark 6.3.1. Note that (6.3.10) is a time-varying feedback control law, as can be seen by making $u(t)$ the subject of the formula

$$u(t) = \frac{1}{\hat{\theta}_{n+1}(t)}[-\hat{\theta}_1(t)y(t) - \hat{\theta}_2(t)y(t - 1) - \cdots - \hat{\theta}_n(t)y(t - n + 1)$$

$$- \hat{\theta}_{n+2}(t)u(t - 1) - \cdots - \hat{\theta}_{n+m+d}(t)u(t - m - d + 1) + y^*(t + d)]$$

where $\hat{\theta}_j(t)$ denotes the jth element of $\hat{\theta}(t)$.

Note also that the algorithm requires that the desired trajectory $y^*(t + d)$ be known at time t. This is clearly reasonable for controlling a system with delay.

▼▼▼

Remark 6.3.2

1. In practice, the feedback law (6.3.10) works well but in theory there is the remote possibility of division by zero in finding $u(t)$. This can easily be avoided. For example, if the sign of $\theta_{n+1} = \beta_0$ is known, the constrained parameter estimation techniques of Section 3.7 can be applied to ensure that $\hat{\theta}_{n+1}$ does not go to zero. Alternatively, the gain constant, $a(t)$, can be computed as follows:

$$a(t) = \begin{cases} 1 & \text{if } [(n + 1)\text{th component of right-hand side of (6.3.9)} \\ & \text{evaluated using } a(t) = 1] \neq 0 \\ \gamma & \text{(where } \gamma \text{ is a fixed constant in the interval} \\ & \epsilon < \gamma < 2 - \epsilon, \gamma \neq 1, \text{ and } 0 < \epsilon < 1) \quad \text{otherwise} \end{cases} \tag{6.3.11}$$

2. It is also worth noting that the purpose of the coefficient c in the term $[c + \phi(t - d)^T\phi(t - d)]^{-1}$ of (6.3.9) is solely to avoid division by zero. [See the discussion leading to the projection algorithm of (3.3.19).]

An alternative is to simply augment $\phi(t - d)$ with an additional entry having the value of 1 and to put $c = 0$. This has the added advantage of allowing arbitrary offsets between input and output in (6.3.1); see further discussion in Section 3.3.

▼▼▼

We shall now show that the algorithm above has attractive convergence properties. In particular, we will prove that the algorithm is global convergent in the case of a time-invariant system.

We make the following assumptions.

Assumption 6.3.A

1. The time delay d is known.
2. An upper bound for the orders of the polynomials in (6.3.3) is known.
3. (i) All modes of the inverse of the model (6.3.1) [i.e., the zeros of the polynomial $\bar{B}'(z^{-1})$] lie inside or on the closed unit disk.
   (ii) All controllable modes of the inverse of the model (6.3.1) [i.e., the zeros of the transfer function $\bar{B}'(z^{-1})/\bar{A}(z^{-1})$] lie strictly inside the unit circle.
   (iii) Any modes of the inverse of the model (6.3.1) on the unit circle have a Jordan block size of 1.

▼▼▼

Note that Assumption 6.3.A, part 3 was necessary to achieve perfect tracking and closed-loop stability for the one-step-ahead controller when the parameters were known (see Theorem 5.2.1). Thus this is a natural assumption in the case where the parameters are unknown. Assumption 6.3.A, part 2, is of significance since it allows the true system order to be overestimated. Assumption 6.3.A, part 1, is vital and is necessary to ensure that $\beta_0 \neq 1$. This is needed in the subsequent development.

We can now establish the following *global convergence* result.

Theorem 6.3.1. Subject to Assumption 6.3.A, the one-step-ahead adaptive control algorithm (6.3.9)–(6.3.10) when applied to the system (6.3.1) yields

(1) $\{y(t)\}$ and $\{u(t)\}$ are bounded sequences $\qquad\qquad\qquad\qquad$ (6.3.12)

(2) $\lim_{N \to \infty} y(t) - y^*(t) = 0$ $\qquad\qquad\qquad\qquad\qquad\qquad\qquad$ (6.3.13)

(3) $\lim_{N \to \infty} \sum_{t=d}^{N} [y(t) - y^*(t)]^2 < \infty$ $\qquad\qquad\qquad\qquad\qquad$ (6.3.14)

Proof. Elementary properties of the iteration (6.3.9) have been discussed in Lemma 3.3.2 [where $\phi(t - 1)$ has now been replaced by $\phi(t - d)$]. (The reader should revise Lemma 3.3.2 before proceeding.)

Part 1: We first recall results (a) and (e) of Lemma 3.3.2:

$$\lim_{t \to \infty} \frac{e(t)}{[c + \phi(t - d)^T\phi(t - d)]^{1/2}} = 0 \qquad\qquad (6.3.15)$$

where

$$e(t) = y(t) - \phi(t - d)^T \hat{\theta}(t - 1) \tag{6.3.16}$$

$$\lim_{t \to \infty} \| \hat{\theta}(t) - \hat{\theta}(t - k) \| = 0 \tag{6.3.17}$$

Now, if we define the tracking error $\epsilon(t)$ as

$$\epsilon(t) = y(t) - y^*(t) \tag{6.3.18}$$

then from (6.3.5) and (6.3.10),

$$\begin{aligned} \epsilon(t) &= \phi(t - d)^T \theta_0 - \phi(t - d)^T \hat{\theta}(t - d) \\ &= -\phi(t - d)^T \tilde{\theta}(t - d) \end{aligned} \tag{6.3.19}$$

where

$$\tilde{\theta}(t) = \hat{\theta}(t) - \theta_0 \tag{6.3.20}$$

Thus

$$\begin{aligned} \frac{-\epsilon(t)}{[c + \phi(t - d)^T \phi(t - d)]^{1/2}} &= \frac{\phi(t - d)^T \tilde{\theta}(t - d)}{[c + \phi(t - d)^T \phi(t - d)]^{1/2}} \\ &= \frac{\phi(t - d)^T \tilde{\theta}(t - 1) + \phi(t - d)^T [\tilde{\theta}(t - d) - \tilde{\theta}(t - 1)]}{[c + \phi(t - d)^T \phi(t - d)]^{1/2}} \end{aligned}$$

$$\tag{6.3.21}$$

Now the limit of the right-hand side is clearly zero from (6.3.15) and (6.3.17). Hence

$$\lim_{t \to \infty} \frac{\epsilon(t)^2}{c + \phi(t - d)^T \phi(t - d)} = 0 \tag{6.3.22}$$

Part 2: We now proceed to use the key technical lemma (Lemma 6.2.1).

We have established condition (1) of Lemma 6.2.1 with $s(t) = \epsilon(t)$, $\sigma(t) = \phi(t - d)$, $b_1(t) = c$, and $b_2(t) = 1$. Condition (2) of Lemma 6.2.1 is also clearly satisfied. To establish condition (3) [namely, that $\| \phi(t - d) \|$ is bounded by $\epsilon(t)$] we note from Assumption 6.3.A, part 3 and Lemma B.3.3 of Appendix B that there exists constants $m_3 < \infty$ and $m_4 < \infty$ such that

$$u(k - d) \le m_3 + m_4 \max_{1 \le \tau \le t} | y(\tau) | \qquad \text{for all } 1 \le k \le t$$

Therefore, using the definition of $\phi(t - d)$ as in (6.3.6),

$$\| \phi(t - d) \| \le p \{ m_3 + [\max (1, m_4)] \max_{1 \le \tau \le t} | y(\tau) | \}$$

where p is the dimension of ϕ. But

$$| \epsilon(t) | \ge | y(t) | - | y^*(t) | \ge | y(t) | - m_1; \qquad m_1 < \infty$$

Hence

$$\begin{aligned} \| \phi(t - d) \| &\le p \{ m_3 + [\max (1, m_4)] \max_{1 \le \tau \le t} (| \epsilon(\tau) + m_1) \} \\ &= C_1 + C_2 \max_{1 \le \tau \le t} | \epsilon(\tau) |; \qquad 0 \le C_1 < \infty, \quad 0 < C_2 < \infty \end{aligned}$$

and it follows that the linear boundedness condition is also satisfied. Equations (6.3.12) and (6.3.13) now follow immediately from Lemma 6.2.1.

Finally, from Lemma 3.3.2, part (ii),

$$\lim_{N \to \infty} \sum_{t=1}^{N} \frac{e(t)^2}{c + \phi(t-d)^T \phi(t-d)} < \infty \qquad (6.3.23)$$

$$\lim_{N \to \infty} \sum_{t=k}^{N} \| \hat{\theta}(t) - \hat{\theta}(t-k) \|^2 < \infty \qquad (6.3.24)$$

Hence using (6.3.23), the Schwarz inequality, and the boundedness of $\{\phi(t)\}$ (established above), we conclude (6.3.14).

▼▼▼

The result above is important because it is a simple and rigorous proof of stability for a representative adaptive control algorithm.

The key conclusions are that

1. Closed-loop stability is achieved.
2. The output tracking error asymptotically goes to zero (perfect tracking is achieved).
3. The convergence rate is better than $1/t$.

Note that nothing has been said about the convergence of the parameter estimates.

Theorem 6.3.1 is remarkable in that such a strong result (global convergence) can be established for adaptive control which is a highly nonlinear and time-varying problem due to the presence of the parameter estimator. It also brings out the importance of the judicious choice of parameter estimation algorithm and control law.

Admittedly, Assumption 6.3.A is somewhat restrictive. It is partly a consequence of the simple control law used. Later we shall see that other control laws lead to comparable but different assumptions.

We now present a simple example to illustrate both the advantages and disadvantages of the algorithm in practice.

Example 6.3.1

The algorithm above [with $a(t) = 1$] has been implemented on an on-line computer and tested on the servo kit described in Example 5.2.3.

Figure 6.3.1 shows the *sampled* output response during the initial training of the adaptive controller on startup (with the servo gain set to 50% and $\{y^*(t)\}$ a square wave of period 10 sec). It can be seen that nearly perfect tracking is obtained after one period of $y^*(t)$.

Figure 6.3.2 shows the effect of a sudden large change in the system parameters. The controller was first tuned with the gain set to 10%. Then the gain was increased by a factor of 6 to 60%. It can be seen, from the figure, that the controller tuned at 10% gain is quite unsuitable for operation at 60% gain, but that retuning of the controller to give near perfect tracking occurs in approximately one period of $\{y^*(t)\}$.

The reader's attention is drawn to the results in Section 5.4 which show that the *continuous-time* response resulting from one-step-ahead control of a lightly damped system is rather unsatisfactory! In fact, nearly 50% overshoot is observed on the continuous-time response for their example independent of the sampling interval. Thus one-step-ahead control is not recommended for lightly damped systems such as a servo system.

▼▼▼

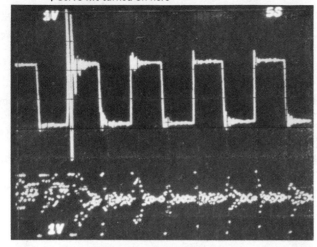

Output

Input

Figure 6.3.1 Initial training of adaptive controller for servo kit.

↓ 6:1 gain change occurs here

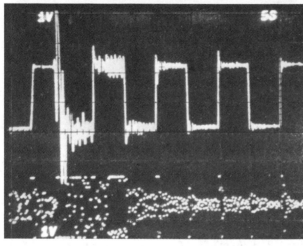

Output

Input

Figure 6.3.2 Returning of the adaptive controller following a 6-to-1 change in system gain.

Remark 6.3.3. An important practical point is that with any large change in system parameters or in $\{y^*(t)\}$, large input signals can be called for by the algorithm. In this case, the control law (6.3.10) can be modified as follows. Put

$$u'(t) = \frac{1}{\hat{\theta}_{n+1}(t)}[-\hat{\theta}_1(t)y(t) \cdots -\hat{\theta}_n(t)y(t-n+1) - \hat{\theta}_{n+2}(t)u(t-1)$$
$$\cdots -\hat{\theta}_{n+m+d}(t)u(t-m-d+1) + y^*(t+d)]$$

If $u_{\min} \leq u'(t) \leq u_{\max}$ put $u(t) = u'(t)$

If $u'(t) < u_{\min}$ put $u(t) = u_{\min}$

If $u'(t) > u_{\max}$ put $u(t) = u_{\max}$

where $u_{\min}$ and $u_{\max}$ are specified minimum and maximum input levels.

It is important that $\{u(t)\}$ be used in $\phi(t-d)$ and not $\{u'(t)\}$! If $u\{(t)\}$ is used, Lemma 3.3.2 remains valid for the algorithm with saturated inputs. The proof of Theorem 6.3.1, however, no longer applies and thus global convergence of the adaptive control algorithm is not assured. However, it is reassuring that the properties of the parameter estimator are maintained.

▼▼▼

Remark 6.3.4. As pointed out earlier, the remote possibility of division by zero in (6.3.10) can be avoided by using the constrained algorithms of Section 3.7 to ensure that

$$\hat{\theta}_{n+1}(t) \neq 0 \tag{6.3.25}$$

We assume knowledge of the sign and lower bound on the magnitude of β_0 in the predictor model (6.3.3). This leads to the following slightly modified form of (6.3.9)–(6.3.10):

$$\hat{\theta}'(t) = \hat{\theta}(t-1) + \frac{\phi(t-d)}{c + \phi(t-d)^T\phi(t-d)}[y(t) - \phi(t-d)^T\hat{\theta}(t-1)]; \qquad c > 0 \tag{6.3.26}$$

If

$$[\hat{\theta}'_{n+1}(t)]\, \text{sign}\, \beta_0 \geq |\beta_0|_{\min} \tag{6.3.27}$$

then

$$\hat{\theta}(t) = \hat{\theta}'(t) \tag{6.3.28}$$

Otherwise,

$$\hat{\theta}_i(t) = \hat{\theta}'_i(t); \qquad i \neq n+1 \tag{6.3.29}$$

$$= |\beta_0|_{\min}\, \text{sign}\, \beta_0; \qquad i = n+1 \tag{6.3.30}$$

The control is then generated from

$$\phi(t-d)^T\hat{\theta}(t-d) = y^*(t) \tag{6.3.31}$$

Note that the algorithm above simply stops the $(n+1)$th component of $\hat{\theta}(t)$ (which is the estimate of β_0) from going outside its known region. A similar idea has been employed by Jones (1973) to ensure bounded parameter estimates in adaptive filtering. The idea has also been used in the adaptive control context by Goodwin, Long, and McInnis (1980) and Elliott (1980).

The algorithm ensures that

$$|\hat{\theta}_{n+1}(t)| \geq |\beta_0|_{\min} \qquad \text{for all } t \tag{6.3.32}$$

Also, as a consequence of the algorithm,

$$|\hat{\theta}_{n+1}(t) - \beta_0| \leq |\hat{\theta}'_{n+1}(t) - \beta_0| \tag{6.3.33}$$

it follows that

$$\|\hat{\theta}(t+1) - \theta_0\| \leq \|\hat{\theta}'(t+1) - \theta_0\| \tag{6.3.34}$$

and also that

$$\|\hat{\theta}(t+1) - \hat{\theta}(t)\| \leq \|\hat{\theta}'(t+1) - \hat{\theta}(t)\| \tag{6.3.35}$$

Subject to Assumptions 6.3.A and 6.3.B, the properties of the constrained algorithm above are exactly as in Lemma 3.3.2 and Theorem 6.3.1, with the additional result (6.3.32). (See Exercise 6.1.)

▼▼▼

In the discussion above we have used the projection algorithm as the parameter estimator. In view of the results in Chapter 3, it should be clear that the least-squares algorithm can be used instead. This leads to the following least-squares form of the algorithm (6.3.9)–(6.3.10):

One-Step-Ahead Adaptive Controller (Least-Squares Iteration)

$$\hat{\theta}(t) = \hat{\theta}(t-1) + \frac{P(t-d-1)\phi(t-d)}{1 + \phi(t-d)^T P(t-d-1)\phi(t-d)}[y(t) - \phi(t-d)^T\hat{\theta}(t-1)]$$

(6.3.36)

$$P(t-d) = P(t-d-1) - \frac{P(t-d-1)\phi(t-d)\phi(t-d)^T P(t-d-1)}{1 + \phi(t-d)^T P(t-d-1)\phi(t-d)}$$
(6.3.37)

$$P(0) = \epsilon I; \quad 0 < \epsilon < \infty$$
(6.3.38)

The control law is

$$\phi(t)^T\hat{\theta}(t) = y^*(t+d)$$
(6.3.39)

Subject to Assumption 6.3.A, the algorithm above can be shown to be globally convergent *exactly* as in Theorem 6.3.1 except that Lemma 3.3.6 is used for the underlying properties of the parameter estimator. (The reader is encouraged to carry out the steps; see Exercise 6.2.)

Remark 6.3.5. It is apparent that we can slightly modify the algorithm to avoid the possibility of division by zero in calculating the control. As for the projection form we assume that the sign of β_0 and a lower bound for $|\beta_0|$ are known. We can then apply the algorithm given in (3.7.12) to (3.7.17) to constrain $\hat{\beta}_0 \neq 0$.

▼▼▼

Example 6.3.2

We now give an example of the least-squares form of the one-step-ahead adaptive controller. We also use the example to illustrate how the one-step-ahead controller performs in the presence of deterministic disturbances. Consider the system

$$y(t) = bu(t-1) + d(t)$$

where $d(t)$ is a deterministic disturbance given by

$$d(t) = D\sin(\theta t + \phi)$$

The appropriate DARMA model for the system is readily seen to be

$$A(q^{-1})y(t) = B(q^{-1})u(t)$$

where

$$A(q^{-1}) \triangleq 1 - (2\cos\theta)q^{-1} + q^{-2} \triangleq 1 + a_1 q^{-1} + a_2 q^{-2}$$

$$B(q^{-1}) \triangleq q^{-1}(b - (2b\cos\theta)q^{-1} + bq^{-2}) \triangleq q^{-1}(b_0 + b_1 q^{-1} + b_2 q^{-2})$$

The true value of the parameter vector, θ_0, and the initial parameter estimates, $\hat{\theta}(0)$, were as follows:

$$\theta_0 = (-1.4, 1, 1, -1.4, 1)$$

$$\hat{\theta}(0) = (-1.0, 1.2, 1.0, -1.5, 1.5)$$

The desired output sequence $\{y^*(t)\}$, was taken to be a square wave of amplitude 1 and period 100 samples. The disturbance was chosen as a sine wave of amplitude 1 and frequency 0.8 radian per unit time.

The performance of the least-squares adaptive control algorithm (6.3.36) to (6.3.39) is shown in Figs. 6.3.3 to 6.3.5. Figures 6.3.3, 6.3.4, and 6.3.5 show the output and reference signal, input signal, and the estimated parameter, respectively. It can be seen that asymptotic perfect tracking is achieved as predicted by the theory.

▼▼▼

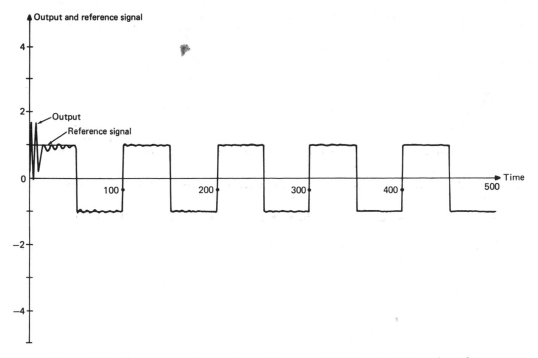

Figure 6.3.3 Output and reference signal of Example 6.3.2 (showing asymptotic perfect tracking).

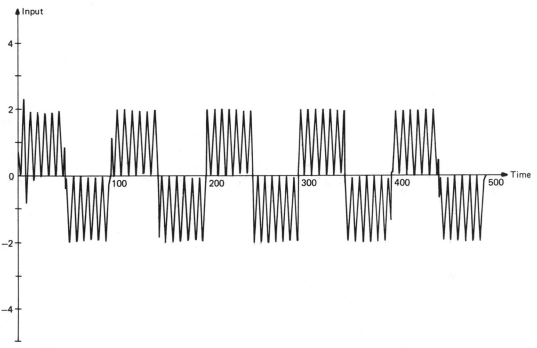

Figure 6.3.4 Input of Example 6.3.2.

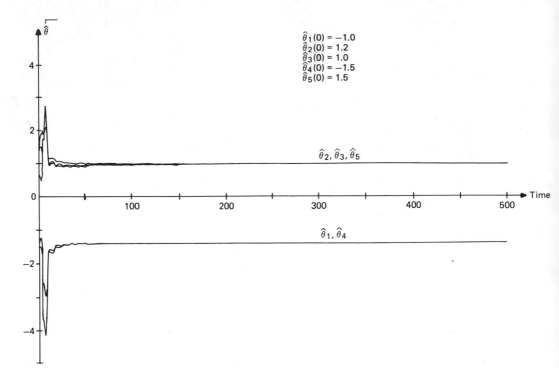

Figure 6.3.5 Estimated parameters of Example 6.3.2.

In the discussion above, we have shown how the projection algorithm and the least-squares algorithm can be employed in adaptive control. However, it should now be clear to the reader that any parameter estimation with suitable convergence properties can be used.

An important aspect that is common to all the algorithms discussed so far is the fact that the predictor form has been used to parameterize the model. This results in a convenient way of generating the control law. Other closely related parameterization can also achieve the same purpose as we illustrate below.

One-Step-Ahead Adaptive Control in Linear Control Form

The preceding algorithms in this chapter generated $u(t)$ from the equation

$$\phi(t)^T \hat{\theta}(t) = y^*(t + d) \tag{6.3.40}$$

This equation, as it stands, is linear in $\hat{\theta}(t)$. However, when rearranged to give an explicit expression for $u(t)$, it results in a control form that is nonlinear in $\hat{\theta}(t)$:

$$u(t) = \frac{1}{\hat{\theta}_n(t)} [-\hat{\theta}_1(t)y(t) \cdots -\hat{\theta}_n(t)y(t - n + 1) - \hat{\theta}_{n+2}(t)u(t - 1)$$
$$\cdots -\hat{\theta}_{n+m+d}(t)u(t - m - d + 1) + y^*(t + d)] \tag{6.3.41}$$

It is also possible to reparameterize the system so that $u(t)$ is a linear function of the estimated parameters, $\hat{\theta}(t)$:

$$u(t) = \phi'(t)^T \hat{\theta}(t) \tag{6.3.42}$$

The essential idea here is to rewrite the predictor (6.3.3) in the form

$$\frac{1}{\beta_0}y(t+d) = \frac{1}{\beta_0}[\alpha(q^{-1})]y(t) + \frac{1}{\beta_0}[\beta(q^{-1})]u(t) \qquad (6.3.43)$$

where $\beta_0 \neq 0$ as before.

Equation (6.3.43) can be rewritten as

$$\frac{1}{\beta_0}y(t+d) = \alpha'(q^{-1})y(t) + [\beta'(q^{-1}) + 1]u(t) \qquad (6.3.44)$$

where

$$\alpha'(q^{-1}) = \frac{1}{\beta_0}\alpha(q^{-1}) = \frac{\alpha_0}{\beta_0} + \frac{\alpha_1}{\beta_1}q^{-1} + \cdots + \frac{\alpha_{n-1}}{\beta_0}q^{-(n-1)}$$

$$= \alpha'_0 + \alpha'_1 q^{-1} + \cdots + \alpha'_{n-1}q^{-(n-1)} \qquad (6.3.45)$$

$$\beta'(q^{-1}) = \frac{1}{\beta_0}[\beta(q^{-1}) - \beta_0]$$

$$= \frac{1}{\beta_0}[\beta_1 q^{-1} + \cdots + \beta_{m+d-1}q^{-(m+d-1)}] \qquad (6.3.46)$$

$$= \beta'_1 q^{-1} + \cdots + \beta'_{m+d-1}q^{-(m+d-1)}$$

Thus the predictor (6.3.44) can be rewritten as

$$u(t) = \bar{\phi}(t)^T \theta_0 \qquad (6.3.47)$$

where

$$\theta_0^T = \left[\alpha'_0, \ldots, \alpha'_{n-1}, \beta'_1, \ldots, \beta'_{m+d-1}, \frac{1}{\beta_0}\right] \qquad (6.3.48)$$

$$\bar{\phi}(t)^T = [-y(t), \ldots, -y(t-n+1), -u(t-1), \ldots, -u(t-m-d+1), y(t+d)] \qquad (6.3.49)$$

The predictor (6.3.47) is simply a reparameterization of the system model and is valid regardless of how $\{u(t)\}$ is generated.

If θ_0 is known, $y(t+d)$ can be brought to $y^*(t+d)$ by replacing $y(t+d)$ by $y^*(t+d)$ in the reformulated model (6.3.47) to give the required control:

$$u^*(t) = \phi'(t)^T \theta_0 \qquad (6.3.50)$$

$$\phi'(t)^T = [-y(t), \ldots, -y(t-n+1), -u(t-1), \ldots,$$
$$-u(t-m-d+1), y^*(t+d)] \qquad (6.3.51)$$

(We assume that the sign of $|1/\beta_0|$, and the lower bound of $|1/\beta_0|$, are known.) We are thus led to the following adaptive control algorithm.

One-Step-Ahead Adaptive Control (Linear Control Form)

$$\hat{\theta}'(t) = \hat{\theta}(t-1) + \frac{\bar{\phi}(t-d)}{c + \bar{\phi}(t-d)^T\bar{\phi}(t-d)}[u(t-d) - \bar{\phi}(t-d)^T\hat{\theta}(t-1)]; \quad c > 0 \qquad (6.3.52)$$

where $\bar{\phi}(t-d)$ is as in (6.3.49). The control is generated from

$$u(t) = \phi'(t)^T\hat{\theta}(t) \qquad (6.3.53)$$

where $\phi'(t)$ is as in (6.3.51). The estimate of $1/\beta_0$ is kept away from zero using the standard constrained estimator; that is, if

$$[\hat{\theta}'_{n+m+d}(t)] \text{ sign} \left(\frac{1}{\beta_0}\right) > \left|\frac{1}{\beta_0}\right|_{\min}$$

then

$$\hat{\theta}(t) = \hat{\theta}'(t) \tag{6.3.54}$$

Otherwise,

$$\hat{\theta}_i(t) = \hat{\theta}'_i(t); \qquad i = 1, \ldots, n+m+d-1 \tag{6.3.55}$$

$$= \left|\frac{1}{\beta_0}\right|_{\min} \text{sign} \left(\frac{1}{\beta_0}\right); \qquad i = n+m+d \tag{6.3.56}$$

Note that the error term driving the parameter estimator (6.3.52) is

$$e(t) \triangleq u(t-d) - \bar{\phi}(t-d)^T\hat{\theta}(t-1) \tag{6.3.57}$$

$$= \phi'(t-d)^T\theta(t-d) - \bar{\phi}(t-d)^T\hat{\theta}(t-1) \qquad \text{using (6.3.53)}$$

$$= \phi'(t-d)^T\theta(t-d) - \phi'(t-d)^T\hat{\theta}(t-1)$$
$$\qquad\qquad\qquad + [\phi'(t-d) - \bar{\phi}(t-d)]^T\hat{\theta}(t-1) \tag{6.3.58}$$

$$= \hat{\theta}_{n+m+d}(t-1)[y^*(t) - y(t)] + \phi'(t-d)^T[\hat{\theta}(t-d) - \hat{\theta}(t-1)]$$

$$= -\hat{\theta}_{n+m+d}(t-1)\epsilon(t) + \tilde{\epsilon}(t)$$

where $\epsilon(t)$ is the output tracking error, $y(t) - y^*(t)$, and $\tilde{\epsilon}(t)$ is an *auxiliary signal*.

In practice, it is always preferable to use the form (6.3.57) for $e(t)$ rather than the derived form (6.3.58). This is especially true when there is input saturation since with input saturation (6.3.57) remains valid, whereas (6.3.53) may not be strictly true and thus (6.3.58) may no longer hold.

Global convergence can be established as before on noting that:

1. $|\hat{\theta}_{n+m+d}(t)| \geq \left|\frac{1}{\beta_0}\right|_{\min} > 0$ \hfill (6.3.59)

2. $e(t) = u(t-d) - \bar{\phi}(t-d)^T\hat{\theta}(t-1)$
$\qquad = -\bar{\phi}(t-d)^T\tilde{\theta}(t-1) \qquad$ using (6.3.47)
(This allows us to establish the counterpart of Lemma 3.3.2.)

3. $e(t) = -\hat{\theta}_{n+m+d}(t-1)\epsilon(t) + \tilde{\epsilon}(t)$ \hfill (6.3.60)
[see (6.3.58)] and thus

$$\frac{e(t)}{[c + \bar{\phi}(t-d)^T\bar{\phi}(t-d)]^{1/2}} = \frac{-\hat{\theta}_{n+m+d}(t-1)\epsilon(t)}{[c + \bar{\phi}(t-d)^T\bar{\phi}(t-d)]^{1/2}}$$
$$+ \frac{\phi'(t-d)^T[\hat{\theta}(t-d) - \hat{\theta}(t-1)]}{[c + \bar{\phi}(t-d)^T\bar{\phi}(t-d)]^{1/2}} \tag{6.3.61}$$

Then we argue as before to conclude

$$\lim_{t\to\infty} \frac{\hat{\theta}_{n+m+d}(t-1)\epsilon(t)}{[c + \bar{\phi}(t-d)^T\bar{\phi}(t-d)]^{1/2}} = 0 \tag{6.3.62}$$

or, using (6.3.59),

$$\lim_{t\to\infty} \left|\frac{1}{\beta_0}\right|_{\min} \frac{\epsilon(t)}{[c + \bar{\phi}(t-d)^T\bar{\phi}(t-d)]^{1/2}} = 0 \tag{6.3.63}$$

We can then apply Lemma 6.2.1 as before.

Remark 6.3.6. A closely related technique to the linear control form algorithms that has appeared in the literature is the *input matching* algorithms (Johnson and Tse, 1978; Goodwin, Johnson, and Sin, 1981). We have seen that, in the case of the linear control forms, the system model is rearranged so that the coefficient of $u(t)$ is 1. In the input matching algorithms, however, the model is left in its original form but the coefficient of $u(t)$ is set to a fixed value $\hat{\beta}_0$ (which is not estimated in the algorithm). This clearly requires that $\hat{\beta}_0$ be a good estimate of the true coefficient β_0. In fact, the theory shows that we require $0 < \beta_0/\hat{\beta}_0 < 2$ for convergence [see Goodwin, Ramadge, and Caines (1978a)]. Simulation experience (Redman, 1980) indicates that a practical range for $\beta_0/\hat{\beta}_0$ is in the range (0.5, 1.5). Some typical input matching algorithms are described in the exercises. We will not go into further details since these algorithms are minor variants of the basic design philosophy discussed above.

Weighted One-Step-Ahead Adaptive Control

The adaptive control algorithms developed above can be generalized to the case where a penalty is associated with the control effort by use of the following cost function:

$$J(t + d) = \frac{1}{2}[y(t + d) - y^*(t + d)]^2 + \frac{\lambda}{2}u(t)^2 \qquad (6.3.64)$$

The motivation for the cost function above is as in Section 5.2.2.

For a model of the system, we return to the predictor of (6.3.3) and rewrite it as

$$y(t + d) - \alpha(q^{-1})y(t) - \beta'(q^{-1})u(t - 1) - \beta_0 u(t) = 0 \qquad (6.3.65)$$

where

$$\beta'(q^{-1}) = q[\beta(q^{-1}) - \beta_0] \qquad (6.3.66)$$

We aim to manipulate (6.3.65) into a form such that the control law $u(t)$ minimizing (6.3.64) can be expressed as a linear function of the model parameters appearing in the resulting form (i.e., we are after a *direct* adaptive control algorithm). Multiplying (6.3.65) by $\beta_0/(\beta_0^2 + \lambda)$ gives

$$\frac{\beta_0}{\beta_0^2 + \lambda}[y(t + d) - \alpha(q^{-1})y(t) - \beta'(q^{-1})u(t - 1) - \beta_0 u(t)] = 0$$

Then adding and subtracting $[\lambda/(\beta_0^2 + \lambda)]u(t)$, we have

$$\frac{\beta_0}{\beta_0^2 + \lambda}\left[y(t + d) + \frac{\lambda}{\beta_0}u(t)\right] - \frac{\beta_0}{\beta_0^2 + \lambda}\alpha(q^{-1})y(t) - \frac{\beta_0\beta'(q^{-1})}{\beta_0^2 + \lambda}u(t - 1) - u(t) = 0 \qquad (6.3.67)$$

The foregoing special form of the output predictor (6.3.65) can be expressed as

$$u(t) = \bar{\phi}(t)^T \theta_0 \qquad (6.3.68)$$

where

$$\theta_0^T = \left[\frac{\beta_0}{\beta_0^2 + \lambda}, \frac{\beta_0\alpha_0}{\beta_0^2 + \lambda}, \cdots, \frac{\beta_0\alpha_{n-1}}{\beta_0^2 + \lambda}, \frac{\beta_0\beta_1}{\beta_0^2 + \lambda}, \cdots, \frac{\beta_0\beta_{m+d-1}}{\beta_0^2 + \lambda}\right] \qquad (6.3.69)$$

$$\bar{\phi}(t)^T = [y(t + d) + \gamma u(t), -y(t), \ldots, -y(t - n + 1),$$
$$-u(t - 1), \ldots, -u(t - m - d + 1)] \qquad (6.3.70)$$

$$\gamma \triangleq \frac{\lambda}{\beta_0} \qquad (6.3.71)$$

Equation (6.3.68) will be used as the system model for the parameter estimator (i.e., θ_0 will be estimated).

Now from Theorem 5.2.2, the control law minimizing (6.3.64) can be expressed in the form [see (5.2.13)]

$$u^*(t) = \phi(t)^T \theta_0 \tag{6.3.72}$$

$$\phi(t)^T = [y^*(t + d), -y(t), \ldots, -y(t - n + 1), -u(t - 1),$$
$$\ldots, -u(t - m - d + 1)] \tag{6.3.73}$$

As before, we impose mild restrictions on β_0. In particular, we assume that the sign of β_0 is known, and a lower bound is available for $|\beta_0/(\beta_0^2 + \lambda)|$. We denote the lower bound by $(\theta_1)_{\min} > 0$. We then have the following algorithm (for a given γ):

Weighted One-Step-Ahead Adaptive Controller

$$\hat{\theta}'(t) = \hat{\theta}(t - 1) + \frac{\bar{\phi}(t - d)}{c + \bar{\phi}(t - d)^T \bar{\phi}(t - d)}[u(t - d) - \bar{\phi}(t - d)^T \hat{\theta}(t - 1)]; \quad c > 0 \tag{6.3.74}$$

where $\bar{\phi}(t)$ is as in (6.3.70). If

$$[\hat{\theta}'_1(t)] \text{ sign } \beta_0 \geq (\theta_1)_{\min}$$

then

$$\hat{\theta}(t) = \theta'(t) \tag{6.3.75}$$

Otherwise,

$$\hat{\theta}_i(t) = \hat{\theta}'_i(t); \quad i = 2, \ldots, n + m + d \tag{6.3.76}$$

$$= (\theta_1)_{\min} \text{ sign } \beta_0; \quad i = 1 \tag{6.3.77}$$

The control law is generated from

$$u(t) = \phi(t)^T \hat{\theta}(t) \tag{6.3.78}$$

where $\phi(t)$ is as in (6.3.73).

▼▼▼

Convergence properties for the parameter estimator (6.3.74) can be established immediately as in Lemma 3.3.2 on substituting (6.3.68) into (6.3.74). We note that these properties are valid irrespective of the form of the control law.

The use of the control law above allows us to relax Assumption 6.3.A to the following:

Assumption 6.3.B

1. The time delay d is known.
2. An upper bound for the orders of the polynomials in (6.3.67) is known.
3. (i) The zeros of the polynomial $[B'(z^{-1}) + (\lambda/\beta_0)A(z^{-1})]$ lie inside or on the unit circle.
   (ii) The zeros of the following two transfer functions lie strictly inside the unit circle.

$$T_1(z) = \frac{B'(z^{-1}) + (\lambda/\beta_0)A(z^{-1})}{B'(z^{-1})}$$

$$T_2(z) = \frac{B'(z^{-1}) + (\lambda/\beta_0)A(z^{-1})}{A(z^{-1})}$$

(iii) Any modes of $[B'(q^{-1}) + (\lambda/\beta_0)A(q^{-1})]$ on the unit circle have a Jordan block size of 1.

The assumptions above are very natural since they correspond to the conditions needed to ensure stability in the nonadaptive case. (See Theorem 5.2.2.)

We can now establish the following global convergence result:

Theorem 6.3.2. Let γ be a specified scalar (with the same sign as β_0). Subject to Assumption 6.3.B, the algorithm (6.3.74) to (6.3.78) when applied to the system (6.3.3) leads to a stable closed-loop system with the properties:

(a) $u(t)$, $y(t)$ bounded for all t (6.3.79)

(b) $\lim_{t \to \infty} [u(t) - u^*(t)] = 0$ (6.3.80)

where $u^*(t)$ is the one-step-ahead optimal input for the system when the cost function is chosen as in (6.3.64) with $\lambda = \gamma\beta_0$.

Proof. We proceed much as in the proof of Theorem 6.3.1. We first recall results (a) and (e) of Lemma 3.3.2:

$$\lim_{t \to \infty} \frac{\bar\phi(t-d)^T\tilde\theta(t-1)}{[c + \bar\phi(t-d)^T\bar\phi(t-d)]^{1/2}} = 0 \tag{6.3.81}$$

and

$$\lim_{t \to \infty} \|\hat\theta(t) - \hat\theta(t-k)\| = 0 \tag{6.3.82}$$

Thus

$$\lim_{t \to \infty} \frac{\bar\phi(t-d)^T\tilde\theta(t-d)}{[c + \bar\phi(t-d)^T\bar\phi(t-d)]^{1/2}} = 0 \tag{6.3.83}$$

Now

$$\begin{aligned}
\bar\phi(t-d)^T\tilde\theta(t-d) &= \bar\phi(t-d)^T\hat\theta(t-d) - \bar\phi(t-d)^T\theta_0 \\
&= \bar\phi(t-d)^T\hat\theta(t-d) - u(t-d) \quad \text{using (6.3.68)} \\
&= \bar\phi(t-d)^T\hat\theta(t-d) - \phi(t-d)^T\hat\theta(t-d) \quad \text{using (6.3.78)} \\
&= \hat\theta_1(t-d)[y(t) + \gamma u(t-d) - y^*(t)]
\end{aligned} \tag{6.3.84}$$

where $|\hat\theta_1(t-d)| > (\theta_1)_{\min}$ from (6.3.75) to (6.3.77). Hence it immediately follows that

$$\left| \frac{\bar\phi(t-d)^T\tilde\theta(t-d)}{\hat\theta_1(t-d)[c + \bar\phi(t-d)^T\bar\phi(t-d)]} \right| \leq \left| \frac{\bar\phi(t-d)\tilde\theta(t-d)}{(\theta_1)_{\min}[c + \bar\phi(t-d)^T\bar\phi(t-d)]} \right|$$

and using (6.3.83) and (6.3.84),

$$\lim_{t \to \infty} \frac{v(t)}{[c + \bar\phi(t-d)^T\bar\phi(t-d)]^{1/2}} = 0$$

where

$$v(t) = y(t) + \gamma u(t-d) - y^*(t) \tag{6.3.85}$$

Operating on the left of (6.3.85) by $A(q^{-1})$ leads to

$$A(q^{-1})v(t) = A(q^{-1})y(t) + \gamma A(q^{-1})u(t-d) - A(q^{-1})y^*(t)$$
$$= [B'(q^{-1}) + \gamma A(q^{-1})]u(t-d) - A(q^{-1})y^*(t)$$

Similarly operating on (6.3.85) by $B'(q^{-1})$ gives

$$B'(q^{-1})v(t) = [B'(q^{-1}) + \gamma A(q^{-1})]y(t) - B'(q^{-1})y^*(t)$$

Since $y^*(t)$ is a bounded sequence, it follows from Assumption 6.3.B, part 3 and the equations above that the growth of $\{y(t)\}$ and $\{u(t-d)\}$ is stably related to the growth of $v(t)$. Hence we can apply the key technical lemma (Lemma 6.2.1) to conclude (6.3.79) and

$$\lim_{t \to \infty} v(t) = 0$$

But from (6.3.68) and (6.3.72),

$$u(t) - u^*(t) = (\theta_0)_1[y(t+d) + \gamma u(t) - y^*(t+d)]$$
$$= (\theta_0)_1 v(t+d)$$

Hence (6.3.80) follows.

▼▼▼

A simple generalization of Theorem 6.3.2 can be made to the case where we use the cost function (see Section 5.2.1)

$$J_3 = \frac{1}{2}[y(t+d) - y^*(t+d)]^2 + \frac{\lambda}{2}\bar{u}(t)^2 \qquad (6.3.86)$$

where $\{\bar{u}(t)\}$ is related to $\{u(t)\}$ by a given rational transfer function

$$P(q^{-1})\bar{u}(t) = R(q^{-1})u(t) \qquad (6.3.87)$$

with $P(q^{-1})$ and $R(q^{-1})$ monic.

As might be expected, following the discussion in Section 5.2, the conditions for global convergence of the appropriate one-step-ahead control law now relate to the zeros of

$$P(q^{-1})B'(q^{-1}) + \gamma R(q^{-1})A(q^{-1}) \qquad (6.3.88)$$

The appropriate adaptive control law is as in (6.3.74) to (6.3.77), with the control signal now being generated from

$$u(t) = \phi'(t)^T \hat{\theta}(t) \qquad (6.3.89)$$

where

$$\phi'(t) = \{y^*(t+d) + \gamma[P'(q^{-1})\bar{u}(t-1) - R'(q^{-1})u(t-1)], -y(t), \ldots, \qquad (6.3.90)$$
$$-y(t-n+1), -u(t-1), \ldots, -u(t-m-d+1)\}$$

with

$$P'(q^{-1}) = q[P(q^{-1}) - 1] \qquad (6.3.91)$$
$$R'(q^{-1}) = q[R(q^{-1}) - 1] \qquad (6.3.92)$$
$$\bar{u}(t) = -P'(q^{-1})\bar{u}(t-1) + R'(q^{-1})u(t-1) + u(t) \qquad (6.3.93)$$

Global convergence follows as before (Exercise 6.5) on noting from (6.3.68), (6.3.70) (6.3.89), and (6.3.90) that

$$\bar{\phi}(t-d)^T\hat{\theta}(t-d) = \hat{\theta}_1(t-d)\{y(t) + \gamma u(t-d) - y^*(t)$$
$$- \gamma[P'(q^{-1})\bar{u}(t-d-1) - R'(q^{-1})u(t-d-1)]\} \quad (6.3.94)$$
$$= \hat{\theta}_1(t-d)v(t)$$

where

$$v(t) = y(t) + \gamma\bar{u}(t-d) - y^*(t) \quad (6.3.95)$$

Remark 6.3.7. Note that γ is assumed specified in the algorithm above. Then the corresponding value of λ is given by (6.3.71) as $\lambda = \gamma\beta_0$. We require that $\lambda > 0$. and hence γ must have the same sign as β_0.

▼▼▼

Remark 6.3.8. As we have seen in Remark 5.2.1, part 2, a special case of the generalization above that has practical significance is when we take

$$P(q^{-1}) = 1 - \gamma q^{-1}; \quad 0 < \gamma < 1 \quad (6.3.96)$$
$$R(q^{-1}) = 1 - \gamma^{-1}$$

This choice of $P(q^{-1})$ and $R(q^{-1})$ leads to a controller with integral action. Thus there will be no steady-state errors for constant desired sequences $\{y^*(t)\}$. Of course, care must be taken with integral wind-up as usual.

▼▼▼

Closely related algorithms to that described above have been described by Clarke et al. (1973), Clarke and Gawthrop (1975), Johnson and Tse (1978), and Goodwin, Johnson, and Sin (1981).

Following our discussion in Section 5.2.2, it is straightforward to introduce a reference model and hence give more flexibility in the assignment of the closed-loop poles. Similar results to the algorithms above can be established. We shall look at these in more detail in the next section.

6.3.2 Model Reference Adaptive Control

In this section we consider a slightly different problem formulation in which the desired output sequence is generated by a linear reference model. (Previously, the desired output could be an arbitrary sequence.) We shall find that the reference model may be used to advantage in the design of the adaptive controller.

Essentially what we will be doing here is developing adaptive forms of the model reference control laws described in Section 5.2.2. We shall see that this approach fits naturally into the general framework within which we have discussed the previous algorithms.

As before, the system model is described by (6.3.1):

$$\bar{A}(q^{-1})y(t) = q^{-d}\bar{B}'(q^{-1})u(t) \quad (6.3.97)$$

where

$$\bar{A}(q^{-1}) = 1 + \bar{a}_1q^{-1} + \cdots + \bar{q}_\bar{n}q^{-\bar{n}}$$
$$\bar{B}'(q^{-1}) = \bar{b}_0 + \bar{b}_1q^{-1} + \cdots + \bar{b}_\bar{m}q^{-\bar{m}}; \quad \bar{b}_0 = b_0 \neq 0$$

As usual we do not assume that $\bar{A}(q^{-1})$ and $\bar{B}'(q^{-1})$ are relatively prime since we wish to include uncontrollable disturbances in our model.

We introduce the usual stable inverse and order assumptions as in Assumption 6.3.A. In particular, we shall make use of the overparameterized model from (6.3.2),

$$A(q^{-1})y(t) = q^{-d}B'(q^{-1})u(t) \tag{6.3.98}$$

where $A(q^{-1})$ and $B'(q^{-1})$ are of order n and m, respectively, where $n \geq \bar{n}$ and $m \geq \bar{m}$. We recall the reference model assumptions from Section 5.2.2:

Reference Model Assumption

1. The desired output $y^*(t)$ satisfies the following reference model:

$$E(q^{-1})y^*(t) = q^{-d'}gH(q^{-1})r(t) \tag{6.3.99}$$

with associated transfer function $G(z) = z^{-d'}H(z^{-1})g/E(z^{-1})$, where g is a constant gain and

$$H(z^{-1}) = h_0 + h_1 z^{-1} + \cdots + h_l z^{-l}; \qquad h_0 = 1 \tag{6.3.100}$$

$$E(z^{-1}) = e_0 + e_1 z^{-1} + \cdots + e_l z^{-l}; \qquad e_0 = 1 \tag{6.3.101}$$

subject to

2. $E(z^{-1})$ is stable.
3. $d' = d$

We also recall that the system model (6.3.98) can be expressed in predictor form for $E(q^{-1})y(t + d)$ as [see (5.2.31)]

$$E(q^{-1})y(t + d) = \alpha(q^{-1})y(t) + \beta(q^{-1})u(t) \tag{6.3.102}$$

where as in Lemma 5.2.1,

$$\alpha(q^{-1}) = G(q^{-1}), \qquad \beta(q^{-1}) = F(q^{-1})B'(q^{-1}) \tag{6.3.103}$$

and

$$E(q^{-1}) = F(q^{-1})A(q^{-1}) + q^{-d}G(q^{-1}) \tag{6.3.104}$$

and the corresponding (optimal) model reference control law [see (5.2.33)]

$$\alpha(q^{-1})y(t) + \beta(q^{-1})u^*(t) = gH(q^{-1})r(t) \tag{6.3.105}$$

The feedback control law (6.3.105) depends on knowledge of the system parameters. We shall now see how the controller can be made adaptive. Any of the adaptive control algorithms of the previous sections can be employed. We illustrate by using the one-step-ahead adaptive controller (linear control form) of (6.3.52)–(6.3.56).

As before, the essential idea is to rewrite the predictor (6.3.102) so that it is linear in $u(t)$. We do this by dividing by β_0 to give

$$\frac{1}{\beta_0}E(q^{-1})y(t + d) = \frac{1}{\beta_0}\alpha(q^{-1})y(t) + \frac{1}{\beta_0}\beta(q^{-1})u(t) \tag{6.3.106}$$

$$= \alpha'(q^{-1})y(t) + [\beta'(q^{-1}) + 1]u(t) \tag{6.3.107}$$

where

$$\alpha'(q^{-1}) = \frac{\alpha_0}{\beta_0} + \frac{\alpha_1}{\beta_0}q^{-1} + \cdots + \frac{\alpha_{n-1}}{\beta_0}q^{-(n-1)} \tag{6.3.108}$$

$$= \alpha'_0 + \alpha'_1 q^{-1} + \cdots + \alpha'_{n-1}q^{-(n-1)}$$

Similarly,

$$\beta'(q^{-1}) = \beta'_1 q^{-1} + \cdots + \beta'_{m+d-1} q^{-(m+d-1)} \tag{6.3.109}$$

Thus the predictor (6.3.106) can be written as

$$u(t) = \bar{\phi}(t)^T \theta_0 \tag{6.3.110}$$

where

$$\theta_0^T = \left[\alpha'_0, \ldots, \alpha'_{n-1}, \beta'_1, \ldots, \beta'_{m+d-1}, \frac{1}{\beta_0} \right] \tag{6.3.111}$$

$$\bar{\phi}(t)^T = [-y(t), \ldots, -y(t-n+1), -u(t-1), \ldots, \\ -u(t-m-d+1), y_a(t+d)] \tag{6.3.112}$$

$$y_a(t+d) = E(q^{-1}) y(t+d) \tag{6.3.113}$$

When θ_0 is known, the control law (6.3.105) can be written as

$$u^*(t) = \phi'(t)^T \theta_0 \tag{6.3.114}$$

where

$$\phi'(t) = [-y(t), \ldots, -y(t-n+1), -u(t-1), \ldots, \\ -u(t-m-d+1), r_a(t)] \tag{6.3.115}$$

with

$$r_a(t) = gH(q^{-1}) r(t) \tag{6.3.116}$$

For the purpose of the convergence analysis, we shall require that the sign of $1/\beta_0$ and a lower bound for $1/\beta_0$ be known.

We now have the following straightforward generalization of the one-step-ahead adaptive controller (linear control form):

Model Reference Adaptive Controller (Linear Control Form)

$$\hat{\theta}'(t) = \hat{\theta}(t-1) + \frac{\bar{\phi}(t-d)}{c + \bar{\phi}(t-d)^T \bar{\phi}(t-d)} [u(t-d) - \bar{\phi}(t-d)^T \hat{\theta}(t-1)] \tag{6.3.117}$$

where $\bar{\phi}(t-d)$ is as in (6.3.112). The control is then generated from

$$u(t) = \phi'(t)^T \hat{\theta}(t) \tag{6.3.118}$$

where $\phi'(t)$ is as in (6.3.115). The usual constrained estimation technique can be used to ensure that the estimate of $1/\beta_0 \neq 0$.

[Alternative (more complicated algorithms) for model reference adaptive control may be found in Morse (1980), Narendra and Lin (1980), and Narendra, Lin, and Valavani (1980). The additional complications arise from the fact that various filtering operations are included as is necessary in the continuous-time case but *not* in the discrete-time case.]

Global convergence follows as before using the arguments presented for the one-step adaptive controllers in the preceding subsection. It is also straightforward to develop a least-squares iteration for the algorithm above. We leave it to the reader to supply the details.

6.3.3 One-Step-Ahead Adaptive Controllers
for Multi-input Multi-output Systems

All of the adaptive algorithms discussed so far can be extended in a straightforward fashion to the multi-input multi-output case by combining the parameter estimation algorithm of Section 3.8 with the multivariable control laws of Section 5.2. The multivariable case of course has a much richer structure but the essential simplicity of the adaptive algorithms is retained, as we shall presently see. We shall describe the one-step-ahead controller to illustrate the principles involved.

Consider a multi-input multi-output system described in DARMA form:

$$A(q^{-1})y(t) = B(q^{-1})u(t) \qquad (6.3.119)$$

where $\{u(t)\}$ and $\{y(t)\}$ denote the $r \times 1$ and $m \times 1$ input and output vectors, respectively. In the sequel we will take $r = m$ since, in the light of Lemma 5.2.2, output tracking is possible only when $r \geq m$, and if $r > m$, some of the inputs can be discarded without loss of generality. The reader will recall that a consequence of Lemma 5.2.2 is that the following assumption is necessary to achieve output tracking.

Assumption 6.3.M(1). The transfer function $T(z) \triangleq A(z^{-1})^{-1}B(z^{-1})$ has rank m.

▼▼▼

It then follows from Lemma 5.2.3, that associated with $T(z)$ there is an interactor matrix $\xi(z)$ such that

1. $\xi(z) = H(z) \operatorname{diag}[z^{f_1} \cdots z^{f_m}]$ \qquad (6.3.120)

$$H(z) = \begin{bmatrix} 1 & & & \\ h_{21}(z) & & & 0 \\ & \ddots & & \\ & & \ddots & \\ h_{m1}(z) & h_{m2}(z) & \cdots & 1 \end{bmatrix} \qquad (6.3.121)$$

and $h_{ij}(z)$ is divisible by z or is zero.

2. $\lim\limits_{z \to \infty} \xi(z)T(z) = K$ nonsingular \qquad (6.3.122)

Also from Lemma 5.2.4 we know that $[\xi(z)]^{-1}$ is a stable operator and provided that $T(z)$ is strictly proper, then $f_i \geq 1, i = 1, \ldots, m$.

Now from Theorem 5.2.4, there exists a positive integer d [= maximum order of any polynomial in $\xi(z)$] such that, with

$$\bar{y}(t) \triangleq \xi(q)y(t) \qquad (6.3.123)$$

$\bar{y}(t)$ is a known function of $y(t + d), y(t + d - 1), \ldots,$ and can be described by

$$\bar{y}(t) = \alpha(q^{-1})y(t) + \beta(q^{-1})u(t) \qquad (6.3.124)$$

where

$$\alpha(q^{-1}) = \alpha_0 + \alpha_1 q^{-1} + \cdots + \alpha_{n_i-1} q^{-n_i-1} \qquad (6.3.125)$$

$$\beta(q^{-1}) = \beta_0 + \beta_1 q^{-1} + \cdots + \beta_{n_i-1} q^{-n_i-1} \qquad (6.3.126)$$

$$\beta_0 \text{ nonsingular [see (5.2.67)]} \qquad (6.3.127)$$

We recall that the interactor matrix is the multivariable equivalent of the delay in the single-input single-output case. Since in the single-input single-output case our standing assumption was that the delay was known, it is natural in the multivariable case to introduce the following assumption:

Assumption 6.3.M(2). $\xi(z)$ is known.

▼▼▼

As before, we also introduce the following assumption, which essentially ensures that the adaptive controller has sufficient degrees of freedom:

Assumption 6.3.M(3). Upper bounds are available for the orders of the polynomials in (6.3.125)–(6.3.126).

We have seen in Section 5.2.1 that one-step-ahead controllers lead to bounded inputs only when the system is "stably invertible." It is therefore natural to introduce the following assumptions for adaptive control.

Assumption 6.3.M(4). The system is stably invertible in the sense described in Theorem 5.2.5, part (d).

▼▼▼

As usual, the control objective is to maintain $\{y(t)\}$ and $\{u(t)\}$ bounded and to asymptotically achieve output tracking: that is, we require that

$$\lim_{t \to \infty} [y_i(t) - y_i^*(t)] = 0; \qquad i = 1, \ldots, m \qquad (6.3.128)$$

where $y^*(t)$ denotes an $m \times 1$ bounded reference sequence. As in Section 5.2.3 we define

$$\bar{y}^*(t) = \xi(q) y^*(t) \qquad (6.3.129)$$

and note that $\bar{y}^*(t)$ can be computed from knowledge of $\{y^*(t)\}$ and the interactor matrix. In the single-input single-output case, $\bar{y}^*(t)$ is simply $y^*(t + d)$. Clearly, $\bar{y}^*(t)$ must be known, or be computable, at time t to achieve tracking.

We shall now describe the multi-input multi-output equivalent of the one step ahead adaptive controller of Section 6.3.1.

Equation (6.3.124) can be written in the form

$$\bar{y}(t) = \theta_0^T \phi(t) \qquad (6.3.130)$$

where θ_0 is an $m \times (n_1' + n_2')$ matrix consisting of the coefficients of $[\alpha(q^{-1}), \beta(q^{-1})]$. $\phi(t)$ is an $(n_1' + n_2') \times 1$ column vector defined by

$$\phi(t) = [y(t)^T, y(t - 1)^T, \ldots, u(t)^T, u(t - 1)^T, \ldots]^T$$

these components being available at time t.

Referring to the parameter estimation algorithm described in (3.8.19), we are led to the following adaptive control algorithm.

MIMO One-Step-Ahead Adaptive Controller

$$\hat{\theta}(t) = \hat{\theta}(t - 1) + \frac{a(t)\phi(t - d)[y(t)^T - \phi(t - d)^T\hat{\theta}(t - 1)]}{c + \phi(t - d)^T\phi(t - d)}; \qquad c > 0 \qquad (6.3.131)$$

$$\bar{y}^*(t) = \hat{\theta}(t)^T\phi(t) \qquad (6.3.132)$$

Note that (6.3.132) represents a feedback control law, as can be seen by rearranging the equation so that $u(t)$ is the subject of the formula. Solvability of this equation requires that the matrix of coefficients multiplying $u(t)$ be nonsingular for all t. Singularity of the matrix is a zero probability event and $a(t)$ can thus be chosen as 1 in almost all cases as in Section 6.3.1. Alternatively, we have the following technical result:

Lemma 6.3.1. Provided that $\hat{\theta}(0)$ is chosen so that the coefficient matrix of $u(0)$ is nonsingular, then $a(t)$ can be chosen in the range $\epsilon < a(t) < 2 - \epsilon, 0 < \epsilon < 1$, so that the coefficient of $u(t)$ is nonsingular for all t.

Proof. The proof of this result in rather involved and since it is not really of interest in practice, we refer the reader to the proof given in Goodwin, Ramadge, and Caines (1978a).

▼▼▼

Global convergence of the algorithm (6.3.131)–(6.3.132) is established in the following theorem [see Goodwin and Long (1980)]:

Theorem 6.3.3. Subject to Assumptions 6.3.M(1) to 6.3.M(4), if the algorithm (6.3.131)–(6.3.132) is used, then $\{y(t)\}$ and $\{u(t)\}$ are bounded for all time and each output asymptotically tracks the corresponding desired output sequence, that is,

$$\lim_{t \to \infty} |y_i(t) - y_i^*(t)| = 0 \qquad \text{for } i = 1, \ldots, m \qquad (6.3.133)$$

Proof. From Lemma 3.8.1 we have the following fundamental property of the iteration (6.3.131):

$$\lim_{t \to \infty} \frac{e_i(t)}{[c + \phi(t - d)^T \phi(t - d)]^{1/2}} = 0; \qquad c > 0 \qquad (6.3.134)$$

where

$$e_i(t) = -\tilde{\theta}_{i.}(t - d)^T \phi(t - d) \qquad (6.3.135)$$
$$= \bar{y}_i(t - d) - \bar{y}_i^*(t - d) \qquad (6.3.136)$$

where

$$\tilde{\theta}_{i.}(t) = \hat{\theta}_{i.}(t) - [\theta_0]_i. \qquad (6.3.137)$$

and $\theta_{i.}$ denotes the ith row of θ^T and $\bar{y}_i$ denotes the ith element of $\bar{y}$.

The proof now follows that of Theorem 6.3.1 except in the case that the vector $\{\bar{y}(t)\}$ may be unbounded. In this case there exists a subsequence $\{t_n\}$ such that

$$\lim_{t_n \to \infty} \|\bar{y}(t_n)\| = \infty$$

and

$$|\bar{y}_i(t)| \leq |\bar{y}_{J(t_n)}(t_n)| \qquad \text{for some } 1 \leq j(t_n) \leq m$$

and for all $1 \leq i \leq m$ and all $t \leq t_n$. [The times t_n denote the points of increase of the vector sequence $\{\|\bar{y}(t)\|\}$ and at time t_n it is the jth component of $|\bar{y}(t_n)|$ which is the largest. Note that j depends on t_n.] It then follows from the inverse stability of the system and the stability of $\xi(z)$ that there exist constants $0 \leq C_1 < \infty$ and $0 < C_2 < \infty$ such that

$$\|\phi(t_n)\| \leq C_1 + C_2 |\bar{y}_{J(t_n)}(t_n)| \qquad \text{for some } 1 \leq j(t_n) \leq m$$

Since m is finite, there exists a further subsequence $\{t_{n'}\}$ of the subsequence $\{t_n\}$ such that

$$\|\phi(t_{n'})\| \leq C_1 + C_2 |\bar{y}_i(t_{n'})| \qquad \text{for at least one } i, \quad 1 \leq i \leq m$$

and $\{\bar{y}_i(t_{n'})\}$ is unbounded. The remainder of the proof then follows that of Theorem 6.3.1 using the key technical lemma (Lemma 6.2.1), where we note that

$$|e_i(t)| = |\bar{y}_i(t-d) - \bar{y}_i^*(t-d)|$$

giving $\{y(t)\}$ and $\{u(t)\}$ bounded and

$$\lim_{t \to \infty} |\bar{y}_i(t) - \bar{y}_i^*(t)| = 0 \qquad \text{for } i = 1, \ldots, m$$

Finally, (6.3.133) follows from (6.3.123) and (6.3.129) and the stability of $\xi(z)^{-1}$.

▼▼▼

A least-squares iteration can also be used to replace the parameter estimator (6.3.131), giving the following algorithm:

MIMO One-Step-Ahead Adaptive Controller (Least-Squares Iteration)

$$\hat{\theta}(t) = \hat{\theta}(t-1) + \frac{P(t-d-1)\phi(t-d)[y(t)^T - \phi(t-d)^T\hat{\theta}(t-1)]}{1 + \phi(t-d)^T P(t-d-1)\phi(t-d)} \tag{6.3.138}$$

$$P(t) = P(t-1) - \frac{P(t-1)\phi(t)\phi(t)^T P(t-1)}{1 + \phi(t)^T P(t-1)\phi(t)} \tag{6.3.139}$$

$$P(0) = \epsilon I; \qquad \epsilon > 0 \tag{6.3.140}$$

$$\bar{y}^*(t) = \hat{\theta}(t)^T\phi(t) \tag{6.3.141}$$

Global convergence follows for this algorithm using the techniques which should by now be familiar to the reader.

▼▼▼

As in the scalar case, other extensions of the basic algorithm are possible. For example, weighted one-step-ahead and model reference control laws can be developed. We will not go into details since the results follow naturally from the single-input single-output case. Some of the extensions are explored in the exercises.

The analysis above has assumed knowledge of the system interactor matrix. This is reasonable when the interactor has a simple structure; for example, when it is diagonal. However, in the general case, the interactor contains real variables and it is difficult to see how these variables could be known a priori. In the following we present a straightforward extension of the result above in which the real variables appearing in the interactor matrix are estimated along with the other parameters appearing in the system model. This approach follows a suggestion originally made by Johansson (1982).

Consider the predictor model (6.3.124). We write this as

$$H(q)D(q)y(t) = \alpha(q^{-1})y(t) + \beta(q^{-1})u(t) \tag{6.3.142}$$

where $D(q) = \text{diag}\,[q^{f_1} \cdots q^{f_m}]$. Using the structure of $H(q)$

$$D(q)y(t) = L(q)D(q)y(t) + \alpha(q^{-1})y(t) + \beta(q^{-1})u(t) \tag{6.3.143}$$

where

$$L(q) = [I - H(q)]$$

$$= \begin{bmatrix} 0 & & & & \\ & \cdot & \cdot & & \\ & & \cdot & \cdot & \\ -h_{21}(q) & & & \cdot & \cdot \\ & & & & \cdot \\ -h_{m1}(q) & \cdots & -h_{mm-1}(q) & & 0 \end{bmatrix} \qquad (6.3.144)$$

Equation (6.3.143) can be written in regression form as:

$$i\text{th element of } [D(q)y(t)] = \phi_i(t + \ell_i)^T \theta_i^0; \qquad i = 1, \ldots, m \qquad (6.3.145)$$

Using the definition of $D(q)$, we have

$$i\text{th element of } D(q)y(t) = y_i(t + f_i) = \phi_i(t + \ell_i)^T \theta_i^0; \qquad i = 1, \ldots, m \qquad (6.3.146)$$

In equation (6.3.145), $\phi_i(t + \ell_i)^T$ is a vector of values of $y(\cdot)$ up to time $(t + \ell_i)$ and of $u(\cdot)$ up to time t where ℓ_i is the maximum forward shift in the ith row of $L(q)D(q)$. θ_i^0 is a vector of coefficients in the ith row of $L(q), \alpha(q^{-1})$ and $\beta(q^{-1})$. (Note, in general $\ell_i \geq f_i; i = 2, \ldots, m$.)

The adaptive control law can now be written as follows: (We illustrate by using the projection algorithm but naturally the same results would apply to least squares and other related algorithms).

$$\hat{\theta}_i(t) = \hat{\theta}_i(t - 1) + \frac{a(t)\phi_i(t)}{c + \phi_i(t)^T\phi_i(t)}[y_i(t + f_i - \ell_i) - \phi_i(t)^T\hat{\theta}_i(t - 1)] \qquad (6.3.147)$$
$$c > 0; \qquad i = 1, \ldots, m$$

The estimated vectors $\{\hat{\theta}_i(t)\}$ are now regrouped to form

$$\hat{\alpha}(t, q^{-1}), \hat{\beta}(t, q^{-1}) \text{ and } \hat{\xi}(t, q) \triangleq (I - \hat{L}(t, q)D(q)).$$

The feedback control signal $u(t)$ is then generated by solving

$$\hat{\beta}(t, q^{-1})u(t) + \hat{\alpha}(t, q^{-1})y(t) = \hat{\xi}(t, q^{-1})y^*(t) \qquad (6.3.148)$$

The scalar coefficient $a(t)$ in equation 6.3.147 is chosen as in Lemma 6.3.1 to ensure that $\hat{\beta}_0(t)$ (the estimate of the leading coefficient matrix of $\beta(q^{-1})$) is nonsingular for all finite t. This ensures that (6.3.148) can be solved for $u(t)$.

We then have the following global convergence result.

Theorem 6.3.4. Subject to Assumptions 6.3.M(1), 6.3.M(3), 6.3.M(4), then the adaptive control algorithm (6.3.147), (6.3.148) with $a(t)$ chosen as in Lemma 6.3.1 when applied to the system (6.3.119) ensures that

(a) $[u(t)], [y(t)]$ are bounded for all t; and
(b) $\lim_{t \to \infty} [y(t) - y^*(t)] = 0$

Proof. [See Dugard, Goodwin and de Souza (1983)]
The reader is encouraged to supply the details which follow closely the proof of theorem 6.3.3.

▼▼▼

In conclusion, we see that the minimum prediction error adaptive control algorithms extend in a very natural and straightforward way to the multi-input multi-output case provided that a suitable parameterization of the model is used.

6.4 MINIMUM PREDICTION ERROR ADAPTIVE CONTROLLERS (INDIRECT APPROACH)

The basic idea here is to fit a standard model (not necessarily the predictor) to the data and then to determine the control law by carrying out the design calculations as in Section 5.2. We shall develop an indirect form of the model reference adaptive controller for the SISO case to illustrate the procedure.

As before, we assume that the system is described by a DARMA model of the form

$$A(q^{-1})y(t) = q^{-d}B'(q^{-1})u(t) = B(q^{-1})u(t) \tag{6.4.1}$$

We then estimate the parameters in $A(q^{-1})$ and $B(q^{-1})$ using any of the standard on-line algorithms. Given $\hat{\theta}(t)$, we form $\hat{A}(t, q^{-1})$ and $\hat{B}'(t, q^{-1})$ from the coefficients as follows:

$$\hat{A}(t, q^{-1}) = 1 + \hat{a}_1(t)q^{-1} + \cdots + \hat{a}_n(t)q^{-n} \tag{6.4.2}$$

$$\hat{B}'(t, q^{-1}) = \hat{b}_0(t) + \cdots + \hat{b}_m(t)q^{-m} \tag{6.4.3}$$

Next we determine $\hat{F}(t, q^{-1})$ and $\hat{G}(t, q^{-1})$ by solving the usual prediction equality [see (5.2.32)]

$$\hat{F}(t, q^{-1})\hat{A}(t, q^{-1}) + q^{-d}\hat{G}(t, q^{-1}) = E(q^{-1}) \tag{6.4.4}$$

where $E(q^{-1})$ is the denominator polynomial in the reference model [see (5.2.28)].

Then $\alpha(t, q^{-1})$ and $\beta(t, q^{-1})$ are evaluated as follows:

$$\alpha(t, q^{-1}) = \hat{G}(t, q^{-1}) \tag{6.4.5}$$

$$\beta(t, q^{-1}) = \hat{F}(t, q^{-1})\hat{B}'(t, q^{-1}) \tag{6.4.6}$$

Finally, the feedback control signal is generated from

$$\beta(t, q^{-1})u(t) + \alpha(t, q^{-1})y(t) = E(q^{-1})y^*(t + d) \tag{6.4.7}$$

where $\{y^*(t + d)\}$ is the output of the reference model.

The indirect adaptive control algorithm above is globally convergent, as shown below.

Theorem 6.4.1. Provided that the projection or least-squares algorithm is used to generate $\hat{\theta}(t)$ and provided the system is stably invertible in the usual sense [see Theorem 5.2.1, part (c)], then the indirect model reference adaptive control algorithm (6.4.2) to (6.4.7) is globally convergent in the sense that

(i) $\{u(t)\}, \{y(t)\}$ are bounded for all time

(ii) $\lim_{t \to \infty} [y(t) - y^*(t)] = 0$ \hfill (6.4.8)

where $y^*(t)$ is the output of the reference model.

Proof. As in the proof of Lemma 4.3.2, we define $\hat{A}\hat{B}$ and $\hat{A}\cdot\hat{B}$ as follows:

$$\hat{A}\hat{B} = \sum_i \sum_j \hat{a}_i(t)\hat{b}_j(t)q^{-i-j} = \hat{B}\hat{A} \qquad (6.4.9)$$

$$\hat{A}\cdot\hat{B} = \sum_i \sum_j \hat{a}_i(t)\hat{b}_j(t-i)q^{-i-j} \neq \hat{B}\cdot\hat{A} \qquad (6.4.10)$$

We also define

$$\hat{\bar{B}} = \hat{B}(t-1, q^{-1}) \qquad (6.4.11)$$

$$\bar{B} = \hat{B}(t+d-1, q^{-1}) \qquad (6.4.12)$$

The key equations for future reference are (we suppress q^{-1})

$$Ay(t) = Bu(t) \qquad (6.4.13)$$

$$\hat{G}y(t) + \hat{F}\hat{B}'u(t) = Ey^*(t+d) \qquad (6.4.14)$$

$$\hat{F}\hat{A} + q^{-d}\hat{G} = E \qquad (6.4.15)$$

$$\begin{aligned} e(t) &= y(t) - \phi(t-1)^T\hat{\theta}(t-1) \\ &= \hat{\bar{A}}y(t) - \hat{\bar{B}}'u(t-d) \end{aligned} \qquad (6.4.16)$$

where

$$\phi(t-1)^T = [-y(t-1), \ \ldots, \ u(t-d), \ \ldots]; \qquad \hat{\theta}(t)^T = [\hat{a}_1(t), \ \ldots, \ \hat{b}_0(t), \ \ldots]$$

We shall derive a model for the closed-loop system. Recall that when the system parameters are known, the closed-loop system is given by $Ey(t+d) = Ey^*(t+d)$ and $EB'u(t) = EAy^*(t+d)]$. From (6.4.16),

$$e(t+d) = \bar{A}y(t+d) - \bar{B}u(t+d)$$

Thus

$$\begin{aligned} \hat{F}e(t+d) &= \hat{F}\cdot\bar{A}y(t+d) - \hat{F}\cdot\bar{B}u(t+d) \\ &= \hat{F}\hat{A}y(t+d) + [\hat{F}\cdot\bar{A} - \hat{F}\hat{A}]y(t+d) - \hat{F}\cdot\bar{B}u(t+d) \\ &= (E - q^{-d}\hat{G})y(t+d) + [\hat{F}\cdot\bar{A} - \hat{F}\hat{A}]y(t+d) - \hat{F}\cdot\bar{B}u(t+d) \\ &= Ey(t+d) - \hat{G}y(t) - \hat{F}\hat{B}'u(t) + [\hat{F}\cdot\bar{A} - \hat{F}\hat{A}]y(t+d) \\ &\qquad\qquad\qquad\qquad\qquad\qquad - [\hat{F}\cdot\bar{B}' - \hat{F}\hat{B}']u(t) \end{aligned}$$

or

$$\begin{aligned} Ey(t+d) + [\hat{F}\cdot\bar{A} - \hat{F}\hat{A}]y(t+d) &- [\hat{F}\cdot\bar{B}' - \hat{F}\hat{B}']u(t) \\ &= Ey^*(t+d) + \hat{F}e(t+d) \end{aligned} \qquad (6.4.17)$$

Operating on (6.4.17) by A gives

$$\begin{aligned} EAy(t+d) + A\cdot[\hat{F}\cdot\bar{A} - \hat{F}\hat{A}]y(t+d) &- A\cdot[\hat{F}\cdot\bar{B}' - \hat{F}\hat{B}']u(t) \\ &= EAy^*(t+d) + A\cdot\hat{F}e(t+d) \end{aligned} \qquad (6.4.18)$$

Using (6.4.13) in (6.4.18), then (6.4.17)–(6.4.18) can be summarized as

$$\begin{aligned} &\begin{bmatrix} E + [\hat{F}\cdot\bar{A} - \hat{F}\hat{A}] & -[\hat{F}\cdot\bar{B}' - \hat{F}\hat{B}'] \\ A\cdot[\hat{F}\cdot\bar{A} - \hat{F}\hat{A}] & EB' - A\cdot[\hat{F}\cdot\bar{B}' - \hat{F}\hat{B}'] \end{bmatrix} \begin{bmatrix} y(t+d) \\ u(t) \end{bmatrix} \\ &\qquad\qquad = \begin{bmatrix} Ey^*(t+d) + \hat{F}e(t+d) \\ EAy^*(t+d) + A\cdot\hat{F}e(t+d) \end{bmatrix} \end{aligned} \qquad (6.4.19)$$

Adaptive Control of Linear Deterministic Systems   Chap. 6

Equation (6.4.19) can be regarded as a linear time-varying system having inputs $\{e(t)\}$ and $\{y^*(t + d)\}$ and outputs $\{u(t)\}$ and $\{y(t)\}$. The terms in $e(t)$ arise due to the modeling error and the terms in square brackets, for example, $[\hat{F} \cdot \bar{B}' - \hat{F}\hat{B}']$, and so on, arise due to the time-varying nature of the parameter estimates.

Now the standard parameter estimation algorithms have the following properties (see, e.g., Lemma 3.3.2):

1. $\hat{A}, \hat{B}$ bounded for all t
2. $\|\hat{\theta}(t + k) - \hat{\theta}(t)\|^2 \rightarrow 0$ for all finite k
3. $\dfrac{e(t)^2}{c + \phi(t - 1)^T\phi(t - 1)} \rightarrow 0$

Property 1 also implies that $\hat{F}$ and $\hat{G}$ are bounded since (6.4.15) is solvable for any $\hat{A}$.

Property 2 ensures that the model (6.4.19) is asymptotically time invariant and stable provided that E^{-1} and B'^{-1} are both stable. (Desoer and Vidyasagar (1975).)

Thus, from (6.4.19), $\{u(t - d)\}$ and $\{y(t)\}$ are asymptotically bounded by $\{e(t)\}$. Thus we can apply the key technical lemma (Lemma 6.2.1) to show using property 3 that $e(t)$ converges to zero and that $\{\phi(t)\}$ is bounded. This establishes the theorem.

▼▼▼

The advantage of the indirect approach over the direct approach is that one need not reparameterize to the d-step-ahead predictor form before applying the parameter estimation algorithm. Thus one has more flexibility and, in the case of large delay, fewer parameters have to be estimated in the system model. This completes our discussion of minimum prediction error and model reference control.

The reader will recall that model reference control is a special case of pole assignment in which the observer is chosen to have dynamics given by E and the closed-loop poles are assigned to B'. In the next section we take up the more general question of arbitrary pole assignment.

6.5 ADAPTIVE ALGORITHMS FOR CLOSED-LOOP POLE ASSIGNMENT

As we have pointed out in Chapter 5, control laws based on the one-step-ahead principle require restrictive assumptions on the zeros of the system. As we have seen above, these restrictions carry over to.the adaptive case. In Chapter 5 the pole assignment algorithm was introduced to overcome this difficulty.

In this section we develop an adaptive form of the closed-loop pole assignment algorithm (for simplicity we treat the single-input single-output case). Algorithms of this type have been described by Åström and Wittenmark (1974), Wittenmark (1977), Wouters (1977), Wellstead, Prager, and Zanker (1979), Elliott and Wolovich (1979), Kreisselmeier (1980), Goodwin and Sin (1981), and Åström and Wittenmark (1980).

As before, we consider a linear time-invariant finite-dimensional system having the DARMA model

$$A(q^{-1})y(t) = B(q^{-1})u(t) \tag{6.5.1}$$

where

$$A(q^{-1}) = 1 + a_1^0 q^{-1} + \cdots + a_n^0 q^{-n} \qquad (6.5.2)$$

$$B(q^{-1}) = b_1^0 q^{-1} + \cdots + b_m^0 q^{-m} \qquad (6.5.3)$$

[Note that the time delay has been incorporated in $B(q^{-1})$ so that the leading coefficients b_1^0 and so on may be zero.]

The following assumptions about the system will be used:

Assumption 6.5.A

1. $r = \max(n, m)$ is known.
2. $A(q^{-1})$ and $B(q^{-1})$ are relatively prime (but having unknown coefficients).

▼▼▼

Assumption 6.5.A should be contrasted with the previous assumptions of an upper bound on n and m and knowledge of the system time delay.

A consequence of Assumption 6.5.A, part 2, and Theorem 5.3.1 is that there exists unique polynomials $L(q^{-1})$ and $P(q^{-1})$, both of order $(r - 1)$ with $L(0) = 1$, such that

$$A(q^{-1})L(q^{-1}) + B(q^{-1})P(q^{-1}) = A^*(q^{-1}) \qquad (6.5.4)$$

where $A^*(q^{-1})$ is an arbitrary polynomial of order $2r - 1$ with $r = \max(n, m)$. (Note that the minimal degrees for L, P, and A^* are, m, $n - 1$, and $n + m$, respectively, but here we assume knowledge of r only and not n or m.)

The implication of the above is that if the input $\{u(t)\}$ is generated by the causal feedback control law,

$$L(q^{-1})u(t) = P(q^{-1})\{y^*(t) - y(t)\} \qquad (6.5.5)$$

where $y^*(t)$ is an arbitrary but bounded set-point sequence, the resulting closed-loop system has characteristic polynomial $A^*(q^{-1})$. Thus if $A^*(q^{-1})$ is a stable polynomial, that is, $A^*(z) \neq 0$ for $|z| \leq 1$, the closed-loop system is asymptotically stable. (See Theorem 5.3.1.)

Here we shall be concerned with the situation when the coefficients in $A(q^{-1})$ and $B(q^{-1})$ are unknown. We first express the system model (6.5.1) in our usual regression form,

$$y(t) = \phi(t - 1)^T \theta_0 \qquad (6.5.6)$$

where

$$\phi(t - 1)^T = [y(t - 1), \ldots, y(t - r), u(t - 1), \ldots, u(t - r)] \qquad (6.5.7)$$

$$\theta_0^T = [-a_1^0, \ldots, -a_r^0, b_1^0, \ldots, b_r^0] \qquad (6.5.8)$$

where $r = \max(n, m)$; $a_i^0 = 0$ for $i > n$: $b_i^0 = 0$ for $i > m$.

Then combining the least-squares parameter estimation algorithms with the pole assignment algorithm of Chapter 5, we are led to the following indirect adaptive pole assignment algorithm.

Adaptive Pole Assignment Controller

$$\hat{\theta}(t) = \hat{\theta}(t - 1) + \frac{P(t - 2)\phi(t - 1)}{1 + \phi(t - 1)^T P(t - 2)\phi(t - 1)}[y(t) - \phi(t - 1)^T \hat{\theta}(t - 1)] \qquad (6.5.9)$$

$$P(t) = P(t-1) - \frac{P(t-1)\phi(t)\phi(t)^T P(t-1)}{1 + \phi(t)^T P(t-1)\phi(t)} \qquad (6.5.10)$$

with $P(0) = \epsilon I$, $\epsilon > 0$. In (6.5.9), the symbols have the following meaning:

$$\hat{\theta}(t)^T = (-\hat{a}_1(t), \ldots, -\hat{a}_r(t), \hat{b}_1(t), \ldots, \hat{b}_r(t)) \qquad (6.5.11)$$

$$\hat{A}(t, q^{-1}) = \hat{a}_0 + \hat{a}_1(t)q^{-1} + \cdots + \hat{a}_r(t)q^{-r}; \qquad \hat{a}_0 = 1 \qquad (6.5.12)$$

$$\hat{B}(t, q^{-1}) = \hat{b}_1(t)q^{-1} + \cdots + \hat{b}_r(t)q^{-r} \qquad (6.5.13)$$

The input $\{u(t)\}$ is then determined by solving the pole assignment equation at each time instant:

$$\hat{A}(t, q^{-1})\hat{L}(t, q^{-1}) + \hat{B}(t, q^{-1})\hat{P}(t, q^{-1}) = A^*(q^{-1}) \qquad (6.5.14)$$

and then using $\hat{L}(t, q^{-1})$ and $\hat{P}(t, q^{-1})$ in the following feedback control law:

$$\hat{L}(t, q^{-1})u(t) = \hat{P}(t, q^{-1})[y^*(t) - y(t)] \qquad (6.5.15)$$

with $\hat{L}(t, q^{-1})$ and $\hat{P}(t, q^{-1})$ of order $(r-1)$.

In the algorithm above, $A^*(q^{-1})$ is an arbitrary stable monic polynomial of order $(2r - 1)$ and $\{y^*(t)\}$ is an arbitrary bounded set-point sequence.

Properties of the least-squares parameter estimator (6.5.9)–(6.5.10) have been discussed in Lemma 3.3.6. We shall use these properties here.

We first investigate the solvability of (6.5.14). It is clear that the values of $\hat{\theta}(t)$ that give rise to an exact pole–zero cancellation [and hence to exact singularity of (6.5.14)] are on a set of measure zero. Thus (6.5.14) is solvable with probability 1. However, in the analysis to follow, we require $\hat{L}(t, q^{-1})$ and $\hat{P}(t, q^{-1})$ to have bounded coefficients and hence near singularity of (6.5.14) must also be avoided. This is clearly also necessary for numerical stability in practical applications. This contrasts with model reference adaptive control discussed in the preceding section. For model reference control, the special form of the pole assignment equation (i.e., the prediction equality) is always solvable.

The solvability of the pole assignment equation is a key question. Difficulties arise when $\hat{A}(t, q^{-1})$ and $\hat{B}(t, q^{-1})$ are not relatively prime. However, stable pole–zero cancellations do not present a problem since one can then simply include the common roots in the closed-loop polynomial, $A^*(q^{-1})$, and thus cancel these roots from the pole assignment equation.

The possibility of an unstable pole–zero cancellation must be avoided. If the system parameters are known sufficiently accurately so that the initial estimate lies in a convex region in parameter space surrounding θ_0 such that, within this region, $\hat{A}$ and $\hat{B}$ are relatively prime, the constrained parameter estimation algorithms of Section 3.7 can be used to ensure that $\hat{A}$ and $\hat{B}$ will remain inside this region.

The convergence properties of the adaptive control algorithm above are explored in the following theorem.

Theorem 6.5.1. Subject to Assumption 6.5.A and provided that the estimates are constrained as described above, the algorithm (6.5.9) to (6.5.15) leads to

(i) $\{u(t)\}$ bounded $\qquad\qquad\qquad\qquad\qquad\qquad\qquad$ (6.5.16)

(ii) $\{y(t)\}$ bounded $\qquad\qquad\qquad\qquad\qquad\qquad\qquad$ (6.5.17)

(iii) The closed-loop characteristic polynomial tends to $A^*(q^{-1})$ in the sense that

$$\lim_{t \to \infty} [A^*(q^{-1})y(t) - G(t-1, q^{-1})y^*(t)] = 0 \tag{6.5.18}$$

where

$$G(t-1, q^{-1}) = \sum_{j=1}^{r} \hat{b}_j(t-1) \sum_{k=0}^{r-1} \hat{p}_k(t-1)q^{-j-k}$$

Proof. Before embarking on the proof we shall outline the approach to be followed to make the principal ideas more transparent. We first obtain a closed-loop model for the system by introducing two quantities $w(t)$ and $z(t)$ which are filtered versions of $y^*(t)$ using the time-varying filters $\hat{A}\hat{P}$ and $\hat{B}\hat{P}$, respectively. We then show that $\{u(t)\}$ and $\{y(t)\}$ are stably related to $w(t)$, $z(t)$, and $e(t) \triangleq y(t) - \phi(t-1)^T\hat{\theta}(t-1)$, the modeling error. Hence we can argue that $\{u(t)\}$ and $\{y(t)\}$ cannot grow faster than $\{e(t)\}$. We are then able to use the key technical lemma (Lemma 6.2.1) to show that $\{u(t)\}$ and $\{y(t)\}$ remain bounded and $\{e(t)\}$ converges to zero.

In the following proof we shall need to manipulate time-varying operators. This is facilitated by the following notation. Given time-varying polynomials $\hat{A}(t, q^{-1})$ and $\hat{B}(t, q^{-1})$, define the following:

$$\hat{A}\hat{B} = \sum_i \sum_j \hat{a}_i(t)\hat{b}_j(t)q^{-i-j} = \hat{B}\hat{A} \tag{6.5.19}$$

$$\hat{A} \cdot \hat{B} = \sum_i \sum_j \hat{a}_i(t)\hat{b}_j(t-i)q^{-i-j} \neq \hat{B} \cdot \hat{A} \tag{6.5.20}$$

(Note that $\hat{A}\hat{B} = \hat{A} \cdot \hat{B} = \hat{B} \cdot \hat{A}$ when $\hat{A}$ and $\hat{B}$ are time invariant.) We also define

$$\hat{\bar{B}} = \hat{B}(t-1, q^{-1}) \tag{6.5.21}$$

The key equations for future reference are

$$Ay(t) = Bu(t) \quad \text{from (6.5.1)} \tag{6.5.22}$$

$$\hat{L}u(t) = \hat{P}y^*(t) - \hat{P}y(t) \quad \text{from (6.5.15)} \tag{6.5.23}$$

$$\hat{A}\hat{L} + \hat{B}\hat{P} = A^* \quad \text{from (6.5.14)} \tag{6.5.24}$$

$$
\begin{aligned}
e(t) &= y(t) - \phi(t-1)^T\hat{\theta}(t-1) \\
&= y(t) - \{(1 - \hat{\bar{A}})y(t) + \hat{\bar{B}}u(t)\} \\
&= \hat{\bar{A}}y(t) - \hat{\bar{B}}u(t)
\end{aligned} \tag{6.5.25}
$$

Now we shall attempt to derive a model for the closed-loop system. (Recall that when the system parameters are known, the closed-loop system is given by $A^*u(t) = APy^*(t)$ and $A^*y(t) = BPy^*(t)$. We therefore define $w(t) \triangleq \hat{A} \cdot \hat{P}y^*(t)$ and $z(t) \triangleq \hat{B} \cdot \hat{P}y^*(t)$. Then

$$
\begin{aligned}
w(t) &= \hat{A} \cdot \hat{P}y^*(t) \\
&= \hat{A} \cdot \hat{L}u(t) + \hat{A} \cdot \hat{P}y(t) \\
&= \hat{A}\hat{L}u(t) + [\hat{A} \cdot \hat{L} - \hat{A}\hat{L}]u(t) + \hat{A}\hat{P}y(t) + [\hat{A} \cdot \hat{P} - \hat{A}\hat{P}]y(t) \\
&= \hat{A}\hat{L}u(t) + \hat{B}\hat{P}u(t) + \hat{P}e(t) + [\hat{A} \cdot \hat{L} - \hat{A}\hat{L}]u(t) + [\hat{A} \cdot \hat{P} - \hat{A}\hat{P}]y(t) \\
&\quad + [\hat{P} \cdot \hat{\bar{B}} - \hat{P}\hat{B}]u(t) - [\hat{P} \cdot \hat{\bar{A}} - \hat{P}\hat{A}]y(t) \quad \text{using (6.5.25)} \\
&= A^*u(t) + \hat{P}e(t) + [\hat{A} \cdot \hat{L} - \hat{A}\hat{L}]u(t) + [\hat{A} \cdot \hat{P} - \hat{A}\hat{P}]y(t) \\
&\quad + [\hat{P} \cdot \hat{\bar{B}} - \hat{P}\hat{B}]u(t) + [\hat{P} \cdot \hat{\bar{A}} - \hat{P}\hat{A}]y(t) \quad \text{using (6.5.24)}
\end{aligned} \tag{6.5.27}
$$

Also define

$$z(t) = \hat{B} \cdot \hat{P} y^*(t) \tag{6.5.28}$$

and using an argument similar to the one above, we have

$$z(t) = A^* y(t) - \hat{P} e(t) + [\hat{B} \cdot \hat{L} - \hat{B}\hat{L}]u(t) + [\hat{B} \cdot \hat{P} - \hat{B}\hat{P}]y(t) \\ - [\hat{L} \cdot \hat{\hat{B}} - \hat{L}\hat{B}]u(t) + [\hat{L} \cdot \hat{\hat{A}} - \hat{L}\hat{A}]y(t) \tag{6.5.29}$$

Combining (6.5.27) and (6.5.29), we obtain the following closed-loop model:

$$\begin{bmatrix} A^* + [\hat{A} \cdot \hat{L} - \hat{A}\hat{L}] + [\hat{P} \cdot \hat{\hat{B}} - \hat{P}\hat{B}] & [\hat{A} \cdot \hat{P} - \hat{A}\hat{P}] - [\hat{P} \cdot \hat{\hat{A}} - \hat{P}\hat{A}] \\ [\hat{B} \cdot \hat{L} - \hat{B}\hat{L}] - [\hat{L} \cdot \hat{B} - \hat{L}\hat{B}] & A^* + [\hat{B} \cdot \hat{P} - \hat{B}\hat{P}] + [L \cdot \hat{\hat{A}} - \hat{L}\hat{A}] \end{bmatrix} \begin{bmatrix} u(t) \\ y(t) \end{bmatrix}$$
$$= \begin{bmatrix} w(t) - \hat{P} e(t) \\ z(t) + \hat{P} e(t) \end{bmatrix} \tag{6.5.30}$$

Equation (6.5.30) can be regarded as a linear time-varying dynamical system having inputs $\{w(t)\}$, $\{z(t)\}$, and $\{e(t)\}$ and outputs $\{u(t)\}$ and $\{y(t)\}$. The terms in $e(t)$ arise due to the modeling error and the terms in brackets, for example, $[\hat{A} \cdot \hat{L} - \hat{A}\hat{L}]$, arise due to the time-varying nature of the parameters estimates.

Now, it follows from Lemma 3.3.6, part (i), that $\hat{A}(t, q^{-1})$ and $\hat{B}(t, q^{-1})$ have bounded coefficients for all t. Also, since the estimates are constrained, the pole assignment equation is solvable and hence $\hat{L}(t, q^{-1})$ and $\hat{P}(t, q^{-1})$ have bounded coefficients for all t. Then, from Lemma 3.3.6, part (e), and the continuity of (6.5.14), it follows that the coefficients of all terms in square brackets in (6.5.30) approach zero as t tends to infinity. Thus for t sufficiently large, but finite, the system (6.5.30) is arbitrarily close to an asymptotically exponentially stable system having characteristic polynomial $[A^*(z)]^2$.

Also, in view of the boundedness of the coefficients of $\hat{A}(t, q^{-1})$, $\hat{B}(t, q^{-1})$, $\hat{L}(t, q^{-1})$, and $\hat{P}(t, q^{-1})$, it follows from the definitions (6.5.26) and (6.5.28) and the boundedness of $\{y^*(t)\}$ that $\{w(t)\}$ and $\{z(t)\}$ are bounded. We can then conclude from (6.5.30) that the sequences $\{u(t)\}$ and $\{y(t)\}$ and hence $\{\|\phi(t)\|\}$ grow no faster than linearly with $e(t)$.

Thus we are able to apply the key technical lemma (Lemma 6.2.1) to conclude from part (a) of Lemma 3.3.6 that

$$\lim_{t \to \infty} e(t) = 0 \tag{6.5.31}$$

and $\{\|\phi(t)\|\}$ is bounded. Hence from the definition of $\phi(t)$, it follows that $\{y(t)\}$ and $\{u(t)\}$ are bounded. This establishes parts (i) and (ii) of the theorem. To establish part (iii) we proceed as follows. From (6.5.25) and (6.5.31),

$$\lim_{t \to \infty} \hat{A} y(t+1) - \hat{B} u(t+1) = 0 \tag{6.5.32}$$

Arguing as in the derivation of (6.5.27), using the boundedness of $\{u(t)\}$, $\{y(t)\}$, and $\{y^*(t)\}$ and of the boundedness of the coefficients of $\hat{A}$, $\hat{B}$, $\hat{L}$, and $\hat{P}$ plus Lemma 3.3.6, we can conclude that for given ϵ arbitrarily small there exists an N_3 such that

$$|\hat{L}\hat{A} y(t+1) - \hat{B}\hat{P}[y^*(t+1) - y(t+1)]| < \epsilon, \quad t \geq N_3$$

Finally, using an argument similar to that used above, we can establish part (iii) of the theorem.

▼▼▼

We shall now make a few comments on the result above:

1. In the case of systems having a "stable inverse," it has been shown in previous sections that adaptive control algorithms exist that cause $\{y(t)\}$ to track $\{y^*(t)\}$. It is clear that this will be impossible when the system does not have a stable inverse. Thus in the case of the pole assignment algorithms special consideration has to be given to the question of steady-state errors in the output sequence. All of the discussion in Section 5.3.5 applies *mutatis mutandis* to the adaptive case.

2. An alternative to solving (6.5.14) at every step would be to use an iterative algorithm aimed at solving the equation asymptotically. This is the approach advocated in Elliott and Wolovich (1979) and Kreisselmeier (1980). An advantage is that it simplifies the computation at each step.

We now present a simulated example of showing the performance of the algorithm.

Example 6.5.1

This example is the deterministic part of the system discussed in Wouters (1977). The system is given by

$$A(q^{-1}) = 1 - 2.0q^{-1} + 0.99q^{-2} \tag{6.5.33}$$

$$B(q^{-1}) = q^{-1}(0.5 + 1.0q^{-1}) \tag{6.5.34}$$

The system has one stable pole at 0.9 and one unstable pole at 1.1. The unstable zero is at -2.0. The following conditions were used:

$$A^*(q^{-1}) = 1.0 \tag{6.5.35}$$

$$y^*(t) = 1.0; \qquad 0 \leq t \leq 30 \tag{6.5.36}$$

$$= -1.0; \qquad 31 \leq t \leq 60$$

Table 6.5.1 sets out the initial parameter estimates.

TABLE 6.5.1

Parameters	a_1	a_2	b_1	b_2
True value	−2.0	0.99	0.5	1.0
Initial value	0.0	0.5	1.0	0.0

Figure 6.5.1a and b show the output and input of the system using the adaptive control algorithm (actually with the projection algorithm used in place of least squares in the parameter estimator; see Exercise 6.6. Figure 6.5.1c and d show the convergence of $\hat{A}$, $\hat{B}$ and $\hat{L}$, $\hat{P}$, respectively, for this example.

A shortcoming of the result in Theorem 6.5.1 (as it stands) is that it requires that a convex region be known in which pole–zero cancellations do not occur. If this is not the case, then a possible strategy is to add a persistently exciting external perturbation so that after a finite time the parameter estimates, $\hat{\theta}(t)$, are close enough to the true value θ_0 so that pole–zero cancellation will not occur. This can be achieved

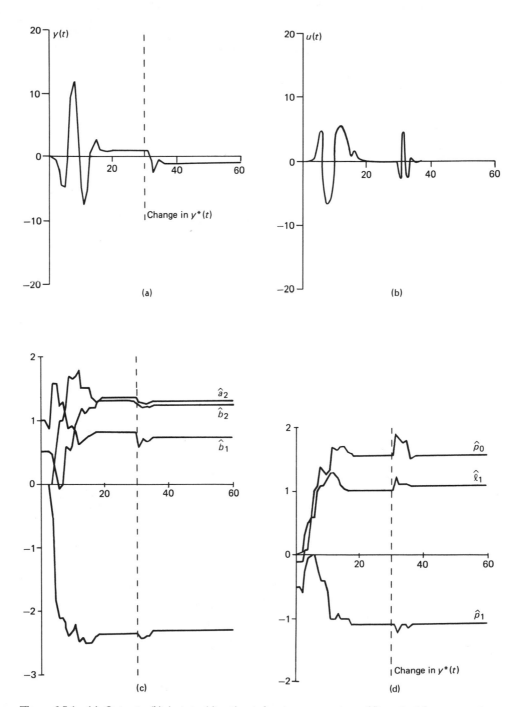

Figure 6.5.1 (a) Output; (b) input; (c) estimated system parameters; (d) control law parameters.

in view of the result in Lemma 3.4.9. We illustrate this procedure below using the covariance resetting algorithm of section 3.3.

We shall consider an indirect pole-assignment adaptive control law using a two-time frame estimator such that the control law is updated every N samples.

The system will be taken to be of the form (6.5.1) subject to assumption 6.5.A.

Adaptive Pole Assignment Controller with Persistent Excitation

Parameter Estimation Update (Least Squares)

$$\hat{\theta}(t) = \hat{\theta}(t-1) + \frac{P(t-2)\phi(t-1)}{1 + \phi(t-1)^T P(t-2)\phi(t-1)}[y(t) - \phi(t-1)^T \hat{\theta}(t-1)]$$

$$t = 1, 2, \ldots \; ; \hat{\theta}(0) \text{ given} \qquad (6.5.37)$$

Covariance Update with Resetting (see also Section 3.3)

$$P'(t-1) = P(t-2) - \frac{P(t-2)\phi(t-1)\phi(t-1)^T P(t-2)}{1 + \phi(t-1)^T P(t-2)\phi(t-1)} \qquad (6.5.38)$$

If t/N is an integer, then resetting occurs as follows

$$P(t-1) = \frac{1}{k_0}I; \qquad 0 < k_0 < \infty \qquad (6.5.39)$$

or else

$$P(t-1) = P'(t-1) \qquad (6.5.40)$$

Control Law Update (ever N samples)

(a) If t/N is an integer, then evaluate

$$\hat{A}(t, q^{-1}) = 1 + \hat{\theta}_0(t)q^{-1} + \cdots + \hat{\theta}_r(t)q^{-r} \qquad (6.5.41)$$

$$\hat{B}(t, q^{-1}) = \hat{\theta}_{r+1}(t)q^{-1} + \cdots + \hat{\theta}_{2r}(t)q^{-r} \qquad (6.5.42)$$

Solve the following equation for $\hat{L}(t, q^{-1})$, $\hat{P}(t, q^{-1})$ each of order $(r-1)$:

$$\hat{A}(t, q^{-1})\hat{L}(t, q^{-1}) + \hat{B}(t, q^{-1})\hat{P}(t, q^{-1}) = A^*(q^{-1}) \qquad (6.5.43)$$

where $A^*(q-1)$ is an arbitrary stable polynomial.

(In the event $\hat{A}(t, q^{-1})$, $\hat{B}(t, q^{-1})$ are not relatively prime, then $\hat{L}(t, q^{-1})$ and $\hat{P}(t, q^{-1})$ can be chosen arbitrarily. Relative primeness can be tested by use of a prespecified lower bound on the magnitude of the determinant of the Sylvester matrix for the true system polynomials and thus sufficient a priori knowledge is implicitly assumed to ensure this.)

(b) Otherwise

$$\hat{L}(t, q^{-1}) = \hat{L}(t-1, q^{-1}); \hat{P}(t, q^{-1}) = \hat{P}(t-1, q^{-1}) \qquad (6.5.44)$$

Evaluation of the Input

$$\hat{L}(t, q^{-1})u(t) = -\hat{P}(t, q^{-1})y(t) + v(t) \qquad (6.5.45)$$

where $\{v(t)\}$ is a persistently exciting external input of the form given in (3.4.63), that is,

$$v(t) = \sum_{k=1}^{s} \Gamma_k \sin(\omega_k t + \sigma_k) \tag{6.5.46}$$

where Γ_k denote the Fourier coefficients—possibly different for each interval, and

$$\omega_k \epsilon(0, \pi); \ \Gamma_k \neq 0 \text{ and } \omega_j \neq \omega_k; \ k = 1, \ldots, s; j = 1, \ldots, s$$

such that the length of the interval N and the number, s of non-zero Γ_k satisfy

(a) $N \geq 10r$ (6.5.47)

(b) $s \geq 4r$ (6.5.48)

Theorem 6.5.2 (Global Convergence for Adaptive Pole Assignment). Consider the algorithm (6.5.37) to (6.5.48) applied to the system (6.5.1) subject to assumption 6.5.A, then $\hat{\theta}(t)$ approaches the true value, θ_0, exponentially fast and $[u(t)], [y(t)]$ remain bounded for all time.

Proof. It is readily shown as in Section 3.3 that

$$\tilde{\theta}((k+1)N)^T P'((k+1)N - 1)^{-1} \tilde{\theta}((k+1)N) \leq \tilde{\theta}(kN)^T P(kN-1)^{-1} \tilde{\theta}(kN)$$
$$k = 0, 1, 2, \ldots \tag{6.5.49}$$

where

$$\tilde{\theta}(t) = \hat{\theta}(t) - \theta_0 \tag{6.5.50}$$

Now, from (6.5.38) to (6.5.40) we have

$$P(kN - 1)^{-1} = k_0 I \tag{6.5.51}$$

$$P'((k+1)N - 1)^{-1} = k_0 I + \sum_{t=kN}^{(k+1)N-1} \phi(t)\phi(t)^T \tag{6.5.52}$$

$$= k_0 I + x(kN)x(kN)^T \tag{6.5.53}$$

where

$$x(t) = [\phi(t), \phi(t+1), \ldots, \phi(t+N-1)] \tag{6.5.54}$$

Since (6.5.47), (6.5.48) are satisfied and $v(t)$ is chosen as in (6.5.46), we can immediately apply Lemma 3.4.9 to conclude there exists an $\epsilon_1 > 0$ such that

$$(k_0 + \epsilon_1) \|\tilde{\theta}((k+1)N)\|^2 \leq \lambda_{\min}[P'((k+1)N - 1)^{-1}] \|\tilde{\theta}((k+1)N)\|^2$$
$$\leq \tilde{\theta}((k+1)N)^T P'((k+1)N - 1)^{-1}\tilde{\theta}((k+1)N)$$
$$\leq \tilde{\theta}(kN)^T P(kN - 1)^{-1}\tilde{\theta}(kN) \tag{6.5.55}$$
$$= k_0 \|\tilde{\theta}(kN)\|^2$$

Hence

$$\|\tilde{\theta}((k+1)N)\|^2 \leq \left(\frac{k_0}{k_0 + \epsilon_1}\right) \|\tilde{\theta}(kN)\|^2 \tag{6.5.56}$$

Hence, since $\epsilon_1 > 0$, we can conclude the subsequence $\{\|\tilde{\theta}(kN)\|^2; k = 0, 1, \ldots\}$ is exponentially convergent to zero. Also, it can be readily seen that

$$\|\tilde{\theta}(kN + \tau)\|^2 \leq \|\tilde{\theta}(kN)\|^2 \quad \text{for } \tau = 1, \ldots, N \tag{6.5.57}$$

Thus we have that $\{\|\tilde{\theta}(t)\|^2; t = 1, 2, \ldots\}$ converges exponentially fast to zero.

In view of assumption 6.5A and continuity there exists an ϵ-neighborhood of θ_0 in which $A(\theta, q^{-1})$ and $B(\theta, q^{-1})$ are relatively prime. Hence, from the result above, there exists a finite time N_ϵ such that $\hat{\theta}(t)$ belongs to this neighborhood and hence for

$t \geq N_\epsilon$, the system is asymptotically stable at each control law setting. Moreover, from the triangle inequality $\|\hat{\theta}(t) - \hat{\theta}(t-1)\|$ approaches zero exponentially fast, and hence from the small gain theorem (Desoer and Vidyasagar (1975)) it follows $\{u(t)\}, \{y(t)\}$ remain bounded.

▼▼▼

The above algorithm uses iterative least squares with covariance resetting. Four points can be made about this procedure:

(i) If resetting is not used, then the algorithm reduces to ordinary recursive least squares. In this case it can still be shown that $\hat{\theta}(t)$ converges to θ_0 but not exponentially fast.

(ii) In the above analysis, we have reset to a scaled value of the identity matrix. However, it can be seen that an identical result is achieved if the resetting is made to any matrix, P^{-1}, satisfying:

$$\lambda_{\max}(P^{-1}) < [\lambda_{\min}(P^{-1}) + \epsilon_1] \qquad (6.5.58)$$

In particular, if one knows an upper bound $(\bar{\epsilon}_1)$ on ϵ_1, one could reset to

$$\frac{\bar{\epsilon}_1}{\text{trace } P'(t-1)^{-1}}[P'(t-1)^{-1}]$$

This satisfies (6.5.50) and has the advantage that the directional information built up in $P(t-1)^{-1}$ is retained.

(iii) We have chosen to analyze the above algorithm which uses least squares with covariance resetting as in Section 3.3 since the gain of this algorithm does not go to zero and thus it is potentially useful for time varying problems. Note that this is not true of ordinary recursive least squares since the gain goes to zero in this case. Moreover, since we have established exponential convergence, then certain robustness properties are sutomatically guaranteed (see Anderson, 1982). In particular, certain types of time variations can be accommodated.

(iv) In the algorithm above, we have chosen to adjust the control law every N samples. This was necessary to be able to apply Lemma 3.4.9 but it also has the advantage that it separates the bandwidth of the control law adjustment from that of the system. Intuitively, this would seem to enhance the robustness of the overall system to unmodeled high frequency dynamics, and so on, since the effect of such modes will be smoothed out by the time averaging introduced by the interval update of the control law.

6.6 ADAPTIVE CONTROL OF NONLINEAR SYSTEMS

We pointed out in Chapter 5 the important fact that the better one is able to predict the output of a system, the better one is able to control it. Thus if the system response is dominated by nonlinear characteristics, it is probably better to use a nonlinear model rather than an approximate linear model.

So far in this chapter we have discussed only linear systems and for these systems it suffices to use linear functions of the data to predict the system output response.

However, in general, it may be desirable, or even necessary, to consider the use of nonlinear functions to get good predictions and hence good control. For example, industrial control situations, measurements are usually corrected for temperature, flow, and so on, and this can be thought of as producing nonlinear functions of the raw output measurements. One can take this idea a step further and explicitly use a nonlinear model of the system under study.

In Chapter 2 it was pointed out that many nonlinear systems can be described by a regression model of the form

$$y(t + 1) = \phi(t)^T \theta_0 \qquad (6.6.1)$$

For example, in the case of discrete-time bilinear systems, we have seen that the model above is appropriate (see Section 2.4 and associated discussion). In the bilinear case, the vector $\phi(t)$ has components linear in $y(t), \ldots, y(t + 1 - n)$, *linear* in $u(t)$, and multilinear (nonlinear) in $u(t - 1), \ldots, u(t + 2 - n)$.

We have also pointed out in Remark 5.2.2 that the one-step-ahead control idea can be extended to nonlinear systems by simply solving the following equation for $u(t)$:

$$\phi(t)^T \theta_0 = y^*(t + 1) \qquad (6.6.2)$$

Again in the bilinear case, solution of the equation above is relatively straightforward since $\phi(t)$ is a linear function of $u(t)$. For general nonlinear systems we would need to check that the equation above can indeed be solved for $u(t)$.

Now adaptive control algorithms for nonlinear systems can be obtained by combining the control law (6.6.2) (or any other control law for that matter) with a suitable parameter estimation scheme. We will not go into details since this is an obvious extension of the linear case.

Of course, in the nonlinear case, it will be more difficult to prove convergence of the algorithm and it is often necessary to exploit special features of the problem under study. For example, in Goodwin, McInnis, and Long (1981) global convergence is established for adaptive control algorithms applied to certain bilinear systems of practical interest. We shall not go into details since the proof techniques largely parallel the linear case. Instead, we shall present some examples to illustrate the ideas involved.

Dissolved Oxygen Controller for Wastewater Treatment

In this section we deal with one particular process, the dissolved oxygen-activated sludge process. A simplified but realistic model for this process results in the first-order *bilinear* equation (6.6.3) below. The objective is to bring the level of oxygen concentration in the processed water to a specified level, by controlling the rate of injection of oxygen via compressed air.

We use the following continuous-time model (McInnis, Lin, and Butler, 1979; Olsson, 1977, 1980):

$$\frac{d\bar{y}(t)}{dt} = \frac{\bar{Q}(t)}{V}[\bar{c}_i(t) - \bar{y}(t)] + K_1 \bar{u}(t)[\bar{c}_s - \bar{y}(t)] - K_2 f(\bar{y}(t)) \qquad (6.6.3)$$

where

$$f(x) = 1 \text{ for } x > 0, f(x) = 0 \text{ for } x \leq 0 \qquad (6.6.4)$$

$$\bar{y}(t) = \text{dissolved oxygen concentration}$$
$$\bar{u}(t) = \text{airflow rate}$$
$$\bar{Q}(t) = \text{influent flow rate}$$
$$\bar{c}_i(t) = \text{influent dissolved oxygen concentration}$$
$$\bar{c}_s = \text{maximum dissolved oxygen concentration (saturation level)}$$
$$V = \text{reactor volume}$$
$$K_1 = \text{mass transfer coefficient}$$
$$K_2 = \text{oxygen uptake rate}$$

The factor $f(\bar{y}(t))$ ensures that the oxygen uptake becomes zero when the dissolved oxygen content falls to zero. This guarantees that $\bar{y}(t)$ cannot become negative, in accordance with the physical situation. In the sequel we shall take $V = 1$ by choice of units.

Equation (6.6.3) can be discretized by use of a first-order Euler expansion to give

$$\bar{y}(t+1) \simeq \bar{y}(t) + \frac{T\bar{Q}(t)}{V}[\bar{c}_i(t) - \bar{y}(t)] + K_1 T\bar{u}(t)[\bar{c}_s - \bar{y}(t)] - K_2 T f(\bar{y}(t)) \qquad (6.6.5)$$

where T is the sampling interval.

Clearly, the equation above can be written as in (6.6.1) and subject to certain reasonable assumptions (Goodwin, Long, and McInnis, 1980), a one-step-ahead control law can be derived as in (6.6.2). An adaptive control algorithm can then be derived by combining the one-step-ahead control law with a particular parameter estimation scheme.

Computer simulations of the algorithms have been carried out (Redman, 1980; Ko, 1980). We present a typical result in which the set point was held fixed at 8 mg/ℓ and the flow rate varied as in Fig. 6.6.1. Figure 6.6.2 shows the output of the system with the adaptive controller working in the presence of additive white measurement noise. It can be seen from the figure that the adaptive controller works well and the dissolved oxygen content differs from the set point by the added noise, which is as good as can be achieved with white output noise. It is particularly significant that the flow-rate variations do not affect the output concentration.

Control of pH in Acidic Wastewater

We consider the problem of controlling acidity in a continuous flow of industrial wastewater using a nonlinear adaptive controller. Other techniques of controlling pH neutralization systems include time optimal controllers (McAvoy, 1972) and linear

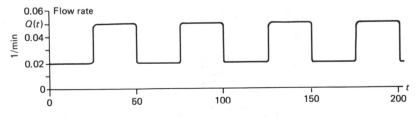

Figure 6.6.1 Flow-rate variation.

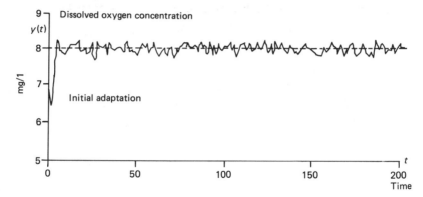

Figure 6.6.2 Performance of nonlinear adaptive controller with measurement noise and flow-rate variations. Dashed line, set point.

adaptive controllers (Burholt and Kummel, 1979; Bergmann and Lachmann, 1980; Shinskey, 1978).

The model that we consider is that described by McAvoy, Hsu, and Lowenthal (1972). A strong acid flows into a tank and is there thoroughly mixed with a strong base whose inward rate of flow is controlled in such a way as to produce a neutral outward flow from the tank. Because the acid and the base are strong, each is completely dissociated, and also the dissociation of the water can be disregarded. The equation describing this model is

$$V \frac{d\bar{y}}{dt} = \bar{F}(t)(a - \bar{y}(t)) - \bar{u}(t)(\bar{b} + \bar{y}(t)) \tag{6.6.6}$$

where $\bar{y}(t) = [H^+] - [OH^-]$ is the distance from neutrality
$\quad V =$ volume of the tank
$\quad \bar{F}(t) =$ rate of flow of the acid
$\quad a =$ concentration of the acid
$\quad \bar{u}(t) =$ rate of flow of the base
$\quad \bar{b} =$ concentration of the base

Note that $\bar{y}(t)$ can be determined from the pH value, $p(t)$ by the following nonlinear transformation (Quagliano, 1958):

$$\bar{y}(t) = 10^{-p(t)} - 10^{p(t)}K_w \tag{6.6.7}$$

where $K_w =$ water equilibrium constant $\simeq 10^{-14}$ (using gram-ion/liter units).

We suppose that $\bar{b}$ is fixed and known, that a is fixed but unknown, that $\bar{F}(t)$ can be measured on line, and that $\bar{u}(t)$ can be given assigned values within certain limits. In general, a will be time dependent and the current value will be estimated by the parameter estimation algorithm.

We can now develop an approximate discrete-time model, incorporating measurement and input actuator errors, as follows:

$$y(t + 1) \simeq y(t) + \frac{T}{V}[F(t)(a - y(t)) - u(t)(b + y(t))] \tag{6.6.8}$$

$$= \phi(t)^T \theta_0 \tag{6.6.9}$$

where $y(t)$ denotes the distance from neutrality, T denotes the sampling interval, and

$$\phi(t) = [y(t), -F(t)y(t) - u(t)(b + y(t)), F(t)]^T \tag{6.6.10}$$

$$\theta_0 = [\theta_1 \quad \theta_2 \quad \theta_3] = \left[1 \quad \frac{T}{V} \quad \frac{aT}{V}\right] \tag{6.6.11}$$

We now estimate θ_0 by use of a modified least-squares algorithm ensuring that $\hat{\theta}_2(t)$ remains positive (see Section 3.7).

For a desired output sequence $\{y^*(t); t = t_0 + nT\}$, we define the input $u(t)$ as follows:

$$u^0(t) = \frac{\hat{\theta}_1(t)y(t) - \hat{\theta}_2(t)F(t)y(t) + \hat{\theta}_3(t)F(t) - y^*(t)}{\hat{\theta}_2(t)(b + y(t))} \tag{6.6.12}$$

and choose

$$u(t) = \begin{cases} u^0(t) & \text{if } 0 < u^0(t) \geq u_{max} & \text{(6.6.13)} \\ u_{max} & \text{if} & u^0(t) \geq u_{max} & \text{(6.6.14)} \\ 0 & \text{if} & u^0(t) \leq 0 & \text{(6.6.15)} \end{cases}$$

We present some typical simulation results below [see Redman (1980)]. The following values were adopted for the various quantities of interest:

$$0.1 \leq F(t) \leq 0.125 \; \ell/\text{min} \qquad 0 \leq u(t) \leq 0.2 \; \ell/\text{min}$$

$$a = 10^{-3} \; \text{mol}/\ell \qquad V = 2 \; \ell$$

$$b = 10^{-3} \; \text{mol}/\ell \qquad T = 1 \; \text{min}$$

In simulations, we have taken the acid and base concentrations to be measured quantities, leaving only θ_1 and θ_2 to be estimated in the bilinear adaptive controller.

To compare the nonlinear adaptive controller with other controllers, a test was conducted where, once steady state was reached, the acid flow rate was varied in a slow sine wave given by

$$F(t) = 0.1125 \; \ell/\text{min}, \qquad\qquad\qquad t < 100$$

$$F(t) = 0.1125 + 0.0125 \sin \frac{t\pi}{25} \; \ell/\text{min}, \qquad t \geq 100$$

Figure 6.6.3 shows the output and input resulting from a classical proportional + integral controller tuned manually to the steady-state conditions. Figure 6.6.4 shows the output and input for an adaptive *linear* controller with four parameters. Figure 6.6.5 shows the output and input for the nonlinear adaptive controller with least-squares parameter estimator.

The figures show only the outputs and inputs after $t = 90$. Prior to this time, the linear and nonlinear adaptive controllers were tuned with constant acid flow. For $t \geq 100$, when the acid flow changes, it is clear from Figs. 6.6.3 to 6.6.5 that the nonlinear adaptive controller gives vastly superior control to either the classical controller or linear adaptive controller primarily because the bilinear adaptive controller uses the true nonlinear model and this gives feedforward action following an acid flow change.

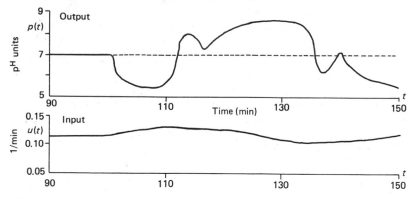

Figure 6.6.3 Output and input under proportional + integral control with acid flow change. Dashed line, set point.

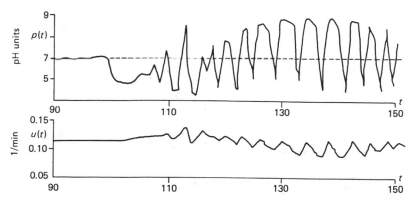

Figure 6.6.4 Output and input under adaptive linear control with acid flow change.

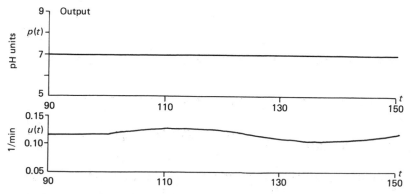

Figure 6.6.5 Output and input under adaptive bilinear control with acid flow change.

6.7 ADAPTIVE CONTROL OF TIME-VARYING SYSTEMS

So far in our analysis of adaptive control algorithms we have restricted attention to time-invariant systems. However, often in practice, the real motivation for adaptive control is to provide a mechanism for dealing with time-varying systems. The basic philosophy of combining an on-line parameter estimator with on-line control system synthesis is all the more natural in the time-varying case. The only major difference is that now the parameter estimator must be capable of continuously tracking time-varying parameters.

We have discussed two basic forms of parameter estimation algorithm: gradient (or projection) and least squares. The gradient algorithm has the property that the gain of the estimator does not go to zero and thus it can, in principle, be applied to time-varying problems. For time-varying systems, the algorithm is often implemented in the following form:

$$\hat{\theta}(t) = \hat{\theta}(t-1) + \frac{\mu \phi(t-1)}{c + \phi(t-1)^T \phi(t-1)}[y(t) - \phi(t-1)^T \hat{\theta}(t-1)] \qquad (6.7.1)$$

The gain constant, μ, allows one to trade off the rate of convergence versus the reduction of noise through smoothing. More will be said about the latter trade-off in Section 9.4.2. The main advantage of the algorithm is its simplicity. However, in practice, it often exhibits slow convergence and thus is generally suitable only for very slowly time-varying problems.

On the other hand, least squares is known to provide extremely rapid initial convergence but then the algorithm gain approaches zero and the estimator switches off. Thus, in its simplest form, the sequential least-squares algorithm is unsuitable for time-varying problems. Several modifications have been proposed which aim to retain the rapid initial convergence of the algorithm while ensuring that the gain does not go to zero. Three possibilities are:

1. Keep the P matrix in the least squares bounded away from zero by adddition of a positive term to the P update equation.
2. Use a finite data window.
3. Use exponential data weighting.

We discuss each of these possibilities briefly below.

Least Squares with Covariance Modification

One form of this algorithm was given in (3.3.106) to (3.3.108), that is,

$$\hat{\theta}(t) = \hat{\theta}(t-1) + \frac{P(t-2)\phi(t-1)}{1 + \phi(t-1)^T P(t-2)\phi(t-1)}[y(t) - \phi(t-1)^T \hat{\theta}(t-1)]$$
$$(6.7.2)$$

$$\bar{P}(t-1) = P(t-2) - \frac{P(t-2)\phi(t-1)\phi(t-1)^T P(t-2)}{1 + \phi(t-1)^T P(t-2)\phi(t-1)} \qquad (6.7.3)$$

$$P(t-1) = \bar{P}(t-1) + Q(t-1) \qquad \text{where } 0 \leq Q(t-1) < \infty \qquad (6.7.4)$$

By suitable choice of the matrix sequence $\{Q(t)\}$ the matrix $P(t)$ can be bounded away from zero. However, it is now theoretically possible that $P(t)$ could become very large and thus it may be desirable to monitor the trace of $P(t)$ and to set $Q(t)$ temporarily to zero whenever prespecified upper bound is reached.

Those readers who are familiar with Kalman filtering theory will appreciate that the algorithm above can be motivated by assuming that the parameters undergo a random walk, that is,

$$\theta(t) = \theta(t-1) + \omega(t) \tag{6.7.5}$$

where $\{\omega(t)\}$ is a white noise sequence having covariance $Q(t)$. (We will discuss the Kalman filter in some detail in Chapter 7.)

Finite Data Window

The basic idea here is to discard old data to keep the algorithm "alive." There are several alternative ways of achieving this. For example:

1. The data prior to some point can be periodically discarded.
2. A moving window of fixed length can be used by throwing away an old data point each time a new one is added.
3. The memory length can be varied by adding and discarding data in some pattern (e.g., the effective data length can be varied from N to $2N$ by using an oscillating memory filter).
4. The covariance matrix P can be periodically reset to some suitable value.

All of these methods achieve roughly the same end result.

Methods 1, 2, and 3 require data to be discarded. This can be achieved as follows. Consider the usual least-squares cost function

$$J_N(\theta) = \tfrac{1}{2} \sum_{t=1}^{N} [y(t) - \phi(t-1)^T \theta]^2 + \tfrac{1}{2}(\theta - \hat{\theta}_0)^T P_0^{-1}(\theta - \hat{\theta}_0) \tag{6.7.6}$$

By completing the square this cost function can be written as

$$J_N(\theta) = \tfrac{1}{2}[\theta - \hat{\theta}(N)]^T P(N-1)^{-1}[\theta - \hat{\theta}(N)] + \beta_N \tag{6.7.7}$$

where $\hat{\theta}(N)$ is the usual least-squares estimate and $P(N-1)$ the usual covariance matrix. β_N is the residual given by

$$\beta_N = \tfrac{1}{2} \sum_{t=1}^{N} [y(t) - \phi(t-1)^T \hat{\theta}(N)]^2 + \tfrac{1}{2}[\hat{\theta}(N) - \theta_0]^T P_0^{-1}[\hat{\theta}(N) - \hat{\theta}_0] \tag{6.7.8}$$

Now, if a data point is to be deleted (say the jth), the cost function becomes

$$\bar{J}_N(\theta) = J_N(\theta) - \tfrac{1}{2}[y(j) - \phi(j-1)^T \theta]^2$$
$$= \tfrac{1}{2}(\theta - \hat{\theta}(N))^T P(N-1)^{-1}(\theta - \hat{\theta}(N)) + \beta_N - \tfrac{1}{2}[y(j) - \phi(j-1)^T \theta]^2 \tag{6.7.9}$$

By differentiating and using the matrix inversion lemma (Lemma 3.3.4), the new minimum can be shown to be given by

$$\hat{\theta}(N)' = \hat{\theta}(N) - \frac{P(N-1)\phi(j-1)}{1 - \phi(j-1)^T P(N-1)\phi(j-1)}[y(j) - \phi(j-1)^T \hat{\theta}(N)] \tag{6.7.10}$$

and the P matrix is modified to

$$P(N-1)' = P(N-1) + \frac{P(N-1)\phi(j-1)\phi(j-1)^T P(N-1)}{1 - \phi(j-1)^T P(N-1)\phi(j-1)} \tag{6.7.11}$$

Given the expressions above together with the usual least-squares update it is clearly possible to add or subtract data points whenever appropriate.

The related algorithm of resetting P has been discussed in some detail in Section 3.3. From (3.3.85) to (3.3.88) the least-squares algorithm with covariance resetting is

$$\hat{\theta}(t) = \hat{\theta}(t-1) + \frac{P(t-2)\phi(t-1)}{1 + \phi(t-1)^T P(t-2)\phi(t-1)}[y(t) - \phi(t-1)^T \hat{\theta}(t-1)]$$
(6.7.12)

$$P(-1) = k_0 I; \qquad k_0 > 0$$
(6.7.13)

Let $\{Z_s\} = \{t_1, t_2, \ldots\}$ be the times at which resetting occurs; then for $t \in \{Z_s\}$ an ordinary sequential least-squares update is used, that is,

$$P(t-1) = P(t-2) - \frac{P(t-2)\phi(t-1)\phi(t-1)^T P(t-2)}{1 + \phi(t-1)^T P(t-2)\phi(t-1)}$$
(6.7.14)

Otherwise, for $t = t_i \in \{Z_s\}$, $P(t_i - 1)$ is reset as follows:

$$P(t_i - 1) = k_i I \qquad \text{where } 0 < k_{\min} \leq k_i \leq k_{\max} < \infty$$
(6.7.15)

To determine when resetting should be applied, one can monitor the magnitude of the P matrix (e.g., trace P). One simple scheme is to keep P above a threshold by resetting when the threshold is crossed.

Exponential Data Weighting

This algorithm was described in Section 3.3 [equations (3.3.81) to (3.3.84)]. The algorithm takes the form

$$\hat{\theta}(t) = \hat{\theta}(t-1) + \frac{P(t-2)\phi(t-1)}{\alpha(t-1) + \phi(t-1)^T P(t-2)\phi(t-1)}$$
$$\times [y(t) - \phi(t-1)^T \hat{\theta}(t-1)]$$
(6.7.16)

$$P(t-1) = \frac{1}{\alpha(t-1)}\left[P(t-2) - \frac{P(t-2)\phi(t-1)\phi(t-1)^T P(t-2)}{\alpha(t-1) + \phi(t-1)^T P(t-2)\phi(t-1)}\right]$$
(6.7.17)

$$0 < \alpha(t) \leq 1$$
(6.7.18)

With this algorithm the data are exponential weighted and the time constant of the data weighting (i.e., roughly the number of significant data points) is $1/(1 - \alpha)$.

It has sometimes been suggested that a fixed value of α can be used, but problems can occur if the data are not persistently exciting. This leads to exponential growth of the P matrix. In an adaptive control situation this can result in "burst" phenomena, as explained below. Initially, with poor parameter estimates, the resulting feedback will lead to bad regulation and hence the data will be rich in information. Then, as the estimates converge, the system under feedback tends to settle down, but simultaneously the P matrix begins to grow due to the loss of persistent excitation. After some time, the parameter estimator can go unstable since P appears as a gain in the algorithm. This can give rise to poor estimates and the resulting feedback controller will begin to perform badly. Thus the cycle will repeat itself (Fortescue, Kershenbaum, and Ydstie, 1981).

The obvious way to overcome the problem above is to vary the exponential data weighting constant, depending on the information content of the data. This leads to the idea of *exponential data weighting with variable foregetting factor*. A simple algorithm of this type (Åström, 1980) is achieved by using the following data-dependent forgetting factor:

$$\alpha(t) = 1 - \alpha \frac{\epsilon(t)^2}{\bar{\epsilon}^2} \tag{6.7.19}$$

where $\epsilon(t)^2$ is the prediction error, $\bar{\epsilon}^2$ is the mean value of $\epsilon(t)^2$ over a certain period, and α is a small constant (say 1/1000). It can be shown that $\bar{\epsilon}(t)$ is proportional to $1 + \phi(t-1)^T P(t-2)\phi(t-1)$ (see Exercise 6.20) and hence (6.7.19) is approximately the same as the following algorithm due to Fortescue et al. (1981):

$$\alpha(t) = 1 - \alpha' \frac{\epsilon(t)^2}{1 + \phi(t-1)^T P(t-2)\phi(t-1)} \tag{6.7.20}$$

The effect of the choice above on the algorithm can be seen as follows. If a sudden change in the plant occurs, $\epsilon(t)^2$ increases; this reduces $\alpha(t)$ temporarily but increases $P(t)$ quickly so that rapid adaptation can occur. After adaptation $\epsilon(t)^2$ decreases and $\alpha(t)$ returns to a value near 1.

Neither of the algorithms (6.7.19) or (6.7.20) guarantees that P will not get excessively large. However, this can be avoided by monitoring the trace of P and temporarily reverting to ordinary least squares whenever trace P goes beyond the preset threshold (Cordero and Mayne, 1981).

An interesting industrial application of the Fortescue et al. (1981) variable forgetting factor algorithm has been described in Dumont (1982).

Further Discussion

Simulation experience suggests that, in practice, there is not that much difference between the performance characteristics of each of these schemes. The underlying principle of each method is to try to retain the fast initial convergence of least squares while preventing the algorithm from turning off. One can distinguish two broad classes of parameter variation with time: jump parameters and drifting parameters.

In the case of *jump parameters*, we consider the parameter to be piecewise constant with infrequent changes. The key problem then is to detect when a change has occurred and to take appropriate action. In order that one can be sure of detecting a parameter change, it is theoretically necessary to probe the system continuously (or periodically) using a persistently exiciting external signal. Once a parameter change has been detected, the obvious strategy is to revitalize (or reinitialize) the least-squares algorithm at the point where the parameters change. This is automatically achieved by the variable forgetting factor algorithm. The same effect can also be ahieved by the least-squares algorithm with covariance resetting if the prediction error is monitored to detect parameter changes and subsequently to initiate resetting of the covariance. In practice, both alternatives have been found to work well.

In the case of *drifting parameters*, the algorithm must be continuously kept alive to keep abreast of the current parameter values. There are many ways of achieving this: for example, by (1) periodic resetting of P (which is equivalent to having ordinary least squares over consecutive finite data windows during which the parameters are approximately constant), (2) exponential data weighting, or (3) covariance modification (which attempts to counteract continuous random parameter variations).

If covariance resetting is used with drift parameters, an obvious practical question that arises is how long the resetting interval should be. In the presence of measurement noise, it can be argued that the mean-square parameter estimation error due to the effect of the noise decreases inversely with the window length (because of averaging), whereas the error due to the effect of parameter drift increases proportionally to the window length (because of the accumulation effect). Thus, in general, the mean-square error, $\bar{e}^2$, has the form $\bar{e}^2 = K_1/L + K_2L$, where L is the window length (or reset interval), K_1 is proportional to the noise variance, and K_2 is proportional to the rate of parameter drift. Hence there exists an optimum window length, L^*, which achieves a balanced compromise between the two competing effects; specifically $L^* = \sqrt{K_1/K_2}$ giving $\bar{e}^2 = 2\sqrt{K_1K_2}$. Obviously, the product of the noise variance and rate of parameter drift sets a lower bound on the achievable performance.

Theoretical Convergence Considerations

It is difficult to analyze rigorously the performance of the algorithms above without making precise assumptions about the nature of the parameter time variations and noise. One highly desirable property is that the algorithms should at least be globally convergent if the parameters happen to be time invariant! Most of the algorithms described above have this property. For example, the algorithm with covariance modification [equations (6.7.2) to (6.7.4)] and the algorithm with covariance resetting [equations (6.7.12) to (6.7.15)] can easily be shown to lead to globally convergent one-step-ahead adaptive controllers by using the properties of these algirithms derived in Section 3.3. (See Exercises 6.17 and 6.18.) Similarly, Cordero and Mayne (1981) have proven global convergence of a one-step-ahead adaptive controller based on the variable forgetting factor algorithm (6.7.20).

With a precise description of the time variations, one may carry out a rigorous theoretical analysis as in Goodwin and Teoh (1983). In practice, it is probably preferable to take a pragmatic approach and simply to tailor the algorithm to the expected nature of the parameter variations following the general guidelines suggested above.

6.8 SOME IMPLEMENTATION CONSIDERATIONS

In this section we give a brief account of some implementation considerations in adaptive control design. This is intended to raise some issues that are relevant to practical applications.

At the outset we would like to make three observations:

1. Adaptive control is a useful way of approaching control problems, but in practice one must always keep in mind the practical realities of the problem under study and include as much physical insight as possible.

2. Adaptive control complements rather than competes with other design techniques.

3. It is sometimes advantageous to combine an adaptive controller with nonadaptive controllers when implementing the final control law (e.g., by implementing an adaptive controller around an existing conventional controller).

We shall discuss the implementation considerations under the following headings:

- Choice of sampling interval
- Choice of control law
- Choice of parameter estimation algorithm
- Use of additional outputs

Choice of Sampling Period

1. One of the first decisions that must be made in digital control is how fast to sample. This will vary greatly depending on the application (from milliseconds to hours). Roughly the sampling period should be approximately one-fifth of the fastest time constant of interest but will also depend on computation speed and other factors.

2. In model reference adaptive control it is desirable that the sampling period be an even multiple of the time delay. If it is not, a nonstably invertible discrete-time model may result from a stably invertible continuous-time system (Wellstead, Prager, and Zanker, 1979a, 1979b). As an example, consider a first-order linear continuous-time system

$$G(s) = \frac{e^{-s\tau}\alpha}{s - \alpha} \qquad \text{where } \tau \text{ is a pure time delay} \qquad (6.8.1)$$

The corresponding discrete-time model (with sampling interval Δ, $\tau = k\Delta + \tau'$; $0 \leq \tau' \leq \Delta$) is

$$G(z) = \frac{z^{-(k+1)}(b_1 + b_2 z^{-1})}{1 - az^{-1}} \qquad (6.8.2)$$

where

$$b_1 = 1 - \exp\left[-(\Delta - \tau')\alpha\right] \qquad (6.8.3)$$

$$b_2 = \exp\left[-(\Delta - \tau')\alpha\right] - \exp\left(-\Delta\alpha\right) \qquad (6.8.4)$$

$$a = \exp\left(-\Delta\alpha\right) \qquad (6.8.5)$$

Note that when τ is an integer multiple of Δ, then $|b_2|$ is zero. Otherwise, $|b_2|$ is nonzero and may be greater than $|b_1|$, giving a nonstably invertible discrete-time system.

3. When a continuous-time system is sampled, the poles, p, are transformed to e^{pT}, where T is the sampling period. However, there is no simple transformation for the zeros. For example, it is not true that a continuous-time system with zeros in the left half-plane will transform to a discrete-time system with zeros inside the unit disk. On the other hand, it is possible to obtain a discrete-time system with zeros all inside the unit disk from a continuous-time system with zeros in the right half-plane (Åström, Hagander, and Sternby, 1980). It turns out that all continuous-time systems with pole excess larger than 2 will always give sampled systems with unstable zeros if the sampling period is sufficiently small. Conversely, for a stable strictly proper system a large sampling period tends to produce stable zeros for the discrete system (Åström, Hagander, and Sternby, 1980). Thus one must exercise caution in the choice of sampling frequency if one uses a control law that relies on inverse stability of the system (e.g., one-step-ahead controllers).

4. It may appear that the minimum sampling period is restricted by the time taken to update the parameters and output the control. A possible strategy to overcome this is to update the parameter estimates over several sampling periods while computing a new control in every sampling instance based on the most recent parameter estimates. [This idea is discussed further in Elliott (1980).]

5. An advantage of keeping the sampling period reasonably long is that the bandwidth of the controller is thereby limited, and hence unmodeled high-frequency dynamics will not be inadvertently excited by the controller.

Further insight into the effect of various sampling strategies may be obtained by studying the example given in Section 5.4. The reader is strongly advised to consider the various implications of the results presented in that section.

Choice of Control Law

Adaptive control provides a way of automatically tuning a control law, but the choice of the actual form of the control law is up to the designer.

1. The range of possible control laws is very wide. We have given emphasis to:

a. *Minimum prediction error controllers* (*including model reference*). These controllers aim to achieve good output tracking. The system must be stably invertible, but steady-state errors are not a problem. With weighted one-step-ahead control a compromise is made between output tracking and the size of the control effort. It can apply to a limited class of nonstably invertible systems. In general, steady-state tracking errors will result, but this can be avoided by weighting the change in control rather than the control itself. This introduces integral action into the controller.

b. *Pole assignment*. Here the closed-loop poles are shifted to desired locations. The algorithm is more complicated than those in item a but applies to any linear system. Steady-state errors can be substantial and it is usually necessary to incorporate integral action in the controller. With this modification it is necessary to guard against integral wind-up. (See Chapter 5.)

We have analyzed each of the algorithms above under somewhat idealized assumptions. This is of course necessary to have a comprehensive theoretical treatment. However, in practice one should be aware of the degree of robustness of the algorithms to conditions that do not strictly comply with the assumptions. An example is the choice of time delay in single-input single-output one-step-ahead control. We have argued in Section 5.2.4 that common sense in the choice of the time delay can improve the robustness of the algorithm. This is also true in other situations. For example, if the order of a system is grossly underestimated, then the robustness of any control law will be jeopardized (Rohrs, et al., 1981). Alternatively, even a good model can only represent the system adequately over a certain bandwidth and thus one has to be careful that a high-performance control law does not extend the system beyond the range of validity of the model. This was mentioned in Section 5.2.2 as an advantage of model reference type of adaptive control over one-step-ahead control since the choice of model dynamics allows the bandwidth of the control system to be automatically restricted.

2. In systems having long sampling periods, the time required to carry out the on-line control computations will be negligible compared with the time between successive samples. Thus one may ignore the computation time without serious consequences. However, in systems having short sampling periods, the time required to carry out the on-line control computations can be a significant fraction of the time between samples. In this case it is often best to synchronize the sampling of the output with the change in input (i.e., make the processing time one sampling period). If this is done, the delay of the system is effectively increased by one unit. This fact should be taken account of in the design of the control system. An illustration of this procedure was given in Example 5.2.3. The reader is also referred to Tsuchiya (1982) for further discussion.

Choice of Parameter Estimation Algorithm

It is possible to use any sensible parameter estimation algorithm. We have concentrated on projection and least-squares algorithms (and minor variants thereof). We make the following observations:

1. Least squares is generally *much* superior in terms of convergence rate compared with projection (sometimes a factor of 1000 or more), although the projection scheme does have the advantage of relative simplicity.

2. The projection algorithm gain does not go to zero and therefore it can automatically track time-varying parameters. For the least-squares algorithm, possible modifications for time-varying problems have been discussed in Section 6.7. These modified least-squares-based algorithms are in general preferable to the projection-type algorithms due to the faster initial convergence rate.

3. In theory, any initial parameter estimates can be used. However, in practice poor initial estimates will lead to large transients during initial training. Thus it is generally advantageous to obtain good initial estimates. This can be done in a number of ways:
   a. Using physical modeling.
   b. To carry out off-line identification.
   c. To initially run the parameter estimator using inputs and outputs from the plant with some existing control law. (In this case it is highly desirable to fluctuate the set points to avoid lack of parameter identifiability.) The adaptive controller can then be switched in after a suitable learning period.

4. If operating conditions can be separated into a number of regions, it may be sensible to store different initial parameter estimates corresponding to each region. This is a form of "gain scheduling."

5. It is always desirable to include as much a priori information as possible in the structure of the model. In some cases, this may mean transformation of data (possibly using nonlinear functions) or it may mean using nonlinear models (see Section 6.6).

6. It is a straightforward matter to constrain the parameter estimates to lie in a certain region (see Section 3.7). Thus if a priori information is available, this can be incorporated into the algorithm. Also, many adaptive control laws require

that certain parameter estimates do not approach zero. This can be achieved by constraining the parameter estimates.

7. It is highly desirable to scale variables so that they have approximately the same order of magnitude. This is crucial in the scalar gain algorithm (projection) to avoid slow convergence and is advisable under all circumstances to avoid numerical problems.

8. In some cases it may be desirable to add an external perturbation to enhance parameter convergence. In theory, this is not necessary with time-invariant parameters. However, with time-varying parameters it is important so that parameters drift is detected at an early stage. In the case of low external excitation, it may be helpful to turn off the parameter estimator when the prediction error is small to avoid drift in the estimated parameters.

9. Sometimes unmodeled nonlinearities can lead to difficulties. For example, a dead zone in electromechanical systems with gears can lead to underestimation of the system gain when the change in input level is small. This can then lead to an excessive input signal when a larger input is called for [see Zanker and Wellstead (1979)]. This kind of difficulty requires a problem specific remedy. Some ideas that could be tried would be to estimate the dead zone or turn off the parameter estimator when the change in input is small.

10. We have basically ignored noise and disturbances in our discussion in this part of the book. However, in practice there will always be errors due to unmodeled dynamics, measurement noise, small disturbances, discretization approximations, and so on. In these cases, the robustness of the algorithms can be improved by introducing a dead zone into the parameter estimation algorithms as in Section 3.6. The heuristic motivation for this is that when the prediction error is small, one is more likely to be influenced by the noise than by the parameters errors. In fact, it is possible to establish global convergence of adaptive control laws using the results of Section 3.6 when the noise is bounded (see Exercise 6.7).

11. When the input goes into saturation, it is highly desirable to use the actual input signal in the parameter estimator since this retains the convergence properties. Thus input saturation should be done in the software (see Remark 6.3.3).

Use of Additional Outputs

So far in this chapter we have assumed that the number of outputs is equal to the number of inputs and that each output is required to track a given desired output sequence. However, in many practical situations other output measurements may be available in addition to those outputs that are required to track the reference values.

Two possibilities come to mind:

1. It may be possible to directly measure certain disturbances acting on the system and this may enhance the predictability of the output of interest.

2. It may be possible to measure certain "internal" variables which although not strictly necessary in predicting the output may vastly improve the prediction if they are incorporated in the model.

An appropriate formalization of this is shown in Fig. 6.8.1, where $y(t)$ is a $r \times 1$ vector of outputs required to track $y^*(t)$, $d(t)$ is a vector of measured disturbances, $z(t)$ is an $l \times 1$ vector of additional output measurements, and $u(t)$ is an $r \times 1$ input vector.

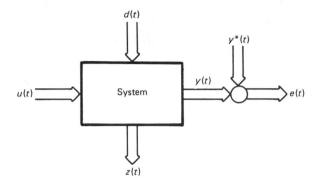

Figure 6.8.1 System with additional outputs.

In order that a good predictor for the output, $\{y(t)\}$, be obtained, it may be highly desirable, if not essential, to incorporate $d(t)$ and $z(t)$ in the model. This can be done by simply writing the predicted model output (assuming a common delay for illustration) as

$$y(t + d) = \theta_0 \phi(t) \tag{6.8.6}$$

where $\phi(t)$ is now made a function of $y(t)$, $y(t - 1)$, ..., $u(t)$, $u(t - 1)$, ..., $z(t)$, $z(t - 1)$, ..., $d(t)$, $d(t - 1)$, ...

The adaptive control algorithms of previous sections can now be applied to the model (6.8.6). The proofs of convergence of the algorithms follow exactly as before with similar assumptions and provided that the system states are observable from $\{y(t)\}$.

The inclusion of $\{d(t)\}$ in the model gives *feedforward action* in as much as the input is instantaneously adjusted to compensate for changes in the disturbances. The reader may wonder what could be done if the disturbances were not directly available. In this case, one tries to predict the disturbances from observations of $\{y(t)\}$ (and $\{z(t)\}$).

There are basically two choices open. One can assume that the disturbances are exactly precictable and can be modeled by observable but uncontrollable linear finite-dimensional systems (as can be done for steps, ramps, sine waves, etc.). These can then be incorporated into the DARMA predictor without difficulty, as we have seen in Chapter 2. Alternatively, one can assume that the disturbances are predictable up to a white noise residual. This leads to the stochastic adaptive control considerations to be discussed in the next part of the book.

To illustrate how the model of (6.8.6) might be obtained in practice, consider the simple cascade system of Fig. 6.8.2. If the data $\{z(t)\}$ is ignored, the input–output model becomes

$$A_1(q^{-1})A_2(q^{-1})y(t) = q^{-(d_1+d_2)}B_1(q^{-1})B_2(q^{-1})u(t) \tag{6.8.7}$$

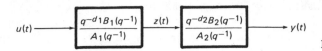

Figure 6.8.2 Simple cascade system.

where A_1, A_2, B_i, and B_2 are taken to be of order n_1, n_2, m_1, and m_2, respectively. The $(d_1 + d_2)$-step-ahead predictor for $\{y(t)\}$ then has the form

$$y(t + d_1 + d_2) = \alpha(q^{-1})y(t) + \beta(q^{-1})u(t) \tag{6.8.8}$$

where $\alpha(q^{-1})$ has order $(n_1 + n_2 - 1)$.
$\beta(q^{-1})$ has order $(m_1 + m_2 + d_1 + d_2 - 1)$ (a total of $n_1 + n_2 + m_1 + m_2 + d_1 + d_2$ coefficients).
Alternatively, using $\{z(t)\}$ we have

$$A_1(q^{-1})z(t) = q^{-d_1}B_1(q^{-1})u(t) \tag{6.8.9}$$

$$A_2(q^{-1})y(t) = q^{-d_2}B_2(q^{-1})z(t) \tag{6.8.10}$$

From (6.8.10) we can write

$$y(t + d_1 + d_2) = \alpha_2(q^{-1})y(t) + \beta_2(q^{-1})z(t + d_1) \tag{6.8.11}$$

where $\alpha_2(q^{-1})$ has order $(n_2 - 1)$ and $\beta_2(q^{-1})$ has order $(m_2 + d_2 - 1)$. Now $\beta_2(q^{-1})$ can be written as

$$\beta_2(q^{-1}) = \beta_2^F(q^{-1}) + \beta_2^L(q^{-1})q^{-d_1} \tag{6.8.12}$$

where

$$\beta_2^F(q^{-1}) = \beta_{2,0}\beta_{2,1}q^{-1} + \cdots + \beta_{2,d-1}q^{-(d_1-1)}$$

$$\beta_2^L(q^{-1}) = \beta_{2,d_1} + \beta_{2,d_1+1}q^{-1} + \cdots + \beta_{2,m_2+d_2-1}q^{-(m_2+d_2-1)+d_1}$$

Thus (6.8.11) can be expressed as

$$y(t + d_1 + d_2) = \alpha_2(q^{-1})y(t) + \beta_2^L(q^{-1})z(t) + \beta_2^F(q^{-1})z(t + d_1) \tag{6.8.13}$$

Then using (6.8.9), we can write

$$z(t + d_1) = \alpha_1(q^{-1})z(t) + \beta_1(q^{-1})u(t) \tag{6.8.14}$$

where $\alpha_1(q^{-1})$ is of order $(n_1 - 1)$ and $\beta_1(q^{-1})$ is of order $(m_1 + d_1 - 1)$. Substituting (6.8.14) into (6.8.13) gives

$$y(t + d_1 + d_2) = \bar{\alpha}(q^{-1})y(t) + \bar{\beta}(q^{-1})u(t) + \bar{\gamma}(q^{-1})z(t) \tag{6.8.15}$$

where

$\bar{\alpha}(q^{-1}) = \alpha_2(q^{-1})$ [with n coefficients]

$\bar{\beta}(q^{-1}) = \beta_2^F(q^{-1})\beta_1(q^{-1})$ [with $m + 2d_1 - 1$ coefficients]

$\bar{\gamma}(q^{-1}) = \beta_2^L(q^{-1}) + \beta_2^F(q^{-1})\alpha_1(q^{-1})$ [with max $(m_2 + d_2 - d_1, d_1 + n_1 - 1)$ coefficients]

Equation (6.8.15) has approximately the same number of coefficients as (6.8.8). However, (6.8.15) has the practical advantage that recent values of $z(t)$ have been used to replace distant values of $y(t)$ in (6.8.8). Thus, although (6.8.8) and (6.8.15) are theoretically equivalent, (6.8.15) may give better predictions of the output in practice. (We shall find that this is definitely the case for stochastic systems discussed later.)

Further Reading

Further discussion on the practical aspects of adaptive control may be found in Clarke, Cope, and Gawthrop (1975), Zanker and Wellstead (1979), Unbehauen, Schmid, and Klein (1978), Jacobs and Saratchandran (1980), Carvalhal, Wellstead, and Pereira (1976), Keviczky and Kumar (1979), Wellstead and Zanker (1978), Narendra and Monopoli (1980), and Harris and Billings (1981). (Additional references are also given in the latter two books.)

As we have remarked earlier, adaptive control algorithms are typically analyzed under ideal assumptions. Clearly, this is a desirable first step since one would have little confidence in an algorithm that could not be shown to work well under ideal conditions. A number of results are also available concerning the robustness properties of adaptive control algorithms in nonideal situations. For further information the reader is referred to Johnson and Goodwin (1982), Rohrs et al. (1981, 1982), Anderson (1982), Kosut and Friedlander (1982), Ioannou and Kokotovic (1982), Balas (1982), and Wittenmark and Åström (1982).

This ends our discussion of adaptive filtering, prediction, and control for deterministic systems. If you have enjoyed the material so far, please read on. Alternatively, if you have not enjoyed the book so far, then, alas, you definitely will not enjoy what is to come. Unfortunately, we do not offer refunds.

EXERCISES

6.1. Establish the convergence properties of the adaptive control algorithm given in (6.3.26) to (6.3.31).

6.2. Establish the convergence properties of the algorithm given in (6.3.36) to (6.3.39) [one-step-adaptive controller (least-squares iteration)].

6.3. Modify the algorithm of Exercise 6.2 so that the estimate of β_0 is constrained away from zero. Establish convergence of the resulting algorithm.

6.4. (a) Extend the one-step-ahead adaptive controller linear control form algorithm (6.3.52) to (6.3.56) to include a least-squares parameter estimator.
  (b) Establish convergence properties for the resulting algorithm.

6.5. Establish convergence for the weighted one-step-ahead control algorithm described in (6.3.88) to (5.3.93).

6.6. Establish the convergence properties of an adaptive pole assignment algorithm based on the projection algorithm for parameter estimation. (*Hint:* Follow the approach of Theorem 6.5.1.)

6.7. Consider the following linear system with bounded errors:

$$y(t + 1) = \phi(t)^T\theta_0 + \omega(t + 1)$$

where

$$\phi(t)^T = \{y(t), \ldots, u(t), \ldots\}$$
$$|\omega(t)| < \Delta$$

We propose use of the projection algorithm with dead zone [equations (3.6.2) and (3.6.3)].

$$\hat{\theta}(t) = \hat{\theta}(t-1) + \frac{a(t-1)\phi(t-1)}{c + \phi(t-1)^T\phi(t-1)}[y(t) - \phi(t-1)^T\hat{\theta}(t-1)]$$

$$a(t-1) = \begin{cases} 1 & \text{if } |y(t) - \phi(t-1)^T\hat{\theta}(t-1)| > 2\Delta \\ 0 & \text{otherwise} \end{cases}$$

$$\phi(t-1)^T\hat{\theta}(t-1) = y^*(t)$$

Assume that the system is stably invertible and use Lemma 3.6.1 to show (by using a contradiction argument as in Lemma 6.2.1):

(a) $\{y(t)\}$, $\{u(t)\}$ are bounded

(b) $\limsup_{t\to\infty} |y(t) - y^*(t)| \le 2\Delta$

6.8. (a) Which adaptive control algorithm would you propose to use on the following system?

$$A(q^{-1})y(t) = B(q^{-1})u(t)$$

where

$A(q^{-1})$ is nominally $1 - 1.2q^{-1}$

$B(q^{-1})$ is nominally $q^{-1}(1 - 3.1q^{-1} + 2.2q^{-2})$

(b) Show that the controller must be unstable so that the closed-loop system is stable. (*Hint:* Use root locus.)

6.9. Consider a model

$$A(q^{-1})y(t) = q^{-1}B(q^{-1})u(t)$$

Say that a weighted one-step-ahead controller is designed using the following cost function:

$$J = \frac{1}{2}[y(t+1) - y^*(t+1)]^2 + \frac{\lambda}{2}\bar{u}(t)^2$$

where

$$P(q^{-1})\bar{u}(t) = R(q^{-1})u(t)$$

(a) Show that if the controller is implemented on a system

$$A_0(q^{-1})y(t) = q^{-1}B_0'(q^{-1})u(t)$$

where

$$A_0(q^{-1}) = A(q^{-1}); \qquad B_0'(q^{-1}) = B(q^{-1})$$

then zero steady-state error is achieved for constant $y^*(t)$ provided that $(1 - q^{-1})$ is a factor of $R(q^{-1})$.

(b) Repeat part (a) for $A_0(q^{-1}) \ne A(q^{-1})$ and $B_0'(q^{-1}) \ne B'(q^{-1})$ (i.e., the model used for design does not correspond exactly with the actual system). Show that the steady-state error is, in general, *not* zero in this case.

(c) Explain why zero-state error is always achieved if an integrator is used in the pole assignment algorithms irrespective of whether $A(q^{-1}) = A_0(q^{-1})$ and $B'(q^{-1}) = B_0'(q^{-1})$.

(d) Why can adaptive controllers overcome the problem in part (b)?

6.10. A particular input matching algorithm in which a fixed value of β_0 is used is described below (for definition of input matching see Remark 6.3.6). We factor β_0 from the predictor (6.3.3) to obtain

$$y(t+d) = \beta_0(\alpha_0'y(t) + \cdots + \alpha_{n-1}'y(t-n+1) + u(t) + B_1'u(t-1)$$
$$+ \cdots + \beta_{m+d-1}'u(t-m-d+1))$$

As before, we define the tracking error as $\epsilon(t + d)$, where

$$\epsilon(t + d) \triangleq y(t + d) - y^*(t + d)$$

$$= \beta_0\Big(u(t) + \alpha'_0 y(t) + \cdots + \alpha'_{n-1} y(t - n + 1) + \beta'_1 u(t - 1) + \cdots$$

$$+ \beta'_{m+d-1} u(t - m - d + 1) - \frac{1}{\beta_0} y^*(t + d)\Big)$$

$$= \beta_0(u(t) - \phi(t)^T \theta'_0)$$

where

$$\phi(t)^T = (-y(t) - \cdots - y(t - n + 1), -u(t - 1) - \cdots$$

$$- u(t - m - d + 1), y^*(t + d))$$

$$\theta'^T_0 = \Big(\alpha'_0, \ \ldots, \ \alpha'_{n-1}, \ \beta'_1 \ \ldots, \ \beta'_{m+d-1}, \ \frac{1}{\beta_0}\Big)$$

Show that the tracking error can be made identically zero by choosing $u(t)$ such that

$$u(t) = u^*(t) \triangleq \phi(t)^T \theta'_0$$

6.11. Consider the following input matching adaptive control algorithm using the same notation and setup as Exercise 6.10.

$$\hat{\theta}(t) = \hat{\theta}(t - d) - \frac{1}{\hat{\beta}_0} \phi(t - d)[c + \phi(t - d)^T \phi(t - d)]^{-1} \epsilon(t); \qquad c > 0$$

$$\epsilon(t) = y(t) - y^*(t)$$

$$u(t) = \phi(t)^T \hat{\theta}(t)$$

where $\hat{\beta}_0$ is a *fixed* constant and $\hat{\theta}(t)$ is a vector of $(n + m + d)$ reals depending on d initial values $\hat{\theta}(0), \ \ldots, \ \hat{\theta}(d - 1)$ and on $y(\tau), 0 < \tau \leq t, u(t), 0 < \tau \leq t - d - 1$. Note that there are actually d separate recursions interlaced. Show that the algorithm above yields

$$\frac{1}{\hat{\beta}_0} \epsilon(t) = \frac{\beta_0}{\hat{\beta}_0}[u(t - d) - u^*(t - d)]$$

where $u^*(t - d)$ can be regarded as the unknown optimal input. (This justifies the name "input matching.")

6.12. Analyze the convergence properties of the algorithm in Exercise 6.11.
[*Hint:* Show that

$$\tilde{\theta}(t) = \tilde{\theta}(t - d) - \frac{\beta_0}{\hat{\beta}_0} \frac{\phi(t - d)\phi(t - d)^T \tilde{\theta}(t - d)}{c + \phi(t - d)^T \phi(t - d)}; \qquad \tilde{\theta}(t) = \hat{\theta}(t) - \theta_0$$

and consider the d-interlaced algorithms separately.]

6.13. Consider a multi-input multi-output system with interactor matrix

$$\xi(q) = qI$$

Factor the matrix β_0 from the right-hand side of the predictor to give

$$y(t + 1) = \beta_0\{u(t) + D(q^{-1})u(t - 1) + C(q^{-1})y(t)\}$$

where

$$D(q^{-1}) = \beta_0^{-1}\beta(q^{-1}) - I$$

$$C(q^{-1}) = \beta_0^{-1}\alpha(q^{-1})$$

(a) Show that the tracking error can be written as follows, (*Hint:* See Exercise 6.10.)

$$\epsilon(t) \triangleq y(t + 1) - y^*(t + 1)$$
$$= \beta_0\{u(t) - \theta_0^T\phi(t)\}$$

What is the form of θ_0^T and $\phi(t)$?

(b) Show that the following input matching adaptive control algorithm will converge provided that

$$K^T + K - K^T K$$

is positive definite where $K = \Gamma\beta_0$.

MIMO Input Matching Algorithm

$$\hat{\theta}(t + 1) = \hat{\theta}(t) - [c + \phi(t)^T\phi(t)]^{-1}\phi(t)\epsilon(t + 1)^T\Gamma^T; \quad c > 0$$
$$u(t) = \hat{\theta}(t)^T\phi(t)$$
$$\epsilon(t) = \bar{y}(t) - \bar{y}^*(t)$$

and Γ is a fixed $r \times r$ matrix.

(c) Comment on the relationship between the condition given in part (b) and that found in Exercise 6.10(c).

6.14. Extend Exercise 6.13 to the case of general interactor matrices.

6.15. Let us presume that a system has p parameters $(\theta_0)_i$, $i = 1, \ldots, p$, and that a range of values is known for each $(\theta_0)_i$, that is,

$$(\theta_0)_i^l \leq (\theta_0)_i \leq (\theta_0)_i^u$$

(a) Show how the ideas described in Section 3.7 can be used to constrain the estimated parameters as above. (Consider both projection and least squares separately.)

(b) Develop an adaptive control algorithm and establish global convergence for it.

6.16. *Model Reference Control with Filtered Errors*

The control law (6.3.105) can be expressed in a slightly different form as

$$H(q^{-1})u^*(t) = -M(q^{-1})u^*(t) - P(q^{-1})y(t) + \frac{g}{b_0}H(q^{-1})r(t)$$

where

$$M(q^{-1}) = \frac{1}{\beta_0}\beta(q^{-1}) - H(q^{-1})$$
$$= m_1 q^{-1} + \cdots + m_{s-1}q^{-(s-1)}$$
$$P(q^{-1}) = \frac{1}{\beta_0}\alpha(q^{-1})$$

Let's assume $H(q^{-1})$ is stable and b_0 is known. With this assumption, the above control law is asymptotically equivalent to

$$u^*(t) = -M(q^{-1})u_F(t) - P(q^{-1})y_F(t) + \frac{g}{b_0}r(t)$$

where $\{u_F(t)\}$, $\{y_F(t)\}$ are filtered versions of $\{u(t)\}$, $\{y(t)\}$ satisfying:

$$H(q^{-1})u_F(t) = u(t)$$
$$H(q^{-1})y_F(t) = y(t)$$

Thus the optimal control law can be expressed as

$$u^*(t) = -\psi(t)^T\bar{\theta}_0 + \frac{g}{b_0}r(t)$$

where

$$\psi(t)^T = [u_F(t-1), \ldots, u_F(t-s+1), y_F(t), \ldots, y_F(t-s+1)]$$
$$\theta_0^T = [m_1, \ldots, m_{s-1}, p_0, \ldots, p_{s-1}]$$

Show that irrespective of what control law is used, the system model can be expressed as

$$\frac{1}{b_0} y_a'(t+d) = P(q^{-1}) y_F(t) + M(q^{-1}) u_F(t) + u(t)$$

where

$$H(q^{-1}) y_a'(t) = E(q^{-1}) y(t)$$

Hence $(1/b) y_a'(t+d)$ can be expressed as

$$\frac{1}{b_0} y_a'(t+d) = u(t) + \psi(t)^T \bar{\theta}_0$$

Now subtracting $g/b_0\, r(t)$ from both sides of the above equation, we obtain

$$\frac{1}{b_0}[y_a(t+d) - gr(t)] = u(t) + \psi(t)^T \bar{\theta}_0 - \frac{g}{b_0} r(t)$$

(Note the right hand side represents $u(t) - u^*(t)$). Define the tracking error as usual as

$$\epsilon(t) = y(t) - y^*(t)$$

and show that

$$E(q^{-1})\epsilon(t+d) = b_0 H(q^{-1})\left[u(t) + \psi(t)^T \bar{\theta}_0 - \frac{g}{b_0} r(t)\right]$$

which is true independently of how we choose $u(t)$. Now let $\hat{\theta}(t)$ denote an estimate of $\bar{\theta}_0$. Consider the following adaptive control law:

$$u(t) = -\psi(t)^T \hat{\theta}(t) + \frac{g}{b_0} r(t)$$

Show that the corresponding tracking error $\{\epsilon(t)\}$ is given by

$$E(q^{-1})\epsilon(t+d) = b_0 H(q^{-1})[\psi(t)^T(\bar{\theta}_0 - \hat{\theta}(t))]$$

In the light of this equation and the discussion of estimation with filtered errors in Exercises 3.23 and 3.24, consider the following algorithm.

Adaptive Control Algorithm with Filtered Errors (b_0 Known)

$$\hat{\theta}(t) = \hat{\theta}(t-1) + \frac{\bar{\psi}(t-d)}{c + \bar{\psi}(t-d)^T \bar{\psi}(t-d)} \bar{e}(t); \quad c > 0$$

where

$$\bar{e}(t) = \frac{1}{b_0}\epsilon(t) + \tilde{e}(t)$$

$\tilde{e}(t)$ is an auxiliary signal given by

$$\tilde{e}(t) = G'(q^{-1})\{\psi(t)^T \hat{\theta}(t)\} - [G'(q)\psi(t)]^T \hat{\theta}(t-1)$$
$$= G'(q)\{\psi(t)^T \hat{\theta}(t)\} - \bar{\psi}(t-d)^T \hat{\theta}(t-1)\}$$
$$\bar{\psi}(t-d) = G'(q)\psi(t)$$
$$G'(q) = \frac{1}{g} G(q) \quad \text{where } G(q) \text{ is as in (6.3.99)}$$

Finally, the control law is given by

$$u(t) = -\psi(t)^T \hat{\theta}(t) + \frac{g}{b_0} r(t)$$

Using the properties of the algorithm above (established in Exercises 3.23 and 3.24), establish the following global convergence result. Subject to the usual stable invertible assumptions, the reference model assumptions, and $H(q^{-1})$ stable, the algorithm when applied to the system (6.3.97) yields

1. $\{u(t)\}, \{y(t)\}$ bounded
2. $\lim\limits_{t \to \infty} [y(t) - y^*(t)] = 0$

6.17. (a) Develop a one-step-ahead adaptive controller using the least-squares algorithm with covariance modification [equations (6.7.2) to (6.7.4)].

(b) Establish global convergence of the resulting algorithm when applied to a linear *time-invariant* system.

6.18. Repeat Exercise 6.17 for the least-squares algorithm with covariance resetting [equations (6.7.12) to (6.7.15)].

6.19. Repeat Exercise 6.17 for the least-squares algorithm with variable forgetting factor chosen as in (6.7.20) [see Cordero and Mayne (1981)].

6.20. Assuming white measurement noise of variance σ^2, that is, $y(t) = \phi(t-1)^T \theta_0 + \omega(t)$, where $E\{\omega(t)^2\} = \sigma^2$, verify that the mean-square prediction error is given by

$$E\{\epsilon(t)^2\} = [1 + \phi(t-1)P(t-2)\phi(t-1)]\sigma^2$$

where

$$\epsilon(t) = y(t) - \phi(t-1)^T \hat{\theta}(t-1)$$

(This exercise relies on some knowledge of Kalman filtering theory; see Chapter 7.)

6.21. Extend the model reference adaptive controller (linear control form) [equations (6.3.117) and (6.3.118)] to include a least-squares parameter estimator.

6.22. Develop a model reference adaptive control algorithm using the same approach as in the prototype one-step-ahead adaptive controller [equations (6.3.9) and (6.3.10)].

6.23. (a) Assuming that β_0 is known, develop adaptive versions of the one-step-ahead control laws described in Section 5.2.3 for nonsquare systems (consider the case $m > r$ and $m < r$).

(b) Establish global convergence for the adaptive control algorithms described above under the same assumptions as those which give closed-loop stability in the non-adaptive case (assume that β_0 is known).

(c) Discuss the role played by the matrix β_0 in the algorithms above.

6.24. Supply details of the steps leading to (6.7.10) and (6.7.11), that is, the data discarding algorithm.

6.25. (a) Show that if the parameters change, then with a persistently exciting external signal, this change will be detected in the prediction error in at most $2n$ steps, where n is the order of the system.

(b) Show that in the absence of persistent excitation, a parameter change can go unnoticed in the prediction error.

6.26. Verify the responses shown in Figs. 6.3.3 to 6.3.5 by simulation.

6.27. (a) Explain why it is desirable to reset the covariance in the least-squares algorithm even in the case of time-invariant systems.

(b) Simulate a system with sinusoidal disturbances and compare the performance of a one-step-ahead controller using ordinary least squares and least squares with periodic resetting of the covariance.

6.28. **(a)** Develop a direct model reference adaptive control for a multivariable system having known interactor matrix. [*Hint:* Consider the predictor for $E(q^{-1})\xi(q)y(t)$.]

(b) Prove global convergence of your algorithm.

6.29. Repeat Exercise 6.28 using an indirect adaptive control algorithm.

6.30. Investigate the relationship between the pole assignment adaptive controllers and the weighted one-step-adaptive controllers with cost function as in (6.3.64) by considering the constraint on the pole locations for the latter algorithm. Illustrate your answer by reference to the system discussed in Section 5.4.

Part II

STOCHASTIC SYSTEMS

In the second part of the book we treat adaptive filtering, prediction, and control in the presence of random or stochastic disturbances. We shall find that most of the ideas from Part I carry over in a natural way to the stochastic case.

The main difference here is that, whereas in Part I we implicitly assumed that the system output could be predicted perfectly, we will now assume that the output can be predicted up to a "white noise" residual (the innovations). The nature of the problem is still basically the same, although the details of the algorithms may differ slightly to handle the random components.

The presence of random components means that the problems are best formulated in a probabilistic setting. This will require slightly more sophisticated mathematics than Part I, but the underlying approach follows essentially the same line.

The second part of the book relies on an elementary knowledge of probability theory and stochastic processes, as covered for example in Burrill (1973) and at a more advanced level in Loeve (1963). Adequate background material with an engineering bias is covered in Åström (1970) and Cox and Miller (1965).

7

Optimal Filtering and Prediction

7.1 INTRODUCTION

In this chapter we give an introduction to optimal filtering for systems having random disturbances. We also discuss the relationship between optimal filtering and the modeling of stochastic processes.

Filtering is concerned with the extraction of signals from noise. If the signal and noise spectra are essentially nonoverlapping, it is possible to design a filter that passes the desired signal but attenuates the unwanted noise component. The resulting filter would either be of a low-pass, band-pass, or high-pass type, depending on the relative frequencies of the desired signal and noise. There are standard procedures for the design of such filters; see, for example, the books by Stanley (1975), Rabiner and Gold (1975), Chen (1979), Bogner and Constantinides (1975), Oppenheim and Schafer (1975), and Oppenheim (1978).

When the signal and noise spectra are overlapping, the question arises as to what is the best filter characteristic to pass the signal while suppressing the noise. This problem was first studied by Wiener (1949) and Kolmogorov (1941), who formulated the filter design problem using statistical and frequency-domain ideas.

We shall restrict attention to linear causal filters of the type shown in Fig. 7.1.1. We shall denote by $s(t + \alpha)$ the signal that is to be estimated by the "filter" output, $g(t)$. We shall require that the filter be causal in the sense that $g(t)$ may only be a function of $y(t)$, $y(t - 1)$, ... We shall distinguish the following types of estimates (Brown, 1962):

1. If $\alpha > 0$, $g(t)$ is called a *predicted* estimate of $s(t)$.

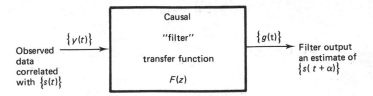

Observed data correlated with $\{s(t)\}$ $\xrightarrow{\{y(t)\}}$

Causal "filter" transfer function $F(z)$

$\xrightarrow{\{g(t)\}}$ Filter output an estimate of $\{s(t+\alpha)\}$

Figure 7.1.1 Generalized filtering problem.

2. If $\alpha = 0$, $g(t)$ is called a *filtered* estimate of $s(t)$.

3. If $\alpha < 0$, $g(t)$ is called a *smoothed* estimate of $s(t)$.

In general, $\{s(t)\}$ may be any signal that is correlated with $\{y(t)\}$. For example, $\{s(t)\}$ may be a signal that is corrupted by noise in forming $\{y(t)\}$. A special case of some interest is when $s(t + \alpha)$ corresponds to the future value of the sequence $\{y(t)\}$. The filtering problem then corresponds to *prediction*.

Prediction was described in Chapter 4 for deterministic systems. In the deterministic case, it was possible to predict the future response exactly. In the case of stochastic systems, it is not possible to obtain a perfect prediction. Instead, the predictor aims to minimize the prediction error variance. Thus an optimal predictor is a special form of optimal filter.

We show below that if $\{y(t)\}$ and $\{s(t)\}$ are modeled by linear stochastic models, a solution to the general filtering problem (prediction, filtering or smoothing) can be obtained by use of the Kalman filter.

We shall also show that the idea of an optimal linear filter leads naturally to the idea of an innovations sequence. Roughly speaking, the innovations correspond to the new information that is contained in each successive observation. This idea leads us in turn to develop innovations models for stochastic systems and finally to describe stochastic autoregressive moving-average models (ARMA models). These are discussed in Section 7.4.1.

In Section 7.5 we describe optimal restricted complexity filters and predictors. These are filters whose structure is prespecified and hence does not necessarily correspond to the true underlying model of the data generating mechanism. A very important class of optimal restricted complexity filters and predictors is described in Section 7.6, namely those based on a lattice structure. These filters have found considerable application in the area of adaptive filtering.

We begin our discussion of optimal filtering in the next section by describing state-space models for stochastic systems.

7.2 STOCHASTIC STATE-SPACE MODELS

State-space models for linear deterministic systems were discussed in Section 2.2. In the discrete-time case, these models have the following form:

$$x(t + 1) = Ax(t) + Bu(t); \qquad x(0) = x_0 \qquad (7.2.1)$$

$$y(t) = Cx(t) \qquad (7.2.2)$$

where $t \in \{0, 1, 2, \ldots\}$.

We have seen in Chapter 2 that the model above can describe the internal and external behaviour of a general linear finite-dimensional system for arbitrary initial state. This can include a description of various purely deterministic disturbances (sinusoids, etc.). These appear as observable but uncontrollable modes in the model.

Stochastic components can be added to the model above in a very natural way by simply including additional "white noise" inputs to the model. This leads us to the following stochastic state-space model:

$$x(t + 1) = Ax(t) + Bu(t) + v_1(t); \qquad x(t_0) = x_0 \qquad (7.2.3)$$

$$y(t) = Cx(t) + v_2(t); \qquad\qquad t \geq t_0 \qquad (7.2.4)$$

where $\{v_1(t)\}$ and $\{v_2(t)\}$ denote "white noise" sequences defined on some probability space $(\Omega, \mathcal{C}, P)$. We shall call $\{v_1(t)\}$ the *process noise* and $\{v_2(t)\}$ the *observation noise*. It will subsequently become clear that the model (7.2.3)–(7.2.4) allows a very general description of the stochastic and deterministic components of the noise processes as they appear at the system output. (See also Exercise 7.1.)

We have seen in Chapter 2 that the state of a system represents that minimal amount of information about the past history of the system, which is necessary to predict the future response. In the deterministic case, the model (7.2.1)–(7.2.2) uniquely determines the future state and output response in terms of the present state and the present and future inputs. In the stochastic case, the model (7.2.3)–(7.2.4) has a slightly different interpretation (expressed in a probabilistic setting). In fact, the model (7.2.3)–(7.2.4) describes the evolution of the probability distribution of the system state and output. The model is a *Markov model* since if $t_1 < t_2 < \cdots < t_m < t$, the probability density of $x(t)$ conditioned on $x(t_1), x(t_2), \ldots, x(t_m)$ is simply the probability density of $x(t)$ conditioned on $x(t_m)$:

$$p(x(t)\,|\,x(t_1), \ldots, x(t_m)) = p(x(t)\,|\,x(t_m)) \qquad (7.2.5)$$

We shall work with the following assumptions concerning the model (7.2.3)–(7.2.4).

Assumption 7.2.A

(1) $\{v_1(t)\}$ and $\{v_2(t)\}$ are zero mean stationary white noise processes with covariance given by

$$E\left\{\begin{bmatrix} v_1(t) \\ v_2(t) \end{bmatrix} [v_1(s)^T v_2(s)^T]\right\} = \begin{bmatrix} Q & S \\ S^T & R \end{bmatrix} \delta(t - s) \qquad (7.2.6)$$

where $\delta(t - s)$ is the Kronecker delta.

(2) The initial state x_0 is a random variable of mean $\bar{x}_0$ and covariance P_0. (Further $\{v_1(t)\}$ and $\{v_2(t)\}$ are assumed to be uncorrelated with x_0.)

A consequence of Assumption 7.2.A is that the mean, $\bar{x}(t)$, and covariance, $P(t)$, of the probability distribution of the state satisfy the following recursions:

$$\bar{x}(t + 1) = A\bar{x}(t) + Bu(t); \qquad \bar{x}(t_0) = \bar{x}_0 \qquad (7.2.7)$$

$$P(t + 1) = AP(t)A^T + Q; \qquad P(t_0) = P_0 \qquad (7.2.8)$$

where

$$\bar{x}(t) \triangleq E\{x(t)\} \tag{7.2.9}$$

$$P(t) \triangleq E\{(x(t) - \bar{x}(t))(x(t) - \bar{x}(t))^T\} \tag{7.2.10}$$

The corresponding mean and covariance of the output process, $\{y(t)\}$, are readily seen to be

$$\bar{y}(t) \triangleq E\{y(t)\} = C\bar{x}(t) \tag{7.2.11}$$

$$E\{[y(t) - \bar{y}(t)][y(s) - \bar{y}(s)]^T\} = CA^{t-s}P(s)C^T + R\delta(t - s)$$
$$+ CA^{t-s-1}S[\delta(t - s) - 1]; \qquad t \geq s \geq t_0 \tag{7.2.12}$$

In the special case when Assumption 7.2.A is strengihened to *gaussian* random processes, the joint distribution of the states is gaussian with mean and covariance as given above. (In the gaussian case, the first two moments describe the complete probability distribution.)

In the next section we describe an optimal filter for estimating the states of the model (7.2.3)–(7.2.4).

7.3 LINEAR OPTIMAL FILTERING AND PREDICTION

We have seen in the preceding section that a general model for a linear finite-dimensional stochastic system is

$$x(t + 1) = Ax(t) + Bu(t) + v_1(t) \tag{7.3.1}$$

$$y(t) = Cx(t) + v_2(t); \qquad t \geq t_0 \tag{7.3.2}$$

$x(t_0)$ has mean $\bar{x}_0$ and covariance Σ_0 and

$$E\left\{\begin{bmatrix} v_1(t) \\ v_2(t) \end{bmatrix} [v_1(\tau)^T v_2(\tau)^T]\right\} = \begin{bmatrix} Q & S \\ S^T & R \end{bmatrix} \delta(t - \tau);$$
$$Q \geq 0, \quad R > 0 \tag{7.3.3}$$

(Note that we have used the symbol Σ_0 instead of P_0.)

We show below that the state, $x(t)$, of the model above can be estimated in a simple fashion by use of the Kalman filter (Kalman, 1960, 1963; Kalman and Bucy, 1961).

7.3.1 The Kalman Filter

The Kalman filter provides a way of estimating the state $x(t)$ of the model (7.3.1) to (7.3.3). The filter has the following two interpretations.

1. If the noise is *gaussian*, the filter gives the minimum variance estimate of the state; that is, it evaluates the conditional mean of $x(t)$ given the past data $\{y(t - 1), y(t - 2), \ldots\}$.
2. If the gaussian assumption is removed, the filter gives the *linear* minimum variance estimate (LMVE) of the state (i.e., having the smallest unconditional

error covariance among all linear estimates), but this will not, in general, be the conditional mean.

The connection between properties 1 and 2 above is developed in Section D.7 of Appendix D. For simplicity we will derive the filter under the assumption that the signals are gaussian. We then have the following result:

Theorem 7.3.1 (The Kalman Filter). Consider the signal model of (7.3.1) to (7.3.3) and assume that the initial state and noise sequences are jointly gaussian. Let $\hat{x}(t + 1)$ denote the conditional mean of $x(t + 1)$ given observations of $\{y(t)\}$ up to and including time t. Then $\hat{x}(t + 1)$ satisfies the following recursion (the Kalman filter):

$$\hat{x}(t + 1) = A\hat{x}(t) + K(t)[y(t) - C\hat{x}(t)] + Bu(t) \tag{7.3.4}$$

$$\hat{x}(t_0) = \bar{x}_0 \tag{7.3.5}$$

where $K(t)$ is the filter gain given by

$$K(t) = [A\Sigma(t)C^T + S][C\Sigma(t)C^T + R]^{-1} \tag{7.3.6}$$

$\Sigma(t)$ is the state error covariance, that is,

$$\Sigma(t) \triangleq E\{[\hat{x}(t) - x(t)][\hat{x}(t) - x(t)]^T \mid y(t - 1), y(t - 2), \ldots, y(t_0)\} \tag{7.3.7}$$

$\Sigma(t)$ satisfies the following *Riccati difference equation* (RDE):

$$\Sigma(t + 1) = A\Sigma(t)A^T + Q - K(t)[C\Sigma(t)C^T + R]K(t)^T \tag{7.3.8}$$

$$\Sigma(t_0) = \Sigma_0 \tag{7.3.9}$$

Proof. We use induction: We first note that the result is true at t_0 since $\hat{x}(t_0) = \bar{x}_0, \Sigma(t_0) = \Sigma_0$.

Then assuming that the conditional distribution for $x(t)$ given $y(t - 1) \cdots y(t_0)$ is gaussian with mean $\hat{x}(t)$ and covariance $\Sigma(t)$, we determine the conditional distribution for $x(t + 1)$ given $y(t) \cdots y(t_0)$.

From (7.3.1) and (7.3.2),

$$\begin{bmatrix} x(t + 1) \\ y(t) \end{bmatrix} = \begin{bmatrix} A & I & 0 \\ C & 0 & I \end{bmatrix} \begin{bmatrix} x(t) \\ v_1(t) \\ v_2(t) \end{bmatrix} + \begin{bmatrix} B \\ 0 \end{bmatrix} u(t)$$

Hence using the result for transformation of gaussian random variables (see Appendix D, Lemma D.6.2), the joint distribution of $\begin{bmatrix} x(t + 1) \\ y(t) \end{bmatrix}$ given $(y(t - 1), \ldots, y(t_0))$ is gaussian with mean

$$\begin{bmatrix} A\hat{x}(t) \\ C\hat{x}(t) \end{bmatrix} + \begin{bmatrix} B \\ 0 \end{bmatrix} u(t)$$

and covariance

$$\begin{bmatrix} A\Sigma(t)A^T + Q & A\Sigma(t)C^T + S \\ C\Sigma(t)A^T + S^T & C\Sigma(t)C^T + R \end{bmatrix}$$

The theorem then follows immediately by application of Lemma D.6.3.

▼▼▼

Properties of the Kalman Filter

The key properties of the Kalman filter are:

1. In the case of gaussian noise, $\hat{x}(t)$ is the conditional mean of $x(t)$, that is,

$$\hat{x}(t) = E\{x(t) \,|\, \mathcal{Y}(t - 1)\} \tag{7.3.10}$$

where $\mathcal{Y}(t - 1) \triangleq \{y(t - 1), \ldots, y(t_0)\}$.

2. A consequence of property 1 is that $\hat{x}(t)$ is the minimum variance estimator (MVE) for $x(t)$, that is,

$$E\{\hat{x}(t) - x(t) \,|\, \mathcal{Y}(t - 1)\} = 0 \tag{7.3.11}$$

and

$$\Sigma(t) \triangleq E\{[\hat{x}(t) - x(t)][\hat{x}(t) - x(t)]^T \,|\, \mathcal{Y}(t - 1)\} \leq \Sigma_F \tag{7.3.12}$$

where Σ_F is the error covariance given by any other filter. (For a brief discussion of minimum variance estimators, see Appendix D.)

3. It can be seen from (7.3.6) and (7.3.7) that the Kalman gain $K(t)$, and the conditional error covariance $\Sigma(t)$, are independent of $\mathcal{Y}(t - 1)$ provided that A, C, R, Q, and S are all independent of $\mathcal{Y}(t - 1)$. In this case, $K(t)$ and $\Sigma(t)$ are precomputable and $\Sigma(t)$ is also the unconditional error covariance.

4. If the gaussian assumption is removed, the Kalman filter given by (7.3.4) to (7.3.9) becomes the linear minimum variance estimator (LMVE) of $x(t)$.

   In this case, the error $(x(t) - \hat{x}(t))$ is *uncorrelated* with $y(t_0) \cdots y(t - 1)$, that is,

$$E\{x(t) - \hat{x}(t))y(t - i)\} = 0; \qquad i = 1, \ldots, t - t_0 \tag{7.3.13}$$

and the error covariance $\Sigma(t)$ is minimal among the set of *linear* estimators. (For a discussion of the relationship between LMVE estimators and MVE estimators, see Appendix D.)

5. We shall give a special symbol, namely $\omega(t)$, to the error term $y(t) - C\hat{x}(t)$ appearing in the Kalman filter (7.3.4). We shall call the sequence $\{\omega(t)\}$ the *innovations sequence*. It is defined by

$$\omega(t) \triangleq y(t) - C\hat{x}(t) \tag{7.3.14}$$

where $\hat{x}(t)$ is the state estimate provided by the Kalman filter.

   By using (7.3.10) we see that the innovations have the following property:

$$E\{\omega(t) \,|\, y(t - 1), \ldots, y(t_0)\} = 0 \tag{7.3.15}$$

We also see from (7.3.14) that $\omega(t)$ represents the new information contained in $y(t)$ which is not contained in $\{y(t - 1), \ldots, y(t_0)\}$. This is because

$$y(t) = C\hat{x}(t) + \omega(t) \tag{7.3.16}$$
$$= E\{y(t) \,|\, y(t - 1), \ldots, y(t_0)\} + \omega(t)$$

This motivates the name *innovations sequence*.

6. Using the sequence $\{\omega(t)\}$, the Kalman filter (7.3.4) can be rewritten as

$$\hat{x}(t + 1) = A\hat{x}(t) + Bu(t) + K(t)\omega(t) \tag{7.3.17}$$
$$y(t) = C\hat{x}(t) + \omega(t) \tag{7.3.18}$$

Equations (7.3.17) and (7.3.18) can be thought of as an alternative model for $\{y(t)\}$ driven by the innovations sequence. The model is known as the *innovations model*. Equations (7.3.17) and (7.3.18) show how $\{y(t)\}$ can be generated from $\{\omega(t)\}$. This calculation can also be reversed such that $\omega(t)$ is expressed in terms of $\{y(t)\}$. This leads to the following "*whitening filter*," which is a simple rearrangement of (7.3.17)–(7.3.18):

$$\hat{x}(t + 1) = [A - K(t)C]\hat{x}(t) + Bu(t) + K(t)y(t) \tag{7.3.19}$$

$$\omega(t) = y(t) - C\hat{x}(t) \tag{7.3.20}$$

We thus note that the Kalman filter, innovations model, and whitening filter are simply different ways of writing the optimal linear filter. In general, the filters will be time varying (in spite of the time invariance of the signal model) with $K(t)$ given as in (7.3.6) and (7.3.8). However, we will see below, that subject to mild assumptions, $K(t)$ is asymptotically time invariant.

▼▼▼

Remark 7.3.1. In discussing the properties of the Kalman filter it is possible to take $S = 0$ in (7.3.3) without loss of generality. The case $S \neq 0$ can be handled by noting that (7.3.1) can be written as

$$\begin{aligned} x(t + 1) - Ax(t) + Bu(t) + v_1(t) &- SR^{-1}[y(t) - y(t)] \\ = Ax(t) + Bu(t) + v_1(t) &- SR^{-1}[Cx(t) + v_2(t) - y(t)] \tag{7.3.21} \\ = A'x(t) + Bu(t) &+ SR^{-1}y(t) + v_1'(t) \end{aligned}$$

where

$$A' = A - SR^{-1}C \tag{7.3.22}$$

$$E\left\{ \begin{bmatrix} v_1'(t) \\ v_2(t) \end{bmatrix} [v_1'(\tau)^T v_2(\tau)^T] \right\} = \begin{bmatrix} Q' & 0 \\ 0 & R \end{bmatrix} \tag{7.3.23}$$

$$Q' = Q - SR^{-1}S^T \tag{7.3.24}$$

Since the term $SR^{-1}y(t)$ is known, the optimal filter for the system (7.3.21) and (7.3.2) can be obtained as for the system (7.3.1)–(7.3.2) with $S = 0$.

Thus in the sequel we will take $S = 0$ without further comment.

▼▼▼

For the case $S = 0$, the *Kalman filter* can be summarized as

$$\hat{x}(t + 1) = A\hat{x}(t) + Bu(t) + K(t)[y(t) - C\hat{x}(t)]; \qquad \hat{x}(t_0) = \bar{x}_0 \tag{7.3.25}$$

or, equivalently,

$$\hat{x}(t + 1) = \bar{A}(t)\hat{x}(t) + Bu(t) + K(t)y(t) \tag{7.3.26}$$

where $\bar{A}(t)$ and $K(t)$ are the *filter state transition* and *filter gain* matrices, respectively, given by

$$\bar{A}(t) = A - K(t)C \tag{7.3.27}$$

$$K(t) = A\Sigma(t)C^T(C\Sigma(t)C^T + R)^{-1} \tag{7.3.28}$$

$$\Sigma(t + 1) = A\Sigma(t)A^T - A\Sigma(t)C^T(C\Sigma(t)C^T + R)^{-1}C\Sigma(t)A^T + Q;$$

$$\Sigma(t_0) = \Sigma_0 \tag{7.3.29}$$

Remark 7.3.2. The reader should note that for $A = I$, $B = 0$, $Q = 0$, $C = \phi(t-1)^T$, and $\theta(t) \triangleq x(t)$, the system becomes

$$\theta(t+1) = \theta(t)$$

$$y(t) = \phi(t-1)^T\theta(t) + v_2(t)$$

For this special case, the Kalman filter reduces to the least-squares algorithm discussed in Chapter 3.

▼▼▼

In many cases the error covariance $\Sigma(t)$ and hence the Kalman gain $K(t)$ converge to steady-state values as $t \to \infty$. If $\Sigma(t)$ converges as $t \to \infty$, the limiting solution Σ will satisfy the following *algebraic Riccati equation* (ARE), obtained from (7.3.29) by putting $\Sigma(t+1) = \Sigma(t) \triangleq \Sigma$:

$$\Sigma - A\Sigma A^T + A\Sigma C^T(C\Sigma C^T + R)^{-1}C\Sigma A^T - Q = 0 \tag{7.3.30}$$

By analogy with (7.3.27)–(7.3.28) we also define the *steady-state filter state transition matrix*, $\bar{A}$, and *steady-state filter gain matrix*, K, by

$$\bar{A} \triangleq A - KC \tag{7.3.31}$$

$$K \triangleq A\Sigma C^T(C\Sigma C^T + R)^{-1} \tag{7.3.32}$$

For future convenience, we also factorize R and Q as

$$R \triangleq (R^{1/2})(R^{1/2})^T \tag{7.3.33}$$

$$Q \triangleq DD^T \tag{7.3.34}$$

and define

$$\bar{C} \triangleq R^{-1/2}C \tag{7.3.35}$$

Then the ARE becomes

$$\Sigma - A\Sigma A^T + A\Sigma\bar{C}^T(\bar{C}\Sigma\bar{C}^T + I)^{-1}\bar{C}\Sigma A^T - DD^T = 0 \tag{7.3.36}$$

with $\bar{A}$ given by

$$\bar{A} = A - A\Sigma\bar{C}^T(\bar{C}\Sigma\bar{C}^T + I)^{-1}\bar{C} \tag{7.3.37}$$

Stability of the Kalman Filter

We shall be particularly interested in those solutions of the ARE which are real, symmetric, positive semidefinite, and which give a steady-state filter having roots on or inside the unit circle. We theorefore introduce the following definitions.

Definition 7.3.A. A real symmetric positive semidefinite solution of the ARE is said to be a *stabilizing solution* if the corresponding filter state transition matrix, $\bar{A}$, has all its eigenvalues *inside* the unit circle.

Definition 7.3.B. A real symmetric positive semidefinite solution of the ARE is said to be a *strong solution* if the corresponding filter state transition matrix, $\bar{A}$, has all its eigenvalues *inside or on* the unit circle.

Appendix E gives a summary of the properties of the solution of the ARE. In particular, conditions are given for the existence and uniqueness of both stabilizing and strong solutions. Some of the key properties of the ARE are summarized in the following lemma (see Appendix E for details):

Lemma 7.3.1. Provided that (C, A) is detectable:

(i) The strong solution of the ARE exists and is unique.
(ii) If (A, D) is stabilizable, the strong solution is the only positive semidefinite solution of the ARE.
(iii) If (A, D) has no uncontrollable modes on the unit circle, the strong solution coincides with the stabilizing solution.
(iv) If (A, D) has an uncontrollable mode on the unit circle, then, although the strong solution exists, there is no stabilizing solution.
(v) If (A, D) has an uncontrollable mode inside, or on, the unit circle, the strong solution is not positive definite.
(vi) If (A, D) has an uncontrollable mode outside the unit circle, then as well as the strong solution, there is at least one other positive semidefinite solution of the ARE.

Proof. See Martensson (1971), Kucera (1972a, 1972b, 1972c), and Chan, Goodwin, and Sin (1983).

▼▼▼

The lemma above generalizes standard results (Anderson and Moore, 1979) on the solutions of the ARE to the case where (A, D) is not necessarily stabilizable.

We also have the following three results on the convergence of the solution of the matrix Riccati difference equation (7.3.29) to the stabilizing or strong solutions of the algebraic Riccati equation:

Theorem 7.3.2. Subject to

(i) (A, D) is stabilizable [with D as in (7.3.34)].
(ii) (C, A) is detectable.
(iii) $\Sigma_0 \geq 0$

then

$$\lim_{t \to \infty} \Sigma(t) = \Sigma_s \text{ (exponentially fast)}$$

$$\lim_{t \to \infty} K(t) = K_s \text{ (exponentially fast)}$$

$$\lim_{t \to \infty} \bar{A}(t) = \bar{A}_s \text{ (exponentially fast)}$$

where

$\Sigma(t)$ = solution of the Riccati difference equation with initial condition Σ_0

Σ_s = unique stabilizing solution of the ARE

K_s = steady-state Kalman gain, i.e., $K_s \triangleq A\Sigma_s C^T (C\Sigma_s C^T + R)^{-1}$

$\bar{A}_s$ = steady-state filter state transition matrix, i.e., $\bar{A}_s = A - K_s C$

Proof. See Appendix E.

▼▼▼

Theorem 7.3.3. Subject to:

  (i) there are no uncontrollable modes of (A, D) on the unit circle,

 (ii) (C, A) is detectable,

(iii) $\Sigma_0 > 0$,

then

$$\lim_{t \to \infty} \Sigma(t) = \Sigma_s \qquad \text{(exponentially fast)}$$

$$\lim_{t \to \infty} K(t) = K_s \qquad \text{(exponentially fast)}$$

$$\lim_{t \to \infty} \bar{A}(t) = \bar{A}_s \qquad \text{(exponentially fast)}$$

where $\Sigma(t)$, Σ_s, K_s, and $\bar{A}_s$ are as in Theorem 7.3.2.

Proof. See Appendix E.

▼▼▼

Theorem 7.3.4. Subject to:

 (i) (C, A) is observable,

(ii) $(\Sigma_0 - \Sigma_s) > 0$, or $\Sigma_0 = \Sigma_s$,

then

$$\lim_{t \to \infty} \Sigma(t) = \Sigma_s$$

$$\lim_{t \to \infty} K(t) = K_s$$

$$\lim_{t \to \infty} \bar{A}(t) = \bar{A}_s$$

where $\Sigma(t)$ is as in Theorem 7.3.2, but now Σ_s is the (unique) *strong solution* of the ARE and K_s and $\bar{A}_s$ are the corresponding steady-state filter gain and state transition matrix. [Note that $\bar{A}_s$ will have roots inside the unit circle unless (A, D) has uncontrollable modes on the unit circle, in which case $\bar{A}_s$ will also have the same roots on the unit circle.]

Proof. See Appendix E.

▼▼▼

The theorems above give the limiting properties of the Kalman filter under a variety of conditions. It is clear that these properties are of central importance in optimal filtering since they give sufficient conditions for the asymptotic time invariance and stability of the filter. We shall also see later (Section 7.4) that the properties are important in developing simplified models for stochastic systems.

Theorem 7.3.2 is a standard result in the theory of optimal filtering [see, e.g., Anderson and Moore (1979)]. Theorems 7.3.3 and 7.3.4 extend the result to nonstabilizable system, in particular to those systems having uncontrollable roots on the unit circle. The later results are of importance since they allow us to greatly extend the class of system to which adaptive filtering, prediction, and control can be applied. This shall become evident later. Some elementary applications of the results are discussed below.

Example 7.3.1

Consider the following *scalar* system:

$$x(t + 1) = Ax(t) + v_1(t); \quad A = a \quad (a \text{ is a scalar})$$
$$y(t) = Cx(t) + v_2(t); \quad C = 1$$

where

$$E\left\{ \begin{bmatrix} v_1(t) \\ v_2(t) \end{bmatrix} [v_1(s) \quad v_2(s)] \right\} = \begin{bmatrix} Q & 0 \\ 0 & R \end{bmatrix} \delta(t - s); \quad \begin{matrix} Q = D^2 \\ R = 1 \end{matrix}$$

Note that

1. If $Q \neq 0$, then (A, D) is stabilizable for all A.
2. If $|a| < 1$, then (A, D) is stabilizable for all Q.
3. If $|a| = 1$ and $Q = 0$, then (A, D) has an uncontrollable mode on the unit circle.

For the case $Q \neq 0$ and/or $|a| < 1$, we can apply Theorem 7.3.2 to show that the solution of the RDE converges exponentially fast to the stabilizing solution for all $\Sigma_0 \geq 0$.

For the case $Q = 0, |a| > 1$, then (A, D) has an unstable uncontrollable mode. Thus from Lemma 7.3.1(vi) there are at least two positive semidefinite solutions to the ARE (in fact, here there are exactly 2). The system is detectable and thus the stabilizing solution to the ARE exists and is unique. Moreover, Theorem 7.3.3 ensures that $\lim_{t \to \infty} \Sigma(t) = \Sigma_s$ (the stabilizing solution), which gives a stable filter with $\bar{A}_s = 1/a$ (with $|1/a| < 1$) for any $\Sigma_0 > 0$.

For the case $a = 1, Q = 0$ the system becomes

$$x(t + 1) = x(t); \quad x(t_0) = x_0$$
$$y(t) = x(t) + v_2(t)$$

where $\{v_2(t)\}$ is an independent and identically distributed sequence having unit variance. This system is not stabilizable since it has an uncontrollable mode on the unit circle. The initial state estimate is assumed to have mean $\bar{x}_0$ and covariance $P_0 = 1$ at time $t_0 = 1$.

The Kalman filter for this system is

$$\hat{x}(t+1) = \hat{x}(t) + K(t)[y(t) - \hat{x}(t)]; \qquad \hat{x}(1) = \bar{x}_0$$

where

$$K(t) = \frac{\Sigma(t)}{\Sigma(t) + 1}$$

and the associated Riccati difference equation is

$$\Sigma(t+1) = \frac{\Sigma(t)}{\Sigma(t) + 1}; \qquad \Sigma(1) = 1$$

The analytic solution of the RDE is

$$\Sigma(t) = \frac{1}{t}$$

Moreover, from the form of the RDE it is clear that the only possible solution to the algebraic Riccati equation is $\Sigma_s = 0$.[Compare with Lemma 7.3.1(iv) and (v).]

Next observe that $\Sigma(t)$ converges to Σ_s as $1/t$ (compare with Theorem 7.3.4) and the resulting steady-state Kalman filter is

$$\hat{x}(t+1) = \hat{x}(t)$$

Of course, the estimate generated by the (time-varying) Kalman filter converges (in means square) to x_0 as t tends to ∞ and thus asymptotically the correct initial conditions are generated for the steady-state filter $\hat{x}(t+1) = \hat{x}(t)$.

▼▼▼

We next investigate other applications of the Kalman filter.

The Kalman filter can be applied to any linear stochastic filtering problem in which the variables of interest appear as components of the state vector (or linear combinations of the components of state vector). By convention, the Kalman filter actually corresponds to the one-step-ahead state estimator. In filtering applications it is often desirable to obtain a filtered or smoothed estimate. We show below how fixed-lag and fixed-point smoothers can be derived by direct application of the Kalman filter.

7.3.2 Fixed-Lag Smoothing

Consider the usual state-space model for the data-generating mechanism, that is,

$$x(t+1) = Ax(t) + v_1(t) \tag{7.3.38}$$

$$y(t) = Cx(t) + v_2(t) \tag{7.3.39}$$

where $v_1(t)$ and $v_2(t)$ are zero mean uncorrelated noise sequence having covariance Q and R, respectively.

A fixed-lag smoother can be derived by augmenting the state vector by delayed versions of the state to give

$$
\begin{bmatrix} x(t+1) \\ x^1(t+1) \\ \cdot \\ \cdot \\ \cdot \\ x^d(t+1) \end{bmatrix}
=
\begin{bmatrix} A & 0 & \cdots\cdots & 0 \\ I & 0 & & \cdot \\ 0 & \cdot & \cdot & \cdot \\ \cdot & \cdot & \cdot & \cdot \\ \cdot & & \cdot & \cdot \\ 0 & \cdots & 0 & I & 0 \end{bmatrix}
\begin{bmatrix} x(t) \\ x^1(t) \\ \cdot \\ \cdot \\ \cdot \\ x^d(t) \end{bmatrix}
+
\begin{bmatrix} I \\ 0 \\ \cdot \\ \cdot \\ \cdot \\ 0 \end{bmatrix} v_1(t) \tag{7.3.40}
$$

$$y(t) = [C \quad 0 \quad \cdots \quad 0] \begin{bmatrix} x(t) \\ \vdots \\ \vdots \\ x^d(t) \end{bmatrix} + v_2(t) \tag{7.3.41}$$

The Kalman filter for the augmented model above can be immediately obtained to give a one-step-ahead state estimate of the composite state vector which contains smoothed estimates of the original state, $x(t)$, for lags up to d, that is,

$$\hat{x}^d(t) = E\{x^d(t)\,|\,y(0), \cdots, y(t-1)\} = E\{x(t-d)\,|\,y(0), \cdots, y(t-1)\} \tag{7.3.42}$$

The filter for the composite state has the following form:

$$\begin{bmatrix} \hat{x}(t+1) \\ \hat{x}^1(t+1) \\ \vdots \\ \vdots \\ \vdots \\ \hat{x}^d(t+1) \end{bmatrix} = \begin{bmatrix} A & 0 & \cdots\cdots\cdots & 0 \\ I & 0 & & \vdots \\ 0 & \cdot & & \vdots \\ \vdots & \cdot & \cdot & \vdots \\ \vdots & & \cdot & \vdots \\ 0 & \cdots & 0 & I & 0 \end{bmatrix} \begin{bmatrix} \hat{x}(t) \\ \vdots \\ \vdots \\ \vdots \\ \vdots \\ \hat{x}^d(t) \end{bmatrix} + \begin{bmatrix} K(t) \\ K^1(t) \\ \vdots \\ \vdots \\ \vdots \\ K^d(t) \end{bmatrix} [y(t) - C\hat{x}(t)] \tag{7.3.43}$$

where $[K(t), \ldots, K^d(t)]$ are obtained from the standard Kalman filter for the composite state. The corresponding state covariance matrix can be partitioned as follows:

$$\bar{\Sigma}(t) = \begin{bmatrix} \Sigma(t) & \Sigma^1(t)^T & \cdots & \Sigma^d(t)^T \\ \Sigma^1(t) & \Sigma^{11}(t) & & \\ \vdots & & & \\ \Sigma^d(t) & & & \Sigma^{dd}(t) \end{bmatrix} \tag{7.3.44}$$

Due to the special structure of (7.3.40)–(7.3.41), the equation generating $K(t)$, $\ldots$, $K^d(t)$ depends only on the first block column of $\bar{\Sigma}(t)$ as follows:

$$K^i(t) = \Sigma^{i-1}(t)C^T[C\Sigma(t)C^T + R]^{-1}; \qquad i = 1, 2, \ldots, d \tag{7.3.45}$$

where

$$\Sigma^i(t+1) = \Sigma^{i-1}(t)[A - K(t)C]^T \tag{7.3.46}$$

$$\Sigma^{ii}(t+1) = \Sigma^{i-1,i-1}(t) - \Sigma^{i-1}(t)[K^i(t)C]^T$$

with initial conditions $\Sigma^0(t) = \Sigma(t)$, where $\Sigma(t)$ satisfies the standard matrix Riccati equation for (7.3.38)–(7.3.39) and

$$K(t) = A\Sigma(t)C^T[C\Sigma(t)C^T + R]^{-1} \tag{7.3.47}$$

It is clear from the equations above that the fixed-lag smoother is asymptotically time invariant and stable under the same condition as for the Kalman filter for the original system.

As an application of the above, consider a typical filtering problem in which an input signal is passed through a linear system where it is convolved with itself and is corrupted by noise. (See, for example, the setup in Fig. 7.3.1.) We assume that $\{s(t)\}$ can be modeled as the output of a linear finite-dimensional system of the form

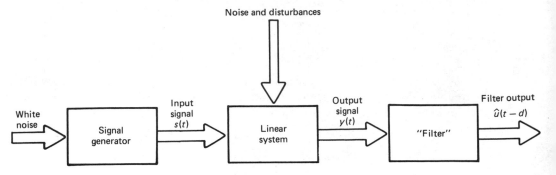

Figure 7.3.1 Fixed-lag smoothing problem.

$$x_1(t+1) = A_1 x_1(t) + w_1(t) \tag{7.3.48}$$

$$s(t) = C_1 x_1(t) + v_1(t) \tag{7.3.49}$$

where $x_1(t)$ is an $n_1 \times 1$ state vector and $w_1(t)$ and $v_1(t)$ are uncorrelated noise sequences with covariance

$$\begin{bmatrix} Q_1 & 0 \\ 0 & R_1 \end{bmatrix}$$

The signal $s(t)$ is assumed to be passed through a linear time-invariant system where it suffers dispersion and is corrupted by white or colored noise. The output of the system is denoted by $y(t)$ and in the light of the assumption above is related to the input $s(t)$ by a model of the form

$$x_2(t+1) = A_2 x_2(t) + B_2 s(t) + w_2(t) \tag{7.3.50}$$

$$y(t) = C_2 x_2(t) + v_2(t) \tag{7.3.51}$$

where $x_2(t)$ is an $n_2 \times 1$ state vector arising from the dispersion of the channel and coloring of the noise. The sequences $\{w_2(t)\}$ and $\{v_2(t)\}$ denote uncorrelated noise sequences with covariance

$$\begin{bmatrix} Q_2 & 0 \\ 0 & R_2 \end{bmatrix}$$

We shall be interested in a fixed-lag smoothed estimate of $s(t)$.

In order that the required signal $(s(t))$ form part of the state vector, we adjoin past values of $\{s(t)\}$ to the model (7.3.38) to (7.3.41) by introducing additional states as follows:

$$x_3^1(t+1) = s(t)$$

$$x_3^2(t+1) = x_3^1(t) = s(t-1)$$

$$\cdot$$
$$\cdot$$
$$\cdot$$

$$x_3^d(t+1) = x_3^{d-1}(t) = s(t-d+1)$$

or putting

$$x_3(t)^T = [x_3^1(t) \quad \cdots \quad x_3^d(t)]$$

we have

Optimal Filtering and Prediction   Chap. 7

$$x_3(t+1) = \begin{bmatrix} 0 & & & 0 \\ 1 & & & \cdot \\ 0 & \cdot & & \cdot \\ & & \cdot & \cdot \\ 0 & \cdots & 0 & 1 & 0 \end{bmatrix} x_3(t) + \begin{bmatrix} 1 \\ 0 \\ \cdot \\ \cdot \\ 0 \end{bmatrix} s(t) \tag{7.3.52}$$

$$\triangleq Sx_3(t) + e_1 s(t)$$

where $x_3(t)$ is a $d \times 1$ vector and e_i denotes a vector of zeros except for the ith entry, which is 1. We note in passing that

$$x_3^j(t) = s(t-j), \qquad 1 \le j \le d \tag{7.3.53}$$

The composite model (7.3.48) to (7.3.53) can be written in compact form as

$$x(t+1) = Ax(t) + \eta_1(t) \tag{7.3.54}$$

$$y(t) = Cx(t) + \eta_2(t) \tag{7.3.55}$$

where $x(t)$ is an $n = (n_1 + n_2 + d)$-dimensional state vector $[x_1(t)^T \ x_2(t)^T \ x_3(t)^T]^T$, and $\eta_1(t)$ and $\eta_2(t)$ denote uncorrelated noise sequences (see Exercise 7.23). Now applying the standard Kalman filter, we can obtain the one-step-ahead estimate, $\hat{x}(t)$, of the composite state.

To recover $s(t-d)$, we observe that the last component of $\hat{x}(t)$ is

$$e_n^T \hat{x}(t) = E\{e_n^T x(t) \mid y(t-1), y(t-2), \ldots\}$$
$$= E\{s(t-d) \mid y(t-1), y(t-2), \ldots\}$$
$$= \hat{s}(t-d \mid t-1)$$

This is the desired fixed-lag smoothed estimate of $s(t-d)$.

Note that in the formulation of the problem above it has only been necessary to augment the state by the variable of interest $\{s(t)\}$ instead of the complete state. This can reduce the dimension of the problem under study. Actually, this reduction is also apparent by inspection of (7.3.45) and (7.3.46).

The reader may wonder what advantage, if any, there is in using a fixed-lag smoother over a simple filter. The point is that as the lag d increases, the estimation error variance decreases due to the information provided by the additional data. As the lag length approaches infinity, the performance approaches that of the noncausal *Wiener filter*. However, in practice, nearly optimal performance may be achieved by making d equal to two or three times the dominant time constants of the system.

7.3.3 Fixed-Point Smoothing

As a further application of Kalman filtering ideas, consider the problem of obtaining a *fixed-point* smoothed estimate of a state vector. The key idea here is that we fix the time at which we want to estimate the state. We then continue to refine the state estimate at this fixed time as additional data become available.

We shall use the standard state-space model for our data, namely

$$x(t+1) = Ax(t) + v_1(t) \tag{7.3.56}$$

$$y(t) = Cx(t) + v_2(t) \tag{7.3.57}$$

where $\{v_1(t)\}$ and $\{v_2(t)\}$ are uncorrelated white noise sequences having covariance Q and R, respectively. We also define another state vector $x'(t)$ as follows:

$$x'(t) = x(t) \qquad \text{for } t \leq t_f \tag{7.3.58}$$

$$x'(t+1) = x'(t) \qquad \text{for } t \geq t_f \tag{7.3.59}$$

where t_f is some fixed time. Note that $x'(t) = x(t_f)$ for all $t \geq t_f$.

We now use the standard Kalman filter of Section 7.3.1 to obtain a filtered estimate of the composite state vector $[x(t)^T \quad x'(t)^T]^T$. From Theorem 7.3.1 for $t \geq t_f$:

$$\begin{bmatrix} \hat{x}(t+1) \\ \hat{x}'(t+1) \end{bmatrix} = \begin{bmatrix} A & 0 \\ 0 & I \end{bmatrix} \begin{bmatrix} \hat{x}(t) \\ \hat{x}'(t) \end{bmatrix} + \begin{bmatrix} K(t) \\ K^a(t) \end{bmatrix} [y(t) - C\hat{x}(t)] \tag{7.3.60}$$

$$\begin{bmatrix} K(t) \\ K^a(t) \end{bmatrix} = \begin{bmatrix} A & 0 \\ 0 & I \end{bmatrix} \begin{bmatrix} \Sigma_{11}(t) & \Sigma_{12}(t) \\ \Sigma_{21}(t) & \Sigma_{22}(t) \end{bmatrix} \begin{bmatrix} C^T \\ 0 \end{bmatrix} [C\Sigma_{11}(t)C^T + R]^{-1} \tag{7.3.61}$$

$$\begin{bmatrix} \Sigma_{11}(t+1) & \Sigma_{12}(t+1) \\ \Sigma_{21}(t+1) & \Sigma_{22}(t+1) \end{bmatrix} = \begin{bmatrix} A & 0 \\ 0 & I \end{bmatrix} \begin{bmatrix} \Sigma_{11}(t) & \Sigma_{12}(t) \\ \Sigma_{21}(t) & \Sigma_{22}(t) \end{bmatrix} \begin{bmatrix} A^T & 0 \\ 0 & I \end{bmatrix}$$
$$+ \begin{bmatrix} Q & 0 \\ 0 & 0 \end{bmatrix} - \begin{bmatrix} K(t) \\ K^a(t) \end{bmatrix} [C\Sigma_{11}(t)C^T + R] \begin{bmatrix} K(t) \\ K^a(t) \end{bmatrix} \tag{7.3.62}$$

with

$$\begin{bmatrix} \Sigma_{11}(t_f) & \Sigma_{12}(t_f) \\ \Sigma_{21}(t_f) & \Sigma_{22}(t_f) \end{bmatrix} = \begin{bmatrix} \Sigma_{11}(t_f) & \Sigma_{11}(t_f) \\ \Sigma_{11}(t_f) & \Sigma_{11}(t_f) \end{bmatrix} \tag{7.3.63}$$

It can be seen from (7.3.60) that

$$\hat{x}(t+1) = A\hat{x}(t) + K(t)[y(t) - C\hat{x}(t)]$$
$$= A\hat{x}(t) + K(t)\tilde{y}(t) \tag{7.3.64}$$

where $\tilde{y}(t)$ is equal to $y(t) - C\hat{x}(t)$.

Also, from (7.3.59)–(7.3.60) for $t \geq t_f$:

$$E\{x(t_f)\,|\,y(t), y(t-1), \ldots\} = E\{x'(t+1)\,|\,y(t), y(t-1), \ldots\}$$
$$= \hat{x}'(t+1) \tag{7.3.65}$$
$$= \hat{x}'(t) + K^a(t)\tilde{y}(t)$$

with

$$\hat{x}'(t_f) = \hat{x}(t_f) \tag{7.3.66}$$

The gain $K^a(t)$ is called the fixed-point smoothing gain and from (7.3.61)–(7.3.62) satisfies

$$K^a(t) = \Sigma_{21}(t)C^T[C\Sigma_{11}(t)C^T + R]^{-1} \tag{7.3.67}$$

where $\Sigma_{11}(t)$ is the usual state estimate covariance satisfying

$$\Sigma_{11}(t+1) = A\Sigma_{11}(t)A^T + Q - K(t)[C\Sigma_{11}(t)C^T + R]K(t)^T \tag{7.3.68}$$

$K(t)$ is the usual Kalman gain:

$$K(t) = A\Sigma_{11}(t)C^T[C\Sigma_{11}(t)C^T + R]^{-1} \tag{7.3.69}$$

Finally from (7.3.62), $\Sigma_{21}(t)$ satisfies

$$\Sigma_{21}(t+1) = \Sigma_{21}(t)A^T - \Sigma_{21}(t)C^T K(t)^T$$
$$= \Sigma_{21}(t)[A - K(t)C]^T \tag{7.3.70}$$

with

$$\Sigma_{21}(t_f) = \Sigma_{11}(t_f) \tag{7.3.71}$$

We can summarize the discussion above in the following lemma:

Lemma 7.3.2. The fixed-point smoothed estimate of $x(t_f)$ is given by the following recursions for $t \geq t_f$:

$$E\{x(t_f) | y(t), y(t-1), \ldots \} = \hat{x}'(t+1) \tag{7.3.72}$$

$$\hat{x}'(t+1) = \hat{x}'(t) + K^a \tilde{y}(t); \qquad \hat{x}'(t_f) = \hat{x}(t_f) \tag{7.3.73}$$

$$K^a(t) = \Sigma_{21}(t) C^T [C\Sigma_{11}(t)C^T + R]^{-1} \tag{7.3.74}$$

$$\Sigma_{21}(t+1) = \Sigma_{21}(t)[A - K(t)C]^T; \qquad \Sigma_{21}(t_f) = \Sigma_{11}(t_f) \tag{7.3.75}$$

$$\tilde{y}(t) = y(t) - C\hat{x}(t) \tag{7.3.76}$$

where $\hat{x}(t)$, $\Sigma_{11}(t)$, and $K(t)$ are state estimate, state estimation error covariance, and Kalman gain appearing in the usual Kalman filter (7.3.64), (7.3.68), and (7.3.69) for estimating $x(t)$ given $y(t-1)$, $y(t-2)$, ...

Proof. See the reasoning as outlined above.

▼▼▼

In the next subsection we shall show how the Kalman filter leads naturally to an optimal predictor for a linear stochastic system.

7.3.4 Optimal Prediction

Consider, the usual state-space model of (7.2.3)–(7.2.4):

$$x(t+1) = Ax(t) + v_1(t); \qquad x(t_0) = x_0 \tag{7.3.77}$$

$$y(t) = Cx(t) + v_2(t); \qquad t \geq t_0 \tag{7.3.78}$$

where $\{v_1(t)\}$ and $\{v_2(t)\}$ are gaussian white noise sequences. Then as shown in (7.3.17) and (7.3.18), the output $\{y(t)\}$ can be described by the following innovations model:

$$\hat{x}(t+1) = A\hat{x}(t) + K(t)\omega(t) \tag{7.3.79}$$

$$y(t) = C\hat{x}(t) + \omega(t) \tag{7.3.80}$$

where $K(t)$ denotes the Kalman filter gain sequence and $\omega(t)$ is the innovations process satisfying

$$E\{\omega(t) | \mathcal{F}_{t-1}\} = 0 \tag{7.3.81}$$

where $\mathcal{F}_{t-1}$ denotes the sigma algebra generated by $\{y(t-1), \ldots, y(0)\}$. $\mathcal{F}_0$ denotes the initial condition. Note also that $\hat{x}(t)$ is measurable with respect to $\mathcal{F}_{t-1}$ and

$$E\{x(t) | \mathcal{F}_{t-1}\} = \hat{x}(t) \tag{7.3.82}$$

Now the *optimal d-step-ahead predictor* of $\{\hat{x}(t)\}$ is the conditional mean, that is, $E\{x(t+d) | \mathcal{F}_t\}$. Using the smoothing property of conditional expectations, we have

$$E\{x(t+d) | \mathcal{F}_t\} = E\{E\{x(t+d) | \mathcal{F}_{t+d-1}\} | \mathcal{F}_t\} \tag{7.3.83}$$

Now from (7.3.79), we have

$$\hat{x}(t + d) = A^{d-1}\hat{x}(t + 1) + \sum_{j=t+1}^{t+d-1} A^{t+d-j-1}K(j)\omega(j) \qquad (7.3.84)$$

Taking conditional expectations and using (7.3.81) yields

$$E\{\hat{x}(t + d)|\mathcal{F}_t\} = A^{d-1}\hat{x}(t + 1) \qquad (7.3.85)$$

Combining (7.3.83) and (7.3.85) gives

$$E\{x(t + d)|\mathcal{F}_t\} = A^{d-1}\hat{x}(t + 1) \qquad (7.3.86)$$

The equation above gives a very simple means of computing the optimal d-step-ahead prediction of $\{x(t)\}$. The corresponding optimal d-step-ahead prediction of $\{y(t)\}$ is then obtained by combining (7.3.80) and (7.3.86) as

$$E\{y(t + d)|y(0) \cdots y(t)\} = CA^{d-1}\hat{x}(t + 1) \qquad (7.3.87)$$

The derivation of the predictor above is greatly facilitated by the use of the innovations model of (7.3.79)–(7.3.80).

We note that it is straightforward to incorporate a known input, $\{u(t)\}$, into the system model (7.3.77)–(7.3.78) to give

$$x(t + 1) = Ax(t) + Bu(t) + v_1(t) \qquad (7.3.88)$$

$$y(t) = Cx(t) + v_2(t) \qquad (7.3.89)$$

The corresponding expressions for the optimal d-step-ahead predictors for $\{x(t)\}$ and $\{y(t)\}$ then become

$$E\{x(t + d)|\mathcal{F}_t\} = A^{d-1}\hat{x}(t + 1)$$
$$+ \sum_{j=t+1}^{t+d-1} A^{t+d-j-1}B(j)u(j) \qquad (7.3.90)$$

$$E\{y(t + d)|\mathcal{F}_t\} = CA^{d-1}\hat{x}(t + 1)$$
$$+ \sum_{j=t+1}^{t+d-1} CA^{t+d-j-1}B(j)u(j) \qquad (7.3.91)$$

7.4 FILTERING AND PREDICTION USING STOCHASTIC ARMA MODELS

All the results in the preceding section have been described in state-space form. In the section we will see how the same results can be expressed in terms of input–output models of high order difference equation type. We begin by describing the appropriate input–output model for a stochastic system.

7.4.1 The Stochastic ARMA Model

We have seen in Chapter 2 that the input–output characteristics of a general *deterministic* linear system can be described by a DARMA model of the form

$$A(q^{-1})y(t) = B(q^{-1})u(t) \qquad (7.4.1)$$

Optimal Filtering and Prediction   Chap. 7

A natural generalization of this to the stochastic case is to add an independent noise input. This leads to the following stochastic *autoregressive moving-average model with auxiliary input*:

$$A(q^{-1})y(t) = B(q^{-1})u(t) + C(q^{-1})\omega(t) \tag{7.4.2}$$

where $\{\omega(t)\}$ is a white noise sequence and $C(q^{-1})$ is a filter of the form

$$C(q^{-1}) = I + c_1 q^{-1} + \cdots + c_n q^{-n} \tag{7.4.3}$$

We have used the shorthand notation DARMA for the model (7.4.1). We introduce a similar notation ARMA for the model (7.4.2) when $u(t) \equiv 0$, that is,

$$A(q^{-1})y(t) = C(q^{-1})\omega(t) \tag{7.4.4}$$

The full model (7.4.2) will be called the stochastic *autoregressive moving-average* model with *auxiliary* input, or ARMAX model for short.

The ARMAX model (7.4.2) can simply be taken as the valid description of the input–output properties of the system under study. Alternatively, the model can be motivated via Kalman filtering theory beginning with the state-space model (7.2.3)–(7.2.4). In particular, we shall show that the ARMAX model (7.4.2) can be thought of as a compact way of writing the innovations model (7.3.17)–(7.3.18).

To make the derivation more transparent, we consider the single-input single-output case. In this case, if we express the stochastic state-space model in observer form, the innovations model (7.3.17)–(7.3.18) will have the following structure:

$$\hat{x}(t+1) = \begin{bmatrix} -a_1 & 1 & & \\ \vdots & & \ddots & \\ \vdots & & & 1 \\ -a_n & & & 0 \end{bmatrix} \hat{x}(t) + \begin{bmatrix} b_1 \\ \vdots \\ \vdots \\ b_n \end{bmatrix} u(t) + \begin{bmatrix} k_1(t) \\ \vdots \\ \vdots \\ k_n(t) \end{bmatrix} \omega(t) \tag{7.4.5}$$

$$y(t) = [1 \quad 0 \quad \cdots \quad 0]\hat{x}(t) + \omega(t) \tag{7.4.6}$$

By successive substitution, the state-space model above can be written as the following time-varying input–output model:

$$A(q^{-1})y(t) = B(q^{-1})u(t) + C(t, q^{-1})\omega(t) \tag{7.4.7}$$

where

$$A(q^{-1}) = 1 + a_1 q^{-1} + \cdots + a_n q^{-n} \tag{7.4.8}$$

$$B(q^{-1}) = b_1 q^{-1} + \cdots + b_n q^{-n} \tag{7.4.9}$$

$$C(t, q^{-1}) = 1 + [k_1(t-1) + a_1]q^{-1} + \cdots + [k_n(t-n) + a_n]q^{-n} \tag{7.4.10}$$

Thus

$$y(t) + a_1 y(t-1) + \cdots + a_n y(t-n) = b_1 u(t-1) + \cdots + b_n u(t-n)$$
$$+ \omega(t) + [a_1 + k_1(t-1)]\omega(t-1) + \cdots + [a_n + k_n(t-n)]\omega(t-n)$$

The time-varying nature of $C(t, q^{-1})$ in (7.4.7) arises from the fact that the Kalman gain $K(t)$ in (7.4.5) is time varying. However, subject to the conditions of Theorem 7.3.2,

7.3.3, or 7.3.4, we have that $K(t)$ is asymptotically time invariant and thus $C(t, q^{-1})$ is asymptotically time invariant. Thus we can often replace the time-varying model (7.4.7) by a steady-state form, that is,

$$A(q^{-1})y(t) = B(q^{-1})u(t) + C(q^{-1})\omega(t) \qquad (7.4.11)$$

We see from the discussion above that the model (7.4.11) is simply an alternative way of describing the steady-state innovations model. With this in mind we are able to be more specific about the form of $C(q^{-1})$. We see from (7.4.11) that $C(q^{-1})$ is the denominator polynomial matrix in the (whitening) model expressing $\{\omega(t)\}$ in terms of $\{y(t)\}$. However, we have seen in (7.3.17) and (7.3.19) that this steady state "whitening filter" has state transition matrix, $(A - K_sC)$. Moreover, Theorems 7.3.2, 7.3.3, and 7.3.4 show that $(A - K_sC)$ generally has eigenvalues that are on or inside the unit circle. Further, the eigenvalues will be inside the unit circle provided that there are no uncontrollable modes of (A, D) on the unit circle.

Thus in the absence of uncontrollable modes on the unit circle, we may assume without loss of generality that $C(q^{-1})$ has all roots strictly inside the unit circle. This will be a standing assumption in much of our subsequent work.

In some cases we will purposefully introduce uncontrollable modes on the unit circle to describe deterministic disturbances, for example sine waves, as in Chapter 2. In this case the appropriate assumption on $C(q^{-1})$ will be that it has roots on or inside the unit circle. Further, any modes that are uncontrollable form $u(t)$ will appear in $A(q^{-1})$, $B(q^{-1})$, and $C(q^{-1})$.

We illustrate the discussion above by a simple example.

Example 7.4.1

The output of a system is given by

$$y(t) = d + v_2(t) + u(t - 1)$$

where d = constant

$v_2(t)$ = white noise with variance 1

$u(t)$ = a known input

The corresponding state-space model is

$$x(t + 1) = Ax(t) + Bu(t)$$
$$y(t) = Cx(t) + v_2(t)$$

where

$$A = \begin{bmatrix} 1 & 0 \\ 0 & 0 \end{bmatrix}; \quad B = \begin{bmatrix} 0 \\ 1 \end{bmatrix}; \quad C^T = \begin{bmatrix} 1 \\ 1 \end{bmatrix}$$

The associated innovations model is

$$\hat{x}_1(t + 1) = x_1(t) + k_1(t)\omega(t)$$
$$\hat{x}_2(t + 1) = u(t) + k_2(t)\omega(t)$$
$$y(t) = \hat{x}_1(t) + \hat{x}_2(t) + \omega(t)$$

The ARE and steady-state filter gain for this problem are, respectively,

$$\Sigma - A\Sigma A^T + A\Sigma C^T(C\Sigma C^T + 1)^{-1}C\Sigma A = 0$$
$$K = A\Sigma C^T(C\Sigma C^T + 1)^{-1}$$

We note that the original system is observable but has an uncontrollable mode on the unit circle. Thus the strong solution to the ARE exists, is unique, and gives a steady-state filter with a root on the unit circle. The strong solution to the ARE is $\Sigma_s = 0$, giving $K_s = 0$. Also, we see from Theorem 7.3.4 that $\Sigma(t)$ and $K(t)$ (the time-varying quantities in the innovations model) converge to Σ_s and K_s for any $\Sigma(t_0) > \Sigma_s$ [i.e., for any $\Sigma(t_0) > 0$]. The resulting steady-state Kalman filter is

$$\hat{x}(t+1) = A\hat{x}(t) + Bu(t)$$
$$y(t) = C\hat{x}(t) + \omega(t)$$

Finally, the steady-state ARMAX model is

$$A(q^{-1})y(t) = q^{-1}B'(q^{-1})u(t) + C(q^{-1})\omega(t)$$

where

$$A(q^{-1}) = B'(q^{-1}) = C(q^{-1}) = 1 - q^{-1}$$

▼▼▼

Example 7.4.2

Consider the MA equation

$$y(t) = \omega(t) + c_1\omega(t-1) + \cdots + c_n\omega(t-n)$$

An appropriate state-space model is

$$x(t+1) = \begin{bmatrix} 0 & 1 & & & \\ & & \ddots & & \\ & & & \ddots & \\ & & & & 1 \\ 0 & & & & 0 \end{bmatrix} x(t) + \begin{bmatrix} 0 \\ \vdots \\ \vdots \\ 0 \\ 1 \end{bmatrix} \omega(t)$$

$$y(t) = [c_n \quad c_{n-1} \quad \cdots \quad c_1]x(t) + \omega(t)$$

Applying the transformation given in (7.3.22) to (7.3.24), we see that the resulting system is uncontrollable. In fact, it is not stabilizable unless the polynomial $C(z^{-1}) = 1 + c_1 z^{-1} + \cdots + c_n z^{-n}$ is stable, but it is observable provided that $c_n \neq 0$. Thus, given $c_n \neq 0$:

1. When $C(z^{-1})$ has roots on the unit circle, we can apply Theorem 7.3.4 to conclude that $\Sigma(t)$ will converge to Σ_s giving all roots of the filter on or inside the unit circle provided that $\Sigma_0 - \Sigma_s > 0$.
2. When $C(z^{-1})$ has no root on the unit circle, we can apply Theorem 7.3.3 to conclude that $\Sigma(t)$ will converge to Σ_s (exponentially fast) giving all roots of the filter inside the unit circle for any $\Sigma_0 > 0$.

Thus in steady state we may assume that the MA model has all zeros insider or on the unit circle without loss of generality.

▼▼▼

*Remark 7.4.1 (Rapprochement with the Theory of Stationary Stochastic Processes). Those readers who are familiar with the theory of *stationary stochastic processes* will appreciate that the innovations model and ARMAX model have a nice interpretation in terms of the Wold decomposition (Wold, 1938, p. 89; Rozanov, 1967,

p. 56). In the Wold decomposition, a stationary stochastic process is decomposed into the sum of a deterministic and nondeterministic (or purely random component). The deterministic component is perfectly predictable from the infinite past and is usually taken to be a finite combination of sinusoids. The purely nondeterministic component is represented as the output of a linear system driven by a white noise sequence (the innovations). We see that the innovations and ARMAX models allow both the deterministic and nondeterministic components of a stationary stochastic process to be described in a simple and unified fashion.

Further, those readers who are familiar with the theory of stationary stochastic processes will recall that the Fourier transform of the autocorrelation of these processes is known as the *spectral distribution function*. This function is a bounded nondecreasing function and can be represented as the sum of a discrete component and an absolutely continuous component. The discrete component always corresponds to purely deterministic components in the original process (in fact, the discrete steps in the spectral distribution function occur at the frequencies of any sinusoidal components that are present). On the other hand, the absolutely continuous part of the spectral distribution can correspond to either a deterministic or nondeterministic component. In particular, if $f(\lambda)$ is the *spectral density* corresponding to an absolutely continuous spectral distribution function, the corresponding process is purely deterministic if

$$\int_{-\pi}^{\pi} \log f(\lambda)\, d\lambda = -\infty \qquad (7.4.12)$$

and is nondeterministic otherwise. This condition is known as the *Wiener–Paley condition* (Rozanov, 1967). A consequence of condition (7.4.12) is that the spectral density of a nondeterministic process must be nonzero almost everywhere. Also, the condition implies that any *rational spectral density* (i.e., one that is expressible as a ratio of finite polynomials) always corresponds to a nondeterministic component.

The form of the autocorrelation function guarantees that the spectral distribution function and spectral density functions have certain properties. In particular, if the spectral density is rational, the roots appear in mirror-image pairs, i.e., if ρ is a pole (or zero) of the spectral density, then so is ρ^{-1} a pole (or zero)]. This implies that a spectral density can be factored as follows:

$$f(\lambda) = H(\lambda)H(\lambda^{-1})\sigma^2 \qquad (7.4.13)$$

where $H(\lambda)$ has all poles inside the unit circle and all zeros inside or on the unit circle, with the additional property that $\lim_{\lambda \to \infty} H(\lambda) = 1$. This suggests that the corresponding purely nondeterministic process can be represented as the output of a linear system having zeros on or inside the unit circle and driven by a white noise process of variance σ^2. Of course, this is precisely the format of the ARMAX model, which has zeros [roots of $C(q^{-1})$] on or inside the unit circle. Note, however, that the development on the ARMAX model given earlier in this section does *not* require that the process be stationary; in particular, the model can be unstable and thus $A(q^{-1})$ can have roots anywhere in the complex plane.

▼▼▼

Remark 7.4.2 (Nonlinear Stochastic Models). So far in this chapter, we have considered only linear models. However, we wish to remind the reader of the observation made in Chapter 2 that, in many cases, a nonlinear model may be more realistic than a linear model.

In this regard the Wold decomposition (see Remark 7.4.1) is restrictive since only linear regressions on past data are considered. For processes generated in some nonlinear way, the Wold decomposition may be artificial and disguise the essential simplicity of the process. In fact, it is possible to have processes that are purely nondeterministic in the sense of the Wold decomposition, but which can be predicted exactly by a suitable nonlinear predictor [see Cox and Miller (1965, p. 288)].

Thus, in some situations, there may be advantages in considering models involving nonlinear regressions on past data. For example, one might consider a nonlinear model of the form

$$y(t) = f(y(t-1), y(t-2), \ldots, u(t-1), \ldots) + \omega(t) \tag{7.4.14}$$

where f is a nonlinear function and $\{\omega(t)\}$ denotes a white noise (or nonlinear innovations sequence) with the assumed property

$$E\{\omega(t) | \mathcal{F}_{t-1}\} = 0 \tag{7.4.15}$$

where $\mathcal{F}_{t-1}$ is the sub-sigma algebra generated by data up to time $t-1$.

A further generalization is to consider models of the form

$$y(t) = g(t-1) + \eta(t) \tag{7.4.16}$$

where

$$g(t-1) = f(y(t-1), \ldots, u(t-1), \ldots) \tag{7.4.17}$$

and $\eta(t)$ is a stationary stochastic process represented as

$$A(q^{-1})\eta(t) = C(q^{-1})\omega(t) \tag{7.4.18}$$

where $\omega(t)$ is a white noise process.

If we operate on (7.4.16) by $A(q^{-1})$ and use (7.4.18), we obtain the following nonlinear ARMAX model:

$$A(q^{-1})y(t) = A(q^{-1})g(t-1) + C(q^{-1})\omega(t) \tag{7.4.19}$$

$$\triangleq h(t-1) + C(q^{-1})\omega(t) \tag{7.4.20}$$

Of course, there are many other different forms of nonlinear models. The main point that we want to stress here is that the reader should keep in mind that a judicious choice of model can often lead to enahanced modeling capability and thus greater chance of success in a given application. Of course, the choice of the form of the nonlinear functions will, in general, depend upon a priori knowledge about the system dynamics; see, for example, the discussion of bilinear models in Chapter 2.

▼▼▼

7.4.2 Optimal Filters and Predictors in ARMA Form

Given the ARMAX representation (7.4.11), it is possible to derive optimal steady-state filters and predictors. Of course, (7.4.11) is nothing more than a compact way of writing the Kalman filter, as we have seen above. We will show below how optimal steady-state predictors can also be expressed in ARMAX form.

Consider the ARMAX model described in Section 7.4.1. For simplicity we temporarily restrict attention to the scalar case where we can write the model as

$$A(q^{-1})y(t) = q^{-d}B'(q^{-1})u(t) + C(q^{-1})\omega(t) \tag{7.4.21}$$

where

$$A(q^{-1}) = 1 + a_1q^{-1} + \cdots + a_nq^{-n}$$
$$B'(q^{-1}) = b'_0 + b'_1q^{-1} + \cdots + b'_mq^{-m}$$
$$C(q^{-1}) = 1 + c_1q^{-1} + \cdots + c_lq^{-l}$$
$$E\{\omega(t)|\mathcal{F}_{t-1}\} = 0, \qquad E\{\omega(t)^2|\mathcal{F}_{t-1}\} = \sigma^2 \tag{7.4.22}$$

where $\mathcal{F}_{t-1}$ is the sigma algebra generated by $\{y(t-1), \ldots, y(0)\}$ together with initial condition data (we assume that $\{u(t)\}$ is a known sequence).

It has been shown above that a model of the form (7.4.21) can be used to describe the output of linear systems having both deterministic and nondeterministic disturbances. In the general case, $C(z^{-1})$ will have zeros on or inside the unit circle. For the moment we will restrict our attention to those cases in which $C(z^{-1})$ has its roots strictly inside the unit circle. We shall treat the more general case presently.

We first present the following result:

Lemma 7.4.1. For the system (7.4.21), provided that $C(q^{-1})$ is asymptotically stable, the optimal d-step-ahead prediction, $y^0(t+d|t)$, of $y(t)$ satisfies

$$C(q^{-1})y^0(t+d|t) = \alpha(q^{-1})y(t) + \beta(q^{-1})u(t) \tag{7.4.23}$$

where

$$y^0(t+d|t) \triangleq E\{y(t+d)|\mathcal{F}_t\} = y(t+d) - F(q^{-1})\omega(t+d) \tag{7.4.24}$$
$$\alpha(q^{-1}) = G(q^{-1}) \tag{7.4.25}$$
$$\beta(q^{-1}) = F(q^{-1})B'(q^{-1}) \tag{7.4.26}$$

and $G(q^{-1})$ and $F(q^{-1})$ are the unique polynomials satisfying

$$C(q^{-1}) = F(q^{-1})A(q^{-1}) + q^{-d}G(b^{-1}) \tag{7.4.27}$$
$$F(q^{-1}) = 1 + f_1q^{-1} + \cdots + f_{d-1}q^{-d+1} \tag{7.4.28}$$
$$G(q^{-1}) = g_0 + g_1q^{-1} + \cdots + g_{n-1}q^{-n+1} \tag{7.4.29}$$

Also,

$$E\{[y(t+d) - y^0(t+d|t)]^2\} = E\{E\{[F(q^{-1})\omega(t)]^2|\mathcal{F}_t\} = \sum_{j=0}^{d-1} f_j^2\sigma^2$$

Proof. From the division algorithm of algebra [see Burton (1967, p. 200)], there exists unique polynomials such that (7.4.27) is satisfied. Multiplying (7.4.21) on the left by $F(q^{-1})$ gives

$$F(q^{-1})A(q^{-1})y(t) = q^{-d}F(q^{-1})B'(q^{-1})u(t) + F(q^{-1})C(q^{-1})\omega(t) \tag{7.4.30}$$

Using (7.4.27), we have

$$[C(q^{-1}) - q^{-d}G(q^{-1})]y(t) = q^{-d}F(q^{-1})B'(q^{-1})u(t) + F(q^{-1})C(q^{-1})\omega(t)$$

or

$$C(q^{-1})[y(t) - F(q^{-1})\omega(t)] = q^{-d}G(q^{-1})y(t) + q^{-d}F(q^{-1})B'(q^{-1})u(t) \tag{7.4.31}$$

Now define
$$y^0(t \mid t - d) = y(t) - F(q^{-1})\omega(t) \tag{7.4.32}$$
Then from (7.4.31)
$$C(q^{-1})y^0(t + d \mid t) = G(q^{-1})y(t) + F(q^{-1})B'(q^{-1})u(t) \tag{7.4.33}$$
Finally, we note from (7.4.33) that $y^0(t + d \mid t)$ is $\mathfrak{F}_t$ measurable and hence
$$
\begin{aligned}
y^0(t + d \mid t) &= E\{y^0(t + d \mid t) \mid \mathfrak{F}_t\} \\
&= E\{y(t + d) - F(q^{-1})\omega(t + d) \mid \mathfrak{F}_t\} \\
&= E\{y(t + d) - \sum_{j=0}^{d-1} f_j\omega(t + d - j) \mid \mathfrak{F}_t\} \\
&= E\{y(t + d) \mid \mathfrak{F}_t\} \quad \text{using (7.4.22)}
\end{aligned}
$$
This establishes the optimality of $y^0(t + d \mid t)$.

▼▼▼

Remark 7.4.3. Note that if we require $y^0(t + d \mid t)$ satisfying (7.4.23) to be the optimal predictor for all t, then the initial conditions must be chosen appropriately. However, since $C(q^{-1})$ is asymptotically stable (by assumption), the effect of arbitrary initial conditions will diminish exponentially. Thus asymptotically there is no error incurred if the predictor (7.4.23) is used with any initial conditions.

We can now see one source of difficulty in the general case when $C(z^{-1})$ has zeros on the unit circle. In this case the effect of incorrect initial conditions will not die away. We shall show later that this can be overcome by using time-varying predictors or restricted complexity predictors.

▼▼▼

We pause briefly to prove the following interesting concatenation property for optimal predictors. This result shows that optimal d-step-ahead predictions can be obtained from optimal one-step-ahead predictors.

Lemma 7.4.2 (Concatenation Property for Optimal Predictors). Let
$$y^0(t + j \mid t) \triangleq E\{y(t + j) \mid \mathfrak{F}_t\} \tag{7.4.34}$$
Then $y^0(t + j \mid t)$ satisfies the following recursive formula:
$$
\begin{aligned}
y^0(t + j \mid t) = {}& -a_1 y^0(t + j - 1 \mid t) - \cdots - a_{j-1} y^0(t + 1 \mid t) \\
& - a_j y(t) - \cdots - a_n y(t + j - n) \\
& + c_j\omega(t) + \cdots + c_l\omega(t + j - l) + B(q^{-1})u(t + j)
\end{aligned} \tag{7.4.35}
$$
where $B(q^{-1}) = q^{-d}B'(q^{-1})$ and
$$\omega(t) = y(t) - y^0(t \mid t - 1) \tag{7.4.36}$$
   Proof. From (7.4.21),
$$
\begin{aligned}
y(t + j) = {}& -a_1 y(t + j - 1) - \cdots - a_n y(t + j - n) + B(q^{-1})u(t + j) \\
& + \omega(t + j) + c_1\omega(t + j - 1) + \cdots + c_l\omega(t + j - l)
\end{aligned} \tag{7.4.37}
$$

Taking expectations given $\mathcal{F}_t$ and using (7.4.22), we obtain

$$y^0(t+j|t) = -a_1 y^0(t+j-1|t) - \cdots - a_{j-1} y^0(t+1|t)$$
$$- a_j y(t) - \cdots - a_n y(t+j-n) + B(q^{-1})u(t+j) \qquad (7.4.38)$$
$$+ c_j E\{\omega(t)|\mathcal{F}_t\} + \cdots + c_l E\{\omega(t+j-l)|\mathcal{F}_t\}$$

However, from (7.4.36),

$$\omega(t) = y(t) - y^0(t|t-1) \qquad (7.4.39)$$

So $\omega(t)$ is $\mathcal{F}_t, \mathcal{F}_{t+1}, \mathcal{F}_{t+2}, \ldots$ measurable and the result follows from (7.4.38)–(7.4.39).

▼▼▼

The implication of the lemma above is that one-step-ahead predictors can be concatenated to produce a d-step-ahead prediction. This can be seen by writing (7.4.35) for $j = 1, 2, \ldots$ as follows:

$$y^0(t+1|t) = -a_1 y(t) - \cdots - a_n y(t-n+1) + B(q^{-1})u(t+1)$$
$$+ c_1 \omega(t) + \cdots + c_l \omega(t+1-l) \qquad (7.4.40)$$

$$y^0(t+2|t) = -a_1 y^0(t+1|t) - a_2 y(t) - \cdots - a_n y(t-n+2)$$
$$+ B(q^{-1})u(t+2) + c_2 \omega(t) + \cdots + c_l \omega(t+2-l) \qquad (7.4.41)$$

$$y^0(t+3|t) = -a_1 y^0(t+2|t) - a_2 y^0(t+1|t) - a_3 y(t) - \cdots$$
$$- a_n y(t+3-n) + B(q^{-1})u(t+3) + c_3 \omega(t) + \cdots + c_l \omega(t+3-l) \qquad (7.4.42)$$

and so on. Thus we note that Lemma 7.4.2 gives the ARMAX equivalent of the property given in (7.3.91) for state-space models.

The result of Lemma 7.4.2 has a nice intuitive interpretation. Basically, it says: given an optimal one-step-ahead predictor, one can generate an optimal d-step-ahead predictor by replacing unknown data in the model with the appropriate conditional estimate. Thus future outputs are replaced by the previously obtained predictions and future innovations are replaced by zero.

Lemmas 7.4.1 and 7.4.2 can also be extended in a straightforward manner to the multioutput case. To illustrate, we present the multivariable version of Lemma 7.4.1.

Lemma 7.4.3. In the multioutput case where the m-dimensional process $\{(y(t)\}$ is modeled by

$$A(q^{-1})y(t) = B(q^{-1})u(t) + C(q^{-1})\omega(t) \qquad (7.4.43)$$

where

$$A(q^{-1}) = m \times m \text{ polynomial matrix}; \quad A_0 = I$$
$$C(q^{-1}) = m \times m \text{ polynomial matrix}; \quad C_0 = I$$
$$B(q^{-1}) = m \times r \text{ polynomial matrix}$$
$$\{\omega(t)\} = \text{vector white noise process}$$

Then provided that $\det C(q^{-1})$ is stable (has roots strictly inside the unit circle), the optimal d-step-ahead predictor satisfies

$$\bar{C}(q^{-1})y^0(t+d|t) = \bar{G}(q^{-1})y(t) + q^d\bar{F}(q^{-1})B(q^{-1})u(t) \qquad (7.4.44)$$

where $\bar{G}(q^{-1})$ and $\bar{F}(q^{-1})$ are defined as follows: $F(q^{-1})$ and $G(q^{-1})$ are the unique polynomial matrices such that

$$C(q^{-1}) = A(q^{-1})F(q^{-1}) + q^{-d}G(q^{-1}) \qquad (7.4.45)$$

$$F(q^{-1}) = I + F_1q^{-1} + \cdots + F_{d-1}q^{-d+1} \qquad (7.4.46)$$

$$G(q^{-1}) = G_0 + G_1q^{-1} + \cdots + G_{n-1}q^{-n+1} \qquad (7.4.47)$$

Then define polynomial matrices $\bar{F}(q^{-1})$, $\bar{G}(q^{-1})$, and $\bar{C}(q^{-1})$ (not necessarily unique) such that

$$\bar{F}(q^{-1})C(q^{-1}) = \bar{C}(q^{-1})F(q^{-1}) \qquad (7.4.48)$$

with $\bar{F}_0 = I$ and such that

$$\det C(z) = \det \bar{C}(z) \qquad (7.4.49)$$

Finally, define

$$\bar{G}(q^{-1}) = q^d[\bar{C}(q^{-1}) - \bar{F}(q^{-1})A(q^{-1})] \qquad (7.4.50)$$

Proof. [Here we basically follow Borisson (1979).] The choice of $\bar{F}(q^{-1})$ and $\bar{C}(q^{-1})$ in (7.4.48) corresponds to converting the right difference operator representation $[F(q^{-1})C(q^{-1})^{-1}]$ into a left difference operator representation $[\bar{C}(q^{-1})^{-1}\bar{F}(q^{-1})]$. This can always be done so that (7.4.49) is satisfied; see Chapter 2 and Wolovich (1974).

The matrix $\bar{G}(q^{-1})$ defined in (7.4.50) is a polynomial matrix in q^{-1}, as can be verified by multiplying (7.4.50) on the right by $F(q^{-1})$ and using (7.4.48) to yield

$$\bar{G}(q^{-1})F(q^{-1}) = q^d[\bar{C}(q^{-1})F(q^{-1}) - \bar{F}(q^{-1})A(q^{-1})F(q^{-1})]$$
$$= q^d\bar{F}(q^{-1})[C(q^{-1}) - A(q^{-1})F(q^{-1})]$$

or using (7.4.45),

$$\bar{G}(q^{-1})F(q^{-1}) = \bar{F}(q^{-1})G(q^{-1}) \qquad (7.4.51)$$

Then operating with $\bar{F}(q^{-1})$ on the left of (7.4.43) gives

$$\bar{F}(q^{-1})A(q^{-1})y(t) = \bar{F}(q^{-1})B(q^{-1})u(t) + \bar{F}(q^{-1})C(q^{-1})\omega(t) \qquad (7.4.52)$$

or using (7.4.48) and (7.4.50), we have

$$[\bar{C}(q^{-1}) - q^{-d}\bar{G}(q^{-1})]y(t) = \bar{F}(q^{-1})B(q^{-1})u(t) + \bar{C}(q^{-1})F(q^{-1})\omega(t) \qquad (7.4.53)$$

Hence

$$\bar{C}(q^{-1})[y(t) - F(q^{-1})\omega(t)] = q^{-d}\bar{G}(q^{-1})y(t) + \bar{F}(q^{-1})B(q^{-1})u(t) \qquad (7.4.54)$$

The remainder of the proof now exactly parallels the proof of Lemma 7.4.1 from (7.4.32) onward.

▼▼▼

Note that Remark 7.4.3 also applies to the multivariable case.

Predictors for Systems with Zeros on the Unit Circle

In the development above we found it necessary to restrict attention to ARMAX models having zeros [roots of $C(z^{-1})$] strictly inside the unit circle. This was necessary so that the effect of arbitrary initial conditions in the predictor would decay (recall Remark 7.4.3). In this section we return to the more general case in which the $C(q^{-1})$ polynomial can have roots on or inside the unit circle. In this case, an optimal *time-varying* predictor can still be obtained from the Kalman filter. Moreover, subject to weak assumptions, as in Theorem 7.3.4, the optimal predictor will become asymptotically time invariant. However, whereas previously the time-invariant predictor could be used from $t = 0$ with little error (since the effect of incorrect initial conditions decayed), this is not possible in the case when there are roots on the unit circle (since the effect of erroneous initial conditions will not decay). The reader is encouraged to look at Exercise 7.16, which highlights the distinction between the two situations for a simple case [see also Hannan (1970, pp. 127–136) for other insights].

One way out of this difficulty is to use a time-invariant predictor which is restricted in such a way that the effect of incorrect initial conditions decays. In the next section we describe the Levinson predictor, which falls into this category. The Levinson predictor is based on a moving average of past data and thus is guaranteed stable. Another alternative, discussed below, uses a time-invariant predictor very much like the ones used above, except that we constrain the predictor to be asymptotically stable. This gives optimal performance when the C polynomial has roots inside the unit circle and *nearly optimal* performance otherwise.

Consider again the stochastic model (7.4.21)

$$A(q^{-1})y(t) = q^{-d}B'(q^{-1})u(t) + C(q^{-1})\omega(t) \tag{7.4.55}$$

where, as before, the innovations sequence satisfies

$$E\{\omega(t)\,|\,\mathcal{F}_{t-1}\} = 0; \qquad E\{\omega(t)^2\,|\,\mathcal{F}_{t-1}\} = \sigma^2 \tag{7.4.56}$$

The polynomial $z^l C(z^{-1})$ will be taken to have roots on or inside the unit circle. We therefore express $C(q^{-1})$ in the following form:

$$C(q^{-1}) = C_s(q^{-1}) + C_R(q^{-1}) \tag{7.4.57}$$

where $C_s(q^{-1})$ is stable and $C_R(q^{-1})$ is a remainder polynomial. Note that $C_R(q^{-1})$ can be made as small as desired by appropriate choice of $C_s(q^{-1})$.

As an example, say that $C(q^{-1}) = 1 - q^{-1}$; then we can choose $C_s(q^{-1}) = 1 - \alpha q^{-1}$, with $\alpha = 1 - \epsilon$, $0 < \epsilon \ll 1$, and $C_R(q^{-1}) = -\epsilon q^{-1}$. Some comments on the best choice of $C_s(q^{-1})$ will be made later. We summarize the idea in the following Lemma:

Lemma 7.4.4. Given the system (7.4.55), there exists a time-invariant d-step-ahead predictor of the form

$$C_s(q^{-1})\hat{y}(t + d) = G(q^{-1})y(t) + \beta(q^{-1})u(t) \tag{7.4.58}$$

where

$$\beta(q^{-1}) = F(q^{-1})B'(q^{-1}) \tag{7.4.59}$$

and $F(q^{-1})$ and $G(q^{-1})$ are the unique polynomials of order $d - 1$ and $n - 1$, satisfying

$$C_s(q^{-1}) = F(q^{-1})A(q^{-1}) + q^{-d}G(q^{-1}) \tag{7.4.60}$$

The predictor (7.4.58) has the following properties:

(i) It is asymptotically stable.
(ii) The unconditional prediction error variance, σ_p^2, is bounded and asymptotically converges to

$$\sigma_p^2 = \frac{\sigma^2}{2\pi} \int_{-\pi}^{\pi} |F(e^{jw})|^2 \left| 1 + \frac{C_R(e^{jw})}{C_s(e^{jw})} \right|^2 dw \tag{7.4.61}$$

and when $C(q^{-1})$ has no roots on the unit circle then the unconditional prediction error variance is

$$\sigma_p^2 = \frac{\sigma^2}{2\pi} \int_{-\pi}^{\pi} |F(e^{jw})|^2 \, dw \tag{7.4.62}$$

(iii) The predictor gives the optimal prediction when $C_s(q^{-1}) = C(q^{-1})$.

Proof. Multiplying (7.4.55) by $F(q^{-1})$ and using (7.4.57) and (7.4.60) gives

$$F(q^{-1})A(q^{-1})y(t) = q^{-d}F(q^{-1})B'(q^{-1})u(t) + F(q^{-1})[C_s(q^{-1}) + C_R(q^{-1})]\omega(t) \tag{7.4.63}$$

Then using (7.4.60),

$$C_s(q^{-1})[y(t) - F(q^{-1})\omega(t) - \eta(t)] = q^{-d}G(q^{-1})y(t) + q^{-d}F(q^{-1})B'(q^{-1})u(t) \tag{7.4.64}$$

where

$$C_s(q^{-1})\eta(t) = F(q^{-1})C_R(q^{-1})\omega(t) \tag{7.4.65}$$

Now if we let $e(t + d)$ denote the prediction error; that is,

$$e(t + d) = y(t + d) - \hat{y}(t + d) \tag{7.4.66}$$

then we see from (7.4.58) and (7.4.64) that $e(t + d)$ satisfies

$$C_s(q^{-1})e(t + d) = C_s(q^{-1})F(q^{-1})\omega(t + d) + C_R(q^{-1})F(q^{-1})\omega(t + d) \tag{7.4.67}$$

Since $C_s(q^{-1})$ is stable (by choice), it follows from (7.4.67) that the unconditional prediction error variance is as in (7.4.61).

When $C_s(q^{-1}) = C(q^{-1})$, then $C_s(q^{-1}) = 0$ and the result (7.4.62) is obtained. We note from Lemma 7.4.1 that this is the minimum prediction error variance.

▼▼▼

A natural question that arises in connection with the (suboptimal) predictor (7.4.58) is how one should choose $C_s(q^{-1})$. If $C_s(q^{-1})$ is chosen with roots well inside the unit circle, the predictor will have "short memory" and this will enhance its *robustness* in the presence of nonideal factors (e.g., phase changes in the deterministic components). On the other hand, it is clear from (7.4.61)–(7.4.62) that the closer $C_s(q^{-1})$ is to $C(q^{-1})$, the better the asymptotic performance will be *in the ideal case.* Thus, in general, a compromise has to be made. It is felt that even in cases where $C(q^{-1})$ is stable but having roots near the unit circle, the robustness of the stochastic predictor could be improved by using a predictor of the form (7.4.58) with roots a "reasonable distance" from the unit circle.

Example 7.4.3

Consider the following simple process:

$$y(t) = d + u(t - 1) + \omega(t) \qquad (7.4.68)$$

where $\omega(t)$ is "white noise" and d is a constant disturbance.

The corresponding ARMAX model is

$$A(q^{-1})y(t) = q^{-1}B'(q^{-1})u(t) + C(q^{-1})\omega(t) \qquad (7.4.69)$$

where

$$A(q^{-1}) = B'(q^{-1}) = C(q^{-1}) = 1 - q^{-1} \qquad (7.4.70)$$

Note that $A(q^{-1})$, $B'(q^{-1})$, and $C(q^{-1})$ all have roots on the unit circle (at 1) corresponding to the purely deterministic component in the model.

We propose that $C_s(q^{-1})$ be chosen as follows:

$$C_s(q^{-1}) = 1 - \alpha q^{-1}; \qquad \alpha = 1 - \epsilon; \quad 0 < \epsilon \ll 1 \qquad (7.4.71)$$

Then the predictor (7.4.58) has the following form for $d = 1$:

$$(1 - \alpha q^{-1})\hat{y}(t + 1) = \epsilon y(t) + u(t) - u(t - 1) \qquad (7.4.72)$$

The unconditional prediction error variance is

$$\sigma_p^2 = \left(1 + \frac{\epsilon}{1 + \alpha}\right)\sigma^2 \qquad (7.4.73)$$

which can be made as close as desired to σ^2 by choosing ϵ sufficiently small.

Note that even if α is chosen as zero, the unconditional variance is only $2\sigma^2$!

▼▼▼

The discussion above has established the existence of a linear *time-invariant* predictor which gives nearly optimal performance in the presence of both deterministic and nondeterministic disturbances. This is as good as can be hoped for when both types of disturbances are present.

The predictor described in Lemma 7.4.4 can also be extended to the multivariable case. We leave the details to the reader; see Exercise 7.19.

Remark 7.4.4 (Predictors for Nonlinear Systems). So far we have discussed predictors for linear dynamic systems only. However, as we have said before, one should always be on the lookout for nonlinear models which may better represent the dynamics of a system than a linear model; see, for example, the discussion in Section 4.2.1 and Remark 7.4.2.

The models in Remark 7.4.2 have been judiciously chosen so that *optimal nonlinear predictors* can be readily constructed. This is explored in the following lemma:

Lemma 7.4.5. Consider a nonlinear model of the form given in (7.4.20):

$$A(q^{-1})y(t) = h(t - d) + C(q^{-1})\omega(t) \qquad (7.4.74)$$

where $\omega(t)$ is a white noise sequence satisfying $E\{\omega(t)\,|\,\mathcal{F}_{t-1}\} = 0$:

$$A(q^{-1}) = 1 + a_1 q^{-1} + \cdots + a_n q^{-n}$$
$$C(q^{-1}) = 1 + c_1 q^{-1} + \cdots + C_n q^{-n}$$

$C(q^{-1})$ is stable and $h(t - d)$ is a nonlinear function of $\{y(t - d), \ldots, u(t - d), \ldots\}$, that is, $h(t - d)$ is $\mathcal{F}_{t-d}$ measurable. Then the optimal d-step-ahead prediction $y^0(t + d|t)$ of $y(t + d)$ satisfies

$$C(q^{-1})y^0(t + d|t) = G(q^{-1})y(t) + F(q^{-1})h(t) \tag{7.4.75}$$

where

$$y^0(t + d|t) = E\{y(t + d)|\mathcal{F}_t\} \tag{7.4.76}$$

and $G(q^{-1})$ and $F(q^{-1})$ are the unique polynomials satisfying

$$C(q^{-1}) = F(q^{-1})A(q^{-1}) + q^{-d}G(q^{-1}) \tag{7.4.77}$$

$$F(q^{-1}) = 1 + f_1 q^{-1} + \cdots + f_{d-1}q^{-d+1} \tag{7.4.78}$$

$$G(q^{-1}) = g_0 + g_1 q^{-1} + \cdots + g_{n-1}q^{-n+1} \tag{7.4.79}$$

Proof. Exactly as for Lemma 7.4.1 on noting that $h(t - d)$ is $\mathcal{F}_{t-d}$ measurable.

▼▼▼

Note that the model in Remark 7.4.2 has been specifically engineered so that the optimal prediction has the simple form described in the result above. In general, it is not possible to obtain a simple closed-form expression for the optimal predictor for the output of a nonlinear system. As explored in the next section, a sensible approach in these cases may be to seek the best *predictor of a given structure* (i.e., a *restricted complexity predictor*).

7.5 RESTRICTED COMPLEXITY FILTERS AND PREDICTORS

The optimal filters and predictors of Sections 7.3 and 7.4 relied on the assumption that the signal-generating mechanism was linear and finite-dimensional. In some cases this may be an unrealistic assumption. In such cases a sensible approach may be to consider a filter (or predictor) of prespecified structure which may not correspond to the structure of the optimal filter. We shall call this class of filters *restricted complexity filters*.

7.5.1 General Filters

Suppose that we wish to estimate a signal $s(t + \alpha)$, given an observation $\{y(t)\}$, which is presumed to contain information about $s(t + \alpha)$ in some sense. [Note that the relationship between $y(t)$ and $s(t + \alpha)$ can be arbitrary, including nonlinear, infinite-dimensional, etc.] We assume only that the signals $\{s(t + \alpha)\}$ and $\{y(t)\}$ are stationary.

As an illustration of a restricted complexity filter, let us consider the following *transversal filter*:

$$\hat{s}(t + \alpha) = F(q^{-1})y(t)$$

where $F(q^{-1}) = f_0 + f_1 q^{-1} + \cdots + f_n q^{-n}$.

We now show that $F(q^{-1})$ can be chosen to minimize the mean-square filtering error, J, where $J \triangleq E\{[s(t + \alpha) - \hat{s}(t + \alpha)]^2\}$. Differentiating with respect to the coefficients in $F(q^{-1})$ leads to the following set of normal equations for the coefficients in the optimal filter:

$$\theta_0 = \Gamma^{-1}\beta$$

with corresponding optimal mean-square error $J^* = E\{s(t + \alpha)^2\} - \theta_0^T \Gamma \theta_0$, where

$$\theta_0 = (f_0, f_1, \cdots, f_n)^T$$

$$\Gamma = \begin{bmatrix} R_{yy}(0) & \cdots & R_{yy}(n) \\ & \cdot & \\ & \cdot & \\ & \cdot & \\ R_{yy}(n) & \cdots & R_{yy}(0) \end{bmatrix}$$

$$\beta = \begin{bmatrix} R_{sy}(-\alpha) \\ \cdot \\ \cdot \\ \cdot \\ R_{sy}(-\alpha - n) \end{bmatrix}$$

$$R_{yy}(j) = E\{y(t)y(t + j)\}$$

$$R_{sy}(j) = E\{s(t)y(t + j)\}$$

A very special case of the above is explored in Exercise 7.11.

In some applications a transversal filter with many coefficients may be required to give a good estimate. In such cases it may be desirable to introduce poles into the filter. Thus another common class of restricted complexity filter is the following filter, which is called a *recursive filter*:

$$H(q^{-1})\hat{s}(t + \alpha) = F(q^{-1})y(t)$$

$$H(q^{-1}) = 1 + h_1 q^{-1} + \cdots + h_m q^{-m}$$

$$F(q^{-1}) = f_0 + f_1 q^{-1} + \cdots + f_n q^{-n}$$

The recursive filter above can be expressed in transfer function form as

$$\hat{s}(t + \alpha) = G(q^{-1})y(t)$$

where

$$G(q^{-1}) = \frac{F(q^{-1})}{H(q^{-1})}$$

The mean-square filtering error is given by

$$J \triangleq E\{[s(t + \alpha) - \hat{s}(t + \alpha)]^2\}$$

$$= E\left\{\left[s(t + \alpha) - \frac{F(q^{-1})}{H(q^{-1})}y(t)\right]^2\right\}$$

A difficulty with J, as given above, is that it is a nonquadratic function of the coefficients in $H(q^{-1})$ and thus there is, in general, no closed-form solution to the problem of minimizing J. (We will see in Chapter 9 that there actually exist interative on-line procedures for finding a local minimum of J.) An alternative approach is to replace J by a *mean-square filtered error*, J_f, defined as follows:

$$J_f = E\{[H(q^{-1})(s(t + \alpha) - \hat{s}(t + \alpha))]^2\}$$

$$= E\{[H(q^{-1})s(t + \alpha) - F(q^{-1})y(t)]^2\}$$

Note that minimization of J_f is *not* equivalent to minimization of J. However, the advantage of working with J_f is that it is quadratic in the coefficients of $F(q^{-1})$ and $H(q^{-1})$ and thus a simple analytic expression can be obtained for the global minimum. By setting the derivative of J_f to zero, we obtain the following expression for the optimal coefficients:

$$\theta_0 = \bar{\Gamma}^{-1}\bar{\beta}$$

where

$$\theta_0 = (h_1 \cdots h_m, f_0 \cdots f_n)^T$$

$$\bar{\Gamma} = \begin{bmatrix} R_{ss}(0) & \cdots & R_{ss}(m-1) & -R_{sy}(1-\alpha) & \cdots & -R_{sy}(1-\alpha+n) \\ & \ddots & \vdots & & \ddots & \vdots \\ & & R_{ss}(0) & -R_{sy}(-\alpha+m) & \cdots & -R_{sy}(-\alpha+m+n) \\ \hline & & & R_{yy}(0) & \cdots & R_{yy}(n) \\ & & & & \ddots & \vdots \\ & & & & & R_{yy}(0) \end{bmatrix}$$

$$(\bar{\Gamma} \text{ is symmetric})$$

$$\bar{\beta} = [R_{ss}(1) \cdots R_{ss}(m), -R_{sy}(-\alpha), \cdots, -R_{sy}(n-\alpha)]^T$$

Remark 7.5.1 (The Finite-Alphabet Case). We wish to make a very brief comment about the case where it is known that a signal is drawn from a finite alphabet (i.e., has a finite number of possible values). This commonly arises in digital communication filtering applications, where it is known that the input has only one of two possible values (say ± 1). All of the linear filters described so far do not take this into account and, in fact, give an estimate with a continuous amplitude distribution.

The simplest way to construct an estimate of a signal drawn from a finite alphabet is to simply take the nearest discrete value to the estimate provided by a linear filter. If the estimates are used in forming future estimates (i.e., they are fed back into the filters), then it makes sense to feedback the discretized estimates. Thus if we consider the recursive filter

$$H(q^{-1})\hat{s}(t + \alpha) = F(q^{-1})y(t)$$

and we replace $\hat{s}(t + \alpha - 1)$, ... by its discretized values $\bar{s}(t + \alpha - 1)$ ..., we obtain the following filter

$$\bar{s}(t + \alpha) = \frac{\text{quantized}}{\text{value of}} \{[-H(q^{-1}) + 1]q\bar{s}(t + \alpha - 1) + F(q^{-1})y(t)\}$$

This kind of filter is commonly called a decision feedback filter and is as shown in Fig. 7.5.1. [See also George, Bowen, and Storey (1971).]

The procedure above leads to suboptimal restricted complexity estimates in the finite-alphabet case. A slightly more fundamental approach is to use *maximum likelihood* estimation and choose those discrete values that maximize the likelihood func-

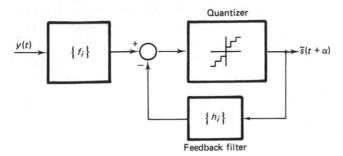

Quantizer

$y(t)$ → $\{f_i\}$ → + ⊖ − → [Quantizer] → $\bar{s}(t + \alpha)$

$\{h_i\}$

Feedback filter

Figure 7.5.1 Decision feedback equalizer.

tion. *Dynamic programming* can be used to simplify the computation. An example of this type of approach is the well-known *Viterbi algorithm* [see Viterbi (1966, 1967), Ungerboeck (1971), Omura (1971), and Forney (1972)].

These kinds of problems are very important in certain applications, but we will not persue this topic in detail since this would be outside the general scope of this book.

▼▼▼

7.5.2 Whitening Filters

We recall from Section 7.3 that for a *linear finite-dimensional* system, the innovations sequences, $\{\omega^0(t)\}$, can be obtained from the output process, $\{y(t)\}$, via a *whitening filter* which is a simple rearrangement of the Kalman filter. We recall from (7.3.19) and (7.3.20) that this (unrestricted optimal) whitening filter has the form

$$\hat{x}(t + 1) = [A - K(t)C]\hat{x}(t) + Bu(t) + K(t)y(t) \tag{7.5.1}$$

$$\omega^0(t) = -C\hat{x}(t) + y(t) \tag{7.5.2}$$

[Note that here we have used $\omega^0(t)$ to denote the innovations.]

In the gaussian cause, $\hat{x}(t)$ is the conditional mean of $x(t)$ given $y(0)$, $\cdots$, $y(t - 1)$ and the initial conditions. Thus in the gaussian case, $\omega^0(t)$ has the following property (as developed in Theorem 7.3.1):

$$E\{\omega^0(t)\,|\,\mathcal{F}_{t-1}\} = 0$$

where $\mathcal{F}_{t-1}$ denotes the sigma algebra generated by $y(0)$, $\ldots$, $y(t - 1)$ plus initial conditions.

In the nongaussian case $\hat{x}(t)$, given by the Kalman filter, is the linear minimum variance estimator. A consequence of this is that

$$E\{(x(t) - \hat{x}(t)y(t - i)\} = 0; \qquad i = 1, \ldots, t$$

and this implies that

$$E\{\omega^0(t)y(t - i)\} = 0; \qquad i = 1, \ldots, t$$

In this case, the innovations should strictly be called the *linear innovations*. If we introduce the Hilbert space, H, spanned by the random process $y(t)$, and with norm $\|\cdot\|$, where $\|v(t)\| \triangleq E\{v(t)^2\}$, $v(t) \in H$, then $\omega^0(t)$ is orthogonal to the observations generating the estimates [i.e., $\omega^0(t) \perp y(0)$, $\cdots$, $y(t - 1)$]. Equivalently, the estimate $\hat{y}(t)$ is the orthogonal projection of $y(t)$ onto the past data.

A remarkable feature of the above (unrestricted optimal) whitening filter is that

Optimal Filtering and Prediction   Chap. 7

$y(t)$ can be projected onto all past data (which may extend into the infinite past) by using a linear finite-dimensional filter [namely, (7.5.1) and (7.5.2)]. This is a consequence of the finite-dimensional nature of the hypothesized signal model (7.2.3)–(7.2.4). In general, this may not hold, and then there is no simple way of projecting $y(t)$ onto the infinite past. In this case it makes sense to simplify the problem by projecting $y(t)$ onto a finite window of the past data, as in Section 7.5.1. Thus consider $\hat{y}(t)$ formed by a finite projection as follows:

$$\hat{y}(t) = \sum_{j=1}^{n} -a_j y(t-j) \qquad (7.5.3)$$

The corresponding residue, $\omega(t)$, is by definition

$$\omega(t) \triangleq y(t) - \hat{y}(t) \qquad (7.5.4)$$

and from (7.5.3), $\omega(t)$ is given by the following *restricted complexity whitening filter*:

$$\omega(t) = A(q^{-1})y(t) \qquad (7.5.5)$$

where $A(q^{-1}) = 1 + a_1 q^{-1} + \cdots + a_n q^{-n}$. Filters of the form (7.5.5) are commonly called *transversal* or *tapped delay line filters*.

The coefficients of $A(q^{-1})$ in (7.5.3) are determined by minimizing the mean-square error,

$$J = E\{[y(t) - \hat{y}(t)]^2\} \qquad (7.5.6)$$

Differentiation with respect to a_j gives

$$\frac{\partial J}{\partial a_j} = E\{[y(t) - \hat{y}(t)]y(t-j)\} = 0; \qquad j = 1, \ldots, n \qquad (7.5.7)$$

Thus the coefficients are determined from the following set of equations:

$$\sum_{i=1}^{n} E\{y(t-i)y(t-j)\}a_i^* = -E\{y(t)y(t-j)\}; \qquad j = 1, \ldots, n \qquad (7.5.8)$$

(where the superscript * denotes the minimum mean-square error solution).

The set of equations above can be solved provided that the process $\{y(t)\}$ is of full rank. The set of equations (7.5.7) [or equivalently (7.5.8)] are called the *normal equations*. Substituting (7.5.7) into (7.5.6) gives the minimum mean-square error as

$$J_n^* = \sum_{j=0}^{n} a_j^* E\{y(t)y(t-j)\}; \qquad a_0^* = 1 \qquad (7.5.9)$$

Also from (7.5.4) and (7.5.7), we have

$$E\{\omega(t)y(t-j)\} = 0; \qquad j = 1, \ldots, n \qquad (7.5.10)$$

Note that the determination of a_j from (7.5.8) depends on knowledge of the *covariance* of the process $\{y(t)\}$, in contrast to the unrestricted optimal filter for the finite-dimensional case (7.5.1)–(7.5.2), which was developed from knowledge of the underlying *state-space signal model*. We will explore an alternative way of implementing the restricted complexity whitening filter (7.5.5) using a lattice structure in a later section.

In the next subsection we show how the same ideas used in developing the transversal whitening filter can be used to form a restricted complexity predictor (called the Levinson predictor).

7.5.3 Levinson Predictors

The Levinson predictor forms a prediction of the future output as a simple moving average of past data; that is, the predicted output, $\hat{y}(t)$, is given by

$$\hat{y}(t) = -\sum_{j=1}^{n} a_j y(t-j) \tag{7.5.11}$$

This predictor is guaranteed stable and can be applied to nonlinear systems or systems having roots on the unit circle (see Section 7.4.2). However, it will not in general correspond to the optimal unrestricted predictor. Instead, we seek the values of the coefficients $a_1, \ldots, a_n$ so that the prediction given by the pre-specified form (7.5.11) is as good as possible (in the mean square sense). The reader will note that this is a special case of the transversal restricted complexity filter described in Section 7.5.1. We shall further develop the idea here and show that a simple method exists for increasing the window length (the size of n) without completely redoing all the calculations. Also, the method leads to a way of choosing the window length so that a specified mean-square prediction error is achieved. The development in this section is based on the work of Levinson (1947) and thus we will call the resulting predictor the *Levinson predictor*.

We are interested in choosing the coefficients $a_1 \cdots a_n$ so that the mean-square prediction error is minimized. Thus consider the following cost function:

$$J_n = E\{[\hat{y}(t) - y(t)]^2\} \tag{7.5.12}$$

Substituting (7.5.11) into (7.5.12) and differentiating with respect to a_j gives the following equations for the restricted complexity optimal predictor coefficients, a_j^n, when the window length is n:

$$E\{y(t-j)(\hat{y}(t) - y(t))\} = 0; \qquad j = 1, \ldots, n \tag{7.5.13}$$

or

$$E\{y(t-j)[\sum_{k=1}^{n} -a_k^n y(t-k) - y(t)]\} = 0; \qquad j = 1, \ldots, n \tag{7.5.14}$$

or

$$\sum_{k=1}^{n} a_k^n C_{|j-k|} = -C_j; \qquad j = 1, \ldots, n \tag{7.5.15}$$

where $\{C_j\}$ is the covariance of the process $\{y(t)\}$ and is given by

$$C_j = E\{y(t)y(t+j)\} \tag{7.5.16}$$

Substituting (7.5.13) into (7.5.12) gives the following expression for the mean-square prediction error:

$$\begin{aligned}
J_n &= E\{(y(t) - \hat{y}(t))^2\} \\
&= E\{[y(t) + \sum_{j=1}^{n} a_j^n y(t-j)][y(t) - \hat{y}(t)]\} \\
&= E\{y(t)[y(t) - \hat{y}(t)]\} \qquad \text{using (7.5.13)} \\
&= C_0 + \sum_{j=1}^{n} a_j^n C_j
\end{aligned} \tag{7.5.17}$$

Combining (7.5.15) and (7.5.17), we have

$$
\begin{bmatrix}
C_0 & C_1 & \cdots & C_n \\
C_1 & \cdot & & \\
\cdot & & \cdot & \\
\cdot & & & \cdot \\
\cdot & & & \\
C_n & \cdot & \cdot & C_0
\end{bmatrix}
\begin{bmatrix}
1 \\
a_1^n \\
\cdot \\
\cdot \\
\cdot \\
a_n^n
\end{bmatrix}
=
\begin{bmatrix}
J_n \\
0 \\
\cdot \\
\cdot \\
\cdot \\
0
\end{bmatrix}
\tag{7.5.18}
$$

Now consider the situation when the window length is increased to $(n+1)$; then the optimal coefficients satisfy

$$
\begin{bmatrix}
C_0 & C_1 & \cdots & C_n & C_{n+1} \\
\cdot & \cdot & & & \cdot \\
\cdot & & \cdot & & \cdot \\
\cdot & & & \cdot & \cdot \\
\cdot & & & & \cdot \\
C_n & C_{n-1} & \cdots & C_0 & C_1 \\
C_{n+1} & \cdot & \cdot & C_1 & C_0
\end{bmatrix}
\begin{bmatrix}
1 \\
a_1^{n+1} \\
\cdot \\
\cdot \\
\cdot \\
a_n^{n+1} \\
a_{n+1}^{n+1}
\end{bmatrix}
=
\begin{bmatrix}
J_{n+1} \\
0 \\
\cdot \\
\cdot \\
\cdot \\
0
\end{bmatrix}
\tag{7.5.19}
$$

Now if we replace the solution $(1, a^{n+1}, \ldots, a_n^{n+1}, a_{n+1}^{n+1})^T$ on the left-hand side of the equation above by $(1, a_1^n, \ldots, a_n^n, 0)^T$, then we see from (7.5.18) that

$$
\begin{bmatrix}
C_0 & C_1 & & & C_{n+1} \\
\cdot & & & & \cdot \\
\cdot & & & & \cdot \\
\cdot & & & & \cdot \\
\cdot & & & & \cdot \\
C_n & C_{n-1} & \cdots & C_0 & C_1
\end{bmatrix}
\begin{bmatrix}
1 \\
a_1^n \\
\cdot \\
\cdot \\
a_n^n \\
0
\end{bmatrix}
=
\begin{bmatrix}
J_n \\
0 \\
\cdot \\
\cdot \\
\cdot \\
0
\end{bmatrix}
\tag{7.5.20}
$$

but

$$
[C_{n+1} \quad C_n \quad \cdots \quad C_0]
\begin{bmatrix}
1 \\
a_1^n \\
\cdot \\
\cdot \\
a_n^n \\
0
\end{bmatrix}
\triangleq \alpha_n \neq 0 \qquad \text{in general}
\tag{7.5.21}
$$

Equation (7.5.21) can be recognized as

$$
E\{y(t-n-1)[y(t) + \sum_{k=1}^{n} a_k^n y(t-k)]\} = \alpha_n
\tag{7.5.22}
$$

where is nonzero, in general, since $y(t-n-1)$ is an additional data point not included in the nth-order regression. Equations (7.5.20) and (7.5.21) can be combined to give

$$
\begin{bmatrix} C_0 & C_1 & \cdots & C_{n+1} \\ C_1 & & & \\ \cdot & \cdot & & \cdot \\ \cdot & & \cdot & \cdot \\ \cdot & & & \cdot \\ C_{n+1} & \cdots & & C_0 \end{bmatrix} \begin{bmatrix} 1 \\ a_1^n \\ \cdot \\ \cdot \\ a_n^n \\ 0 \end{bmatrix} = \begin{bmatrix} J_n \\ 0 \\ \cdot \\ \cdot \\ 0 \\ \alpha_n \end{bmatrix} \tag{7.5.23}
$$

Note from (7.5.19) and (7.5.23) that if α_n turned out to be zero, this would imply that $(a_1^n, \ldots, a_n^n, 0)$ would be the optimal predictor coefficients for window of length $n + 1$. We shall thus manipulate (7.5.23) to reduce α_n to zero.

Since the matrix on the left-hand side of (7.5.23) is Toeplitz [a matrix with elements a_{ij} is Toeplitz if $a_{ij} = a_{pq}$ whenever $i - j = p - q$; see also Grenander and Szego (1958)], we can rearrange (7.5.23) into the following equivalent form:

$$
\begin{bmatrix} C_0 & C_1 & \cdots & C_{n+1} \\ C_1 & & & \\ \cdot & & & \\ \cdot & & & \\ \cdot & & & \\ C_{n+1} & \cdots & & C_0 \end{bmatrix} \begin{bmatrix} 0 \\ a_n^n \\ \cdot \\ \cdot \\ a_1^n \\ 1 \end{bmatrix} = \begin{bmatrix} \alpha_n \\ 0 \\ \cdot \\ \cdot \\ 0 \\ J_n \end{bmatrix} \tag{7.5.24}
$$

Now multiplying (7.5.24) by $-\alpha_n/J_n$ and adding the result to (7.5.23) gives

$$
\begin{bmatrix} C_0 & C_1 & \cdots & C_{n+1} \\ C_1 & & & \\ \cdot & & & \\ \cdot & & & \\ \cdot & & & \\ C_{n+1} & \cdots & & C_0 \end{bmatrix} \begin{bmatrix} 1 \\ a_1^n - \dfrac{\alpha_n}{J_n} a_n^n \\ \cdot \\ \cdot \\ a_n^n - \dfrac{\alpha_n}{J_n} a_1^n \\ -\dfrac{\alpha_n}{J_n} \end{bmatrix} = \begin{bmatrix} J_n - \dfrac{\alpha_n^2}{J_n} \\ 0 \\ \cdot \\ \cdot \\ 0 \\ 0 \end{bmatrix} \tag{7.5.25}
$$

It is clear from (7.5.19) and (7.5.25) that the restricted complexity optimal predictor of length $n + 1$ has coefficients $\{a_i^{n+1}\}$ given by

$$
a_i^{n+1} = a_i^n - \frac{\alpha_n}{J_n} a_{n+1-i}^n; \qquad i = 1, \ldots, n \tag{7.5.26}
$$

$$
a_{n+1}^{n+1} = -\frac{\alpha_n}{J_n} \tag{7.5.27}
$$

where α_n is given by (7.5.23) as

$$
\alpha_n = C_{n+1} + \sum_{j=1}^{n} a_j^n C_{n+1-j} \tag{7.5.28}
$$

Also from (7.5.25)

$$
J_{n+1} = J_n - \frac{\alpha_n^2}{J_n} \tag{7.5.29}
$$

Equations (7.5.26) to (7.5.29) provide a simple means of deriving the
of length $(n + 1)$ given the predictor of length n. The equations are initializ'

$$a_1^1 = \frac{-C_1}{C_0}$$

$$J_1 = C_0 - \frac{C_1^2}{C_0} \tag{7.5.31}$$

We note from (7.5.29) that $\{J_n\}$ is a nonincreasing function of n. We can thus
increase n until a specified mean-square prediction error is reached or n reaches some
maximum value. This facility is one of the prime advantages of this approach.

The procedure above has found wide applications; see, for example, Robinson
(1967) and Durbin (1960). The algorithm can also be extended to the multivariable
case; see Whittle (1963) and Wiggins and Robinson (1965).

In the next section we describe lattice filters, which turn out to be closely related
to the Levinson predictor.

7.6 LATTICE FILTERS AND PREDICTORS

7.6.1 Lattice Filters

In this section we shall describe the lattice filter introduced by Itakura and Saito
(1971). We shall find that this filter has many features that make it attractive in applications as an alternative to the transversal filter described in Section 7.5.1. The main
feature of the lattice filter is that it is an orthogonalization device which replaces the
original signal process by a sequence of orthogonal residuals generating the same
space.

We shall see later that the lattice filter has a number of interesting properties,
including: (1) the successive stages are decoupled in a certain sense; (2) the stability
of the inverse of the filter can be checked by inspection; and (3) a simple procedure
exists for upgrading from a filter of order n to a filter of order $n + 1$.

We shall consider a (wide-sense) *stationary* random process $\{y(t)\}$ and the Hilbert
space, H, spanned by $y(t)$. We denote by Y_t^{t+k} for $k \geq 0$ the closed linear subspace of
H spanned by $\{y(t), \ldots, y(t + k)\}$, Y_t^{t+k} for $k < 0$ will denote the empty space, and
finally, for every element $y(i)$ in H, $\hat{y}(i \mid Y_t^{t+k})$ will denote the orthogonal projection of
$y(i)$ onto Y_t^{t+k}; that is,

$$E\{y(i) - \hat{y}(i \mid Y_t^{t+k}))y(j) = 0; \qquad j = t, \ldots, t + k \tag{7.6.1}$$

In terms of the notation of the preceding section $\hat{y}(t \mid Y_{t-n}^{t-1})$ would have been
simply denoted by $\hat{y}(t)$. Here however, we are interested in determining $\hat{y}(t \mid Y_{t-n}^{t-1})$
for a range of values of n. In particular, we will be interested in the conversion of
$\hat{y}(t \mid Y_{t-n}^{t-1})$ to $\hat{y}(t \mid Y_{t-n-1}^{t-1})$ (i.e., increasing the order of the restricted complexity filter).

If we use (7.5.8) to determine the coefficients in the filter in (7.5.5), we see that
we have to solve n simultaneous equations. However, these equations are greatly
simplified if the *past data* are *orthogonal*. In this case we have

$$E\{y(t - i)y(t - j)\} = \beta_i \delta_{i-j}; \qquad i \geq 1, \; j \geq 1 \tag{7.6.2}$$

where $\beta_i \triangleq E\{y(t-i)^2\}$. Then (7.5.8) simplifies to

$$\beta_i a_i^n = -\gamma_i; \qquad i = 1, \ldots, n \tag{7.6.3}$$

where

$$\gamma_i \triangleq E\{y(t)y(t-i)\} \tag{7.6.4}$$

(The superscript n in a_i^n denotes the nth-order filter.) Clearly, we now have

$$a_i^{n+1} = a_i^n; \qquad i = 1, \ldots, n \tag{7.6.5}$$

$$a_{n+1}^{n+1} = \frac{-\gamma_{n+1}}{\beta_{n+1}} \tag{7.6.6}$$

Thus, in this special case, the determination of a_i^{n+1} is very simple as is the conversion from order n to order $n+1$.

The discussion above suggests that determination of the filter coefficients can be simplified if we first orthogonalize the past data. This can be readily achieved by using the *Gram–Schmidt orthogonalization* procedure (in the Hilbert space defined above). Thus define the following orthogonal basis for the past data:

$$b_0(t-1) \triangleq y(t-1) \tag{7.6.7}$$

$$b_1(t-1) \triangleq y(t-2) - \hat{y}(t-2 \mid y(t-1)) \tag{7.6.8}$$

$$b_{n-1}(t-1) \triangleq y(t-n) - \hat{y}(t-n \mid Y_{t-n+1}^{t-1}) \tag{7.6.9}$$

[Check that $b_0(t-1), \ldots, b_{n-1}(t-1)$ are orthogonal as required].

Now $\hat{y}(t-j-1 \mid Y_{t-j}^{t-1})$ is a linear function of $y(t-j), \ldots, y(t-1)$. Thus (7.6.7) to (7.6.9) can be written as

$$\begin{bmatrix} b_0(t-1) \\ b_1(t-1) \\ \cdot \\ \cdot \\ \cdot \\ b_{n-1}(t-1) \end{bmatrix} = \begin{bmatrix} 1 & & & \\ w_1^1 & 1 & & \\ \cdot & & & \\ \cdot & & 1 & \\ \cdot & & & \\ w_{n-1}^{n-1} & \cdots & w_1^{n-1} & 1 \end{bmatrix} \begin{bmatrix} y(t-1) \\ y(t-2) \\ \cdot \\ \cdot \\ \cdot \\ y(t-n) \end{bmatrix} \tag{7.6.10}$$

Since $b_0(t-1), \ldots, b_{n-1}(t-1)$ span the same space as do $y(t-1), \ldots, y(t-n)$, we can compute $\hat{y}(t \mid Y_{t-n}^{t-1})$ by projecting onto $b_0(t-1), \ldots, b_{n-1}(t-1)$. This can be achieved as follows:

$$\hat{y}(t \mid Y_{t-n}^{t-1}) = - \sum_{j=1}^{n} K_j^b b_{j-1}(t-1) \tag{7.6.11}$$

Also, in view of the arguments leading to (7.6.3) we immediately have that K_j^b can be simply computed as follows [see the normal equations (7.5.8) and the orthogonality of the b's]:

$$K_j^b = \frac{-E\{y(t)b_{j-1}(t-1)\}}{E\{b_{j-1}(t-1)^2\}}; \qquad j = 1, \ldots, n \tag{7.6.12}$$

Thus we have achieved the desired result of simplifying the evaluation of the filter coefficients and of proceeding from a filter of length n to $n+1$.

Now, we are actually interested in forming a "whitening filter," and therefore we look at the residuals $f_n(t)$ defined as follows:

$$f_n(t) \triangleq y(t) - \hat{y}(t \mid Y_{t-n}^{t-1}) \tag{7.6.13}$$

$$w_i^n = a_i^n; \qquad i = 1, \ldots, n$$

and

$$R_n^b = R_n^f$$

The proof of part (c) is immediate from part (a) and (7.6.10) and (7.6.20).

(iv) Equation (7.6.37) follows from (7.6.12), (7.6.22), (7.6.29), (7.6.30), and (7.6.32). Equation (7.6.38) is then a combination of (7.6.12) and (7.6.22).

(v) From (7.6.25), we have

$$A_{n+1}(q^{-1})y(t) = A_n(q^{-1})y(t) + K_{n+1}\bar{A}_n(q^{-1})q^{-1}y(t)$$

The result (7.6.39) then follows since $E\{y(t)^2\} \neq 0$. Similarly, (7.6.41) follows from (7.6.26).

(vi) From (7.6.25) we have

$$J_{n+1}^* = J_n^* + 2K_{n+1}E\{f_n(t)b_n(t-1)\} + K_{n+1}^2 J_n^*$$

The result then follows by using (7.6.38).

(vii) We first show that (a) $\Rightarrow$ (b) $\Rightarrow$ (c). We recall that

$$f_n(t) = A_n(q^{-1})y(t)$$

From this, we can construct a state-space model for $y(t)$:

$$x(t+1) = \begin{bmatrix} 0 & 1 & & \\ & & \ddots & \\ & & & 1 \\ -a_n & \cdots & & -a_1 \end{bmatrix} x(t) + \begin{bmatrix} 0 \\ \vdots \\ 0 \\ 1 \end{bmatrix} f_n(t+1)$$

$$y(t) = [0 \quad \cdots \quad 0 \quad 1]x(t)$$

where $x_{n-i}(t) = y(t-i)$, $\quad 0 \le i \le n-1$.

The stationary covariance for $x(t)$ satisfies

$$P_n = AP_nA^T + e_nJ_n^*e_n^T$$

with P_n as in (7.6.44) and where

$$A = \begin{bmatrix} 0 & 1 & & \\ & & \ddots & \\ & & & 1 \\ -a_n & \cdots & & -a_1 \end{bmatrix}; \quad e_n = \begin{bmatrix} 0 \\ \vdots \\ 0 \\ 1 \end{bmatrix}$$

We can now apply the discrete-time lemma of Lyapunov (Section B.3 of Appendix B) to conclude (b).

Now from (7.6.39) it follows that K_n is a_n^n for all n. Now, since $A_n(q^{-1})$ is stable, a_n^n must be less than 1 since it is the product of the roots of $A_n(q^{-1})$. Hence $|K_n| < 1$ for all n and (c) follows.

Now conversely, (c) $\Rightarrow$ (a) via (vi).

(viii) Equation (7.6.45) follows from (7.6.19)–(7.6.20) since the set $\{f_n(t), \cdots, f_n(t+n)\}$ is an orthogonal basis for $\{y(t), \cdots, y(t+n)\}$. Similarly, (7.6.46) follows from (7.6.10) and (7.6.47) follows from the definition of $b_n(t)$. Equation (7.6.48) follows from (7.6.16).

▼▼▼

We have seen above that the lattice filter is nothing more than an interesting and effective way of implementing the restricted complexity transversal whitening filter. Of course, its importance arises because of the orthogonalization procedure used in the implementation. An alternative filter structure based on orthogonalization is presented in Exercise 7.13. This particular structure has been described as an *escalator* realization in the literature (Ahmed and Youn, 1980).

This completes our discussion of lattice filters. For further discussion of lattice structures, see Makhoul (1977, 1978), Burg (1975), Gray and Markel (1973), Griffiths (1977), Griffiths and Medaugh (1978), Morf, Vieira, and Lee (1977), and Morf and Lee (1979).

The basic lattice structure can also be generalized in several ways. One possibility discussed in (Messerschmitt, 1980) is to replace the delay elements by more general all-pass filters. However, the basic idea remains as described above.

7.6.2 Lattice Predictors

We now show that the Lattice whitening filter can be used to give a restricted complexity optimal one-step-ahead prediction.

From (7.6.25) of Theorem 7.6.1 and using (7.6.37), we have

$$f_i(t) = f_{i-1}(t) + K_i b_{i-1}(t-1); \qquad i = 1, \ldots, n \qquad (7.6.49)$$

Summing the equation above from 1 to n gives us

$$f_n(t) = f_0(t) + \sum_{i=1}^{n} K_i b_{i-1}(t-1); \qquad f_0(t) = y(t) \qquad (7.6.50)$$

However, from the definition of $f_n(t)$, we have

$$f_n(t) = y(t) - \hat{y}_n(t) \qquad (7.6.51)$$

Combining (7.6.50) and (7.6.51), we have that $\hat{y}_n(t)$ can be computed from the lattice as follows:

$$\begin{aligned} \hat{y}_n(t) &= y(t) - f_n(t) \\ &= -\sum_{i=1}^{n} K_i b_{i-1}(t-1) \end{aligned} \qquad (7.6.52)$$

Thus, the one-step-ahead predictor using the lattice structure is as shown in Fig. 7.6.3.

The ideas above can be extended to the case of *d-step-ahead prediction* [see Gevers and Wertz (1980a) and Reddy, Egardt, and Kailath (1981)].

The Relationship between the Lattice and Levinson Predictors

We have derived the lattice predictor shown in Fig. 7.6.3 from the lattice whitening filter. We will now show, however, that the lattice predictor is actually nothing more than a particular way of implementing the Levinson predictor.

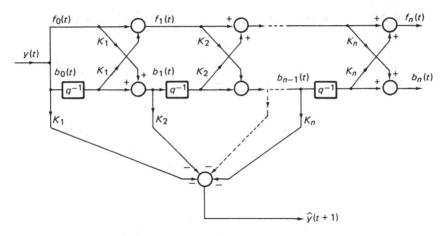

Figure 7.6.3 Lattice form of the one-step-ahead predictor.

We first define the following two polynomials from the optimal predictor coefficients in the Levinson predictor [see (7.5.11)]:

$$A_n(z^{-1}) = \sum_{j=0}^{n} a_j^n z^{-j}; \qquad a_0^n = 1 \tag{7.6.53}$$

$$\begin{aligned} \bar{A}_n(z^{-1}) &= z^{-n} A_n(z) \\ &= \sum_{j=0}^{n} a_{n-j}^n z^{-j} \end{aligned} \tag{7.6.54}$$

Note that $\bar{A}_n(z)$ is actually the reverse (or reciprocal) polynomial corresponding to $A_n(z)$.

We also note that the quantity $-\alpha_{n-1}/J_{n-1}$ has special significance in the Levinson procedure [see (7.5.26), (7.5.27), and (7.5.29)]. We therefore give a special symbol to this quantity. We define

$$K_n' \triangleq \frac{-\alpha_{n-1}}{J_{n-1}} \tag{7.6.55}$$

We also note from (7.5.27) that K_n' is the last coefficient in $A_n(z^{-1})$, that is,

$$K_n' = a_n^n \tag{7.6.56}$$

We now see that the Levinson algorithm is closely related to the lattice filter. This is shown in the following lemma:

Lemma 7.6.1. The coefficient K_n' defined in (7.6.55) is identical to the reflection coefficient (or partial correlation coefficient) in the lattice filter.

Proof. Note that the definition of $\hat{y}(t)$ in (7.5.11) is identical to $\hat{y}(t \mid Y_{t-1}^{t-n})$ in the lattice filter. Thus from the definition of $f_n(t)$ in Section 7.6.1 and J_n in (7.5.12) it is clear that

$$J_n = E\{f_n(t)^2\} \tag{7.6.57}$$

Also, from (7.5.22) we note that

$$\alpha_n = E\{y(t - n - 1)f_n(t)\} \tag{7.6.58}$$

Hence

$$K'_{n+1} \triangleq \frac{-\alpha_n}{J_n} = -\frac{E\{y(t - n - 1)f_n(t)\}}{E\{f_n(t)^2\}}$$

$$= K_{n+1} \quad \text{from (7.6.22)}$$

▼▼▼

We can now rewrite the Levinson algorithm of (7.5.26) to (7.5.28) as

$$A_0(z^{-1}) = \bar{A}_0(z^{-1}) = 1 \tag{7.6.59}$$

$$A_n(z^{-1}) = A_{n-1}(z^{-1}) + K_n z^{-1} \bar{A}_{n-1}(z^{-1}) \tag{7.6.60}$$

Also, we have

$$\bar{A}_n(z^{-1}) = K_n A_{n-1}(z^{-1}) + z^{-1} \bar{A}_{n-1}(z^{-1}) \tag{7.6.61}$$

We can reverse the procedure above and calculate all lower-order polynomials from $A_n(z^{-1})$. This and its consequences are described in

Lemma 7.6.2

(a) Given $A_n(z^{-1})$, we can compute $A_{n-1}(z^{-1})$, ..., $A_0(z^{-1})$ as follows:

$$A_{n-1}(z^{-1}) = \frac{A_n(z^{-1}) - K_n \bar{A}_n(z^{-1})}{1 - K_n^2} \tag{7.6.62}$$

where

$$K_n = a_n^n \tag{7.6.63}$$

(b) The polynomial $A_n(z^{-1})$ of order n is asymptotically stable if and only if the values of $K_j, j = n, \ldots, 1$, generated by the procedure above satisfy $|K_j| < 1$.

Proof. (a) Immediate from (7.6.60)–(7.6.61) by rearrangement.
(b) Follows immediately from Lemma 7.6.1 and Theorem 7.6.2, part (vii).

▼▼▼

Remark 7.6.1. Part (b) of Lemma 7.6.2, when generalized to arbitrary polynomials, is known as Cohn's stability test [see Cohn (1922)].

Remark 7.6.2. We can now see that the lattice filter may be viewed as a particular way of implementing the Levinson algorithm. In fact, the lattice structure immediately follows by defining

$$f_n(t) \triangleq A_n(q^{-1})y(t) \tag{7.6.64}$$

$$b_n(t) \triangleq \bar{A}_n(q^{-1})y(t) \tag{7.6.65}$$

The equations above can be rewritten in the conventional lattice form of Section 7.6.1 by using (7.6.59), (7.6.60), and (7.6.61), that is,

$$f_0(t) = b_0(t) = y(t) \tag{7.6.66}$$

$$f_n(t) = f_{n-1}(t) + K_n b_{n-1}(t - 1) \tag{7.6.67}$$

$$b_n(t) = K_n f_{n-1}(t) + b_{n-1}(t - 1) \tag{7.6.68}$$

7.7 THE EXTENDED KALMAN FILTER

We conclude this chapter by briefly considering state estimation for nonlinear systems. We have seen in Section 7.3 that the Kalman filter provides a simple and elegant solution to the problem of estimating the states in a *linear* finite-dimensional stochastic dynamical system. The question arises as to whether or not this method can be extended to nonlinear systems.

For nonlinear problems the Kalman filter is not strictly applicable since linearity plays an important role in its derivation and performance as an optimal filter. The extended Kalman filter attempts to overcome this difficulty by using a linearized approximation where the linearization is performed about the current state estimate. The approach seems to have been first suggested by Kopp and Orford (1963) and Cox (1964). There have been many papers on the topic of extended Kalman filtering, including Sage and Wakefield (1972), Leung and Padmanabhan (1973), Nelson and Stear (1976), Farison, Graham, and Shelton (1967), and many others.

Consider a general nonlinear model of the form

$$z(t + 1) = f(z(t), u(t), v_1(t)) \tag{7.7.1}$$

$$y(t) = h(z(t), u(t), v_2(t)) \tag{7.7.2}$$

with

$$E\left\{ \begin{bmatrix} v_1(s) \\ v_2(s) \end{bmatrix} [v_1(t)^T v_2(t)^T] \right\} = \begin{bmatrix} Q & S \\ S & R \end{bmatrix} \delta(t - s) \tag{7.7.3}$$

Let $\hat{z}(t)$ denote an estimate of the state at time t and linearize (7.7.1)–(7.7.2) about $z(t) = \hat{z}(t)$, $v_1(t) = 0$, and $v_2(t) = 0$. This leads to

$$z(t + 1) \simeq f(\hat{z}(t), u(t), 0) + F(t)[z(t) - \hat{z}(t)] + G(t)v_1(t) \tag{7.7.4}$$

$$y(t) \simeq h(\hat{z}(t), u(t), 0) + H(t)[z(t) - \hat{z}(t)] + J(t)v_2(t) \tag{7.7.5}$$

where

$$F(t) = \left. \frac{\partial f(z(t), u(t), v_1(t))}{\partial z(t)} \right|_{z(t) = \hat{z}(t), \, v_1(t) = 0} \tag{7.7.6}$$

$$G(t) = \left. \frac{\partial f(z(t), u(t), v_1(t))}{\partial v_1(t)} \right|_{z(t) = \hat{z}(t), \, v_1(t) = 0} \tag{7.7.7}$$

$$H(t) = \left. \frac{\partial h(z(t), u(t), v_2(t))}{\partial z(t)} \right|_{z(t) = \hat{z}(t), \, v_2(t) = 0} \tag{7.7.8}$$

$$J(t) = \left. \frac{\partial h(z(t), u(t), v_2(t))}{\partial v_2(t)} \right|_{z(t) = \hat{z}(t), \, v_2(t) = 0} \tag{7.7.9}$$

It is now a straightforward matter to write down the Kalman filter for the linearized equations (7.7.4) and (7.7.5). This leads to the *extended Kalman filter*:

$$\hat{z}(t + 1) = f(\hat{z}(t), u(t), 0) + L(t)[y(t) - h(\hat{z}(t), u(t), 0)]; \quad \hat{z}(0) = \hat{z}_0 \tag{7.7.10}$$

where $L(t)$ is the Kalman gain satisfying

$$L(t) = [F(t)\Sigma(t)H(t)^T + \bar{S}(t)][H(t)\Sigma(t)H(t)^T + \bar{R}(t)]^{-1} \tag{7.7.11}$$

and

$$\Sigma(t + 1) = F(t)\Sigma(t)F(t)^T + \bar{Q}(t) - L(t)[H(t)\Sigma(t)H(t)^T + \bar{R}(t)]L(t)^T;$$
$$\Sigma(0) = \Sigma_0 \tag{7.7.12}$$

with $F(t)$, $G(t)$, $H(t)$, and $J(t)$ as in (7.7.6) to (7.7.9) and

$$\bar{Q}(t) = G(t)QG(t)^T \qquad (7.7.13)$$

$$\bar{S}(t) = G(t)SJ(t)^T \qquad (7.7.14)$$

$$\bar{R}(t) = J(t)RJ(t)^T \qquad (7.7.15)$$

The extended Kalman filter is certainly not optimal in general. In fact, it can be thought of as a *restricted complexity filter* which is constrained to have a similar format as that used for linear systems. Due to the linear approximation, it is quite possible that the filter may diverge, and thus care must be exercised in using this method. Various improvements have been proposed, including iteratively refining the linearization.

We shall show later in Chapter 9 that the extended Kalman filter is one possible way of deriving a state estimator for a linear system with unknown parameters.

EXERCISES

7.1. Write down the stochastic state-space model corresponding to the following data sequence:

$$y(t) = d(t) + n(t) + z(t) + v(t)$$

where

$d(t) = A \sin(\omega t + \phi), \qquad -\pi < \omega < \pi$

$n(t)$ is a colored noise sequence generated as follows:

$n(t) = -a_1 n(t-1) + \omega(t); \qquad \{\omega(t)\}$ white noise

$z(t) = -a_2 z(t-1) + b_1 u(t); \qquad \{u(t)\}$ a known input

$\{v(t)\}$ white measurement noise

7.2. Consider the following system:

$$x(t+1) = ax(t) + bu(t)$$

$$y(t) = x(t) + v_2(t); \qquad \{v(t)\}$$ white noise

Determine the steady-state innovations model for the following cases:

(a) $|a| < 1$

(b) $|a| = 1$

(c) $|a| > 1$

7.3. Refer to (7.3.21) in Remark 7.3.1. Write down the form of the Kalman filter for this model using (7.3.26) to (7.3.29). Compare your answer with Theorem 7.3.1.

7.4. Consider the following state-space model (which is assumed to be observable):

$$x(t+1) = Ax(t) + Lv(t); \qquad \{v(t)\}$$ white noise
$$y(t) = Cx(t) + v(t)$$

(a) Transform this model to the form (7.3.21) and (7.3.24).

(b) With D defined as in (7.3.34) for the transformed model, show that (A', D) is always completely uncontrollable.

(c) Show that $\Sigma = 0$ is always one possible solution of the ARE.

(d) If A is stable, show that the RDE will converge to $\Sigma = 0$ for all initial conditions $\Sigma(t_0) > 0$.

(e) If A has unstable roots, will $\Sigma = 0$ be the strong solution of the ARE?

(f) If A has no eigenvalues with modulus equal to 1, show that the solutions of the RDE will converge to the stabilizing solution for all $\Sigma(t_0) > 0$.

7.5. Consider the following moving-average process:

$$y(t) = v_1(t) + c_1 v_1(t-1) + \cdots + c_n v_1(t-n); \qquad \{v_1(t)\} \text{ white noise}$$

(a) Show that the corresponding observer-form state-space model is

$$x(t+1) = \begin{bmatrix} 0 & 1 & & \\ & & \cdot & \\ & & & \cdot \\ & & & 1 \\ 0 & & & 0 \end{bmatrix} x(t) + \begin{bmatrix} c_1 \\ \cdot \\ \cdot \\ \cdot \\ c_n \end{bmatrix} v_1(t)$$

$$y(t) = [1 \quad 0 \quad \cdots \quad 0] x(t) + v_1(t)$$

(b) Transform the state-space model above as in Exercise 7.4 and show that (A', D) will have uncontrollable modes on the unit circle if and only if $(1 + c_1 q^{-1} + \cdots + c_n q^{-n})$ has roots on the unit circle.

(c) Show that $\{y(t)\}$ can be represented in the form

$$y(t) = C'(q^{-1})\omega(t); \qquad \{\omega(t)\} \text{ white noise}$$

where $C'(q^{-1})$ has roots on or inside the unit circle. When does $C'(q^{-1})$ have roots on the unit circle?

7.6. Show that the fixed-lag and fixed-point smoothers of Sections 7.3.2 and 7.3.3 are stable if and only if the Kalman filter is stable.

7.7. Consider a special fixed-lag smoothing problem where a signal $\{u(t)\}$ is observed in the presence of white measurement noise, that is,

$$y(t) = u(t) + n(t); \qquad n(t) \text{ white with variance } \sigma_n^2$$

$u(t)$ is a stationary stochastic process modeled by

$$A(q^{-1})u(t) = E(q^{-1})v(t); \qquad v(t) \text{ white with variance } \sigma_v^2, \quad A(q^{-1}) \text{ stable}$$

(a) Show that in the steady state $y(t)$ can also be described by an ARMA model of the form

$$A(q^{-1})y(t) = G(q^{-1})\omega(t); \qquad \omega(t) \text{ white with variance } \sigma_\omega^2$$

(b) Show that the optimal fixed-lag smoother for estimating $u(t)$ given data $\{y(t)\}$ up to time $t + d$ is given by

$$\hat{u}(t \mid t+d) = y(t) - \frac{\sigma_n^2}{\sigma_\omega^2} F_d(q)\omega(t); \qquad \frac{\sigma_n^2}{\sigma_\omega^2} = \frac{g_n}{a_n}$$

where

$$F_d(q) = \sum_{i=0}^{d} f_i q^i$$

and $F_d(q)$ is the first of $(d+1)$ terms from

$$F(q) = \sum_{i=0}^{\infty} f_i q^i$$

where $A(q) = G(q)F(q)$ [see Hagander and Wittenmark (1977)].

7.8. Let Σ_s denote the stabilizing solution of the ARE and let Σ_t^d denote the covariance of a fixed-point state estimate at time t for data $\{y(t_0), \ldots, y(t), \ldots, y(t+d)\}$. Show that $(\Sigma_s - \Sigma_t^d)$ satisfies the following equation:

$$(\Sigma_s - \Sigma_t^d) - \Sigma_s \bar{A}_s^T \Sigma_s^{-1} (\Sigma_s - \Sigma_t^d) \Sigma_s^{-1} \bar{A}_s \Sigma_s$$
$$= \Sigma_s C^T [C\Sigma_s C^T + R]^{-1} C\Sigma_s - \Sigma_s [\bar{A}_s^T]^{d-1} C^T [C\Sigma_s C^T + R]^{-1} C \bar{A}_s^{d-1} \Sigma_s$$

where $\bar{A}_s$ is the stabilizing state transition matrix for the Kalman filter. Hence show that as $d \to \infty$, so the improvement due to smoothing, $(\Sigma_s - \Sigma_t^d)$, approaches a limit $\tilde{\Sigma}$ which satisfies the following equation:

$$\tilde{\Sigma} - \Sigma_s A_s^{-T} \Sigma_s^{-1} \tilde{\Sigma} \Sigma_s^{-1} \bar{A}_s \Sigma_s = \Sigma_s C^T [C\Sigma_s C^T + R]^{-1} C\Sigma_s$$

7.9. Determine the steady-state ARMAX model for the system given in Exercise 7.1.

7.10. Write down the nonlinear ARMAX model for the following system:

$$y(t + 1) = -\tfrac{1}{3} y(t)^3 + \eta(t + 1)$$

where $\eta(t)$ is a colored noise process satisfying

$$\eta(t + 1) = -\tfrac{1}{2} \eta(t) + \omega(t); \qquad \{\omega(t)\} \text{ white}$$

7.11. Consider the problem set up in Section 7.5.1 specializing to the following case:

$$y(t) = G(q^{-1}) s(t) + v(t)$$
$$G(q^{-1}) = g_0 + g_1 q^{-1} + \cdots + g_N q^{-N}$$

and where both $\{s(t)\}$ and $\{v(t)\}$ are white noise sequences of variance σ_s^2 and σ_v^2, respectively, and uncorrelated with each other. Consider a restricted complexity fixed-lag smoother of transversal type, that is,

$$\hat{s}(t - d) = F(q^{-1}) y(t)$$
$$d > 0; \qquad F(q^{-1}) = f_0 + f_1 q^{-1} + \cdots + f_n q^{-n}; \qquad n \ll N$$

Show that the restricted complexity optimal filter of the above form has coefficients given by

$$\theta_0 = [GG^T \sigma_s^2 + \sigma_v^2 I]^{-1} G e_{d+1} \sigma_s^2 \,; \text{ where } \qquad \theta_0 \triangleq [f_0 \, \ldots, \, f_n]^T$$

$$G = \begin{bmatrix} g_0 & \cdots & g_N & & & \\ & g_0 & \cdots & g_N & & \\ & & \cdot & & \cdot & \\ & & & \cdot & & \cdot \\ & & & & \cdot & \\ & & g_0 & \cdots & & g_N \end{bmatrix} \begin{matrix} \uparrow \\ n+1 \\ \text{rows} \\ \downarrow \end{matrix}$$

e_{d-1} = zero vector except for 1 in the $(d - 1)$ row

7.12. Consider a process

$$y(t) = v(t) - v(t - 1); \qquad \{v(t)\} \text{ white with variance } \sigma_v^2$$

Say that we use a zeroth-order transversed filter to predict $y(t + 1)$; that is, we form

$$\hat{y}(t + 1) = f_0 y(t)$$

Show that the optimal value of f_0 is $-\tfrac{1}{2}$, giving a mean-square prediction error of $\tfrac{3}{2} \sigma_v^2$. What is the variance of $\{y(t)\}$?

7.13. Consider again the orthogonalization presented in (7.6.20).
(a) Show that the transformation can be factored as follows:

$$\begin{bmatrix} 1 & & & & \\ a_1^1 & \cdot & & & \\ \cdot & & \cdot & & \\ \cdot & & & 1 & \\ \cdot & & & & \\ a_n^n & \cdots & a_1^n & 1 \end{bmatrix} = M_n \cdots M_1 M_0 \qquad (1)$$

where $M_0 = I$ and $M_j (j \geq 1)$ has the form

$$M_j = \begin{bmatrix} 1 & & & & & & & & \\ 0 & \cdot & & & & & & & \\ & & \cdot & & & & & & \\ & & 0 & 1 & & & & & \\ \cdot & & & \alpha_1^j & \cdot & & & & \\ \cdot & & & & 0 & \cdot & & & \\ & & & & & & 1 & & \\ 0 & \cdots & 0 & \alpha_{n-j}^j & 0 & \cdots & 0 & 1 \end{bmatrix} \qquad (2)$$

(b) By substituting (7.6.69) into (7.6.20), show that $f_0(t-n), \ldots, f_n(t)$ can be computed in $n+1$ stages as follows, with each stage generating an additional orthogonal component. The zeroth stage gives $f_0(t-n) = y(t-n)$. The first stage gives

$$\begin{bmatrix} f_0(t-n) \\ f_1(t-n+1) \\ * \\ * \\ \vdots \\ * \end{bmatrix} = \begin{bmatrix} 1 & & & & \\ \alpha_1^1 & 1 & & & \\ & \cdot & 0 & \cdot & \\ \cdot & \cdot & \cdot & \cdot & \\ \cdot & \cdot & \cdot & \cdot & \\ \cdot & \cdot & & 1 & \\ \alpha_n^1 & 0 & \cdots & 0 & 1 \end{bmatrix} \begin{bmatrix} f_0(t-n) \\ y(t-n+1) \\ \cdot \\ \cdot \\ \cdot \\ y(t) \end{bmatrix}$$

where $*$ denotes an intermediate variable. Similarly, the jth stage gives

$$\begin{bmatrix} f_0(t-n) \\ f_1(t-n+1) \\ \cdot \\ \cdot \\ f_j(t-n+j) \\ * \\ \cdot \\ \cdot \\ * \end{bmatrix} = \begin{bmatrix} 1 & & & \\ 0 & \cdot & & \\ & \cdot & & \\ & & 1 & \\ & & \alpha_1^i & \cdot \\ & & \cdot & \cdot \\ & & \cdot & \cdot \\ 0 & & \alpha_{n-j}^n & 1 \end{bmatrix} \begin{bmatrix} f_0(t-n) \\ f_1(t-n+1) \\ \cdot \\ \cdot \\ f_{j-1}(t-n+j-1) \\ * \\ \cdot \\ \cdot \\ * \end{bmatrix}$$

(c) Show that the filter above can be realized by the escalator structure shown in Fig. 7.A (illustrated for the case $n = 3$).

(d) Show that the escalator structure allows the order of the filter to be updated from n to $n+1$ by simply adding an additional row to the top of the structure shown in Fig. 7.A without changing the existing filter.

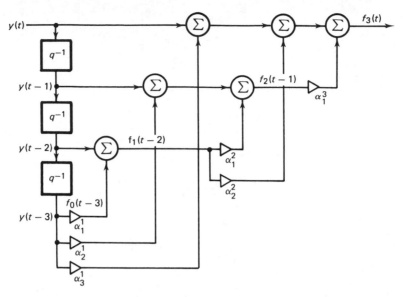

Figure 7.A Escalator realization of whitening filter.

7.14. Consider the state-space model of (7.3.1) to (7.3.3). Derive an expression for the "at observation" filter for $x(t)$; that is, evaluate $E\{x(t)\,|\,y(0),\,\ldots,\,y(t)\}$. Express your answer as a linear function of $\hat{x}(t)$ [i.e., the estimate given $\{y(0),\,\ldots,\,y(t-1)\}$] and $y(t)$. (*Hint:* Follow the derivation of Theorem 7.3.1.)

7.15. Consider the d-step-ahead predictor of (7.4.23). Note that the predicted output $y^0(t+d\,|\,t)$ is a function of $y^0(t+d-1\,|\,t-1)$, $y^0(t+d-2\,|\,t-2)$, $\ldots$. Express the predictor in the form

$$y^0(t+d\,|\,t) = R(q^{-1})y^0(t\,|\,t-d) + S(q^{-1})y(t) + L(q^{-1})u(t)$$

where

$$R(q^{-1}) = r_0 + r_1 q^{-1} + \cdots$$
$$S(q^{-1}) = s_0 + s_1 q^{-1} + \cdots$$
$$L(q^{-1}) = l_0 + l_1 q^{-1} + \cdots$$

That is, express $y^0(t+d\,|\,t)$ in terms of $y^0(t\,|\,t-d)$, $y^0(t-1\,|\,t-d-1)$, $\ldots$. (*Hint:* Apply the division algorithm of algebra again.)

7.16. Consider the following system:

$$y(t) = \omega(t) - \omega(t-1); \qquad \omega(t) \text{ white noise of variance } \sigma^2$$

(Note that the noise model has a zero on the unit circle at 1.)

(a) Write down the corresponding state-space model in observer form. Hence show that the optimal *time-varying* one-step-ahead prediction, $\hat{y}(t)$, for $y(t)$ satisfies

$$\hat{x}(t+1) = \frac{1}{[\Sigma(t)+1]}\hat{x}(t) + \frac{1}{[\Sigma(t)+1]}y(t)$$
$$\hat{y}(t) = -\hat{x}(t)$$

where

$$\Sigma(t+1) = \frac{\Sigma(t)}{\Sigma(t)+1}$$

Optimal Filtering and Prediction   Chap. 7

Hence show that $\Sigma(t)$ converges to 0 for all $\Sigma_0 \geq 0$ and thus show that the state error covariance converges to zero and the output prediction error variance converges to σ^2. Also verify that this leads to the following steady-state time-invariant predictor:

$$\hat{x}(t+1) = \hat{x}(t) + y(t); \qquad \hat{y}(t) = -\hat{x}(t)$$

Why can't this time-invariant predictor be used from $t = 0$?

(b) Explain why the time-varying Kalman filter in this case has the property that $\hat{x}(t) - x(t)$ approaches zero in mean square.

(c) Using part (b), define

$$\tilde{x}(t) \triangleq \hat{x}(t) - x(t)$$

Show that $\tilde{x}(t)$ satisfies

$$\tilde{x}(t+1) = \frac{1}{[\Sigma(t)+1]}\tilde{x}(t) - \frac{\Sigma(t)}{\Sigma(t+1)+1}\omega(t)$$

For $\Sigma(0) = 1$, show that

$$\tilde{x}(t+1) = \frac{1}{t+1}\tilde{x}(0) - \frac{1}{t+1}\sum_{j=0}^{t}\omega(j)$$

Hence show that $\tilde{x}(t)$ approaches zero (in mean square) for all initial states. [This shows that the time-varying predictor converges to the time-invariant predictor but in the process it "discards" the erroneous initial conditions so that the asymptotically time-invariant state estimator gives zero estimation error (in the mean-square sense).]

7.17. Establish (7.4.61) and (7.4.62).

7.18. Consider the following restricted complexity predictor for the system discussed in Exercise 7.16.

$$\hat{y}(t+1) = (1-\epsilon)\hat{y}(t) - (1-\epsilon)\hat{y}(t); \qquad 0 < \epsilon \ll 1$$

Establish the fact that this predictor gives nearly optimal performance as compared with the time-varying predictor by showing that the resulting prediction error variance is $[2/(2-\epsilon)]\sigma^2$. Hence show that optimality is achieved as $\epsilon \to 0$. Discuss the relative advantages of the predictors discussed in Exercises 7.16 and 7.18.

7.19. Extend Lemma 7.4.4 to the multivariable case.

7.20. Consider the following nonlinear system:

$$y(t) = y(t-1)y(t-2)^2 + \eta(t)$$

where $\eta(t)$ is colored noise satisfying

$$\eta(t) = \tfrac{1}{2}\eta(t-1) + \omega(t); \qquad \omega(t) \text{ white noise}$$

(a) Write down the optimal predictor for $y(t)$ given $y(t-1), y(t-2), \ldots$.

(b) Why is it difficult to repeat part (a) for the following model?

$$x_1(t) = x_1(t-1) + v_1(t)$$
$$x_2(t) = x_2(t-1)^2 + v_2(t)$$
$$y(t) = x_1(t)x_2(t) + v_3(t); \qquad v_1(t), v_2(t), v_3(t) \text{ white noise}$$

7.21. Consider the following restricted complexity predictor:

$$\hat{y}(t) = -a_1 y(t-1) - a_2 y(t-2)$$

Determine a_1 and a_2 so that the mean-square prediction error is minimized for the system in Exercise 7.16. Show that the minimum prediction error variance is $\tfrac{4}{3}\sigma^2$. Compare your result with that obtained in Exercises 7.16 and 7.18. Discuss.

7.22. Derive a d-step-ahead form of the lattice predictor.

7.23. Verify that in (7.3.54) and (7.3.55), A, C, $\eta_1(t)$, and $\eta_2(t)$ have the form

$$
A = \begin{bmatrix} A_1 & 0 & 0 \\ B_2 C_1 & A_2 & 0 \\ e_1 C_1 & 0 & S \end{bmatrix}
$$

$$
C = [\, 0 \quad C_2 \quad 0\,]
$$

$$
\eta_1(t) = \begin{bmatrix} w_1(t) \\ w_2(t) + B_2 v_1(t) \\ e_1 v_1(t) \end{bmatrix}; \qquad \eta_2(t) = v_2(t)
$$

and hence show that the covariance of $\eta_1(t)$ and $\eta_2(t)$ is

$$
E\left\{ \begin{bmatrix} \eta_1(t) \\ \eta_2(t) \end{bmatrix} [\eta_1(s)^T \eta_2(s)^T] \right\} = \begin{bmatrix} Q_1 & 0 & 0 & 0 \\ 0 & B_2^T R_1 B_2 + Q_2 & B_2 R_1 e_1^T & 0 \\ 0 & e_1 R_1 B_2^T & e_1 R_1 e_1^T & 0 \\ \hdashline 0 & 0 & 0 & R_2 \end{bmatrix} \delta(t - s)
$$

$$
\triangleq \begin{bmatrix} Q & 0 \\ \hdashline 0 & R \end{bmatrix} \delta(t - s)
$$

8

Parameter Estimation for

Stochastic Dynamic Systems

8.1 INTRODUCTION

In this chapter we discuss parameter estimation schemes for estimating the parameters in stochastic models (such as those introduced in Section 7.4). Much of the background to parameter estimation given in Chapter 3 is pertinent here and the reader is encouraged to briefly review Chapter 3.

There is much in common between our discussion here and that in Chapter 3. One main depature is that here we work in a probabilistic framework so that we can adequately handle the stochastic nature of the problem. Most of the algorithms discussed here will not be radically different from those described in Chapter 3, although the properties and details will differ to reflect the stochastic nature of the models.

We first introduce some basic notions involved in stochastic parameter estimation.

Description versus Application

One can distinguish two basic motivations for parameter estimation in stochastic systems. These can be regarded as being *description oriented* or *applications orientated*. In the description-orientated approach the main emphasis is placed on obtaining an understanding of the process, whereas in the applications-orientated approach the main emphasis is on achieving a certain objective (such as prediction or control of the underlying system). This distinction may influence the choice of algorithms and the performance criteria. For example, in the description-orientated approach it is natural to focus attention on the properties of the estimated parameters. On the other hand, in an applications-orientated approach one focuses attention principally on the ultimate objective (such as *d*-step-ahead prediction), and thus what happens to the parameter estimates will usually be of secondary importance. Of course, the distinction is somewhat arbitrary. For example, in the description-orientated approach it will generally be impossible in practice to describe *all* properties of the system and thus one must be selective and aim to describe those properties most relevant to the kind of application one has in mind. Vice versa, in the applications-orientated approach, the ability to achieve a specified objective often depends implicitly on having a "good" description of the process within a range of operating conditions.

Choice of Model Structure

As we have seen in Chapter 3, one of the first choices that must be made in parameter estimation is the model structure. Specifically, one must choose between using a linear or nonlinear model and one must choose the order of the model. As we have seen in Part I, depending on the nature of the process, there may be significant advantages in using nonlinear models. Indeed, there is a growing interest in parameter estimation for nonlinear systems [see, e.g., Sage and Melsa (1971), Eykhoff (1974), and Priestley (1978, 1980)]. We discuss parameter estimation for nonlinear systems briefly in Sections 8.2 and 8.3. However, our main emphasis will be on linear models since these have been predominately used in applications, and since a richer theoretical treatment is currently possible for this case.

The question of choosing the model order (or structure) in the linear case has received much attention in the literature. Knowledge of the correct model order is clearly necessary to achieve unique parameter identifiability. A wide range of algorithms have been proposed for choosing the model order, such as *F*-tests [see Goodwin and Payne (1977)], information criteria [see Akaike (1976)], prediction error criteria [see Akaike (1974)], various properties of the information matrix [see Ljung and Rissanen (1976)], and minimum data-length descriptions [see Rissanen (1981)], etc. In practice, it must be borne in mind that every model is at best an approximation. Thus the question of choosing the model order is really a question of choosing that particular model structure which on the one hand is simple and yet gives an adequate description of the system for the intended purpose without being excessively complicated. In this chapter we generally assume that the model structure is prespecified and we concentrate on estimating the parameters within a given model. We shall analyze the performance of the algorithms under two types of assumptions: (1) that the model structure encompasses the true system (i.e., has adequate degrees of freedom), or (2) that the model is essentially arbitrarily chosen (restricted complexity) and is not therefore necessarily related in any way to the underlying system structure.

Estimation Criteria

In parameter estimation one usually uses some objective criterion to measure the relative merits of one set of parameter estimates compared with another. For example, the criterion could be the sum of squares of the difference between the model output and system output over some period of time. In some cases, the criterion may be directly related to the intended application. For example, if one is interested in *d*-step-ahead prediction, the criterion could be chosen as the sample mean-square *d*-step-ahead prediction error. In other cases, the criterion may be motivated by statistical considerations in an effort to gain a good description of the system. Examples of this type of criterion are maximum likelihood and Bayesian criteria (Goodwin and Payne, 1977; Åström and Eykhoff, 1971; Box and Jenkins, 1970). In this book we concentrate on applications-orientated criteria. In particular we emphasize quadratic criteria since this leads to relatively simple algorithms such as ordinary least squares and its variants.

Off-Line Estimation

In general, the criterion function can be a highly nonlinear function of the parameters and thus its minimization requires off-line optimization techniques of an iterative nature. We discuss these techniques briefly in Section 8.2.

On-Line Algorithms

Our prime concern will be with on-line techniques for parameter estimation since they will form the basis of our subsequent development of adaptive filtering, prediction, and control for stochastic systems. We shall initially motivate these algorithms as on-line implementations of particular off-line schemes. Subsequently, we will give alternative interpretations of the algorithms. Our discussion will cover in detail most of the commonly used algorithms. This will be adequate for our treatment of adaptive techniques for filtering, prediction, and control. Those readers who wish to explore the subject of recursive parameter estimation algorithms in greater detail are referred to books dealing specifically with this topic, for example the one by Ljung and Söderström (1982).

Properties of Algorithms

In discussing the various algorithms we shall focus attention on their convergence properties under a range of different assumptions. In brief we shall distinguish three alternative situations. First, we shall explore such questions as: Do the estimated parameters converge to the true values (*consistency*)? And: What is the nature of the asymptotic distribution of the parameter estimates (*efficiency*)? This will require us to make relatively strong assumptions about the nature of the underlying system and signals. Second, we shall explore such applications orientated questions as: Does the prediction error variance converge to the optimal prediction error variance? And: Does the estimated control law asymptotically yield optimal performance? This will require moderate assumptions on the structure of the system and weak assumptions about the nature of the signals. Third, we shall ask such questions as: Does the algorithm asymptotically give the best possible performance within a restricted class (the restricted complexity optimum)? This will require only very weak assumptions on the nature of the system and signals.

8.2 OFF-LINE PREDICTION ERROR ALGORITHMS

In this section we briefly review off-line algorithms for parameter estimation. It is true that these algorithms are not directly used in adaptive applications. However, we wish to give the reader an appreciation for off-line schemes, for two reasons.

First, these schemes can be employed to give initial parameter estimates for use in an on-line scheme. Second, many of the on-line schemes in current use can be thought of as sequential implementations of off-line algorithms.

Our discussion of off-line algorithms will follow the standard approach as in other texts, for example Eykhoff (1974), Sage and Melsa (1971), and Goodwin and Payne (1977).

In this section and the next one we use a very general "model" as the basis of our discussion. This will include both linear and nonlinear systems. In Section 8.3.2 we consider the specific case of linear systems to further embelish the discussion.

For the moment, we shall assume that the "model" for the system can be described via a *prediction* of the system output at time t of the form $\hat{y}(t, \theta, \mathcal{U}(t-1), \mathcal{Y}(t-1))$, where

$$\mathcal{Y}(t) \triangleq \{y(0), y(1), \ldots, y(t)\}$$

$$\mathcal{U}(t) \triangleq \{u(0), u(1), \ldots, u(t)\}$$

and θ is a parameter vector. For notational simplicity, we will suppress the dependence on the past data, $\mathcal{U}(t-1)$ and $\mathcal{Y}(t-1)$. Thus $\hat{y}(t, \theta, u(t-1), y(t-1))$ will be denoted as $\hat{y}(t, \theta)$.

A very special case of the model above would be the linear finite-dimensional models discussed in Chapter 7. For example, if we use the innovations model, (7.3.17)–(7.3.18) with $K(t)$ replaced by its steady-state value (see Theorems 7.3.2 to 7.3.4), the predicted output has the form

$$\hat{x}(t+1) = A\hat{x}(t) + Bu(t) + K(y(t) - \hat{y}(t)) \tag{8.2.1}$$

$$\hat{y}(t) = C\hat{x}(t) \tag{8.2.2}$$

where A, B, K, and C are parameterized by θ. The model above can be thought of as giving an explicit expression for a one-step-ahead output prediction.

We shall not assume that there necessarily exists a value of θ (say θ_0) for which the model output, $\hat{y}(t, \theta)$, "corresponds" to the true system output. Instead, we shall ask what value of θ causes the model output to best approximate the system output. Our measure of the "quality" of the approximation will be a scalar criterion of the following general form:

$$V_N(\theta) = \frac{1}{N} \sum_{t=1}^{N} l(\hat{y}(t, \theta), y(t)) \tag{8.2.3}$$

where $l(\cdot, \cdot)$ is a suitably chosen scalar measure of model fit.

A typical choice for $l(\hat{y}, y)$ in the case of a scalar output is

$$l(\hat{y}, y) = \tfrac{1}{2}[\hat{y}(t, \theta) - y(t)]^2 \tag{8.2.4}$$

With this choice, the value of θ is chosen so that the sum of the squares of the errors between the true system output and the predicted output is minimized. This motivates the term *prediction error algorithm*.

Further motivation, from a statistical point of view, for the criterion (8.2.3) is given in Appendix D, where the criterion is related to maximum likelihood estimation.

When the input–output data $\mathcal{Y}(N)$ and $\mathcal{U}(N)$ have been recorded, the criterion (8.2.3) can be seen to be a scalar-valued function of the parameter vector θ. With the exception of some special cases, it is not possible to minimize $V_N(\theta)$ by analytic means. A notable exception is when $\hat{y}(t, \theta)$ is a linear function of θ, that is, when we can write

$$\hat{y}(t, \theta) = \phi(t - 1)^T\theta \tag{8.2.5}$$

[where $\phi(t - 1)$ is a vector function of $\mathcal{Y}(t - 1)$ and $\mathcal{U}(t - 1)$], and the quadratic criterion (8.2.4) is used.

The resulting algorithm (least squares) was discussed in Section 3.3; see, in particular, Lemma 3.3.5. It is also possible to derive an on-line version of the algorithm [equations (3.3.46) and (3.3.47)] which produces exactly the same estimates as those generated by the off-line algorithm. For other cases, (8.2.3) has to be minimized by *iterative*, numerical *search procedures*. A very common form (Gupta and Mehra, 1974) for such a procedure is as follows:

$$\hat{\theta}(i + 1) = \hat{\theta}(i) + \alpha(i)g(i) \tag{8.2.6}$$

where $\hat{\theta}(i)$ denotes the estimate at the ith iteration based on data up to time N, $g(i)$ is a search direction based on information about V_N (typically the negative gradient), and $\alpha(i)$ is a search constant chosen so that a large decrease in the criterion (8.2.3) is obtained.

In the case of gradient search, $g(i)$ is taken to be the negative gradient, that is,

$$g(i) = -\dot{V}_N(\hat{\theta}(i)) \triangleq -\frac{d}{d\theta}V_N(\theta)\bigg|_{\theta=\theta(i)} \tag{8.2.7}$$

For example, when the quadratic criterion (8.2.4) is used, $\dot{V}_N(\theta)$ is as follows:

$$\dot{V}_N(\theta) = \sum_{t=1}^{N}\frac{1}{N}[\hat{y}(t, \hat{\theta}(i)) - y(t)]\hat{y}^{\cdot}(t, \hat{\theta}(i)) \tag{8.2.8}$$

where $\hat{y}^{\cdot}(t, \hat{\theta}(i))$ denotes the derivative of $\hat{y}(t, \theta)$ with respect to θ evaluated at $\hat{\theta}(i)$.

An improved algorithm can be obtained by considering a second-order Taylor's series expansion of $V_N(\theta)$ about $\hat{\theta}(i)$, that is,

$$V_N(\theta) \simeq V_N(\hat{\theta}(i)) + \dot{V}_N(\hat{\theta}(i))^T[\theta - \hat{\theta}(i)] + \frac{1}{2}[\theta - \hat{\theta}(i)]^T\frac{d^2V_N}{d\theta^2}[\theta - \hat{\theta}(i)]$$

Minimization of the expression above with respect to θ gives

$$\theta \simeq \hat{\theta}(i) + \left[\frac{d^2V_N}{d\theta^2}\right]^{-1}\dot{V}_N(\hat{\theta}(i))$$

This motivates the following *Newton iteration* (Luenberger, 1973):

$$\hat{\theta}(i + 1) = \hat{\theta}(i) + \alpha(i)\left[\frac{d^2V_N}{d\theta^2}\right]^{-1}\dot{V}_N(\hat{\theta}(i)) \tag{8.2.9}$$

where

$$\frac{d^2V_N}{d\theta^2} = \frac{1}{N}\sum_{t=1}^{N}\hat{y}^{\cdot}(t, \hat{\theta}(i))\hat{y}^{\cdot}(t, \hat{\theta}(i))^T + \frac{1}{N}\sum_{t=1}^{N}[\hat{y}(t, \hat{\theta}(i)) - y(t)]\frac{d}{d\theta}\hat{y}^{\cdot}(t, \hat{\theta}(i)) \tag{8.2.10}$$

A difficulty with the algorithm (8.2.9) is that $d^2V_N/d\theta^2$ may not be positive definite and thus $[d^2V_N/d\theta^2]^{-1}\dot{V}_N(\hat{\theta}(i))$ may not point in a "downhill" direction; see, for example, Speedy, Brown, and Goodwin (1970). A sensible remedy is to neglect the second term on the right-hand side of (8.2.10) so as to ensure that the matrix used in (8.2.9) is nonnegative definite. Further motivation for neglecting this term is provided by the observation that this term has asymptotically zero mean value in many cases due to the independence of the residual, $[\hat{y}(t, \hat{\theta}) - y(t)]$, in the limit.

This gives the so-called *Gauss–Newton iteration*:

$$\hat{\theta}(i + 1) = \hat{\theta}(i) + \alpha(i)P(i)\dot{V}_N(\hat{\theta}(i)) \qquad (8.2.11)$$

where

$$P(i)^{-1} \triangleq R(i) \triangleq \frac{1}{N}\sum_{t=1}^{N} \hat{y}'(t, \hat{\theta}(i))\hat{y}'(t, \hat{\theta}(i))^T \qquad (8.2.12)$$

In practice, it is often desirable to do a linear search for $\alpha(i)$ in the direction $P(i)\dot{V}_N(\hat{\theta}(i))$ so that $V_N(\theta)$ is minimized in the search direction, that is,

$$\alpha(i) = \arg\min_{\alpha} V[\hat{\theta}(i) + \alpha P(i)\dot{V}_N(\hat{\theta}(i))] \qquad (8.2.13)$$

In practice, it is usually necessary to constrain the estimated parameter values $\hat{\theta}(i)$ so that they remain within some prespecified region (for example, a region in which the model is known to be stable).

Subject to regularity conditions, the sequence of estimates $\hat{\theta}(i)$ generated iteratively by the algorithms above converges to θ_N^*, where θ_N^* is either a local minimum of $V_N(\theta)$ or a boundary point of the constrained region.

The discussion above is essentially "deterministic." This is because given the data $\mathcal{Y}(N)$ and $\mathcal{U}(N)$ together with the model structure, minimization of the criterion function V_N becomes a deterministic problem. However, $\mathcal{Y}(N)$ [and maybe also $\mathcal{U}(N)$] are random variables and thus V_N and θ_N^* are also random variables. Note that θ_N^* minimizes the criterion function V_N only for the given realization $\mathcal{Y}(N)$ and $\mathcal{U}(N)$ (of length N). The question of how the random variable θ_N^* behaves as N tends to infinity is a question of *convergence* and *consistency* of the algorithm.

To remove sample-path dependent considerations, we introduce the following criterion function, which is basically the expected value of V_N:

$$E\{V_N(\theta)\} = \frac{1}{N}\sum_{t=1}^{N} E\{l[\hat{y}(t, \theta), y(t)]\} \qquad (8.2.14)$$

In Ljung (1978a, 1981), the following result is established for the case of linear models:

Theorem 8.2.1. Subject to the following regularity conditions and assumptions: Provided:

(i) $\lim_{N\to\infty} E\{V_N(\theta)\}$ exists almost surely $\qquad (8.2.15)$
and is denoted by $\bar{V}(\theta)$.

(ii) The system generating the data is exponentially stable.

(iii) The model output $\hat{y}(t, \theta)$ is twice differentiable with respect to θ and the predictor model generating $\hat{y}(t, \theta)$ is stable for all θ in the constrained region, which we denote by $D_{\mathfrak{M}}$.

Then $V_N(\theta)$ converges uniformly to $\bar{V}(\theta)$ almost surely (a.s.); that is,

$$\lim_{N\to\infty} \sup_{\theta\in D_{\mathfrak{M}}} |V_N(\theta) - \bar{V}(\theta)| = 0 \quad \text{a.s.} \qquad (8.2.16)$$

and this, in turn, implies that

$$\lim_{N \to \infty} \theta_N^* = \theta^* \quad \text{a.s.} \tag{8.2.17}$$

where θ^* is either a local minimum of $\bar{V}(\theta)$ or a boundary point of $D_{\mathfrak{M}}$.

Proof. See Ljung (1978a, 1981).

▼▼▼

Heuristically, the result above can be viewed as saying that the value θ_N^* which minimizes the sample criterion $V_N(\theta)$ [as given in (8.2.3)] converges as N tends to infinite to a value which minimizes $\bar{V}(\theta)$, which is sample path independent. Clearly, this result relies on some sort of ergodicity condition on the data.

It is also possible to investigate the asymptotic distribution of θ_N^* around θ^*. This is explored in the following result:

Theorem 8.2.2. Subject to the assumptions of Theorem 8.2.1 and the additional assumption that the matrix

$$\frac{d^2}{d\theta^2} \bar{V}(\theta^*) \triangleq \ddot{\bar{V}}(\theta^*)$$

is invertible, then $\sqrt{N}(\theta_N^* - \theta^*)$ converges in distribution to a normal random variable having zero mean and covariance, P, where

$$P = [\ddot{\bar{V}}(\theta^*)]^{-1} [\lim_{N \to \infty} NE\{\dot{V}_N(\theta^*)\dot{V}_N(\theta^*)^T\}]\ddot{\bar{V}}(\theta^*)]^{-1} \tag{8.2.18}$$

Proof. See Ljung and Caines (1979).

▼▼▼

An interesting feature of the results above is that the model need not "match" the system [i.e., there need not exist a value of $\theta \in D_{\mathfrak{M}}$ such that $\hat{y}(t, \theta)$ is the unrestricted optimal prediction for $y(t)$]. The result is important since it shows that asymptotically the procedures leads to a "good" model in the sense that the prediction errors are minimized for the given class of predictors. Further discussion and amplification on this point will be made in a subsequent chapter on adaptive prediction.

So far we have not discussed how one may choose the structure of the predictor for a given problem. However, it seems reasonable that two factors should be taken into account: (1) the a priori knowledge about the system, and (2) the intended application. Bearing these two factors in mind, one can essentially choose the predictor structure arbitrarily since the algorithms have been shown to yield the "best" predictor within the specified class. In the following we illustrate how one might generate a suitable predictor structure using heuristic arguments.

Example 8.2.1

Consider the following nonlinear model:

$$x(t + 1) = f(x(t), u(t), \theta) + v_1(t) \tag{8.2.19}$$

$$y(t) = h(x(t), u(t), \theta) + v_2(t) \tag{8.2.20}$$

with

$$E\left\{\begin{bmatrix} v_1(t) \\ v_2(t) \end{bmatrix} [v_1(s)^T \quad v_2(s)^T]\right\} = \begin{bmatrix} Q(\theta) & S(\theta) \\ S(\theta)^T & R(\theta) \end{bmatrix} \delta(t - s) \tag{8.2.21}$$

Let us presume that $\hat{x}(t)$ is a reasonable estimate of $x(t)$; then (8.2.19)–(8.2.20) can be linearized about $\hat{x}(t)$ as [in the equation, we have again dropped the explicit dependence on $u(t)$ for notational simplicity]

$$x(t + 1) \simeq f(\hat{x}(t), \theta) + F(t, \theta)[x(t) - \hat{x}(t)] + v_1(t) \qquad (8.2.22)$$

$$y(t) \simeq h(\hat{x}(t), \theta) + H(t, \theta)[x(t) - \hat{x}(t)] + v_2(t) \qquad (8.2.23)$$

where

$$F(t, \theta) = \left.\frac{\partial f(x, \theta)}{\partial x}\right|_{x = \hat{x}(t)} \qquad (8.2.24)$$

$$H(t, \theta) = \left.\frac{h(x, \theta)}{\partial x}\right|_{x = \hat{x}(t)} \qquad (8.2.25)$$

Given the (approximate) time-varying linear model (8.2.22)–(8.2.23), we can use the Kalman filter to write down the "optimal" one-step-ahead predictor:

$$\hat{x}(t + 1, \theta) = f(\hat{x}(t, \theta), \theta) + K(t, \theta)[y(t) - h(\hat{x}(t), \theta)] \qquad (8.2.26)$$

$$\hat{y}(t) = h(\hat{x}(t, \theta), \theta) \qquad (8.2.27)$$

where

$$K(t, \theta) = [F(t, \theta)\Sigma(t, \theta)H(t, \theta)^T + S(\theta)][H(t, \theta)\Sigma(t, \theta)H(t, \theta)^T + R(\theta)]^{-1} \qquad (8.2.28)$$

$$\begin{aligned}\Sigma(t + 1) = {} & F(t, \theta)\Sigma(t, \theta)F(t, \theta)^T + Q(\theta) \\ & - K(t, \theta)[H(t, \theta)\Sigma(t, \theta)H(t, \theta)^T + R(\theta)]K(t, \theta)^T\end{aligned} \qquad (8.2.29)$$

Given the particular predictor structure (8.2.26) to (8.2.29) (whether it is, in fact, optimal or not), we can apply a prediction error algorithm to estimate the value of θ that gives the best predictor of the given structure. This will involve taking derivatives of each of the expressions above. It can be seen that the major source of computational complexity will arise from the differentiation of $K(t, \theta)$ using (8.2.28) and (8.2.29). Thus a sensible approach in practice may be to simply take $K(t, \theta)$ as a time-invariant function parameterized directly in terms of θ. This can be justified on the grounds that (8.2.26)–(8.2.27) is but an approximation to the "optimal" predictor; moreover, the prediction error algorithm finds the "best" prediction of the given structure.

▼▼▼

We show in the next section how the prediction error algorithm discussed above can be turned into a sequential form.

8.3 SEQUENTIAL PREDICTION ERROR METHODS

In this section we show how the prediction error algorithms can be implemented in a sequential form whereby the current parameter estimate is updated when a new data point is added. Let us assume that $\hat{\theta}(N)$ has been found such that the criterion function, $V_N(\theta)$, is minimized at $\theta = \hat{\theta}(N)$. Then let us pressume that another observation becomes available. We ask the question: Is there a simple way to obtain $\hat{\theta}(N + 1)$ that minimizes $V_{N+1}(\theta)$ using $\hat{\theta}(N)$ and the additional observation without having to start from scratch? It may be heuristically argued that this is at least plausible since for large N, $V_{N+1}(\theta)$ will not be very different from $V_N(\theta)$ and thus $\hat{\theta}(N + 1)$ should be close to $\hat{\theta}(N)$. We shall study this question below. For simplicity we shall consider the use of a quadratic error criterion.

8.3.1 General Systems

Development of the Algorithm

Consider the scalar criterion of goodness of fit (8.2.3)–(8.2.4), that is,

$$V_N(\theta) = \frac{1}{N} \sum_{t=1}^{N} l(\hat{y}(t, \theta), y(t)) \qquad (8.3.1)$$

where $y(t)$ is assumed to be a scalar and $l(\hat{y}, y)$ is chosen as the quadratic criterion

$$l(\hat{y}, y) \triangleq \tfrac{1}{2}[\hat{y}(t, \theta) - y(t)]^2 \qquad (8.3.2)$$

It follows from (8.3.1)–(8.3.2) that we can express V_{N+1} as

$$(N + 1)V_{N+1}(\theta) = NV_N(\theta) + \tfrac{1}{2}[\hat{y}(N + 1, \theta) - y(N + 1)]^2 \qquad (8.3.3)$$

We now approximate $\hat{y}(N + 1, \theta)$ by using a first-order Taylor's series expansion about $\hat{\theta}(N)$. This gives

$$\hat{y}(N + 1, \theta) \simeq \hat{y}(N + 1, \hat{\theta}(N)) + \hat{y}\,'(N + 1)^T[\theta - \hat{\theta}(N)] \qquad (8.3.4)$$

where

$$\hat{y}\,'(N + 1) \triangleq \left.\frac{d\hat{y}(N + 1, \theta)}{d\theta}\right|_{\theta = \hat{\theta}(N)} \qquad (8.3.5)$$

Substituting (8.3.4) into (8.3.3) gives

$$(N + 1)V_{N+1}(\theta)$$
$$\simeq NV_N(\theta) + \tfrac{1}{2}[\hat{y}(N + 1, \hat{\theta}(N)) + \hat{y}\,'(N + 1)^T(\theta - \hat{\theta}(N)) - y(N + 1)]^2 \qquad (8.3.6)$$
$$= NV_N(\theta) + \tfrac{1}{2}[\psi(N)^T\theta - z(N + 1)]^2$$

where

$$\psi(N) \triangleq \hat{y}\,'(N + 1) \qquad (8.3.7)$$
$$z(N + 1) \triangleq y(N + 1) - \hat{y}(N + 1, \hat{\theta}(N)) + \psi(N)^T\hat{\theta}(N) \qquad (8.3.8)$$

Note that (8.3.6) appears in the form of the usual linear least-squares problem as discussed in Chapter 3. (More will be said about this presently; see Remark 8.3.1.)

Differentiating (8.3.6) with respect to θ gives

$$(N + 1)\dot{V}_{N+1}(\theta) \simeq N\dot{V}_N(\theta) - \psi(N)[z(N + 1) - \psi(N)^T\theta] \qquad (8.3.9)$$

where

$$\dot{V}_{N+1}(\theta) \triangleq \frac{d}{d\theta}V_{N+1}(\theta) \qquad (8.3.10)$$

Expanding $\dot{V}_N(\theta)$ in a first-order Taylor's series about $\hat{\theta}(N)$ gives

$$\dot{V}_N(\theta) \simeq \dot{V}_N(\hat{\theta}(N)) + \ddot{V}_N[\theta - \hat{\theta}(N)] \qquad (8.3.11)$$

Substituting (8.3.11) into (8.3.9) and noting that $\dot{V}_N(\theta_N)$ is zero in view of the optimality of $\hat{\theta}(N)$ gives

$$(N + 1)\dot{V}_{N+1}(\theta) \simeq N\ddot{V}_N[\theta - \hat{\theta}(N)] - \psi(N)[z(N + 1) - \psi(N)^T\theta] \qquad (8.3.12)$$

Now the value, $\hat{\theta}(N + 1)$, of θ optimizing $V_{N+1}(\theta)$ gives $\dot{V}_{N+1}(\theta) = 0$ and hence from (8.3.12) satisfies

$$N\ddot{V}_N[\hat{\theta}(N + 1) - \hat{\theta}(N)] - \psi(N)[z(N + 1) - \psi(N)^T\hat{\theta}(N + 1)] = 0 \qquad (8.3.13)$$

that is,

$$[N\ddot{V}_N + \psi(N)\psi(N)^T]\hat{\theta}(N+1)$$

$$\simeq N\ddot{V}_N\hat{\theta}(N) + \psi(N)z(N+1) \qquad (8.3.14)$$

$$= [N\ddot{V}_N + \psi(N)\psi(N)^T]\hat{\theta}(N) + \psi(N)[z(N+1) - \psi(N)^T\hat{\theta}(N)]$$

Hence

$$\hat{\theta}(N+1) \simeq \hat{\theta}(N) + [N\ddot{V}_N + \psi(N)\psi(N)^T]^{-1}\psi(N)[z(N+1) - \psi(N)^T\hat{\theta}(N)] \qquad (8.3.15)$$

This is basically the final form of the algorithm. For computational reasons, we next show how $\ddot{V}_N$ can be computed iteratively. Thus differentiating (8.3.9) with respect to θ gives

$$(N+1)\ddot{V}_{N+1} \simeq N\ddot{V}_N + \psi(N)\psi(N)^T \qquad (8.3.16)$$

Now define

$$P(N)^{-1} \triangleq (N+1)\ddot{V}_{N+1} \qquad (8.3.17)$$

Then using the matrix inversion lemma (Lemma 3.3.4), we have from (8.3.16)

$$P(N) = P(N-1) - \frac{P(N-1)\psi(N)\psi(N)^T P(N-1)}{1 + \psi(N)^T P(N-1)\psi(N)} \qquad (8.3.18)$$

Finally, using (8.3.8) and (8.3.16) in (8.3.15), the algorithm can be summarized as follows:

Sequential Nonlinear Least-Squares Algorithm

$$\hat{\theta}(t) = \hat{\theta}(t-1) + P(t-1)\psi(t-1)[y(t) - \hat{y}(t, \hat{\theta}(t-1))] \qquad (8.3.19)$$

$$P(t-1) = P(t-2) - \frac{P(t-2)\psi(t-1)\psi(t-1)^T P(t-2)}{1 + \psi(t-1)^T P(t-2)\psi(t-1)} \qquad (8.3.20)$$

$$\psi(t-1) = \frac{d\hat{y}(t, \theta)}{d\theta}\bigg|_{\theta=\theta(t-1)} = \frac{-d\epsilon(t, \theta)}{d\theta}\bigg|_{\theta=\theta(t-1)} \qquad (8.3.21)$$

where $\epsilon(t, \theta) = y(t) - \hat{y}(t, \theta)$.

The algorithm above can be seen to be very close to the sequential least-squares algorithm of Chapter 3. In fact, we can see that in the special case where $\hat{y}(t, \theta)$ is expressed as

$$\hat{y}(t, \theta) = \phi(t-1)^T\theta \qquad (8.3.22)$$

where $\phi(t-1)$ is some function of $\mathcal{U}(t-1)$ and $\mathcal{Y}(t-1)$, then

$$\psi(t-1) = \phi(t-1) \qquad (8.3.23)$$

and the algorithm (8.3.19) to (8.3.21) becomes identical to the least-squares algorithm of Chapter 3.

Remark 8.3.1. Many standard programs exist for solving the usual least-squares problem [see, e.g., Bierman and Nead (1977)]. These programs frequently use superior numerical techniques such as UD factorization or square-root algorithm to implement the least-squares recursion. They can be used to implement the nonlinear least-squares algorithm (8.3.19) to (8.3.21) if the regression vector that is fed to the least-squares update subroutine is $\psi(t-1)$ and if the observation is changed to $z(t)$

as in (8.3.8), that is,

$$z(t) = y(t) - \hat{y}(t, \hat{\theta}(t-1)) + \psi(t-1)^T\hat{\theta}(t-1) \qquad (8.3.24)$$

The reason that this works is that the least-squares programs use the following error term to update the parameters:

$$\begin{aligned}
e(t) &= z(t) - \psi(t-1)^T\hat{\theta}(t-1) \\
&= y(t) - \hat{y}(t, \hat{\theta}(t-1)) \qquad \text{as required in (8.3.19)}
\end{aligned}$$

▼▼▼

The development above can be extended to more general criteria as in (8.2.3); see Exercise 8.1.

Alternative Model Formats

The algorithm (8.3.19) to (8.3.21) is a genuine sequential algorithm for nonlinear models in which $\hat{y}(t, \hat{\theta}(t-1))$ is computed from past data $u(j), y(j), j = 1, \ldots, t-1$, via an algebraic function of the form

$$\hat{y}(t, \theta) = f(\mathcal{Y}(t-1), \mathcal{U}(t-1), \theta) \qquad (8.3.25)$$

where f is a differentiable function.

However, in some cases, the model can take the following form:

$$\hat{y}(t, \theta) = h(\mathcal{Y}(t-1), \mathcal{U}(t-1), \theta, \hat{y}(t-1, \theta), \ldots) \qquad (8.3.26)$$

An example of this kind of model is the predictor model given in (8.2.1) and (8.2.2). We shall see below that models of the form (8.3.26) are very common in the linear case.

For models of the form (8.3.26) we can see that evaluation of $\hat{y}(t, \hat{\theta}(t-1))$ requires that the following equation be solved from $j = 1, \ldots, t$ at each iteration [using $\hat{\theta}(t-1)$]:

$$\hat{y}(j, \hat{\theta}(t-1)) = h(\mathcal{Y}(j-1), \mathcal{U}(j-1), \hat{\theta}(t-1), \hat{y}(j-1, \hat{\theta}(t-1)), \ldots) \qquad (8.3.27)$$

Similarly, evaluation of $\psi(t-1)$ requires that the following equation be solved from $j = 1, \ldots, t$ at each iteration:

$$\frac{d\hat{y}(j, \hat{\theta}(t-1))}{d\theta} = \left[\frac{\partial h(j-1)}{\partial \theta} + \frac{\partial h(j-1)}{\partial \hat{y}(j-1, \theta)} \frac{d\hat{y}(j-1, \theta)}{d\theta} \right]_{\theta = \hat{\theta}(t-1)} \qquad (8.3.28)$$

In these cases the algorithm (8.3.19) to (8.3.21) is not truely sequential. However, since we expect $\hat{\theta}(t-1)$ to be close to $\hat{\theta}(t-2), \ldots$ in the limit, a reasonable approximation is to replace (8.3.27) by

$$\hat{y}(t, \hat{\theta}(t-1)) \simeq h(\mathcal{Y}(t-1), \mathcal{U}(t-1), \hat{\theta}(t-1), \hat{y}(t-1, \hat{\theta}(t-2)), \ldots) \qquad (8.3.29)$$

Similarly, (8.3.28) may be replaced by

$$\psi(t-1) \simeq \frac{\partial h(t-1)}{\partial \theta} + \frac{\partial h}{\partial \hat{y}(t-1, \theta)} \psi(t-2) \cdots \qquad (8.3.30)$$

Using (8.3.29) and (8.3.30), the algorithm (8.3.19) to (8.3.21) again becomes truly sequential. This sequential form of the algorithm will form the basis of much of our discussion below.

Remark 8.3.2. Those readers who are familiar with stochastic approximation algorithms will recognize the close connection between the algorithms discussed above and the stochastic approximation algorithms (Nevel'son and Khasminskii, 1973). We briefly explore this connection below.

The *Robbins–Monro* scheme (Robbins and Monro, 1951) is the prototype stochastic approximation scheme. This algorithm aims to solve an equation of the form

$$E\{Q[\theta, \eta(t)]\} = M(\theta) = 0 \qquad (8.3.31)$$

where $\{\eta(t)\}$ is a sequence of random variables and E denotes expectation over $\eta(t)$. For each value θ it is presumed that it is possible to observe the random vector $Q[\theta, \eta(t)]$ at time t. The Robbins–Monro algorithm to sequentially solve (8.3.31) is as follows:

Robbins–Monro Algorithm

$$\hat{\theta}(t) = \hat{\theta}(t + 1) + \alpha(t)Q[\hat{\theta}(t - 1), \eta(t)] \qquad (8.3.32)$$

where $\{\alpha(t)\}$ is a sequence of positive constants satisfying certain conditions, typically

$$\sum_{t=1}^{\infty} \alpha(t) = \infty, \qquad \lim_{t \to \infty} \alpha(t) = 0 \qquad (8.3.33)$$

Blum (1954) has shown that, subject to certain assumptions, $\hat{\theta}(t)$ given by (8.3.32) will converge almost surely to the solution of (8.3.31).

To show how this might be applied to parameter estimation in stochastic dynamic systems, we consider again the quadratic error criterion given in (8.2.4):

$$l(\hat{y}, y) = \tfrac{1}{2}[\hat{y}(t, \theta) - y(t)]^2 \qquad (8.3.34)$$

Now, rather than using the sample sum-of-squares criterion as in (8.2.3), we use an ensemble average computed as follows:

$$\bar{V}(\theta) \triangleq E\{l(\hat{y}(t, \theta), y(t)) \qquad (8.3.35)$$

The value of θ minimizing (8.3.35) can be found by solving

$$-\frac{d\bar{V}(\theta)}{d\theta} = 0 \qquad (8.3.36)$$

where, from (8.3.34),

$$-\frac{d\bar{V}(\theta)}{d\theta} = E\left\{\frac{d\hat{y}(t, \theta)}{d\theta}(y(t) - \hat{y}(t, \theta))\right\} \qquad (8.3.37)$$

Application of the Robbins–Monro algorithm to (8.3.36) leads to the following scheme:

$$\hat{\theta}(t) = \hat{\theta}(t - 1) + \alpha(t)\hat{y}^{\cdot}(t)[y(t) - \hat{y}(t, \hat{\theta}(t - 1))] \qquad (8.3.38)$$

where

$$\hat{y}^{\cdot}(t) = \frac{d\hat{y}(t, \theta)}{d\theta}\bigg|_{\theta = \hat{\theta}(t-1)} \qquad (8.3.39)$$

In essence, the algorithm (8.3.38) can be viewed as a parameter update in the "estimated" negative gradient direction.

It can be seen that the algorithm (8.3.38)–(8.3.39) is basically the same as the algorithm (8.3.19) except that the matrix algorithm gain, $P(t-1)$, in (8.3.19) has been replaced by the scalar gain $\alpha(t)$ in (8.3.38). Of course, we might expect that, in practice, faster convergence would be obtained by using the algorithm (8.3.19) rather than the scalar gain algorithm (8.3.38)–(8.3.39). On the other hand, the algorithm (8.3.38)–(8.3.39) has the advantage of simplicity.

An alternative form of (8.3.38) is obtained where the gradient $\hat{y}'(t)$ cannot be evaluated analytically but must be estimated. This leads to the *Kiefer–Wolfowitz* algorithm (Kiefer and Wolfowitz, 1952), which generates a sequence of candidate solutions as follows:

$$\hat{\theta}(t) = \hat{\theta}(t-1) - \frac{\alpha(t)}{\beta(t)}[l(\hat{\theta}(t-1) + \beta(t)) - l(\hat{\theta}(t-1) - \beta(t))] \qquad (8.3.40)$$

[Stochastic approximation has been extensively studied. For example, Blum (1953) has proved almost sure convergence for both the Robbins–Monro and Kiefer–Wolfowitz procedures in the multidimensional case. Dvoretzky (1956) has proved both mean-square and almost-sure convergence in the scalar case for a general algorithm incorporating the two procedures above as special cases. A simple proof of convergence of the multidimensional version of Dvoretzky's general algorithm has also been presented by Derman and Sacks (1959). A central limit theorem for the estimates provided by the algorithm has been established by Sacks (1958). For a more complete account of the theory of stochastic approximation, the reader is referred to the book by Albert and Gardner (1967). A very general convergence result for systems with dependent observations has been presented by Ljung (1978b). An analysis of the transient convergence rate has been given by Polyak (1976, 1977). General discussions of stochastic approximation particularly in relation to system identification are given by Young (1968), Saridis (1977), Ljung (1977a), Tsypkin (1971), and Mendel (1973).]

▼▼▼

We next consider the application of the sequential prediction algorithms to the case of linear stochastic systems.

8.3.2 Linear Systems

In this section we specialize the prediction error method of the preceding section to the case where the predictor is a linear dynamical system.

State-Space Models

To motivate the form of the predictor, let us presume for the moment that the system generating the data is described by a linear state-space model:

$$x(t+1) = A(\theta)x(t) + B(\theta)u(t) + v_1(t) \qquad (8.3.41)$$

$$y(t) = C(\theta)x(t) + v_2(t) \qquad (8.3.42)$$

where

$$E\left\{\begin{bmatrix} v_1(s) \\ v_2(s) \end{bmatrix}[v_1(t)v_2(t)]^T\right\} = \begin{bmatrix} Q & S \\ S & R \end{bmatrix}\delta(t-s) \qquad (8.3.43)$$

We have shown in Section 7.3 that the optimal one-step-ahead prediction of $y(t)$ then satisfies

$$\hat{x}(t+1) = A(\theta)\hat{x}(t) + B(\theta)u(t) + K(t, \theta)[y(t) - C(\theta)\hat{x}(t)] \qquad (8.3.44)$$

$$\hat{y}(t, \theta) = C(\theta)\hat{x}(t) \qquad (8.3.45)$$

where $K(t, \theta)$ is obtained from solving the usual matrix Riccati equation.

The equations above give the optimal predictor for given θ. Of course, as we have said in Section 8.2, it is possible to hypothesize (8.3.44)–(8.3.45) as the predictor structure *irrespective* of the true nature of the system. We now ask the question: What value of θ causes the predicted output to be close to the output of the true system?

As in Section 8.2, we shall use a scalar criterion, $V_N(\theta)$, to measure the goodness of fit. In the scalar output case, a suitable choice for V_N is

$$V_N(\theta) = \frac{1}{N} \sum_{t=1}^{N} \epsilon(t, \theta)^2 \qquad (8.3.46)$$

where $\epsilon(t, \theta)$ is the prediction error given by

$$\epsilon(t, \theta) = y(t) - \hat{y}(t, \theta) \qquad (8.3.47)$$

In the multioutput case, a suitable criterion is

$$V_N(\theta) = \log \det \frac{1}{N} \sum_{t=1}^{N} \epsilon(t, \theta)\epsilon(t, \theta)^T \qquad (8.3.48)$$

It is shown in Appendix D that the criteria (8.3.46)–(8.3.48) actually corresponds to the negative log-likelihood function in the case of gaussian noise.

We can now apply the sequential nonlinear least-squares algorithm of (8.3.19) to (8.3.21) to generate $\hat{\theta}(t)$ to minimize (8.3.46). For simplicity we shall consider the case where the steady-state Kalman gain, $K(\theta)$, is directly parameterized in terms of θ. We may then summarise the algorithm in the single-input single-output case as follows:

Sequential Prediction Error Algorithm (Linear Innovations Model)

$$\hat{\theta}(t) = \hat{\theta}(t - 1) + P(t - 1)\psi(t - 1)[y(t) - \hat{y}(t)] \qquad (8.3.49)$$

$$P(t - 1) = P(t - 2) - \frac{P(t - 2)\psi(t - 1)\psi(t - 1)^T P(t - 2)}{1 + \psi(t - 1)^T P(t - 2)\psi(t - 1)} \qquad (8.3.50)$$

$$\hat{y}(t) = C_{t-1}\hat{x}(t) \qquad (8.3.51)$$

$$\hat{x}(t + 1) = A_t\hat{x}(t) + B_t u(t) + K_t[y(t) - \hat{y}(t)] \qquad (8.3.52)$$

where we have used the notation

$$A_t = A(\hat{\theta}(t)), \quad B_t = B(\hat{\theta}(t)), \quad \text{etc.}$$

In (8.3.49) and (8.3.50), $\psi(t - 1)$ has its ith component given by

$$[\psi(t - 1)]_i = \left\{\frac{\partial}{\partial \theta_i} C(\theta)\right\} \hat{x}(t) + C_{t-1}[\hat{x}^{\cdot}(t)]_i \qquad (8.3.53)$$

and

$$[\hat{x}^{\cdot}(t)]_i = \left\{\frac{\partial}{\partial \theta_i} A(\theta)\right\} \hat{x}(t - 1) + A_{t-1}[\hat{x}^{\cdot}(t - 1)]_i$$

$$+ \left\{\frac{\partial}{\partial \theta_i} B(\theta)\right\} u(t) + \frac{\partial}{\partial \theta_i} K(\theta)[y(t - 1) - \hat{y}(t - 1)] - K_{t-1}[\psi(t - 2)]_i \qquad (8.3.54)$$

where all derivatives are evaluated at $\theta = \hat{\theta}(t - 1)$.

A projection facility as in Section 3.7 would generally be required to ensure that the estimated predictor remains in the stability region $D_{\mathfrak{M}}$.

Our discussion so far has been essentially deterministic. As noted in Section 8.2, $\hat{\theta}(t)$ is a random variable. It is therefore relevant to ask if the counterparts of Theorems 8.2.1 and 8.2.2 will apply to the sequential form of the prediction error algorithm. We quote without proof, the following result due to Ljung (1978a, 1981).

Theorem 8.3.1. Consider the algorithm (8.3.49) to (8.3.54) applied to the linear predictor (8.3.44)–(8.3.45) with a projection facility to ensure that the estimated predictor remains stable (i.e., $\theta \in D_{\mathfrak{M}}$). Subject to the following conditions:

(i) Assumptions (i), (ii), and (iii) of Theorem 8.2.1 hold.
(ii) $P(t)$ satisfies $\delta_1 I \leq (1/t)P(t)^{-1} \leq \delta_2 I$; $\delta_1 > 0$, $\delta_2 > 0$ [i.e., $P(t)^{-1}$ grows linearly with t and has bounded condition number].

Then $\hat{\theta}(t)$ converges almost surely to either the set

$$D_c = \left\{ \theta \left| \frac{d}{d\theta} \bar{V}(\theta) = 0 \right. \right\} \tag{8.3.55}$$

where $\bar{V}(\theta)$ is as in (8.2.15) or to a boundary point of $D_{\mathfrak{M}}$.

Proof. See Ljung (1978a, 1981), Hannan (1978), and Ljung and Söderström (1982).

▼▼▼

Asymptotic normality results similar to those presented in Theorem 8.2.2 have also been investigated [see Solo (1978, 1981)].

In summary, the properties of the off-line algorithm carry over to the on-line version. This is a remarkable and important result.

The convergence analysis of the algorithm above is complicated because of its time-varying nonlinear stochastic nature. In Sections 8.5.1 and 8.5.2 we shall present a rigorous analysis of a more restricted class of algorithms using the Martingale convergence theorem. Later, in Section 8.5.4, we shall return to the analysis of more general sequential algorithms, for example the sequential prediction error algorithms discussed in this section.

ARMAX Models

Here we specialize the sequential prediction error algorithms to the case of ARMAX models. Recall that ARMAX models have the form

$$A(q^{-1})y(t) = q^{-1}B'(q^{-1})u(t) + C(q^{-1})\omega(t) \tag{8.3.56}$$

where $A(q^{-1})$, $B'(q^{-1})$, and $C(q^{-1})$ are of order n, m, and l, respectively.

For simplicity, we shall treat only the single-input single-output case. It has been shown in Chapter 7 (Lemma 7.4.1) that the optimal one-step-ahead predictor corresponding to the model (8.3.56) is

$$C(q^{-1})y^0(t+1|t) = \alpha(q^{-1})y(t) + B'(q^{-1})u(t+1-d) \tag{8.3.57}$$

where in this case

$$\alpha(q^{-1}) = q[C(q^{-1}) - A(q^{-1})]$$
$$= \alpha_1 + \alpha_2 q^{-1} + \cdots + \alpha_{\bar{n}} q^{-\bar{n}+1} \tag{8.3.58}$$
$$\bar{n} = \max(n, l)$$

Now, as before, we can use (8.3.57) to define a prediction of $y(t+1)$ whether or not the true system is described by a model of the form (8.3.56). Thus in the sequel we shall use the following predictor structure without further comment:

$$C(q^{-1}, \theta)\hat{y}(t+1, \theta) = \alpha(q^{-1}, \theta)y(t) + B'(q^{-1}, \theta)u(t+1-d) \tag{8.3.59}$$

The sequential prediction error algorithm then takes the following form:

Sequential Prediction Error Algorithm (ARMAX Model)

$$\hat{\theta}(t) = \hat{\theta}(t-1) + P(t-1)\psi(t-1)[y(t) - \hat{y}(t)] \tag{8.3.60}$$

$$P(t-1) = P(t-2) - \frac{P(t-2)\psi(t-1)\psi(t-1)^T P(t-2)}{1 + \psi(t-1)^T P(t-2)\psi(t-1)} \tag{8.3.61}$$

$$\hat{y}(t) = [1 - \hat{C}(q^{-1}, \hat{\theta}(t-1))]\hat{y}(t) + \hat{\alpha}(q^{-1}, \hat{\theta}(t-1))y(t-1)$$
$$+ B'(q^{-1}, \hat{\theta}(t-1))u(t-d) \tag{8.3.62}$$

where

$$\hat{C}(q^{-1}, \hat{\theta}(t-1)) = 1 + \hat{c}_1(t-1)q^{-1} + \cdots + \hat{c}_l(t-1)q^{-l}$$

and so on; also,

$$\hat{\theta}(t) = [\hat{\alpha}_1(t), \ldots, \hat{\alpha}_{\bar{n}}(t), \hat{b}_0(t), \ldots, \hat{b}_m(t), \hat{c}_1(t), \ldots, \hat{c}_l(t)] \tag{8.3.63}$$

Equation (8.3.62) can be expressed alternatively as

$$\hat{y}(t) = \phi(t-1)^T \hat{\theta}(t-1) \tag{8.3.64}$$

where

$$\phi(t-1)^T = [y(t-1), \ldots, y(t-\bar{n}), u(t-d), \ldots, u(t-d-m),$$
$$-\hat{y}(t-1), \ldots, -\hat{y}(t-l)] \tag{8.3.65}$$

Finally, the gradient vector $\psi(t-1)$ in (8.3.60) satisfies (8.3.30), which now takes the specific form

$$\hat{C}[q^{-1}, \hat{\theta}(t-1)]\psi(t-1) = \phi(t-1) \tag{8.3.66}$$

▼▼▼

There are several alternative ways of expressing the algorithm above. For example, by adding and subtracting the term $[1 - \hat{C}(q^{-1}, \hat{\theta}(t-1))]y(t)$ from the right-hand side of (8.3.62) and noting (8.3.58), $\hat{y}(t)$ can be expressed as

$$\hat{y}(t) = [\hat{C}(q^{-1}, \hat{\theta}(t-1)) - 1]\epsilon(t) + [1 - \hat{A}(q^{-1}, \hat{\theta}(t-1))]y(t)$$
$$+ \hat{B}'(q^{-1}, \hat{\theta}(t-1))u(t-d) \tag{8.3.67}$$

where $\epsilon(t) = y(t) - \hat{y}(t)$. Or, alternatively,

$$\hat{C}(q^{-1}, \hat{\theta}(t-1))\epsilon(t) = \hat{A}(q^{-1}, \hat{\theta}(t-1))y(t) - \hat{B}'(q^{-1}, \hat{\theta}(t-1))u(t-d) \tag{8.3.68}$$

which can be written as

$$\epsilon(t) = y(t) - \phi'(t-1)^T \hat{\theta}'(t-1) \tag{8.3.69}$$

where

$$\phi'(t-1)^T = [-y(t-1), \ldots, -y(t-n), u(t-d), \ldots, u(t-d-m),$$
$$\epsilon(t-1), \ldots, \epsilon(t-l)] \tag{8.3.70}$$

$$\hat{\theta}'(t-1)^T = [\hat{a}_1(t-1), \ldots, \hat{a}_n(t-1), \hat{b}_0(t-1), \ldots, \hat{b}_m(t-1),$$
$$\hat{c}_1(t-1), \ldots, \hat{c}_l(t-1)] \tag{8.3.71}$$

The algorithm can then be written as

$$\hat{\theta}'(t) = \hat{\theta}'(t-1) + P'(t-1)\psi'(t-1)\epsilon(t) \tag{8.3.72}$$

$$P'(t-1) = P'(t-2) - \frac{P'(t-2)\psi'(t-1)\psi'(t-1)^T P'(t-2)}{1 + \psi'(t-1)^T P(t-2)\psi'(t-1)} \tag{8.3.73}$$

$$\hat{C}(q^{-1}, \hat{\theta}(t-1))\epsilon(t) = \hat{A}(q^{-1}, \hat{\theta}(t-1))y(t) - \hat{B}'(q^{-1}, \hat{\theta}(t-1))u(t-d) \tag{8.3.74}$$

$$\hat{C}(q^{-1}, \hat{\theta}(t-1))\psi'(t-1) = \phi'(t-1) \tag{8.3.75}$$

The algorithm above is commonly known in the identification literature as RML2 (recursive maximum likelihood, version 2).

Remark 8.3.3. It is important to note that in the algorithm (8.3.60)–(8.3.61), the gradient vector $\psi(t)$ is obtained by passing the input–output data $\phi(t)$ through a time-varying filter, $(1/\hat{C}(q^{-1}, \hat{\theta}(t)))$ based on the estimated parameters $\hat{\theta}(t)$. Divergence problems can occur unless this filter is kept within the stability region. This is usually done by projecting the parameter estimates into a set such that the roots of the estimated $\hat{C}(q^{-1})$ polynomial lie strictly inside the unit circle. Programs implementing this method usually incorporate a stability test [e.g., Jury's test; see Kuo (1980, p. 278)] so that this projection can be carried out.

Clearly, if there is a good a priori estimate of $C(q^{-1})$, it seems reasonable to use a fixed filter in determining $\psi(t)$. We will investigate algorithms based on this principle in Section 8.4.

Models Closely Related to ARMAX Models

In the literature, there are many algorithms that can be considered as variants of the algorithm above, as has been shown by Ljung (1979a). In order that we can describe these algorithms, it is convenient to replace the ARMAX model (8.3.56) by a slightly more general model:

$$A(q^{-1})y(t) = \frac{q^{-d}B'(q^{-1})}{E(q^{-1})}u(t) + \frac{C(q^{-1})}{D(q^{-1})}\omega(t) \tag{8.3.76}$$

where $E(0) = D(0) = C(0) = 1$.

For the model above, the optimal one-step-ahead predictor has the form (see Exercise 8.2)

$$\hat{y}(t) = \left[1 - \frac{D(q^{-1})A(q^{-1})}{C(q^{-1})}\right]y(t) + \frac{D(q^{-1})B'(q^{-1})}{C(q^{-1})E(q^{-1})}u(t-d) \tag{8.3.77}$$

We introduce the auxiliary variables

$$u'(t-d) = \frac{B'(q^{-1})}{E(q^{-1})}u(t-d) \tag{8.3.78}$$

$$v'(t) = A(q^{-1})y(t) - u'(t-d) \tag{8.3.79}$$

Then from (8.3.77), we have

$$\epsilon(t) = \frac{D(q^{-1})}{C(q^{-1})}\left[A(q^{-1})y(t) - \frac{B'(q^{-1})}{E(q^{-1})}u(t-d) \right]$$

$$= \frac{D(q^{-1})}{C(q^{-1})}v'(t) \qquad (8.3.80)$$

where

$$\epsilon(t) = y(t) - \hat{y}(t) \qquad (8.3.81)$$

Now (8.3.80) can be written as

$$\epsilon(t) = [1 - C(q^{-1})]\epsilon(t) + D(q^{-1})v'(t)$$

$$= [1 - C(q^{-1})]\epsilon(t) + [D(q^{-1}) - 1]v'(t) + A(q^{-1})y(t) - u'(t-d)$$

$$= y(t) + [1 - C(q^{-1})]\epsilon(t) + [D(q^{-1}) - 1]v'(t) + [A(q^{-1}) - 1]y(t)$$

$$\qquad + [E(q^{-1}) - 1]u'(t-d) - B'(q^{-1})u(t-d) \qquad (8.3.82)$$

and this can be written as

$$\epsilon(t) = y(t) - \phi(t)^T\theta \qquad (8.3.83)$$

where

$$\phi(t)^T = [-y(t-1), \ldots, -y(t-n), u(t-d), \ldots, u(t-d-m)$$
$$\epsilon(t-1), \ldots, \epsilon(t-l), -v'(t-1), \ldots, -v'(t-n_d) \qquad (8.3.84)$$
$$- u'(t-d-1), \ldots, -u'(t-d-n_e)]$$

$$\theta = [a_1, \ldots, a_n, b_0, \ldots, b_m, c_1, \ldots, c_l, d_1, \ldots, d_{n_d}, e_1, \ldots e_{n_e}] \qquad (8.3.85)$$

Simple algebra shows that

$$\frac{\partial}{\partial a_i}\epsilon(t) = \frac{D(q^{-1})}{C(q^{-1})}y(t-i) \qquad (8.3.86)$$

$$\frac{\partial}{\partial b_i}\epsilon(t) = -\frac{D(q^{-1})}{C(q^{-1})E(q^{-1})}u(t-i-d) \qquad (8.3.87)$$

$$\frac{\partial}{\partial c_i}\epsilon(t) = -\frac{1}{C(q^{-1})}\epsilon(t-i) \qquad (8.3.88)$$

$$\frac{\partial}{\partial d_i}\epsilon(t) = \frac{1}{C(q^{-1})}v'(t-i) \qquad (8.3.89)$$

$$\frac{\partial}{\partial e_i}\epsilon(t) = \frac{D(q^{-1})}{C(q^{-1})E(q^{-1})}u'(t-d-i) \qquad (8.3.90)$$

The equations above can be used in an obvious way to define the gradient vector $\psi(t)$. The resulting sequential algorithm is quite general and specializes to many commonly used algorithms:

1. $E(q^{-1}) = C(q^{-1}) = D(q^{-1}) = 1$ gives ordinary least squares (equation error formulation).
2. $E(q^{-1}) = D(q^{-1}) = 1$ gives the RML2 algorithm described in the preceding section; see Söderström (1973) and Furht (1973).

3. $E(q^{-1}) = C(q^{-1}) = 1$ gives the identification algorithm of Gertler and Banyasz (1974).

4. $A(q^{-1}) = 1$ gives the method of Young and Jakeman (1979).

8.4 ALGORITHMS BASED ON PSEUDO LINEAR REGRESSIONS

In this section we investigate a class of algorithms that can be thought of as approximations to the sequential prediction error algorithms of Section 8.3. We begin by returning to the sequential prediction error algorithm for ARMAX models described in Section 8.3.2 [equations (8.3.60) to (8.3.66)]. As pointed out in Remark 8.3.3, it is usually necessary to include a projection facility in this algorithm so that the filter $[1/\hat{C}(q^{-1}, \hat{\theta}(t))]$ used in generating the gradient vector, $\psi(t)$, is kept inside the stability domain. It was suggested in Remark 8.3.3 that if a good initial estimate of $C(q^{-1})$ is available, it may be possible to use a fixed filter based on available a priori knowledge. The use of a fixed filter will then lead to an approximate form of the sequential prediction error algorithm. We investigate this idea in this section.

Our discussion in this section is again based on the ARMAX model of (8.3.56):

$$A(q^{-1})y(t) = q^{-d}B'(q^{-1})u(t) + C(q^{-1})\omega(t) \tag{8.4.1}$$

Let $D(q^{-1})$ be a fixed a priori estimate of $C(q^{-1})$.

Introducing the fixed filter $D(q^{-1})$ into the algorithm (8.3.60) to (8.3.66) now leads to

Pseudo Linear Regression Algorithm (with Filtering)

$$\hat{\theta}(t) = \hat{\theta}(t-1) + P(t-1)\psi(t-1)[y(t) - \hat{y}(t)] \tag{8.4.2}$$

$$P(t-1) = P(t-2) - \frac{P(t-2)\psi(t-1)\psi(t-1)^T P(t-2)}{1 + \psi(t-1)^T P(t-2)\psi(t-1)} \tag{8.4.3}$$

$$\hat{y}(t) = \phi(t-1)^T \hat{\theta}(t-1) \tag{8.4.4}$$

$$\phi(t-1)^T = [y(t-1), \ldots, y(t-\bar{n}), u(t-d), \ldots, u(t-d-m), \\ -\hat{y}(t-1), \ldots, -\hat{y}(t-l)] \tag{8.4.5}$$

$$D(q^{-1})\psi(t-1) = \phi(t-1) \tag{8.4.6}$$

In many cases, the algorithm above can be simplified even further by putting $D(q^{-1}) = 1$. The algorithm is then known by a number of names, including *extended least squares*, *Panuska's method* (Panuska, 1968, 1969), *RML1* (Söderström, Ljung, and Gustavsson 1974), and *extended matrix method* (Talmon and van der Boom, 1973).

Remark 8.4.1. When viewed from the prediction error approach, the vector $\psi(t-1)$ in (8.4.6) is an approximation to the gradient vector $\psi(t-1)$ in (8.3.66). It is natural that there will be restrictions on the relationship between $D(q^{-1})$ and $C(q^{-1})$ for the approximation to be good enough for the method to still converge. An appreciation of this point is important in analyzing or modifying algorithms to ensure good performance. Roughly the condition is that $D(q^{-1})/C(q^{-1})$ should be "positive" (in a sense to be made precise later) so that the $\psi(t)$ makes an acute angle with the gradient vector.

▼▼▼

Remark 8.4.2. We have motivated the algorithm above as an approximation to the sequential prediction error algorithm. However, it has been traditionally motivated in the literature as an extension of the standard least-squares algorithm to accommodate disturbances other than white noise.

To see how this is done, we return to the optimal predictor structure of (8.3.57):

$$C(q^{-1})y^0(t + 1 \mid t) = \alpha(q^{-1})y(t) + B'(q^{-1})u(t + 1 - d) \qquad (8.4.7)$$

or

$$y^0(t \mid t - 1) = \alpha(q^{-1})y(t - 1) + B'(q^{-1})u(t - d) + [1 - C(q^{-1})]y^0(t \mid t - 1)$$

$$= \phi_0(t - 1)^T\theta \qquad (8.4.8)$$

where

$$\phi_0(t - 1)^T = [y(t - 1), \ldots, y(t - \bar{n}), u(t - d), \ldots, u(t - d - m),$$
$$-y^0(t - 1 \mid t - 2), \ldots, -y^0(t - l \mid t - l - 1)] \qquad (8.4.9)$$

Now if $y^0(t - 1 \mid t - 2)$, $y^0(t - 2 \mid t - 3)$, and so on, were known, we could estimate θ by using ordinary least squares in conjunction with the following predictor:

$$\hat{y}(t) = \phi_0(t - 1)^T\hat{\theta}(t - 1) \qquad (8.4.10)$$

This would lead to an ordinary *linear regression* problem. However, since $y^0(t - 1 \mid t - 2)$, $y^0(t - 2 \mid t - 3)$, and so on, are not available it makes sense to replace them by previously estimated values in a bootstrap fashion; that is, we define

$$\hat{y}(t) = \phi(t - 1)^T\hat{\theta}(t - 1) \qquad (8.4.11)$$

where

$$\phi(t - 1)^T = [y(t - 1), \ldots, y(t - \bar{n}), u(t - d), u(t - d - m),$$
$$-\hat{y}(t - 1), \ldots, -\hat{y}(t - l)] \qquad (8.4.12)$$

We then apply the ordinary least-squares algorithm and note that this leads directly to (8.4.2) to (8.4.6) with $D(q^{-1}) = 1$.

Note that if $\phi(t)$ were not a function of $\hat{\theta}(t - 1)$, we would have an ordinary linear regression problem. However, here $\hat{y}(t)$ and hence $\phi(t)$ depend on $\hat{\theta}(t - 1)$ and thus (8.4.11) is not really a linear regression. For this reason, we often use the term *pseudo linear regression* for this type of algorithm. Clearly, ignoring the dependence of $\phi(t)$ on θ leads to an approximation.

▼▼▼

Remark 8.4.3. In a sense, one can pinpoint the source of difficulty in parameter estimation in stochastic systems as lack of knowledge of the complete state vector. For example, the reader may consider the $\phi_0(t - 1)$ vector in (8.4.9) as containing unknown states: $y^0(t - 1 \mid t - 2)$, $y^0(t - 2 \mid t - 3)$, and so on. This suggests that an algorithm may be able to be generated by using a state estimator—say the Kalman filter, in conjunction with ordinary least squares. We show below that this leads again to the pseudo linear regression algorithm.

We will represent the system in observer state-space form (since we know from Chapter 2 that this ties up with ARMAX models). Therefore, consider the following state-space model:

$$x(t+1) = \begin{bmatrix} -a_1 & 1 & & \\ \vdots & & \ddots & \\ & & & 1 \\ -a_n & 0 & \cdots & 0 \end{bmatrix} x(t) + \begin{bmatrix} b_1 \\ \vdots \\ \vdots \\ b_n \end{bmatrix} u(t) + v_1(t) \tag{8.4.13}$$

$$y(t) = [1 \quad 0 \quad \cdots \quad 0]x(t) + v_2(t) \tag{8.4.14}$$

If the parameters are known, the Kalman filter has the form

$$\hat{x}(t+1) = \begin{bmatrix} -a_1 & 1 & & \\ \vdots & & \ddots & \\ & & & 1 \\ -a_n & 0 & \cdots & 0 \end{bmatrix} \hat{x}(t) + \begin{bmatrix} b_1 \\ \vdots \\ \vdots \\ b_n \end{bmatrix} u(t) + \begin{bmatrix} k_1 \\ \vdots \\ \vdots \\ k_n \end{bmatrix} \{y(t) - \hat{y}(t)\} \tag{8.4.15}$$

$$\hat{y}(t) = \hat{x}_1(t) \tag{8.4.16}$$

[i.e., the first component of $\hat{x}(t)$] or

$$\hat{x}(t+1) = \begin{bmatrix} -c_1 & 1 & & \\ \vdots & & \ddots & \\ & & & 1 \\ -c_n & 0 & \cdots & 0 \end{bmatrix} \hat{x}(t) + \begin{bmatrix} b_1 \\ \vdots \\ \vdots \\ b_n \end{bmatrix} u(t) + \begin{bmatrix} k_1 \\ \vdots \\ \vdots \\ k_n \end{bmatrix} y(t)$$

where

$$\hat{y}(t) = \hat{x}_1(t) \tag{8.4.17}$$

$$c_i \triangleq a_i + k_i; \quad i = 1, \ldots, n \tag{8.4.18}$$

Now by successive substitution

$$\hat{y}(t) = \hat{x}_1(t)$$
$$= -c_1\hat{x}_1(t-1) + \hat{x}_2(t-1) + b_1u(t-1) + k_1y(t-1)$$
$$= -c_1\hat{x}_1(t-1) - c_2\hat{x}_1(t-2) + \hat{x}_3(t-2) + b_1u(t-1) + b_2u(t-2)$$
$$+ k_1y(t-1) + k_2y(t-2)$$

Finally, we can write

$$\hat{y}(t) = \phi(t-1)^T\theta \tag{8.4.19}$$

where

$$\theta^T = [k_1 \cdots k_n, b_1 \cdots b_n, c_1 \cdots c_n] \tag{8.4.20}$$

$$\phi(t-1)^T = [y(t-1), \ldots, y(t-n), u(t-1), \ldots, u(t-n), \\ -\hat{x}_1(t-1), \ldots, -\hat{x}_1(t-n)] \tag{8.4.21}$$

The vector $\phi(t-1)$ depends on $\hat{x}(t-1)$, $\hat{x}(t-2)$, and so on. However, from

(8.4.15), these are in turn functions of θ. This suggests a bootstrap scheme in which $\hat{x}(t)$ is generated using a Kalman filter in which θ is replaced by $\hat{\theta}(t-1)$. Then if the resulting $\hat{x}(t)$ is used in (8.4.19), we are lead to the usual least-squares algorithm for estimating θ. This leads to the following algorithm:

$$\hat{x}(t+1) = \begin{bmatrix} -\hat{c}_1(t) & 1 & & \\ \vdots & & \ddots & \\ \vdots & & & 1 \\ -\hat{c}_n(t) & \cdots & & 0 \end{bmatrix} \hat{x}(t) + \begin{bmatrix} \hat{b}_1(t) \\ \vdots \\ \vdots \\ \hat{b}_n(t) \end{bmatrix} u(t) + \begin{bmatrix} \hat{k}_1(t) \\ \vdots \\ \vdots \\ \hat{k}_n(t) \end{bmatrix} y(t) \quad (8.4.22)$$

$$\hat{y}(t) = \hat{x}_1(t) \quad (8.4.23)$$

$$\hat{\theta}(t) = \hat{\theta}(t-1) + P(t-1)\phi(t-1)[y(t) - \hat{y}(t)] \quad (8.4.24)$$

where $P(t-1)$ is as in (8.4.3) [where $\psi(t-1) = \phi(t-1)$], and $\phi(t-1)$ is as in (8.4.21) but with $\hat{x}(t)$ defined by (8.4.22). This algorithm is shown diagramatically in Fig. 8.4.1.

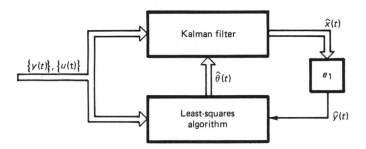

Figure 8.4.1 Boot-strap parameter estimator for stochastic systems.

However, in view of the fact that only $\hat{x}_1(t) = \hat{y}(t)$ appears in the regression vector, we can use successive substitution in (8.4.22) to express $\hat{y}(t)$ in terms of $\hat{y}(t-1)$, and so on. This gives

$$\begin{aligned} \hat{x}_1(t) &= \hat{y}(t) \\ &= -\hat{c}_1(t-1)\hat{y}(t-1) - \hat{c}_2(t-2)\hat{y}(t-2) - \cdots - \hat{c}_n(t-n)\hat{y}(t-n) \\ &\quad + \hat{b}_1(t-1)u(t-1) + \cdots + \hat{b}_n(t-n)u(t-n) \\ &\quad + \hat{k}_1(t-1)y(t-1) + \cdots + \hat{k}_n(t-n)y(t-n) \end{aligned} \quad (8.4.25)$$

Thus (8.4.25) is almost identical to (8.4.4) except that the latter vector is based on the most recent parameter estimate $\hat{\theta}(t-1)$. With this in mind it makes sense to replace (8.4.25) by

$$\begin{aligned} \hat{x}_1(t) = \hat{y}(t) &= [1 - \hat{C}(q^{-1}, t-1)]\hat{y}(t) + \hat{B}'(q^{-1}, t-1)u(t-1) \\ &\quad + \hat{\alpha}(q^{-1}, t-1)y(t-1) \end{aligned} \quad (8.4.26)$$

in which case the algorithm can be seen to be identical to the pseudo linear regression algorithm [with $D(q^{-1}) = 1$]. [In (8.4.26), $\hat{B}'(q^{-1}, t-1) = \hat{b}_1(t-1) + \hat{b}_2(t-1)q^{-1} + \cdots$.]

Note that the simple linear regression given in (8.4.19) results from the use of the observer canonical form, which in a certain sense is linear in the parameters (as is apparent from the corresponding ARMAX form). If alternative state-space models are used, this linearity property will not hold and the simplicity of the pseudo linear regression schemes will be lost. In such cases, the more general sequential prediction error methods will have to be used since these do not rely on any particular linear model structure.

▼▼▼

Remark 8.4.4. The comments made immediately prior to (8.4.26) suggest a further improvement to the algorithm, that is, whenever possible, to make use of the most recent parameter estimate. Now we see from (8.4.2) that $\hat{\theta}(t)$ depends on $\psi(t-1)$ and $\hat{y}(t)$. Further $\hat{y}(t)$ is a function of $\phi(t-1)$ and $\hat{\theta}(t-1)$. Now if we focus attention on the "gradient" vector $\psi(t-1)$ and $\phi(t-1)$, we see that they depend on $\hat{y}(t-1)$, $\hat{y}(t-2)$, ..., which are in turn a function of $\hat{\theta}(t-2)$ and $\hat{\theta}(t-3)$. Thus a simple modification to the algorithm is to replace the a priori estimates $\hat{y}(t-1)$ and $\hat{y}(t-2)$ in the definition of $\phi(t-1)$ and $\psi(t-1)$ by a posteriori estimate $\bar{y}(t-1)$, $\bar{y}(t-2)$, ..., where

$$\bar{y}(t) \triangleq \phi(t-1)^T \hat{\theta}(t) \tag{8.4.27}$$

Intuitively this should lead to an improved algorithm since a more recent estimate of θ is used in generating the regression vector. We shall call $\bar{y}(t)$ an *a posteriori prediction* of $y(t)$. Clearly, it is preferable to use a posteriori predictions in this general class of algorithms.

As discussed in Section 3.5, the notion of an a posteriori output was first introduced by Young (1974). Since then, algorithms incorporating this idea have also been studied by Moore and Ledwich (1977), Solo (1979), Kumar and Moore (1979), and Sin and Goodwin (1982).

▼▼▼

In the following subsections we shall present a rigorous analysis of the class of pseudo linear regression algorithms. We shall use the Martingale approach to the convergence analysis. To illustrate the ideas involved, we shall begin by looking at the stochastic gradient variant of the algorithm. Later we shall generalize this to the least-squares-based algorithms. Many of the ideas used in the convergence analysis have counterparts in the analysis of convergence of deterministic algorithms presented in Chapter 3.

8.5 CONVERGENCE ANALYSIS OF SEQUENTIAL ALGORITHMS

8.5.1 The Stochastic Gradient Algorithm

Consider again the ARMAX model

$$A(q^{-1})y(t) = q^{-d}B'(q^{-1})u(t) + C(q^{-1})\omega(t) \tag{8.5.1}$$

where $\{y(t)\}$, $\{u(t)\}$, and $\{\omega(t)\}$ denote the output, input, and "white noise" sequence, respectively, and

$$A(q^{-1}) = 1 + a_1 q^{-1} + \cdots + a_n q^{-n} \tag{8.5.2}$$

$$B'(q^{-1}) = b_0 + b_1 q^{-1} + \cdots + b_m q^{-m} \tag{8.5.3}$$

$$C(q^{-1}) = 1 + c_1 q^{-1} + \cdots + c_l q^{-l} \tag{8.5.4}$$

We shall be concerned with analyzing a stochastic gradient variant of the pseudo linear regression method introduced in the preceding section for estimating the coefficients in $A(q^{-1})$, $B'(q^{-1})$, and $C(q^{-1})$.

The convergence analysis presented here will necessarily use some ideas drawn from the theory of stochastic processes. A summary of some of these ideas is given in Appendix D. However, since the analysis technique used here parallels the corresponding analysis for the deterministic case presented in Chapter 3, the reader should be able to grasp the key ideas without necessarily becoming embroiled in some of the finer technicalities of the proofs.

In line with the corresponding work in Chapter 3, we shall perform the convergence analysis under relative weak assumptions. The reason for doing this is that we will subsequently be able to utilize the results in adaptive filtering, prediction, and control. The algorithms will be a pseudo linear regression type and, as has been shown in Section 8.4, are closely related to the prediction error algorithms. Thus we shall pay particular attention in the convergence analysis to the convergence of the prediction error rather than the convergence of the parameter estimates.

In particular:

1. We will assume that only an upper bound is known for the orders of the polynomials $A(q^{-1})$, $B(q^{-1})$, and $C(q^{-1})$.
2. We make no prior assumptions about the stability of $A(z^{-1})$ or the nature of the signals $\{y(t)\}$ and $\{u(t)\}$.
3. We will not impose any "persistently exciting" conditions on the input.

In the model (8.5.1), q^{-d} represents a pure time delay. The scalar sequence $\{\omega(t)\}$ can be thought of as a *white noise sequence*. We shall express this in a technically precise way: The sequence $\{\omega(t)\}$ is a real stochastic process defined on a probability space $(\Omega, \mathcal{F}, P)$ adapted to the sequence of increasing sub-sigma algebras $(\mathcal{F}_n, n \in \mathfrak{N})$, where $\mathcal{F}_n$ is generated by the observations up to time n, and such that $\{\omega(t)\}$ satisfies:

Noise Assumptions

$$E\{\omega(t) \mid \mathcal{F}_{t-1}\} = 0 \quad \text{a.s.} \tag{8.5.5}$$

$$E\{\omega(t)^2 \mid \mathcal{F}_{t-1}\} = \sigma^2 \quad \text{a.s.} \tag{8.5.6}$$

$$\sup_N \frac{1}{N} \sum_{t=1}^{N} \omega(t)^2 < \infty \quad \text{a.s.} \tag{8.5.7}$$

It is assumed that the coefficients in $A(q^{-1})$, $B(q^{-1})$, $C(q^{-1})$, and σ^2 are unknown and that only the input, $\{u(t)\}$, and output, $\{y(t)\}$, sequences are directly available.

The following prior knowledge regarding the system (8.5.1) will be required:

System Assumptions

1. An upper bound for n, m, and l is known.
2. $C(z^{-1})$ is a stable polynomial.

The algorithm that we shall be concerned with here is the following:

Stochastic Gradient Algorithm

$$\hat{\theta}(t) = \hat{\theta}(t-1) + \frac{\phi(t-1)}{r(t-2) + \phi(t-1)^T\phi(t-1)}[y(t) - \hat{y}(t)] \qquad (8.5.8)$$

where

$$\hat{y}(t) = \phi(t-1)^T\hat{\theta}(t-1) \qquad (8.5.9)$$

$$\phi(t-1)^T = (y(t-1), \ldots, y(t-\bar{n}), u(t-d), \ldots, u(t-d-m),$$
$$-\bar{y}(t-1), \ldots, -\bar{y}(t-l)) \qquad (8.5.10)$$

and $\bar{y}(t)$ is the a posteriori predicted output given by

$$\bar{y}(t) = \phi(t-1)^T\hat{\theta}(t) \qquad (8.5.11)$$

and $\hat{\theta}(t)$ is an estimate of θ_0 at time t, and

$$\theta_0^T = (\alpha_1, \ldots, \alpha_{\bar{n}}, b_0, \ldots, b_m, c_1, \ldots, c_l) \qquad (8.5.12)$$

as in (8.3.57). In (8.5.8), $r(t-1)$ is given sequentially by

$$r(t-1) = r(t-2) + \phi(t-1)^T\phi(t-1); \qquad r(-1) = r_0 > 0 \qquad (8.5.13)$$

▼▼▼

We have presented the algorithm above in its most basic form in order to simplify the presentation. However, the basic convergence analysis technique is applicable to a wide range of algorithms. We shall describe presently a more practical form of the algorithm having improved convergence rate.

Preliminary properties of the algorithm are summarized in the following lemma:

Lemma 8.5.1. The following results hold for the algorithm (8.5.8) to (8.5.12):

(i) $\lim\limits_{N} \sum\limits_{t=1}^{N} \frac{\phi(t-1)^T\phi(t-1)}{r(t-1)r(t-2)} < \infty$ \qquad (8.5.14)

(ii) $e(t) = \frac{r(t-1)}{r(t-2)}\eta(t)$ \qquad (8.5.15)

where $e(t)$ and $\eta(t)$ are the *a priori* and *a posteriori errors*, respectively, and are given by

$$e(t) \triangleq y(t) - \hat{y}(t) = y(t) - \phi(t-1)^T\hat{\theta}(t-1) \qquad (8.5.16)$$
$$\eta(t) \triangleq y(t) - \bar{y}(t) = y(t) - \phi(t-1)^T\hat{\theta}(t) \qquad (8.5.17)$$

(iii) $C(q^{-1})z(t) = b(t)$ \qquad (8.5.18)

where

$$z(t) \triangleq \eta(t) - \omega(t) \qquad (8.5.19)$$
$$b(t) \triangleq -\phi(t-1)^T\tilde{\theta}(t) \qquad (8.5.20)$$
$$\tilde{\theta}(t) \triangleq \hat{\theta}(t) - \theta_0 \qquad (8.5.21)$$

(iv) $E\{b(t)\omega(t)|\mathcal{F}_{t-1}\} = -\dfrac{\phi(t-1)^T\phi(t-1)}{r(t-1)}\sigma^2$ $\qquad\qquad$ (8.5.22)

Proof. (i) $\displaystyle\sum_{t=1}^{\infty}\frac{\phi(t-1)^T\phi(t-1)}{r(t-1)r(t-2)} = \sum_{t=1}^{\infty}\frac{r(t-1)-r(t-2)}{r(t-1)r(t-2)}$

$$= \sum_{t=1}^{\infty}\frac{1}{r(t-2)} - \frac{1}{r(t-1)} \qquad (8.5.23)$$

$$\leq \frac{1}{r(-1)} < \infty; \qquad r(-1) = r_0$$

[Actually, this is a special case of the Pringsheim theorem (Knopp, 1956), which states that

$$\sum_{t=1}^{\infty}\frac{d_t}{D_t D_{t-1}^{\delta}} < \infty \qquad \text{for every } \delta > 0$$

where $d_t > 0$ and $D_t = \sum_{j=1}^{t} d_j$ (see Exercise 8.4).]

(ii) Multiplying (8.5.8) by $\phi(t-1)^T$ and then subtracting both sides from $y(t)$ gives the result.

(iii) Equation (8.5.1) can be written as

$$C(q^{-1})(y(t) - \omega(t)) = \alpha(q^{-1})y(t-1) + q^{-d}B(q^{-1})u(t) \qquad (8.5.24)$$

where $G(q^{-1}) = \alpha_1 + \alpha_2 q^{-1} + \cdots + \alpha_{\bar{n}}q^{-\bar{n}+1}$, $\bar{n} = \max(n, l)$ is given by the relation

$$C(q^{-1}) = A(q^{-1}) + q^{-1}\alpha(q^{-1}) \qquad (8.5.25)$$

Subtracting $C(q^{-1})\bar{y}(t)$ from both sides of (8.5.22) gives

$$C(q^{-1})(y(t) - \bar{y}(t) - \omega(t)) = \alpha(q^{-1})y(t-1) + q^{-d}B(q^{-1})u(t) - C(q^{-1})\bar{y}(t)$$

$$= \phi(t-1)^T\theta_0 - \bar{y}(t) \qquad (8.5.26)$$

$$= -\phi(t-1)^T\tilde{\theta}(t)$$

where we have used (8.5.10).

(iv) Subtracting θ_0 from (8.5.8) and multiplying by $\omega(t)\phi(t-1)^T$ gives

$$\omega(t)\phi(t-1)^T\tilde{\theta}(t) = \omega(t)\phi(t-1)^T\tilde{\theta}(t-1) + \frac{\phi(t-1)^T\phi(t-1)}{r(t-1)}$$

$$\times [e(t) - \omega(t)) + \omega(t)]\omega(t)$$

Taking conditional expectations, we immediately have (8.5.22).

$\blacktriangledown\blacktriangledown\blacktriangledown$

We can now establish the following properties of the algorithm (8.5.8) to (8.5.13).

Theorem 8.5.1. Subject to the noise assumptions, the system assumptions, and provided that

$$C(z^{-1}) \text{ is input strictly passive} \qquad \text{(see Appendix C)} \qquad (8.5.27)$$

the algorithm (8.5.8) ensures that with probability 1:

(i) Parameter norm convergence, that is,

$$\limsup_{t\to\infty} \|\hat{\theta}(t) - \theta_0\|^2 < \infty \qquad (8.5.28)$$

(ii) Parameter difference convergence, that is,

$$\lim_{N \to \infty} \sum_{t=1}^{N} \| \hat{\theta}(t) - \hat{\theta}(t - k) \|^2 < \infty \qquad \text{for any finite } k \qquad (8.5.29)$$

(iii) Normalized prediction error convergence, that is,

$$\lim_{N \to \infty} \sum_{t=1}^{N} \frac{(e(t) - \omega(t))^2}{r(t - 1)} < \infty \qquad (8.5.30)$$

Proof. The result is an elementary consequence of the Martingale convergence theorem (Appendix D). We make use of the following nonnegative Lyapunov function:

$$V(t) \triangleq \tilde{\theta}(t)^T \tilde{\theta}(t)$$

From (8.5.8) and using (8.5.15) and (8.5.16), we have

$$\tilde{\theta}(t) - \frac{\phi(t - 1)}{r(t - 2)} \eta(t) = \tilde{\theta}(t - 1)$$

Hence taking squares and using (8.5.20), we have

$$V(t) + \frac{2b(t)\eta(t)}{r(t - 2)} + \frac{\phi(t - 1)^T \phi(t - 1)}{r(t - 2)^2} \eta(t)^2 = V(t - 1)$$

or

$$V(t) = V(t - 1) - \frac{2b(t)z(t)}{r(t - 2)} - \frac{2b(t)\omega(t)}{r(t - 2)} - \frac{\phi(t - 1)^T \phi(t - 1)}{r(t - 2)^2} \eta(t)^2$$

Taking conditional expectations and using Lemma 8.5.1 yields

$$E\{V(t) \mid \mathfrak{F}_{t-1}\} = V(t - 1) - \frac{2}{r(t - 2)} E\{b(t)z(t) \mid \mathfrak{F}_{t-1}\}$$

$$- E\left\{\frac{\phi(t - 1)^T \phi(t - 1)}{r(t - 2)^2} \eta(t)^2 \mid \mathfrak{F}_{t-1}\right\} \qquad (8.5.31)$$

$$+ \frac{2\phi(t - 1)^T \phi(t - 1)}{r(t - 2)r(t - 1)} \sigma^2 \quad \text{a.s.}$$

Now define

$$S(t) = 2 \sum_{j=1}^{t} \left[b(j)z(j) - \frac{\rho}{2} z(j)^2 \right] + K$$

where $0 < \rho < \infty$, $K < \infty$. Now $b(t)$ is related to $z(t)$ by the transfer function $C(z)$, which is input strictly passive (Appendix C); it therefore follows that $\rho > 0$ can be chosen such that $S(t) \geq 0$.

Also define

$$X(t) = V(t) + \frac{S(t)}{r(t - 2)} + \rho \sum_{j=1}^{t} \frac{z(j)^2}{r(j - 2)} + \sum_{j=1}^{t} \frac{\phi(j - 1)^T \phi(j - 1)}{r(j - 2)^2} \eta(j)^2 \qquad (8.5.32)$$

Hence from (8.5.31) and the monotonicity of $r(t)$,

$$E\{X(t) \mid \mathfrak{F}_{t-1}\} \leq X(t) + \frac{2\phi(t - 1)^T \phi(t - 1)}{r(t - 2)r(t - 1)} \sigma^2 \quad \text{a.s.} \qquad (8.5.33)$$

Noting (8.5.14), we can immediately apply the Martingale convergence theorem to conclude that

$$X(t) \longrightarrow X < \infty \quad \text{a.s.} \qquad (8.5.34)$$

Using (8.5.32) and (8.5.34), this establishes (i) and also we may conclude that

$$\lim_{N \to \infty} \sum_{t=1}^{N} \frac{z(t)^2}{r(t-2)} < \infty \quad \text{a.s.} \tag{8.5.35}$$

$$\lim_{N \to \infty} \sum_{t=1}^{N} \frac{\phi(t-1)^T \phi(t-1)}{r(t-2)^2} \eta(t)^2 < \infty \quad \text{a.s.} \tag{8.5.36}$$

From (8.5.8),

$$\|\hat{\theta}(t) - \hat{\theta}(t-1)\|^2 = \frac{\phi(t-1)^T \phi(t-1)}{r(t-2)^2} \eta(t)^2$$

(ii) This follows from (8.5.36) using the Schwarz inequality.

(iii) From (8.5.8) and using Lemma 8.5.1, part (ii), we have

$$\hat{\theta}(t) - \frac{\phi(t-1)}{r(t-2)} \eta(t) = \hat{\theta}(t-1)$$

Hence

$$[y(t) - \phi(t-1)^T \hat{\theta}(t) - \omega(t)] + \frac{\phi(t-1)^T \phi(t-1)}{r(t-2)} \eta(t)$$

$$= [y(t) - \phi(t-1)^T \hat{\theta}(t-1) - \omega(t)]$$

or

$$z(t) + \frac{\phi(t-1)^T \phi(t-1)}{r(t-2)} \eta(t) = e(t) - \omega(t)$$

Using the Schwarz inequality we conclude that

$$[e(t) - \omega(t)]^2 \leq 2z(t)^2 + \frac{2[\phi(t-1)^T \phi(t-1)]^2}{r(t-2)^2} \eta(t)^2$$

Thus, using (8.5.13),

$$\frac{[e(t) - \omega(t)]^2}{r(t-1)} \leq \frac{2z(t)^2}{r(t-1)} + \frac{2\phi(t-1)^T \phi(t-1)}{r(t-2)^2} \eta(t)^2$$

Equation (8.5.30) then follows from (8.5.35) and (8.5.36).

▼▼▼

The proof technique above is very general and can be applied to other algorithms, as we shall see presently.

Note that the parameter norm convergence and parameter difference convergence established in the theorem above are weaker than parameter convergence with probability 1. However, they suffice to show that the estimates remain bounded with probability 1 and that the estimates at successive time points approach each other asymptotically. The normalized prediction error convergence result will later allow us to establish the performance of the algorithms in adaptive filtering, prediction, and control under relatively weak assumptions. This property is the stochastic analog of the property given in Lemma 3.3.2 [equation 3.3.23)] for the projection algorithm in the deterministic case (Chapter 3).

The term $e(t) - \omega(t)$ appearing in the normalized predictor error is given by [using (7.4.24) of Lemma 7.4.1]

$$e(t) - \omega(t) = y(t) - \hat{y}(t) - \omega(t)$$
$$= y^0(t \,|\, t-1) - \hat{y}(t)$$

Thus $[e(t) - \omega(t)]$ is a measure of the difference between the optimal prediction and the prediction computed by the sequential algorithm.

We shall repeatedly generate results on parameter norm convergence, parameter difference convergence, and normalized prediction error convergence for the various algorithms that we study. We shall subsequently apply these results in adaptive filtering, prediction, and control.

The convergence proof above has been structured so as to highlight the common ground between the stochastic and deterministic cases. It can be seen that there is a close analogy between Theorem 3.5.1 and Theorem 8.5.1 in terms of the results obtained and the proof technique employed.

Remark 8.5.1. The crucial role of positive realness (passivity) for pseudo linear regression algorithms was first pointed out by Ljung (1977b). The use of the normalization factor, $r(t - 1)$, to accommodate possibly unbounded data is also crucial for the convergence analysis presented. The importance of such a normalization was first pointed out by Goodwin, Ramadge, and Caines (1978b). In their work a priori predictions were used in the regression vector. (The reader is asked to establish corresponding results for this algorithm in Exercise 8.5. Here we have used a posteriori prediction in the regression vector. This allows us to develop a theory within a common framework for both deterministic and stochastic cases. We shall also see later that this common framework can also be extended in a natural way to cover least-squares algorithms.

8.5.2 The Least-Squares Form of the Pseudo Linear Regression Algorithm

Consider again the ARMAX model of (8.5.1) to (8.5.4) with the noise assumptions (8.5.5) to (8.5.7) and system assumptions. Our objective here will be to establish convergence properties for the following algorithm (see Section 8.4):

Pseudo Linear Regression Algorithm (with A Posteriori Predictions)

$$\hat{\theta}(t) = \hat{\theta}(t - 1) + \frac{P(t - 2)\phi(t - 1)}{1 + \phi(t - 1)^T P(t - 2)\phi(t - 1)}[y(t) - \phi(t - 1)^T \hat{\theta}(t - 1)]$$

(8.5.37)

where $\phi(t - 1)$ is as in (8.5.10) and $P(t)$ is given recursively by

$$P(t - 1) = P(t - 2) - \frac{P(t - 2)\phi(t - 1)\phi(t - 1)^T P(t - 2)}{1 + \phi(t - 1)^T P(t - 2)\phi(t - 1)}, \qquad P(-1) > 0$$

(8.5.38)

Preliminary properties of the algorithm are summarized in:

Lemma 8.5.2

(i) $\displaystyle \lim_{N} \sum_{t=1}^{N} \frac{\phi(t - 1)^T P(t - 1)\phi(t - 1)}{r(t - 2)} < \infty$ (8.5.39)

where

$$r(t - 1) = r(t - 2) + \phi(t - 1)^T \phi(t - 1), \ r(-1) = \text{trace } P(-1)^{-1}$$ (8.5.40)

(ii) $\eta(t) = \dfrac{e(t)}{1 + \phi(t-1)^T P(t-2)\phi(t-1)}$ (8.5.41)

where

$$e(t) \triangleq y(t) - \hat{y}(t) = y(t) - \phi(t-1)^T \hat{\theta}(t-1) \tag{8.5.42}$$

$$\eta(t) \triangleq y(t) - \bar{y}(t) \tag{8.5.43}$$

$$\bar{y}(t) = \phi(t-1)^T \hat{\theta}(t) \tag{8.5.44}$$

(iii) $C(q^{-1})z(t) = b(t)$ (8.5.45)

where

$$z(t) \triangleq \eta(t) - \omega(t) \tag{8.5.46}$$

$$b(t) \triangleq -\phi(t-1)^T \tilde{\theta}(t) \tag{8.5.47}$$

$$\tilde{\theta}(t) \triangleq \hat{\theta}(t) - \theta_0 \tag{8.5.48}$$

(iv) $E\{b(t)\omega(t)\,|\,\mathfrak{F}_{t-1}\} = -\dfrac{\phi(t-1)^T P(t-2)\phi(t-1)}{1 + \phi(t-1)^T P(t-2)\phi(t-1)}\sigma^2$

$$= -\phi(t-1)^T P(t-1)\phi(t-1)\sigma^2 \tag{8.5.49}$$

Proof. (Along the same lines as Lemma 8.5.1). Part (i) is proved as follows:

$$\frac{\phi(t-1)^T P(t-1)\phi(t-1)}{r(t-2)} = \frac{\phi(t-1)^T P(t-2)\phi(t-1)}{r(t-2)[1 + \phi(t-1)^T P(t-2)\phi(t-1)]}$$

$$\leq \frac{\phi(t-1)^T P(t-2)^2\phi(t-1)}{1 + \phi(t-1)^T P(t-2)\phi(t-1)}$$

$$= \phi(t-1)^T P(t-1)P(t-2)\phi(t-1)$$

$$= \text{trace } P(t-2)\phi(t-1)\phi(t-1)^T P(t-1)$$

$$= \text{trace } P(t-2) - \text{trace } P(t-1) \qquad \text{using (8.5.38)}$$

Summing both sides gives the result.

Parts (ii), (iii), and (iv) are as in Lemma 8.5.1.

▼▼▼

We can now establish the following properties of the algorithm (8.5.37)–(8.5.38).

Theorem 8.5.2. Subject to the noise assumptions, the system assumptions, and provided that:

1. *Passivity assumption:*

$$\left[\frac{1}{C(z)} - \frac{1}{2}\right] \text{ is very strictly passive} \qquad \text{(see Appendix C)} \tag{8.5.50}$$

then the algorithm (8.5.37)–(8.5.38) ensures that with probability 1:
(i) Parameter norm convergence

(a) $\lim\limits_{t\to\infty} \sup \dfrac{\tilde{\theta}(t)^T P(t-1)^{-1}\tilde{\theta}(t)}{r(t-1)} < \infty$

(b) $\lim\limits_{N\to\infty} \sum\limits_{t=1}^{N} \dfrac{\phi(t-1)^T\phi(t-1)}{r(t-2)} \dfrac{\tilde{\theta}(t)^T P(t-1)^{-1}\tilde{\theta}(t)}{r(t-1)} < \infty$

If, in addition,

2. Bounded condition number assumption

$$\lim_{N\to\infty} \sup \frac{\lambda_{max}P(N)}{\lambda_{min}P(N)} < \infty \tag{8.5.51}$$

then

(ii) Parameter difference convergence

$$\lim_{N\to\infty} \sum_{t=1}^{N} \|\hat{\theta}(t) - \hat{\theta}(t-k)\|^2 < \infty; \qquad t \geq k \tag{8.5.52}$$

for any finite k.

(iii) Normalized prediction error convergence

$$\lim_{N\to\infty} \sum_{t=1}^{N} \frac{(e(t) - \omega(t))^2}{r(t-1)} < \infty \tag{8.5.53}$$

Proof. (Along the same lines as the proof of Theorem 8.5.1). From (8.5.37) and using (8.5.41), (8.5.42), and (8.5.48), we have

$$\tilde{\theta}(t) - P(t-2)\phi(t-1)\eta(t) = \tilde{\theta}(t-1) \tag{8.5.54}$$

Thus

$$\tilde{\theta}(t)^T P(t-2)^{-1}\tilde{\theta}(t) - 2\phi(t-1)^T\tilde{\theta}(t)\eta(t) + \phi(t-1)^T P(t-2)\phi(t-1)\eta(t)^2$$
$$= \tilde{\theta}(t-1)^T P(t-2)^{-1}\tilde{\theta}(t-1)$$

Define $V(t) \triangleq \tilde{\theta}(t)^T P(t-1)^{-1}\tilde{\theta}(t)$, and using (8.5.38) and (8.5.47),

$$V(t) = V(t-1) + b(t)^2 - 2b(t)\eta(t) - \phi(t-1)^T P(t-2)\phi(t-1)\eta(t)^2$$
$$= V(t-1) + b(t)^2 - 2b(t)z(t) - 2b(t)\omega(t) - \phi(t-1)^T P(t-2)\phi(t-1)\eta(t)^2$$

Taking conditional expectation and using Lemma 8.5.2 gives us

$$E\{V(t)|\mathcal{F}_{t-1}\} = V(t-1) + E\{b(t)^2 - 2b(t)z(t)|\mathcal{F}_{t-1}\}$$
$$+ 2\phi(t-1)^T P(t-1)\phi(t-1)\sigma^2 \tag{8.5.55}$$
$$- E\{\phi(t-1)^T P(t-2)\phi(t-1)\eta(t)^2|\mathcal{F}_{t-1}\}$$

Define

$$g(t) \triangleq z(t) - \tfrac{1}{2}b(t) \tag{8.5.56}$$

We then note from (8.5.45) that $g(t)$ is related to $b(t)$ by the transfer function $[1/C(z) - \tfrac{1}{2}]$. Dividing (8.5.55) by $r(t-2)$ and using (8.5.56) gives

$$\frac{r(t-1)}{r(t-2)}E\left\{\frac{V(t)}{r(t-1)}\bigg|\mathcal{F}_{t-1}\right\} = \frac{V(t-1)}{r(t-2)} - \frac{2}{r(t-2)}E\{b(t)g(t)|\mathcal{F}_{t-1}\}$$
$$- E\left\{\frac{\phi(t-1)^T P(t-2)\phi(t-1)}{r(t-2)}\eta(t)^2|\mathcal{F}_{t-1}\right\} \tag{8.5.57}$$
$$+ \frac{2\phi(t-1)^T P(t-1)\phi(t-1)}{r(t-2)}\sigma^2 \quad \text{a.s.}$$

Now define

$$S(t) = 2\sum_{j=1}^{t}\left[b(j)g(j) - \frac{\rho_1}{2}g(j)^2 - \frac{\rho_2}{2}b(j)^2\right] + K$$

where $0 < \rho_1 < \infty, 0 < \rho_2 < \infty$, and $0 \leq K < \infty$. Now, as we have observed above, $g(t)$ is related to $b(t)$ by the transfer function $[1/C(z) - \tfrac{1}{2}]$ and (by assumption) this transfer function is very strictly passive. It follows that p_1 and p_2 can be chosen (non zero) such that $S(t) \geq 0$.

Also define

$$X(t) = \frac{V(t)}{r(t-1)} + \frac{S(t)}{r(t-2)} + p_1 \sum_{j=1}^{t} \frac{g(j)^2}{r(j-2)} + p_2 \sum_{j=1}^{t} \frac{b(j)^2}{r(j-2)}$$

$$+ \sum_{j=1}^{t} \frac{\phi(j-1)^T P(j-2)\phi(t-1)}{r(j-2)} \eta(j)^2 \qquad (8.5.58)$$

$$+ \sum_{j=1}^{t} \frac{\phi(j-1)^T \phi(j-1)}{r(j-2)} \frac{V(j)}{r(j-1)}$$

Hence from (8.5.57),

$$E\{X(t)|\mathcal{F}_{t-1}\} \leq X(t-1) + \frac{\phi(t-1)^T P(t-1)\phi(t-1)}{r(t-2)} \sigma^2 \quad \text{a.s.}$$

Noting (8.5.39), we can again immediately apply the Martingale convergence theorem (Appendix D) to conclude that

$$X(t) \longrightarrow X < \infty \quad \text{a.s.}$$

Then using (8.5.58) this establishes parts (a) and (b) and

$$\lim_{N \to \infty} \sum_{t=1}^{N} \frac{g(t)^2}{r(t-2)} < \infty \quad \text{a.s.} \qquad (8.5.59)$$

$$\lim_{N \to \infty} \sum_{t=1}^{N} \frac{b(t)^2}{r(t-2)} < \infty \quad \text{a.s.} \qquad (8.5.60)$$

$$\lim_{N \to \infty} \sum_{t=1}^{N} \frac{\phi(t-1)^T P(t-2)\phi(t-1)}{r(t-2)} \eta(t)^2 < \infty \quad \text{a.s.} \qquad (8.5.61)$$

From (8.5.56) and using (8.5.59) and (8.5.60) and the Schwarz inequality, we have

$$\lim_{N \to \infty} \sum_{t=1}^{N} \frac{z(t)^2}{r(t-2)} < \infty \quad \text{a.s.} \qquad (8.5.62)$$

(ii) From (8.5.37),

$$\hat{\theta}(t) = \hat{\theta}(t-1) + P(t-2)\phi(t-1)\eta(t)$$

Hence

$$\|\hat{\theta}(t) - \hat{\theta}(t-1)\|^2 = \phi(t-1)^T P(t-2)^2 \phi(t-1)\eta(t)^2 \qquad (8.5.63)$$

$$\sum_{t=1}^{N} \|\hat{\theta}(t) - \hat{\theta}(t-1)\|^2 \leq K_1 \sum_{t=1}^{N} \frac{\phi(t-1)^T P(t-2)\phi(t-1)}{r(t-2)} \eta(t)^2 \qquad \text{using (8.5.51)}$$

where $0 \leq K_1 < \infty$.

Thus (8.5.52) follows by using (8.5.61) and the Schwarz inequality.

(iii) From (8.5.54), we have [after multiplying by $\phi(t-1)^T$ and then subtracting from $(y(t) - \omega(t))$]

$$[y(t) - \phi(t-1)^T \hat{\theta}(t) - \omega(t)] + \phi(t-1)^T P(t-2)\phi(t-1)\eta(t)$$
$$= [y(t) - \phi(t-1)^T \hat{\theta}(t-1) - \omega(t)]$$

or

$$z(t) + \phi(t-1)^T P(t-2)\phi(t-1)\eta(t) = e(t) - \omega(t)$$

Using Schwarz inequality,

$$(e(t) - \omega(t))^2 \leq 2z(t)^2 + 2[\phi(t-1)^T P(t-2)\phi(t-1)]^2 \eta(t)^2 \qquad (8.5.64)$$

Now $[\phi(t-1)^T P(t-2)\phi(t-1)]^2 \leq [\phi(t-1)^T\phi(t-1)][\phi(t-1)^T P(t-2)^2\phi(t-1)]$
using the triangle inequality. Hence from (8.5.64),

$$\frac{[e(t)-\omega(t)]^2}{r(t-1)} \leq \frac{2z(t)^2}{r(t-1)} + \frac{2\phi(t-1)^T\phi(t-1)}{r(t-1)}\phi(t-1)^T P(t-2)^2\phi(t-1)\eta(t)^2$$

Thus

$$\sum_{t=1}^{N}\frac{[e(t)-\omega(t)]^2}{r(t-1)} \leq \sum_{t=1}^{N}\frac{2z(t)^2}{r(t-1)} + 2K_1\sum_{t=1}^{N}\frac{\phi(t-1)^T\phi(t-1)}{r(t-1)}$$

$$\frac{\phi(t-1)^T P(t-2)\phi(t-1)}{r(t-2)}\eta(t)^2 \qquad \text{using (8.5.51)}$$

Hence the result follows from (8.5.61) and (8.5.62).

▼▼▼

Remark 8.5.2. Convergence properties (8.5.59) to (8.5.62) for the *a posteriori error* quantities are immediate from the analysis. However, to establish convergence for the *a priori error* is more difficult and requires the stronger assumption of bounded condition number. Theorem 8.5.2 therefore reveals a potential problem with the least-squares-type algorithm. In practice, of course, numerical difficulty will also arise when the condition number gets large. There are a number of schemes for overcoming the possibility of unboundedness of the condition number of the P matrix. One such scheme (Sin and Goodwin, 1982) is to replace the standard P update (8.5.38) by the following *condition number monitoring* scheme:

$$P'(t-1) = P(t-2) - \frac{P(t-2)\phi(t-1)\phi(t-1)^T P(t-2)}{1+\phi(t-1)^T P(t-2)\phi(t-1)}, \qquad P(-1) > 0$$
(8.5.65)

$$\bar{r}(t-1) = \bar{r}(t-2)(1+\phi(t-1)^T P(t-2)\phi(t-1)), \qquad \bar{r}(-1) > 0 \qquad (8.5.66)$$

If

$$(\bar{r}(t-1)\lambda_{max}P'(t-1) \leq K \qquad 0 < K < \infty \qquad (8.5.67)$$

Then

$$P(t-1) = P'(t-1) \qquad (8.5.68)$$

Otherwise,

$$P(t-1) = \frac{K}{\bar{r}(t-1)\lambda_{max}P'(t-1)}P'(t-1) \qquad (8.5.69)$$

[Note that $\lambda_{max}P'(t-1)$ is bounded by trace $P(t-1)$, which is easily computed.]
   With the scheme above, one can establish the conclusions of Theorem 8.5.2 without the need for the additional assumption (8.5.51). (See Exercises 8.16 to 8.20.)

▼▼▼

8.5.3 The Stochastic Key Technical Lemma

In the subsections above it has been shown that a key property of many sequential parameter estimation algorithms is the following normalized prediction error convergence result:

$$\lim_{N\to\infty}\sum_{t=1}^{N}\frac{[e(t)-v(t)]^2}{r(t-1)} < \infty \qquad \text{a.s.} \qquad (8.5.70)$$

where for some increasing sequence of σ-algebras $\mathfrak{F}_{t-1}$

1. $E\{v(t)|\mathfrak{F}_{t-1}\} = 0$ a.s. $\hspace{5cm}$ (8.5.71)
2. $E\{v(t)^2|\mathfrak{F}_{t-1}\} = \gamma^2$ a.s. $\hspace{4cm}$ (8.5.72)
3. $\limsup\limits_{N\to\infty} \dfrac{1}{N}\sum\limits_{t=1}^{N} v(t)^2 < \infty$ a.s. $\hspace{3cm}$ (8.5.73)
4. $e(t) = y(t) - \hat{y}(t)$ $\hspace{5.5cm}$ (8.5.74)
   where $\hat{y}(t)$ and $(y(t) - v(t))$ is $\mathfrak{F}_{t-1}$ measurable.
5. $\{r(t-1)\}$ is a nondecreasing nonnegative sequence such that $r(t-1)$ is $\mathfrak{F}_{t-1}$ measurable.

In many applications it is desirable to replace the normalization factor $r(t-1)$ in (8.5.70) by a simple function of time. This is often possible using the following stochastic version of the key technical lemma of Section 6.2.

Lemma 8.5.3 (The Stochastic Key Technical Lemma).   If condition (8.5.70) holds (together with properties 1 to 5), and if there exists constants K_1, K_2, and $\bar{N}$ $[0 \le K_1 < \infty, 0 < K_2 < \infty, 0 < \bar{N} < \infty]$ such that

$$\frac{1}{N}r(N-1) \le K_1 + \frac{K_2}{N}\sum_{t=1}^{N}[e(t)-v(t)]^2, \qquad N \ge \bar{N} \quad \text{a.s.} \qquad (8.5.75)$$

then

(1) $\lim\limits_{N\to\infty} \dfrac{1}{N}\sum\limits_{t=1}^{N}[e(t)-v(t)]^2 = 0$ a.s. $\hspace{3cm}$ (8.5.76)

(2) $\limsup\limits_{N\to\infty} \dfrac{1}{N}r(N-1) < \infty$ a.s. $\hspace{3.5cm}$ (8.5.77)

(3) $\lim\limits_{N\to\infty} \dfrac{1}{N}\sum\limits_{t=1}^{N} E\{[y(t)-\hat{y}(t)]^2|\mathfrak{F}_{t-1}\} = \gamma^2$ a.s. $\hspace{1.5cm}$ (8.5.78)

If assumption (8.5.73) is strengthened to

$$E\{v(t)^4|\mathfrak{F}_{t-1}\} < \infty \quad \text{a.s.} \qquad (8.5.79)$$

then, in addition,

(4) $\lim\limits_{N\to\infty} \dfrac{1}{N}\sum\limits_{t=1}^{N}[y(t)-\hat{y}(t)]^2 = \gamma^2$ a.s. $\hspace{2.5cm}$ (8.5.80)

Proof. (1) If $r(t-1) < K_3 < \infty$, then (8.5.70) implies

$$\lim_{N\to\infty} \frac{1}{K_3}\sum_{t=1}^{N}[e(t)-v(t)]^2 < \infty \quad \text{a.s.} \qquad (8.5.81)$$

and (8.3.76) follows trivially.

Alternatively, if $r(t-1)$ is unbounded, then since the sum in (8.5.70) is nondecreasing, we can apply Kronecker's lemma (Appendix D) to conclude that

$$\lim_{N\to\infty} \frac{N}{r(N-1)}\frac{1}{N}\sum_{t=1}^{N}[e(t)-v(t)]^2 = 0 \quad \text{a.s.} \qquad (8.5.82)$$

Substituting (8.5.75) into (8.5.82) gives

$$\lim_{N \to \infty} \frac{\frac{1}{N} \sum_{t=1}^{N} [e(t) - v(t)]^2}{K_1 + \frac{K_2}{N} \sum_{t=1}^{N} [e(t) - v(t)]^2} = 0 \quad \text{a.s.} \tag{8.5.83}$$

Equation (8.3.76) again follows immediately.

(2) Equation (8.5.77) follows from (8.5.75) and (8.5.76).

(3) Note that

$$
\begin{aligned}
E\{[y(t) - \hat{y}(t)]^2 \,|\, \mathcal{F}_{t-1}\} &= E\{[(y(t) - \hat{y}(t) - v(t)) + v(t)]^2 \,|\, \mathcal{F}_{t-1}\} \\
&= E\{[y(t) - \hat{y}(t) - v(t)]^2 + 2[y(t) - \hat{y}(t) - v(t)]v(t) \\
&\quad + v(t)^2 \,|\, \mathcal{F}_{t-1}\}
\end{aligned} \tag{8.5.84}
$$

Since $y(t) - v(t)$ and $\hat{y}(t)$ are $\mathcal{F}_{t-1}$ measurable and using (8.5.71), we have

$$E\{(y(t) - \hat{y}(t))^2 \,|\, \mathcal{F}_{t-1}\} = [e(t) - v(t)]^2 + E\{v(t)^2 \,|\, \mathcal{F}_{t-1}\} \tag{8.5.85}$$

Equation (8.5.78) now follows from (8.5.72), (8.5.76), and (8.5.85).

(4) Here we argue as in Lafortune (1982):

$$
\begin{aligned}
\lim_{N \to \infty} \frac{1}{N} \sum_{t=1}^{N} [y(t) - \hat{y}(t)]^2 \\
= \lim_{N \to \infty} \frac{1}{N} \sum_{t=1}^{N} [y(t) - \hat{y}(t) - v(t) + v(t)]^2 \\
= \lim_{N \to \infty} \frac{1}{N} \sum_{t=1}^{N} \{[y(t) - \hat{y}(t) - v(t)]^2 + v(t)^2 + 2[y(t) - \hat{y}(t) - v(t)]v(t)\} \\
= \gamma^2 + \lim_{N \to \infty} \frac{2}{N} \sum_{t=1}^{N} [y(t) - \hat{y}(t) - v(t)]v(t)
\end{aligned} \tag{8.5.86}
$$

using (8.5.72), (8.5.76), and (8.5.79) together with Lemma D.5.2 of Appendix D. Now consider

$$\sum_{t=1}^{N} \frac{1}{t} [y(t) - \hat{y}(t) - v(t)]v(t)$$

Clearly from (8.5.70), (8.5.72), and (8.5.77), we have

$$\sum_{t=1}^{N} \frac{1}{t^2} E\{[y(t) - \hat{y}(t) - v(t)]^2 v(t)^2 \,|\, \mathcal{F}_{t-1}\} < \infty$$

Hence from Lemma D.5.1,

$$\lim_{N \to \infty} \frac{1}{N} \sum_{t=1}^{N} [y(t) - \hat{y}(t) - v(t)]v(t) = 0 \tag{8.5.87}$$

Substituting (8.5.87) into (8.5.86) gives (8.5.80).

▼▼▼

8.5.4 The ODE Approach to the Analysis of Sequential Algorithms

In Sections 8.5.1 and 8.5.2 we have used the Martingale convergence theorem to establish the convergence properties. This is an extremely powerful method and relies on relatively weak assumptions. For example, in a later chapter, when we discuss stochastic adaptive control, we will show that the Martingale approach allows a simultaneous proof of closed-loop stability and convergence. However, as the name suggests, the Martingale approach relies on a Martingale-type property and it is not always clear how this property can be extracted from a general sequential algorithm. Therefore, in this subsection we present an alternative approach to the analysis of sequential algorithms. This approach relies on relating the asymptotic trajectories of the algorithm to the solutions of an ordinary differential equation (ODE). The method has been called the *ODE approach*.

The ODE approach is very widely applicable, but its very generality is not without limitations. The key restriction is that it relies on stringent regularity conditions, including boundedness of the data. Thus it does not strictly apply to certain problems (e.g., adaptive control), where boundedness cannot be a priori verified. Not withstanding these limitations, the ODE approach is a powerful tool which can give important insights into the performance of iterative algorithms. It can often distinguish between good and bad algorithms and can suggest improvements to existing algorithms. An example of this is the *improved extended Kalman filter* discussed later.

Also, the ODE analysis can often be used as a substitute for stochastic simulation studies.

The proof of the ODE approach is rather formidable. However, in view of its importance, we shall give a heuristic discussion of the result together with a statement, without proof, of the formal theorem.

Heuristic Discussion

In many sequential algorithms, the updating formula typically has the form

$$\hat{\theta}(t) = \hat{\theta}(t-1) + \gamma(t)R(t)^{-1}\psi(t)\epsilon(t) \tag{8.5.88}$$

where $\hat{\theta}(t)$ denotes the estimate at time t

$\gamma(t)$ denotes a gain sequence [typically, $\gamma(t) = 1/t$]

$\psi(t)$ denotes a regression vector (or alternatively a gradient search direction)

$\epsilon(t)$ denotes a prediction error

The matrix $R(t)$ allows for the possibility of a Newton step, in which case $R(t)$ is chosen as

$$R(t) = R(t-1) + \gamma(t)[\psi(t)\psi(t)^T - R(t-1)] \tag{8.5.89}$$

A typical example of the algorithm above is given in (8.3.8), (8.3.15)–(8.3.16), where

$$\gamma(t) = \frac{1}{t} \tag{8.5.90}$$

$$R(t) = \ddot{V}_t \tag{8.5.91}$$

$$\epsilon(t) = y(t) - \hat{y}(t, \hat{\theta}(t-1)) \tag{8.5.92}$$

with $\hat{y}(t, \hat{\theta}(t-1))$ being the predicted output based on $\hat{\theta}(t-1)$.

The variables $\psi(t)$, $\epsilon(t)$, and $R(t)$ are in general functions of $\hat{\theta}(t-1)$, $\hat{\theta}(t-2)$, ... via a set of dynamic equations. A property of algorithms of the form (8.5.88) is ideally $\hat{\theta}(t)$ approaches $\hat{\theta}(t-1)$ asymptotically (subject to regularity conditions such as stability, stationarity, etc.). Thus for asymptotic analysis, it makes sense to approximate $\psi(t)$, $\epsilon(t)$ by $\psi(t, \theta)$, and $\epsilon(t, \theta)$, where θ is a fixed nominal value of $\hat{\theta}(t)$. Similarly, it makes sense to replace $R(t)$ by a nominal value $R(\theta)$ (actually, R is the average information matrix and would asymptotically be expected to be a function of θ and independent of t).

Then, on average, the asymptotic updating direction in (8.5.88) is given by

$$h(\theta) = R(\theta)^{-1}f(\theta) \tag{8.5.93}$$

where

$$f(\theta) = E\{\psi(t, \theta)\epsilon(t, \theta)\} \tag{8.5.94}$$

In the expression above, we have assumed that the expectation does not depend on t. A more general definition of $f(\theta)$ is

$$f(\theta) = \bar{E}\{\psi(t, \theta)\epsilon(t, \theta)\} \triangleq \lim_{N \to \infty} \frac{1}{N} \sum_{t=1}^{N} E\{\psi(t, \theta)\epsilon(t, \theta)\} \tag{8.5.95}$$

(existence of the limit requires stationarity conditions on the underlying stochastic process).

Similarly, from (8.5.89) the average updating direction for R is

$$H(\theta) = G(\theta) - R(\theta) \tag{8.5.96}$$

where

$$G(\theta) = \bar{E}\{\psi(t, \theta)\psi(t, \theta)^T\} \tag{8.5.97}$$

Now returning to (8.5.88), we look at the adjustment of $\hat{\theta}$ over an interval of length s in order that we can replace the individual updating directions by an average value (in some sense)

$$
\begin{aligned}
\hat{\theta}(t + s) &= \hat{\theta}(t) + \sum_{k=t}^{t+s-1} \gamma(k)R(k)^{-1}\psi(k)\epsilon(k) \\
&\simeq \hat{\theta}(t) + \sum_{k=t}^{t+s-1} \gamma(k)R(\theta)^{-1}[E\{\psi(k, \theta)\epsilon(k, \theta)\} + \omega(k)]
\end{aligned}
\tag{8.5.98}
$$

where $\omega(k)$ denotes a zero mean random variable. From (8.5.98) we have

$$
\begin{aligned}
\hat{\theta}(t + s) &\simeq \hat{\theta}(t) + \left[\sum_{k=t}^{t+s-1} \gamma(k)\right] R(\theta)^{-1}f(\theta) + \sum_{k=t}^{t+s-1} \gamma(k)R(\theta)^{-1}\omega(k) \\
&\simeq \hat{\theta}(t) + \left[\sum_{k=t}^{t+s-1} \gamma(k)\right] R(\theta)^{-1}f(\theta)
\end{aligned}
\tag{8.5.99}
$$

Now defining the compressed time scales:

$$\tau = \sum_{k=1}^{t} \gamma(k) \tag{8.5.100}$$

$$\Delta\tau = \sum_{k=t}^{t+s-1} \gamma(k) \tag{8.5.101}$$

and mapping $\hat{\theta}(t)$ into $\theta(\tau)$, we obtain

$$\theta(\tau + \Delta\tau) \simeq \theta(\tau) + \Delta\tau R(\theta)^{-1}f(\theta) \tag{8.5.102}$$

Asymptotically, when $\Delta\tau$ becomes small, we have

$$\frac{d}{d\tau}\theta(\tau) \simeq R(\tau)^{-1}f(\theta(\tau)) \tag{8.5.103}$$

Similarly, from (8.5.89), we obtain

$$\frac{d}{d\tau}R(\tau) \simeq G(\theta(\tau)) - R(\tau) \tag{8.5.104}$$

Thus it can be heuristically seen that the asymptotic trajectories of the algorithm (8.5.88)–(8.5.89) are related to the solutions of the ordinary differential equations (8.5.103) and (8.5.104). In fact, it can be shown (Ljung, 1977a, 1977b) that stability of the differential equations will imply convergence a.s. for the algorithm (8.5.88)–(8.5.89). Moreover, only stable, stationary points of (8.5.103) and (8.5.104) are possible convergence points for the algorithm.

Note that in many applications, the ODE need not be formally derived. Since the right-hand side of the ODE is the average updating direction, this can sometimes be derived by evaluating directly certain covariances between inputs and outputs.

Formal Results

Subject to certain smoothness assumptions (Ljung, 1977a), we have

Theorem 8.5.3. Let D_s denote the stability domain for $\hat{\theta}(t)$ such that the dynamical systems giving rise to $\psi(t)$ and $\epsilon(t)$ are stable. Subject to:

(a) *Boundedness condition:* There exists a random variable C such that

$$\hat{\theta}(t) \in D_s \text{ and } |\psi(t)| < C \text{ infinitely often a.s.} \qquad (8.5.105)$$

(b) *Lyapunov condition:* There exists a positive twice differentiable function V (a function of Q and R) whose time derivative along the solutions of (8.5.103)–(8.5.104) satisfies

$$\frac{d}{d\tau} V \le 0; \quad \theta \in D_s, \quad R > 0 \quad \left[\text{where } \frac{d}{d\tau} V = \frac{\partial V}{\partial \theta} R^{-1} f + \frac{\partial V}{\partial R} (G - R) \right]$$
$$(8.5.106)$$

then either

(i) $\hat{\theta}(t) \longrightarrow D_c$ a.s. $\qquad (8.5.107)$
where

$$D_c = \left\{ \theta, R \mid \theta \in D_s \text{ and } \frac{d}{d\tau} V = 0 \right\} \qquad (8.5.108)$$

or

(ii) $\{\hat{\theta}(t)\}$ has a cluster point on the boundary of D_s.

Proof. See Ljung (1977a).

▼▼▼

Theorem 8.5.4. If $\hat{\theta}(t)$ converges to θ^* and $R(t)$ converges to R^* (positive definite) with nonzero probability, then

$$f(\theta^*) = 0 \quad \text{and} \quad G(\theta^*) = R^* \qquad (8.5.109)$$

and the matrix

$$H(\theta^*) = (R^*)^{-1} \frac{d}{d\theta} f(\theta) \Big|_{\theta = \theta^*} \qquad (8.5.110)$$

must have all its eigenvalues in the closed left half-plane.

Proof. See Ljung (1977a).

▼▼▼

The theorem above states that the algorithm can only converge to locally stable stationary points of the ODE. This result can sometimes be used to prove failure of convergence.

Theorem 8.5.5. The trajectories of the ODE (8.5.103)–(8.5.104) are the asymptotic paths of the estimates generated by the algorithms (8.5.88) and (8.5.89).

Proof. See Ljung (1977a).

▼▼▼

The result above says that the ODE effectively describes the asymptotic expected value of the trajectories of the sequential algorithm. Numerical solutions of (8.5.103)–(8.5.104) may therefore be an alternative to simulating the algorithm. An advantage of this approach is that time scaling (8.5.100) allows the ODE to reveal more rapidly the asymptotic properties of the algorithm under study.

8.6 PARAMETER CONVERGENCE

In Sections 8.5.1 and 8.5.2 we have developed properties of various sequential algorithms under very weak assumptions. These properties fall short of establishing convergence of the estimated parameters to the "true" system parameters. However, the properties so obtained are useful in diverse applications because of the weak assumptions imposed.

In this section we show what additional assumptions are necessary in order that parameter convergence can be established. In summary, the additional assumptions are: (1) the model order must be known exactly (previously on upper bound on the model order sufficed); (2) the input signal is required to be persistently exciting (previously no restrictions were placed on the input signal); and (3) the data must be mean-square bounded (previously, no boundedness assumptions were made).

We shall begin by establishing convergence of the estimated parameters generated by the sequential least-squares algorithm. Later we shall establish the corresponding result for the pseudo linear regression algorithm of Section 8.4.

The reader should note the connection between the results presented here for stochastic systems and those presented in Section 3.4 for deterministic systems.

8.6.1 The Ordinary Least-Squares Algorithm

In this section we investigate the performance of the least-squares algorithm introduced in Section 3.3 when the data are noisy.

In particular, consider the usual ARMAX model

$$A(q^{-1})y(t) = q^{-d}B'(q^{-1})u(t) + C(q^{-1})\omega(t) \tag{8.6.1}$$

where $\{\omega(t)\}$ is a white noise sequence. For the moment, we shall take $C(q^{-1}) = 1$. The more general case will be treated in the next subsection. With $C(q^{-1}) = 1$, the model (8.6.1) can be written in our standard form as

$$y(t) = \phi(t-1)^T\theta_0 + \omega(t) \tag{8.6.2}$$

where

$$\phi(t-1)^T = [-y(t-1), \ldots, -y(t-n), u(t-d), \ldots, u(t-d-m)] \qquad (8.6.3)$$

$$\theta_0 = [a_1, \ldots a_n, b_0, \ldots b_m] \qquad (8.6.4)$$

$$A(q^{-1}) = 1 + a_1 q^{-1} + \cdots + a_n q^{-n} \qquad (8.6.5)$$

$$B'(q^{-1}) = b_0 + b_1 q^{-1} + \cdots + b_m q^{-m} \qquad (8.6.6)$$

The least-squares algorithm now takes the form

Least-Squares Algorithm

$$\hat{\theta}(t) = \hat{\theta}(t-1) + \frac{P(t-2)\phi(t-1)}{1 + \phi(t-1)^T P(t-2)\phi(t-1)}[y(t) - \phi(t-1)^T\hat{\theta}(t-1)]$$
$$(8.6.7)$$

where

$$P(t-1) = P(t-2) - \frac{P(t-2)\phi(t-1)\phi(t-1)^T P(t-2)}{1 + \phi(t-1)^T P(t-2)\phi(t-1)} \qquad (8.6.8)$$

$$P(-1) = P_0 \text{ any positive definite matrix}$$

As in Sections 8.5.1 and 8.5.3, we define $\{\mathcal{F}_t\}$ to be the increasing sequence of sub-sigma algebras generated by $\{y(t)\}$ and we introduce the following technical assumptions regarding the sequence $\{\omega(t)\}$:

$$E\{\omega(t)\,|\,\mathcal{F}_{t-1}\} = 0 \quad \text{a.s.} \qquad (8.6.9)$$

$$E\{\omega(t)^2\,|\,\mathcal{F}_{t-1}\} = \sigma^2 < \infty \quad \text{a.s.} \qquad (8.6.10)$$

We then have the following convergence result:

Theorem 8.6.1. Consider the algorithm (8.6.7)–(8.6.8) applied to the system (8.6.2) and subject to (8.6.9)–(8.6.10). Then provided that:

(i) "Persistent excitation"

$$\lim_{t\to\infty} \lambda_{\min}[P(t)^{-1}] = \infty \qquad (8.6.11)$$

(ii) "Order condition"

$$\limsup_{t\to\infty} \frac{\lambda_{\max}[P(t)^{-1}]}{\lambda_{\min}[P(t)^{-1}]} < \infty \qquad (8.6.12)$$

we may conclude parameter convergence, that is,

$$\hat{\theta}(t) \xrightarrow{\text{a.s.}} \theta_0 \qquad (8.6.13)$$

Proof. We can argue precisely as in Theorem 8.5.2, part (i), to conclude [actually the proof is simpler here because $C(q^{-1}) = 1$]:

$$\frac{V(t)}{r(t-1)} \text{ converges a.s. to a finite random variable} \qquad (8.6.14)$$

and

$$\sum_{t=1}^{\infty} V(t)\left[\frac{1}{r(t-2)} - \frac{1}{r(t-1)}\right] < \infty \quad \text{a.s.} \qquad (8.6.15)$$

where $V(t) = \tilde{\theta}(t)^T P(t-1)^{-1} \tilde{\theta}(t)$ and $r(t) = \text{trace } P(t)^{-1}$. From (8.6.15),

$$\sum_{t=1}^{\infty} \frac{V(t)}{r(t-1)} [r(t-1) - r(t-2)] \frac{1}{r(t-2)} < \infty \qquad (8.6.16)$$

However, we now show by contradiction that

$$\sum_{t=1}^{\infty} [r(t-1) - r(t-2)] \frac{1}{r(t-2)} = \infty \qquad (8.6.17)$$

By (8.6.11), $\lim_{t \to \infty} r(t) = \infty$ and thus if we assume that

$$\sum_{t=1}^{\infty} [r(t-1) - r(t-2)] \frac{1}{r(t-2)} \text{ converges}$$

then by Kronecker's lemma (Appendix D),

$$\lim_{N \to \infty} \frac{1}{r(N-2)} \sum_{t=1}^{N} [r(t-1) - r(t-2)] = 0$$

or, since $r(N-1) \geq r(N-2)$,

$$\lim_{N \to \infty} \frac{1}{r(N-1)} [r(N-1) - r(-1)] = 0 \qquad (8.6.18)$$

However, this contradicts $\lim_{N \to \infty} r(N) = \infty$; thus (8.6.17) is established. Then from (8.6.14), (8.6.16), and (8.6.17), we can conclude that

$$\lim_{t \to \infty} \frac{V(t)}{r(t-1)} = 0 \quad \text{a.s.} \qquad (8.6.19)$$

Now from the definition of $V(t)$,

$$\frac{V(t)}{r(t-1)} \geq \frac{\lambda_{\min} P(t-1)^{-1} \tilde{\theta}(t)^T \tilde{\theta}(t)}{r(t-1)}$$

$$\geq \frac{\lambda_{\min} P(t-1)^{-1} \tilde{\theta}(t)^T \tilde{\theta}(t)}{p \lambda_{\max} P(t-1)^{-1}} \qquad (8.6.20)$$

where P is the number of parameters. From (8.6.19) and (8.6.20) we have, using (8.6.12),

$$\lim_{t \to \infty} \tilde{\theta}(t)^T \tilde{\theta}(t) = 0 \quad \text{a.s.} \qquad (8.6.21)$$

and (8.6.13) follows.

▼▼▼

  The theorem above is based on the work of Ljung and Wittenmark (1974), Ljung (1976), Sternby (1977), and Solo (1978).

  Assumption (i) in the theorem statement is a form of "persistently exciting" condition (see Definition 3.4.A). Assumption (ii) is a constraint on the condition number of $P(t)^{-1}$ matrix. Heuristically, this constraint implies that the information in the data applies equally to all linear combinations of the parameters, and requires that the system order not be overestimated.

8.6.2 The Pseudo Linear Regression Algorithm

In this subsection we establish parameter convergence for the pseudo linear regression algorithm introduced in Section 8.4. In fact, we shall generalize slightly and analyze a version of the algorithm incorporating a prespecified filter $D(q^{-1})$ which is used in the generation of the regression vector. We shall see that the presence of this filter allows us to weaken the condition for convergence.

The various filtering operations are organized with a view to the subsequent convergence analysis. As suggested in Remark 8.4.4, we use a posteriori predictions in the regression vector.

We consider again the ARMAX model:

$$A(q^{-1})y(t) = q^{-d}B'(q^{-1})u(t) + C(q^{-1})\omega(t) \qquad (8.6.22)$$

$$A(q^{-1}) = 1 + a_1 q^{-1} + \cdots + a_n q^{-n} \qquad (8.6.23)$$

$$B'(q^{-1}) = b_0 + b_1 q^{-1} + \cdots + b_m q^{-m} \qquad (8.6.24)$$

$$C(q^{-1}) = 1 + c_1 q^{-1} + \cdots + c_l q^{-l} \qquad (8.6.25)$$

with the following usual assumption on the noise:

Noise Assumptions

$$E\{\omega(t) \mid \mathcal{F}_{t-1}\} = 0 \quad \text{a.s.} \qquad (8.6.26)$$

$$E\{\omega(t)^2 \mid \mathcal{F}_{t-1}\} = \sigma^2 \quad \text{a.s.} \qquad (8.6.27)$$

$$\sup_N \frac{1}{N} \sum_{t=1}^N \omega(t)^2 < \infty \quad \text{a.s.} \qquad (8.6.28)$$

Our objective here is to estimate the parameters θ_0 in the polynomials $A(q^{-1})$, $B'(q^{-1})$, and $C(q^{-1})$:

$$\theta_0 = [a_1, \ldots, a_n, b_0, \ldots, b_m, c_1, \ldots, c_l] \qquad (8.6.29)$$

The algorithm is described as follows:

Pseudo Linear Regression Algorithm (Incorporating Filtering and Using A Posteriori Predictions)

$$\hat{\theta}(t) = \hat{\theta}(t-1) + P(t-1)\psi(t-1)\bar{v}(t) \qquad (8.6.30)$$

where

$$P(t-1) = P(t-2) - \frac{P(t-2)\psi(t-1)\psi(t-1)^T P(t-2)}{1 + \psi(t-1)^T P(t-2)\psi(t-1)} \qquad (8.6.31)$$

$P(-1)$ any positive definite matrix.

The various quantities required by the algorithm above are described as follows. We introduce a stably invertible moving-average filter $D(q^{-1})$, where

$$D(q^{-1}) = 1 + d_1 q^{-1} + \cdots + d_l q^{-l} \qquad (8.6.32)$$

We then define:

Filtered output $[y_F(t)]$

$$D(q^{-1})y_F(t) = y(t) \qquad (8.6.33)$$

Filtered input $[u_F(t)]$

$$D(q^{-1})u_F(t) = u(t) \qquad (8.6.34)$$

Regression vector $[\psi(t-1)]$

$$\psi(t-1)^T = [-y_F(t-1), \ldots, -y_F(t-n),$$
$$u_F(t-d), \ldots, u_F(t-d-m), \qquad (8.6.35)$$
$$\eta(t-1), \ldots, \eta(t-l)]$$

A posteriori filtered output $[\bar{y}_F(t)]$

$$\bar{y}_F(t) = \psi(t-1)^T \hat{\theta}(t) \tag{8.6.36}$$

A posteriori filtered output error $[\eta(t)]$

$$\eta(t) = y_F(t) - \bar{y}_F(t) \tag{8.6.37}$$

A priori filtered output $[\hat{y}_F(t)]$

$$\hat{y}_F(t) = \psi(t-1)^T \hat{\theta}(t-1) \tag{8.6.38}$$

A priori filtered output error $[e(t)]$

$$e(t) = y_F(t) - \hat{y}_F(t) \tag{8.6.39}$$

Generalized a priori error $[\bar{v}(t)]$

$$\bar{v}(t) = e(t) + [D(q^{-1}) - 1]\eta(t) \tag{8.6.40}$$

In the subsequent analysis we shall also require the following quantity:

Generalized a posteriori error

$$\bar{\eta}(t) = D(q^{-1})\eta(t) \tag{8.6.41}$$

Remark 8.6.1. The algorithm above is basically the same as the extended least-squares method with filtering described in (8.4.2) to (8.4.6). All that has been done is the filter operations have been arranged in a different order and we have introduced a posteriori prediction. The full interrelationships between the algorithms are discussed in Exercise 8.9.

▼▼▼

We now turn to the convergence analysis. We will require the following assumptions:

Assumption 8.6.A (Stability Assumption)

$$A(z^{-1}) \text{ and } C(z^{-1}) \text{ are both stable polynomials} \tag{8.6.42}$$

Assumption 8.6.B (Persistently Exciting Condition)

$$\lim_{N\to\infty} \frac{1}{N} \sum_{t=1}^{N} \overset{*}{\phi}(t-1)\overset{*}{\phi}(t-1)^T = R \tag{8.6.43}$$

exists and is positive definite, where

$$\overset{*}{\phi}(t-1)^T = [-y(t-1), \ \ldots, \ -y(t-n),$$
$$u(t-d), \ \ldots, \ u(t-d-m), \tag{8.6.44}$$
$$\omega(t-1), \ \ldots, \ \omega(t-l)]$$

Assumption 8.6.C (Passivity Condition). The transfer function

$$\left\{ \frac{D(z^{-1})}{C(z^{-1})} - \frac{1}{2} \right\} \text{ is very strictly passive} \tag{8.6.45}$$

where $D(z^{-1})$ is the a priori chosen filter.

We then have the following global convergence result:

Theorem 8.6.2. The algorithm (8.6.30) to (8.6.41) when applied to the system (8.6.22) and subject to the noise assumptions (8.6.26) to (8.6.28), the stability assumption (8.6.42), the persistently exciting condition (8.6.43), and the passivity condition (8.6.45) yields with probability 1

$$\lim_{t \to \infty} \hat{\theta}(t) = \theta_0 \tag{8.6.46}$$

Proof. The result is an extension of Theorem 8.6.1 using the ideas of data prefiltering and error filtering.

We present an outline proof.

We first note the following relationships:

$$C(q^{-1})[\eta(t) - \omega_F(t)] = -\psi(t-1)^T \tilde{\theta}(t) \triangleq b(t) \tag{8.6.47}$$

$$D(q^{-1})[\eta(t) - \omega_F(t)] = \bar{\eta}(t) - \omega(t) \tag{8.6.48}$$

where $\{\omega_F(T)\}$ is defined by $D(q^{-1})\omega_F(t) = \omega(t)$. Thus $[\bar{\eta}(t) - \omega(t)]$ (the "deterministic component" of the generalized a posteriori error) is related to $b(t)$ (the parameter error) by the transfer function $H(z) = D(z)/C(z)$ such that $[H(z) - \frac{1}{2}]$ is very strictly passive. (Note in this case the important role played by data prefiltering in developing the error relationship above.)

$$\bar{\eta}(t) = \frac{\bar{v}(t)}{1 + \psi(t-1)^T P(t-2)\psi(t-1)} \tag{8.6.49}$$

$$\hat{\theta}(t) = \hat{\theta}(t-1) + P(t-2)\psi(t-1)\bar{\eta}(t) \tag{8.6.50}$$

We can now introduce

$$\frac{V(t)}{r(t-1)} \triangleq \frac{\tilde{\theta}(t)^T P(t-1)^{-1}\tilde{\theta}(t)}{r(t-1)} \tag{8.6.51}$$

where $r(t-1) = r(t-2) + \psi(t-1)^T\psi(t-1)$, $r(-1) > 0$, and proceed essentially as in the proof of Theorem 8.6.1 to conclude:

$$\frac{V(t)}{r(t-1)} \xrightarrow{\text{a.s.}} \text{a finite random variable} \tag{8.6.52}$$

$$\sum_{t=1}^{\infty} V(t)\left[\frac{1}{r(t-2)} - \frac{1}{r(t-1)}\right] < \infty \quad \text{a.s.} \tag{8.6.53}$$

$$\sum_{t=1}^{\infty} \frac{b(t)^2}{r(t-2)} < \infty \quad \text{a.s.} \tag{8.6.54}$$

Now noting that $y(t)$, $u(t)$, and $\omega(t)$ and hence $y_F(t)$, $u_F(t)$, and $\omega_F(t)$ are mean-squared bounded, we have from (8.6.47) and the stability of $C(q^{-1})^{-1}$ (Assumption 8.6.A) that

$$\frac{1}{N}\sum_{t=1}^{N} \eta(t)^2 \le K_1 \frac{1}{N}\sum_{t=1}^{N} b(t)^2 + K_2 \quad \text{a.s.} \qquad \begin{matrix} 0 < K_1 < \infty \\ 0 \le K_2 < \infty \end{matrix} \tag{8.6.55}$$

Hence from the definition of $r(t)$ and using (8.6.55), we have for some $\bar{N}$

$$\frac{r(N)}{N} \le K_3\left[\frac{1}{N}\sum_{t=1}^{N} b(t)^2\right] + K_4 \qquad \text{for } N \ge \bar{N} \tag{8.6.56}$$

where $0 < K_3 < \infty$, $0 \le K_4 < \infty$. Thus we can apply the stochastic key technical lemma (Lemma 8.5.3) to conclude from (8.6.54) that

$$\lim_{N \to \infty} \sup \frac{1}{N}\sum_{t=1}^{N} b(t)^2 = 0 \quad \text{a.s.} \tag{8.6.57}$$

$$\lim_{N \to \infty} \sup \frac{1}{N} \sum_{t=1}^{N} \eta(t)^2 < \infty \quad \text{a.s.} \qquad (8.6.58)$$

$$\lim_{N \leftarrow \infty} \sup \frac{r(N)}{N} < \infty \quad \text{a.s.} \qquad (8.6.59)$$

Now from (8.6.53) we have

$$\sum_{t=1}^{\infty} \frac{V(t)}{r(t-1)} \left[\frac{r(t-1) - r(t-2)}{r(t-2)} \right] < \infty \quad \text{a.s.} \qquad (8.6.60)$$

and since, from Assumption 8.6.B, $r(t-1) \to \infty$ so that as in Theorem 8.6.1

$$\sum \frac{r(t-1) - r(t-2)}{r(t-1)} = \infty$$

we conclude that

$$\frac{V(t)}{r(t-2)} \xrightarrow{\text{a.s.}} 0$$

or using (8.6.51) and (8.6.59),

$$\frac{\tilde{\theta}(t)^T P(t-1)^{-1} \tilde{\theta}(t)}{t} \xrightarrow{\text{a.s.}} 0 \qquad (8.6.61)$$

We now proceed to prove $\lim \inf_{t \to \infty} P(t-1)^{-1}/t > 0$ a.s. This is equivalent to proving positive definiteness of the correlation matrix whose typical elements are of the form

$$\frac{1}{N} \sum_{1}^{N} \eta(t)\eta(t-k) \qquad (8.6.62)$$

$$\frac{1}{N} \sum_{1}^{N} \eta(t)y_F(t-k) \qquad (8.6.63)$$

$$\frac{1}{N} \sum_{1}^{N} \eta(t)u_F(t-k) \qquad (8.6.64)$$

$$\frac{1}{N} \sum_{1}^{N} y_F(t)u_F(t-k) \qquad (8.6.65)$$

$$\frac{1}{N} \sum_{1}^{N} y_F(t)y_F(t-k) \qquad (8.6.66)$$

$$\frac{1}{N} \sum_{1}^{N} u_F(t)u_F(t-k) \qquad (8.6.67)$$

We show that this correlation matrix converges to the limit that is obtained if $\eta(t)$ were to be replaced by $\omega_F(t)$. The resulting matrix is then positive definite in view of the persistently exciting condition (8.6.43) and the fact that $\{y_F(t)\}$, $\{u_F(t)\}$, and $\{\omega_F(t)\}$ are obtained from $\{y(t)\}$, $\{u(t)\}$, and $\{\omega(t)\}$, respectively, by the same filtering operation. The result is immediate for (8.6.65), (8.6.66), and (8.6.67). For (8.6.63) consider

$$\frac{1}{N} \sum_{1}^{N} \eta(t)y_F(t-k) = \frac{1}{N} \sum_{1}^{N} [\eta(t) - \omega_F(t)]y_F(t-k) + \frac{1}{N} \sum_{1}^{N} \omega_F(t)y_F(t-k) \quad (8.6.68)$$

Now from (8.6.47) and (8.6.57) and the stability of $C(q^{-1})^{-1}$,

$$\lim_{N \to \infty} \sup \frac{1}{N} \sum_{1}^{N} [\eta(t) - \omega_F(t)]^2 = 0 \quad \text{a.s.} \qquad (8.6.69)$$

Hence applying the Schwarz inequality to the first term on the right-hand side of (8.6.68), we conclude that it converges to zero. The second term on the right-hand side of (8.6.68) converges to the limit consistent with (8.6.43). Similar arguments apply to (8.6.62) and (8.6.64). This completes the proof.

▼▼▼

Remark 8.6.2. The idea of using prefiltering $[D(q^{-1}) \neq 1]$ of the data to weaken the positive real condition has been studied by Ljung (1977b) via the ODE approach (see Section 8.5.4) for the extended least-squares algorithm with a priori prediction. For this case, monitoring is required to ensure boundedness of the regression vector. Solo (1978, 1979) showed, for the case $D(q^{-1}) = 1$, that monitoring was unnecessary provided that a posteriori predictions were used in the regression vector.

As recognized by Ljung (1977b) the extension to the case $D(q^{-1}) \neq 1$ has practical importance since if the algorithm fails in a particular instance, then one has the flexibility of trying again with a different $D(q^{-1})$ polynomial. Moreover, it is known that if $D(q^{-1})$ is approximately $C(q^{-1})$, then (8.6.45) is automatically satisfied.

Remark 8.6.3. The alert reader will have recognized that if $A(q^{-1}) = C(q^{-1})$, then the algorithm analyzed above reduces to the output error method of Section 3.5. In fact, the basic steps in the proof above have much in common with the proof of convergence of the output error method in the deterministic case given in Chapter 3. The distinction is that the stochastic proof replaces the deterministic Lyapunov argument by the corresponding Martingale convergence argument. Thus we have now established the result stated in Section 3.5 that the output error scheme is globally convergent in the presence of white output noise. This is explored further in Exercise 8.10.

Remark 8.6.4. The results of this section indicate that the filtering of data may be important for convergence and also has practical importance. Moreover, the filter should be something like the inverse of the noise polynomial $C(q^{-1})$. In our discussion a fixed estimate of this inverse was used, namely $D(q^{-1})$. The convergence result then depends on the very strict passivity of $[D(z^{-1})/C(z^{-1}) - \frac{1}{2}]$. This suggests that a useful approach may be to replace $D(q^{-1})$ by the on-line estimate of the $C(q^{-1})$ polynomial. This leads to a stochastic form of Landau's scheme (Landau, 1978b) aimed at eliminating the positive real condition in output error methods. It is clear that the resulting algorithm then resembles the sequential prediction error algorithm of (8.3.60) to (8.3.66) (Söderström, Ljung, and Gustavsson, 1978).

▼▼▼

8.7 CONCLUDING REMARKS

In this chapter two broad classes of related algorithms have been studied: sequential prediction error and pseudo linear regression methods. In this section we summarize some of the key conclusions and discuss some of the practical aspects of parameter estimation in stochastic systems.

We have also introduced two forms of parameter update: the stochastic gradient and the least-squares interations. In practice, the least-squares-based algorithms have much faster convergence than the gradient-type algorithms. Thus whenever possible, the least-squares-type algorithms should be used. The advantage of the gradient algorithms is their relative simplicity.

The underlying theme in our discussion has been the prediction error approach. This is a very general idea and applies to a wide range of problems. A key point is that the model need not match the system, and yet the algorithms lead to the "best" predictor from a specified class.

Within the prediction error framework, the pseudo linear regression algorithms can be viewed as approximate prediction error algorithms. However, we have also shown that the pseudo linear regression schemes can be alternatively motivated as an extension of ordinary least squares. A key point here, however, is that the convergence analysis of the pseudo linear regression algorithms depends on the system having the same structure as the model. Nothing definite can be said about the performance of these algorithms when the model does not match the system.

We have also discussed the use of a priori and a posteriori predictions in the regression vector. The latter are based on a more recent parameter estimates and thus should lead to improved performance. Certainly, as we have seen in this chapter stronger theoretical results can be proved for the algorithms incorporating a posteriori predictions. In practice the distinction is probably not important except perhaps during the transient phase, but it is obviously sensible to use the most recent parameter estimates when forming the regression vector.

Another important point is that many of the algorithms require that the estimated C polynomial corresponding to the noise dynamics be monitored and projected to remain in the stability region. If this is not done, the algorithms may diverge.

The passivity condition (Appendix C) plays an important role in those algorithms based on pseudo linear regressions. These algorithms may fail if the passivity condition is violated. This has been shown in simulation studies. In practice, this is probably less of a problem than it seems at first sight since the noise model is likely to be quite simple. However, the potential failure of the algorithms should be borne in mind together with the additional flexibility offered by incorporating filtering in the algorithm to modify the passivity condition.

So far in this chapter we have considered algorithms whose gains go to zero asymptotically. This is a necessary requirement for convergence. However, in practice it is desirable to modify the algorithms so that they can continuously track the parameters in time-varying systems. The gain must then be kept from going to zero so that the algorithms will continuously track the parameters being estimated. This has been discussed in considerable detail in Sections 3.3 and 6.7.

If the least-squares algorithm is used, the algorithm gain can be prevented from going to zero by one of the following techniques:

1. Covariance resetting
2. Covariance modification
3. Exponentially weighted least squares (with variable forgetting factor)
4. Finite data windows

These methods were discussed in detail in Section 6.7 for the deterministic case. Similar observations apply to the stochastic case and thus we will not go into further detail. Our personal preference from experience is covariance resetting.

In some cases it is not feasible to utilize the least-squares-based algorithms due to the computational effort involved in updating and storing the $P(t)$ matrix. This is especially so when the number of parameters is large. In this case it is possible to use the stochastic gradient algorithms or variants thereof. Several algorithms of this type have appeared in the literature. We summarize some of these below.

Normalized Least-Mean-Square Algorithm (NLMS)

$$\hat{\theta}(t+1) = \hat{\theta}(t) + \frac{\mu(t)\phi(t)}{\phi(t)^T\phi(t) + c}[y(t+1) - \phi(t)^T\hat{\theta}(t)] \qquad (8.7.1)$$

where $\mu(t)$ is a scalar gain, usually taken to be a small positive constant.

This algorithm was suggested by Albert and Gardner (1967), Nagumo and Noda (1967), and others. The algorithm can be seen to be identical to the projection algorithm of Section 3.3. We have introduced the new name NLMS for the reason that this is the name more commonly used in the context of adaptive filtering.

It is possible to simplify the algorithm even further by removing the normalization all together. This leads to

Least-Mean-Square Algorithm (LMS)

$$\hat{\theta}(t+1) = \hat{\theta}(t) + \mu(t)\phi(t)[y(t+1) - \phi(t)^T\hat{\theta}(t)] \qquad (8.7.2)$$

The algorithm above was advocated and extensively studied by Widrow and Hoff (1960) and others.

Continuing in the same vein, we can simplify even further by quantizing the values in the $\phi(t)$ vector. This leads to:

Quantized State Least-Mean-Square Algorithm

$$\hat{\theta}(t+1) = \hat{\theta}(t) + \mu(t)\bar{\phi}(t)[y(t+1) - \phi(t)^T\hat{\theta}(t)] \qquad (8.7.3)$$

where

$$\bar{\phi}_i(t) = \text{sign } \phi_i(t) \qquad (8.7.4)$$

The algorithm above was described by Moschner (1970) and later studied by Kumar and Moore (1980).

There are also other algorithms, but the list above represents those most commonly used.

Various properties for the algorithms above are presented in the following references for:

1. The stationary case: Widrow (1970), Widrow et al. (1975), Gersho (1969), Senne (1970), Nagumo and Noda (1967), Polyak (1976, 1977), Davisson (1970), Daniell (1970), Kim and Davisson (1975), Jones (1973), Farden, Goding, and Saywood (1979), Sondhi and Mitra (1976), Weiss and Mitra (1979), Bitmead (1979b)
2. The nonstationary case: Widrow et al. (1976), Bitmead (1979a)

EXERCISES

8.1. Develop a recursive prediction error algorithm for the general cost function given in (8.2.3).

8.2. Develop the optimal one-step-ahead predictor for the general ARMAX model given in (8.3.76).

8.3. Reformulate the pseudo linear regression algorithm with filtering [equations (8.4.2) to (8.4.6)] so that the regression vector contains $[y(t-1), \ldots, u(t-d), \ldots, \epsilon(t-1)]$, where $\epsilon(t-1) = y(t-1) - \hat{y}(t-1)$. (This is the form most frequently discussed in the literature.)

8.4. Establish Pringsheim's theorem [see equation (8.5.23) and sequel].

8.5. Consider the following stochastic gradient algorithm:

$$\hat{\theta}(t) = \hat{\theta}(t-1) + \frac{\bar{a}\phi(t-1)}{r(t-1)}[y(t) - \phi(t-1)^T\hat{\theta}(t-1)]$$

$$r(t-1) = r(t-2) + \phi(t-1)^T\phi(t-1); \qquad r(-1) = 1$$

and $\phi(t-1)$ contains the a priori predictions, that is,

$$\phi(t-1)^T = [y(t-1), \ldots, y(t-n), u(t-1), \ldots, u(t-m),$$
$$- \hat{y}(t-1), \ldots, -\hat{y}(t-l)])$$

where $\hat{y}(t) = \phi(t-1)^T\hat{\theta}(t-1)$. Develop a convergence theory for the algorithm above. In particular, show that properties (i), (ii), and (iii) of Theorem 8.5.1 apply to the algorithm above provided that $C(q^{-1}) - \bar{a}/2$ is input strictly passive.

8.6. Examination of the proof of Theorem 8.5.1 reveals that the sequence $r(t-2)$ has to satisfy:

1. $r(t-2) \geq r(t-j), \quad j \geq 2$

2. $\lim_{N} \sum_{t=1}^{N} \dfrac{\phi(t-1)^T\phi(t-1)}{r(t-2)[r(t-2) + \phi(t-1)^T\phi(t-1)]} < \infty$

From simulation experience with this kind of algorithm, it is found that in order to improve the convergence rate of the algorithm, we have to reduce the rate of increase of $r(t-2)$ (and yet still satisfy conditions 1 and 2). Based on the Pringsheim theorem (Knopp, 1956), a plausible choice is to replace (8.5.13) by

$$r(t-1) = r(t-2) + a(t-1)\phi(t-1)^T\phi(t-1); \qquad r(-1) > 0$$

where $0 < a(t-1) \ll 1$. A simple example of such a choice is

$$r(t-1) = \begin{cases} r(t-2) + 1 & \text{when } \|\phi(t-1)\| \leq \dfrac{1}{\epsilon} \\ r(t-2) + (\epsilon\|\phi(t-1)\|)^2 & \text{otherwise.} \end{cases}$$

▼▼▼

Show that the convergence properties given in Theorem 8.5.1 are retained with the foregoing choice of the sequence $\{r(t)\}$.

8.7. Consider the following weighted least-squares algorithm [a similar algorithm has been studied by Kumar and Moore (1979)]:

$$\hat{\theta}(t) = \hat{\theta}(t-1) + \frac{\bar{P}(t-2)\phi(t-1)}{\bar{r}(t-2) + \phi(t-1)^T\bar{P}(t-2)\phi(t-1)}[y(t) - \phi(t-1)^T\hat{\theta}(t-1)]$$

where $\phi(t-1)$ is as in (8.5.10) and

$$\bar{r}(t-1) = \bar{r}(t-2) + \phi(t-1)^T \bar{P}(t-2)\phi(t-1), \qquad \bar{r}(-1) > 0$$

$$\bar{P}(t-1) = \bar{P}(t-2) - \frac{\bar{P}(t-2)\phi(t-1)\phi(t-1)^T\bar{P}(t-2)}{\bar{r}(t-2) + \phi(t-1)^T\bar{P}(t-2)\phi(t-1)}, \qquad \bar{P}(-1) > 0$$

or equivalently,

$$\bar{P}(t-1)^{-1} = \bar{P}(t-2)^{-1} + \frac{\phi(t-1)\phi(t-1)^T}{\bar{r}(t-2)}$$

Observe the similarity of the algorithm above to the stochastic gradient algorithm of (8.5.8) and (8.5.13). This has the consequence that much of the convergence analysis of the algorithm above parallels that of the stochastic gradient in Theorem 8.5.1. Establish the following convergence properties of the algorithm above.

Theorem. Subject to the usual noise and system assumptions and provided that the system having transfer function

$$\left[\frac{1}{C(z^{-1})} - \frac{1}{2}\right]$$

is very strictly passive, the algorithm above ensures that with probability 1:

   (i) Parameter norm convergence:

$$\limsup_{t\to\infty} \tilde{\theta}(t)^T \bar{P}(t-1)^{-1}\tilde{\theta}(t) < \infty$$

  (ii) Parameter difference convergence:

$$\lim_{N\to\infty} \sum_{t=1}^{N} \|\hat{\theta}(t) - \hat{\theta}(t-k)\|^2 < \infty$$

     for any finite k.

  (iii) Normalized prediction error convergence:

$$\lim_{N\to\infty} \sum_{t=1}^{N} \frac{(e(t) - \omega(t))^2}{\bar{r}(t-1)} < \infty$$

[*Hint:* Follow the proof of Theorem 8.5.1 using the matrix inversion lemma (Lemma 3.3.4) and the Lyapunov function $V(t) \triangleq \tilde{\theta}(t)^T\bar{P}(t-1)^{-1}\tilde{\theta}(t)$. Note that

$$E\{b(t)^2 - 2b(t)z(t)\,|\,\mathcal{F}_{t-1}\} = -2E\{b(t)g(t)\,|\,\mathcal{F}_{t-1}\}$$

such that $g(t) = (z(t) - b(t)/2)$ and is related to $b(t)$ by $[1/C(z) - \frac{1}{2}]$. The condition of $[1/C(z) - \frac{1}{2}]$ being input strictly passive allows one to establish

$$\lim_{N\to\infty} \sum_{t=1}^{N} \frac{b(t)^2}{\bar{r}(t-2)} < \infty \quad \text{a.s.}$$

whereas the condition $[1/C(z) - \frac{1}{2}]$ being output strictly passive allows one to conclude that

$$\lim_{N\to\infty} \sum_{t=2}^{N} \frac{g(t)^2}{\bar{r}(t-2)} < \infty \quad \text{a.s.}$$

The two passivity conditions are implied by very strict passivity (Appendix C).]

8.8. Consider the pseudo linear regression algorithm [equations (8.4.2) to (8.4.6)]. Develop an alternative form of this algorithm in which the prediction *error* (rather than the predicted output) appears in the regression vector, that is,

$$\phi(t-1) = [-y(t-1), \ldots, u(t-d), \ldots, e(t-1), \ldots]^T$$

where $e(t-1) = y(t-1) - \hat{y}(t-1)$.

8.9. Consider the pseudo linear regression algorithm (incorporating filtering and using a priori errors) (8.6.30) to (8.6.40).

(a) Show that

    (i) $\psi(t)$ is "something like"

$$\frac{1}{D(q^{-1})}\phi(t-1)$$

with $\phi(t-1)$ as in Exercise 8.8. [*Hint:*

$$\eta(t) \simeq \frac{1}{D(q^{-1})}e(t-1).]$$

    (ii) Show that $\bar{v}(t)$ is "something like"

$$y(t) - \phi(t-1)^T\hat{\theta}(t-1)$$

    [*Hint:* Ignore the distinction between a priori and a posteriori predictions; that is, put $\bar{v}(t) \simeq D(q^{-1})\eta(t)$.]

(b) Combine Exercises 8.8 and 8.9(a) to establish a link between the pseudo linear regression algorithm given in (8.4.2) to (8.4.6) with the algorithm given in (8.6.30) to (8.6.40).

8.10. Consider the following ARMAX model:

$$A(q^{-1})z(t) = q^{-d}B'(q^{-1})u(t)$$

$$y(t) = z(t) + \omega(t); \qquad \{\omega(t)\} \text{ white noise}$$

(a) Show that the system can be expressed as

$$A(q^{-1})y(t) = q^{-d}B'(q^{-1})u(t) + C(q^{-1})\omega(t)$$

with $A(q^{-1}) = C(q^{-1})$.

(b) Show that the model can be written as

$$y(t) = \overset{*}{\phi}(t-1)^T\theta_0 + \omega(t)$$

$$\overset{*}{\phi}(t-1)^T = [-y(t-1) + \omega(t-1), \ \ldots \ -y(t-n) + \omega(t-n), u(t-d), \ \ldots$$

$$u(t-d-m)]$$

(c) Develop a pseudo linear regression algorithm using a posteriori prediction as in (8.6.30) to (8.6.41) for the model above by defining the following quantities. [Take $D(q^{-1}) = 1$.]

    (i) Regression vector:

$$\phi(t-1)^T = [-y(t-1) + \eta(t-1), \ \ldots, \ -y(t-n) + \eta(t-n),$$

$$u(t-d), \ \ldots, \ u(t-d-m)]$$

    (ii) A posteriori output and error:

$$\bar{y}(t) = \phi(t-1)^T\hat{\theta}(t)$$

$$\eta(t) = y(t) - \bar{y}(t)$$

    (iii) A priori output and error:

$$\hat{y}(t) = \phi(t-1)^T\hat{\theta}(t-1)$$

$$e(t) = y(t) - \hat{y}(t)$$

    Then noting that $-y(t-i) + \eta(t-i) = -\bar{y}(t-i)$, show that the resulting algorithm is identical to the output error scheme discussed in Section 3.5. Hence show that the *output error* scheme causes $\hat{\theta}(t) \longrightarrow \theta_0$ provided that $[1/A(z^{-1}) - \frac{1}{2}]$ is very strictly passive.

8.11. Prove that

$$\left(\sum_{1}^{N} a_i\right)^2 \le N \sum_{1}^{N} a_i^2 \qquad \text{for any sequence } \{a_i\}$$

8.12. Show that

$$|2ab| \le \frac{a^2}{x} + xb^2 \qquad \text{for any } x > 0$$

8.13. Use the Schwarz inequality to show that

$$\left[\frac{1}{N} \sum_{t=1}^{N} [\eta(t) - \omega_F(t)] y_F(t-k)\right]^2 \le \frac{1}{N} \sum_{t=1}^{N} [\eta(t) - \omega_F(t)]^2 \frac{1}{N} \sum_{t=1}^{N} y_F(t-k)^2$$

8.14. Show that the vector $\psi(t-1)$ in the recursive prediction error algorithm for ARMAX models can be evaluated via

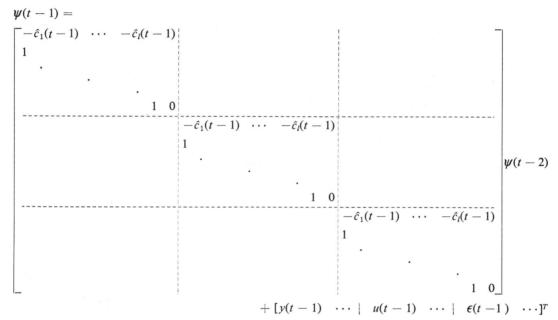

$$+ [y(t-1) \quad \cdots \mid u(t-1) \quad \cdots \mid \epsilon(t-1) \quad \cdots]^T$$

8.15. Based on the geometric interpretation of deterministic algorithms in Chapter 3, one can take the view that stochastic gradient and least-squares update represent but two special (perhaps extreme) cases. In between these extremes there are infinitely many other plausible directions of update. By a proper choice of direction *and* magnitude of adjustment, it should be possible to reach a compromise that is "optimal." As an illustration of this principle, consider the least-squares update of (8.5.37):

$$\hat{\theta}(t) = \hat{\theta}(t-1) + \frac{P(t-2)\phi(t-1)}{1 + \phi(t-1)^T P(t-2)\phi(t-1)} [y(t) - \phi(t-1)^T \hat{\theta}(t-1)]$$

(a) By defining

$$\bar{P}(t-2) = \bar{r}(t-2)P(t-2)$$

$$\bar{r}(t-1) = \bar{r}(t-2) + \phi(t-1)^T \bar{P}(t-2)\phi(t-1)$$

show that the algorithm above can be expressed in the following alternative form:

$$\hat{\theta}(t) = \hat{\theta}(t-1) + \frac{\bar{P}(t-2)\phi(t-1)}{\bar{r}(t-2) + \phi(t-1)^T \bar{P}(t-2)\phi(t-1)}$$

$$\times [y(t) - \phi(t-1)^T \hat{\theta}(t-1)]$$

where

$$P(t - 1) = \frac{\bar{r}(t - 1)}{\bar{r}(t - 2)}\left[\bar{P}(t - 2) - \frac{\bar{P}(t - 2)\phi(t - 1)\phi(t - 1)^T\bar{P}(t - 2)}{\bar{r}(t - 2) + \phi(t - 1)^T\bar{P}(t - 2)\phi(t - 1)}\right]$$

[Note that this equation is identical to that given for the algorithm in Exercise 8.7 save for the additional gain factor $\bar{r}(t - 1)/\bar{r}(t - 2)$.]

(b) Show that the equation for $\bar{P}(t - 1)$ above is equivalent to

$$\bar{P}(t - 1)^{-1} = \frac{\bar{r}(t - 2)}{\bar{r}(t - 1)}\left[\bar{P}(t - 2)^{-1} + \frac{1}{\bar{r}(t - 2)}\phi(t - 1)\phi(t - 1)^T\right]$$

(c) Define

$$\bar{V}(t) = \tilde{\theta}(t)^T\bar{P}(t - 1)^{-1}\tilde{\theta}(t)$$

Hence show that

$$\frac{\bar{r}(t - 1)}{\bar{r}(t - 2)}E\{\bar{V}(t)\,|\,\mathcal{F}_{t-1}\} = \bar{V}(t - 1) + \frac{1}{\bar{r}(t - 2)}E\{b(t)^2 - 2b(t)z(t)\,|\,\mathcal{F}_{t-1}\}$$

$$+ \frac{2\phi(t - 1)^T\bar{P}(t - 2)\phi(t - 1)}{\bar{r}(t - 2)\bar{r}(t - 1)}\sigma^2$$

$$- E\left\{\frac{\phi(t - 1)^T\bar{P}(t - 2)\phi(t - 1)}{\bar{r}(t - 2)^2}\eta(t)^2\,|\,\mathcal{F}_{t-1}\right\}$$

[*Hint:* Show that

$$\frac{\bar{r}(t - 1)}{\bar{r}(t - 2)}E\{\tilde{\theta}(t)^T\bar{P}(t - 1)^{-1}\tilde{\theta}(t)\,|\,\mathcal{F}_{t-1}\} = \tilde{\theta}(t - 1)^T\bar{P}(t - 2)^{-1}\tilde{\theta}(t - 1)$$

$$+ \frac{1}{\bar{r}(t - 2)}E\{b(t)^2 - 2b(t)z(t)\,|\,\mathcal{F}_{t-1}\}$$

$$+ \frac{2\phi(t - 1)^T\bar{P}(t - 1)\phi(t - 1)}{\bar{r}(t - 2)\bar{r}(t - 1)}\sigma^2$$

$$- E\left\{\frac{\phi(t - 1)^T\bar{P}(t - 2)\phi(t - 1)}{\bar{r}(t - 2)^2}\eta(t)^2\,|\,\mathcal{F}_{t-1}\right\}$$

and

$$\phi(t - 1)^T\bar{P}(t - 1)\phi(t - 1) = \phi(t - 1)^T\bar{P}(t - 2)\phi(t - 1)]$$

(d) Show that $\bar{P}(t)$, although ≥ 0, may not necessarily be monotone and can be increasing or decreasing depending on the time instant.

(e) By comparing the result in part (c) with the corresponding result for the algorithm of Exercise 8.7, show that the factor $\bar{r}(t - 1)/\bar{r}(t - 2)$ plays a critical role in establishing convergence of the algorithm.

8.16. Consider the condition number monitoring scheme given in (8.5.65) to (8.5.69). For this algorithm, show that

(a) $P(t - 1) = \beta(t - 1)\left[P(t - 2) - \frac{P(t - 2)\phi(t - 1)\phi(t - 1)^TP(t - 2)}{1 + \phi(t - 1)^TP(t - 2)\phi(t - 1)}\right]$

or equivalently

$$P(t - 1)^{-1} = \beta(t - 1)^{-1}[P(t - 2)^{-1} + \phi(t - 1)\phi(t - 1)^T]$$

where

$$\beta(t - 1) \triangleq \frac{K}{\max\{K, \bar{r}(t - 1)\lambda_{max}P'(t - 1)\}}$$

and $\beta(t - 1)$ can be expressed in the form

$$\beta(t - 1) = \frac{1 + \alpha(t - 1)\phi(t - 1)^TP(t - 2)\phi(t - 1)}{1 + \phi(t - 1)^TP(t - 2)\phi(t - 1)},$$

$$\text{for some } \alpha(t - 1); \quad 0 \leq \alpha(t - 1) \leq 1$$

(b) $\bar{r}(t-1)\lambda_{\max}P(t-1)\leq K, \quad t\geq 0$

(c) $\bar{r}(t-1)\leq Kr(t-1)$

where

$$r(t-1)=r(t-2)+\phi(t-1)^T\phi(t-1), \qquad r(-1)\geq\frac{1}{K}\bar{r}(-1)$$

[Note that the algorithm ensures that the gain factor $\beta(t-1)$ is equal to 1 whenever this is consistent with convergence and otherwise is set as close to 1 as is possible.]

[*Hint*:

1. If $\bar{r}(t-1)\lambda_{\max}P'(t-1)\leq K$, show that

$$\beta(t-1)=1 \qquad \text{and} \qquad \alpha(t-1)=1$$

Otherwise, it is true that

$$\bar{r}(t-1)\lambda_{\max}P'(t-1)>K$$

or

$$\bar{r}(t-2)[1+\phi(t-1)^TP(t-2)\phi(t-1)]\lambda_{\max}P'(t-1)>K$$

But

$$\bar{r}(t-2)\lambda_{\max}P'(t-1)\leq\bar{r}(t-2)\lambda_{\max}P(t-2)$$

Hence

$$\bar{r}(t-2)\lambda_{\max}P'(t-1)$$
$$\leq K<\bar{r}(t-2)[1+\phi(t-1)^TP(t-2)\phi(t-1)]\lambda_{\max}P'(t-1)$$

or

$$K=\bar{r}(t-2)[1+\alpha(t-1)\phi(t-1)^TP(t-2)\phi(t-1)]\lambda_{\max}P'(t-1)$$

for some $\alpha(t-1)$; $0\leq\alpha(t-1)<1$. Now

$$\beta(t-1)=\frac{K}{\bar{r}(t-1)\lambda_{\max}P'(t-1)}$$

and so

$$\beta(t-1)=\frac{\bar{r}(t-2)[1+\alpha(t-1)\phi(t-1)^TP(t-2)\phi(t-1)]\lambda_{\max}P'(t-1)}{\bar{r}(t-2)[1+\phi(t-1)^TP(t-2)\phi(t-1)]\lambda_{\max}P'(t-1)}$$

$$=\frac{1+\alpha(t-1)\phi(t-1)^TP(t-2)\phi(t-1)}{1+\phi(t-1)^TP(t-2)\phi(t-1)}$$

2. Use induction, assume that the result is true for $t-2$, and note that the result is true for $t=1$, using

$$\bar{r}(t-1)=\bar{r}(t-2)+\bar{r}(t-2)\phi(t-1)^TP(t-2)\phi(t-1)$$
$$\leq\bar{r}(t-2)+\bar{r}(t-2)\lambda_{\max}P(t-2)\phi(t-1)^T\phi(t-1)$$
$$\leq\bar{r}(t-2)+K\phi(t-1)^T\phi(t-1)$$
$$\leq Kr(t-2)+K\phi(t-1)^T\phi(t-1) \qquad \text{by induction hypothesis}$$
$$=Kr(t-1)]$$

8.17. Consider again the scheme of Exercise 8.16. Show that the following relationships hold between the various error quantities:

(a) $\eta(t)=\dfrac{e(t)}{1+\phi(t-1)^TP(t-2)\phi(t-1)}$

where

$$e(t)\triangleq y(t)-\hat{y}(t)=y(t)-\phi(t-1)^T\hat{\theta}(t-1)$$
$$\eta(t)\triangleq y(t)-\bar{y}(t)$$
$$\bar{y}(t)=\phi(t-1)^T\hat{\theta}(t)$$

(b) $C(q^{-1})z(t) = b(t)$

where

$$z(t) \triangleq \eta(t) - \omega(t)$$
$$b(t) \triangleq -\phi(t-1)^T \tilde{\theta}(t)$$
$$\tilde{\theta}(t) \triangleq \hat{\theta}(t) - \theta_0$$

(c) $E\{b(t)\omega(t)\,|\,\mathcal{F}_{-1}\} = \dfrac{-\phi(t-1)^T P(t-2)\phi(t-1)}{1 + \phi(t-1)^T P(t-2)\phi(t-1)}\sigma^2$

(*Hint:* See Lemma 8.5.1.)

8.18. Establish the following convergence result for the algorithm of Exercises 8.16 and 8.17.

Theorem. Subject to the usual assumptions, and provided that $[1/C(z) - \frac{1}{2}]$ is very strictly passive, the algorithm ensures that with probability 1:

(i) Parameter norm convergence:

(a) $\displaystyle \limsup_{t\to\infty} \frac{\tilde{\theta}(t)^T P(t-1)^{-1}\tilde{\theta}(t)}{r(t-1)} < \infty$

(b) $\displaystyle \lim_{N} \sum_{t=1}^{N} \alpha(t-1)\phi(t-1)^T P(t-2)\phi(t-1)\frac{\tilde{\theta}(t)^T P(t-1)^{-1}\tilde{\theta}(t)}{r(t-1)} < \infty$

(ii) Parameter difference convergence:

$$\lim_{N\to\infty} \sum_{t=1}^{N} \|\hat{\theta}(t) - \hat{\theta}(t-k)\|^2 < \infty \qquad \text{for any finite } k$$

(iii) Normalized prediction error convergence:

$$\lim_{N\to\infty} \sum_{t=1}^{N} \frac{[e(t) - \omega(t)]^2}{r(t-1)} < \infty$$

(*Hint:* Follow closely the proof of Theorem 8.5.2 and Exercises 8.16 and 8.17.)
(a) Show that

$$\tilde{\theta}(t) - P(t-2)\phi(t-1)\eta(t) = \tilde{\theta}(t-1)$$

and hence

$$\tilde{\theta}(t)^T P(t-2)^{-1}\tilde{\theta}(t) - 2\phi(t-1)^T \tilde{\theta}(t)\eta(t) + \phi(t-1)^T P(t-2)\phi(t-1)\eta(t)^2$$
$$= \tilde{\theta}(t-1)^T P(t-2)^{-1}\tilde{\theta}(t-1)$$

Then define $V(t) \triangleq \tilde{\theta}(t)^T P(t-1)^{-1}\tilde{\theta}(t)$ and show that

$$\beta(t-1)V(t) = V(t-1) + b(t)^2 - 2b(t)\eta(t) - \phi(t-1)^T P(t-2)\phi(t-1)\eta(t)^2$$
$$= V(t-1) + b(t)^2 - 2b(t)z(t) - 2b(t)\omega(t)$$
$$\quad - \phi(t-1)^T P(t-2)\phi(t-1)\eta(t)^2$$

Taking conditional expectations, show that

$$\beta(t-1)E\{V(t)\,|\,\mathcal{F}_{t-1}\} = V(t-1) + E\{b(t)^2 - 2b(t)z(t)\,|\,\mathcal{F}_{t-1}\}$$
$$+ \frac{2\phi(t-1)^T P(t-2)\phi(t-1)}{1 + \phi(t-1)^T P(t-2)\phi(t-1)}\sigma^2$$
$$- E\{\phi(t-1)^T P(t-2)\phi(t-1)\eta(t)^2\,|\,\mathcal{F}_{t-1}\} \quad \text{a.s.}$$

Divide by $\bar{r}(t-2)$ and using $g(t) \triangleq z(t) - \frac{1}{2}b(t)$ (note that $g(t)$ is related to $b(t)$ by the transfer function $[1/C(z) - \frac{1}{2}]$), show that

$$\beta(t-1)\frac{\bar{r}(t-1)}{\bar{r}(t-2)}E\left\{\frac{V(t)}{\bar{r}(t-1)}\bigg|\mathfrak{F}_{t-1}\right\} = \frac{V(t-1)}{\bar{r}(t-2)} - \frac{2}{\bar{r}(t-2)}E\{b(t)g(t)|\mathfrak{F}_{t-1}\}$$

$$+ E\left\{\frac{\phi(t-1)^T P(t-2)\phi(t-1)}{\bar{r}(t-2)}\eta(t)^2\bigg|\mathfrak{F}_{t-1}\right\}$$

$$+ \frac{2\phi(t-1)^T P(t-2)\phi(t-1)}{(1+\phi(t-1)^T P(t-2)\phi(t-1))\bar{r}(t-2)}\sigma^2 \quad \text{a.s.}$$

Hence show that

$$(1 + \alpha(t-1)\phi(t-1)^T P(t-2)\phi(t-1))E\left\{\frac{V(t)}{\bar{r}(t-1)}\bigg|\mathfrak{F}_{t-1}\right\}$$

$$= \frac{V(t-1)}{\bar{r}(t-2)} - \frac{2}{\bar{r}(t-2)}E\{b(t)g(t)|\mathfrak{F}_{t-1}\}$$

$$- E\left\{\frac{\phi(t-1)^T P(t-2)\phi(t-1)}{\bar{r}(t-2)}\eta(t)^2\bigg|\mathfrak{F}_{t-1}\right\}$$

$$+ \frac{2\phi(t-1)^T P(t-2)\phi(t-1)}{\bar{r}(t-1)}\sigma^2$$

Now define

$$\bar{P}(t-2) = \bar{r}(t-2)P(t-2)$$

and show that

$$\bar{r}(t-1) = \bar{r}(t-2) + \phi(t-1)\bar{P}(t-2)\phi(t-1)$$

and thus that

$$\sum_{t=1}^{\infty}\frac{\phi(t-1)^T P(t-2)\phi(t-1)}{\bar{r}(t-1)}\sigma^2 = \sum_{t=1}^{\infty}\frac{\phi(t-1)^T \bar{P}(t-2)\phi(t-1)}{\bar{r}(t-2)\bar{r}(t-1)}\sigma^2$$

$$= \sum_{t=1}^{\infty}\frac{\bar{r}(t-1) - \bar{r}(t-2)}{\bar{r}(t-2)\bar{r}(t-1)}\sigma^2$$

$$= \sum_{t=1}^{\infty}\left[\frac{1}{\bar{r}(t-2)} - \frac{1}{\bar{r}(t-1)}\right]\sigma^2$$

$$\leq \frac{1}{\bar{r}(-1)}\sigma^2 < \infty$$

Now apply the Martingale convergence theorem to show that

$$\limsup_{t\to\infty}\frac{\tilde{\theta}(t)^T P(t-1)^{-1}\tilde{\theta}(t)}{\bar{r}(t-1)} < \infty \quad \text{a.s.}$$

$$\lim_{N\to\infty}\sum_{t=1}^{N}\alpha(t-1)\phi(t-1)^T P(t-2)\phi(t-1)\left(\frac{\tilde{\theta}(t)^T P(t-1)^{-1}\tilde{\theta}(t)}{\bar{r}(t-1)}\right) < \infty \quad \text{a.s.}$$

$$\lim_{N\to\infty}\sum_{t=1}^{N}\frac{g(t)^2}{\bar{r}(t-2)} < \infty \quad \text{a.s.}$$

$$\lim_{N\to\infty}\sum_{t=1}^{N}\frac{b(t)^2}{\bar{r}(t-2)} < \infty \quad \text{a.s.}$$

$$\lim_{N\to\infty}\sum_{t=1}^{N}\frac{\phi(t-1)^T P(t-2)\phi(t-1)}{\bar{r}(t-2)}\eta(t)^2 < \infty \quad \text{a.s.}$$

and hence

$$\lim_{N\to\infty}\sum_{t=1}^{N}\frac{z(t)^2}{\bar{r}(t-2)} < \infty \quad \text{a.s.}$$

Result (i) then follows easily.

(b) Show that

$$\| \hat{\theta}(t) - \hat{\theta}(t-1) \|^2 = \phi(t-1)^T P(t-2)^2 \phi(t-1) \eta(t)^2$$

$$\leq \lambda_{\max} P(t-2) \phi(t-1)^T P(t-2) \phi(t-1) \eta(t)^2$$

$$= [\bar{r}(t-2) \lambda_{\max} P(t-2)] \frac{\phi(t-1)^T P(t-2) \phi(t-1)}{\bar{r}(t-2)} \eta(t)^2$$

$$\leq K \frac{\phi(t-1)^T P(t-2) \phi(t-1)}{\bar{r}(t-2)} \eta(t)^2$$

(c) Show that

$$[y(t) - \phi(t-1)^T \hat{\theta}(t) - \omega(t)] + \phi(t-1)^T P(t-2) \phi(t-1) \eta(t)$$
$$= [y(t) - \phi(t-1)^T \hat{\theta}(t-1) - \omega(t)]$$

or

$$z(t) + \phi(t-1)^T P(t-2) \phi(t-1) \eta(t) = e(t) - \omega(t)$$

and hence that

$$[e(t) - \omega(t)]^2 \leq 2z(t) + 2[\phi(t-1) P(t-2) \phi(t-1)]^2 \eta(t)^2$$

and

$$\frac{[e(t) - \omega(t)]^2}{\bar{r}(t-1)} \leq \frac{2z(t)^2}{\bar{r}(t-1)} + \frac{2\bar{r}(t-2) \phi(t-1)^T P(t-2) \phi(t-1)}{\bar{r}(t-1)}$$

$$\left[\frac{\phi(t-1)^T P(t-2) \phi(t-1)}{\bar{r}(t-2)} \eta(t)^2 \right]$$

$$\leq \frac{2z(t)^2}{\bar{r}(t-1)} + \frac{2\phi(t-1)^T P(t-2) \phi(t-1)}{\bar{r}(t-2)} \eta(t)^2$$

8.19. Define

$$\bar{P}(t-2) = \bar{r}(t-2) P(t-2)$$

and hence show that the algorithm of Exercise 8.16 can be represented as

$$\hat{\theta}(t) = \hat{\theta}(t-1) + \frac{\bar{P}(t-2) \phi(t-1)}{\bar{r}(t-2) + \phi(t-1)^T \bar{P}(t-2) \phi(t-1)} [y(t) - \phi(t-1)^T \hat{\theta}(t-1)]$$

where

$$\bar{r}(t-1) = \bar{r}(t-2) + \phi(t-1)^T \bar{P}(t-2) \phi(t-1), \qquad \bar{r}(-1) > 0$$

$$\bar{P}(t-1) = \lambda(t-1) \left[\bar{P}(t-2) - \frac{\bar{P}(t-2) \phi(t-1) \phi(t-1)^T \bar{P}(t-2)}{\bar{r}(t-2) + \phi(t-1)^T \bar{P}(t-2) \phi(t-1)} \right]$$

$$\bar{P}(-1) > 0$$

or equivalently,

$$\bar{P}(t-1)^{-1} = \lambda(t-1)^{-1} \left[\bar{P}(t-2)^{-1} + \frac{\phi(t-1) \phi(t-1)^T}{\bar{r}(t-2)} \right]$$

with

$$\lambda(t-1) \triangleq 1 + \alpha(t-1) \phi(t-1)^T P(t-2) \phi(t-1), \qquad 0 \leq \alpha(t-1) \leq 1$$

Also, show that

$$\lambda_{\max} \bar{P}(t-1) \leq K, \qquad t \geq 0$$

8.20. Develop an alternative proof for the result in Exercise 8.18, using the alternative form of the algorithm given in Exercise 8.19.

[*Hint:* First note that:

1. $\displaystyle\lim_{N} \sum_{t=1}^{N} \frac{\phi(t-1)^T \bar{P}(t-2)\phi(t-1)}{\bar{r}(t-2)\bar{r}(t-1)} < \infty$

2. $\displaystyle \eta(t) = \frac{\bar{r}(t-2)}{\bar{r}(t-1)} e(t)$

3. Lemma 8.5.2(iii) holds.

4. $\displaystyle E\{b(t)\omega(t) \,|\, \mathcal{F}_{t-1}\} = \frac{-\phi(t-1)^T \bar{P}(t-2)\phi(t-1)}{\bar{r}(t-1)}$

Now proceed as in the proof of Theorem 8.5.1. Define

$$\bar{V}(t) = \tilde{\theta}(t)^T \bar{P}(t-1)^{-1}\tilde{\theta}(t)$$

Show that

$$\tilde{\theta}(t) - \frac{\bar{P}(t-2)\phi(t-1)}{\bar{r}(t-2)}\eta(t) = \tilde{\theta}(t-1)$$

and

$$\lambda(t-1)\bar{V}(t) - \frac{b(t)^2}{\bar{r}(t-2)} + \frac{2b(t)\eta(t)}{\bar{r}(t-2)} + \frac{\phi(t-1)^T\bar{P}(t-2)\phi(t-1)}{\bar{r}(t-2)}\eta(t)^2 = \bar{V}(t-1)$$

Take conditional expectations and show that

$$\lambda(t-1)E\{\bar{V}(t)\,|\,\mathcal{F}_{t-1}\} = \bar{V}(t-1) + \frac{1}{\bar{r}(t-2)}E\{b(t)^2 - 2b(t)z(t)\,|\,\mathcal{F}_{t-1}\}$$
$$- E\left\{\frac{\phi(t-1)^T\bar{P}(t-2)\phi(t-1)}{\bar{r}(t-2)^2}\eta(t)^2\,\bigg|\,\mathcal{F}_{t-1}\right\}$$
$$+ \frac{2\phi(t-1)^T\bar{P}(t-2)\phi(t-1)}{\bar{r}(t-2)\bar{r}(t-1)}\sigma^2 \quad \text{a.s.}$$

Define

$$g(t) \triangleq z(t) - \frac{b(t)}{2}$$

$$S(t) \triangleq 2\sum_{j=1}^{t}\left[b(j)g(j) - \frac{\rho_1}{2}g(j)^2 - \frac{\rho_2}{2}b(j)^2\right] + K_1$$

Note from the very strict passivity assumption that ρ_1 and ρ_2 can be chosen (nonzero) such that $S(t) \geq 0$. Also define

$$X(t = \bar{V}(t) + \frac{S(t)}{\bar{r}(t-2)} + \rho_1\sum_{j=1}^{t}\frac{g(j)^2}{\bar{r}(j-2)} + \rho_2\sum_{j=1}^{t}\frac{b(j)^2}{\bar{r}(j-2)}$$
$$+ \sum_{j=1}^{t}\frac{\phi(j-1)^T\bar{P}(j-2)\phi(j-1)}{\bar{r}(j-2)^2}\eta(j)^2$$

Hence show that

$$E\{X(t)\,|\,\mathcal{F}_{t-1}\} \leq X(t) + \frac{2\phi(t-1)^T\bar{P}(t-2)\phi(t-1)}{\bar{r}(t-2)\hat{r}(t-1)}\sigma^2 \quad \text{a.s.}$$

Use the Martingale convergence theorem to conclude that

$$X(t) \longrightarrow X < \infty \quad \text{a.s.}$$

Hence, with probability 1 we have

$$\limsup_t \bar{V}(t) < \infty$$

$$\lim_N \sum_{t=1}^N \frac{b(t)^2}{\bar{r}(t-2)} < \infty$$

$$\lim_N \sum_{t=1}^N \frac{g(t)^2}{\bar{r}(t-2)} < \infty$$

$$\lim_N \sum_{t=1}^N \frac{\phi(t-1)^T \bar{P}(t-2)\phi(t-1)}{\bar{r}(t-2)^2}\eta(t)^2 < \infty$$

Hence show that

$$\lim_N \sum_{t=1}^N \frac{z(t)^2}{\bar{r}(t-2)} < \infty$$

Also show that

$$\lim_N \|\hat{\theta}(t) - \hat{\theta}(t-1)\|^2 \le K \lim_N \sum_{t=1}^N \frac{\phi(t-1)^T \bar{P}(t-2)\phi(t-1)}{\bar{r}(t-2)^2}\eta(t)^2 < \infty$$

Note that

$$\hat{\theta}(t) - \frac{\bar{P}(t-2)\phi(t-1)}{\bar{r}(t-2)}\eta(t) = \hat{\theta}(t-1)$$

Hence show that

$$[e(t) - \omega(t)]^2 \le 2z(t)^2 + 2\frac{[\phi(t-1)^T \bar{P}(t-2)\phi(t-1)]^2}{\bar{r}(t-2)^2}\eta(t)^2$$

Thus establish that

$$\lim_N \sum_{t=1}^N \frac{[e(t) - \omega(t)]^2}{\bar{r}(t-1)} \le \lim_N \sum_{t=1}^N \frac{2z(t)^2}{\bar{r}(t-1)}$$

$$+ \lim_N \sum_{t=1}^N \frac{\phi(t-1)^T \bar{P}(t-2)\phi(t-1)}{\bar{r}(t-2)^2}\eta(t)^2$$

$$< \infty.]$$

9

Adaptive Filtering
and Prediction

9.1 INTRODUCTION

In this chapter we show how parameter estimation techniques can be combined with filter design methods to yield adaptive filters and predictors. In this way the information necessary to design the filter or predictor is obtained directly by analyzing data from the system. This is a very attractive alternative, especially in those cases where there is little or no prior knowledge of the signal model, for example when the system is time varying or too complex for a precise signal model to be obtained.

We have seen in Chapter 7 that when the signal model is known, an optimal filter can be designed using Kalman filtering techniques. We have also seen that the Kalman filter corresponds exactly to the optimal one-step-ahead predictor for the observed data. This suggests that the Kalman filter can be fitted to on-line data using the prediction error estimation algorithms of Chapter 8. This idea leads to a class of adaptive filtering algorithms. We shall study these in Section 9.2.

It has been pointed out in Chapter 7 that restricted complexity filters and predictors of arbitrary structure can be designed to optimize a given performance criterion. Again the use of parameter estimation algorithms allows us to turn these into adaptive restricted complexity filters and predictors.

Because of their in-built flexibility, adaptive filters and predictors have found application in a number of diverse problems, including:

1. Antenna systems: see Griffiths (1969), Frost (1972), Gabriel (1976), Monzingo and Miller (1980), and Cantoni (1980).
2. Sonar arrays: see Change and Tuteur (1971) and Owsley (1980).

3. Echo canceling: see Lucky (1966), Gersho (1969), Sondhi and Berkley (1980), and Duttweiler (1980).

4. Noise canceling: see Widrow et al. (1975) and Glover (1977).

5. Speech processing: see Griffiths (1975) and Makhoul and Cosell (1981).

6. Forecasting: see Wittenmark (1974), and Holst (1977).

7. Plant operator guidance: see De Keyzer and Van Cauwenberghe (1979).

The majority of applications fall into the four categories of adaptive state estimation, adaptive deconvolution, adaptive noise canceling, and adaptive prediction. Techniques specific to these applications will be studied in some detail in subsequent sections. However, before going into these details, we present below a brief outline of the motivation for each of these problems as a guide to the reader:

1. Adaptive state estimation arises in several contexts. For example, the state vector may be of interest in its own right. An example of this is flight path reconstruction from observed data in aircraft trials. In other cases, the state may simple be an intermediate variable in a larger design problem such as control system design. In the former case, it is important that the basis for the state space be chosen in such a way that the states have physical significance. In the latter case, the basis for the state space can be chosen arbitrarily (e.g., to facilitate the subsequent design calculations).

2. In adaptive deconvolution, a signal, $s(t)$, is passed through a system, where it is convolved with itself and is also corrupted by noise. The problem then is to "filter" the output from the system so that a good estimate of the original signal, $s(t)$, is obtained. To determine appropriate parameter values for the filter it is usually necessary first to send a known *training sequence* through the system. The filter is then adjusted so that its output is close to the known sequence. Then, when unknown signals are sent through the system, the filter is used without further adjustment. However, periodic retraining of the filter may be desirable to adjust for time variations. This is achieved by retransmitting a known sequence. A major application of this scheme is in the equalization of communication channels.

3. In adaptive noise canceling the basic problem is to remove an unwanted "noise" component from a signal. It is usually assumed that a reference signal is available which contains information about the noise process. A example would be where measurements are corrupted by mains frequency interference. The reference signal can then be drawn from a power outlet. A feature of this kind of adaptive filter is that the adjustment of the filter parameters can be made continuously since the reference signal is always available. There are many applications for noise canceling filters, including echo cancelers, line enhancers, noise suppressors, sidelobe cancelers in antenna arrays, and so on.

4. Prediction is concerned with the extrapolation of a time series into the future. In a sense, one may regard an adaptive predictor as a special kind of adaptive filter in which the filter output is an estimate of the future values of the given series. Alternatively a predictor can be regarded as a particular representation,

or model, of a system. Viewed in this way the problem of adaptive prediction reduces to a simple problem in parameter estimation as in Chapter 8. We will contrast these two alternative viewpoints.

Techniques applicable to each of the problems above will be described in the remainder of the chapter.

9.2 ADAPTIVE OPTIMAL STATE ESTIMATION

In this section we consider adaptive implementation of optimal filters for state estimation. We consider two basic approaches to the problem of constructing an adaptive state estimator. In the first, the state vector is augmented by addition of the unknown, but constant parameters. This leads to a nonlinear filtering problem due to the occurrence of products between parameters and states. The extended Kalman filter of Section 7.7 can then, in principle, be applied to estimate the composite state comprising the original state and the parameters. We shall study this approach in Section 9.2.1. In Section 9.2.2 we describe an alternative approach based on recognizing that the Kalman filter for a linear system coincides with the one-step-ahead predictor. Thus the prediction error algorithm of Chapter 8 can be used to estimate the unknown parameters from input–output data and hence to generate an adaptive filter.

We will discuss the extended Kalman filter and prediction error formulations of the adaptive state estimation problem below.

9.2.1 The Extended Kalman Filter Approach

To set the scene, let us consider a linear state-space model of the form

$$x(t + 1) = Ax(t) + Bu(t) + v_1(t) \tag{9.2.1}$$

$$y(t) = Cx(t) + v_2(t) \tag{9.2.2}$$

where

$$E\left\{ \binom{v_1(s)}{v_2(s)} (v_1(t)^T v_2(t))^T \right\} = \begin{bmatrix} Q & S \\ S^T & R \end{bmatrix} \delta(t - s) \tag{9.2.3}$$

In general, A, B, C, Q, R, and S will all depend on a parameter vector θ. The unknown parameters of the system can be included in an augmented state space model as follows. Define

$$\theta(t + 1) = \theta(t) \tag{9.2.4}$$

and write (9.2.1)–(9.2.2) as

$$x(t + 1) = A(\theta(t))x(t) + B(\theta(t))u(t) + v_1(t) \tag{9.2.5}$$

$$y(t) = C(\theta(t))x(t) + v_2(t) \tag{9.2.6}$$

$$E\left\{ \binom{v_1(s)}{v_2(s)} (v_1(t)^T v_2(t)^T) \right\} = \begin{bmatrix} Q(\theta(t)) & S(\theta(t)) \\ S(\theta(t))^T & R(\theta(t)) \end{bmatrix} \delta(t - s) \tag{9.2.7}$$

We can think of the equations above as being a *nonlinear* stochastic state-space model in terms of the composite vector $(x(t), \theta(t))$. Thus in principle the extended Kalman filter of Section 7.7 can be used.

Hence consider the following augmented state vector:

$$z(t) = \begin{pmatrix} x(t) \\ \theta(t) \end{pmatrix} \tag{9.2.8}$$

We then have the following nonlinear state equations:

$$z(t+1) = f(z(t), u(t), v_1(t)) \tag{9.2.9}$$

$$y(t) = h(z(t), u(t), v_2(t)) \tag{9.2.10}$$

where

$$f(z(t), u(t), v_1(t)) = \begin{bmatrix} A(\theta(t))x(t) + B(\theta(t))u(t) + v_1(t) \\ \theta(t) \end{bmatrix} \tag{9.2.11}$$

$$h(z(t), u(t), v_2(t)) = C(\theta(t))x(t) + v_2(t) \tag{9.2.12}$$

The extended Kalman filter for the model above is then as in (7.7.10) to (7.7.15), with $F(t)$, $H(t)$, $Q(t)$, $\bar{S}(t)$, and $\bar{R}(t)$ taking the following specific values:

$$F(t) = \begin{bmatrix} A(\hat{\theta}(t)) & M(\hat{\theta}(t), \hat{x}(t), u(t)) \\ \hline 0 & I \end{bmatrix} \tag{9.2.13}$$

$$H(t) = [C(\hat{\theta}(t)) \quad D(\hat{\theta}(t), \hat{x}(t))] \tag{9.2.14}$$

$$Q(t) = \begin{bmatrix} Q & 0 \\ 0 & 0 \end{bmatrix} \tag{9.2.15}$$

$$\bar{S}(t) = \begin{bmatrix} S \\ 0 \end{bmatrix} \tag{9.2.16}$$

$$\bar{R}(t) = R \tag{9.2.17}$$

where

$$M(\hat{\theta}(t), \hat{x}(t), u(t)) = \frac{\partial}{\partial \theta}(A(\theta)\hat{x}(t) + B(\theta)u(t))\Big|_{\theta = \hat{\theta}(t)} \tag{9.2.18}$$

$$D(\hat{\theta}(t), \hat{x}(t)) = \frac{\partial}{\partial \theta}(C(\theta)\hat{x}(t))\Big|_{\theta = \hat{\theta}(t)} \tag{9.2.19}$$

It is helpful to introduce the following notation:

$$M_t \triangleq M(\hat{\theta}(t), \hat{x}(t), u(t)) \tag{9.2.20}$$

$$D_t \triangleq D(\hat{\theta}(t), \hat{x}(t)) \tag{9.2.21}$$

$$A_t \triangleq A(\hat{\theta}(t)) \tag{9.2.22}$$

$$B_t \triangleq B(\hat{\theta}(t)) \tag{9.2.23}$$

$$C_t \triangleq C(\hat{\theta}(t)) \tag{9.2.24}$$

$$P_t \triangleq [H(t)\Sigma(t)H(t)^T + \bar{R}(t)] \tag{9.2.25}$$

and the natural block structure

$$\begin{bmatrix} L_1(t) \\ L_2(t) \end{bmatrix} \triangleq L(t) \tag{9.2.26}$$

$$\begin{bmatrix} \Sigma_{11}(t) & \Sigma_{12}(t) \\ \Sigma_{12}(t) & \Sigma_{22}(t) \end{bmatrix} \triangleq \Sigma(t) \tag{9.2.27}$$

Then the extended Kalman filter can be written explicitly as

Extended Kalman Filter (Linear State-Space Model with Unknown Parameters)

$$\hat{x}(t+1) = A_t\hat{x}(t) + B_t u(t) + L_1(t)[y(t) - C_t\hat{x}(t)]; \qquad \hat{x}(0) = \hat{x}_0 \tag{9.2.28}$$

$$\hat{\theta}(t+1) = \hat{\theta}(t) + L_2(t)[y(t) - C_t\hat{x}(t)]; \qquad \hat{\theta}(0) = \hat{\theta}_0 \tag{9.2.29}$$

$$L_1(t) = [A_t\Sigma_{11}(t)C^T + M_t\Sigma_{12}(t)^T C_t^T + A_t\Sigma_{12}(t)D_t^T$$
$$+ M_t\Sigma_{12}(t)D_t^T + S]P_t^{-1} \tag{9.2.30}$$

$$L_2(t) = [\Sigma_{12}(t)^T C_t^T + \Sigma_{22}(t)D_t^T]P_t^{-1} \tag{9.2.31}$$

$$P_t = C_t\Sigma_{11}(t)C_t^T + C_t\Sigma_{12}(t)D_t^T + D_t\Sigma_{12}(t)^T C_t^T + D_t\Sigma_{22}(t)D_t^T + R \tag{9.2.32}$$

$$\Sigma_{11}(t+1) = A_t\Sigma_{11}(t)A_t^T + A_t\Sigma_{12}(t)M_t^T + M_t\Sigma_{12}(t)^T A_t^T + M_t\Sigma_{12}(t)M_t^T$$
$$- L_1(t)P_t L_1(t)^T + Q; \qquad \Sigma_{11}(0) = \Sigma_{11}^0 \tag{9.2.33}$$

$$\Sigma_{12}(t+1) = A_t\Sigma_{12}(t) + M_t\Sigma_{22}(t) - L_1(t)P_t L_2(t)^T; \qquad \Sigma_{12}(0) = \Sigma_{12}^0 \tag{9.2.34}$$

$$\Sigma_{22}(t+1) = \Sigma_{22}(t) - L_2(t)P_t L_2(t)^T; \qquad \Sigma_{22}(0) = \Sigma_{22}^0 \tag{9.2.35}$$

It is to be noted that the algorithm above depends on knowledge of the noise covariances Q, S, and R. This may be unreasonable, especially as we do not assume here knowledge of the system dynamics. It has been observed from simulations and practical applications that the algorithm above may give biased estimates, and may sometimes diverge. The convergence behavior has been investigated by Ljung (1979b), who has demonstrated that the convergence difficulties arise from a combination of factors. These include incorrect specification of noise covariances, and the dependence of the Kalman gain on the parameter estimates. Taking these factors into account, Ljung (1979b) has shown that the algorithm may be improved by the following modification:

Improved Extended Kalman Filter (for Linear Systems with Unknown Parameters)

1. Replace M_t in (9.2.18) and (9.2.20) by M_t^*, whose ith column is given by

$$[M_t^*]_i = [M_t]_i + [K_t]_i(y(t) - C_t x(t)) \tag{9.2.36}$$

where $[K_t]_i$ is the (asymptotic) sensitivity of the Kalman gain, $L_1(t)$, with respect to θ and is defined by

$$[K_t]_i = \left[\frac{\partial}{\partial\theta_i} A(\theta)\Sigma_{11}(t)C_t^T + A_t[\dot{\Sigma}(t)]_i C_t^T \right.$$
$$\left. + A_t\Sigma_{11}(t)\frac{\partial}{\partial\theta_i}C(\theta)^T + \frac{\partial}{\partial\theta_i}S(\theta)\right]P_t^{-1} \tag{9.2.37}$$
$$- L_1(t)P_t^{-1}[\dot{P}_t]_i P_t^{-1}|_{\theta=\hat{\theta}(t)}$$

$$[\dot{P}_t]_i = \left[\frac{\partial}{\partial\theta_i}C(\theta)\Sigma_{11}(t)C_t^T + C_t[\dot{\Sigma}(t)]_i C_t^T \right.$$
$$\left. + C_t\Sigma_{11}(t)\frac{\partial}{\partial\theta_i}C(\theta) + \frac{\partial}{\partial\theta_i}R(\theta)]\right|_{\theta=\hat{\theta}(t)} \tag{9.2.38}$$

$$[\dot{\Sigma}(t+1)]_i = \left[\frac{\partial}{\partial\theta_i}A(\theta)\Sigma_{11}(t)A_t^T + A_t[\dot{\Sigma}(t)]_iA_t^T + A_t\Sigma_{11}(t)\frac{\partial}{\partial\theta_i}A(\theta)^T\right.$$

$$+ \frac{\partial}{\partial\theta_i}Q(\theta) - [K_t]_iP_tL_1(t)^T - L_1(t)[\dot{P}_t]_iL_1(t)^T \qquad (9.2.39)$$

$$\left. - L_1(t)P_t[K_t]_i\right] \qquad \text{evaluated at} \quad \theta = \hat{\theta}(t)$$

2. Replace P_t^{-1} in (9.2.31) by the identity matrix, I.
3. "Regularize" $\Sigma_{22}(t)$ by replacing (9.2.35) by

$$\Sigma_{22}(t+1) = \Sigma_{22}(t) - L_2(t)P_tL_2(t)^T - \delta\Sigma_{22}(t)\Sigma_{22}(t) \qquad (9.2.40)$$

for some positive small δ.

4. Introduce a projection facility to keep $\hat{\theta}(t)$ in a compact subset of

$$D_s = \{\theta \mid A(\theta), C(\theta)) \text{ detectable and } (A(\theta), Q^{1/2}) \text{ stabilizable}\} \qquad (9.2.41)$$

With the modification above, Ljung (1979b) has proved that the estimate, $\hat{\theta}(t)$, converges with probability to a stationary point of the expected prediction error variance, and among isolated stationary points, only local minima are possible convergence points.

Unfortunately, the expressions (9.2.36) to (9.2.39) for the derivatives are rather difficult to implement, especially for high-order systems. A significant simplification is achieved if the system (9.2.5)–(9.2.6) is initially represented in an innovations model format:

$$x(t+1) = A(\theta)x(t) + B(\theta)u(t) + K(\theta)\omega(t)$$

$$y(t) = C(\theta)X(t) + \omega(t) \qquad (9.2.42)$$

with

$$E[\omega(t)\omega(s)^T] = \Lambda(\theta)\,\delta(t-s) \qquad (9.2.43)$$

It then follows that we may take $\Sigma_{11}(t) = 0$, $P_t = \Lambda(\hat{\theta}(t))$, $L_1(t) = K(\hat{\theta}(t)) \triangleq K_t$, and

$$M_t = \frac{\partial}{\partial\theta}(A(\theta)\hat{x}(t) + B(\theta)u(t) + K(\theta)\epsilon(t))\bigg|_{\theta=\hat{\theta}(t)} \qquad (9.2.44)$$

$$\epsilon(t) = y(t) - C_t\hat{x}(t) \qquad (9.2.45)$$

For the model above the extended Kalman filter becomes significantly simpler, as shown below (see also Exercise 9.3).

Extended Kalman Filter (for Innovations Model with Unknown Parameters)

$$\hat{x}(t+1) = A_t\hat{x}(t) + B_tu(t) + K_t\epsilon(t) \qquad (9.2.46)$$

$$\hat{\theta}(t+1) = \hat{\theta}(t) + L_2(t)\epsilon(t) \qquad (9.2.47)$$

$$L_2(t) = [\Sigma_{12}(t)^TC_t^T + \Sigma_{12}(t)D_t^T]\hat{\Lambda}(t)^{-1} \qquad (9.2.48)$$

$$\Sigma_{12}(t+1) = A_t\Sigma_{12}(t) + M_t\Sigma_{22}(t) - K_t\hat{\Lambda}(t)L_2(t)^T \qquad (9.2.49)$$

$$\Sigma_{22}(t+1) = \Sigma_{22}(t) - L_2(t)\hat{\Lambda}(t)L_2(t)^T - \delta\Sigma_{22}(t)\Sigma_{22}(t) \qquad (9.2.50)$$

$$\hat{\Lambda}(t) = \hat{\Lambda}(t-1) + \frac{1}{t}[\Sigma(t)\Sigma(t)^T - \hat{\Lambda}(t-1)] \qquad (9.2.51)$$

Again, it can be shown that $(\hat{\theta}(t), \hat{\Lambda}(t))$ converges with probability to a stationary point of the function

$$V(\theta, \Lambda) = E[\epsilon(t, \theta)^T \Lambda^{-1} \Sigma(t, \theta)] + \log \det \Lambda \qquad (9.2.52)$$

which is the negative log-likelihood function in the gaussian case (see Appendix D).

Remark 9.2.1. In the development above we have considered a linear model. However, it is a simple matter to extend the same idea to nonlinear models. The basic idea is to augment the state vector by adding $\theta(t + 1) = \theta(t)$. Then, in principle, the nonlinear extended Kalman filtering algorithms of Section 7.7 can again be immediately applied. However, the reader is warned of the difficulties inherent in this approach (bias, divergence, etc.) due to the nonlinear nature of the problem. An alternative approach which appears to overcome many of the disadvantages of the brute-force extended Kalman filter approach is described in the next subsection.

▼▼▼

9.2.2 The Prediction Error Approach

As pointed out in the Introduction, the Kalman filter for linear systems actually coincides with a one-step-ahead predictor. This immediately suggests that an *adaptive Kalman filter* can be obtained by direct application of the sequential prediction error algorithm of Chapter 8. An adaptive state estimator can then be constructed as shown in Fig. 9.2.1.

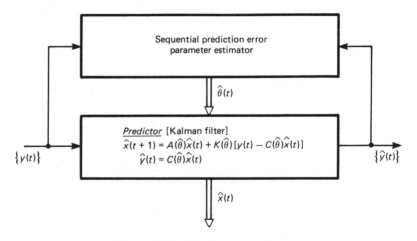

Figure 9.2.1 Adaptive state estimator.

Subject to the conditions discussed in Chapter 8, the sequential prediction error algorithm will converge to a (local) minimum of the mean-square prediction error. Thus this appears to be an attractive way of generating an adaptive filter.

In using the general adaptive state estimator shown in Fig. 9.2.1, one must decide how the system is to be parameterized, that is, what coordinate basis to use for the state, and how the Kalman gain $K(\theta)$ is to be derived. Regarding the latter point, it is highly desirable to parameterize $K(\theta)$ directly in terms of θ rather than indirectly via the Riccati equation. This eliminates a great deal of complexity in fitting the predictor to the data. As to the choice of coordinate basis, it depends on the application. For example, in some applications it may be necessary to use particular coordinate bases so that the state estimates obtained have physical significance (e.g., velocities, temperature, etc.). In other applications, the coordinate basis may be arbitrary and it may be possible to choose the basis so as to simplify subsequent design calculations. For example, the controller state-space from simplifies the determination of state variable feedback controllers. This immediately raises the question of the uniqueness of the state estimates obtained. Since the variable $\{u(t)\}$ and $\{y(t)\}$ are the only measurable data, the best one can hope for is that the predictor output, $\{\hat{y}(t)\}$, approaches the true system output, $\{y(t)\}$. However, this does not necessarily imply that the corresponding state estimate, $\hat{x}(t)$, is a good estimate of states having physical significance. To ensure the latter, additional constraints are required. For example, if it can be shown that $\hat{\theta}(t)$, generated by the parameter estimator, is a good estimate of the true system parameters in a particular parameterization of the model, then the corresponding state estimates will also be good.

The setup in Fig. 9.2.1 is a very natural and simple way to implement an adaptive state estimator and seems to overcome many of the disadvantages associated with the extended Kalman filter approach.

Remark 9.2.2. The prediction error approach to adaptive state estimation can also be applied to nonlinear systems. As we have stated in Section 7.7, there is, in general, no simple optimal predictor for nonlinear systems. However, an advantage of the prediction error algorithms is that they will provide (locally) the best predictor of a given arbitrary structure.

Thus, consider the following nonlinear system:

$$x(t+1) = f(x(t), u(t), \theta_1, v_1(t))$$

$$y(t) = h(x(t), u(t), \theta_2, v_2(t))$$

where $\{v_1(t)\}$ and $\{v_2(t)\}$ are white noise sequences.

Then by reference to the development of the extended Kalman filter in Section 7.7, it may be inferred that a reasonable structure for a one-step-ahead state estimator for this system would have the form

$$\hat{x}(t+1) = f(\hat{x}(t), u(t), \theta_1, 0) + K(\theta_3)[y(t) - \hat{y}(t)]$$

$$\hat{y}(t) = h(\hat{x}(t), u(t), \theta_2, 0)$$

Given the predictor above, the general sequential prediction error methods of Section 8.3.1 are immediately applicable to the estimation of θ_1, θ_2, and θ_3, leading to the adaptive state estimator shown in Fig. 9.2.2. This would seem to be a feasible alternative to the extended Kalman filter in many applications. Two practical points are: (1) a projection facility may again be necessary to ensure that the estimates give rise to a stablefilter; and (2) some provision for parameter time variations is almost certainly necessary, since no single value of θ_3 will give good performance under all conditions due to the restricted complexity nature of the predictor.

▼▼▼

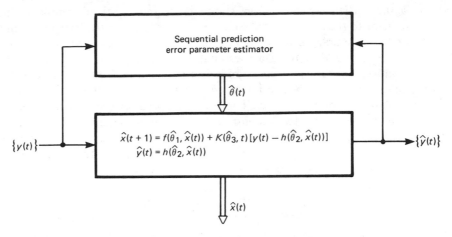

Figure 9.2.2 Adaptive state estimator for nonlinear system.

In the next section we investigate refined forms of adaptive state estimators based on fixed-lag smoothing ideas.

9.2.3 Self-Tuning Fixed-Lag Smoothers

The concept of fixed-lag smoothing was introduced in Section 7.3.2, where it was shown that a lag in the estimation can give improved performance. This happens because additional data are used to generate the estimates.

We recall that many different data-generating mechanisms can give rise to the same output statistics and hence to the same Kalman filter. Thus the state estimate provided by the Kalman filter applies to members of an equivalence class. However, different models in the class will have different state estimation error covariances. In the special case when the data-generating mechanism corresponds to the innovations model, the state estimation error is zero (see Exercise 9.1). In this case there is no benefit to be gained by fixed-lag smoothing. For fixed-lag smoothing to make sense, one has to be more specific about the nature of the data-generating mechanism so that one may distinguish between different members of the equivalence class having the same Kalman filter. We illustrate this by the application of fixed-lag smoothing to a specific problem, adaptive deconvolution. The setup for this problem is as shown in Fig. 9.2.3.

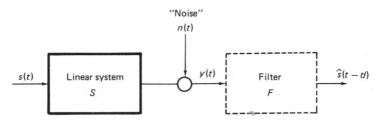

Figure 9.2.3 Signal deconvolution.

We shall require that:

1. The system, S, is linear, finite dimensional, and time invariant.
2. The input signal $\{s(t)\}$ is a stationary stochastic process having rational power density spectrum.

We shall be interested in obtaining an *optimal* fixed-lag smoothed estimate of $s(t - d)$ given observations on $y(t)$ up to time t. We shall investigate two alternative design methods, an indirect approach and a direct approach.

Indirect approach. The determination of the optimal fixed-lag smoother relies on knowledge of the model for $\{s(t)\}$, and the (conditional) model for $\{y(t)\}$ given $\{s(t)\}$. When these models are unknown, they can be estimated using the methods of Chapter 8. Moreover, it suffices to use innovations models for these subsystems as follows:

$$x_1(t + 1) = A_1 x_1(t) + K_1 \omega_1(t) \tag{9.3.53}$$

$$s(t) = C_1 x_1(t) + \omega_1(t) \tag{9.2.54}$$

where

$$E\{\omega_1(t)\omega_1(t)^T\} = \text{prediction error covariance for } s(t)$$

$$= E\{[s(t) - \hat{s}(t)][s(t) - \hat{s}(t)]^T \,|\, s(t-1) \cdots\} \tag{9.2.55}$$

$$\triangleq \Gamma_1$$

$$x_2(t + 1) = A_2 x_2(t) + B_2 s(t) + K_2 \omega_2(t) \tag{9.2.56}$$

$$y(t) = C_2 x_2(t) + \omega_2(t) \tag{9.2.57}$$

where

$$E\{\omega_2(t)\omega_2(t)^T\} = \text{conditional prediction error covariance for } y(t) \text{ given } \{s(t)\}$$

$$= E\{[y(t) - \hat{y}(t)][y(t) - \hat{y}(t)]^T \,|\, y(t-1), \ldots, s(t), \ldots\} \tag{9.2.58}$$

$$\triangleq \Gamma_2$$

Once the models (9.2.53) to (9.2.58) have been estimated by analyzing appropriate data from the system, one can construct a composite state-space model as follows:

$$\begin{bmatrix} x_1(t+1) \\ x_2(t+1) \\ x_3^1(t+1) \\ \vdots \\ x_3^d(t+1) \end{bmatrix} = \begin{bmatrix} A_1 & 0 & 0 & \cdots & 0 \\ B_2 C_1 & A_2 & 0 & \cdots & 0 \\ C_1 & 0 & 0 & \cdots & 0 \\ 0 & 0 & 1 & & \\ & & & \ddots & \\ & & & 1 & 0 \end{bmatrix} \begin{bmatrix} x_1(t) \\ x_2(t) \\ x_3^1(t) \\ \vdots \\ x_3^d(t) \end{bmatrix} + \begin{bmatrix} K_1\omega_1(t) \\ B_2\omega_1(t) + K_2\omega_2(t) \\ \omega_1(t) \\ 0 \\ \vdots \\ 0 \end{bmatrix} \tag{9.2.59}$$

$$y(t) = \begin{bmatrix} 0 & C_2 & 0 & \cdots & 0 \end{bmatrix} \begin{bmatrix} x_1(t) \\ x_2(t) \\ \vdots \\ x_3^d(t) \end{bmatrix} + \omega_2(t) \tag{9.2.60}$$

where
$$x_3^1(t) = s(t-1), \quad \ldots, \quad x_3^d(t) = s(t-d) \qquad (9.2.61)$$

The Kalman filter can then be immediately applied to form a fixed-lag smoothed estimate of $s(t-d)$ given the *output data* $\{y(0), \cdots, y(t)\}$. [Note that the fixed-lag smoother has a slightly different form to that in Section 7.3.2 due to the correlation between the process and measurement noise in (9.2.59)–(9.2.60); see Exercise 9.2.]

In this application there are no restrictions on the coordinate basis and thus one may as well use the observer form. If this is done, a pseudo linear regression algorithm can be used to estimate $A_1, K_1, C_1, \Gamma_1, A_2, B_2, K_2, C_2$, and Γ_2. The estimated parameters can then be used to construct the composite model (9.2.59)–(9.2.60). In order to estimate the parameters in A_2, B_2, K_2, C_2, and Γ_2 it is necessary to send a *known training sequence* through the system. Should the system subsequently change, it is necessary to reuse this training sequence to restimate the parameters.

Note that the special structure of the model in (9.2.59)–(9.2.60) allows one sensibly to use a fixed-lag smoother to estimate $\{s(t)\}$. The point is that (9.2.59)–(9.2.60) is not, in general, the innovations model for $\{y(t)\}$. It is therefore important to estimate the parameters in (9.2.53) to (9.2.58) and not in the innovations model for $\{y(t)\}$. The subsequent determination of the fixed-lag smoother from (9.2.59)–(9.2.60) requires the solution of a matrix Riccati equation and hence the method is an indirect algorithm in the usual sense in which we use this term. In the next subsection we describe a direct method of estimating a fixed-lag smoother which avoids the use of a matrix Riccati equation.

Direct approach. The method used above is an indirect approach in that an appropriate signal model is first obtained and then the smoother is evaluated by solving a matrix Riccati equation. [In very special cases (see Exercise 7.7), the fixed-lag smoother can be obtained by simple algebraic manipulations from the estimated models. However, in general, the determination of the filter requires that a matrix Riccati equation be solved.]

Here we describe an alternative direct method [as originally proposed in Goodwin, Saluja, and Sin (1980)]. The basic idea of this method is that an optimal one-step predictor for $s(t-d)$ based on observation of $y(0), \ldots, y(t-1)$ is constructed as in Section 7.4.2. This predictor will have the following form:

$$C(q^{-1})s^0(t-d\,|\,S_0^{t-d-1}, Y_0^{t-1}) = \alpha(q^{-1})s(t-d-1) + \beta(q^{-1})y(t-1) \qquad (9.2.62)$$

where $s^0(t-d\,|\,S_0^{t-d-1}, Y_0^{t-1})$ denotes the optimal prediction of $s(t-d)$ given $S_0^{t-d-1} \triangleq \{s(0), \ldots, s(t-d-1)\}$ and $Y_0^{t-1} \triangleq \{y(0), \ldots, y(t-1)\}$. The coefficients in the predictor model (9.2.62) can be estimated in a straightforward way by using the techniques of Chapter 8. This requires that a *known training sequence* be used, as is evident from (9.2.62). This is also consistent with the indirect algorithm described above.

A *suboptimal fixed-lag smoother* can then be directly obtained from the estimated predictor (9.2.62) by replacing the unknown past values of $s(t-d)$ by their estimates. This gives the following filter, which is driven by $\{y(t)\}$ alone:

$$C(q^{-1})\hat{s}(t-d) = \alpha(q^{-1})\hat{s}(t-d-1) + \beta(q^{-1})y(t-1) \qquad (9.2.63)$$

or

$$[C(q^{-1}) - q^{-1}\alpha(q^{-1})]\hat{s}(t-d) = \beta(q^{-1})y(t-1) \qquad (9.2.64)$$

It can be shown using fixed-lag and fixed-point smoothing ideas (see Sections 7.3.2 and 7.3.3) that the filter (9.2.64) is stable and that

$$\lim_{d\to\infty} [\hat{s}(t-d) - s^0(t-d|Y_0^{t-1})] = 0 \quad \text{in mean square} \tag{9.2.65}$$

where $s^0(t-d|Y_0^{t-1})$ is the optimal fixed-lag smoothed estimate of $s(t-d)$. Thus for large lag length the performance of the suboptimal filter approaches the performance of the optimal filter. The key point here is that the suboptimal filter (9.2.64) is directly parameterized in terms of the parameters in the special "predictor model" (9.2.62).

In the next section we turn to a related problem, that of predicting the future values of a given time series.

9.3 ADAPTIVE OPTIMAL PREDICTION

We presume we are given a time series $\{y(t)\}$ that we wish to predict into the future. We shall be concerned with predictors of the general form

$$\hat{y}(t+d) = f(\mathcal{Y}(t), \theta) \tag{9.3.1}$$

where $\mathcal{Y}(t) \triangleq \{y(0), \ldots, y(t)\}$, $\{y(t)\}$ is the sequence of interest and $f(\cdot, \cdot)$ is a particular function (possibly nonlinear) which specifies the predictor output.

It is clear that the parameter estimation techniques of Chapter 8 (in particular the prediction error methods) allow us to estimate θ in an on-line fashion by comparing $\hat{y}(t+d)$ with $y(t+d)$ when it becomes available. The estimation criterion of interest is the mean square d-step-ahead prediction error variance, that is,

$$H = E\{[y(t+d) - \hat{y}(t+d)]^2\} \tag{9.3.2}$$

One distinction between adaptive prediction and some of the other adaptive state estimation problems discussed earlier is that in adaptive prediction the basis for the state space is quite arbitrary. Thus one is free to choose the model structure to aid parameter estimation. With this in mind, we will, for simplicity, choose the model in observer state-space form since this model is "linear" in the parameter entries. As pointed out in Chapter 2, the observer form is equivalent to an ARMA model. Thus we will use ARMAX forms for our predictors. Much of the discussion in this section will therefore be based on the following predictor:

$$C(q^{-1})y^0(t+d|t) = \alpha(q^{-1})y(t) + \beta(q^{-1})u(t) \tag{9.3.3}$$

$\{u(t)\}$ is an input sequence.

There are two possible approaches to the problem of adaptive prediction. These are:

1. *Indirect.* A model is fitted to the time series using the on-line parameter estimation techniques of Chapter 8. The on-line estimated model is then used to derive a predictor using the calculations outlined in Chapter 7.

2. *Direct.* The parameters in the optimal predictor are estimated directly.

The advantage of the indirect approach is that there may be fewer parameters to estimate in general. Also, given the model, it would be possible to generate a range of predictors (e.g., one, two steps ahead, etc.).

The advantage of the direct approach is that no additional calculations are required to convert the model into a predictor. Another advantage is that the optimization criterion for parameter estimation is directly expressed in terms of minimization of the prediction error variance. Thus the ultimate objective is achieved regardless of whether or not the assumptions made about the system model are strictly valid.

Both of these approaches will be discussed based on the following ARMAX model for a general linear finite-dimensional system:

$$A(q^{-1})y(t) = q^{-d_1}B'(q^{-1})u(t) + C(q^{-1})\omega(t) \tag{9.3.4}$$

where $\{y(t)\}$, $\{u(t)\}$, and $\{\omega(t)\}$ denote the scalar output, input, and disturbance sequences, respectively. $A(q^{-1})$, $B'(q^{-1})$, and $C(q^{-1})$ are scalar polynomials in the unit delay operator q^{-1}:

$$A(q^{-1}) = 1 + a_1 q^{-1} + \cdots + a_n q^{-n}$$
$$B'(q^{-1}) = b_0 + b_1 q^{-1} + \cdots + b_m q^{-m}$$
$$C(q^{-1}) = 1 + c_1 q^{-1} + \cdots + c_l q^{-l}$$

In (9.3.4), q^{-d_1}, $d_1 \geq 1$, represents a pure time delay. The scalar sequence $\{\omega(t)\}$ satisfies our usual assumption that it is an innovations sequence defined on a probability space $(\Omega, \mathcal{A}, P)$ adapted to the sequence of increasing sub-sigma algebras $(\mathcal{F}_n, n \in N)$, where $\mathcal{F}_n$ is generated by the observations up to time n, and such that $\{\omega(t)\}$ satisfies:

Noise Assumptions

$$E\{\omega(t)|\mathcal{F}_{t-1}\} = 0 \quad \text{a.s.} \tag{9.3.5}$$

$$E\{\omega(t)^2|\mathcal{F}_{t-1}\} = \sigma^2 \quad \text{a.s.} \tag{9.3.6}$$

$$\sup_N \frac{1}{N} \sum_{t=1}^{N} \omega(t)^2 < \infty \quad \text{a.s.} \tag{9.3.7}$$

We will also require the following:

System Assumptions

(i) An upper bound for n, m, and l is known.

(ii) $C(z)$ has all zeros outside the closed unit disk.

Signal Assumptions The data are mean-square bounded in the sense that, with probability 1,

$$\sup_N \frac{1}{N} \sum_{t=1}^{N} y(t)^2 < \infty \tag{9.3.8}$$

$$\sup_N \frac{1}{N} \sum_{t=1}^{N} u(t)^2 < \infty \tag{9.3.9}$$

We observe that we do not require that the polynomial $A(z)$ be stable. Also, our subsequent analysis will allow arbitrary feedback between $\{y(t)\}$ and $\{u(t)\}$. The only requirement is that the data be mean-square bounded in the foregoing sense.

Our objective is to establish that the output of the adaptive predictor converges to the optimal prediction of the output in the sense that, with probability 1,

$$\lim_{N \to \infty} \frac{1}{N} \sum_{t=1}^{N} [y^0(t) - \hat{y}(t)]^2 = 0 \tag{9.3.10}$$

where $\{\hat{y}(t)\}$ is the output of the adaptive predictor and $\{y^0(t)\}$ is the optimal prediction of the output. Note that $\hat{y}(t)$ will be a function of $y(t-1)$, $y(t-2)$, ... and $u(t-1)$, $u(t-2)$, ... only (i.e., we will not assume knowledge of the parameters in the data-generating model).

We discuss direct and indirect methods for achieving this below:

9.3.1 Indirect Adaptive Prediction

In the indirect approach, one first fits a general model to the data, and then for each set of estimated parameters, a d-step-ahead predictor is evaluated using the methods of Chapter 7.

In the case of one-step-ahead prediction, the problem is greatly simplified since it will generally be the case that the model will automatically be a one-step-ahead predictor. Thus the parameter estimation techniques of Chapter 8 can be immediately applied to give an adaptive predictor.

As an illustration, consider the stochastic gradient algorithm described in (8.5.8) to (8.5.13), that is,

$$\hat{\theta}(t) = \hat{\theta}(t-1) + \frac{\phi(t-1)}{r(t-2) + \phi(t-1)^T \phi(t-1)} [y(t) - \hat{y}(t)] \tag{9.3.11}$$

where $\hat{y}(t)$ is the one-step-ahead adaptive prediction, that is,

$$\hat{y}(t) = \phi(t-1)^T \hat{\theta}(t-1) \tag{9.3.12}$$

$$\phi(t-1)^T = (y(t-1), \ldots, y(t-\bar{n}), u(t-d_1), \ldots, u(t-d_1-m),$$
$$- \bar{y}(t-1), \ldots, -\bar{y}(t-l)); \quad \bar{n} = \max(n, l) \tag{9.3.13}$$

$$\bar{y}(t) = \phi(t-1)^T \hat{\theta}(t) \tag{9.3.14}$$

$\hat{\theta}(t)$ is an estimate of the parameters, θ_0, in the model (9.3.4) expressed as a one-step-ahead predictor, that is,

$$C(q^{-1})y^0(t+1 \mid t) = [C(q^{-1}) - A(q^{-1})]y(t) + B'(q^{-1})u(t-d_1+1) \tag{9.3.15}$$

Thus the vector of unknown parameters is

$$\theta_0 = [c_1 - a_1, \ldots, c_{\bar{n}} - a_{\bar{n}}, b_0, \ldots, b_m, c_1, \ldots, c_l] \tag{9.3.16}$$

Finally, the sequence $\{r(t-1)\}$ in (9.3.11) is computed using

$$r(t-1) = r(t-2) + \phi(t-1)^T \phi(t-1); \quad r(-1) = r_0 > 0 \tag{9.3.17}$$

Properties of the algorithm above when applied to the problem of adaptive one-step-ahead prediction are given in

Lemma 9.3.1

(a) Subject to the noise, system, and signal assumptions and provided that $C(z^{-1})$ is input strictly passive, then the adaptive one-step-ahead prediction, $\hat{y}(t)$, given in (9.3.12) converges to the optimal prediction in the following sense:

$$\lim_{N\to\infty} \frac{1}{N} \sum_{t=1}^{N} E\{(y(t) - \hat{y}(t))^2 \,|\, \mathfrak{F}_{t-1}\} = \sigma^2 \quad \text{a.s.} \tag{9.3.18}$$

σ^2 is the optimal prediction error variance.

(b) If the mean-square boundedness assumption for the noise (9.3.7) is strengthened to

$$E\{\omega(t)^4 \,|\, \mathfrak{F}_{t-1}\} < \infty \quad \text{a.s.} \tag{9.3.19}$$

then the result (9.3.17) is strengthened to

$$\lim_{N\to\infty} \frac{1}{N} \sum_{t=1}^{N} (y(t) - \hat{y}(t))^2 = \sigma^2 \quad \text{a.s.} \tag{9.3.20}$$

Proof. From Theorem 8.5.1 we have the following normalized prediction error convergence result:

$$\lim_{N\to\infty} \sum_{t=1}^{N} \frac{[e(t) - \omega(t)]^2}{r(t-1)} < \infty \quad \text{a.s.} \tag{9.3.21}$$

where $e(t) = y(t) - \hat{y}(t)$. Now from Lemma 8.5.1,

$$\eta(t)^2 = \left[\frac{r(t-2)}{r(t-1)}\right]^2 e(t)^2$$
$$\leq e(t)^2 \tag{9.3.22}$$

Hence using (9.3.7), (9.3.8), (9.3.9), and (9.3.22) and the definition of $r(t-1)$, we have (by an application of the Schwarz inequality)

$$\frac{r(N-1)}{N} \leq K_1 + \frac{K_2}{N} \sum_{t=1}^{N} [e(t) - \omega(t)]^2, \qquad N \geq \bar{N} \tag{9.3.23}$$

for constants K_1, K_2, and $\bar{N}$; $0 \leq K_1 < \infty$, $0 < K_2 < \infty$, and $0 < \bar{N} < \infty$.

The result then follows immediately from the stochastic key technical lemma (Lemma 8.5.3).

▼▼▼

Remark 9.3.1. In the result above we have used, for simplicity of analysis, the stochastic gradient algorithm. Of course, least-squares-based algorithms will generally be preferable because of their superior convergence rate. There is no difficulty extending Lemma 9.3.1 to other algorithms that satisfy the normalized prediction error convergence result (9.3.21), for example, the least-squares algorithm applied in the bounded data case (see Exercise 9.6).

Turning now to the case of d-step-ahead prediction, we note that an indirect adaptive d-step-ahead predictor can be obtained as follows. Use the estimated parameters in a one-step-ahead predictor of the type described above to construct a d-step-ahead predictor using the calculations outlined in Section 7.4.2; that is, given $\hat{\theta}(t)$ as in (9.3.11), we form

$$\hat{C}(t, q^{-1}) = 1 + \hat{c}_1(t)q^{-1} + \cdots + \hat{c}_l(t)q^{-l} \tag{9.3.24}$$

$$\hat{A}(t, q^{-1}) = 1 + (\hat{c}_1(t) - \hat{a}_1(t))q^{-1} + \cdots + (\hat{c}_{\hat{n}}(t) - \hat{a}_{\hat{n}}(t))q^{-\hat{n}} \tag{9.3.25}$$

$$\hat{B}(t, q^{-1}) = [\hat{b}_0(t) + \cdots + \hat{b}_m(t)q^{-m}]q^{-d_1} \tag{9.3.26}$$

Then solve the following prediction equality for

$$\hat{F}(t, q^{-1}) = 1 + \hat{f}_1(t)q^{-1} + \cdots + \hat{f}_{d-1}(t)q^{-d+1} \qquad \text{and}$$

$$\hat{G}(t, q^{-1}) = \hat{g}_0(t) + \hat{g}_1(t)q^{-1} + \cdots + \hat{g}_{\hat{n}-1}(t)q^{-\hat{n}+1}:$$

$$\hat{C}(t, q^{-1}) = \hat{F}(t, q^{-1})\hat{A}(t, q^{-1}) + q^{-d}\hat{G}(t, q^{-1}) \tag{9.3.27}$$

Finally, form the adaptive d-step-ahead prediction, $\hat{y}(t + d)$, as

$$\hat{y}(t + d) = q[1 - \hat{C}(t, q^{-1})]\hat{y}(t + d - 1)$$
$$+ \hat{G}(t, q^{-1})y(t) + [\hat{F}(t, q^{-1})\hat{B}(t, q^{-1})]u(t + d) \tag{9.3.28}$$

Properties of the algorithm above are described in:

Lemma 9.3.2. Provided that $C(z^{-1})$ is strictly input passive and that the system input, output, and noise are absolutely bounded almost surely, the above indirect d-step-ahead adaptive predictor is globally convergent in the following sense:

$$\lim_{N \to \infty} \frac{1}{N} \sum_{t=1}^{N} E\{(y(t) - \hat{y}(t))^2 \,|\, \mathcal{F}_{t-d}\} = \gamma^2 \quad \text{a.s.} \tag{9.3.29}$$

where γ^2 is the *optimal d-step-ahead prediction error variance.*

Proof. Follows the same technique as the proof of Lemma 4.3.2 using the parameter norm convergence, parameter difference conference, and normalized prediction error convergence, as in Theorem 8.5.1. (The details are requested in Exercise 9.7.)

▼▼▼

In the next section we describe a direct form of an adaptive d-step-ahead predictor.

9.3.2 Direct Adaptive Prediction

To motivate our adaptive algorithm, we first develop a special form for the optimal d-step-ahead predictor.

Lemma 9.3.3. The optimal d-step-ahead predictor $y^0(t + d | t)$ for the model (9.3.4) satisfies an equation of the form

$$y^0(t + d | t) = \bar{G}(q^{-1})y^0(t | t - d) + \alpha(q^{-1})y(t) + q^d \beta(q^{-1})u(t - d_1) \tag{9.3.30}$$

where

$$y^0(t + d | t) = E\{y(t + d) | \mathcal{F}_t\} \tag{9.3.31}$$

and $\alpha(q^{-1})$, $\beta(q^{-1})$, and $\bar{G}(q^{-1})$ are finite-order polynomials in the unit delay operator q^{-1}. (We take the orders to be $n_1 - 1$, $n_2 - 1$, and $n_3 - 1$, respectively.)

Proof. We recall from Lemma 7.4.1 that there exist unique polynomials such that

$$C(q^{-1}) = F(q^{-1})A(q^{-1}) + q^{-d}G(q^{-1}) \tag{9.3.32}$$

where

$$F(q^{-1}) = 1 + f_1 q^{-1} + \cdots + f_{d-1}q^{-d+1} \tag{9.3.33}$$

$$G(q^{-1}) = g_0 + g_1 q^{-1} + \cdots + g_{n-1}q^{-n+1} \tag{9.3.34}$$

Hence multiplying (9.3.4) by $F(q^{-1})$ and using (9.3.32), we have

$$C(q^{-1})[y(t) - F(q^{-1})\omega(t)] = q^{-d}G(q^{-1})y(t) + q^{-d_1}F(q^{-1})B'(q^{-1})u(t) \quad (9.3.35)$$

or

$$C(q^{-1})y^0(t \mid t - d) = q^{-d}G(q^{-1})y(t) + q^{-d_1}F(q^{-1})B'(q^{-1})u(t) \quad (9.3.36)$$

where

$$y^0(t \mid t - d) \triangleq y(t) - F(q^{-1})\omega(t) \quad (9.3.37)$$

Using the division algorithm again, we write

$$1 = \bar{F}(q^{-1})C(q^{-1}) + q^{-d}\bar{G}(q^{-1}) \quad (9.3.38)$$

Applying this to (9.3.36) gives

$$[1 - q^{-d}\bar{G}(q^{-1})]y^0(t \mid t - d) = q^{-d}G(q^{-1})\bar{F}(q^{-1})y(t) + q^{-d_1}F(q^{-1})\bar{F}(q^{-1})B'(q^{-1})u(t)$$

or

$$y^0(t \mid t - d) = \bar{G}(q^{-1})y^0(t - d \mid t - 2d) + \alpha(q^{-1})y(t - d)$$
$$+ \beta(q^{-1})u(t - d_1)$$

where

$$\alpha(q^{-1}) = G(q^{-1})\bar{F}(q^{-1})$$
$$\beta(q^{-1}) = F(q^{-1})\bar{F}(q^{-1})B'(q^{-1})$$

▼▼▼

Note that in the special form (9.3.30) for the optimal predictor, $y^0(t \mid t - d)$ depends on $y^0(t - d \mid t - 2d) \cdots y^0(t - d - n_3 \mid t - 2d - n_3)$.

Note also that d_1 is the system time delay, whereas d is the number of steps we look ahead in the predictor. These need not necessarily be the same. If $d_1 \geq d$ or the input, $\{u(t)\}$, is known for all time, no difficulty arises. However, in cases where the input in known only up to the present time and $d_1 < d$, an alternative approach is required. For ease of presentation we assume in the sequel that either $d_1 \geq d$ or the input is known for d steps ahead.

The optimal predictor (9.3.30) is a function of a parameter vector θ_0 which contains the coefficients of the polynomials $\alpha(q^{-1})$, $\beta(q^{-1})$, and $\bar{G}(q^{-1})$. In the adaptive predictor presented below, θ_0 is replaced by an estimate $\hat{\theta}(t)$. Also, $y^0(t - d \mid t - 2d)$ $\cdots y^0(t - d - n_3 \mid t - 2d - n_3)$ are replaced by estimates depending on $\hat{\theta}(t - d)$, $\ldots$, $\hat{\theta}(t - d - n_3)$.

The adaptive predictor that we study here is an improved version of the one given in Goodwin, Sin, and Saluja (1980), where interlaced recursions were used. Basically, we shall replace the interlacing of algorithms by the interlacing of a set of Lyapunov functions using a technique due to Moore and Kumar (1980).

The Direct Adaptive d-Step-Ahead Predictor Based on Least Squares

Consider the model of (9.3.4). The adaptive prediction $\hat{y}(t)$ for $t \geq d$ is given by

$$\hat{y}(t) = \phi(t - d)^T\hat{\theta}(t - d) \quad (9.3.39)$$

where $\hat{\theta}(t - d)$ is a sequentially computed parameter estimate given by

$$\hat{\theta}(t) = \hat{\theta}(t - 1) + \frac{P(t - d - 1)\phi(t - d)\epsilon(t)}{1 + \phi(t - d)P(t - d - 1)\phi(t - d)} \quad (9.3.40)$$

where

$$\epsilon(t) = y(t) - \phi(t-d)^T \hat{\theta}(t-1) \tag{9.3.41}$$

$$\phi(t-d)^T = [y(t-d), \ldots, y(t-d-n_1),$$

$$u(t-d_1), \ldots, u(t-d_1-n_2), \tag{9.3.42}$$

$$\bar{y}(t-d), \ldots, \bar{y}(t-d-n_3)]$$

and where $\bar{y}(t-d)$ is the a posteriori prediction given by

$$\bar{y}(t) = \phi(t-d)^T \hat{\theta}(t) \tag{9.3.43}$$

with $\bar{y}(\tau) = 0$ for $\tau \leq d-1$.

Finally, the matrix $P(t-d)$ is computed using the condition number monitoring scheme of Remark 8.5.2, that is,

$$P'(t-d) = P(t-d-1) - \frac{P(t-d-1)\phi(t-d)\phi(t-d)^T P(t-d-1)}{1 + \phi(t-d)^T P(t-d-1)\phi(t-d)},$$

$$P(-1) > 0 \tag{9.3.44}$$

$$\bar{r}(t-d) = \bar{r}(t-d-1)(1 + \phi(t-d)^T P(t-d-1)\phi(t-1)),$$

$$\bar{r}(-1) > 0 \tag{9.3.45}$$

If

$$(\bar{r}(t-d)\lambda_{\max} P'(t-d) \leq K), \qquad 0 < K < \infty \tag{9.3.46}$$

then

$$P(t-d) = P'(t-d) \tag{9.3.47}$$

Otherwise,

$$P(t-d) = \frac{K}{\bar{r}(t-d)\lambda_{\max} P'(t-d)} P'(t-d) \tag{9.3.48}$$

The prediction error for the algorithm above is

$$e(t) = y(t) - \hat{y}(t) \tag{9.3.49}$$

We also define the a posteriori prediction error,

$$\eta(t) = y(t) - \bar{y}(t) \tag{9.3.50}$$

The reader will note that the algorithm above is basically the least-squares algorithm of Chapter 8 generalized slightly to account for the d-step-ahead prediction.

The convergence analysis of the algorithm above employs the same basic technique as used in Lemma 9.3.1. We will not go into details, but we ask readers to explore it themselves, if interested, in the exercises.

Remark 9.3.2 (Adaptive Predictors for Systems with Zeros on the Unit Circle). We have seen in Section 7.4.2 that the time-invariant predictor (assuming it exists) cannot be used in the case when the noise model has zeros on the unit circle since, in this case, the effect of incorrect initial condition will not decay. However, nearly optimal performance can be achieved by a linear time-invariant predictor of the form (see Lemma 7.4.4)

$$C_s(q^{-1})\hat{y}(t+d) = \alpha(q^{-1})y(t) + \beta(q^{-1})u(t) \tag{9.3.51}$$

The predictor above is motivated from the optimal predictor derivation. However, as we have seen in Section 7.4.2, it is, in fact, a suboptimal predictor having a restricted complexity structure to guarantee stability of $C_s(q^{-1})$. This suggests the use of the prediction error parameter estimation algorithm, which gives the restricted complexity optimal solution. This leads to the adaptive predictor described below. [For simplicity we present the algorithm when $u(t) = 0$. The extension to the case $u(t) \neq 0$ is immediate.]

$$\hat{\theta}(t) = \hat{\theta}(t-1) + P(t-d)\phi(t-d)[y(t) - \hat{y}(t)] \tag{9.3.52}$$

$$\phi(t)^T = [y_F(t), \ldots, y_F(t-n+1),$$
$$-\hat{y}_F(t+d-1), \ldots, -\hat{y}_F(t+d-l_s)]^T \tag{9.3.53}$$

$$\bar{\phi}(t) = [y(t), \ldots, y(t-n+1),$$
$$-\hat{y}(t+d-1), \ldots, -\hat{y}(t+d-l_s)]^T \tag{9.3.54}$$

$$\hat{y}(t+d) = \bar{\phi}(t)^T \hat{\theta}(t+d-1) \tag{9.3.55}$$

$$\hat{\theta}(t)^T = [\hat{a}_0(t), \ldots, \hat{a}_{n-1}(t), -\hat{c}_{s1}(t), \ldots, -\hat{c}_{sl_s}(t)]^T \tag{9.3.56}$$

$$y_F(t) = -\sum_{j=1}^{l_s} \hat{c}_{sj}(t+d-1)y_F(t-j) + y(t) \tag{9.3.57}$$

$$\hat{y}_F(t+d) = -\sum_{j=1}^{l_s} \hat{c}_{sj}(t+d-1)\hat{y}_F(t+d-j) + \hat{y}(t+d) \tag{9.3.58}$$

$$P(t-d) = \frac{1}{\alpha(t-d)} \left\{ P(t-d-1) \right.$$
$$\left. - \frac{P(t-d-1)\phi(t-d)\phi(t-d)^T P(t-d-1)}{\alpha(t-d) + \phi(t-d)^T P(t-d-1)\phi(t-d)} \right\} \tag{9.3.59}$$
$$0 < \alpha(t-d) \leq 1$$

The adaptive d-step-ahead prediction is given by (9.3.55).

The algorithm above is basically the sequential prediction error algorithm discussed in Chapter 8. As usual it will be necessary to project $\hat{\theta}(t)$ so that the polynomial $1 + \hat{c}_{s1}(t)q^{-1} + \cdots + \hat{c}_{sl_s}(t)q^{-l_s}$ remains stable.

The weighting coefficient $\alpha(t-d)$ in (9.3.59) allows exponential data weighting to be introduced in the algorithm if required to handle time-varying systems. (See a full discussion in Section 6.7.)

It can be seen from Section 7.4.2 that for $C_s(q^{-1})$ close to $C(q^{-1})$ the prediction error is very nearly equal to the optimal uncorrelated future noise sequence. Thus one would expect that, in practice, other algorithms, including pseudo linear regression algorithms, could be used provided that the estimated $C_s(q^{-1})$ polynomial is kept inside the unit circle. Also, if the deterministic components of the disturbances dominate the nondeterministic components, then $C_s(q^{-1})$ can be chosen more or less arbitrarily to have a simple form without seriously affecting the performance.

▼▼▼

The restricted complexity predictor described above has been applied (Goodwin and Chan, 1982) to a number of the series, including the generation of a 12-month-ahead prediction for the "series C" of monthly soft drink data of Gersovitz and Mac-Kinnon (1978, Table 4). The same series has also been studied by Akaike (1980). The restricted complexity predictor has been found to work very satisfactorily in all cases tested and is accordingly recommended as a feasible alternative to other methods of predicting the series having significant drift and periodic components.

9.4 RESTRICTED COMPLEXITY ADAPTIVE FILTERS

In all of the development above we have aimed to adaptively generate the unrestricted optimal filter or predictor. However, as pointed out in Section 7.5, it is often more practical to use a filter of prespecified structure (i.e., having *restricted complexity*) and then to seek the "best" filter of the given structure. This is especially important in those applications where the optimal filter (even if it exists) is too complex to be of any use.

In this section we briefly explore the idea of adaptive restricted complexity filters, with particular reference to signal recovery and noise canceling.

9.4.1 Adaptive Deconvolution

The basic setup for deconvolution is as shown in Fig. 9.4.1. A data sequence, $s(t)$, is passed through a dynamic system, G, wherein it is convolved with itself and has noise, $n(t)$, added to it. The filtering objective is to design a filter F such that the output from the filter, $\hat{s}(t - d)$, is a good estimate of a delayed version of $s(t)$.

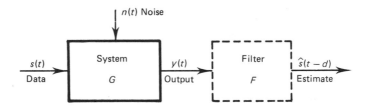

Figure 9.4.1 Deconvolution.

For example, if the system response is linear and is described by a finite impulse response sequence $\{g_k: k = 0, \ldots, N\}$, the system output is given by

$$y(t) = \sum_{k=0}^{N} g_k s(t - k) + n(t) \tag{9.4.1}$$

$$= g_d s(t - d) + \sum_{\substack{k=0 \\ k \neq d}}^{N} g_k s(t - k) + n(t) \tag{9.4.2}$$

The first term in (9.4.2) corresponds to a signal proportional to the transmitted data at time $t - d$. The second term represents the cumulative effect of adjacent data symbols due to the fact that the impulse response is nonzero for $k \neq d$. This term is commonly referred to in communication applications as *intersymbol interference*. The last term in (9.4.2) represents additive noise. A typical nonideal impulse response producing intersymbol interference is shown in Fig. 9.4.2.

We have seen in Section 7.5 that it is possible to design restricted complexity linear filters of both transversal and recursive type to produce an estimate $\hat{s}(t - d)$ which is a good estimate of $s(t - d)$. The filter aims to eliminate or reduce the inter-

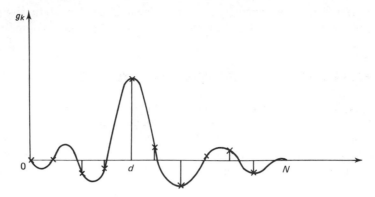

Figure 9.4.2 Nonideal impulse response producing intersymbol interference.

symbol interference. It can be seen from Fig. 9.4.1 that the filter F acts as a kind of inverse to the system. We first treat the case when F is a transversal filter.

Transversal Filters

In the case of a transversal filter, the filter output is described by

$$\hat{s}(t - d) = F(q^{-1})y(t) \tag{9.4.3}$$

where

$$F(q^{-1}) = f_0 + f_1 q^{-1} + \cdots + f_n q^{-n}$$

The corresponding mean-square filtering error is then given by

$$
\begin{aligned}
J &= E\{[s(t - d) - \hat{s}(t - d)]^2\} \\
&= E\{[s(t - d) - F(q^{-1})y(t)]^2\}
\end{aligned} \tag{9.4.4}
$$

We have seen in Chapter 8 that the criterion (9.4.4) can be asymptotically minimized by adjusting the coefficients in the filter on-line by using the following least-squares algorithm:

$$\hat{\theta}(t) = \hat{\theta}(t - 1) + \frac{P(t - 2)\phi(t - 1)}{1 + \phi(t - 1)^T P(t - 2)\phi(t - 1)}[s(t - d) - \phi(t - 1)^T \hat{\theta}(t - 1)] \tag{9.4.5}$$

where

$$\phi(t - 1)^T = [y(t), \ldots, y(t - n)]^T \tag{9.4.6}$$

$$P(t - 1) = \left[P(t - 2) - \frac{P(t - 2)\phi(t - 1)\phi(t - 1)^T P(t - 2)}{1 + \phi(t - 1)^T P(t - 2)\phi(t - 1)} \right] \tag{9.4.7}$$

$$\hat{\theta}(t) = (\hat{f}_0(t), \ldots, \hat{f}_n(t))^T \tag{9.4.8}$$

Of course, the algorithm above can be replaced by any of the algorithms discussed in Chapter 8: LMS, NMLS, and so on.

Recursive Filters

It is possible that a very large number of coefficients may be required in a transversal filter to give satisfactory performance. An *infinite impulse filter* (IIF) can be obtained by using a *recursive filter* (i.e., one having poles in the filter transfer function).

This class of filter has been studied by many authors, including White (1975), Feintuch (1976), Parikh and Ahmed (1978), Kikuchi, Omati, and Soeda (1979), Larimore, Treichler, and Johnson (1980), and Johnson et al. (1981). The general recursive filter has the following form:

$$H(q^{-1})\hat{s}(t - d) = F(q^{-1})y(t) \qquad (9.4.9)$$

where

$$H(q^{-1}) = 1 + h_1 q^{-1} + \cdots + h_m q^{-m}$$
$$F(q^{-1}) = f_0 + f_1 q^{-1} + \cdots + f_n q^{-n}$$

In this case the mean-square filtering error is given by

$$J = E\{[s(t - d) - \hat{s}(t - d)]^2\} \qquad (9.4.10)$$

$$= E\left\{\left[s(t - d) - \frac{F(q^{-1})}{H(q^{-1})}y(t)\right]^2\right\} \qquad (9.4.11)$$

We have seen in Section 7.5.1 that the mean-square filtering error, J, given in (9.4.11) is a nonquadratic function of the filter coefficients. In this case there are three possible approaches. These are described below.

1. We replace J given in (9.4.11) by a related criterion function of the following mean-square filtered error:

$$J_f = E\{[H(q^{-1})(s(t - d) - \hat{s}(t - d))]^2\} \qquad (9.4.12)$$

$$= E\{[H(q^{-1})s(t - d) - F(q^{-1})y(t)]^2\} \qquad (9.4.13)$$

As pointed out in Section 7.5, minimizing J_f is *not* equivalent to minimizing J. However, J_f has the advantage that it is quadratic in the filter coefficients. Thus the filter parameters, $\hat{\theta}(t)$, where $\hat{\theta}(t) \triangleq (\hat{h}_1(t), \ldots, \hat{h}_m(t), \hat{f}_0(t), \ldots, \hat{f}_n(t))^T$, can be estimated by algorithms as for the transversal case but with ϕ now given by

$$\phi(t - 1)^T = [-s(t - d - 1), \ldots, -s(t - d - m), y(t), \ldots, y(t - n)]^T \qquad (9.4.14)$$

2. We can work with the nonquadratic function, J, as in (9.4.11) and simply regard the filter output $\hat{s}(t - d)$ as a prediction of $s(t - d)$. We can then immediately apply the *sequential prediction error* algorithm of Section 8.3 to estimate the filter coefficients. For this particular problem, the prediction error algorithm (8.3.60) to (8.3.66) with data weighting becomes

$$\hat{\theta}(t) = \hat{\theta}(t - 1) + P(t - 1)\psi(t - 1)[s(t - d) - \hat{s}(t - d)] \qquad (9.4.15)$$

$$P(t - 1) = \left[P(t - 2) - \frac{P(t - 2)\psi(t - 1)\psi(t - 1)^T P(t - 2)}{1 + \psi(t - 1)^T P(t - 2)\psi(t - 1)}\right] \qquad (9.4.16)$$

$$\hat{s}(t - d) = [1 - \hat{H}(q^{-1}, \hat{\theta}(t - 1))]\hat{s}(t - d) + \hat{F}(q^{-1}, \hat{\theta}(t - 1))y(t)$$
$$\triangleq \phi(t - 1)^T \hat{\theta}(t - 1) \qquad (9.4.17)$$

where

$$\phi(t - 1)^T = [-\hat{s}(t - d - 1), \ldots, -\hat{s}(t - d - m), y(t), \ldots, y(t - n)]^T \qquad (9.4.18)$$

$$\hat{H}[q^{-1}, \hat{\theta}(t - 1)]\psi(t - 1) = \phi(t - 1) \qquad (9.4.19)$$

It has been shown in Chapter 8 that, subject to reasonable conditions, this algorithm will converge to a local minimum of the criterion function (the restricted complexity optimum). Thus the algorithm (9.4.15)–(9.4.19) immediately gives an adaptive infinite impulse response filter.

3. A simplified form of the algorithm above is obtained if the filtering operation in (9.4.19) is replaced by $\psi(t-1) = \phi(t-1)$. Note that this gives a *pseudo linear regression* algorithm. As we have seen in Chapter 8, an improved form of this algorithm is obtained by replacing $\hat{s}(t - d - 1)$, ... in the regression vector, $\phi(t-1)$, by $\bar{s}(t - d - 1)$, ... where $\bar{s}(t - d - 1)$ is an *a posteriori prediction* given by

$$\bar{s}(t - d - 1) = \phi(t - 2)^T \hat{\theta}(t - 1) \qquad (9.4.20)$$

This algorithm and its gradient form can also be used to obtain infinite impulse response filters. This is a very reasonable approach to parameter estimation, but convergence depends on more restrictive assumptions than the prediction error method described in approach 2. In particular, Theorem 8.6.2 shows that in theory convergence of the algorithm depends on

1. The system G having a structure commensurate with that used to develop the algorithm. [For example, application of Theorem 8.6.2 to the algorithm (9.4.15)–(9.4.19) requires that the system be described as in Fig. 9.4.3, where $\{n(t)\}$ is white noise.]

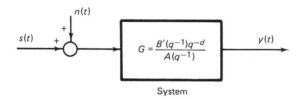

Figure 9.4.3 Model for pseudo linear regression algorithm in adaptive deconvolution.

2. A passivity condition holding for the numerator polynomial in the system transfer function G. (See Exercise 9.14.)

Remark 9.4.1. It is clear that the algorithms above require that $\{s(t)\}$ be known. This implies that a known training sequence must be used to adjust the filter parameters. When the filter is used with an unknown data sequence, the parameter update is usually turned off. This arrangement of training and normal operation is depicted in Fig. 9.4.4.

Of course, a similar approach can be used with finite-alphabet inputs, as discussed in Remark 7.5.1.

9.4.2 Adaptive Noise Canceling

In this section we consider an alternative filtering problem wherein an additional measurement (the reference signal) is available which gives information about the noise interference acting on the observed data. The setup is as shown in Fig. 9.4.5.

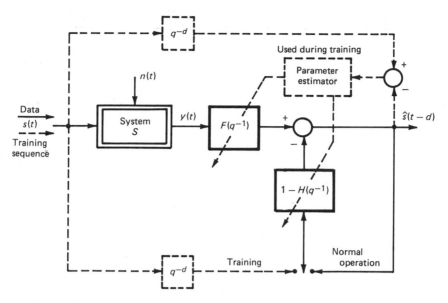

Figure 9.4.4 Training and normal operation of adaptive deconvolution filter.

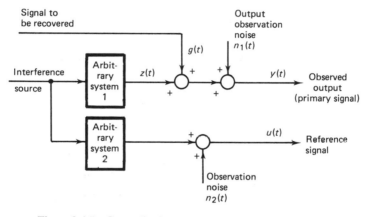

Figure 9.4.5 Generalized signal model for noise cancelling.

Our aim is to design a filter driven by $u(t)$, the reference signal, so that the filter output approximates $z(t)$, the unknown interference in the primary signal, $y(t)$ (which contains the signal to be recovered). If we then subtract the filter output from the primary signal, $y(t)$, we will get an estimate of the signal to be recovered [i.e., $g(t)$]. This motivates the term "noise canceling."

For the moment, let us assume that the noise-canceling filter is of moving average or transversal type. Then the filter output which is an estimate of the unknown interference is given by

$$\hat{z}(t - d) = F(q^{-1})u(t) \qquad (9.4.21)$$

where

$$F(q^{-1}) = f_0 + f_1 q^{-1} + f_2 q^{-2} + \cdots + f_n q^{-n}.$$

The noise-canceler output which provides an estimate of the signal to be recovered is then given by

$$\hat{g}(t - d) = y(t - d) - \hat{z}(t - d) \qquad (9.4.22)$$

The noise canceler is illustrated in Fig. 9.4.6. We now suggest that $F(q^{-1})$ can be estimated by minimizing the power in $\{\hat{g}(t)\}$; that is, we minimize

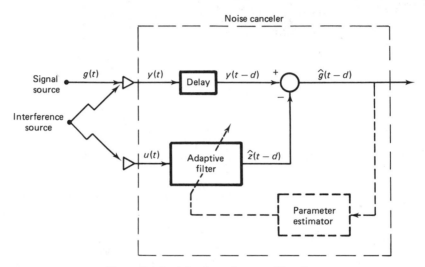

Figure 9.4.6 Adaptive noise-cancelling filter.

$$J = E\{[\hat{g}(t - d)]^2\} \qquad (9.4.23)$$

The rationale behind this choice can be seen by noting from Figs. 9.4.5 and 9.4.6 that

$$J = E\{[y(t - d) - \hat{z}(t - d)]^2\}$$
$$= E\{[g(t - d) + n_1(t - d) + z(t - d) - \hat{z}(t - d)]^2\} \qquad (9.4.24)$$

then provided that $g(t)$, $z(t)$, $n_1(t)$, and $n_2(t)$ are mutually uncorrelated, we have

$$J = E\{n_1(t - d)^2\} + E\{g(t - d)^2\} + E\{[z(t - d) - \hat{z}(t - d)]^2\} \qquad (9.4.25)$$

The first two terms in (9.4.25) are independent of $F(q^{-1})$. Thus minimizing J with respect to $F(q^{-1})$ has the effect of minimizing $E\{[z(t - d) - \hat{z}(t - d)]^2\}$. Note that the noise canceler relies on the reference signal $\{u(t)\}$ being highly correlated with the interference $\{z(t)\}$ but weakly correlated (theoretically zero correlation) with the desired signal $\{g(t)\}$. The performance of the noise canceler depends critically on this assumption.

We can use any parameter estimation algorithm to estimate $F(q^{-1})$ so that (9.4.23) is asymptotically minimized. For Example, the least-squares algorithm is

$$\hat{\theta}(t) = \hat{\theta}(t - 1) + \frac{P(t - 2)\phi(t - 1)}{\alpha + \phi(t - 1)^T P(t - 2)\phi(t - 1)}$$
$$\times [y(t - d) - \phi(t - 1)^T \hat{\theta}(t - 1)] \qquad (9.4.26)$$

where

$$\phi(t-1)^T = [u(t), u(t-1), \ldots, u(t-n)]^T$$
$$0 < \alpha \leq 1 \tag{9.4.27}$$

with $P(t-1)$ satisfying the usual matrix equation.

Alternatively, (9.4.26) can be replaced by a stochastic gradient type of algorithm; for example, the LMS algorithm is

$$\hat{\theta}(t) = \hat{\theta}(t-1) + \mu\phi(t-1)[y(t-d) - \phi(t-1)^T\hat{\theta}(t-1)] \tag{9.4.28}$$

with μ a scalar.

In the description above we have considered the use of a transversal filter in the noise canceler. This can, of course, be generalized to the use of a recursive filter. We then define the filter output via

$$H(q^{-1})\hat{z}(t-d) = F(q^{-1})u(t) \tag{9.4.29}$$

We can again proceed to adjust the filter so as to minimize

$$J = E\{[y(t-d) - \hat{z}(t-d)]^2\} \tag{9.4.30}$$

Note that $\hat{z}(t-d)$ is a nonlinear function of the filter parameters and thus the comments made in Section 9.4.1 apply *mutatis mutandis* here as well.

A number of applications of the adaptive noise-canceling principle to signal processing have been described in the literature. An excellent survey of applications up to 1975 is contained in the paper by Widrow et al. (1975). We shall briefly indicate how the algorithms above can be employed in some of these applications. First, however, it is important to draw a distinction between the deconvolution algorithms of Section 9.4.1 and the noise-canceling algorithm of this section. In the former case, the objective was to estimate $\{s(t)\}$ given observations of $\{y(t)\}$. To estimate the filter parameters a known training sequence $\{s(t)\}$ was required. Thus the usual pattern would be to send a known training sequence initially to estimate the parameters and then to subsequently use the filter to estimate the unknown data $\{s(t)\}$. This makes continuous adaptation of the filter difficult. A common practice is to send the known training sequence periodically to avoid cumulative estimation drift.

By way of contrast, in the noise cancelling case, the reference signal $\{u(t)\}$ is always available and thus continuous adaptation of the filter parameters can be achieved in a straightforward way.

To illustrate how the adaptive noise-canceling principle is applied in practice, we briefly describe three typical applications:

1. *Canceling of mains frequency interference in electrocardiography.* A problem in electrocardiogram (ECG) recording is the appearance of mains frequency components on the output. Figure 9.4.7 shows how an adaptive noise canceler could be used to mitigate this effect.

    Here the adaptive filter requires only two variable weights, one applied to the reference signal directly and one applied to the reference signal delayed by 90°. This allows estimation of the magnitude and phase of the nuisance signal $z(t)$.

2. *Echo cancelers and adaptive hybrids.* A problem in simultaneous two-way data

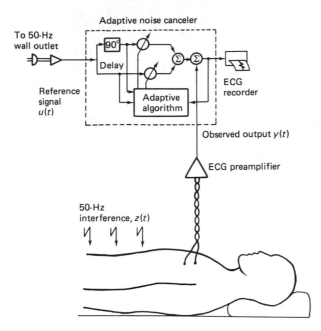

Figure 9.4.7 Canceling 50-Hz interference in electrocardiography.

communication over a single twisted pair is that the received signal contains components (echoes) of the send signal. These components arise from two sources. One is due to imperfect operation of the hybrid at the near end and the other is due to reflections arising from imperfect termination at the far end. This is illustrated in Fig. 9.4.8. An adaptive noise canceler can be used to remove the unwanted echos in the received signal using the send signal as the reference.

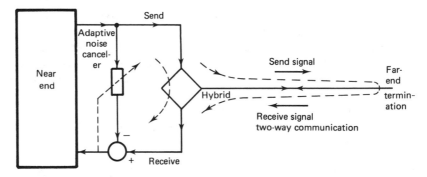

Figure 9.4.8 Adaptive echo canceler.

3. *Adaptive line enhancer.* An adaptive filter configuration known as the adaptive line enhancer (Widrow et al., 1975) can be used for the detection of sinusoidal signals in wideband noise. The key point here is that because of the periodic nature of the sinusoidal signals, they are strongly correlated with their past values. This suggests that a reference signal can be obtained from a delayed version of the input signal (the primary signal). The setup is as shown in Fig. 9.4.9.

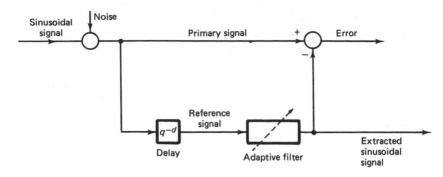

Figure 9.4.9 Adaptive line enhancer.

Remark 9.4.2 (LMS Algorithm). Although it is true that any parameter estimation scheme could be used in the adaptive filters described above, one particular algorithm, the LMS algorithm, has traditionally been applied, mainly because of its simplicity. Because of its popularity, we shall discuss the algorithm in a little more detail below.

The LMS algorithm was described in Section 8.7, where it was shown to be a simple form of stochastic gradient algorithm with nondiminishing adaptation gain.

We recall from (8.7.2) that the LMS algorithm has the form

$$\hat{\theta}(t+1) = \hat{\theta}(t) + \mu(t)\phi(t)[y(t+1) - \phi(t)^T\hat{\theta}(t)] \qquad (9.4.31)$$

Here we shall be particularly interested in the application of this algorithm to the determination of that value of θ which minimizes a mean-square error criterion of the form

$$J = E\{[y(t+1) - \phi(t)^T\theta]^2\} \qquad (9.4.32)$$

We have seen that the criterion above occurs naturally in adaptive filtering when transversal filters are used.

For the case where $\mu(t) = 1/t$, convergence of the algorithm (9.4.31) can be analyzed by the ODE approach discussed in Section 8.5.4, and it can be shown that $\hat{\theta}(t)$ converges to θ^* minimizing (9.4.32) provided that mild regularity conditions are satisfied, including stationarity. Here, however, we are more interested in the case where μ is a constant. We begin by a *heuristic* analysis (under idealized conditions) to expose some of the features of the convergence.

The idea underlying the ODE approach to convergence analysis is that a recursive algorithm can be analyzed by considering the average updating direction. For the algorithm (9.4.31), if we replace the true update direction by its ensemble average, we obtain the following approximation to the algorithm:

$$\theta(t+1) \simeq \theta(t) + \mu E\phi(t)[y(t+1) - \phi(t)^T\theta(t)] \qquad (9.4.33)$$

Differentiating (9.4.32), we obtain the following equation for the value, θ^*, of θ minimizing J:

$$E\phi(t)[y(t+1) - \phi(t)^T\theta^*] = 0 \qquad (9.4.34)$$

Substituting (9.4.34) into (9.4.33) and putting $\tilde{\theta}(t) = \theta(t) - \theta^*$, we have

$$\tilde{\theta}(t+1) \simeq [I - \mu R]\tilde{\theta}(t) \qquad (9.4.35)$$

where
$$R = E\{\phi(t)\phi(t)^T\} \tag{9.4.36}$$

The performance in this idealized case can be more readily seen by diagonalizing R. Thus define $\Lambda \triangleq Q^T R Q$ where $\Lambda = \text{diag}(\lambda_1, \cdots, \lambda_n)$. Then the set of equations (9.4.35) can be written as

$$\tilde{\theta}'(t + 1) \simeq [I - \mu\Lambda]\tilde{\theta}'(t), \qquad \text{where } \tilde{\theta}'(t) \triangleq Q^T\tilde{\theta}(t) \tag{9.4.37}$$

For convergence, it is clearly necessary that

$$0 < \mu\lambda_{\max} < 2 \tag{9.4.38}$$

where $\lambda_{\max} = \max\{\lambda_1 \cdots \lambda_n\} = $ max eigenvalue of R.

The constant μ is generally chosen so that $\mu\lambda_{\max} \ll 1$ to retain tracking and to keep the residual mean-square error small (see below).

If one thinks of (9.4.37) as an Euler solution of a first-order differential equation, it is clear that the "time constant" associated with the jth component of $\tilde{\theta}$ is

$$\tau_j = \frac{1}{\mu\lambda_j} \tag{9.4.39}$$

and hence the slowest rate of convergence is associated with $\lambda_{\min}$. We also see from the above that if the *condition number* of R is large, μ has to be chosen with $\lambda_{\max}$ in mind and this inevitably leads to slow convergence for the components associated with $\lambda_{\min}$.

Ideally, one would like to match the rate of adaptation with the expected rate of change of the parameters. The faster we require the algorithm to track, the larger μ should be. However, if μ is large, then the effect of noise on the fluctuation of $\hat{\theta}(t)$ will be large and this will lead to an increase in the mean-square error. Thus a compromise must be made.

The discussion above gives us a "feel" for the performance of the algorithm (further properties are explored in Exercise 10.12). Further analysis of the performance of the LMS-type filter is given in Widrow et al. (1976), Dupac (1965), Brown (1970), Treichler (1980), Sondhi and Mitra (1976), Weiss and Mitra (1979), Kim and Davisson (1975), Davisson (1970), and Bitmead and Anderson (1980a, 1980b).

The foregoing discussion of convergence of the LMS algorithm shows that slow convergence occurs if the condition number of R is large. However, in the case that R is diagonal, *different* scalar algorithm gain can be attached to different components in the parameter, due to the decoupling of the parameter adjustment. Thus the reader might well wonder if some procedure can be found to orthogonalize the regression vector, thus making the R matrix diagonal. This can be achieved in two ways:

1. The iterative least-squares algorithm amounts to a form of on-line orthogonalization; see the discussion of the orthogonalized projection algorithm and its relationship to least squares in Section 3.3. The least-squares algorithm requires more storage and more computation than the stochastic gradient algorithm. However, it generally has much faster convergence. The argument in favor of the gradient procedures based on their simplicity has been somewhat weakened in recent years due to the advances in microprocessor technology for implementing the algorithms.

2. The iterative lattice filters have a form of in-built orthogonalization use of the backward residuals in the regression vector (see Section assists convergence of the gradient-type algorithm and partially ex these algorithms have attracted so much interest in recent years.

In the next section we describe gradient and least-squares-based algorithms for estimating the parameters in a lattice filter.

9.5 ADAPTIVE LATTICE FILTERS

As explored in detail in Chapter 7, the lattice filter is a special form of restricted complexity filter and incorporates in-built orthogonalization, which leads to decoupling of successive stages in the filter. This can be expected to yield improved convergence rates compared with transversal filters when gradient-type adaptation schemes are used (such as the LMS algorithm). This action has recently spurred considerable interest in the application of adaptive lattice filters; see, for example, Reddy, Egardt, and Kailath (1981).

We describe below three methods of making the lattice filter adaptive.

9.5.1 The Bootstrap Method

We have seen in Section 7.6 that if the reflection coefficient, K_i, are given, then the forward and backward residuals can be computed from (7.6.25)–(7.6.26):

$$f_0(t) = b_0(t) = y(t) \tag{9.5.1}$$

$$f_k(t) = f_{k-1}(t) + K_k b_{k-1}(t-1) \tag{9.5.2}$$

$$b_k(t) = K_k f_{k-1}(t) + b_{k-1}(t-1) \tag{9.5.3}$$

The equations above result in the lattice structure shown in Fig. 7.6.1.

Conversely, if $\{f_k(t)\}$ and $\{b_k(t)\}$ have known properties, K_k can be computed from (7.6.38):

$$K_k = \frac{-2E\{f_{k-1}(t)b_{k-1}(t-1)\}}{E\{f_{k-1}(t)^2\} + E\{b_{k-1}(t-1)^2\}} \tag{9.5.4}$$

Equations (9.5.1)–(9.5.4) can be combined to give a recursive algorithm for computing the reflection coefficients. We show this below.

If the signals are ergodic, it is reasonable to replace the expectations on the right of (9.5.4) by sample means. This leads to

$$\hat{K}_k(t+1) = \frac{-\dfrac{2}{t}\sum\limits_{j=1}^{t} f_{k-1}(j)b_{k-1}(j-1)}{\dfrac{1}{t}\sum\limits_{j=1}^{t}[f_{k-1}(j)^2 + b_{k-1}(j-1)^2]} \tag{9.5.5}$$

where the quantities $\{f_k(j)\}$ and $\{b_k(j)\}$ on the right-hand side of (9.5.5) have been obtained via (9.5.1), (9.5.2), and (9.5.3) using some prior estimate of K_{k-1}.

Equation (9.5.5) suggests that a possible way of forming a recursive estimate for $\hat{K}_k(t+1)$ may be to assume that expression (9.5.5) was obtained via (9.5.1)–(9.5.2) with K_{k-1} set equal to $\hat{K}_{k-1}(t)$. This gives

$$\hat{K}_k(t+1) = \frac{C(t)}{D(t)} \tag{9.5.6}$$

where

$$C(t) = -2 \sum_{j=1}^{t} f_{k-1}(j) b_{k-1}(j-1)$$

$$= C(t-1) - 2f_{k-1}(t) b_{k-1}(t-1) \tag{9.5.7}$$

$$D(t) = \sum_{j=1}^{t} [f_{k-1}(j)^2 + b_{k-1}(j-1)^2]$$

$$= D(t-1) + f_{k-1}(t)^2 + b_{k-1}(t-1)^2 \tag{9.5.8}$$

and from (9.5.2)–(9.5.3),

$$f_k(t) = f_{k-1}(t) + \hat{K}(t) b_{k-1}(t-1) \tag{9.5.9}$$

$$b_k(t) = \hat{K}(t) f_{k-1}(t) + b_{k-1}(t-1) \tag{9.5.10}$$

From (9.5.6)–(9.5.8), we have

$$\hat{K}_k(t+1) = \frac{C(t-1) - 2f_{k-1}(t) b_{k-1}(t-1)}{D(t)}$$

$$= \frac{C(t-1)}{D(t-1)} - \frac{C(t-1)}{D(t-1)}\left[1 - \frac{D(t-1)}{D(t)}\right] - \frac{2f_{k-1}(t) b_{k-1}(t-1)}{D(t)} \tag{9.5.11}$$

$$= \hat{K}_k(t) - \hat{K}_k(t)\frac{f_{k-1}(t)^2 + b_{k-1}(t-1)^2}{D(t)} - \frac{2f_{k-1}(t) b_{k-1}(t-1)}{D(t)}$$

or using (9.5.9)–(9.5.10),

$$K_k(t+1) = \hat{K}_k(t) - \frac{f_{k-1}(t) b_k(t) + f_k(t) b_{k-1}(t-1)}{D(t)} \tag{9.5.12}$$

Notice that $f_k(t)$ and $b_k(t)$ depend on $[y(t-k), y(t-k+1), \ldots, y(t)]$. Thus in the way we have developed the expression above, it is necessary to "rewind" the data at each step in the sense that we regenerate $f(\cdot)$ and $b(\cdot)$ up to time t by passing the data through the lattice with parameters $\hat{K}_k(t)$.

A further simplification would be to ignore the data "rewind" and simply leave the existing values of $f(\cdot)$ and $b(\cdot)$ on the lattice when K is updated. This gives a bootstrap action much the same as was used in Section 8.3 in relation to the prediction error algorithms.

In the case of nonstationary data it is necessary to modify the algorithm (9.5.12) so that the denominator $D(t)$ does not go to infinity. Otherwise, the gain of the algorithm goes to zero and time varying parameters cannot be tracked. Equation (9.5.12) suggests the following algorithm for the nonstationary case:

$$\hat{K}_k(t+1) = \hat{K}_k(t) - \mu[f_{k-1}(t) b_k(t) + f_k(t) b_{k-1}(t-1)] \tag{9.5.13}$$

where μ is a small fixed constant. This is very much like the LMS algorithm. A closely related algorithm which is similar to the NLMS algorithm is as follows:

$$\hat{K}_k(t+1) = \hat{K}_k(t) - \frac{\mu}{D(t)}[f_{k-1}(t) b_k(t) + f_k(t) b_{k-1}(t-1)] \tag{9.5.14}$$

where $D(t)$ involves data weighting and is given by [see (9.5.8)]

$$D(t) = \alpha D(t-1) + f_{k-1}(t)^2 + b_{k-1}(t-1)^2; \quad 0 < \alpha < 1 \tag{9.5.15}$$

Other algorithms suitable for use in the time-varying case are described in Makhoul (1978), Carter (1978), Griffiths (1977), and Griffiths and Medaugh (1978).

9.5.2 The Prediction Error Method

In developing the algorithm of the preceding section, the interdependence of $f(\cdot)$ and $b(\cdot)$ with $\hat{K}_k(\cdot)$ was dealt with in a relaxation manner; that is, in updating $\hat{K}(\cdot)$, $f(\cdot)$ and $b(\cdot)$ were assumed given, and in updating $f(\cdot)$ and $b(\cdot)$, $\hat{K}(\cdot)$ were assumed given. In this section we attempt explicitly to take account of the dependence of $f(\cdot)$ and $b(\cdot)$ on $\hat{K}(\cdot)$ by use of partial derivatives. This will lead us to a stochastic gradient-type prediction error algorithm for estimating the reflection coefficients.

From (7.6.52), the optimal restricted complexity lattice predictor has output given by

$$\hat{y}_N(t + 1) = -\sum_{j=1}^{N} K_j b_{j-1}(t) \tag{9.5.16}$$

where $\{b_k(t)\}$ denotes the kth backward residual and K_k denotes the kth reflection coefficient. To motivate the algorithm we modify the predictor of Fig. 7.6.3 by replacing $K_1 \cdots K_N$ by a parameter vector $\theta^T = (\theta_1 \cdots \theta_N)$. This gives the predictor shown in Fig. 9.5.1 with output $\bar{y}(t + 1, \theta)$.

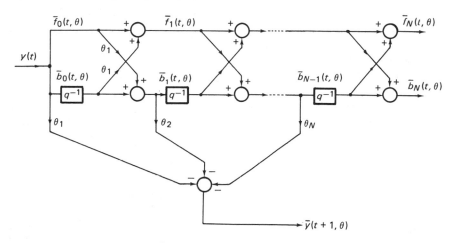

Figure 9.5.1 Lattice predictor.

We note from Fig. 9.5.1 that $\bar{f}_1(t, \theta), \ldots, \bar{f}_N(t, \theta), \bar{b}_1(t, \theta), \ldots, \bar{b}_N(t, \theta)$, $\bar{y}(t + 1, \theta)$ all depend on θ and are equal to $f_1(t), \ldots, f_N(t), b_1(t), \ldots, b_N(t)$, $\hat{y}_N(t + 1)$ if $\theta_i = K_i$, $i = 1, \ldots, N$.

Our objective is to develop a good predictor and thus we shall seek a value of θ so that the mean-square prediction error, J, is minimized where

$$J = E\{\bar{J}(t + 1, \theta)\} \tag{9.5.17}$$

$$\bar{J}(t + 1, \theta) = \bar{\epsilon}(t + 1, \theta)^2$$

$$\bar{\epsilon}(t + 1, \theta) = y(t + 1) - \bar{y}(t + 1, \theta)$$

$$= y(t + 1) + \sum_{j=1}^{N} \theta_j \bar{b}_{j-1}(t, \theta) \tag{9.5.18}$$

The derivative of $\bar{J}(t+1, \theta)$ with respect to θ_i, $i = 1, \ldots, N$, is readily seen to be

$$\frac{\partial \bar{J}(t+1, \theta)}{\partial \theta_i} = 2\bar{\epsilon}(t+1, \theta) \frac{\partial \bar{\epsilon}(t+1, \theta)}{\partial \theta_i}$$

where

$$\frac{\partial \bar{\epsilon}(t+1, \theta)}{\partial \theta_i} = \bar{b}_{i-1}(t, \theta) + \sum_{j=1}^{N} \theta_j \frac{\partial \bar{b}_{j-1}(t, \theta)}{\partial \theta_i} \qquad (9.5.19)$$

We shall now generate expressions for $\partial \bar{b}_{j-1}(t, \theta)/\partial \theta_i$. From Fig. 9.5.1 we note that

$$\bar{f}_0(t, \theta) = y(t); \qquad \bar{b}_0(t, \theta) = y(t) \qquad (9.5.20)$$

$$\bar{f}_i(t, \theta) = \bar{f}_{i-1}(t, \theta) + \theta_i \bar{b}_{i-1}(t-1, \theta);$$

$$\bar{b}_i(t, \theta) = \theta_i \bar{f}_{i-1}(t, \theta) + \bar{b}_{i-1}(t-1, \theta), \qquad i = 1, \ldots, N \qquad (9.5.21)$$

Hence we have

$$\frac{\partial \bar{f}_i(t, \theta)}{\partial \theta_i} = 0; \qquad \frac{\partial \bar{b}_i(t, \theta)}{\partial \theta_j} = 0 \qquad \text{for } i = 0, \ldots, j-1 \qquad (9.5.22)$$

$$\frac{\partial \bar{f}_i(t, \theta)}{\partial \theta_j} = \bar{b}_{i-1}(t-1, \theta); \qquad \frac{\partial \bar{b}_i(t, \theta)}{\partial \theta_j} = \bar{f}_{i-1}(t, \theta) \qquad \text{for } i = j \qquad (9.5.23)$$

and

$$\frac{\partial \bar{f}_i(t, \theta)}{\partial \theta_j} = \frac{\partial \bar{f}_{i-1}(t, \theta)}{\partial \theta_j} + \theta_i \frac{\partial \bar{b}_{i-1}(t-1, \theta)}{\partial \theta_j}$$

$$\frac{\partial \bar{b}_i(t, \theta)}{\partial \theta_j} = \theta_i \frac{\partial \bar{b}_{i-1}(t, \theta)}{\partial \theta_j} + \frac{\partial \bar{b}_{i-1}(t-1, \theta)}{\partial \theta_j} \qquad \text{for } i = j+1, \ldots, N \qquad (9.5.24)$$

It can be seen from the equations above that each component of $\partial \bar{\epsilon}(t+1, \theta)/\partial \theta$ can be obtained from a lattice structure as shown in Figs. 9.5.2 and 9.5.3.

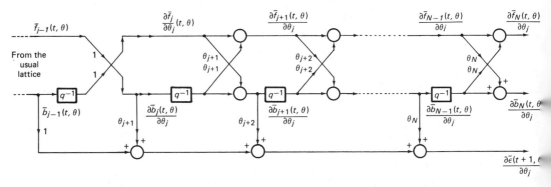

Figure 9.5.2 Detail of lattice structure for generating $\partial \bar{\epsilon}(t+1, \theta)/\partial \theta_j$.

The discussion above suggests the following stochastic gradient prediction error algorithm for estimating the value of θ that minimizes (9.5.2):

$$\hat{\theta}(t+1) = \hat{\theta}(t) - \alpha(t) \left. \frac{\partial \bar{J}(t+1, \theta)}{\partial \theta} \right|_{\hat{\theta}(t)}$$

$$= \hat{\theta}(t) - \alpha(t) 2\phi(t) \bar{\epsilon}(t+1, \hat{\theta}(t)) \qquad (9.5.25)$$

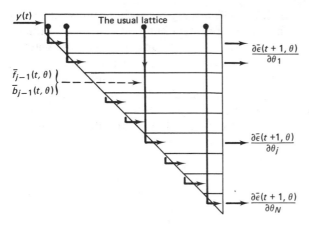

Figure 9.5.3 Schematic of lattice for generating $\partial\bar{\epsilon}(t+1,\theta)/\partial\theta$.

where $\bar{\epsilon}(t+1,\hat{\theta}(t))$ is given by (9.5.18) and

$$\phi(t)^T = \left[\frac{\partial\bar{\epsilon}(t+1,\theta)}{\partial\theta_1} \quad \cdots \quad \frac{\partial\bar{\epsilon}(t+1,\theta)}{\partial\theta_N}\right]\Bigg|_{\hat{\theta}(t)} \tag{9.5.26}$$

The quantities $\{\partial\bar{\epsilon}(t+1,\theta)/\partial\theta_i\}$, $i=1,\ldots,N$, evaluated at $\hat{\theta}(t)$ can be determined from (9.5.19) to (9.5.24) with $\theta - \hat{\theta}(t)$.

To achieve parameter convergence the gain sequence $\{\alpha(t)\}$ in (9.5.25) should converge to zero. A standard choice is (see Chapter 8)

$$\Sigma\alpha(t) = \infty; \qquad \Sigma\alpha(t)^2 < \infty \tag{9.5.27}$$

Alternatively, $\alpha(t)$ can be set to a constant to track time-varying parameters. As we have developed them, (9.5.19) to (9.5.24) should be implemented by "rewinding" the data by the length of the lattice at each step and regenerating $f(\cdot)$, $b(\cdot)$, and their derivatives up to time t. In practice, it may be possible to simply leave the current values of $f(\cdot)$, $b(\cdot)$, and their derivatives in the lattice of Fig. 9.5.2 when the parameters are updated. [See Section 8.3 of Chapter 8.]

The least-squares form of (9.5.25) is as follows:

$$\hat{\theta}(t+1) = \hat{\theta}(t) - P(t)\phi(t)\bar{\epsilon}(t+1,\tilde{\theta}(t)) \tag{9.5.28}$$

$$P(t) = \frac{1}{\lambda}\left[P(t-1) - \frac{P(t-1)\phi(t)\phi(t)^T P(t-1)}{\lambda + \phi(t)^T P(t-1)\phi(t)}\right]; \qquad P(0) = \epsilon I, \tag{9.5.29}$$
$$0 < \lambda \leq, \quad \epsilon > 0$$

($\lambda < 1$ gives an exponential weighted least-squares update.)

The method above is more complicated than the bootstrap method but has the advantage that explicit account has been taken of the dependence of $f(\cdot)$ and $b(\cdot)$ on $K(\cdot)$.

9.5.3 The Exact Least-Squares Method

Finally, we wish to present the "exact least-squares" method of Morf et al. (1977), Morf, Vieira, and Lee (1977), Morf and Lee (1979), Lee and Morf (1980), and Lee, Morf, and Friedlander (1981). This algorithm is closely related to the standard least-squares algorithm save that the solution is converted into a lattice structure by use of the Levinson algorithm (see Section 7.5.3). This gives an algorithm with guaranteed convergence but with more computations per step than that of the bootstrap method. The exact least-squares method does not attempt to find an approximation to the expression (9.5.4) for the reflection coefficient. Instead, the forward and backward residuals are obtained as the solutions of a least-squares minimization problem, for which exact sequential expressions are obtained. Our development will overlap partly with the previous discussion of the Levinson algorithm in Section 7.5.3.

Given a set of data $\{y(j), 0 \leq j \leq t\}$, we define forward and backward residuals of order k.

$$
\begin{aligned}
f_k(t) &= y(t) + \sum_{i=1}^{k} a_{k,i}(t) y(t-i) \\
&= A_k(t)^T \phi_k(t)
\end{aligned}
\tag{9.5.30}
$$

$$
\begin{aligned}
b_k(t) &= y(t-k) + \sum_{i=1}^{k} h_{k,i}(t) y(t-k+i) \\
&= H_k(t)^T \phi_k(t)
\end{aligned}
\tag{9.5.31}
$$

where

$$A_k(t)^T = [1 \quad a_{k,1}(t) \quad \cdots \quad a_{k,k}(t)] \tag{9.5.32}$$

$$H_k(t)^T = [h_{k,k}(t) \quad \cdots \quad h_{k,1}(t) \quad 1] \tag{9.5.33}$$

$$\phi_k(t)^T = [y(t) \quad y(t-1) \quad \cdots \quad y(t-k)] \tag{9.5.34}$$

We also define

$$
Y_k(t) = \begin{bmatrix}
y(0) & y(1) & \cdots & y(k) & \cdots & y(t) \\
0 & y(0) & & & & \\
\vdots & & \ddots & & & \vdots \\
& & & & & \\
0 & 0 & & y(0) & & y(t-k)
\end{bmatrix}
\tag{9.5.35}
$$

and the sample covariance matrix

$$R_k(t) = Y_k(t) Y_k(t)^T \tag{9.5.36}$$

It will be assumed throughout that $R_k(t)$ is nonsingular; in particular, $t \geq k$. The coefficient vectors $A_k(t)$ and $H_k(t)$ are defined at each time t as the solutions of the following minimization problems:

$$A_k(t) = \arg\min A_k(t)^T R_k(t) A_k(t) \tag{9.5.37}$$

$$H_k(t) = \arg\min H_k(t)^T R_k(t) H_k(t) \tag{9.5.38}$$

The least-squares solutions of these minimization problems are obtained via the usual "normal equations."

$$
R_k(t) A_k(t) = \begin{bmatrix} R_k^f(t) \\ \hline 0 \\ \vdots \\ 0 \end{bmatrix} \begin{matrix} \}1 \\ \\ \}k, \end{matrix} \qquad
R_k(t) H_k(t) = \begin{bmatrix} 0 \\ \vdots \\ 0 \\ \hline R_k^b(t) \end{bmatrix} \begin{matrix} \}k \\ \\ \}1 \end{matrix}
\tag{9.5.39}
$$

where $R_k^f(t)$ and $R_k^b(t)$ are the resulting residual variances. Notice that $A_k(t)$ is a

$(k + 1)$-vector containing k unknowns. The last k equations on the left of (9.5.39) determine $a_{k,1}(t), \ldots, a_{k,k}(t)$, while the first equation defines $R_k^f(t)$, the sample covariance of the optimal forward residuals, which is also the minimum value of the criterion (9.5.37):

$$R_k^f(t) = \sum_{j=0}^{t} [y(j) + \sum_{i=1}^{k} a_{k,i}(t)y(j-i)]^2 \qquad (9.5.40)$$

Here the $a_{k,i}(t)$ are the solutions of the last k equations on the left of (9.5.39); similarly for $H_k(t)$ and

$$R_k^b(t) = \sum_{j=0}^{t} [y(j-k) + \sum_{i=1}^{k} h_{k,i}(t)y(j-k+i)]^2 \qquad (9.5.41)$$

The remainder of the work consists in finding recursive expressions for $A_k(t)$ and $H_k(t)$ (recursions both in the order k and in the time t) without having to invert the $(k+1) \times (k+1)$ matrix $R_k(t)$. These recursions will then lead to a recursive implementation in lattice form. All the recursive formulas are derived from the following recursions on $R_k(t)$:

$$R_k(t+1) = R_k(t) + \phi_k(t+1)\phi_k(t+1)^T \qquad (9.5.42)$$

$$R_{k+1}(t) = \begin{bmatrix} \times \times \times \times \times \times \\ \times & & \\ \times & R_k(t-1) & \\ \times & & \\ \times & & \end{bmatrix} = \begin{bmatrix} & & \times \\ & & \times \\ & R_k(t) & \times \\ & & \times \\ \times \times \times \times \times \times \end{bmatrix} \qquad (9.5.43)$$

The $\times$'s are there to indicate that $R_k(t-1)$ and $R_k(t)$ are bordered by one additional row and column.

Finally, we introduce two auxiliary quantities:

a $(k+1)$-vector: $\quad C_k(t) = R_k(t)^{-1}\phi_k(t) \qquad (9.5.44)$

a scalar: $\qquad \gamma_k(t) = \phi_k(t)^T R_k(t)^{-1}\phi_k(t) = \phi_k(t)^T C_k(t) \qquad (9.5.45)$

Note that $0 \le \gamma_k(t) < 1$.

Order-update recursions. We first want to find $A_{k+1}(t)$, the vector of least-squares estimates for the $(k+1)$th-order filter, as a function of $A_k(t)$. By the same argument as before,

$$R_{k+1}(t)A_{k+1}(t) = \begin{bmatrix} R_{k+1}^f(t) \\ \hline 0 \\ \vdots \\ \vdots \\ 0 \end{bmatrix} \begin{matrix} \}1 \\ \\ \}k+1 \\ \\ \end{matrix} \qquad (9.5.46)$$

where the last $k+1$ equations determine $a_{k+1,1}(t), \ldots, a_{k+1,k+1}(t)$ and the first equation defines $R_{k+1}^f(t)$. Using (9.5.43), we also have

$$R_{k+1}(t)\begin{bmatrix} A_k(t) \\ 0 \end{bmatrix} = \begin{bmatrix} R_k(t) & \times \\ & \times \\ & \times \\ \times \times \times \times \end{bmatrix}\begin{bmatrix} A_k(t) \\ 0 \end{bmatrix} = \begin{bmatrix} R_k^f(t) \\ 0 \\ \vdots \\ \vdots \\ 0 \\ S_{k+1}(t) \end{bmatrix} \qquad (9.5.47)$$

where $S_{k+1}(t)$ is defined as

$$S_{k+1}(t) = [\text{last row of } R_{k+1}(t)] \begin{bmatrix} A_k(t) \\ 0 \end{bmatrix} \tag{9.5.48}$$

Similarly,

$$R_{k+1}(t) \begin{bmatrix} 0 \\ H_k(t-1) \end{bmatrix} = \begin{bmatrix} \times \times \times \times \times \times \\ \times \\ \times & R_k(t-1) \end{bmatrix} \begin{bmatrix} 0 \\ H_k(t-1) \end{bmatrix} = \begin{bmatrix} S_{k+1}^*(t) \\ 0 \\ \cdot \\ \cdot \\ \cdot \\ 0 \\ R_k^b(t-1) \end{bmatrix} \tag{9.5.49}$$

with $S_{k+1}^*(t)$ defined analogously. Now, premultiplying (9.5.47) by $[0 \quad H_k(t-1)^T]$ and (9.5.49) by $[A_k(t)^T \quad 0]$ yields

$$S_{k+1}(t) = S_{k+1}^*(t) \tag{9.5.50}$$

Considering (9.5.32), (9.5.33), (9.5.36), (9.5.47), and (9.5.49), and assuming that $R_{k+1}(t)$ is nonsingular, one can write $A_{k+1}(t)$ as

$$A_{k+1}(t) = \begin{bmatrix} A_k(t) \\ 0 \end{bmatrix} + \alpha \begin{bmatrix} 0 \\ H_k(t-1) \end{bmatrix} \tag{9.5.51}$$

In order to satisfy (9.5.46) α must be chosen such that

$$S_{k+1}(t) + \alpha R_k^b(t-1) = 0$$

This yields the following recursions:

$$A_{k+1}(t) = \begin{bmatrix} A_k(t) \\ 0 \end{bmatrix} - \begin{bmatrix} 0 \\ H_k(t-1) \end{bmatrix} \frac{S_{k+1}(t)}{R_k^b(t-1)} \tag{9.5.52}$$

Premultiplying (9.5.52) by $R_{k+1}(t)$ and using (9.5.46), (9.5.47), and (9.5.49) also yields

$$R_{k+1}^f(t) = R_k^f(t) - \frac{S_{k+1}^2(t)}{R_k^b(t-1)} \tag{9.5.53}$$

By exactly the same argument one finds the order-update equations for $H_k(t)$:

$$H_{k+1}(t) = \begin{bmatrix} 0 \\ H_k(t-1) \end{bmatrix} - \begin{bmatrix} A_k(t) \\ 0 \end{bmatrix} \frac{S_{k+1}(t)}{R_k^f(t)} \tag{9.5.54}$$

$$R_{k+1}^b(t) = R_k^b(t-1) - \frac{S_{k+1}^2(t)}{R_k^f(t)} \tag{9.5.55}$$

The procedure for updating $C_k(t)$ is as follows. By definition

$$R_{k+1}(t)C_{k+1}(t) = \begin{bmatrix} y(t) \\ \cdot \\ \cdot \\ \cdot \\ y(t-k-1) \end{bmatrix} \tag{9.5.56}$$

Also, by (9.5.43),

　　　　　　　　　　　　　　Adaptive Filtering and Prediction　　Chap. 9

$$R_{k+1}(t)\begin{bmatrix} C_k(t) \\ 0 \end{bmatrix} = \begin{bmatrix} y(t) \\ \cdot \\ \cdot \\ \cdot \\ y(t-k) \\ x \end{bmatrix} \tag{9.5.57}$$

where x is an unknown value, and by (9.5.39),

$$R_{k+1}(t)H_{k+1}(t) = \begin{bmatrix} 0 \\ \cdot \\ \cdot \\ \cdot \\ 0 \\ R^b_{k+1}(t) \end{bmatrix} \tag{9.5.58}$$

Therefore, using the same argument as for $A_{k+1}(t)$, we can construct

$$C_{k+1}(t) = \begin{bmatrix} C_k(t) \\ 0 \end{bmatrix} + \alpha H_{k+1}(t) \tag{9.5.59}$$

α is determined by premultiplying (9.5.59) by $H_k^T(t+1)R_{k+1}(t)$, using (9.5.38):

$$\begin{aligned}
H_k^t(t+1)R_{k+1}(t)C_{k+1}(t) &= H_k^T(t+1)\phi_{k+1}(t) = b_{k+1}(t) \\
&= [0 \quad \cdots \quad 0 \quad R^b_{k+1}(t)]C_{k+1}(t) \\
&= 0 + \alpha[0 \quad \cdots \quad 0 \quad R^b_{k+1}(t)]H_{k+1}(t) \\
&= \alpha R^b_{k+1}(t)
\end{aligned} \tag{9.5.60}$$

Hence

$$\alpha = \frac{g_{k+1}(t)}{R^b_{k+1}(t)} \tag{9.5.61}$$

and

$$C_{k+1}(t) = \begin{bmatrix} C_k(t) \\ 0 \end{bmatrix} + H_{k+1}(t)\frac{b_{k+1}(t)}{R^b_{k+1}(t)} \tag{9.5.62}$$

Premultiplying (9.5.52), (9.5.54), and (9.5.62) by $\phi_{k+1}(t)^T$ yields

$$f_{k+1}(t) = f_k(t) + K^b_{k+1}(t)b_k(t-1) \tag{9.5.63}$$

$$b_{k+1}(t) = K^f_{k+1}(t)f_k(t) + b_k(t-1) \tag{9.5.64}$$

$$\gamma_{k+1}(t) = \gamma_k(t) + \frac{b^2_{k+1}(t)}{R^b_{k+1}(t)} \tag{9.5.65}$$

with

$$K^b_{k+1}(t) = -\frac{S_{k+1}(t)}{R^b_k(t-1)}; \qquad K^f_{k+1}(t) = -\frac{S_{k+1}(t)}{R^f_k(t)} \tag{9.5.66}$$

and $S_{k+1}(t)$ defined by (9.5.48). K^b and K^f are, as before, called the reflection coefficients. Equations (9.5.63) and (9.5.64) are the least-squares lattice recursions for the forward and backward residuals. Recursions start with $f_0(t) = b_0(t) = y(t)$. $S_{k+1}(t)$ is defined by (9.5.48), while $R^f_k(t)$ and $R^b_k(t-1)$ are updated by (9.5.53) and (9.5.55). Assuming that $t > k$, these recursions start with

$$R_0^f(t) = R_0^b(t) = \sum_{j=0}^{t} y_j^2$$

Time-update recursions. We now derive recursive (in time) equations for A_k, H_k, C_k, R_k^f, R_k^b, γ_k, and S_k. We first express $A_k(t+1)$ as a function of $A_k(t)$. By (9.5.42) and (9.5.39),

$$R_k(t+1)A_k(t) = \begin{bmatrix} R_k^f(t) \\ \hline 0 \\ \vdots \\ \vdots \\ 0 \end{bmatrix}\Big\}1 \\ \Big\}k + \begin{bmatrix} y(t+1) \\ \vdots \\ \vdots \\ y(t+1-k) \end{bmatrix}[y(t+1) \cdots y(t+1-k)]A_k(t) \tag{9.5.67}$$

We now define

$$[y(t+1) \cdots y(t+1-k)]A_k(t) = \hat{f}_k(t+1) \tag{9.5.68}$$

Note that $\hat{f}_k(t+1)$ is different from $f_k(t+1)$ because it uses the non-updated estimates $A_k(t)$ rather than the new estimates $A_k(t+1)$. Now by (9.5.43),

$$R_k(t+1)\begin{bmatrix} 0 \\ C_{k-1}(t) \end{bmatrix} = \begin{bmatrix} \times \times \times \times \times \\ \times \\ \times & R_{k-1}(t) \end{bmatrix}\begin{bmatrix} 0 \\ C_{k-1}(t) \end{bmatrix} = \begin{bmatrix} x \\ y(t) \\ \vdots \\ \vdots \\ y(t+1-k) \end{bmatrix} \tag{9.5.69}$$

Using (9.5.46), (9.5.47), and (9.5.48) and remembering that $A_k(t+1)$ is defined by the last k equations of (9.5.46), we have

$$A_k(t+1) = A_k(t) - \begin{bmatrix} 0 \\ C_{k-1}(t) \end{bmatrix}\hat{f}_k(t+1) \tag{9.5.70}$$

Premultiplying by $\phi_k(t+1)^T$ yields

$$f_k(t+1) = (1 - \gamma_{k-1}(t))\hat{f}_k(t+1) \tag{9.5.71}$$

$$A_k(t+1) = A_k(t) - \begin{bmatrix} 0 \\ C_{k-1}(t) \end{bmatrix}\frac{f_k(t+1)}{1 - \gamma_{k-1}(t)} \tag{9.5.72}$$

Finally, premultiplying (9.5.70) by $A_k(t+1)^T R_k(t+1)$, using (9.5.46)–(9.5.67) and the fact that the first element of $A_k(\cdot)$ is 1, leads to

$$\begin{aligned} R_k^f(t+1) &= R_k^f(t) + f_k(t+1)\hat{f}_k(t+1) \\ &= R_k^f(t) + \frac{f_k^2(t+1)}{1 - \gamma_{k-1}(t)} \end{aligned} \tag{9.5.73}$$

Following exactly the same method, one obtains a time-update recursion for H_k and R_k^b:

$$H_k(t+1) = H_k(t) - \begin{bmatrix} C_{k-1}(t+1) \\ 0 \end{bmatrix}\frac{b_k(t+1)}{1 - \gamma_{k-1}(t+1)} \tag{9.5.74}$$

$$R_k^b(t+1) = R_k^b(t) + \frac{b_k^2(t+1)}{1 - \gamma_{k-1}(t+1)} \tag{9.5.75}$$

Following the same procedure again for C_k yields

$$C_k(t+1) = \left[\begin{array}{c} 0 \\ \hline C_{k-1}(t) \end{array}\right]\begin{array}{l}\}1 \\ \}k\end{array} + \alpha A_k(t+1) \tag{9.5.76}$$

where α must be determined. We use the same idea as for the order update of C_k. Premultiplying (9.5.76) by $A_k^T(t+1)R_k(t+1)$, yields, using (9.5.44) and (9.5.46),

$$A^k(t+1)^T R_k(t+1)C_k(t+1) = A_k(t+1)^T \phi_k(t+1) = f_k(t+1)$$
$$= [R_k^f(t+1) \quad 0 \quad \cdots \quad 0]C_k(t+1)$$
$$= 0 + \alpha[R_k^f(t+1) \quad 0 \quad \cdots \quad 0]A_k(t+1)$$

Hence $\alpha = f_k(t+1)/R_k^f(t+1)$ and

$$C_k(t+1) = \left[\begin{array}{c} 0 \\ \hline C_{k-1}(t) \end{array}\right] + A_k(t+1)\frac{f_k(t+1)}{R_k^f(t+1)} \tag{9.5.77}$$

Premultiplying by $\phi_k(t+1)$ gives

$$\gamma_k(t+1) = \gamma_{k-1}(t) + \frac{f_k^2(t+1)}{R_k^f(t+1)} \tag{9.5.78}$$

Finally we obtain a time-update recursion for $S_k(t)$, the numerator of the reflection coefficients $K_k^b(t)$ and $K_k^f(t)$. Recall that

$$S_{k+1}(t+1) = [l.r. \quad R_{k+1}(t+1)]\left[\begin{array}{c} A_k(t+1) \\ 0 \end{array}\right] \tag{9.5.79}$$

where $l.r.$ stands for "last row of." Using (9.5.70),

$$S_{k+1}(t+1) = [l.r. \quad R_{k+1}(t+1)]\left\{\left[\begin{array}{c} A_k(t) \\ 0 \end{array}\right] - \left[\begin{array}{c} 0 \\ C_{k-1}(t) \\ 0 \end{array}\right]\hat{f}_k(t+1)\right\} \tag{9.5.80}$$

The first term on the right-hand side yields, using (9.5.42) and (9.5.68),

$$[l.r. \quad R_{k+1}(t+1)]\left[\begin{array}{c} A_k(t) \\ 0 \end{array}\right] = [l.r. \quad R_{k+1}(t)]\left[\begin{array}{c} A_k(t) \\ 0 \end{array}\right]$$
$$+ y(t-k)\phi_{k+1}(t+1)^T\left[\begin{array}{c} A_k(t) \\ 0 \end{array}\right] \tag{9.5.81}$$
$$= S_{k+1}(t) + y(t-k)\hat{f}_k(t+1)$$

The second term gives, by (9.5.43) and (9.5.62),

$$[l.r. \quad R_{k+1}(t+1)]\left[\begin{array}{c} 0 \\ C_{k-1}(t) \\ 0 \end{array}\right]\hat{f}_k(t+1)$$
$$= [l.r. \quad R_k(t)]\left[\begin{array}{c} C_{k-1}(t) \\ 0 \end{array}\right]\hat{f}_k(t+1)$$
$$= [l.r. \quad R_k(t)]\left[C_k(t) - H_k(t)\frac{b_k(t)}{R_k^b(t)}\right]\hat{f}_k(t+1) \tag{9.5.82}$$
$$= y(t-k)\hat{f}_k(t+1) - b_k(t)\hat{f}_k(t+1)$$

Inserting (9.5.81)–(9.5.82) into (9.5.80) and using (9.5.71) yields

$$S_{k+1}(t+1) = S_{k+1}(t) + \frac{b_k(t) f_k(t+1)}{1 - \gamma_{k-1}(t)} \qquad (9.5.83)$$

Equations (9.5.73), (9.5.75), (9.5.78), and (9.5.83) together with the lattice equations (9.5.63) and (9.5.64) constitute a complete set of recursions that are required for the adaptive implementation of the least-squares lattice filter. Assuming that $k_{\max}$ is the maximum order to be considered for the lattice filter, and that the filter has been adapted up to time t, then, when $y(t+1)$ is observed, the following quantities must be updated for $k = 1, 2, \ldots, k_{\max}$: $f_k(t+1)$, $b_k(t+1)$, $S_k(t+1)$, $R^b_{k-1}(t+1)$, $R^f_{k-1}(t+1)$, and $\gamma_{k-2}(t+1)$.

If $k_{\max}$ is the maximum order of the lattice, the time recursions can only start after a time t_1 such that $R_{k_{\max}}(t_1)$ is nonsingular (i.e., $t_1 \geq k$) [see (9.5.35) and (9.5.36)]. The initial conditions for $S_k(t)$, $R^b_k(t)$, $R^f_k(t)$, and $\gamma_k(t)$ are then computed using the nonrecursive (in time) expressions of the beginning of this section.

We recall that this third adaptive implementation of the lattice filter is a least-squares solution, which therefore converges to the optimal lattice filter of Section 7.6 in the case of stationary statistics. The amount of computations is higher than the bootstrap method or the prediction error method, but notice that no matrix inversion is required. The recursive least-squares lattice filter described here has also been generalized to the case of d-step-ahead prediction [see Gevers and Wertz (1980b)]. An alternative derivation of the least-squares lattice filter using the geometric approach has been described in Lee, Morf, and Friedlander (1981).

Remark 9.5.1 (Adaptive Lattice Predictors).   We have noted in Section 7.6.2 that there is a close relationship between a whitening filter and the one-step-ahead prediction. Thus an adaptive lattice predictor can be obtained in an analogous fashion to the adaptive (whitening) filter described above. This gives an adaptive one-step-ahead lattice predictor [see also Vizwanathan and Makhoul (1976)]. The idea can be extended to the case of d-step-ahead adaptive lattice predictors; see Reddy, Egardt, and Kailath (1981) and Gevers and Wertz (1980a), (1980b).

▼▼▼

EXERCISES

9.1. Consider a system described by a state-space model in the innovations form

$$x(t+1) = Fx(t) + L\omega(t)$$
$$y(t) = Hx(t) + \omega(t)$$

where $(F - LH)$ is stable.
(a) Show that the state estimation error covariance, $\Sigma(t)$, asymptotically converges to zero.
(b) Using the results of part (a), verify that fixed-lag and fixed-point smoothing offer no advantages (in the steady state) for a system modeled in innovations form.

9.2. Use Kalman filtering theory to write down a fixed-lag smoother for the state $x_3^d(t) = s(t - d)$ using the model (9.2.59)–(9.2.61). Note that the answer differs from that given in Section 7.3.2 due to the coupling between the process and measurement noise vectors.

9.3. (a) Show that the extended Kalman filter [equations (9.2.28)–(9.2.30)] simplifies to the form given in (9.2.46)–(9.2.51) if the system is described in innovations form with the Kalman gain K being directly parameterized.

(b) Comment on the desirability of using innovations representations in adaptive optimal filtering.

9.4. Consider the parameter estimation algorithm described in Exercise 8.5 using a priori predictions in the regression vector. Note that $\hat{y}(t)$, so obtained, is an adaptive one-step-ahead predictor for $y(t)$. Show that, subject to

$$\lim \sup \frac{1}{N} \sum_{t=1}^{N} y(t)^2 < \infty \quad \text{a.s.}$$

$$\lim \sup \frac{1}{N} \sum_{t=1}^{N} u(t)^2 < \infty \quad \text{a.s.}$$

$$C(z^{-1}) - \frac{\bar{a}}{2} \quad \text{input strictly passive}$$

then the adaptive prediction, $\hat{y}(t)$, converges to the optimal prediction, $y^0(t + 1 \mid t)$, in the sense that

$$\frac{1}{N} \sum_{t=1}^{N} [\hat{y}(t + 1) - y^0(t + 1 \mid t)]^2 = 0 \quad \text{a.s.}$$

9.5. Consider the following d-step-ahead adaptive predictor

$$\hat{\theta}(t) = \hat{\theta}(t - d) + \frac{\bar{a}}{r(t - d)} \phi(t - d)[y(t) - \hat{y}(t)]$$

$$r(t - d) = r(t - d - 1) + \phi(t - d)^T \phi(t - d); \quad r(0) = 1$$

$$\hat{y}(t) = \phi(t - d)^T \hat{\theta}(t - d)$$

Note that the algorithm above incorporates d interlaced parameter estimators. Also note that the prediction $\hat{y}(t + d)$ depends on $\{y(\tau): \tau \le t\}$. Show that the algorithm is globally convergent in the sense described in Exercise 9.4 (Goodwin, Sin, and Saluja, 1980).

9.6. Develop an adaptive one-step-ahead predictor using the pseudo linear regression least-squares algorithm of Section 8.4. Prove that the algorithm is globally convergent in the sense described in Exercise 9.4 provided that $[C(z^{-1}) - \frac{1}{2}]$ is strictly input and output passive. [*Hint:* See Section 8.6.2 to establish $\lim \sup r(N - 1)/N < \infty$.]

9.7. Complete the proof of Lemma 9.3.2. (*Hint:* See the proof of Lemma 4.3.2.)

9.8. Consider the direct adaptive d-step-ahead predictor given in Section 9.3.2. Establish the following preliminary properties of the algorithm:

(a) Show that

$$\eta(t) = \frac{\epsilon(t)}{1 + \phi(t - d)^T P(t - d - 1)\phi(t - d)}$$

where $\eta(t)$ and $\epsilon(t)$ are as in (9.3.41) and (9.3.50).

(b) Show that $\{\eta(t)\}$ is related to the parameter estimation error $\tilde{\theta}(t) = \hat{\theta}(t) - \theta_0$ as follows:

$$\bar{C}(q^{-1})[\eta(t) - v(t)] = b(t) \quad \text{[see (9.3.38)]}$$

where

$$\bar{C}(q^{-1}) = 1 - q^{-d}\bar{G}(q^{-1})$$
$$v(t) = y(t) - y^0(t\,|\,t - d) = F(q^{-1})\omega(t)$$
$$b(t) = -\phi(t - d)^T\tilde{\theta}(t)$$
$$\tilde{\theta}(t) = \hat{\theta}(t) - \theta_0$$

*9.9. Using the results of Exercise 9.8, establish the following convergence result for the direct adaptive d-step-ahead predictor of Section 9.3.2.

Theorem. Subject to the usual noise, system, and signal assumptions, and provided that the transfer function $[(1/\bar{C})(z) - \frac{1}{2}]$ is very strictly passive where $\bar{C}(q^{-1})$ is a function of $C(q^{-1})$ as defined in Exercise 9.8, the algorithm (9.3.39) to (9.3.48) yields

$$\limsup_{t\to\infty} \frac{1}{N} \sum_{t=1}^{N} [\hat{y}(t + d) - y^0(t + d\,|\,t)]^2 = 0 \quad \text{a.s.}$$

where $y^0(t + d\,|\,t)$ is the optimal linear d-step-ahead prediction of $y(t + d)$.

[*Hint*:

Step 1: Show that

$$\tilde{\theta}(t) = \tilde{\theta}(t - 1) + P(t - d - 1)\phi(t - d)\eta(t)$$

and then decompose $\tilde{\theta}(t)$ and $\eta(t)$ as follows:

$$\tilde{\theta}(t) = \sum_{i=1}^{d} \tilde{\theta}^i(t)$$

$$\eta(t) = \sum_{i=1}^{d} \eta^i(t)$$

So that

$$\sum_{i=1}^{d} \tilde{\theta}^i = \sum_{i=1}^{d} \tilde{\theta}^i(t - 1) + P(t - d - 1)\phi(t - d)\sum_{i=1}^{d} \eta^i(t)$$

$$\bar{C}(q^{-1})\left[\sum_{i=1}^{d} \eta^i(t) - \sum_{i=1}^{d} f_{i-1}\omega(t - i + 1)\right] = \sum_{i=1}^{d} b^i(t) = b(t)$$

where $b^i(t) \triangleq -\phi(t - d)^T\tilde{\theta}^i(t)$.]

Step 2: Consider the equations above component by component. For $i = 1, \ldots, d$, define $\tilde{\theta}^i(t)$ and $b^i(t)$ by the following partitioned form of the algorithm:

$$\tilde{\theta}^i(t) = \tilde{\theta}^i(t - 1) + P(t - d - 1)\phi(t - d)\eta^i(t)$$
$$\bar{C}(q^{-1})[\eta^i(t) - f_{i-1}\omega(t - i + 1)] = b^i(t)$$

Show that $\phi(t - d)$ is $\mathscr{F}_{t-d}$ measurable and that

$$\mathscr{F}_{t-d} \subset \mathscr{F}_{t-i}, \qquad i = 1, \ldots, d$$
$$\tilde{\theta}^i(t - 1) \text{ is } \mathscr{F}_{t-i} \text{ measurable}$$
$$\eta^i(t - 1) \text{ is } \mathscr{F}_{t-i} \text{ measurable}$$

Now define

$$z^i(t) \triangleq \eta^i(t) - v^i(t); \qquad v^i(t) \triangleq f_{i-1}\omega(t - i + 1)$$
$$g^i(t) \triangleq z^i(t) - \tfrac{1}{2}b^i(t)$$

(Note that $g^i(t)$ is related to $b^i(t)$ by the transfer function $[1/\bar{C}(z^{-1}) - \frac{1}{2}]$.) Now show

that
$$v^i(t)\phi(t-d)^T\tilde{\theta}^i(t) = v^i(t)\phi(t-d)^T\tilde{\theta}^i(t-1)$$
$$+ v^i(t)\phi(t-d)^TP(t-d-1)\phi(t-d)\eta^i(t)$$
$$= v^i(t)\phi(t-d)^T\tilde{\theta}^i(t-1) + v^i(t)\phi(t-d)^TP(t-d-1)$$
$$\times \phi(t-d)\{b^i(t) - (\bar{C}(q^{-1}) - 1)(\eta^i(t) - v^i(t)) + v^i(t)\}$$

Take conditional expectations to show that
$$-\{1 + \phi(t-d)^TP(t-d-1)\phi(t-d)\}E\{v^i(t)b^i(t)|\mathcal{F}_{t-i}\}$$
$$= \phi(t-d)^TP(t-d-1)\phi(t-d)f^2_{t-1}\sigma^2$$

Hence show that
$$E\{v^i(t)b^i(t)|\mathcal{F}_{t-i}\} = \frac{-\phi(t-d)^TP(t-d-1)\phi(t-d)}{1 + \phi(t-d)^TP(t-d-1)\phi(t-d)}f^2_{t-1}\sigma^2$$

Step 3: Introduce a positive Lyapunov-type function associated with *each* $\tilde{\theta}^i$ as follows:
$$V^i(t) \triangleq \tilde{\theta}^i(t)P(t-d)^{-1}\tilde{\theta}^i(t)$$
where
$$\tilde{\theta}^i(t) = \tilde{\theta}^i(t-1) + P(t-d-1)\phi(t-d)\eta^i(t)$$

Then use the standard Martingale convergence argument to establish for $i = 1, \ldots, d$,
$$\lim_{N\to\infty} \sum_{j=d}^N \frac{\phi(j-d)^TP(j-d-1)\phi(j-d)\eta^i(j)^2}{\bar{r}(j-d-1)} < \infty \quad \text{a.s.}$$

$$\lim_{N\to\infty} \sum_{j=d}^N \frac{z^i(j)^2}{\bar{r}(j-d-1)} < \infty \quad \text{a.s.}$$

and (using the Schwarz inequality) that
$$\lim_{N\to\infty} \sum_{t=d}^N \frac{\phi(t-d)^TP(t-d-1)\phi(t-d)}{\bar{r}(t-d-1)}\eta(t)^2 < \infty \quad \text{a.s.}$$

$$\lim_{N\to\infty} \sum_{t=d}^N \frac{[\eta(t) - v(t)]^2}{\bar{r}(t-d-1)} < \infty \quad \text{a.s.}$$

Thus show that
$$\lim_{N\to\infty} \sum_{t=d}^N \frac{[\epsilon(t) - v(t)]^2}{\bar{r}(t-d)} < \infty \quad \text{a.s.}$$

Then show that
$$e(t) - v(t) = \epsilon(t) - v(t) + \phi(t-d)^T[\hat{\theta}(t-1) - \hat{\theta}(t-d)]$$
$$= (\epsilon(t) - v(t)) + \sum_{k=2}^d \phi(t-d)^TP(t-d-k)\phi(t-d-k+1)$$
$$\times \eta(t-k+1)$$

and hence that
$$[e(t) - v(t)]^2 \leq d[\epsilon(t) - v(t)]^2$$
$$+ d\sum_{k=2}^d [\phi(t-d)^TP(t-d-k)^{1/2}P(t-d-k)^{1/2}$$
$$\times \phi(t-d-k+1)]^2\eta(t-k+1)^2$$
$$\leq d[\epsilon(t) - v(t)]^2$$
$$+ d\sum_{k=2}^d [\phi(t-d)^TP(t-d-k)\phi(t-d)]$$
$$\times [\phi(t-d-k+1)^T P(t-d-k)\phi(t-d-k+1)]\eta(t-k+1)^2$$

using the triangle inequality. Then argue that the following results hold:

$$\lim_{N\to\infty} \sum_{t=d}^{N} \frac{[e(t) - v(t)]^2}{\bar{r}(t-d)} < \infty \quad \text{a.s.}$$

$$\lim_{N\to\infty} \sum_{t=d}^{N} \frac{[e(t) - v(t)]^2}{r(t-d)} < \infty \quad \text{a.s.}$$

where

$$r(t - d) = r(t - d - 1) + \phi(t-d)^T\phi(t-d)$$

Step 4: By similar reasoning show that

$$\lim_{N\to\infty} \sum_{t=d}^{N} \frac{[\eta(t) - v(t)]^2}{r(t-d-1)} < \infty \quad \text{a.s.}$$

and hence from Kronecker's lemma that

$$\lim_{N\to\infty} \left[\frac{r(N-d-1)}{N}\right]^{-1} \frac{1}{N} \sum_{t=d}^{N} [\eta(t) - v(t)]^2 = 0$$

Using the definition of $\{r(t)\}$ and $\{\phi(t)\}$ together with Lemma B.3.3 of Appendix B, show that there exist constants $0 < K_{11}, K_{12} < \infty$ such that

$$\frac{r(N-d-1)}{N} \leq \frac{K_{11}}{N} \sum_{t=d}^{N} [\eta(t) - v(t)]^2 + K_{12}$$

Then use the stochastic key technical lemma to show that

$$\limsup_{N\to\infty} \frac{r(N-d-1)}{N} < \infty$$

$$\lim_{N\to\infty} \sum_{t=d}^{N} \frac{[e(t) - v(t)]^2}{t} < \infty$$

which by Kronecker's lemma gives

$$\lim_{N\to\infty} \frac{1}{N} \sum_{t=d}^{N} [y(t) - v(t) - \hat{y}(t)]^2 = 0 \quad \text{a.s.}$$

Finally, combine the results above to establish the theorem.]

9.10. (a) Show that $C(z^{-1}) = 1 - \alpha z^{-1}$ is strictly input passive if and only if $|\alpha| < 1$.

(b) Comment on the use of the pseudo linear regression algorithm when the noise model has zeros on the unit circle.

9.11. Consider a purely deterministic disturbance of period $N = 12$ samples. Each harmonic in the Fourier series of the waveform has the form

$$d_i(t) = A_i \sin[\rho_i t + \phi_i]; \qquad \rho_i = \frac{\pi i}{6}$$

and has the associated model

$$[1 - (2 \cos \rho_i)q^{-1} + q^{-2}]d_i(t) = 0; \qquad i \neq 0, 6$$

$$[1 - q^{-1}]d_0(t) = 0; \qquad i = 0$$

$$[1 + q^{-1}]d_0(t) = 0; \qquad i = 6$$

(a) Show that $y(t) = \sum_{i=0}^{11} d_i(t)$ can be modeled by $(1 - q^{-12})y(t) = 0$. [*Hint:* Show that

$$(1 - q^{-1})(1 + q^{-1}) \prod_{i=1}^{5} [1 - (2 \cos \rho_i)q^{-1} + q^{-2}] = 1 - q^{-12}.]$$

(b) Use the reasoning above to motivate the predictor given in (9.3.51).

9.12. Consider the LMS algorithm for large t (i.e., assume that the initial transients have disappeared). Let $\hat{\theta}(t)$ denote the estimate provided by the LMS algorithm

$$J(\theta) \triangleq E\{[y(t+1) - \phi(t)^T\theta]^2\}$$

$$\theta^* \triangleq \arg\min J(\theta)$$

$$J^* \triangleq J(\theta^*)$$

$$J_{\text{LMS}} \triangleq J_{\text{LMS}} - J^*$$

$$J_{\text{excess}} = J_{\text{LMS}} - J^*$$

(a) Assume that $\{\phi(t-1)\}$ is an uncorrelated sequence [i.e., assume that $\hat{\theta}(t)$ is uncorrelated with $\phi(t)$]. Show that (using the notation of Remark 9.4.2)

$$J_{\text{excess}} = E\{\tilde{\theta}(t)^T\phi(t)\phi(t)^T\tilde{\theta}(t)\}$$
$$= E\{\tilde{\theta}(t)^T R\tilde{\theta}(t)\}$$
$$= E\{\tilde{\theta}'(t)^T \Lambda \tilde{\theta}'(t)\}$$
$$= \text{trace } \Lambda V$$

where $V = E\{\tilde{\theta}'(t)\tilde{\theta}'(t)^T\}$.

(b) Show that

$$J_{\text{excess}} \simeq \frac{\mu}{2}J^* \text{ trace } \Lambda$$

[*Hint:* Show in order:

1. $\tilde{\theta}(t+1) = [I - \mu R]\tilde{\theta}(t) + \mu g(t)$
   where $g(t) \triangleq \phi(t)e(t) - E\{\phi(t)e(t)\}$
   $e(t) \triangleq y(t+1) - \phi(t)^T\hat{\theta}(t)$
2. $\tilde{\theta}'(t+1) = [I - \mu\Lambda]\tilde{\theta}'(t) + \mu g'(t)$
   where $g'(t) \triangleq Q^T g(t)$
3. $E\{g'(t)g'(t)^T\} \simeq (Q^T R Q)J^*$ [close to θ^* when $E\{\phi(t)^T e(t)\} \simeq 0$]
   $= J^*\Lambda$
4. $V \simeq [I - \mu\Lambda]V[I - \mu\Lambda]^T + \mu^2 J^*\Lambda$
5. trace $V = $ trace $[V[I - \mu\Lambda]^2] + \mu^2 J^*$ trace Λ
   2μ trace $\Lambda V \simeq \mu^2 J^*$ trace Λ (for $\mu \ll 1$)
6. trace $\Lambda V \simeq \frac{1}{2}\mu J^*$ trace Λ.]

(c) Note that J_{excess} is small for $\mu\lambda_{\max}$ small. Comment on the relationship between this result and the rate of convergence discussed in the text, particularly when the condition number of R is large.

9.13. Consider a process, $\{y(t)\}$, with covariance $C_\tau = E\{y(t)y(t-\tau)\}$.

(a) Consider a first-order transversal predictor of the form

$$\hat{y}(t+1) = ay(t)$$

Determine the optimal value of a to minimize the mean-square prediction error.

(b) Assuming that the predictor in part (a) is the unrestricted optimal predictor, show that the two-step-ahead predictor is given by

$$\hat{y}(t+2) = a^2 y(t)$$

(c) Consider a restricted complexity two-step-ahead predictor for a general process $\{y(t)\}$ of the form

$$\hat{y}(t+2) = by(t)$$

Determine the optimal value of b to minimize the mean-square two-step-ahead prediction error.

(d) Show that the two-step-ahead predictors in parts (b) and (c) are the same if and only if

$$c_2 = \frac{c_1^2}{c_0}$$

(e) Comment on the relatively advantages of direct and indirect methods for obtaining two-step-ahead predictors.

9.14. Consider the system shown in Fig. 9.4.3. Show that this model can be written as

$$A(q^{-1})y(t) = B'(q^{-1})[s(t - d) + n(t - d)]; \qquad \{n(t)\} \text{ white noise}$$

Show that the algorithm (9.4.15)–(9.4.19) using pseudo linear regressions leads to consistent estimation of $A(q^{-1})$ and $B'(q^{-1})$ provided that $[1/B(q^{-1}) - \frac{1}{2}]$ is very strictly passive.

9.15. Consider the following first-order linear system:

$$x(t + 1) = \theta x(t) + v_1(t)$$
$$y(t) = x(t) + v_2(t); \qquad v_1(t), v_2(t) \text{ white noise}$$

(a) Write down the steady-state Kalman filter for this problem.
(b) Develop an adaptive state estimator using the recursive prediction error algorithm to estimate θ and K.

9.16. Show how the fixed-lag smoother described in Exercise 7.7 can be made adaptive. [*Hint:* Fit a one-step-ahead predictor to $y(t)$ of the form

$$A(q^{-1})y(t) = G(q^{-1})\omega(t)$$

using any parameter estimation algorithm.]

10

Control of Stochastic Systems

10.1 INTRODUCTION

In this chapter we turn our attention to the control of stochastic systems and consider both deterministic and nondeterministic disturbances. This will be a natural generalization of Chapter 5, where the disturbances were assumed to be deterministic.

We shall show that control of linear stochastic systems follows very closely the corresponding deterministic problem. The stochastic nature of the problem arises in relation to uncertainty about the system state and does not appear otherwise in the control system design. The key distinction between deterministic and stochastic control is that in the deterministic case the state can be reconstructed exactly, whereas in the stochastic case there will, in general, be a residual state estimation error due to the presence of noise. We demonstrate that each of the deterministic design methods discussed in Chapter 5 has a stochastic counterpart.

We begin by investigating the applicability of designs based on purely deterministic considerations. We show that, in some cases, this leads to an acceptable solution to the stochastic control problem. We point to its advantages and limitations.

We are then lead to consider designs that incorporate models of the stochastic components. In particular, we explore how explicit models of the noise dynamics can be used in state estimation and disturbance prediction. This underlines the basic distinction between the deterministic and stochastic problems.

We describe a range of stochastic control algorithms beginning with the stochastic version of the minimum prediction error algorithms of Chapter 5. A case of practical importance is when the system has a significant time delay. In this case, the ability to predict the effect of control actions and disturbances on future response can be critical. A similar idea is inherent in the well-known classical design technique based on the Smith predictor, which also uses a predictor or model of the system output to counter the effect of the system time delay.

Finally, we extend the minimum prediction error algorithms by taking into account the control effort via a linear quadratic performance criterion. We show that in the case of gaussian disturbances the optimal design incorporates a Kalman filter together with state-variable feedback. This result is summarized in the separation theorem.

10.2 THE APPLICATION OF DETERMINISTIC DESIGN METHODS

We have seen in Chapter 5 that many of the methods for deterministic control system design can be viewed as closed-loop pole assignment algorithms. In this section we investigate the applicability of these methods to cases where there are stochastic disturbances.

We shall use a stochastic autoregressive moving-average model with auxiliary input (ARMAX model) as the basis of our discussion. In the single-input single-output case, this model has the form

$$A(q^{-1})y(t) = q^{-d}B'(q^{-1})u(t) + C(q^{-1})\omega(t) \tag{10.2.1}$$

where $\{y(t)\}$ and $\{u(t)\}$ denote the output and input, respectively.

$$A(q^{-1}) = 1 + a_1 q^{-1} + \cdots + a_n q^{-n}$$
$$B'(q^{-1}) = b_0 + b_1 q^{-1} + \cdots + b_m q^{-m}; \quad b_0 \neq 0; \quad B(q^{-1}) \triangleq q^{-d}B'(q^{-1})$$
$$C(q^{-1}) = 1 + c_1 q^{-1} + \cdots + c_l q^{-l}$$

[We have seen in Chapter 7 that, without loss of generality, $C(z^{-1})$ can be taken to have roots on or inside the unit circle.]

The noise sequence, $\{\omega(t)\}$, will be taken to satisfy our usual assumption:

$$E\{\omega(t) \mid \mathcal{F}_{t-1}\} = 0 \quad \text{a.s.} \tag{10.2.2}$$

$$E\{\omega(t)^2 \mid \mathcal{F}_{t-1}\} = \sigma^2 \quad \text{a.s.} \tag{10.2.3}$$

where $\mathcal{F}_t$ denotes data up to time t.

To simplify the discussion, we shall assume that there are no deterministic disturbances. (If deterministic disturbances are present, we can simply apply the methods outlined in Section 5.3.5.) Thus we assume that $A(q^{-1})$ and $B'(q^{-1})$ are relatively prime or, at most, have common roots strictly inside the unit circle. [These common roots can arise due to modes (associated with the stochastic disturbance) which are not in the input–output transfer function. These modes are required to be stable to ensure stabilizability of the overall system.]

Following the discussion in Section 5.3, we introduce the following general feedback control law:

$$L(q^{-1})u(t) = -P(q^{-1})y(t) + M(q^{-1})y^*(t) \tag{10.2.4}$$

We shall also adopt the error-actuated structure shown in Fig. 5.3.2; that is, we take

$M(q^{-1}) = P(q^{-1})$. The resulting closed-loop system is then given by

$$[A(q^{-1})L(q^{-1}) + B(q^{-1})P(q^{-1})]y(t)$$
$$= B(q^{-1})P(q^{-1})y^*(t) + L(q^{-1})C(q^{-1})\omega(t) \qquad (10.2.5)$$

The closed-loop response can thus be considered as the superposition of the response due to $\{y^*(t)\}$ and the response due to the noise $\{\omega(t)\}$. The response due to $\{y^*(t)\}$ obeys the standard conditions pertaining in the deterministic case. Thus we can ensure good transient response and good tracking performance (in relation to $\{y^*(t)\}$) by using the design techniques outlined in Chapter 5.

Turning to the random part of the response, we see that the white noise $\{\omega(t)\}$ passes through the following closed-loop transfer function

$$H(z) = \frac{N(z)}{1 + G(z)H(z)} \qquad (10.2.6)$$

where $N(z)$ is the open-loop noise transfer function, $G(z)$ is the open-loop plant transfer function, and $H(z)$ is the transfer function of the controller [i.e., $P(z^{-1})/L(z^{-1})$].

Provided that the closed-loop system is stable (which can always be achieved by proper choice of the feedback law), the variance, σ_c^2, of the noise on the closed-loop output response can be evaluated as

$$\sigma_c^2 = \frac{\sigma^2}{2\pi} \int_{-\pi}^{\pi} \left| \frac{N(e^{j\omega})}{1 + G(e^{j\omega})H(e^{j\omega})} \right|^2 d\omega \qquad (10.2.7)$$

where σ^2 is the variance of $\{\omega(t)\}$.

It is clear from the equation above that the effect of the noise on the closed-loop system will be diminished relative to the open-loop case provided that the loop gain $G(e^{j\omega})H(e^{j\omega})$ is made large over the effective bandwidth of the noise. This is exactly as for the deterministic case, where we aimed to have high loop gain for those frequencies dominant in the disturbances.

One simple way of achieving this is to apply the *internal model principle* as used in the deterministic design. Following the deterministic design methodology, we constrain the denominator of the controller, $L(q^{-1})$, to have the following form:

$$L(q^{-1}) = L'(q^{-1})S(q^{-1})D(q^{-1}) \qquad (10.2.8)$$

and then solve the following pole assignment equation:

$$L'(q^{-1})S(q^{-1})D(q^{-1})\tilde{A}(q^{-1}) + P(q^{-1})\tilde{B}(q^{-1}) = A^*(q^{-1}) \qquad (10.2.9)$$

where $A(q^{-1}) = \tilde{A}(q^{-1})D(q^{-1})$
$\qquad B(q^{-1}) = \tilde{B}(q^{-1})D(q^{-1})$
$\qquad D(q^{-1})$ denotes the uncontrollable modes arising from the disturbance
$\qquad S(q^{-1})$ is the denominator polynomial in the reference model, i.e.,
$\qquad\qquad S(q^{-1})y^*(t) = 0$

If this is done, then (in steady state) the resulting output tracking error, $e(t) \triangleq y(t) - y^*(t)$ satisfies an equation of the following form [obtained from (10.2.1) on multiplying by $L'(q^{-1})S(q^{-1})$] (Exercise 10.1):

$$A^*(q^{-1})e(t) = L'(q^{-1})S(q^{-1})C(q^{-1})\omega(t) \qquad (10.2.10)$$

The expression above allows one to calculate the resulting output variance for a given set of closed-loop pole locations.

A number of practical points arise in relation to the design above (see also Exercise 10.4):

1. If the output noise is predominantly low frequency, its effect on the output can be diminished using the deterministic design considerations outlined above. As in Chapter 5, a good design rule is to make the sampling rate roughly 5 to 10 times faster than the desired closed-loop bandwidth. Then noise having bandwidth up to about one-tenth the sampling rate can be significantly diminished if the loop gain can be made high.

2. Feedback is not helpful in reducing wideband noise on the output. One obvious step that should be taken is to filter the system output prior to sampling so that frequency components above half the sampling rate are eliminated prior to sampling. This avoids aliasing of high-frequency noise components, which leads to large noise variance on the samples.

3. Further reduction in the variance of the disturbances on the output of the closed-loop system can be achieved by explicitly taking account of the disturbance model in the design of the control system. This can be thought of as a stochastic version of the internal model principle. In the deterministic case, it was possible to predict the future values of the disturbance exactly from past data and hence to cancel them perfectly by feedback. In fact, this property is precisely what characterizes a deterministic disturbance. In the stochastic case, it is not possible to achieve perfect prediction. However, the best strategy is now to use the best possible predictor of the future disturbance response obtained by using an optimal predictor (i.e., the Kalman filter). Incorporation of this filter into the control system design will, in general, lead to a reduction of the output variance. In particular, it can be seen from (10.2.10) that if $A^*(q^{-1}) = C(q^{-1})$ then $e(t)$ is simply a moving average of white noise.

In the next section we consider the design of controllers for stochastic systems incorporating an optimal predictor for the disturbances.

10.3 STOCHASTIC MINIMUM PREDICTION ERROR CONTROLLERS

In this section we extend the minimum prediction error controllers of Section 5.2 to the stochastic case.

Initially, we restrict attention to single-input single-output systems. Later we show how the results can be extended to multi-input multi-output case (see Section 10.3.3).

As in Section 10.2 we shall use the ARMAX model (10.2.1):

$$A(q^{-1})y(t) = q^{-d}B'(q^{-1})u(t) + C(q^{-1})\omega(t) \qquad (10.3.1)$$

where $\{u(t)\}$ and $\{y(t)\}$ denote the input and output, respectively, and $\{\omega(t)\}$ is a "white noise" sequence satisfying

$$E\{\omega(t)\,|\,\mathfrak{F}_{t-1}\} = 0 \quad \text{a.s.} \qquad (10.3.2)$$

$$E\{\omega(t)^2\,|\,\mathfrak{F}_{t-1}\} = \sigma^2 \quad \text{a.s.} \qquad (10.3.3)$$

The first controller that we shall develop is the stochastic equivalent of the one-step-ahead controller discussed in Section 5.2.1. In the stochastic case this controller is called a *minimum variance* controller.

A frequent motivation for minimum variance control is that by reducing the variance of a given variable, the set point, y^*, can be put at a less conservative value while still ensuring that a given proportion of the output meets a given acceptance criteria. This can lead to greater throughput or reduced cost. An example is in controlling the feed to a grinding mill. With the use of minimum variance control, the average throughput may be increased for a given probability of an overload trip occurring. This is illustrated in Fig. 10.3.1.

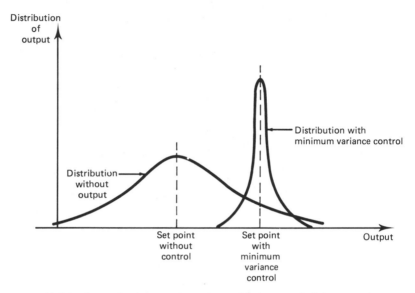

Figure 10.3.1 Increasing average throughput with the use of minimum variance control.

10.3.1 Minimum Variance Control (The SISO Case)

In the case of deterministic systems, the output could be predicted exactly and thus, in the development of the one-step-ahead controller in Section 5.2.1, the input was determined so as to bring the output, $y(t + d)$, at some future time to a desired value $y^*(t + d)$. In the stochastic case, the output cannot be predicted exactly. However, it now makes sense to choose the input so as to minimize the mean-square error between the output and the desired value $y^*(t + d)$. Thus consider the following cost function:

$$J(t + d) \triangleq E\{[y(t + d) - y^*(t + d)]^2\} \tag{10.3.4}$$

We wish to choose $u(t)$ as a function of $y(t), y(t - 1), \ldots, u(t - 1), u(t - 2), \ldots$ so as to minimize $J(t + d)$. Using the smoothing property of conditional expectations (see Appendix D), we have

$$J(t + d) = E\{E\{[y(t + d) - y^*(t + d)]^2 \,|\, \mathcal{F}_t\}\} \tag{10.3.5}$$

Hence the optimal cost is

$$J^*(t + d) = E\{\min_{u(t)} E\{[y(t + d) - y^*(t + d)]^2 | \mathcal{F}_t\}\} \qquad (10.3.6)$$

where $u(t)$ is constrained to be $\mathcal{F}_t$ measurable.

The minimization above is subject to the constraint that (10.3.1) be satisfied. We shall find that the solution to this minimization problem is more transparent if we introduce an optimal d-step-ahead prediction of $y(t + d)$.

As we have seen in Section 7.4.2, a linear time-invariant optimal predictor can be used only when the C-polynomial does not have roots on the unit circle. Thus we shall temporarily assume that the C-polynomial has zeros strictly inside the unit circle. The more general case will be taken up subsequently.

From Lemma 7.4.1 the optimal d-step-ahead prediction of $\{y(t)\}$ for the system (10.3.1) satisfies

$$C(q^{-1})y^0(t + d | t) = \alpha(q^{-1})y(t) + \beta(q^{-1})u(t) \qquad (10.3.7)$$

where

$$y^0(t + d | t) = E\{y(t + d) | \mathcal{F}_t\} \qquad (10.3.8)$$

$$= y(t + d) - F(q^{-1})\omega(t + d) \qquad (10.3.9)$$

$$\alpha(q^{-1}) = G(q^{-1}) \qquad (10.3.10)$$

$$\beta(q^{-1}) = F(q^{-1})B'(q^{-1}) \qquad (10.3.11)$$

and $G(q^{-1})$ and $F(q^{-1})$ are the unique polynomials satisfying

$$C(q^{-1}) = F(q^{-1})A(q^{-1}) + q^{-d}G(q^{-1}) \qquad (10.3.12)$$

$$F(q^{-1}) = f_0 + f_1 q^{-1} + \cdots + f_{d-1}q^{-(d-1)}; \qquad f_0 = 1 \qquad (10.3.13)$$

$$G(q^{-1}) = g_0 + g_1 q^{-1} + \cdots + g_{n-1}q^{-(n-1)} \qquad (10.3.14)$$

We now use (10.3.7) and (10.3.9) to find $u(t)$ such that the cost (10.3.6) is minimized.

Theorem 10.3.1

(a) The minimum variance control satisfying (10.3.6) is given by

$$\beta_0 u(t) = q[\beta_0 - \beta(q^{-1})]u(t - 1) + y^*(t + d) + [C(q^{-1}) - 1]$$
$$y^0(t + d | t) - \alpha(q^{-1})y(t) \qquad (10.3.15)$$

or

$$\phi'(t)^T \theta_0 = y^*(t + d) \qquad (10.3.16)$$

where

$$\phi'(t) = [y(t), \ldots, y(t - n + 1), u(t), \ldots, u(t - m - d + 1),$$
$$-y^0(t + d - 1 | t - 1), \ldots, -y^0(t + d - l | t - l)] \qquad (10.3.17)$$

$$\theta_0 = [\alpha_0, \ldots, \alpha_{n-1}, \beta_0, \ldots, \beta_{m+d-1}, c_1, \ldots, c_l] \qquad (10.3.18)$$

(b) The effect of the control (10.3.15) is to give

$$y^0(t + d | t) = y^*(t + d) \qquad (10.3.19)$$

(i.e., the predicted output is set equal to the desired output).

(c) If (10.3.16) holds for all t, the control law can be written as

$$\beta(q^{-1})u(t) + \alpha(q^{-1})y(t) = C(q^{-1})y^*(t + d) \qquad (10.3.20)$$

(d) The control law (10.3.20) gives a closed-loop system satisfying

$$B'(q^{-1})u(t) = A(q^{-1})y^*(t + d) - G(q^{-1})\omega(t) \qquad (10.3.21)$$

$$y(t + d) = y^*(t + d) + \tilde{y}(t + d) \qquad (10.3.22)$$

where

$$\tilde{y}(t + d) = y(t + d) - y^0(t + d | t) \qquad (10.3.23)$$

$$= F(q^{-1})\omega(t + d) \qquad (10.3.24)$$

Proof. (a) Substituting (10.3.8) into (10.3.6) gives

$$J^*(t + d) = E\{\min_{u(t)} E\{[y(t + d) - y^0(t + d | t) \\ + y^0(t + d | t) - y^*(t + d)]^2 | \mathcal{F}_t\}\} \qquad (10.3.25)$$

$$J^*(t + d) = E\{\min_{u(t)} E\{[y(t + d) - y^0(t + d | t)]^2 \\ + 2[y(t + d) - y^0(t + d | t)][y^0(t + d | t) - y^*(t + d)] \\ + [y^0(t + d | t) - y^*(t + d)]^2 | \mathcal{F}_t\}\} \qquad (10.3.26)$$

$$= E\{\min_{u(t)} \sum_{j=0}^{d-1} f_j^2 \sigma^2 + [y^0(t + d | t) - y^*(t + d)]^2\}\}$$

using (10.3.2), (10.3.3), (10.3.8), and (10.3.9). Substituting for $y^0(t + d | t)$ from (10.3.7) gives

$$J^*(t + d) = E\left\{\sum_{j=0}^{d-1} f_j^2 \sigma^2 + \min_{u(t)} [(1 - C(q^{-1}))y^0(t + d | t) + \beta(q^{-1})u(t) \\ + \alpha(q^{-1})y(t) - y^*(t + d)]^2\right\} \qquad (10.3.27)$$

The input $u(t)$ minimizing (10.3.27) can be seen to be given by

$$\beta(q^{-1})u(t) = y^*(t + d) + [C(q^{-1}) - 1]y^0(t + d | t) - \alpha(q^{-1})y(t) \qquad (10.3.28)$$

This establishes (10.3.15). Equation (10.3.16) follows immediately.
(b) Follows from (10.3.7) and (10.3.28).
(c) Immediate from (10.3.28), since

$$y^0(t + d - j | t - j) = y^*(t + d - j), \qquad j = 1, \ldots, l$$

(d) Equation (10.3.22) follows from (10.3.9) and (10.3.19). Finally, from (10.3.1),

$$B'(q^{-1})u(t) = A(q^{-1})y(t + d) - C(q^{-1})\omega(t + d)$$

$$= A(q^{-1})[y^*(t + d) + F(q^{-1})\omega(t + d)] - C(q^{-1})A(t + d)$$

using (10.3.23)–(10.3.24);

$$= A(q^{-1})y^*(t + d) + G(q^{-1})\omega(t) \qquad (10.3.29)$$

using (10.3.12).
This establishes (10.3.21).

▼▼▼

Remark 10.3.1. In the theorem above we have derived a stochastic controller based on an optimal predictor. The predictor allows us to determine, using past input and output data, the predictable part of the disturbance on the future response and hence to cancel it using the control action. Specifically, we note that result (b) shows that putting the predicted output equal to the desired output achieves minimum variance control. This is in line with the approach for deterministic systems in Section 5.2.1. We see from (10.3.21) the $u(t)$ will be mean-square bounded provided that $B'(q^{-1})$ is stable. This corresponds to the stably invertible condition associated with the deterministic one-step-ahead controller. A more general result will be presented presently.

▼▼▼

Remark 10.3.2. The reader may have noticed that for the predictor (10.3.7) to be optimal for all t, it is necessary that the initial conditions for $y^0(t + d|t)$ be appropriately chosen; see Remark 7.4.3.

In practice, the initial conditions may not be known. In this case, one can simply define a prediction $\hat{y}(t + d)$ by (10.3.7) but with arbitrary initial conditions, that is, define $\hat{y}(t + d)$ by

$$C(q^{-1})\hat{y}(t + d) = \alpha(q^{-1})y(t) + \beta(q^{-1})u(t) \tag{10.3.30}$$

Since $C(q^{-1})$ is asymptotically stable (by assumption), then $\hat{y}(t + d)$ will converge to $y^0(t + d|t)$ exponentially fast. It thus makes sense to implement the control law (10.3.15) by replacing $y^0(t + d|t)$ by $\hat{y}(t + d)$. If this is done, the appropriate generalization of Theorem 10.3.1 is given in the following theorem.

Theorem 10.3.2. Consider the control law

$$\beta_0 u(t) = q[\beta_0 - \beta(q^{-1})]u(t - 1) + y^*(t + d)$$
$$+ [C(q^{-1}) - 1]\hat{y}(t + d) - \alpha(q^{-1})y(t) \tag{10.3.31}$$

where $\hat{y}(t + d)$ is given by (10.3.30) with arbitrary initial conditions.

(a) The effect of the control law (10.3.31) is to give

$$\hat{y}(t + d) = y^*(t + d) \qquad \text{for all } t \tag{10.3.32}$$

(b) If (10.3.31) holds for all t, the control law can be written

$$\beta(q^{-1})u(t) + \alpha(q^{-1})y(t) = C(q^{-1})y^*(t + d) \tag{10.3.33}$$

(c) The control law (10.3.33) gives a closed-loop system satisfying

$$C(q^{-1})B'(q^{-1})u(t) = C(q^{-1})[A(q^{-1})y^*(t + d)$$
$$- G(q^{-1})\omega(t)] \tag{10.3.34}$$
$$C(q^{-1})[y(t + d) - y^*(t + d) - \tilde{y}(t + d)] = 0 \tag{10.3.35}$$

where

$$\tilde{y}(t + d) = y(t + d) - y^0(t + d|t) \tag{10.3.36}$$
$$= F(q^{-1})\omega(t + d) \tag{10.3.37}$$

(d) Provided that $C(q^{-1})$ is asymptotically stable, the output and input of the closed-loop system converge asymptotically to the output and input that would have resulted from the minimum variance controller, (10.3.15).

Proof. Conditions (a) to (c) are as for Theorem 10.3.1. Condition (d) follows by comparing (10.3.34)–(10.3.35) with (10.3.21)–(10.3.22) and noting that $C(q^{-1})$ is asymptotically stable.

▼▼▼

Remark 10.3.3. The minimum variance control strategy discussed above uses an optimal prediction of the system output to generate the feedback control law. The same idea is inherent in the *Smith predictor* (Smith, 1959; Palmer and Shinnar, 1979), which is commonly used for controlling systems with time delays. The basic idea in the Smith predictor is to generate the feedback, in part, by using an open-loop model of the plant with the delay removed. In the minimum variance controller, if the noise happens to be white output noise and the process is asymptotically stable, the ARMAX model has the following special form:

$$A(q^{-1})y(t) = q^{-d}B'(q^{-1})u(t) + A(q^{-1})\omega(t)$$

With this model, the optimal output prediction $y^0(t + d \mid t)$ is readily seen to be

$$A(q^{-1})y^0(t + d \mid t) = B'(q^{-1})u(t)$$

that is, in this case, the minimum variance controller also effectively uses an open-loop model of the plant to generate the feedback.

In general, the minimum variance control relies critically on the accuracy of the predictor for its success and thus, like the one-step-ahead controller in the deterministic case, its performance is very sensitive to the accuracy of the predictor parameters. Therefore, use of this class of algorithm is not recommended unless the parameters are known accurately or unless the algorithm is implemented adaptively so that the parameters are estimated on-line (see Chapter 11).

▼▼▼

Remark 10.3.4 (Control of Stochastic Nonlinear Systems). It is possible to develop stochastic minimum variance controllers for nonlinear systems based on the special model structure described in (7.4.20):

$$A(q^{-1})y(t) = h(t - d) + C(q^{-1})\omega(t)$$

where $\{\omega(t)\}$ is a white noise sequence.

$C(q^{-1})$ is stable and $h(t - d)$ is a nonlinear function of $\{y(t - d), \ldots, u(t - d), \ldots\}$. [Note that this model structure was specifically "engineered" so that the optimal d-step-ahead predictor has a simple structure (see Remark 7.4.4).]

It was shown in Lemma 7.4.5 that the optimal d-step-ahead predictor for the model above satisfies the following equation [see (7.4.75)]:

$$C(q^{-1})y^0(t + d \mid t) = G(q^{-1})y(t) + F(q^{-1})h(t)$$

The minimum variance controller is obtained by setting $y^0(t + d \mid t)$ equal to the desired value $y^*(t + d)$. Thus we immediately have that the minimum variance controller for the nonlinear system above has the form

$$F(q^{-1})h(t) = C(q^{-1})y^*(t + d) - G(q^{-1})y(t)$$

Two points must be made about the control law above:

1. In the general nonlinear case it may be difficult to solve for $u(t)$ in terms of $u(t-1)$, ... etc. This would have to be checked for each specific problem. For example, in the bilinear case $h(t)$ is linear in $u(t)$ and multilinear in $u(t-1)$, and so on, so no problem arises.
2. In the general nonlinear case it may be difficult to establish "nice" properties for the control law above. The nonlinear versions of Theorems 10.3.1 and 10.3.2 would depend on the specific form of the nonlinearity.

▼▼▼

Note that in Theorem 10.3.2, particularly part (d), we have excluded cases in which $C(q^{-1})$ has roots on the unit circle. We therefore now turn to consider the latter case.

Systems with Noise Dynamics Having Zeros on the Unit Circle

We have seen in Section 7.4 that in the general case the C-polynomial may have zeros inside or on the unit circle. The latter case can arise, for example, when purely deterministic disturbances are present.

We have specifically excluded roots on the unit circle in our previous development since we required that erroneous initial conditions in the time-invariant predictor would decay. For the general case, we have seen in Section 7.4.2 that a nearly optimal predictor exists which is linear, time invariant, and has the property that erroneous initial conditions decay. The predictor has the following structure (see Section 7.4.2):

$$C_s(q^{-1})\hat{y}(t+d) = \alpha(q^{-1})y(t) + \beta(q^{-1})u(t) \qquad (10.3.38)$$

where $C_s(q^{-1})$ is asymptotically stable and

$$C(q^{-1}) = C_s(q^{-1}) + C_R(q^{-1}) \qquad (10.3.39)$$

$C_R(q^{-1})$ is a residual which can be made as small as desired.

When deterministic disturbances are present, we can implement the control law (19.3.15) by replacing the optimal prediction, $y^0(t+d\,|\,t)$, by the suboptimal prediction, $\hat{y}(t+d)$, as given by (10.3.38). If this is done, Theorem 10.3.1 has the following counterpart:

Theorem 10.3.3. Consider the control law

$$\beta_0 u(t) = q[\beta_0 - \beta(q^{-1})]u(t-1) + y^*(t+d)$$
$$+ [C_s(q^{-1}) - 1]\hat{y}(t+d) - \alpha(q^{-1})y(t) \qquad (10.3.40)$$

where $\hat{y}(t+d)$ is given by the nearly optimal predictor (10.3.38). Then:

(a) The effect of the control law (10.3.40) is to give

$$\hat{y}(t+d) = y^*(t+d) \qquad (10.3.41)$$

(b) If (10.3.40) holds for all t, the control law can be written

$$\beta(q^{-1})u(t) + \alpha(q^{-1})y(t) = C_s(q^{-1})y^*(t+d) \qquad (10.3.42)$$

(c) The control law (10.3.42) gives a closed-loop system satisfying

$$C_s(q^{-1})B'(q^{-1})u(t) = C_s(q^{-1})A(q^{-1})y^*(t+d) - C(q^{-1})G(q^{-1})\omega(t) \qquad (10.3.43)$$

$$C_s(q^{-1})[y(t+d) - y^*(t+d) - \bar{y}(t+d) - \eta(t+d)] = 0 \qquad (10.3.44)$$

where $\bar{y}(t+d)$ is the optimal tracking error and satisfies

$$\bar{y}(t+d) = y(t+d) - y^0(t+d \,|\, t) = F(q^{-1})\omega(t+d) \qquad (10.3.45)$$

and where $\eta(t)$ is the additional error arising from the use of the nonoptimal predictor (10.3.38) and satisfies [see (7.4.65)]

$$C_s(q^{-1})\eta(t) = F(q^{-1})C_R(q^{-1})\omega(t) \qquad (10.3.46)$$

(d) The control law (10.3.40) will give performance arbitrarily close to optimal performance by choosing $C_R(q^{-1})$ sufficiently small.

(e) $\{y(t)\}$ and $\{u(t)\}$ will be sample mean-square bounded almost surely provided that $\{\omega(t)\}$ is sample mean-square bounded almost surely and provided that:

(i) All zeros of the polynomial $B'(q^{-1})$ lie inside or on the closed unit circle.

(ii) All poles of the transfer function

$$\frac{1}{B'(z^{-1})}[A(z^{-1}) \,|\, C(z^{-1})]$$

lie strictly inside the unit circle.

(iii) Any zeros of the polynomial $B'(q^{-1})$ on the unit circle have a Jordan block size of 1.

Proof. Conditions (a) to (c) are as for Theorem 10.3.1. Condition (d) follows from the properties of the predictor (10.3.38) and the additional error term $\{\eta(t)\}$ discussed in Section 7.4.2. Condition (e): From (10.3.46) and the stability of $C_s(q^{-1})$ it follows that $\{\eta(t)\}$ is mean-square bounded (a.s.).

The result then follows from (10.3.43), (10.3.44), and Lemma B.3.3 of Appendix B.

▼▼▼

Remark 10.3.5. Part (e) of Theorem 10.3.3 provides the stochastic generalization of Theorem 5.2.1. Note that the conditions allow both deterministic and nondeterministic disturbances to be handled.

▼▼▼

Remark 10.3.6. Part (d) of Theorem 10.3.3 shows that the performance of the controller can be made as close to optimal as desired by choosing $C_R(q^{-1})$ sufficiently small. However, as in Section 7.4.2, the "memory" of the control law will depend on how close the roots of $C_s(q^{-1})$ are to the unit circle. The robustness of the controller may therefore be enhanced by keeping the roots of $C_s(q^{-1})$ a reasonable distance from the unit circle.

We present examples showing the performance of these controllers in Chapter 11.

In the next and subsequent sections, we will (for simplicity) assume in our discussion that $C(q^{-1})$ has roots strictly inside the unit circle. Of course, the comments made above apply *mutatis mutandis* to the subsequent development in case there are roots on the unit circle and the extension to the general case is straightforward.

Weighted One-Step-Ahead Stochastic Controllers

A slight generalization of the minimum variance controller is achieved by including a term in the cost function weighting the control effort. This leads to the stochastic equivalent of the weighted one-step-ahead controller (discussed in Section 5.2.1).

Consider the following one-step-ahead cost function:

$$J'(t + d) = E\left\{\frac{1}{2}[y(t + d) - y^*(t + d)]^2 + \frac{\lambda}{2}u(t)^2 \mid \mathcal{F}_t\right\} \tag{10.3.47}$$

We make the same assumptions as above. We then have

Theorem 10.3.4. For the system (10.3.1) having optimal d-step-ahead predictor (10.3.7), we have:

(a) The one-step-ahead control law minimizing (10.3.47) given by

$$u(t) = \frac{\beta_0}{\beta_0^2 + \lambda}\{q[\beta_0 - \beta(q^{-1})]u(t - 1) + y^*(t + d)$$
$$+ [C(q^{-1}) - 1]y^0(t + d \mid t) - \alpha(q^{-1})y(t)\} \tag{10.3.48}$$

which is equivalent to

$$y^0(t + d \mid t) - y^*(t + d) + \frac{\lambda}{\beta_0}u(t) = 0 \tag{10.3.49}$$

(b) When the control law (10.3.48) is used for all t, the closed-loop system is described by

$$\left[B'(q^{-1}) + \frac{\lambda}{\beta_0}A(q^{-1})\right]u(t) = G(q^{-1})\omega(t) + A(q^{-1})y^*(t + d) \tag{10.3.50}$$

$$\left[B'(q^{-1}) + \frac{\lambda}{\beta_0}A(q^{-1})\right]y(t + d) = \left[B'(q^{-1})F(q^{-1}) + \frac{\lambda}{\beta_0}C(q^{-1})\right]$$
$$\times \omega(t + d) + B'(q^{-1})y^*(t + d) \tag{10.3.51}$$

Proof. (a) Using (10.3.7)–(10.3.8), (10.3.47) can be written

$$J'(t + d) = \frac{1}{2}\sum_{j=0}^{d-1} f_j^2\sigma^2 + \frac{1}{2}[y^0(t + d \mid t) - y^*(t + d)]^2 + \frac{\lambda}{2}u(t)^2 \tag{10.3.52}$$

Using (10.3.7) and differentiating (10.3.52) with respect to $u(t)$ gives (10.3.49), which is equivalent to (10.3.48) when (10.3.7) is satisfied.

(b) From (10.3.9) and (10.3.49), we have

$$y(t + d) - F(q^{-1})\omega(t + d) - y^*(t + d) + \frac{\lambda}{\beta_0}u(t) = 0 \tag{10.3.53}$$

Multiplying by $A(q^{-1})$ gives

$$A(q^{-1})y(t + d) - A(q^{-1})F(q^{-1})\omega(t + d) - A(q^{-1})y^*(t + d) + \frac{\lambda}{\beta_0}A(q^{-1})u(t) = 0 \tag{10.3.54}$$

Using (10.3.1) and (10.3.12) yields

$$\left[B'(q^{-1}) + \frac{\lambda}{\beta_0}A(q^{-1})\right]u(t) = [A(q^{-1})F(q^{-1}) - C(q^{-1})]\omega(t + d) + A(q^{-1})y^*(t - d)$$
$$= G(q^{-1})\omega(t) + A(q^{-1})y^*(t + d) \tag{10.3.55}$$

This gives (10.3.50).

Multiplying (10.3.53) by $B'(q^{-1})$ gives

$$B'(q^{-1})y(t+d) - B'(q^{-1})F(q^{-1})\omega(t+d) - B'(q^{-1})y^*(t+d) + \frac{\lambda}{\beta_0}B'(q^{-1})u(t) = 0$$

Using (10.3.1),

$$\left[B'(q^{-1}) + \frac{\lambda}{\beta_0}A(q^{-1})\right]y(t+d) = B'(q^{-1})F(q^{-1})\omega(t+d) + B'(q^{-1})y^*(t+d)$$
$$+ \frac{\lambda}{\beta_0}C(q^{-1})\omega(t+d)$$

This gives (10.3.51).

▼▼▼

Remark 10.3.7. We note from part (b) of Theorem 10.3.4 that $\{u(t)\}$ and $\{y(t)\}$ will be bounded by $\{\omega(t)\}$ and $\{y^*(t)\}$ only if $[B'(q^{-1}) + (\lambda/\beta_0)A(q^{-1})]$ is stable. This is the stochastic equivalent of the corresponding result for deterministic weighted one-step-ahead controllers presented in Theorem 5.2.2.

For later use in developing an adaptive control algorithm (see Chapter 11) we shall express the control law (10.3.48) in a slightly different form.

Theorem 10.3.5

(a) The d-step-ahead predictor (10.3.7) for the system (8.3.1) can be rearranged into the form

$$\frac{\beta_0}{\beta_0^2 + \lambda}C(q^{-1})\left[y^0(t+d\,|\,t) - y^*(t+d) + \frac{\lambda}{\beta_0}u(t)\right] = u(t) + \phi(t)^T\theta_0$$
$$(10.3.56)$$

where

$$\phi(t)^T = [y(t), y(t-1), \ldots, u(t-1), u(t-2), \ldots, \\ y^*(t+d), y^*(t+d-1)\ldots]$$
$$(10.3.57)$$

θ_0 is a finite-length vector of parameters in the polynomials

$$\left[\frac{\beta_0}{\beta_0^2 + \lambda}\alpha(q^{-1}), \frac{\beta_0}{\beta_0^2 + \lambda}\left\{[\beta(q^{-1}) - \beta_0]q + \frac{\lambda}{\beta_0}[C(q^{-1}) - 1]q\right\}, \\ \frac{-\beta_0}{\beta_0^2 + \lambda}C(q^{-1})\right]$$
$$(10.3.58)$$

(b) If the polynomial $C(q^{-1})$ describing the noise dynamics in (10.3.1) is asymptotically stable, and if the input is generated by

$$u(t) = -\phi(t)^T\theta_0 \tag{10.3.59}$$

then the input converges exponentially fast to the one-step-ahead control law (10.3.48)–(10.3.49).

(c) When the control law (10.3.59) is used for all t, the closed-loop system is described by

$$C(q^{-1})\left[B'(q^{-1}) + \frac{\lambda}{\beta_0}A(q^{-1})\right]u(t) = C(q^{-1})[G(q^{-1})\omega(t) \\ + A(q^{-1})y^*(t+d)]$$
$$(10.3.60)$$

$$C(q^{-1})\left[B'(q^{-1}) + \frac{\lambda}{\beta_0}A(q^{-1})\right]y(t+d) = C(q^{-1})\left[B'(q^{-1})F(q^{-1})\right.$$

$$+ \frac{\lambda}{\beta_0}C(q^{-1})\left]\omega(t+d)\right. \tag{10.3.61}$$

$$+ C(q^{-1})B'(q^{-1})y^*(t+d)$$

Proof. (a) From (10.3.7) we have

$$C(q^{-1})y^0(t+d\,|\,t) = \alpha(q^{-1})y(t) + \beta(q^{-1})u(t) \tag{10.3.62}$$

Subtracting $C(q^{-1})[y^*(t+d) - (\lambda/\beta_0)u(t)]$ from both sides of (10.3.62) gives

$$C(q^{-1})\left[y^0(t+d\,|\,t) - y^*(t+d) + \frac{\lambda}{\beta_0}u(t)\right]$$

$$\tag{10.3.63}$$

$$= \alpha(q^{-1})y(t) + \beta(q^{-1})u(t) - C(q^{-1})\left[y^*(t+d) - \frac{\lambda}{\beta_0}u(t)\right]$$

Multiplying (10.3.63) on the left by $\beta_0/(\beta_0^2 + \lambda)$ gives

$$\frac{\beta_0}{\beta_0^2 + \lambda}C(q^{-1})\left[y^0(t+d\,|\,t) - y^*(t+d) + \frac{\lambda}{\beta_0}u(t)\right]$$

$$= u(t) + \frac{\beta_0}{\beta_0^2 + \lambda}\alpha(q^{-1})y(t) + \frac{\beta_0}{\beta_0^2 + \lambda}\{[\beta(q^{-1}) - \beta_0] \tag{10.3.64}$$

$$+ \frac{\lambda}{\beta_0}[C(q^{-1}) - 1]\}qu(t-1) - \frac{\beta_0}{\beta_0^2 + \lambda}C(q^{-1})y^*(t+d)$$

or, using (10.3.57)–(10.3.58),

$$\frac{\beta_0}{\beta_0^2 + \lambda}C(q^{-1})\left[y^0(t+d\,|\,t) - y^*(t+d) + \frac{\lambda}{\beta_0}u(t)\right] = u(t) + \phi(t)^T\theta_0 \tag{10.3.65}$$

This establishes (10.3.56).

(b) In view of the asymptotic stability of $C(q^{-1})$, the control law (10.3.59) will asymptotically yield (10.3.49) as required.

(c) Along the same lines as in the proof of Theorem 10.3.4, part (b).

▼▼▼

Remark 10.3.8. We note from Theorem 10.3.5, part (c) and Theorem 10.3.4, part (b), that the closed-loop system resulting from the control law (10.3.59) is the same as the closed-loop system resulting from the control law (10.3.48) save for a decaying initial transient produced by the stable polynomial $C(q^{-1})$.

The conditions for closed-loop stability are as before; $[B'(q^{-1}) + (\lambda/\beta_0)A(q^{-1})]$ should be stable.

▼▼▼

10.3.2 Model Reference Stochastic Control

The minimum variance controller discussed in the preceding subsection is the stochastic equivalent of the one-step-ahead controller described in Section 5.2.1. In the stochastic case the controller gives minimum variance regulation about the desired output sequence. However, this can often be achieved only by the use of relatively high feedback gains and large input signals.

In Section 5.2.2 for the deterministic case we introduced the concept of model reference control in an effort to overcome these difficulties. The basic idea was that the desired output sequence $\{y^*(t)\}$ was specified by a reference model, and this had the advantage of smoothing the desired output so that the size of the feedback gains and of the control signals would be reduced relative to one-step-ahead control. In this section we explore the extension of the model reference control principle to the stochastic case.

In model reference control, the desired output $y^*(t)$, is specified by a linear dynamical system driven by a reference input, $r(t)$, that is,

$$E(q^{-1})y^*(t) = q^{-d}gH(q^{-1})r(t) \tag{10.3.66}$$

where $E(q^{-1})$ is stable. As in Section 5.2.2, we form a predictor for $E(q^{-1})y(t)$ and then set the predicted value of this quantity equal to the value specified by the reference model. The form of this predictor is established in the following result:

Lemma 10.3.1. Provided that $C(q^{-1})$ is asymptotically stable, then the optimal d-step-ahead prediction $y_F^0(t + d \mid t)$ of $y_F(t + d) \triangleq E(q^{-1})y(t + d)$ satisfies

$$C(q^{-1})y_F^0(t + d \mid t) = \alpha(q^{-1})y(t) + \beta(q^{-1})u(t) \tag{10.3.67}$$

where

$$y_F^0(t + d \mid t) = E\{y_F(t + d) \mid \mathfrak{F}_t\} \tag{10.3.68}$$

(with $\mathfrak{F}_t$ the sigma algebra generated by data up to time t, plus initial condition information):

$$\alpha(q^{-1}) = G(q^{-1}) \tag{10.3.69}$$

$$\beta(q^{-1}) = F(q^{-1})B'(q^{-1}) \tag{10.3.70}$$

and $G(q^{-1})$ and $F(q^{-1})$ are the unique polynomials satisfying

$$C(q^{-1})E(q^{-1}) = F(q^{-1})A(q^{-1}) + q^{-d}G(q^{-1}) \tag{10.3.71}$$

$$F(q^{-1}) = 1 + f_1 q^{-1} + \cdots + f_{d-1}q^{-d+1} \tag{10.3.72}$$

$$G(q^{-1}) = g_0 + g_1 g^{-1} + \cdots + g_{n-1}q^{-n+1} \tag{10.3.73}$$

Proof. Essentially as for Lemma 7.4.1. (See also Lemma 5.2.1.)

▼▼▼

Given the predictor for $E(q^{-1})y(t)$ described above, it is a straightforward matter to extend the minimum variance controller to the case where a reference model is used. This is established in the following theorem:

Theorem 10.3.6

(a) The causal control law minimizing

$$J(t + d) \triangleq E\{[E(q^{-1})[y(t + d) - y^*(t + d)]]^2\} \tag{10.3.74}$$

satisfies

$$\beta_0 u(t) = q[\beta_0 - \beta(q^{-1})]u(t - 1) + gH(q^{-1})r(t)$$
$$+ [C(q^{-1}) - 1]y_F^0(t + d \mid t) - \alpha(q^{-1})y(t) \tag{10.3.75}$$

(b) The effect of the control law above is to give

$$y_F^0(t + d \mid t) = gH(q^{-1})r(t) = E(q^{-1})y^*(t + d) \tag{10.3.76}$$

(c) If (10.3.75) holds for all t, the control law can be written

$$\beta(q^{-1})u(t) + \alpha(q^{-1})y(t) = C(q^{-1})E(q^{-1})y^*(t + d) \tag{10.3.77}$$

(d) The control law (10.3.77) gives a closed-loop system satisfying

$$B'(q^{-1})E(q^{-1})u(t) = gA(q^{-1})H(q^{-1})r(t) - G(q^{-1})\omega(t) \tag{10.3.78}$$

$$E(q^{-1})y(t + d) = gH(q^{-1})r(t) + F(q^{-1})\omega(t + d) \tag{10.3.79}$$

(e) The controller gives minimum variance regulation of $y_F(t + d)$ about $y_F^*(t + d)$ where

$$y_F(t + d) = E(q^{-1})y(t + d) \tag{10.3.80}$$

$$y_F^*(t + d) = E(q^{-1})y^*(t + d) \tag{10.3.81}$$

(f) $\{y(t)\}$ and $\{u(t)\}$ will be sample mean-square bounded provided that $\{\omega(t)\}$ is sample mean-square bounded and that $B'(q^{-1})$ and $E(q^{-1})$ are both asymptotically stable.

Proof. Conditions (a) to (d) as for Theorem 10.3.1. Condition (e) follows from (10.3.74). Note also from (10.3.79) that

$$y_F(t + d) = y_F^*(t + d) + F(q^{-1})\omega(t + d) \tag{10.3.82}$$

Part (f) is Immediate from (10.3.78)–(10.3.79).

▼▼▼

Remark 10.3.9. Note that the controller (10.3.75) does *not* give minimum variance regulation of $y(t + d)$ about $y^*(t + d)$ unless $E(q^{-1}) = 1$. [The reader should reflect on the connection between (10.3.22) and (10.3.82).] Instead, minimum variance regulation of $E(q^{-1})y(t + d)$ about $E(q^{-1})y^*(t + d)$ has been achieved. However, this may require less control effort. The penalty is that the variance of $y(t)$ about $y^*(t)$ will be greater than in the minimum variance case. This is explored further in some of the exercises (see Exercise 10.12).

▼▼▼

Remark 10.3.10. The remarks made in Section 10.3.1 concerning systems with noise dynamics having zeros on the unit circle apply *mutatis mutandis* to model reference control. This is again studied in some of the exercises.

▼▼▼

10.3.3 Control of Multi-input Multi-output Stochastic Systems

We will now examine the extension of the stochastic minimum prediction error control algorithms to the multi-input multi-output case. This extension is straightforward using the tools developed in Section 5.2.3 for the deterministic case.

We describe the system in the multivariable form of (10.3.1):

$$A(q^{-1})y(t) = B(q^{-1})u(t) + C(q^{-1})\omega(t) \tag{10.3.83}$$

where $y(t)$ is $m \times 1$, $u(t)$ is $r \times 1$, and $\omega(t)$ is $m \times 1$.

For the reasons discussed in Section 5.2.3, we introduce the following standing assumptions.

Assumption 10.3.A

(1) The number of outputs, m, is equal to the number of inputs, r.
(2) The input–output transfer function $T(z) = A(z^{-1})^{-1}B(z^{-1})$ is strictly proper and satisfies

$$\det T(z) \neq 0 \qquad \text{almost all } z$$

In addition we assume, as usual, that $C(z^{-1})$ has roots on or inside the unit circle.

We will treat the case of a general delay structure. However, the reader may wish to revise Section 5.2.3, where simple special cases are treated as a preliminary to the general case.

In view of Assumption 10.3.A, part (2), it follows from Lemma 5.2.3 that associated with $T(z)$ there is an interactor matrix $\xi(z)$ such that

1. $\xi(z) = H(z) \, \text{diag} \, [z^{f_1} \quad \cdots \quad z^{f_m}]$ (10.3.84)

$$H(z) = \begin{bmatrix} 1 \, . & & & 0 \\ h_{21}(z) \, \cdot & \cdot & & \\ \cdot & & \cdot & \\ \cdot & & & \cdot \\ \cdot & & & \\ h_{m1}(z) & h_{m2}(z) & \cdots & 1 \end{bmatrix} \qquad (10.3.85)$$

and $h_{ij}(z)$ is divisible by z or is zero.

2. $\lim_{z \to \infty} \xi(z)T(z) = K_T$ nonsingular   (10.3.86)

Also, from Lemma 5.2.4 we know that $[\xi(z)]^{-1}$ is a stable operator, and if $T(z)$ is strictly proper, then $f_i \geq 1, \, i = 1, \, \ldots, \, m$.

We recall from Section 5.2 that the interactor matrix describes the multivariate delay structure of $T(z)$. This is the appropriate generalization of the delay, d, in the scalar case.

We now generalize Theorem 5.2.4 to the stochastic case. This is done in the following theorem:

Theorem 10.3.7. We define a variable $\bar{y}(t)$ by the following difference equation:

$$\bar{y}(t) = \xi(q)y(t) \qquad (10.3.87)$$

Then, subject to Assumption 10.3.A, we have that the optimal prediction $\bar{y}^0(t \,|\, t)$ of $\bar{y}(t)$ given $\{y(t), y(t-1), \, \ldots\}$ satisfies an equation of the form

$$\bar{C}(q^{-1})\bar{y}^0(t \,|\, t) = \bar{\alpha}(q^{-1})y(t) + \bar{\beta}(q^{-1})u(t) \qquad (10.3.88)$$

where

$$\bar{C}(q^{-1}) = I + c_1 q^{-1} + \cdots + c_{n_1} q^{-n_1} \qquad (10.3.89)$$

$$\bar{\alpha}(q^{-1}) = \alpha_0 + \alpha_1 q^{-1} + \cdots + \alpha_{n_2} q^{-n_2} \qquad (10.3.90)$$

$$\bar{B}(q^{-1}) = \beta_0 + \beta_1 q^{-1} + \cdots + \beta_{n_3} q^{-n_3} \tag{10.3.91}$$

with β_0 nonsingular.

Proof. For clarity of exposition, we transform the model (10.3.83) into transfer function form as

$$y(t) = T(q)u(t) + N(q)\omega(t) \tag{10.3.92}$$

where

$$T(q) = A(q^{-1})^{-1}B(q^{-1}); \quad N(q) = A(q^{-1})^{-1}C(q^{-1})$$

Multiplying equation (10.3.92) on the left by $\xi(q)$ gives

$$\xi(q)y(t) = \xi(q)T(q)u(t) + \xi(q)N(q)\omega(t) \tag{10.3.93}$$

The noise term $\xi(q)N(q)$ can be factored into future and past noise by use of the division algorithm of algebra, that is, we write

$$\xi(q)N(q) = F(q) + R(q) \tag{10.4.94}$$

where

$$F(q) = F_d q^d + \cdots F_1 q$$

and $R(q)$ is a proper transfer function. Substituting (10.3.94) into (10.3.93) gives

$$\xi(q)y(t) = \xi(q)T(q)u(t) + F(q)\omega(t) + R(q)\omega(t) \tag{10.3.95}$$

Since $\{\omega(t)\}$ is the innovations sequence it can be causally obtained from $\{y(t)\}$. An appropriate expression for $\omega(t)$ is obtained by inverting (10.3.92), that is

$$\omega(t) = N(q)^{-1}[y(t) - T(q)u(t)] \tag{10.3.96}$$

Substituting (10.3.96) into (10.3.95) gives

$$\begin{aligned}
\xi(q)y(t) &= \xi(q)T(q)u(t) + F(q)\omega(t) + R(q)N(q)^{-1}[y(t) - T(q)u(t)] \\
&= [\xi(q)T(q) - R(q)N(q^{-1})T(q)]u(t) \\
&\quad + R(q)N(q)^{-1}y(t) + F(q)\omega(t)
\end{aligned} \tag{10.3.97}$$

$$\begin{aligned}
[\xi(q)y(t) - F(q)\omega(t)] &= [\xi(q) - R(q)N(q)^{-1}]T(q)u(t) + R(q)N(q)^{-1}y(t) \\
&= [\xi(q)N(q) - R(q)]N(q)^{-1}T(q)u(t) + R(q)N(q)^{-1}y(t) \\
&= F(q)N(q)^{-1}T(q)u(t) + R(q)N(q)^{-1}y(t)
\end{aligned} \tag{10.3.98}$$

where $F(q)\omega(t)$ denotes the future (unpredictable) noise.

From equation (10.3.86) it follows that $F(q)N(q)^{-1}T(q)$ is a proper transfer function. Similarly, $R(q)N(q)^{-1}$ is a proper transfer function by definition of $R(q)$. Hence the composite matrix $[F(q)N(q)^{-1}T(q); R(q)N(q)^{-1}]$ can be described by a causal left difference operator representation as follows:

$$[F(q)N(q)^{-1}T(q); R(q)N(q)^{-1}] = \bar{C}(q^{-1})^{-1}[\bar{B}(q^{-1}); \bar{\alpha}(q^{-1})] \tag{10.3.99}$$

where $\bar{C}(q^{-1})$, $\bar{B}(q^{-1})$, $\bar{\alpha}(q^{-1})$ are polynomial matrices in the unit delay operator q^{-1} such that the zeros of $\bar{C}(q^{-1})$ are the transmission zeros of $N(q)$.

Substituting (10.3.99) into (10.3.98) immediately gives

$$\bar{C}(q^{-1})[\xi(q)y(t) - F(q)\omega(t)] = \bar{B}(q^{-1})u(t) + \bar{\alpha}(q^{-1})y(t)$$

By choosing $\bar{C}(0) = I$, we get $\bar{B}(0) = K_T$ from equation (10.3.96). This establishes the predictor (10.3.88).

▼▼▼

Remark 10.3.11. The optimal steady state predictor given in (10.3.88) can also be derived by simple algebraic manipulation starting from the usual ARMAX model (10.3.83).

In this case, we need the following 3 equalities:

$$A(q^{-1})^{-1}C(q^{-1}) = \tilde{C}(q^{-1})\tilde{A}(q^{-1})^{-1}$$

$$\xi(q)\tilde{C}(q^{-1}) = F(q)\tilde{A}(q^{-1}) + G(q^{-1})$$

$$[F(q)C(q^{-1})^{-1}, G(q^{-1})\tilde{C}(q^{-1})^{-1}] = \bar{C}(q^{-1})^{-1}[\bar{F}(q), \bar{G}(q^{-1})]$$

▼▼▼

As previously, the predictor (10.3.88) can be used to determine a minimum variance control strategy.

Theorem 10.3.8

(a) A minimum variance controller for regulating $\bar{y}(t)$ about a desired output $\bar{y}*(t) = \xi(q)y*(t)$ can be obtained by setting $\bar{y}^0(t\,|\,t) = \bar{y}*(t)$, that is,

$$\bar{\beta}(q^{-1})u(t) = \bar{C}(q^{-1})\bar{y}*(t) - \bar{\alpha}(q^{-1})y(t) \qquad (10.3.100)$$

(b) The control law (10.3.100) results in a closed-loop system satisfying

$$\bar{y}(t) = \bar{y}*(t) + \eta(t) \qquad (10.3.101)$$

where

$$E\{\eta(t)\,|\,y(t), y(t-1), \ldots\} = 0 \qquad (10.3.102)$$

(c) Equation (10.3.101) ensures that the system output asymptotically satisfies

$$y(t) = y*(t) + v(t) \qquad (10.3.103)$$

where

$$v(t) = F'(q^{-1})\omega(t) \qquad (10.3.104)$$

and $F'(q^{-1})$ is a finite moving average polynomial:

$$F'(q^{-1}) = \xi(q)^{-1}F(q)$$

and $F(q)$ is as in (10.3.94)

Proof. (a) Immediate from (10.3.88).
(b) Follows from the optimality of $\bar{y}^0(t\,|\,t)$
(c) Since

$$\bar{y}(t) = \xi(q)y(t)$$

$$\bar{y}*(t) = \xi(q)y*(t)$$

▼▼▼

Note that the comments made in Remark 10.3.6 and Section 10.3.1 regarding systems with noise dynamics having zeros on the unit circle also apply to the multi-input multi-output case.

It is also possible to extend the model reference stochastic controller of Section 10.3.2 and the weighted one-step-ahead stochastic controller of Section 10.3.1 to the multi-input multi-output case. We leave the details to the reader (see Exercises 10.13 and 10.14.)

Finally we remark that all of the results in this chapter refer to the behaviour of the system at the sampling instants. In some cases it may also be desirable to consider the continuous-time response between samples. For example, with minimum variance control of a lightly damped system such as that considered in Section 5.4, the output variance can rise substantially between samples. Thus a suboptimal design (in the sense of the sample variance) may actually give a better overall continuous-time response. Further discussion may be found in de Souza and Goodwin (1983).

10.4 THE LINEAR QUADRATIC GAUSSIAN OPTIMAL CONTROL PROBLEM

We have seen in previous sections that the minimum variance control strategy minimizes the following cost function with respect to admissible control strategies:

$$J_t = E\{(y(t) - y^*(t))^2\} \tag{10.4.1}$$

Similarly, the weighted one-step-ahead controller minimizes a cost function of the form

$$J_t' = E\{(y(t) - y^*(t))^2 + u(t - d)^2\} \tag{10.4.2}$$

A feature of the cost functions above is that they are only concerned with the situation one step ahead. A generalization of the above cost functions is obtained if one looks at the situation N steps ahead by the use of a cost function of the form

$$\bar{J}_N = E\left\{x(N)^T\Omega_N x(N) + \sum_{t=0}^{N-1}(x(t)^T\Omega x(t) + u(t)^T\Gamma u(t))\right\} \tag{10.4.3}$$

where $\{x(t)\}$ denotes the system state in a model of the form (7.3.1)–(7.3.2):

$$x(t + 1) = Ax(t) + Bu(t) + v_1(t); \qquad t \geq t_0 \tag{10.4.4}$$

$$y(t) = Cx(t) + v_2(t) \tag{10.4.5}$$

with initial conditions $x(t_0)$. The sequences $\{x(t)\}$, $\{y(t)\}$, and $\{u(t)\}$ are the state, output, and input vectors, respectively.

A special case of (10.4.3) is obtained if N is chosen as the natural delay, d, of the system and the matrices Γ and Ω_N are chosen as $\Gamma = \lambda$, $\Omega_N = C^TC$. Then in the single-input single-output case, minimization of (10.4.3) is equivalent to minimization of (10.4.2) with $y^*(t) = 0$ (Exercise 10.17).

Use of the cost function (10.4.3) leads to a general quadratic stochastic optimal control problem. If the disturbances are gaussian, the problem is called the *linear quadratic gaussian* (LQG) optimal control problem. We shall study this problem below.

We assume that $x(t_0)$, $v_1(t)$, and $v_2(t)$ are jointly distributed gaussian random variables such that $x(t_0)$ has mean $\bar{x}_0$ and covariance P_0 and is uncorrelated with $\{v_1(t)\}$ and $\{v_2(t)\}$. The latter quantities have zero mean and covariance given by

$$E\left\{\begin{bmatrix} v_1(t) \\ v_2(t) \end{bmatrix}[v_1(s)^T v_2(s)^T]\right\} = \begin{bmatrix} Q & 0 \\ 0 & R \end{bmatrix}\delta(t - s) \tag{10.4.6}$$

We wish to find the control $u(t)$ as a function of $\{y(t - 1), y(t - 2), \ldots, y(0), u(t - 1), u(t - 2), \ldots u(0)\} \triangleq \mathcal{F}_{t-1}$ so that the cost function (10.4.3) is minimized.

10.4.1 The Separation Principle

We shall discuss briefly the solution to the problem above and some of its implications. The solution can be obtained by dynamic programming arguments (Bellman, 1957). The optimal control turns out to comprise a Kalman filter together with state-variable feedback. The result is summarized in the following theorem.

Theorem 10.4.1. The control law minimizing (10.4.3) for the system (10.4.4) to (10.4.6) is given by

$$\hat{x}(t + 1) = A\hat{x}(t) + Bu(t) + K(t)[y(t) - C\hat{x}(t)]; \qquad \hat{x}(t_0) = \bar{x}_0 \qquad (10.4.7)$$

$$u(t) = -L(t)\hat{x}(t) \qquad (10.4.8)$$

where $\hat{x}(t)$ denotes the optimal state estimate given $\mathcal{F}_{t-1}$
$K(t)$ denotes the Kalman gain (as given in Theorem 7.3.1)
$L(t)$ is a linear feedback obtained from

$$L(t) = [\Gamma + B^T S(t + 1)B]^{-1} B^T S(t + 1)A \qquad (10.4.9)$$

and $\{S(t)\}$ satisfies the following matrix Riccati equation:

$$S(t) = \Omega + L(t)^T \Gamma L(t) + (A - BL(t))^T S(t + 1)(A - BL(t)) \qquad (10.4.10)$$

with

$$S(N) = \Omega_N \qquad (10.4.11)$$

Proof. We use dynamic programming. Let V^*_{t+1} be the optimal cost to go from time $t + 1$ to the final time $N - 1$, that is,

$$V^*_{t+1} = \min_{u(t+1)} E\left\{ \min_{u(t+2)} E\left\{ \ldots x(N)^T \Omega_N x(N) + \sum_{j=t+1}^{N-1} x(j)^T \Omega x(j) \right. \right.$$
$$\left. \left. + u(j)^T \Gamma u(j) \cdots | Y_{t+1} \right\} | Y_t \right\}$$

Then by the principle of optimality, the optimal control satisfies the following functional equation:

$$V^*_t = \min_{u(t)} E\{L_t + V^*_{t+1} | Y_{t-1}\} \qquad (10.4.12)$$

where $L_t = x(t)^T \Omega x(t) + u(t)^T \Gamma u(t)$ and Y_{t-1} denotes the data $\{y(0), \ldots, y(t - 1)\}$. The admissible control strategy allows $u(k)$ to be a function of Y_{t-1}.
Boundary conditions for (10.4.12) can be found at $t = N$:

$$V^*_N = \min_{u(N)} E\{x(N)^T \Omega_N x(N) | Y_{N-1}\}$$
$$= \hat{x}(N)^T \Omega_N \hat{x}(N) + \text{trace } \Omega_N P(N) \qquad (10.4.13)$$

where $\hat{x}(N)$ and $P(N)$ denote the conditional mean and covariance of $x(N)$ given Y_{N-1}. [Note that $\hat{x}(N)$ and $P(N)$ can be obtained from the Kalman filter; see Section 7.3.]

We note that V^*_N is a quadratic function of $\hat{x}(N)$. We use induction to prove this is true in general; that is, assume that

$$V^*_{t+1} = \hat{x}(t + 1)^T S(t + 1)\hat{x}(t + 1) + \alpha(t + 1) \qquad (10.4.14)$$

Hence, substituting (10.4.14) into (10.4.12), we have

$$V_t^* = \min_{u(t)} E\{x(t)^T \Omega x(t) + u(t)^T \Gamma u(t) + \hat{x}(t+1)^T S(t+1)\hat{x}(t+1)$$
$$+ \alpha(t+1)\,|\,Y_{t-1}\} \tag{10.4.15}$$

Now from the Kalman filter [equation (7.3.4)],

$$\hat{x}(t+1) = A\hat{x}(t) + Bu(t) + K(t)[y(t) - C\hat{x}(t)] \tag{10.4.16}$$

where $\hat{x}(t)$ is Y_{t-1} measurable and $y(t)$ is as in (10.4.5). Hence the conditional mean and covariance of $\hat{x}(t+1)$ given Y_{t-1} are, respectively, given by

$$E\{\hat{x}(t+1)\,|\,Y_{t-1}\} = A\hat{x}(t) + Bu(t) \tag{10.4.17}$$

$$E\{[\hat{x}(t+1) - A\hat{x}(t) - Bu(t)][\hat{x}(t+1) - A\hat{x}(t) - Bu(t)]^T/Y_{t-1}\}$$
$$= K(t)[CP(t)C^T + R]K(t)^T \tag{10.4.18}$$

Substituting (10.4.17)–(10.4.18) into (10.4.15) gives

$$V_t^* = \min_{u(t)} \{\hat{x}(t)^T \Omega \hat{x}(t) + \text{trace } \Omega P(t) + u(t)^T \Gamma u(t)$$
$$+ [A\hat{x}(t) + Bu(t)]^T S(t+1)[A\hat{x}(t) + Bu(t)]$$
$$+ \text{trace }[S(t+1)K(t)(CP(t)C^T + R)K(t)^T]$$
$$+ \alpha(t+1)\} \tag{10.4.19}$$

Differentiating with respect to $u(t)$ gives

$$u(t) = -[\Gamma + B^T S(t+1)B]^{-1}B^T S(t+1)A\hat{x}(t)$$
$$= -L(t)\hat{x}(t) \tag{10.4.20}$$

Substituting back to V_t^* gives

$$V_t^* = \hat{x}(t)^T S(t)\hat{x}(t) + \alpha(t) \tag{10.4.21}$$

where

$$S(t) = \Omega + L(t)^T \Gamma L(t) + [A - BL(t)]^T S(t+1)[A - BL(t)] \tag{10.4.22}$$

$$\alpha(t) = \text{trace }[\Omega P(t)] + \text{trace }[S(t+1)K(t)(CP(t)C^T + R)K(t)^T]$$
$$+ \alpha(t+1) \tag{10.4.23}$$

Induction now completes the proof.

▼▼▼

An interesting feature of the solution described above is that the linear feedback gain, $L(t)$, is exactly the same as would be obtained if $x(t)$ were directly measured and there were no noise [$v_1(t) \equiv 0, v_2(t) \equiv 0$]. This is known as the *separation principle* (Joseph and Tou, 1961) and is illustrated in Fig. 10.4.1. It is important to note that this result holds only in the linear quadratic gaussian case.

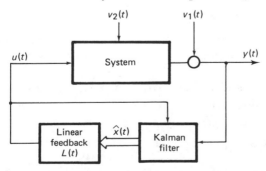

Figure 10.4.1 Solution to the LQG optimal control problem (the separation principle).

Another interesting feature of the solution is that, subject to weak assumptions, the closed-loop system is guaranteed stable provided that the optimization horizon is made infinitely large [i.e., N approaches infinity in (10.4.3)]. This is discussed below.

Theorem 10.4.2

(a) The control law (10.4.7) to (10.4.11) gives a closed-loop system described by

$$\begin{bmatrix} \tilde{x}(t+1) \\ \hat{x}(t+1) \end{bmatrix} = \begin{bmatrix} A - K(t)C & 0 \\ -K(t)C & A - BL(t) \end{bmatrix} \begin{bmatrix} \tilde{x}(t) \\ \hat{x}(t) \end{bmatrix} + \begin{bmatrix} -v_1(t) + K(t)v_2(t) \\ K(t)v_2(t) \end{bmatrix}$$

(10.4.24)

where

$$\tilde{x}(t) = \hat{x}(t) - x(t) \tag{10.4.25}$$

(b) If the optimization horizon is extended to ∞, the feedback is asymptotically time invariant and the resulting closed-loop system is asymptotically stable provided that:
 (i) (C, A) is detectable.
 (ii) (A, D) is stabilizable where $DD^T \triangleq Q$.
 (iii) (A, B) is stabilizable.
 (iv) (E, A) is detectable where $EE^T \triangleq \Omega$.

Proof. (a) The Kalman filter is given by (Theorem 7.3.1)

$$\hat{x}(t+1) = A\hat{x}(t) + Bu(t) + K(t)[y(t) - C\hat{x}(t)] \tag{10.4.26}$$
$$= A\hat{x}(t) + Bu(t) + K(t)[-C\tilde{x}(t) + v_2(t)]$$

Substracting (10.4.4) from (10.4.26) gives

$$\tilde{x}(t+1) = [A - K(t)C]\tilde{x}(t) + K(t)v_2(t) - v_1(t) \tag{10.4.27}$$

Similarly, substituting (10.4.8) into (10.4.26) gives

$$\hat{x}(t+1) = [A - BL(t)]\hat{x}(t) - K(t)C\tilde{x}(t) + K(t)v_2(t) \tag{10.4.28}$$

Combining (10.4.27) and (10.4.28) gives (10.4.24).

(b) Theorem 7.3.2 and its dual [replace A, B, C, Q, R, and $K(t)$ by A^T, C^T, B^T, Ω, Γ, and $L(t)^T$] immediately implies (b) using the fact that the modes of (10.4.24) are the union of the modes of $[A - K(t)C]$ and $[A - BL(t)]$, as is evident from (10.4.24). ▼▼▼

Remark 10.4.1. Using Theorem 7.3.3, it is apparent that condition (ii) of Theorem 10.4.2 can be weakened to

(ii') (A, D) has no uncontrollable modes on the unit circle and $P(0) > 0$.

Similarly, condition (iv) can be weakened to

(iv') (E, A) has no unobservable modes on the unit circle and $\Omega_\infty > 0$. ▼▼▼

Remark 10.4.2. The theorem above demonstrates that the optimal control of a general linear stochastic system makes use of a Kalman filter and state-variable feedback. Note also that the modes of the closed-loop system are the union of the modes of the filter and of the closed-loop system obtained as if there were no noise (see also the equivalent result in Section 5.3.3).

▼▼▼

For further discussion of the LQG optimal control problem, including numerical and practical consideration, the reader is referred to Åström (1970), Anderson and Moore (1971), Kwakernaak and Sivan (1972), Striebel (1975), and Caines (1972).

10.4.2 The Tracking Problem

The optimal control design criterion given in (10.4.3) is concerned with regulation. Here we show how a tracking objective can be incorporated by augmenting the state vector. For simplicity we consider only constant desired outputs and follow the method proposed in Young and Willems (1972).

Let y^* be a multivariable constant desired output vector for $\{y(t)\}$. We are interested in producing a form of "integral action." Thus consider the following variable, $z(t)$, which is the integral (i.e., sum) of the output errors:

$$z(t) = \sum_{j=0}^{t-1} [y^* - y(j)]$$

The equation above can be written in an iterative form as

$$z(t + 1) = z(t) + y^* - y(t); \qquad z(0) = 0$$

We temporarily set $y^* = 0$ and combine the equation above with the state equations (10.4.4) and (10.4.5) to yield

$$\begin{bmatrix} x(t+1) \\ z(t+1) \end{bmatrix} = \begin{bmatrix} A & 0 \\ -C & I \end{bmatrix} \begin{bmatrix} x(t) \\ z(t) \end{bmatrix} + \begin{bmatrix} B \\ 0 \end{bmatrix} u(t) + \begin{bmatrix} v_1(t) \\ -v_2(t) \end{bmatrix}$$

$$\begin{bmatrix} y(t) \\ z(t) \end{bmatrix} = \begin{bmatrix} C & 0 \\ 0 & I \end{bmatrix} \begin{bmatrix} x(t) \\ z(t) \end{bmatrix} + \begin{bmatrix} v_2(t) \\ 0 \end{bmatrix}$$

An infinite-horizon optimal regulation problem can now be formulated in terms of the augmented state vector

$$x'(t) \triangleq \begin{bmatrix} x(t) \\ z(t) \end{bmatrix}$$

The resulting feedback control law will have the following form:

$$u(t) = K\hat{x}'(t) \triangleq [K_1 \quad K_2] \begin{bmatrix} \hat{x}(t) \\ z(t) \end{bmatrix}$$

where $\hat{x}(t)$ is the optimal state estimate provided by the Kalman filter.

Finally, the desired output y^* is reintroduced and the control law is implemented as in Fig. 10.4.2.

Subject to mild assumptions, the design above ensures that the closed-loop system is asymptotically stable and that the mean value of the tracking error is zero.

To verify that the closed-loop system under the feedback is, in fact, asymptoti-

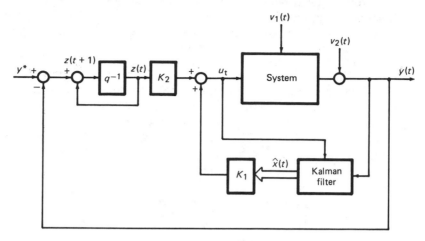

Figure 10.4.2 Tracking form of optimal control.

cally stable, we need to verify conditions (i) to (iv) of Theorem 10.4.2 for the augmented system. Conditions (i) and (iv) are straightforward. Condition (ii) is also immediate since the Kalman filter is applied only to the original system, which is presumed to satisfy condition (ii). Condition (iii) can be shown to be satisfied under reasonable conditions: for example, if (1) the original system is controllable, (2) A is nonsingular, and (3) B has rank equal to the number of outputs (see Exercise 10.18). This implies, among other things, that the number of inputs should at least be equal to the number of outputs to achieve output tracking.

10.4.3 Rapprochement with Minimum Variance Control

It is instructive to interpret the minimum variance controller of Section 10.3.1 as an optimal controller of form shown in Fig. 10.4.1. We recall that the minimum variance controller was based on an ARMAX predictor with the following structure (we assume scalar output and $d = 1$ for illustration):

$$C(q^{-1})y^0(t + 1 \mid t) = G(q^{-1})y(t) + B'(q^{-1})u(t) \tag{10.4.29}$$

This predictor can be written in the equivalent observer state-space form as

$$\hat{x}(t + 1) = \begin{bmatrix} -c_1 & 1 & & \\ -c_2 & & \cdot & \\ \vdots & & & \cdot \\ \vdots & & & & 1 \\ -c_n & 0 & \cdots & 0 \end{bmatrix} \hat{x}(t) + \begin{bmatrix} g_1 \\ \vdots \\ \vdots \\ \vdots \\ g_n \end{bmatrix} y(t) + \begin{bmatrix} b_0 \\ \vdots \\ \vdots \\ \vdots \\ b_{n-1} \end{bmatrix} u(t) \tag{10.4.30}$$

$$\hat{y}(t) = [1 \quad 0 \quad \cdots \quad 0]\hat{x}(t) \tag{10.4.31}$$

Now from Section 7.4.1 the model (10.4.30) is actually the *Kalman filter* for the system and $\hat{x}(t + 1)$ and $\hat{y}(t + 1)$ have the interpretation of being the estimates of $x(t + 1)$ and $y(t + 1)$, respectively, given the data $\{y(0), \ldots, y(t)\}$.

Returning to the minimum variance controller, we recall that this control law is achieved by setting $\hat{y}(t + 1) = y^*(t + 1)$. It follows immediately from (10.4.30)–(10.4.31) that this is equivalent to the following "state-variable" feedback law:

$$u(t) = \left[\frac{c_1}{b_0} \quad -\frac{1}{b_0} \quad 0 \quad \cdots \quad 0\right]\hat{x}(t) + \left[\frac{-g_1}{b_0}\right]y(t) + \left[\frac{1}{b_0}\right]y^*(t + 1) \qquad (10.4.32)$$

The appearance of the term $[-g_1/b_0]y(t)$ on the right-hand side of (10.4.32) takes account of the fact that in the minimum variance controller, $u(t)$ is expressed as a function of $y(t)$, $y(t - 1)$, $\ldots$, whereas $\hat{x}(t)$ (provided by the Kalman filter) is an estimate of $x(t)$ given $[y(t - 1) \quad y(t - 2) \quad \cdots \quad y(0)]$. Thus to get minimum variance control using state-variable feedback we must update $\hat{x}(t)$ by addition of the observation $y(t)$. Using the same method as in the development of the Kalman filter (see Exercise 7.14) we see that the state estimate, $\hat{x}(t \mid t)$, given $[y(t) \quad y(t - 1) \quad \cdots \quad y(0)]$, can be expressed as a linear function of $\hat{x}(t)$ and $y(t)$, that is,

$$\hat{x}(t \mid t) = R\hat{x}(t) + S'y(t) \qquad (10.4.33)$$

We then see that (10.4.32) actually corresponds to a state-variable feedback law of the form

$$u(t) = L\hat{x}(t \mid t) + \left[\frac{1}{b_0}\right]y^*(t + 1) \qquad (10.4.34)$$

In terms of pole assignment, the minimum variance controller gives closed-loop poles at the open-loop zeros plus the state observer (Kalman filter dynamics) given by the $C(q^{-1})$ polynomial. [The reader should check that the closed-loop system obtained by use of the minimum variance controller has roots at the zeros of $C(q^{-1})B'(q^{-1})$.]

EXERCISES

10.1. Establish (10.2.10). In particular, show that the tracking error has zero mean for an arbitrary sequence $\{y^*(t)\}$ satisfying the reference model

$$S(q^{-1})y^*(t) = 0$$

10.2. Consider the following continuous-time system:

$$\dot{x} = -ax + bu + v$$

where v denotes a continuous-time "white noise" process having power density spectrum, σ^2, that is, $E\{v(t)v(t - \tau)\} = \sigma^2 \delta(t - \tau)$, where $\delta(s)$ is the Dirac delta function.

(a) Show that, with a sample hold input, the corresponding discrete-time response can be modeled by

$$x(t + 1) = e^{-a\Delta}x(t) + \frac{b}{a}[1 - e^{-a\Delta}]u(t) + w(t)$$

where $\{w(t)\}$ is a discrete-time white noise sequence having variance

$$\bar{\sigma}^2 = \frac{1 - e^{-2a\Delta}}{2a}\sigma^2$$

(b) If Δ is small compared with $1/a$, show that the discrete-time model above is

approximately given by
$$x(t + 1) = (1 - a\Delta)x(t) + b\Delta u(t) + w(t)$$
where $\{w(t)\}$ has variance $\bar{\sigma}^2 \simeq \Delta\sigma^2$.

10.3. Consider the system described in Exercise 10.2(b). Show that the minimum variance controller for the output $y(t) = x(t)$ has a tracking error variance of $\Delta\sigma^2$.

10.4. Consider again the system of Exercise 10.2(b) and assume that the system is now sampled at 10 times the sampling rate used in Exercise 10.3; that is, put $\Delta' = \Delta/10$.
  (a) Show that the following feedback control law gives a closed-loop bandwidth of approximately $1/\Delta \simeq \frac{1}{10}(1/\Delta')$:
  $$u(t) = -\frac{1}{b}\left(\frac{1}{\Delta} - a\right)[y(t) - y^*]$$

  (b) Explicitly evaluate the performance of the control law given in part (a) for the case $a = \frac{1}{5}$, $\Delta = 1$, $\Delta' = \frac{1}{10}$, and $b = 1$.
  (c) Show that the output variance, σ_0^2, resulting from the control law found in part (a), is approximately given by
  $$\sigma_0^2 \simeq \frac{\sigma^2\Delta}{2}$$

  (d) Why is it possible for the variance found in part (c) to be less than the variance found in Exercise 10.3? (*Hint:* The sampling rate is different.)
  (e) Show that the open-loop output variance, σ_0^2, is approximately
  $$\sigma_0^2 \simeq \frac{\sigma^2}{2a}$$

  Hence if $a = 1/5\Delta$, show that the controller in part (a) reduces the output variance by a factor of 5 to 1.

10.5. Consider the following discrete-time system:
$$(1 - aq^{-1})y(t) = q^{-d}u(t) + w(t)$$
where $\{w(t)\}$ is white noise of variance σ^2.
  (a) Show that the minimum variance controller gives an output variance, σ_0^2, as follows:
  $$\sigma_0^2 = \frac{1 - a^{2d}}{1 - a^2}\sigma^2$$

  (b) Discuss the implications of this result for the cases $a > 1$ and $a < 1$. (Consider, for example, $a = 2$ and $a = 0.9$.)

10.6. Consider the following discrete-time system:
$$(1 - aq^{-1})y(t) = q^{-d}u(t) + (1 - aq^{-1})w(t)$$
where $\{w(t)\}$ is white noise of variance σ^2.
  (a) Consider the case $a < 1$. Show that the minimum variance controller gives an output variance, σ_0^2, given by
  $$\sigma_0^2 = \sigma^2$$

  (b) Draw a block diagram of the minimum variance controller found in part (a). Give a heuristic explanation as to why there is no feedback from the output. Why is the design sensitive to the value of a used in evaluating the feedback controller?

10.7. Consider again the system of Exercise 10.6 but with $a > 1$.
  (a) Show that the minimum variance control strategy now gives an output variance, σ_0^2, as follows:
  $$\sigma_0^2 = \sigma^2(1 - a^{2d-2} + a^{2d})$$

(b) Using the results of part (a), discuss the feasibility and desirability of controlling unstable systems having long time delays.

10.8. Consider the following system:

$$y(t) = q^{-1}u(t) + w(t) + d(t)$$

where disturbance of the form

$$d(t) = G \sin (pt + \phi)$$

(a) Write down the corresponding ARMAX model.

(b) Develop a nearly-optimal minimum variance controller for this system.

10.9. Reformulate the weighted one-step-ahead stochastic controller of Section 10.3 in the form of linear state-variable feedback using the Kalman filter. (*Hint:* See Section 10.4.3.)

10.10. Would you recommend minimum variance control for the following system?

$$A(q^{-1})y(t) = B(q^{-1})u(t) + C(q^{-1})\omega(t)$$
$$A(q^{-1}) = 1 + 5q^{-1} + 7q^{-2} + 8q^{-3} + 9q^{-4}$$
$$B(q^{-1}) = 1 + 3q^{-1} + q^{-2} + 3q^{-3} + 2q^{-4}$$
$$C(q^{-1}) = 1 - 5q^{-1}$$

Why?

10.11. Consider the following reference model:

$$E(q^{-1})y^*(t) = 0$$

where

$$E(q^{-1}) = 1 - (2 \cos p)q^{-1} + q^{-2}$$

Show that this reference model generates a sinusoidal desired output. Will the model reference stochastic controller perform satisfactorily? Why?

10.12. Consider the following discrete-time system:

$$(1 - q^{-1})y(t) = q^{-1}u(t) + \omega(t)$$

where $\{\omega(t)\}$ is a white sequence of variance σ^2.

(a) Show that the minimum variance controller gives an output variance, σ_0^2, given by

$$\sigma_0^2 = \sigma^2$$

(b) If model reference stochastic control is used where the reference model is

$$(1 - eq^{-1})y^*(t) = r(t)$$

show that the output variance is $\sigma^2/(1 - e^2)$.

10.13. Extend the model reference stochastic controller to the multi-input multi-output case.

10.14. Extend the weighted one-step-ahead stochastic controller to the multi-input multi-output case.

10.15. Extend Theorem 10.4.1 to the case where $u(t)$ can be a function of $y(t)$, $y(t - 1)$, ..., $y(0)$. (*Hint:* See Exercise 7.14.)

10.16. Verify the form of the Kalman filter shown in (10.3.95) to (10.3.97).

10.17. Verify that minimization of the cost function (10.4.3) is equivalent to minimizing (10.4.2) in the case where $\Gamma = \lambda$, $\Omega_N = C^T C$, and $N = d$, where d is the natural delay of the system. [*Hint:* $x(t) \cdots x(t + d - 1)$ are not a function of $u(t)$.]

10.18. Verify the conditions for closed-loop stability quoted in Section 10.4.2. [*Hint:* See Young and Willems (1972).]

10.19. Consider the following two-output system:

$$A_1(q^{-1})y_1(t) = C_1(q^{-1})\omega_1(t)$$

$$A_2(q^{-1})y_2(t) = q^{-d}B_2(q^{-1})u(t) + C_2(q^{-1})\omega_2(t)$$

where $\{\omega_1(t)\}$ and $\{\omega_2(t)\}$ are independent white noise sequences of variance σ_1^2 and σ_2^2, respectively. Assuming that $B_2(q^{-1})$ is table, design a minimum variance controller for minimizing the mean-square distance of $y_2(t + d)$ from $y_1(t + d)$.

11

Adaptive Control of
Stochastic Systems

11.1 INTRODUCTION

The notion of adaptive control was introduced in Chapter 6 in the context of deterministic systems. As we have seen, the basic idea is to combine an on-line parameter estimation procedure with some control system design technique to produce a control law with self-tuning capability.

In this chapter we extend the ideas introduced in Chapter 6 to systems having stochastic disturbances. Most of the ideas discussed in Chapter 6 carry over to the stochastic case in a natural and simple way.

Potential applications of stochastic adaptive controllers include:

1. Replacement of manual tuning of control laws by automatic self-tuning in the case of time-invariant plants.
2. Building of a gain schedule. Here the controller is automatically tuned at a number of operating points. The resulting control laws are then stored away for subsequent use whenever the corresponding operating condition is reached.
3. Continuous adjustment of controllers for time-varying plants.

A number of successful industrial feasibility studies and industrial applications of adaptive control have appeared in the literature; see, for example, the books by Narendra and Monopoli (1980) and Unbehauen (1980). Applications have also been described in a number of papers, for example, Borisson and Wittenmark (1974), Borisson and Syding (1976), Van Amerongen and Udink Ten Cate (1975), Källström et al. (1979), Keviczky et al. (1978), Dumont and Bélanger (1978a, 1978b), Cegrell and Hedqvist (1975), Sastry, Seborg, and Wood (1977), and Dumont (1982).

A complication in the stochastic case which is not present in the deterministic case is that the system "state" is not available due to the presence of unknown disturbances. If we view the system parameters as unknown constant states, we can regard the problem of controlling a stochastic system with unmeasured states and unknown parameters as a nonlinear stochastic optimal control problem. In this framework the design criterion is formulated as minimization of a cost function which is a scalar function of states and controls. Unfortunately, the general problem of nonlinear stochastic optimal control is extremely difficult (Fel'dbaum, 1965). For example, conditions for the existence, in general cases, of optimal controls are not known. Under the assumption that a solution exists, a functional equation, called the *Bellman equation*, for the optimal cost can be obtained via *dynamic programming* arguments. However, this equation can be solved numerically only in very special cases. Thus it is desirable to seek simplified formulations of the stochastic adaptive control problem which retain some of the more important features of the general formulation.

The general formulation of the adaptive control problem in the context of optimal control leads to some interesting insights. One of the most important of these is the idea of *dual control*. The idea here is that the system input has a dual role: learning and regulation. With regard to learning, the input introduces perturbations which yield information about the systems dynamics and thus allow the parameter uncertainty to be reduced. With regard to regulation, the input tries to keep the output at the desired value. It is often the case that the two roles of the input are conflicting and thus the controller must achieve an optimal compromise between learning (which may require large perturbations) and regulation (which may only need relatively small signals). The idea of dual control will be discussed in a little more detail in Section 11.2.

At one extreme, by ignoring the uncertainty in the parameter estimates, one can design the control law as if the estimated parameters were the true system parameters. This approach is commonly called *certainty equivalence* (Bar-Shalom and Tse, 1974) and involves the *separation* of the estimation and control problems. Various adaptive control laws based on the certainty equivalence principle will be discussed below.

Perhaps the best known certainty equivalence stochastic adaptive control law is the *self-tuning regulator* of Åström and Wittenmark (1973). This adaptive controller combines the least-squares procedure for parameter estimation with a one-step-ahead (minimum variance) certainty equivalence controller. This algorithm, together with other related algorithms, will be discussed in some detail. The convergence of these algorithms will also be studied. This convergence theory, although requiring idealized assumptions, gives important insights to the operation of adaptive control algorithms for stochastic systems. In particular, we will discuss the role of passivity in the self-tuning regulator-type algorithms and highlight the differences between tracking and regulation.

We will also discuss adaptive control algorithms aimed at cases where the self-tuning regulator can fail and we will briefly discuss adaptive control of nonlinear stochastic systems.

11.2 CONCEPTS OF DUAL CONTROL
AND CERTAINTY EQUIVALENCE CONTROL

Dual Control

As mentioned in the introduction, dual control aims to achieve a compromise between learning and regulation. Dual control has been discussed in detail by many authors, including Fel'dbaum (1965), Jacobs and Patchell (1972), Bar-Shalom and Tse (1974, 1976), Bar-Shalom and Wall (1980), Tse, Bar-Shalom and Meier (1973), and Wenk and Bar-Shalom (1980).

In general, dual control problems are very complex. Thus, to illustrate some of the ideas involved, we shall use a very simple example, suggested by Åström (1981).

Example 11.2.1

Consider the following system:

$$y(t + 1) = y(t) + b(t + 1)u(t) + \omega(t + 1) \tag{11.2.1}$$

where y, u, and ω denote output, input, and noise, respectively. The noise is assumed white and gaussian with zero mean and variance σ_ω^2.

The time-varying parameter $b(t)$ is assumed to satisfy a Markov model of the form

$$b(t + 1) = \alpha b(t) + v(t) \tag{11.2.2}$$

where $|\alpha| < 1$ and $v(t)$ is a gaussian white process with variance σ_v^2. The prior knowledge regarding $b(t)$ is summarized by saying that $b(0)$ has a gaussian distribution with mean b_0 and variance σ_b^2.

A control policy $\{u(t); t = 0, \ldots, N\}$ is sought such that $u(t)$ is a function of $\{y(t), y(t - 1), \ldots, y(0)\}$ and so that the following cost function is minimized:

$$J_{N+1} = \tfrac{1}{2} \sum_{t=1}^{N+1} E\{y(t)^2\}$$

(Note that the problem above has been "engineered" to simplify the determination of the nonlinear dual optimal control.)

We shall use dynamic programming. Let S_{t+1} denote the optimal cost associated with the trajectory from time $t + 1$ to $N + 1$. S_{t+1} is obviously a function of the "state" at time t.

$$S_{t+1} \triangleq \min_{u(t)} E\{\tfrac{1}{2} y(t + 1)^2 + S_{t+2} \mid y(0) \cdots y(t)\}$$

and hence, using (11.2.1),

$$S_{t+1} = \min_{u(t)} \{\tfrac{1}{2}[y(t) + \hat{b}(t + 1)u(t)]^2$$
$$+ \tfrac{1}{2} u(t)^2 P_b(t + 1) + \tfrac{1}{2} \sigma_\omega^2 + E\{S_{t+2} \mid y(0) \cdots y(t)\} \tag{11.2.3}$$

where $\hat{b}(t + 1)$ and $P_b(t + 1)$ are the conditional mean and covariance of $b(t + 1)$, respectively, that is,

$$\hat{b}(t + 1) = E\{b(t + 1) \mid y(0) \cdots y(t)\} \tag{11.2.4}$$

$$P_b(t + 1) = E\{[b(t + 1) - \hat{b}(t + 1)]^2 \mid y(0) \cdots y(t)\} \tag{11.2.5}$$

For $t = N$, (11.2.3) reduces to

$$S_{N+1} = \min_{u(N)} \{\tfrac{1}{2}[y(N) + \hat{b}(N + 1)u(N)]^2 + \tfrac{1}{2}u(N)^2 P_b(N + 1) + \tfrac{1}{2}\sigma_\omega^2\}$$

Hence differentiating with respect to $u(N)$, the optimal control at time N is seen to be

$$u(N) = \frac{\hat{b}(N + 1)\, y(N)}{\hat{b}(N + 1)^2 + P_b(N + 1)} \tag{11.2.6}$$

giving the optimal cost as

$$S_{N+1} = \frac{1}{2} \frac{P_b(N + 1)y(N)^2}{\hat{b}(N + 1)^2 + P_b(N + 1)} + \frac{1}{2}\sigma_\omega^2 \tag{11.2.7}$$

Now we can rewrite the system (11.2.1)–(11.2.2) as a standard linear state-space model as follows:

$$x(t + 1) = \alpha x(t) + v_1(t); \qquad x(0) = b_0 \tag{11.2.8}$$
$$y(t) = y(t - 1) + c(t)x(t) + v_2(t) \tag{11.2.9}$$

where $x(t) \triangleq b(t)$, $v_1(t) \triangleq v(t)$, $v_2(t) \triangleq \omega(t)$, and $c(t) \triangleq u(t - 1)$.

Thus we can use the Kalman filter (Theorem 7.3.1) to evaluate the mean and covariance of $x(t + 1) \triangleq b(t + 1)$ given $\{y(0) \cdots y(t)\}$. This gives

$$\hat{x}(t + 1 | t) = \alpha\hat{x}(t | t - 1) + K(t)[y(t) - y(t - 1) - c(t)\hat{x}(t | t - 1)];$$

$$K(t) = \frac{\alpha \sum(t)c(t)}{[c(t)^2 \sum(t) + \sigma_\omega^2]}$$

$$\sum(t + 1) = \alpha^2 \sum(t) + \sigma_v^2 - \frac{\alpha^2 \sum(t)^2 c(t)^2}{[c(t)^2 \sum(t) + \sigma_\omega^2]}$$

or in terms of the original variables $\hat{b}(t)$ and $P_b(t)$,

$$\hat{b}(t + 1) = \hat{b}(t) + K(t)[y(t) - y(t - 1) - u(t - 1)\hat{b}(t)] \tag{11.2.10}$$

$$K(t) = \frac{\alpha P_b(t)u(t - 1)}{u(t - 1)^2 P_b(t) + \sigma_\omega^2} \tag{11.2.11}$$

$$P_b(t + 1) = \alpha^2 P_b(t) + \sigma_v^2 - \frac{\alpha^2 P_b(t)^2 u(t - 1)^2}{u(t - 1)^2 P_b(t) + \sigma_\omega^2} \tag{11.2.12}$$

$$= \frac{\alpha^2 P_b(t)\alpha_\omega^2}{u(t - 1)^2 P_b(t) + \sigma_\omega^2} + \sigma_v^2; \qquad P_b(0) = \sigma_b^2$$

Proceeding to S_N, we have from (11.2.3) and (11.2.7) that

$$S_N = \min_{u(N-1)} \{\tfrac{1}{2}[y(N - 1) + \hat{b}(N)u(N - 1)]^2 + \tfrac{1}{2}u(N - 1)^2 P_b(N) + \sigma_\omega^2$$

$$+ E\left\{ \frac{\tfrac{1}{2}P_b(N + 1)y(N)^2}{\hat{b}(N + 1)^2 + P_b(N + 1)} \middle| y(0) \cdots y(N - 1)\right\} \tag{11.2.13}$$

▼▼▼

We will not continue with the problem due to the difficulty of evaluating the last term in the equation above. However, the partial solution presented above indicates some general features of the optimal control formulation.

1. The optimal controller takes account of both the parameter uncertainty and the noise. This can readily be seen from (11.2.6), where $u(N)$ depends on the parameter uncertainty as measured by $P_b(N + 1)$. There are two consequences of the inclusion of the parameter uncertainty in generating the control. The first is that the controller exercises *caution* when the parameters are poorly known. This is evident from (11.2.6), where the presence of $P_b(N + 1)$ in the denominator reduces the size of $u(N)$ depending on the size of the uncertainty in $\hat{b}(N + 1)$. The second consequence is that before time $t = N$ (when the controller knows that more observations are to be made), the system

can be *probed* in an effort to learn more about the parameters so that subsequently better control is achieved. Thus the controller may sacrifice some performance at the current step in order that it might reduce the parameter uncertainty for the next step. This can then give better performance when averaged over both steps. This is brought out in expression (11.2.13), where the first term measures the performance as if there were no parameter uncertainty, the second term takes account of the current parameter uncertainty, and the last term looks ahead to see the likely effect of future parameter uncertainty. Now it can be seen from (11.2.12) that the last term in (11.2.13) is reduced if $u(N - 1)$ is made large. However, this obviously conflicts with minimizing the second term (the one giving caution). The optimal control thus makes a compromise between *regulation* [term 1 in (11.2.13)], *caution* [term 2 in (11.2.13)], and *learning* [term 3 in (11.2.13)]. Regulation and learning are often said to be the *dual* role of the input and this is automatically included in the optimal strategy.

2. If we ignore the dependence of the optimal performance on the parameter uncertainty (and hence ignore caution and probing), then (11.2.13) reduces to the first term only, which results in the following control law:

$$u(t) = \frac{- y(t)}{\hat{b}(t + 1)} \qquad (11.2.14)$$

This control law is called the *certainty equivalence* controller because the parameter estimate $\hat{b}(t + 1)$ has been used as if it were the true value in deriving the control law. More will be said about certainty equivalence control below.

3. Some further comments can be made concerning the time-varying nature of the parameters in (11.2.2):

If the system parameters ($b(t)$ in this case) are time invariant, then $\alpha = 1$ and $\sigma_v^2 = 0$. Substituting these into (11.2.12) gives

$$P_b(t + 1)^{-1} = P_b(t)^{-1} + \left[\frac{u(t - 1)}{\sigma_\omega}\right]^2$$

Thus provided that $u(t) > 0$ infinitely often (persistent excitation of order 1), we see that the parameter covariance $P_b(t)$ ultimately converges to zero, and hence $\hat{b}(t)$ converges to the true value. Thus in this case we can asymptotically use the certainty equivalence control law without error.

Also, for $\alpha = 1$, $\alpha_v^2 = 0$, it can be seen from (11.2.10)–(11.2.13) that the parameter estimator reduces to the scalar version of the least-squares update. In this case the gain, $K(t)$, of the algorithm goes to zero asymptotically. However, in the case when $\alpha > 1$, $\alpha_v^2 \neq 0$ (giving a time-varying parameter), $K(t)$ does not converge to zero, as can be seen from (11.2.11)–(11.2.12). Also, we note that in the latter case the form or structure of the parameter estimator is closely related to the model for the time variation of the unknown parameter.

▲▲▲

In summary, the simple example above serves to illustrate a number of ideas:

1. The use of caution when the parameter estimates have large uncertainty
2. The dual action of the control giving learning and regulation
3. The role of the certainty equivalence approximation

4. The simplifications that are possible when the parameters are time invariant

5. The close relationship between the optimal parameter estimator and the model of the time variations of the unknown parameter

Certainty Equivalence Control

As mentioned above, in certainty equivalence control, one bases the control system design on the latest parameter estimates as if they were the true system parameters. Thus the caution and learning aspects of the optimal controller are ignored. In principle any parameter estimation algorithm can be combined with any method for control law synthesis to yield a certainty equivalence adaptive controller. The setup is shown diagrammatically in Fig. 11.2.1.

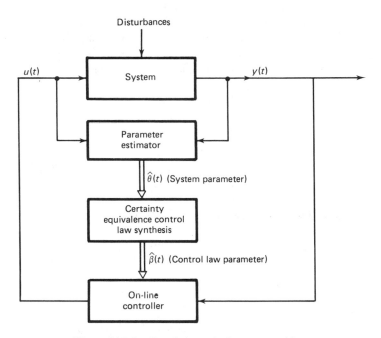

Figure 11.2.1 Certainty equivalence control law.

We have seen in Chapter 6 that the calculations necessary to synthesize the control law can be very simple in certain cases. For example, in the case of minimum variance control (see Chapter 10) the system can be directly parameterized in terms of the control law parameters. In this case the control law synthesis becomes trivial. In the nomenclature of Chapter 6, we then say that the adaptive control law is *implicit* or *direct*. An example of this kind of algorithm is the self-tuning regulator. In other cases, where the control law synthesis requires additional computations, we say that the algorithm is *explicit* or *indirect*. An example of this is the adaptive pole assignment procedure.

Any reasonable estimator can be used in forming an adaptive control algorithm; for example, one could use a prediction error or pseudo linear regression algorithm of the type described in Chapter 8. The latter algorithm has been used extensively in applications of stochastic adaptive control and we will therefore give it emphasis in our subsequent discussion.

11.3 STOCHASTIC MINIMUM PREDICTION ERROR ADAPTIVE CONTROLLERS

In this section we introduce a particular form of adaptive control law which combines a least-squares parameter estimator (see Chapter 8) with the minimum prediction error control laws (see Section 10.3). A special case of this classs of algorithms is the adaptive minimum variance controller [originally called a self-tuning regulator; see Åström, Borisson, Ljung, and Wittenmark (1977)].

The basic idea behind the controller is to form an adaptive prediction of the system output and then to determine the input by setting the predicted output equal to the desired output. This is essentially the same philosphy as was used in the development of the one-step-ahead and model reference adaptive controllers in the deterministic case (see Chapter 6).

As in Chapter 6, one can adopt a direct or indirect approach in formulating the adaptive controller. In the direct approach, the system model is massaged into a form whereby it is directly expressed in terms of the control law parameters. In the indirect approach, one works with a given model of the system and subsequently performs the necessary calculations to evaluate a control law. In this subsection we give most emphasis to the direct approach.

11.3.1 Adaptive Minimum Variance Control

Here we develop an adaptive form of the minimum variance controller described in Section 10.3.1. For simplicity, we initially consider the single-input single-output case.

We assume that the system is described by an ARMAX model as formulated in Section 8.3:

$$A(q^{-1})y(t) = q^{-d}B'(q^{-1})u(t) + C(q^{-1})\omega(t) \tag{11.3.1}$$

where d is an integer time delay (assumed known) and

$$A(q^{-1}) = 1 + a_1 q^{-1} + \cdots + a_n q^{-n}$$
$$B'(q^{-1}) = b_0 + \cdots + b_m q^{-m}; \quad b_0 \neq 0$$
$$C(q^{-1}) = 1 + c_1 q^{-1} + \cdots + c_l q^{-l}$$

$\{y(t)\}$, $\{u(t)\}$, and $\{\omega(t)\}$ are sequences of outputs, inputs, and disturbances, respectively. The sequence $\{\omega(t)\}$ satisfies the usual assumptions.

We initially take the system delay to be unity [$d = 1$ in (11.3.1)]. Later we will show how this can be readily extended to the case $d \geq 1$. (For the case $d = 1$, there is actually no difference between direct and indirect adaptive control, since in this case the ARMAX model coincides with the one-step-ahead predictor and thus the minimum variance controller is directly parameterized in terms of the model parameters).

We recall from Lemma 7.4.1 that the optimal one-step-ahead predictor for the model (11.3.1) with $d = 1$ has the form

$$C(q^{-1})y^\circ(t + 1 \mid t) = \alpha(q^{-1})y(t) + \beta(q^{-1})u(t) \tag{11.3.2}$$

where

$$y^\circ(t + 1 \mid t) = E\{y(t + 1) \mid F_t\} \tag{11.3.3}$$
$$= y(t + 1) - \omega(t + 1)$$

$$\alpha(q^{-1}) = q[C(q^{-1}) - A(q^{-1})] \tag{11.3.4}$$

$$\beta(q^{-1}) = B'(q^{-1}) = b_0 + b_1 q^{-1} + \cdots + b_m q^{-m} \tag{11.3.5}$$

$$(b_0 \neq 0 \text{ by assumption})$$

We have seen in Chapter 8 that many different forms of parameter estimator can be used to estimate the parameters in (11.3.2). To avoid confusion, we will concentrate on one basic type of parameter estimator: the least-squares variant of the pseudo linear regression algorithm. Our reason for choosing this algorithm is that it has been used predominantly in applications of adaptive control.

We now make the predictor (11.3.2) adaptive as in Section 9.3 Then an adaptive minimum variance control law is obtained by simply setting the predicted output, $\hat{y}(t + 1)$, equal to the desired output, $y^*(t + 1)$, to generate the control $u(t)$ at time t. For the purpose of illustration, consider the pseudo linear regression parameter estimator (extended least squares) of Section 8.4. This gives the following adaptive control law. (Later we shall add a few minor modifications to the algorithm, such as replacing a priori predictions in the regression vector by a posteriori predictions. These modifications aid the convergence analysis but are otherwise of secondary importance.)

Adaptive Minimum Variance Controller (Unit Delay)

$$\hat{\theta}(t) = \hat{\theta}(t - 1) + aP(t - 1)\phi(t - 1)e(t): \qquad 0 < a \leq 1 \tag{11.3.6}$$

$$e(t) = y(t) - \hat{y}(t) \tag{11.3.7}$$

$$P(t - 1) = P(t - 2) - \frac{P(t - 2)\phi(t - 1)\phi(t - 1)^T P(t - 2)}{1 + \phi(t - 1)^T P(t - 2)\phi(t - 1)} \tag{11.3.8}$$

$$\hat{y}(t) = \phi(t - 1)^T \hat{\theta}(t - 1) \tag{11.3.9}$$

$$\phi(t - 1) = [y(t - 1), \ldots, y(t - n'), u(t - 1), \ldots, \tag{11.3.10}$$
$$u(t - m - 1), -\hat{y}(t - 1), \ldots, -\hat{y}(t - l)]$$

$$\hat{\theta}(t - 1) = [\hat{\alpha}_1(t - 1), \ldots, \hat{\alpha}_{n'}(t - 1), \hat{\beta}_0(t - 1), \ldots, \hat{\beta}_m(t - 1), \tag{11.3.11}$$
$$\hat{c}_1(t - 1), \ldots, \hat{c}(t - 1)]$$

where $n' = \max(n, l)$.

Finally the input signal $u(t)$ is generated from the following feedback control law:

$$\phi(t)^T \hat{\theta}(t) = y^*(t + 1) \tag{11.3.12}$$

[A minor difficulty with the control law (11.3.12) is that there is a possibility of division by zero if $\hat{\beta}_0(t)$ goes to zero. This is rarely, if ever, a concern in practice unless the system time delay has been chosen too small. However, the problem can be avoided as in Chapter 6 by a number of methods, including:

1. Constraining $\hat{\beta}_0(t)$ to be nonzero. (This requires knowledge of the sign of $\hat{\beta}_0$ and a lower bound on its magnitude.)
2. Using linear control forms.

Remark 11.3.1. The algorithm above is of the direct type since (11.3.12) expresses the control law directly in terms of the system parameters.

▼▼▼

Remark 11.3.2. Convergence of the algorithm above to the minimum variance controller will be discussed in Section 11.3.4. Basically, to ensure convergence, one requires the following assumptions:

1. An upper bound is known for the orders of the polynomials in the system description.
2. The time delay d is known (actually, $d = 1$ here).
3. $[1/C(z^{-1}) - \frac{1}{2}]$ is positive real.
4. $B'(z^{-1})$ is asymptotically stable.

The assumptions above are quite natural. Assumption 1 ensures that the controller has adequate degrees of freedom. Assumption 2 arises due to the look-ahead nature of the control law. Assumption 3 results from the use of a pseudo linear regression type of parameter estimator (see Chapter 8). Assumption 4 is reasonable in the adaptive case since this condition is necessary to ensure a bounded input with minimum variance control in the nonadaptive case.

▼▼▼

For simplicity, we have presented the algorithm for the case $d = 1$. For $d \geq 1$, the predictor (11.3.2) becomes (see Lemma 7.4.1)

$$C(q^{-1})y^0(t + d \,|\, t) = \alpha(q^{-1})y(t) + \beta(q^{-1})u(t) \tag{11.3.13}$$

or

$$y^0(t + d \,|\, t) = \bar{\phi}(t)^T\theta_0 \tag{11.3.14}$$

where

$$y^0(t + d \,|\, t) = y(t + d) - v(t + d)$$

as given in (7.4.32) and $\bar{\phi}(t)$, is as follows:

$$\bar{\phi}(t) = [y(t), \ldots, y(t - n' + 1), u(t), \ldots, u(t - m - d + 1), -y^0(t + d - 1),$$
$$\ldots, -y^0(t + d - l)]^T$$

This leads to a natural extension of the algorithm (11.3.6) to (11.3.12):

Adaptive Minimum Variance Controller (Multiple Delay)

$$\hat{\theta}(t) = \hat{\theta}(t - 1) + P(t - 1)\phi(t - d)[y(t) - \phi(t - d)^T\hat{\theta}(t - 1)] \tag{11.3.15}$$

$$P(t - 1) = P(t - 2) - \frac{P(t - 2)\phi(t - d)\phi(t - d)^T P(t - 2)}{1 + \phi(t - d)^T P(t - 2)\phi(t - d)} \tag{11.3.16}$$

$$\phi(t)^T = [y(t), \ldots, y(t - n' + 1), u(t), \ldots, u(t - m - d + 1),$$
$$- \hat{y}(t + d - 1), \ldots, - \hat{y}(t + d - l)] \tag{11.3.17}$$

$$\hat{y}(t) = \phi(t - d)^T \hat{\theta}(t - d) \tag{11.3.18}$$

Finally, the input $u(t)$ is generated by solving

$$\phi(t)^T \hat{\theta}(t) = y^*(t + d) \tag{11.3.19}$$

[Note that $\phi(t)$ in (11.3.17) is $\bar{\phi}(t)$ in (11.3.14) with $y^0(j \mid j - d)$, and so on, replaced by the corresponding a priori estimates $\hat{y}(j)$, and so on.]

Some Important Special Cases

In the special case when $y^*(t) = 0$, then (11.3.19) gives

$$\phi(t)^T \hat{\theta}(t) = 0 \tag{11.3.20}$$

This ensures that $\hat{y}(t) = 0$ for all t. However, it then makes sense to remove $\hat{y}(t + d - 1), \hat{y}(t + d - 2), \ldots$ from the regression vector $\phi(t)$. Thus in this case (and this case *only*) the parameters in $C(q^{-1})$ may be removed from the problem and ordinary least squares can be used. This important observation was originally made by Åström and Wittenmark (1973).

Another special case is when $y^*(t) = \text{constant} = y^*$ (say). It can be seen that it would then suffice to replace $\hat{y}(t + d - 1) \cdots \hat{y}(t + d - l)$ in the regression vector by a *single* constant (say 1). In this case it can be seen that the last component of the θ vector would become $-y^* \sum_{j=1}^l c_j$.

This is an important observation because it implies that, in the case of regulation about a constant set point, the adaptive minimum variance algorithm developed in a stochastic framework coincides exactly with the deterministic algorithm for one-step-ahead control. (with a 1 in the regression vector to account for dc offsets). Thus in this special case one may simply disregard all the stochastic considerations and approach the problem as if it were purely deterministic and apply the methods of Chapter 6. This is a remarkable result. Of course, for the algorithm to work in the stochastic case it is necessary for there to be a positive real condition on the true C polynomial for the system due to the use of pseudo linear regressions.

However, in the general case, the reader is reminded that it is necessary to take account of the stochastic nature of the problem. For example, if $\{y^*(t)\}$ is not a constant, it is necessary to estimate the $C(q^{-1})$ polynomial explicitly.

The statement above is verified in Figs. 11.3.1 to 11.3.7, which show simulated adaptive control results for a system when $\{y^*(t)\}$ is a square wave (lying between 0 and 10 with period 100 units). Figures 11.3.1 and 11.3.2 show the output and input when ordinary least squares is used [i.e., the coefficients in $C(q^{-1})$ are ignored]. These results are quite unsatisfactory and, in fact, the system can be seen to be unstable. Figures 11.3.3 and 11.3.4 show the corresponding results when $C(q^{-1})$ is estimated. Figures 11.3.5 to 11.3.7 show that in this case the true system parameters are obtained and minimum variance control results, as is evident from Fig. 11.3.5.

It is also evident from Figs. 11.3.5 to 11.3.7 that the C parameters converge slower than the A and B parameters. This is usually the case and results from the fact that C is estimated by the pseudo regression terms in $\phi(t)$.

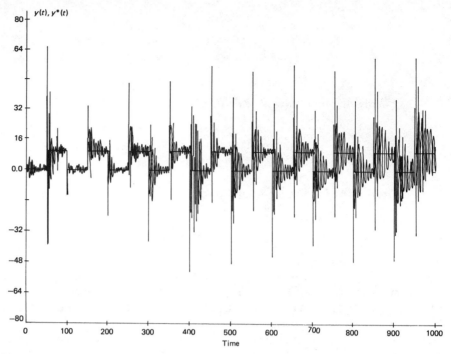

Figure 11.3.1 Output tracking performance of self-tuning regulator using ordinary least squares.

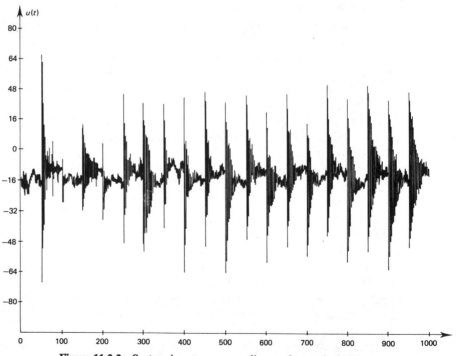

Figure 11.3.2 System input corresponding to the results in Fig. 11.3.1.

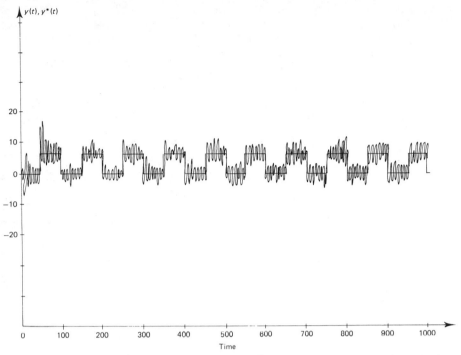

Figure 11.3.3 Output tracking performance of self-tuning regulator in which $C(q^{-1})$ polynomial is estimated.

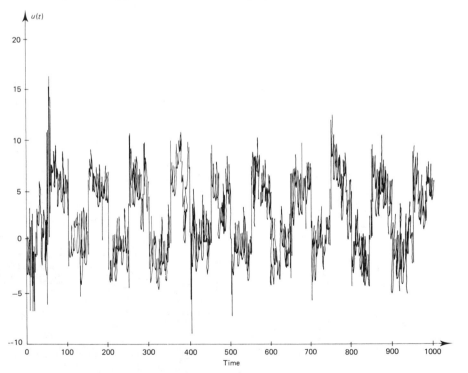

Figure 11.3.4 System input corresponding to the results in Fig. 11.3.3.

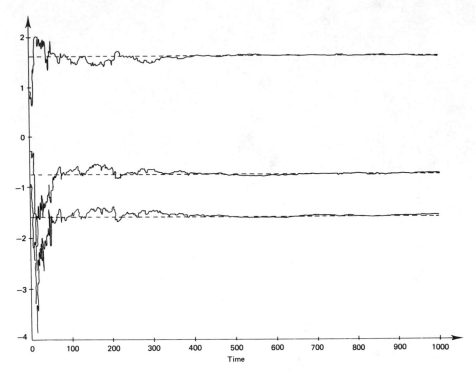

Figure 11.3.5 Convergence of system dynamic parameters.

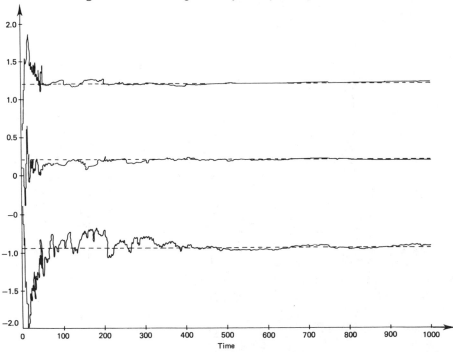

Figure 11.3.6 Convergence of input dynamic parameters.

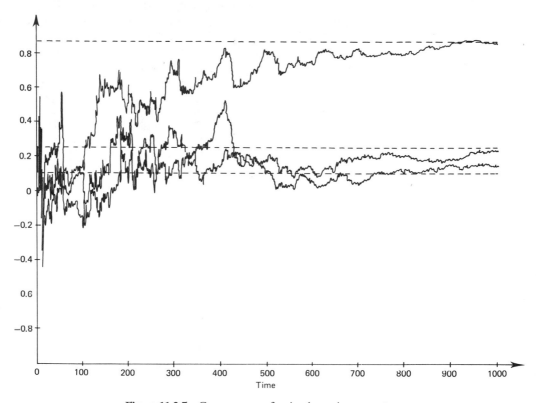

Figure 11.3.7 Convergence of noise dynamic parameters.

Algorithms for Systems with Noise Dynamics Having Zeros on the Unit Circle

We recall from Section 7.4.2 that for systems with noise dynamics having zeros on the unit circle, it is not possible to use the optimal linear time invariant predictor from time $t = 0$. However, a stable time-invariant predictor exists with performance arbitrarily close to the performance of the optimal predictor. This nearly optimal predictor has the form [see (7.4.58)]

$$C_s(q^{-1})\hat{y}(t + d) = G(q^{-1})y(t) + \beta(q^{-1})u(t) \qquad (11.3.21)$$

where $C_s(q^{-1})$ is asymptotically stable. In Section 10.3.1, the predictor above was used to develop a nearly optimal minimum variance control law. We recall from (10.3.42) that this control law can be implemented as

$$\beta(q^{-1})u(t) + \alpha(q^{-1})y(t) = C_s(q^{-1})y^*(t + d) \qquad (11.3.22)$$

The nearly optimal predictor (11.3.21) can also be used to generate stochastic adaptive controllers for systems with noise dynamics having zeros on the unit circle. For the parameter estimation part of these algorithms, it is desirable to use a recursive prediction error method (see Section 8.3). The reason for this choice is that these algorithms have the property that they converge to a local minimum of the prediction error variance even in those cases where the predictor structure does not correspond to the optimal predictor. This is important here, since the predictor (11.3.21) does not correspond to the optimal predictor (although it will give nearly optimal performance).

An adaptive control algorithm can therefore be constructed as follows:

$$\hat{\theta}(t) = \hat{\theta}(t-1) + P(t-d)\phi(t-d)[y(t) - \hat{y}(t)] \tag{11.3.23}$$

$$\phi(t)^T = [y_F(t), \ldots, y_F(t-n+1)$$
$$u_F(t), \ldots, u_F(t-m-d+1) \tag{11.3.24}$$
$$-\hat{y}_F(t+d-1), \ldots, -\hat{y}_F(t+d-l_s)]^T$$

$$\bar{\phi}(t)^T = [y(t), \ldots, y(t-n+1)$$
$$u(t), \ldots, u(t-m-d+1) \tag{11.3.25}$$
$$-\hat{y}(t+d-1), \ldots, -\hat{y}(t+d-l_s)]^T$$

$$\hat{y}(t+d) = \bar{\phi}(t)^T\hat{\theta}(t+d-1) \tag{11.3.26}$$

$$\hat{\theta}(t)^T = [\hat{g}_0(t), \ldots, \hat{g}_{n-1}(t), \hat{\beta}_0(t), \ldots, \hat{\beta}_{m+d-1}(t),$$
$$-\hat{c}_{s1}(t), \ldots, -\hat{c}_{sl_s}(t)]^T \tag{11.3.27}$$

$$y_F(t) = -\sum_{j=1}^{l_s} \hat{c}_{sj}(t+d-1)y_F(t-j) + y(t) \tag{11.3.28}$$

$$u_F(t) = -\sum_{j=1}^{l_s} \hat{c}_{sj}(t+d-1)u_F(t-j) + u(t) \tag{11.3.29}$$

$$\hat{y}_F(t+d) = -\sum_{j=1}^{l_s} \hat{c}_{sj}(t+d-1)\hat{y}_F(t+d-j) + \hat{y}(t+d) \tag{11.3.30}$$

$$P(t-d) = \left\{ P(t-d-1) \right.$$
$$\left. - \frac{P(t-d-1)\phi(t-d)\phi(t-d)^T P(-d-1)}{1 + \phi(t-d)^T P(t-d-1)\phi(t-d)} \right\} \tag{11.3.31}$$

with feedback control law determined from

$$\bar{\phi}(t)^T\hat{\theta}(t) = y^*(t+d) \tag{11.3.32}$$

Equations (11.3.23) to (11.3.31) can be seen to give the iterative prediction error method of Section 8.3. With this parameter estimator it will be necessary to project $\hat{\theta}(t)$ so that the polynomial $(1 + \hat{c}_{1s}(t)q^{-1} + \cdots + c_{sl_s}(t)q^{-l_s})$ remains strictly stable (see Section 8.3).

We illustrate the performance of the algorithms above with a simple example.

Example 11.3.1

We consider the following system with both deterministic and nondeterministic disturbances:

$$y(t) = u(t-1) + d(t) + (1 + \bar{c}_1 q^{-1})\omega(t) \tag{11.3.33}$$

where $\omega(t)$ is a white noise sequence. The unmeasured deterministic disturbance $d(t)$ is taken to be a sine wave of unknown frequency, amplitude, and phase $[d(t) = G\sin(pt + \phi)]$. Thus the ARMAX model for the system is (see Chapter 7)

$$[1 + a_1 q^{-1} + a_2 q^{-2}]y(t) = [b_0 + b_1 q^{-1} + b_2 q^{-2}]u(t-1)$$
$$+ (1 + c_1 q^{-1} + c_2 q^{-2} + c_3 q^{-3})\omega(t) \tag{11.3.34}$$

where the parameters have the following true (but unknown) values:

$$a_1 = -2\cos\rho \qquad b_0 = 1 \qquad\qquad c_1 = \bar{c}_1 - 2\cos\rho$$
$$a_2 = 1 \qquad\qquad b_1 = -2\cos\rho \qquad c_2 = 1 - 2\bar{c}_1\cos\rho \qquad (11.3.35)$$
$$b_2 = 1 \qquad\qquad c_3 = \bar{c}_1$$

It is readily seen from (11.3.35) that $C(q^{-1}) \triangleq (1 + c_1 q^{-1} + c_2 q^{-2} + c_3 q^{-3})$ has zeros on the unit circle (at $\cos\rho \pm j\sin\rho$). Thus, as expected, the noise model has zeros on the unit circle due to the presence of the purely deterministic disturbance.

Figures 11.3.8 to 11.3.10 show the performance of the algorithm (11.3.23)–(11.3.31) on this example. [Here $\rho = 0.8$ radian per unit time; $\omega(t)$ is white noise of standard deviation 0.05; $\bar{c} = 0.9$; the estimated $C_s(q^{-1})$ polynomial was chosen as a first-order polynomial with roots constrained to lie in $|z| \leq 0.98$; the desired output is a square wave lying between ± 1 and with period 100 units.] It can be seen from the figures that the adaptive controller gives excellent tracking performance in spite of the presence of the unmeasured (and unknown) deterministic disturbance.

▼▼▼

An Adaptive Minimum Variance Controller with Weighted Control Effort

As was pointed out in Section 10.3.1 a slight generalization of the minimum variance controller is achieved by including a term in the cost function weighting the control effort. It is probably always a good idea to include this weighting in the adaptive controller since it offers a tuning parameter which can be adjusted on-line

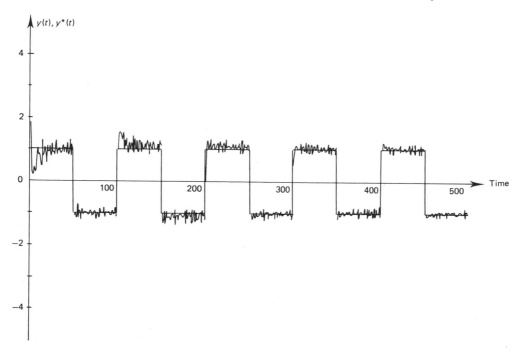

Figure 11.3.8 Output and reference signal for Example 11.3.1. Dashed line, $y^*(t)$; solid line, $y(t)$.

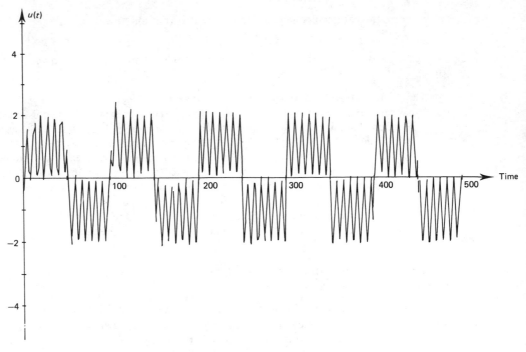

Figure 11.3.9 Input for Example 11.3.1.

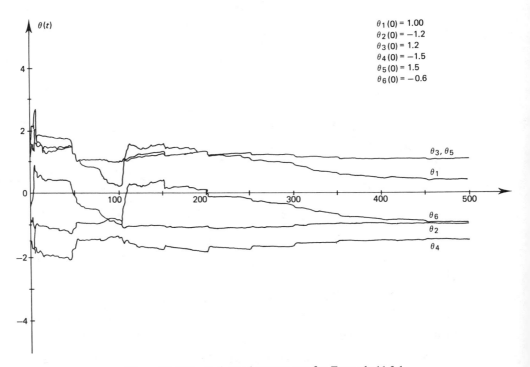

$\theta_1(0) = 1.00$
$\theta_2(0) = -1.2$
$\theta_3(0) = 1.2$
$\theta_4(0) = -1.5$
$\theta_5(0) = 1.5$
$\theta_6(0) = -0.6$

θ_3, θ_5
θ_1
θ_6
θ_2
θ_4

Figure 11.3.10 Estimated parameters for Example 11.3.1.

to limit the amount of control effort used. If appropriate, the weighting can always be set to zero, giving an ordinary minimum variance regulator.

The weighted one-step-ahead cost function is of the form [see (10.3.47)]

$$J'(t + d) = E\left\{\frac{1}{2}[y(t + d) - y^*(t + d)]^2 + \frac{\lambda}{2}u(t)^2 \,|\, F_t\right\} \quad (11.3.36)$$

where $\{y^*(t)\}$ denotes the desired output sequence.

With adaptive minimum variance control in mind it was shown in Theorem 10.3.5 that the optimal d-step-ahead predictor (11.3.2) could be rearranged to the form

$$\frac{\beta_0}{\beta_0^2 + \lambda}C(q^{-1})[y^0(t + d\,|\,t) - y^*(t + d) + \frac{\lambda}{\beta_0}u(t)] = u(t) + \phi(t)^T\theta_0 \quad (11.3.37)$$

where

$$\phi(t)^T = [y(t), y(t - 1), \ldots, u(t - 1), u(t - 2), \ldots, y^*(t + d),$$
$$y^*(t + d - 1), \ldots] \quad (11.3.38)$$

θ_0 are the parameters in the following polynomials:

$$\left[\frac{\beta_0}{\beta_0^2 + \lambda}\alpha(q^{-1}), \frac{\beta_0}{\beta_0^2 + \lambda}\{[\beta(q^{-1}) - \beta_0]q - \frac{\lambda}{\beta_0}[C(q^{-1}) - 1]q\}, -\frac{\beta_0}{\beta_0^2 + \lambda}C(q^{-1})\right] \quad (11.3.39)$$

It was also shown in Theorem 10.3.5 that the one-step-ahead optimal control law could be determined from

$$u^*(t) = -\phi(t)^T\theta_0 \quad (11.3.40)$$

[Notice, however that (11.3.37) is valid *irrespective* of whether or not $u(t)$ is chosen optimally.]

Since the leading coefficient, β_0, associated with the input $u(t)$ in the polynomial $\beta(q^{-1})$ is unknown, we introduce $\bar{\beta}_0$ as a prior fixed estimate of β_0. We then consider a cost function with weighting $\bar{\lambda}$, that is, we aim to minimize $E\{\frac{1}{2}[y(t + d) - y^*(t + d)]^2 + (\bar{\lambda}/2)u(t)^2\,|\,F_t\}$. Due to the possible difference between $\bar{\beta}_0$ and β_0, the algorithm thus developed actually minimizes $J'(t + d)$ as given in (11.3.36), where λ takes the following particular value:

$$\lambda = \frac{\bar{\lambda}\beta_0}{\bar{\beta}_0} \quad (11.3.41)$$

Equations (11.3.37) to (11.3.40) immediately suggest the following adaptive control law (for the case of unit delay):

$$\hat{\theta}(t) = \hat{\theta}(t - 1) + aP(t - 1)\phi(t - 1)e(t) \quad (11.3.42)$$

Note this is exactly as in (11.3.6) except that $e(t)$, $\phi(t - 1)$, and a have modified definitions as below:

$$e(t) = y(t) - y^*(t) + \frac{\bar{\lambda}}{\bar{\beta}_0}u(t - 1) \quad (11.3.43)$$

$\phi(t)$ is as in (11.3.38), and

$$a = \frac{\bar{\beta}_0^2 + \bar{\lambda}}{\bar{\beta}_0} \quad (11.3.44)$$

Finally, the input is generated from

$$u(t) = -\phi(t)^T \hat{\theta}(t) \qquad (11.3.45)$$

The algorithm (11.3.42) to (11.3.45) has been described for $d = 1$. For $d \geq 1$ we simply change to

$$e(t) = y(t) - y^*(t) + \frac{\bar{\lambda}}{\bar{\beta}_0} u(t - d) \qquad (11.3.46)$$

and replace $\phi(t - 1)$ by $\phi(t - d)$ in the algorithm.

Remark 11.3.3. Note that more general forms of the algorithms above can be developed as in Section 5.2.1, where the weighting on $u(t)^2$ is replaced by a weighting on $\bar{u}(t)^2$, where $\bar{u}(t)$ is related to $u(t)$ by a linear dynamic system. This can be used, for example, to give zero steady-state error performance, and so on.

Remark 11.3.4 (Indirect Approach to Minimum Prediction Error Control). In the development above we have used a direct approach to adaptive control. The alternative indirect approach involves first estimating the parameters in a standard model (e.g., a *one*-step-ahead predictor) and then evaluating the control law by on-line calculations.

Thus, let us assume that $\hat{\theta}(t)$ denotes an estimate of the parameters in the one-step-ahead predictor (11.3.2). Then we form $\hat{C}(t, q^{-1})$, $\hat{A}(t, q^{-1})$, and $\hat{B}(t, q^{-1})$ in the corresponding ARMAX model as follows:

$$\hat{C}(t, q^{-1}) = 1 + \hat{c}_1(t)q^{-1} + \cdots + \hat{c}_l(t)q^{-l}$$
$$\hat{A}(t, q^{-1}) = 1 + (\hat{c}_1(t) - \hat{\alpha}_1(t))q^{-1} + \cdots + (\hat{c}_{\bar{n}}(t) - \hat{\alpha}_{\bar{n}}(t))q^{-\bar{n}}$$
$$\hat{B}'(t, q^{-1}) = [\hat{b}_0(t) + \cdots + \hat{b}_m(t)q^{-m}]$$

Then, as in indirect adaptive prediction (Section 9.3.1), we solve the usual prediction equality for $\hat{F}(t, q^{-1})$ and $\hat{G}(t, q^{-1})$, where

$$\hat{C}(t, q^{-1}) = \hat{F}(t, q^{-1})\hat{A}(t, q^{-1}) + q^{-d}\hat{G}(t, q^{-1})$$

Then a d-step-ahead prediction, $\hat{y}(t + d)$, for $y(t - d)$ can be obtained from

$$\hat{y}(t + d) = q[1 - \hat{C}(t, q^{-1})]\hat{y}(t + d - 1) + \hat{G}(t, q^{-1})y(t) + [\hat{F}(t, q^{-1})\hat{B}'(t, q^{-1})]u(t)$$

The d-step-ahead predictor above can then be used to generate an adaptive minimum prediction error control law. For example, an adaptive indirect minimum variance controller is obtained by solving the following equation for $u(t)$:

$$[\hat{F}(t, q^{-1})\hat{B}'(t, q^{-1})]u(t) = \hat{C}(t, q^{-1})y^*(t + d) - \hat{G}(t, q^{-1})y(t)$$

▼▼▼

Remark 11.3.5 (Adaptive Control of Nonlinear Stochastic Systems). We have argued at several points in the book that it is preferable to base control laws on the underlying physical model (which may well be nonlinear) rather than on some linear approximation. This same consideration applies to adaptive control of nonlinear stochastic systems.

A difficulty with general nonlinear stochastic systems is that it may be difficult to determine an optimal predictor for the system output. We present below two ways to get around this difficulty.

A special class of nonlinear systems. If we presume that the input–output response is dominant over the noise response, then a good model for a nonlinear system may be as in (7.4.19):

$$A(q^{-1})y(t) = h(t - d) + C(q^{-1})\omega(t)$$

where $\omega(t)$ is a white sequence,

$$A(q^{-1}) = 1 + a_1 q^{-1} + \cdots + a_n q^{-n}$$
$$C(q^{-1}) = 1 + c_1 q^{-1} + \cdots + c_n q^{-n}$$

$h(t - d)$ is a *nonlinear* function of $\{y(t - d), y(t - d - 1), \ldots, u(t - d), u(t - d - 1), \ldots\}$. The form of the function $h(t - d)$ is assumed known, but it may contain unknown parameters. [A similar idea has recently been described by Svoronos, Stephanopoulos, and Aris (1981).]

It is shown in Lemma 7.4.5 that an optimal *d*-step-ahead predictor can be constructed for the special model above. The predictor has the structure

$$C(q^{-1})y^0(t + d \mid t) = G(q^{-1})y(t) + F(q^{-1})h(t)$$

Now we can write the predictor above in regression form as

$$y^0(t + d \mid t) = \bar{\phi}(t)^T \theta_0$$

where $\bar{\phi}(t)$ will contain $y^0(t + d - 1 \mid t - 1)$ and $y^0(t + d - 2 \mid t - 2)$ together with various *nonlinear* functions of $\{y(t), y(t - 1), \ldots\}$ and $\{u(t), u(t - 1), \ldots\}$. Note that θ_0 represents parameters in $C(q^{-1})$, $G(q^{-1})$, $F(q^{-1})$, and $h(t)$.

Adaptive control algorithms can now be developed exactly as before. For example, we might use the following adaptive control algorithm [with $\phi(t)$ suitably constructed to account for the nonlinear nature of the problem] based on extended least squares:

$$\hat{\theta}(t) = \hat{\theta}(t - 1) + P(t - 1)\phi(t - d)[y(t) - \phi(t - d)^T\hat{\theta}(t - 1)]$$

$$P(t - 1) = P(t - 2) - \frac{P(t - 2)\phi(t - d)\phi(t - d)^T P(t - 2)}{1 + \phi(t - d)^T P(t - 2)\phi(t - d)}$$

with $\phi(t)$ as for $\bar{\phi}(t)$ but with $y^0(t + d - 1 \mid t - 1) \ldots$ replaced by $\hat{y}(t + d - 1), \ldots,$ where

$$\hat{y}(t) = \phi(t - d)^T\hat{\theta}(t - d)$$

Finally, the input $u(t)$ is generated by solving

$$\phi(t)^T\hat{\theta}(t) = y^*(t + d)$$

where $\{y^*(t)\}$ is the desired output sequence.

Discussion

1. The properties of the various parameter estimation algorithms described in Chapter 8 made no particular assumptions about the form of the $\{\phi(t)\}$ sequence. Thus these properties carry over to the nonlinear case! However, the proof of global convergence of the adaptive control algorithm given later in Section 11.3.4 does depend on various properties of the underlying linear model. Thus, while the properties of the parameter estimator are retained, convergence of the nonlinear adaptive control algorithm will depend on the precise nature of the nonlinear system.

2. Solvability of $\phi(t)^T\hat{\theta}(t) = y^*(t + d)$ for $u(t)$ would also have to be verified in particular cases. For example, we know that in the case of bilinear systems $\phi(t)$ is linear in $u(t)$, and this simplifies matters. It is likely that other cases of interest exist in which the control law can be readily solved for $u(t)$.

3. We have described only the one-step-ahead (or minimum variance) form of the nonlinear adaptive control algorithm. However, it would clearly be possible to use other control system synthesis procedures while retaining the certainty equivalence structure.

General nonlinear systems. As was pointed out in Chapter 8, an advantage of the recursive prediction error algorithms is that they lead to the best predictor of a given structure irrespective of whether or not the structure corresponds to the unrestricted optimal predictor for the system. Thus it would be possible to use physical insight, heuristics, informed guesswork, and so on, to come up with a predictor structure. Then a recursive prediction error algorithm from Section 8.3 could be used to fit the predictor to the system. This adaptive predictor could then be used to generate a "minimum variance" controller. Of course, convergence would depend on the precise form of the predictor and on the nature of the underlying system.

▼▼▼

11.3.2 Stochastic Model Reference Adaptive Control

The basic minimum variance controller described above can be readily extended to give an adaptive form of the model reference stochastic controllers described in Section 10.3.2. The approach parallels that presented in Chapter 6 for deterministic model reference adaptive control—see also Landau (1981).

The desired output, $\{y^*(t)\}$, is assumed to satisfy the following reference model:

$$E(q^{-1})y^*(t) = q^{-d}gH(q^{-1})r(t)$$

where $E(q^{-1})$ is stable. In Lemma 10.3.1 it was shown that an optimal predictor could be constructed for the quantity $E(q^{-1})y(t) \triangleq y_F(t)$. This predictor has the form

$$C(q^{-1})y_F^0(t + d \,|\, t) = \alpha(q^{-1})y(t) + \beta(q^{-1})u(t)$$

where $y_F^0(t + d \,|\, t) = E\{y_F(t + d \,|\, t)\}$.

Now we will form an adaptive prediction, $\hat{y}_F(t)$, for $y_F(t)$ and we will then set this predicted value equal to $q^{-d}gH(q^{-1})r(t)$ as suggested by the reference model. This leads to the following algorithm:

Stochastic Model Reference Adaptive Controller

$$\hat{\theta}(t) = \hat{\theta}(t - 1) + P(t - 1)\phi(t - d)[y_F(t) - \phi(t - d)^T\hat{\theta}(t - 1)]$$

$$\phi(t)^T = [y(t), \ldots, u(t), \ldots, -\hat{y}_F(t + d - 1), \ldots]^T$$

$$\hat{y}_F(t) = \phi(t - d)^T\hat{\theta}(t - d)$$

Finally, the input is generated from

$$\phi(t)^T\hat{\theta}(t) = gH(q^{-1})r(t)$$

This algorithm will have identical convergence properties to the ordinary self-tuning regulator; see Section 11.3.4 and the exercises.

11.3.3 Multi-input Multi-output Systems

As in Chapter 6, it is a simple matter in principle to extend the adaptive controllers described above to the multi-input multi-output case. By way of illustration, we will consider the minimum variance control strategy.

Let the system be described by a multivariable ARMAX model [see (10.3.83)]

$$A(q^{-1})y(t) = B(q^{-1})u(t) + C(q^{-1})\omega(t) \tag{11.3.47}$$

where $\{y(t)\}$, $\{u(t)\}$, and $\{\omega(t)\}$ are the $m \times 1$ output, input, and disturbance, respectively. We refer the reader back to Section 10.3.3 for discussion of the one-step-ahead control of the system (11.3.47) when the parameters are known. A key idea introduced in Section 10.3.3 is that of the interactor matrix, $\xi(q)$. This matrix gives the multivariable generalization of the system delay. As in the nonadaptive case, we define $\bar{y}(t)$ as follows:

$$\bar{y}(t) = \xi(q)y(t) \tag{11.3.48}$$

[We assume that $A(z^{-1})^{-1}B(z^{-1})$ is strictly proper and is nonsingular for almost all z.] Now from Theorem 10.3.7, the optimal predictor $\bar{y}^0(t|t)$ for $\bar{y}(t)$ given $\{y(t), y(t-1), \ldots\}$ satisfies an equation of the form

$$\bar{C}(q^{-1})\bar{y}^0(t|t) = \bar{\alpha}(q^{-1})y(t) + \bar{\beta}(q^{-1})u(t) \tag{11.3.49}$$

where $\bar{C}(q^{-1})$, $\bar{\alpha}(q^{-1})$, and $\bar{\beta}(q^{-1})$ are $m \times m$ polynomial matrices as follows:

$$
\begin{aligned}
\bar{C}(q^{-1}) &= I + C_1 q^{-1} + \cdots + C_{n_1} q^{-n_1} \\
\bar{\alpha}(q^{-1}) &= \alpha_0 + \alpha_1 q^{-1} + \cdots + \alpha_{n_2} q^{-n_2} \\
\bar{\beta}(q^{-1}) &= \beta_0 + \beta_1 q^{-1} + \cdots + \beta_{n_3} q^{-n_3}; \qquad \det \beta_0 \neq 0
\end{aligned}
\tag{11.3.50}
$$

The predictor (11.3.49) can be written in regression form as

$$\bar{y}^0(t|t) = \theta_0^T \bar{\phi}(t) \tag{11.3.51}$$

where

$$
\begin{aligned}
\bar{\phi}(t)^T = [y(t)^T, y(t-1)^T, &\ldots, u(t)^T, u(t-1)^T, \ldots, \\
&-\bar{y}^0(t-1|t-1)^T, -\bar{y}^0(t-2|t-2)^T, \ldots]
\end{aligned}
\tag{11.3.52}
$$

We now define a desired output sequence, $\bar{y}^*(t)$, for $\bar{y}(t)$ as follows:

$$\bar{y}^*(t) = \xi(q)y^*(t) \tag{11.3.53}$$

where $y^*(t)$ is the desired value for $y(t)$. Then the minimum variance control algorithm of Theorem 10.3.8 can be expressed in the form

$$\theta_0^T \bar{\phi}(t) = \bar{y}^*(t) \tag{11.3.54}$$

The discussion above motivates the following adaptive control algorithm (again we choose the extended least-squares parameter estimation for illustration):

Adaptive Minimum Variance Control Algorithm (MIMO)

$$\hat{\theta}(t) = \hat{\theta}(t-1) + P(t-1)\phi(t-d)[\bar{y}(t-d)^T - \phi(t-d)^T\hat{\theta}(t-1)] \tag{11.3.55}$$

$$
\begin{aligned}
\phi(t)^T = [y(t)^T, y(t-1)^T, &\ldots, u(t)^T, u(t-1)^T, \\
&-\hat{\bar{y}}(t-1)^T, -\hat{\bar{y}}(t-2)^T, \ldots]
\end{aligned}
\tag{11.3.56}
$$

$$\hat{y}(t) = \hat{\theta}(t)^T \phi(t) \tag{11.3.57}$$

$$P(t-1) = P(t-2) - \frac{P(t-2)\phi(t-d)\phi(t-d)^T P(t-2)}{1 + \phi(t-d)^T P(t-2)\phi(t-d)} \tag{11.3.58}$$

The input $u(t)$ is generated from

$$\hat{\theta}(t)^T \phi(t) = \bar{y}^*(t) \tag{11.3.59}$$

where d is the maximum shift in $\xi(q)$.

In the discussion above, we have assumed that the number of inputs is equal to the number of outputs and that each output is required to track a given desired output sequence $y_i^*(t)$, $i = 1, \ldots, m$. However, as mentioned in Section 6.8, there may be other output measurements available. We denote these additional observations by $\{z(t)\}$. When forming the structure of the predictor upon which the adaptive controller is based, it may be highly desirable to incorporate these additional observations into the information set upon which the prediction is based. Thus we consider predictors of the form

$$C(q^{-1})y^0(t+d\,|\,t) = \alpha(q^{-1})y(t) + \beta(q^{-1})u(t) + \gamma(q^{-1})z(t) \tag{11.3.60}$$

It can be readily seen that the optimal prediction error variance obtained by incorporating $\{z(t)\}$ into the predictor is never worse than if it is left out and will usually be smaller. This statement is heuristically obvious but can also be justified using the Kalman filter development of the optimal predictor (see Exercise 11.5).

The modifications necessary to the adaptive control laws to incorporate $\{z(t)\}$ are straightforward as in Chapter 6 for the deterministic case. We will not comment further other than to stress again that it is highly desirable to use all available information in forming the predictors upon which the adaptive controllers are based.

Remark 11.3.6. The multi-input multi-output stochastic adaptive control law discussed above assumes knowledge of the system interaction matrix. This is a reasonable assumption when the interactor has a simple structure, for example when it is diagonal. In cases where the interaction is unknown, techniques exist for estimating it from on-line data (see Section 6.3.3). Alternatively, one can use the input over an interval rather than at a single point to bring the predicted output to the desired value. This obviates the need to know the interaction matrix but gives suboptimal performance. It has the advantage that it retains the basic simplicity of the minimum variance control law. Further discussion may be found in Goodwin and Dugard (1983).

▾▾▾

11.3.4 Convergence Analysis

In this section we present a rigorous analysis of convergence of several stochastic adaptive control algorithms. The approach to convergence analysis that we describe here is based largely on the work of Goodwin, Ramadge, and Caines (1978b) and Sin and Goodwin (1982). This work, in turn, built on the earlier work of Chang and Rissanen (1968), Åström and Wittenmark (1973), Ljung (1977b), Åström et al. (1977), Ljung and Wittenmark (1976), Gawthrop (1980), and others.

It turns out that a wide class of adaptive control algorithms can be analyzed using the same approach. We shall illustrate this approach by reference to a particular algorithm. Other algorithms are explored in the exercises.

Consider a linear finite-dimensional time-invariant SISO system which is described by an ARMAX model of the following form:

$$A(q^{-1})y(t) = q^{-d}B'(q^{-1})u(t) + C(q^{-1})\omega(t) \tag{11.3.61}$$

where, as usual,

$$A(q^{-1}) = 1 + a_1 q^{-1} + \cdots + a_n q^{-n}$$
$$B'(q^{-1}) = b_0 + b_1 q^{-1} + \cdots + b_m q^{-m}; \qquad b_0 \neq 0$$
$$C(q^{-1}) = 1 + c_1 q^{-1} + \cdots + c_l q^{-l}$$

and $\{y(t)\}, \{u(t)\}$, and $\{\omega(t)\}$ denote the output, input, and noise sequence, respectively. In (11.3.61), q^{-d} represents a pure time delay. As usual the sequence $\{\omega(t)\}$ will be taken to be a real stochastic process defined on a probability space $(\Omega, \mathcal{F}, P)$ and adapted to the sequence of increasing sub-sigma algebras $(\mathcal{F}_t, t \in \mathbb{N})$, where $\mathcal{F}_t$ is generated by the observations up to and including time t. $\mathcal{F}_0$ includes initial condition information. The sequence $\{\omega(t)\}$ is assumed to satisfy the following.

Assumption 11.3.A

1. $E\{\omega(t)\,|\,\mathcal{F}_{t-1}\} = 0$ a.s. $\hspace{3cm}$ (11.3.62)

2. $E\{\omega(t)^2\,|\,\mathcal{F}_{t-1}\} = \sigma^2$ a.s. $\hspace{2.5cm}$ (11.3.63)

3. $\limsup\limits_{N\to\infty} \dfrac{1}{N}\sum\limits_{t=1}^{N} \omega(t)^2 < \infty$ a.s. $\hspace{1.8cm}$ (11.3.64)

▼▼▼

It is assumed that the coefficients in $A(q^{-1})$, $B(q^{-1})$, and $C(q^{-1})$ and σ^2 are unknown and that only the input sequence $\{u(t)\}$ and output sequence $\{y(t)\}$ are directly available.

A feedback control law is to be designed to stabilize the system and to cause the output $\{y(t)\}$ to track a given reference sequence $\{y^*(t)\}$ with minimum (conditional) mean square error. The following assumptions will be made about the system.

Assumption 11.3.B

1. The system time delay, d, is known.
2. An upper bound for n, m, and l is known.
3. $C(z^{-1})$ has all zeros inside the closed unit disk.
4. $B(z^{-1})$ has all zeros inside the closed unit disk.

▼▼▼

Now, as shown in Lemma 7.4.1, the optimal d-step-ahead predictor for the system (11.3.61) is of the form

$$C(q^{-1})[y(t + d) - v(t + d)] = \alpha(q^{-1})y(t) + \beta(q^{-1})u(t) \tag{11.3.65}$$

where $y^0(t + d\,|\,t) = [y(t + d) - v(t + d)]$ is the optimal linear d-step-ahead prediction of $y(t + d)$ given $\mathcal{F}_t$, and $v(t + d)$ is a moving average, of order $d - 1$, driven by the disturbances

$$v(t + d) = \sum_{i=0}^{d-1} f_i \omega(t + d - i); \qquad f_0 = 1 \tag{11.3.66}$$

Then

$$E[v(t + d) | \mathcal{F}_t] = 0 \quad \text{a.s.} \tag{11.3.67}$$

and we write

$$\gamma^2 = E[v(t + d)^2 | \mathcal{F}_t] = \sigma^2 \sum_{i=0}^{d-1} f_i^2 \quad \text{a.s.} \tag{11.3.68}$$

The Control Objective

The control objective is to achieve, with probability 1,

$$\lim_{N} \sup \frac{1}{N} \sum_{t=1}^{N} y(t)^2 < \infty \tag{11.3.69}$$

$$\lim_{N} \sup \frac{1}{N} \sum_{t=1}^{N} u(t)^2 < \infty \tag{11.3.70}$$

$$\lim_{N \to \infty} \frac{1}{N} \sum_{t=1}^{N} E\{y(t) - y^*(t))^2 | \mathcal{F}_{t-d}\} = \gamma^2 \triangleq \lim_{N \to \infty} \frac{1}{N} \sum_{t=1}^{N} E\{v_t^2 | \mathcal{F}_{t-d}\} \tag{11.3.71}$$

where $\{y^*(t)\}$ is a reference sequence. $y^*(t + d)$ represents the desired value of $y(t + d)$ and hence obviously must be known or computable at time t. We also require that $y^*(t)$ be bounded, that is,

$$|y^*(t)| < M < \infty; \qquad t \geq 0 \tag{11.3.72}$$

The key property of the parameter estimator used in the analysis of adaptive control algorithms is that of *normalized prediction error convergence*, that is,

$$\lim_{N \to \infty} \sum_{t=1}^{N} \frac{[e(t) - \omega(t)]^2}{r(t - 1)} < \infty \tag{11.3.73}$$

where

$e(t) =$ prediction error [i.e., $e(t) = y(t) - \hat{y}(t)$]

$\omega(t) =$ noise sequence

$[(e(t) - \omega(t)) =$ deterministic part of the prediction error]

$r(t - 1) =$ cummulative sum of squares of the regression vector, $\phi(t)$; that is,

$r(t)$ satisfies

$$r(t - 1) = r(t - 2) + \phi(t - 1)^T \phi(t - 1) \tag{11.3.74}$$

It can be seen from Chapter 8 that the property above applies to a wide range of estimation algorithms.

Given the property (11.3.73), it is a relatively straightforward matter to establish convergence of an adaptive control algorithm by applying the stochastic key technical lemma (Lemma 8.5.3). The steps in the argument parallel the deterministic case. The main step is to verify assumption (8.5.75) in the stochastic key technical lemma, that is, to verify that

$$\frac{1}{N} r(N - 1) \leq K_1 + \frac{K_2}{N} \sum_{t=1}^{N} [e(t) - \omega(t)]^2; \qquad N \geq \bar{N} \quad \text{a.s.} \tag{11.3.75}$$

$$\text{for } 0 < K_1 \leq \infty, \quad 0 < K_2 < \infty, \quad 0 < \bar{N} < \infty$$

We illustrate the analysis procedure by considering a slightly modified form of the adaptive minimum variance controller with unit delay given in (11.3.6) to (11.3.12). The specific modifications that we introduce are:

1. We use a posteriori predictions in the regression vector in place of a priori predictions.
2. We add a slight modification to the P update (as suggested in Remark 8.5.2) to ensure that the condition number remains bounded.

This leads to the following algorithm:

$$\hat{\theta}(t) = \hat{\theta}(t-1) + \frac{P(t-2)\phi(t-1)}{1 + \phi(t-1)^T P(t-2)\phi(t-1)}[y(t) - \phi(t-1)^T\hat{\theta}(t-1)] \tag{11.3.76}$$

$$\phi(t-1) = [y(t-1), \ldots, y(t-n'), u(t-1), \ldots, u(t-m),$$
$$-\bar{y}(t-1), \ldots, -\bar{y}(t-l)]^T; \qquad n' = \max(n, l) \tag{11.3.77}$$

$$\bar{y}(t) = \phi(t-1)^T\hat{\theta}(t) \tag{11.3.78}$$

$$P'(t-1) = P(t-2) - \frac{P(t-2)\phi(t-1)\phi(t-1)^T P(t-2)}{1 + \phi(t-1)^T P(t-2)\phi(t-1)}; \qquad P(-1) > 0 \tag{11.3.79}$$

(together with the condition number monitoring scheme of Remark 8.5.2). The control is generated from

$$\phi(t-1)^T\hat{\theta}(t-1) = y^*(t) \tag{11.3.80}$$

The property (11.3.75) is verified in the following lemma:

Lemma 11.3.1. There exist finite positive constants K_1 to K_8 such that

(i) $$\eta(t) = \frac{e(t)}{1 + \phi(t-1)^T P(t-2)\phi(t-1)} \tag{11.3.81}$$
where

$$e(t) \triangleq y(t) - \phi(t-1)^T\hat{\theta}(t-1)$$
$$= y(t) - y^*(t) \tag{11.3.82}$$

$$\eta(t) \triangleq y(t) - \bar{y}(t) \tag{11.3.83}$$

(ii) $$\lim_N \frac{1}{N} \sum_{t=1}^{N-1} u(t)^2 \leq \lim_N \frac{K_1}{N} \sum_{t=1}^{N} y(t)^2 + K_2 \tag{11.3.84}$$

(iii) $$\lim_N \frac{1}{N} \sum_{t=1}^{N} y(t)^2 \leq \lim_N \frac{K_3}{N} \sum_{t=1}^{N} [e(t) - \omega(t)]^2 + K_4 \tag{11.3.85}$$

(iv) $$\lim_N \frac{1}{N} \sum_{t=1}^{N} \bar{y}(t)^2 \leq \lim_N \frac{K_5}{N} \sum_{t=1}^{N} [e(t) - \omega(t)]^2 + K_6 \tag{11.3.86}$$

(v) $$\frac{r(N-1)}{N} \leq K_7 \frac{1}{N} \sum_{t=1}^{N} [e(t) - \omega(t)]^2 + K_8 \tag{11.3.87}$$

Proof. (i) Straightforward (see Exercise 8.17).
(ii) Follows immediately from Assumptions 11.3.A, part **3** and 11.3.B, part **4**, (stable invertibility and mean-square bounded noise).

(iii) From (11.3.82), we have

$$y(t) = e(t) + y^*(t)$$
$$= (e(t) - \omega(t)) + \omega(t) + y^*(t)$$

Hence

$$y(t)^2 \leq 3(e(t) - \omega(t))^2 + 3\omega(t)^2 + 3y^*(t)^2$$

Equation (11.3.85) now follows using (11.3.72) and Assumption 11.3.A, part 3 [both noise and $y^*(t)$ mean-square bounded].

(iv) From (11.3.83)

$$\bar{y}(t) = y(t) - \eta(t)$$
$$= e(t) + y^*(t) - \frac{e(t)}{1 + \phi(t-1)^T P(t-2)\phi(t-1)} \qquad \text{using (11.3.81) and (11.3.82)}$$

$$= \frac{\phi(t-1)^T P(t-2)\phi(t-1)}{1 + \phi(t-1)^T P(t-2)\phi(t-1)} [e(t) - \omega(t)]$$

$$+ \frac{\phi(t-1)^T P(t-2)\phi(t-1)}{1 + \phi(t-1)^T P(t-2)\phi(t-1)} \omega(t) + y^*(t)$$

Hence

$$\bar{y}(t)^2 \leq 3(e(t) - \omega(t))^2 + 3\omega(t)^2 + 3y^*(t)^2$$

Equation (11.3.86) follows immediately.

(v) Equation (11.3.87) follows from parts (ii), (iii), and (iv) and the definition of $r(t), \phi(t)$.

▼▼▼

We then have the following global convergence result.

Theorem 11.3.1

(a) Subject to the noise, system, and signal assumption (Assumptions 11.3.A and 11.3.B), and provided that $[1/C(z) - \frac{1}{2}]$ is very strictly passive, the adaptive control algorithm (11.3.76)–(11.3.80) is globally convergent in the following sense:

(i) $\displaystyle \lim_N \sup \frac{1}{N} \sum_{t=1}^N y(t)^2 < \infty$ a.s. $\qquad\qquad$ (11.3.88)

(ii) $\displaystyle \lim_N \sup \frac{1}{N} \sum_{t=1}^N u(t)^2 < \infty$ a.s. $\qquad\qquad$ (11.3.89)

(iii) $\displaystyle \lim_N \frac{1}{N} \sum_{t=1}^N E\{[y(t) - y^*(t)]^2 \,|\, \mathcal{F}_{t-1}\} = \sigma^2$ a.s. $\qquad\qquad$ (11.3.90)

(b) If the mean-square boundedness assumption (11.3.64) for the noise is strengthened to

$$E\{\omega(t)^4 \,|\, \mathcal{F}_{t-1}\} < \infty \quad \text{a.s.} \qquad\qquad (11.3.91)$$

then the result (11.3.90) is strengthened to

$$\lim_{N \to \infty} \frac{1}{N} \sum_{t=1}^N [y(t) - y^*(t)]^2 = \sigma^2 \quad \text{a.s.} \qquad\qquad (11.3.92)$$

Proof. A key property of the algorithm (11.3.76)–(11.3.80) is that it gives normalized prediction error convergence, that is,

$$\lim_{N \to \infty} \sum_{t=1}^{N} \frac{(e(t) - \omega(t))^2}{r(t-1)} < \infty \quad \text{a.s.} \qquad (11.3.93)$$

where

$$r(t-1) = r(t-2) + \phi(t-1)^T \phi(t-1), \qquad r(-1) > 0 \qquad (11.3.94)$$

The proof of the result (11.3.93) is discussed in Chapter 8.

The result then immediately follows from the stochastic key technical lemma using Lemma 11.3.1, part (v).

▼▼▼

The convergence technique above can be applied to a wide class of algorithms used in stochastic adaptive control. The analogy with the deterministic case is evident. Application of the technique to several other algorithms are explored in the exercises.

We have deliberately kept the assumptions relatively weak and thus we have stopped short of proving parameter convergence. However, if the assumptions are strengthened, for example by assuming knowledge of the true system order and by requiring that $\{y^*(t)\}$ be persistently exciting, parameter convergence can be established as in Section 8.6 [see Chen (1982) for specific results of this type]. The essential prerequisite being to first establish boundedness of the system input and output as in Theorem 11.3.1.

Further results relating to parameter convergence in adaptive control can be found in Caines (1981), Caines and Lafortune (1981), Anderson and Johnson (1980), Kumar (1981), Kumar and Moore (1979), Kushner and Kumar (1981), Moore (1980), and Chen (1982).

11.4 ADAPTIVE POLE PLACEMENT AND ADAPTIVE OPTIMAL CONTROLLERS

The adaptive control algorithms discussed above have all relied on some form of stable invertibility assumption. In this section we discuss more general algorithms aimed at adaptive pole placement and adaptive optimal control. Similar algorithms to the ones described here have been suggested by Åström, Westerburg, and Wittenmark (1978), Wellstead, Prager and Zanker (1979a, 1979b), Wellstead and Prager (1981), and Allidina and Hughes (1980).

The basic strategy is again to combine a parameter estimator with a standard control system synthesis method. The exact procedure necessary to adaptively implement any of the algorithms described in Chapter 10 should by now be apparent.

Thus if one aims to use pole assignment, one simply uses the estimated polynomials $\hat{A}(t, q^{-1})$ and $\hat{B}(t, q^{-1})$ to determine the control law by solving the usual pole assignment equation on-line:

$$A(t, q^{-1})\hat{L}(t, q^{-1}) + \hat{B}(t, q^{-1})\hat{P}(t, q^{-1}) = A^*(q^{-1}) \qquad (11.4.1)$$

where $A^*(q^{-1})$ is an arbitrary stable polynomial. We then determine the input signal from the following feedback control law:

$$\hat{L}(t, q^{-1})u(t) = -\hat{P}(t, q^{-1})y(t) + \hat{P}(t, q^{-1})y^*(t) \qquad (11.4.2)$$

Again, as pointed out for the deterministic case in Chapter 6, a potential problem with this algorithm is that estimated parameters may have common unstable roots in the estimated ARMAX model. This clearly leads to difficulty in generating the control via (11.4.1)–(11.4.2). However, practical experience indicates that this is a theoretical problem rather than a real implementation difficulty. With suitable precautions to ensure that unstable pole–zero cancellation do not occur, then convergence can be established for a robust controller to counter both deterministic and stochastic disturbances (see Exercise 11.14), [see, e.g., Hersh and Zarrop (1981)].

Alternatively, if one adopts an optimal control approach, the algorithm will entail on-line solution of a matrix Riccati equation to generate the control law as in Section 10.4. Of course, stabilizability of the estimated model is still an issue as it was in pole assignment, since stabilizability is now necessary to ensure convergence of the matrix Riccati equation of optimal control.

The adaptive optimal controller is now as in Fig. 11.4.1 and comprises: parameter estimator, Kalman filter, control law synthesis (via the Riccati equation), and state-variable feedback. The determination of the Kalman filter is made trivial if the system in parameterized in the form of the one-step-ahead predictor or equivalently the ARMAX model.

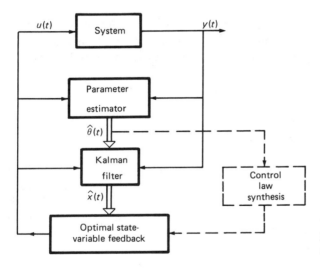

Figure 11.4.1 Adaptive optimal controller.

11.5 CONCLUDING REMARKS

The reader is reminded of all the practical and implementation considerations discussed in Section 6.8. All of these issues are as important in the stochastic case as they are in the deterministic use. In particular, we recall the discussion in Section 6.8 regarding the choice of sampling intervals, choice of control law, choice of parameter estimation algorithm, use of additional output data, and so on.

Regarding the choice of parameter estimator, our recommendation is again to use least squares or one of its variants. In some of the literature on stochastic adaptive control, stochastic gradient-type algorithms are often considered. These algorithms have the advantage of simplicity of analysis, but are generally unsuitable for practical application due to their very slow rate of convergence. The least-squares type of algorithm generally exhibits rapid initial convergence but then slows up after a few steps (say 10 to 20). Possible refinements to overcome this "turn-off" phenomenon have been described in some detail in Section 6.7. For example, one can incorporate periodic covariance resetting into the least-squares algorithm. As pointed out in Chapter 6, this effectively gives least squares over consecutive intervals of chosen length. This has been found to be extremely helpful in both time-invariant and time-varying cases. For the adaptive control of time-varying systems, we do not advocate the use of exponentially slow algorithms such as the projection algorithm, and the LMS algorithm, since these can be up to 1000 times slower than the alternative least-squares-based algorithms. It should also be pointed out that least squares with exponential data weighting, although frequently advocated in some of the literature, is generally unsuitable for industrial applications due to the drift and burst phenomenon discussed in Section 6.7. These difficulties can be overcome in a number of ways (e.g., by the use of a variable forgetting factor or by simply using ordinary least squares with covariance resetting).

In the time-varying case, it is also often desirable to probe the system periodically so that the drift in the parameters will not go unnoticed by the parameter estimator. If parameter drift is not tracked, then when the operating conditions change, large transients can appear.

We wish to point out that adaptive control is a sophisticated approach to control system design and may not be justified or desirable in some cases. If the nature of the problem is such that adaptive control appears desirable, one should still proceed with due caution and regard for the physical reality of the problem under study. For example, it is sometimes helpful, at least as a starting point, to implement the adaptive controller on a process that is already controlled by a conventional fixed controller. This has the advantage that the adaptive controller can be disconnected at any time leaving the fixed controller still present to control the plant. Further, it sometimes allows a low-order adaptive controller to be used since the controlled plant may already be relatively well behaved.

This completes our treatment of adaptive filtering, prediction, and control. Our aim has been to extend, unify, and consolidate various ideas relating to these problems. We hope that the book has helped to improve the reader's appreciation of the utility, and where appropriate, the limitations of adaptive techniques. We would feel that we have succeeded in our aim if the reader now considers adaptive techniques to be feasible alternatives to other methods.

EXERCISES

11.1. Establish convergence properties for the model reference adaptive controller described in Section 11.3.2.

11.2. Consider the case where $y^*(t)$ is a single sinusoid of the form

$$y^*(t) = A^* \sin (p^* t + \phi^*)$$

(a) Show that it is necessary to consider only a second-order $C(q^{-1})$ polynomial in adaptive minimum variance control when the extended least-squares parameter estimator having a priori predictions in the regression vector is used (i.e., show that only two C coefficients are needed).

(b) Does the passivity condition apply for the second-order $C(q^{-1})$ polynomial or the true nth-order $C(q^{-1})$ polynomial?

11.3. Consider the parameter estimator described in Exercise 8.5 and assume that the system delay is 1. Combine this parameter estimator with the following feedback control law:

$$\phi(t-1)^T \hat{\theta}(t-1) = y^*(t)$$

Show that provided that

$$C(z^{-1}) - \frac{\bar{a}}{2}$$

is input strictly passive, the algorithm is globally convergent in the sense that

$$\limsup_{N} \frac{1}{N} \sum_{1}^{N} y(t)^2 < \infty \quad \text{a.s.}$$

$$\limsup_{N} \frac{1}{N} \sum_{1}^{N} u(t)^2 < \infty \quad \text{a.s.}$$

$$\lim_{N} \frac{1}{N} \sum_{1}^{N} E\{[y(t) - y^*(t)]^2 \mid \mathcal{F}_{t-d}\} = \sigma^2 \quad \text{a.s.}$$

where σ^2 is the minimum possible mean-square control error achievable with any casual linear feedback.

11.4. Consider the model obtained by factoring $|\beta_0|$ from the right-hand side of the ARMAX of the d-step-ahead predictor to give (for $d = 1$)

$$C(q^{-1})[y(t+1) - \omega(t+1)] = |\beta_0| [\alpha'(q^{-1})y(t) + (\text{sgn } \beta_0)u(t) + \beta'(q^{-1})u(t)$$

By subtracting $C(q^{-1})y^*(t+1)$ from both sides, show that the tracking error $e(t) = y(t) - y^*(t)$ satisfies an equation of the form

$$C(q^{-1})[e(t+1) - \omega(t+1)] = |\beta_0| [\phi(t)^T \theta_0 + (\text{sgn } \beta_0)u(t)]$$

where

$$\phi(t)^T = [y(t), \ \dots, \ y(t-n+1), u(t-1), \ \dots, \ u(t-m), y^*(t+1), \ \dots, \ y^*(t-l)]$$

Consider the following adaptive control algorithm, which effectively uses $1/\bar{a}$ as an a priori fixed estimate of $|\beta_0|$:

$$\hat{\theta}(t) = \hat{\theta}(t-1) + \bar{a}\phi(t-1)\frac{1}{r(t-1)} [y(t) - y^*(t)]$$

$$r(t) = r(t-1) + \phi(t-1)^T\phi(t-1); \qquad r(0) = 1$$

$$u(t) = -(\text{sgn } \beta_0)\phi(t)^T \hat{\theta}(t)$$

Establish global convergence for the algorithm above subject to the condition that $C(z^{-1}) - \frac{1}{2}\bar{a}|\beta_0|$ is strictly input passive.

11.5. Show, using the Kalman filter, that if additional measurements are included in the predictor, the prediction error covariance decreases [see Section 11.3.3, in particular (11.3.60)].

11.6. Consider the setup of Section 11.3.1 but with $C(q^{-1}) = 1$, $d > 1$. Establish global convergence for the following adaptive control algorithm:

$$\hat{\theta}(t) = \hat{\theta}(t-1) + \frac{\phi(t-d)}{r(t-d)}[y(t) - \phi(t-d)^T\hat{\theta}(t-1)]$$

$$r(t-d) = r(t-d-1) + \phi(t-d)^T\phi(t-d); \qquad r(-1) = 1$$

with $u(t)$ generated by solving

$$\phi(t)^T\hat{\theta}(t) = y^*(t+d)$$

where

$$\phi(t) = [y(t), \ldots, y(t-n), u(t), \ldots, u(t-m-d+1)]$$

[*Hint:* Use the technique described in Exercise 9.9. In particular, define

$$\epsilon(t) \triangleq y(t) - \phi(t-d)^T\hat{\theta}(t-1)$$

$$e(t) \triangleq y(t) - y^*(t) = y(t) - \phi(t-d)^T\hat{\theta}(t-d)$$

$$\eta(t) \triangleq y(t) - \bar{y}(t) = y(t) - \phi(t-d)^T\hat{\theta}(t)$$

$$z(t) \triangleq \eta(t) - v(t)$$

$$= \eta(t) - F(q^{-1})\omega(t)$$

$$= -\phi(t-d)^T\tilde{\theta}(t)$$

$$\tilde{\theta}(t) = \hat{\theta}(t) - \theta_0$$

θ_0 are the coefficients of $\alpha(q^{-1})$ and $\beta(q^{-1})$ in (11.3.13).]

(a) Show that

$$\frac{\eta(t)}{r(t-d-1)} = \frac{\epsilon(t)}{r(t-d)}$$

Decompose $\tilde{\theta}(t), \eta(t)$ as

$$\tilde{\theta}(t) = \sum_{i=1}^{d} \tilde{\theta}^i(t)$$

$$\eta(t) = \sum_{i=1}^{d} \eta^i(t)$$

$$z(t) = \sum_{i=1}^{d} z^i(t) = -\sum_{i=1}^{d} \phi(t-d)^T\tilde{\theta}^i(t)$$

Construct

$$\tilde{\theta}^i(t) = \tilde{\theta}^i(t-1) + \frac{1}{r(t-d)}\phi(t-d)\eta^i(t)$$

$$\eta^i(t) - v^i(t) = -\phi(t-d)^T\tilde{\theta}^i(t) = z^i(t)$$

$$u^i(t) \triangleq f_{i-1}\omega(t+i-1)$$

(b) Show that
   (i) $\tilde{\theta}^i(t-1)$ is $\mathcal{F}_{t-i}$ measurable.
   (ii) $\eta^i(t-1)$ is $\mathcal{F}_{t-i}$ measurable.
(c) For $i = 1, \ldots, d$, show that

$$V^i(t) = V^i(t-1) - \frac{2z^i(t)\eta^i(t)}{r(t-d-1)} - \frac{\phi(t-d)^T\phi(t-d)}{r(t-d-1)^2}\eta^i(t)^2$$

where $V^i(t) = \tilde{\theta}^i(t)^T\tilde{\theta}^i(t)$.
(d) Evaluate

$$E\{v^i(t)z^i(t)|\mathcal{F}_{t-i}\}$$

$$E\{V^i(t)|\mathcal{F}_{t-i}\}$$

(e) Use the Martingale convergence theorem to conclude that

$$\lim_{N\to\infty} \sum_{t=d}^{N} \frac{z^i(t)^2}{r(t-d-1)} < \infty \quad \text{a.s.}$$

$$\lim_{N\to\infty} \sum_{t=d}^{N} \frac{\phi(t-d)^T\phi(t-d)}{r(t-d-1)^2}\eta^i(t)^2 < \infty$$

and hence

$$\lim_{N\to\infty} \sum_{t=d}^{N} \frac{z(t)^2}{r(t-d-1)} < \infty \quad \text{a.s.}$$

$$\lim_{N\to\infty} \sum_{t=d}^{N} \frac{\phi(t-d)^T\phi(t-d)}{r(t-d-1)^2}\eta(t)^2 < \infty$$

(f) Show that

$$e(t) - v(t) = [\eta(t) - v(t)] + \sum_{k=1}^{d} \frac{\phi(t-d)^T\phi(t-d-k+1)}{r(t-d-k)}\eta(t-k+1)$$

Hence conclude using the Schwarz and triangle inequalities plus part (e):

$$\lim_{N\to\infty} \sum_{t=d}^{N} \frac{[e(t) - v(t)]^2}{r(t-d)} < \infty \quad \text{a.s.}$$

(g) Complete the proof using the stochastic key technical lemma.

11.7. Consider a stochastic gradient form of the algorithm given in (11.3.36) to (11.3.45) (adaptive minimum variance controller with weighted control effort):

$$\hat{\theta}(t) = \hat{\theta}(t-1) + \frac{a(t-1)}{r(t-1)}e(t)$$

$$u(t) = -\phi(t)^T\hat{\theta}(t)$$

where a is a positive constant and

$$e(t) = y(t) - y^*(t) + \frac{\bar{\lambda}}{\beta_0}u(t-1)$$

$$\phi(t)^T = [y(t), y(t-1), \ldots, u(t-1), u(t-2), \ldots,$$
$$y^*(t+d), y^*(t+d-1) \ldots]$$

$$r(t-1) = r(t-2) + \phi(t-1)^T\phi(t-1); \qquad r(-1) = r_0$$

Show that the algorithm above is globally convergent provided that:

1. $\left[\dfrac{\beta_0}{\bar{\beta}_0}\right]\left[\dfrac{\bar{\beta}_0^2 + \bar{\lambda}}{\beta_0^2 + \lambda}\right]C(z^{-1}) - a$ is input strictly passive.

2. $B(z^{-1}) + \dfrac{\lambda}{\beta_0}A(z^{-1})$ is asymptotically stable.

[*Hint:* First show that

$$\left[\frac{\beta_0}{\bar{\beta}_0}\right]\left[\frac{\bar{\beta}_0^2 + \bar{\lambda}}{\beta_0^2 + \lambda}\right]C(q^{-1})e(t) = -\phi(t)^T\tilde{\theta}(t)$$

where $\tilde{\theta}(t) = \hat{\theta}(t) - \theta_0$.]

11.8. Here we consider a least-squares form of the algorithm discussed in Exercise 11.7 using a posteriori predictions in the regression vector and condition number monitoring. Define the following a posteriori predicted quantity:

$$\bar{e}(t) \triangleq \phi(t)^T\hat{\theta}(t+1)$$

(a) Consider the usual one-step-ahead predictor:

$$C(q^{-1})[y^0(t+1|t)] = \alpha(q^{-1})y(t) + \beta(q^{-1})u(t)$$

Subtract

$$C(q^{-1})[y^*(t+1) - \frac{\lambda}{\beta}u(t) + \bar{\varepsilon}(t)]$$

from both sides of the equation above and hence show that

$$C(q^{-1})[y^0(t+1|t) - y^*(t+1) + \frac{\lambda}{\beta}u(t) - \bar{\varepsilon}(t)]$$

$$= \alpha(q^{-1})y(t) + [\beta(q^{-1}) + \frac{\lambda}{\beta_0}C(q^{-1})]u(t)$$

$$- C(q^{-1})y^*(t+1) - (C(q^{-1}) - 1)\bar{\varepsilon}(t) - \bar{\varepsilon}(t)$$

$$\triangleq \phi(t)^T\theta_0 - \bar{\varepsilon}(t)$$

where $\phi(t)$ is of the form

$$(y(t), \ldots, u(t), \ldots, y^*(t+1), \ldots, \bar{\varepsilon}(t-1), \ldots)^T$$

(b) Show that the optimal minimum variance control with weighted control effort is achieved when $\bar{\varepsilon}(t) = 0$ and $\phi(t)^T\theta_0 = 0$ since this gives $y^0(t+1|t) - y^*(t+1) + (\lambda/\beta_0)u(t) = 0$.

11.9. Using the setup in Exercise 11.8, consider the following adaptive control algorithm with weighted control effort:

$$\hat{\theta}(t) = \hat{\theta}(t-1) + \frac{P(t-2)\phi(t-1)}{1 + \phi(t-1)^TP(t-2)\phi(t-1)}\left[y(t) - y^*(t) + \frac{\bar{\lambda}}{\beta_0}y(t-1)\right]$$

$$\phi(t-1) = [y(t-1), \ldots, u(t-1), \ldots, y^*(t), \ldots, \bar{\varepsilon}(t-2), \ldots]$$

with $\bar{\varepsilon}(t-2) = -\phi(t-2)^T\hat{\theta}(t-1)$ and so on. The control is generated from the equation

$$\phi(t)^T\hat{\theta}(t) = 0$$

The matrix $P(t)$ is generating using the condition monitoring scheme given in Remark 8.5.2.

[*Hint:* The algorithm above can be motivated by noting that optimality is achieved for $\phi(t)^T\theta_0 = 0$, $\bar{\varepsilon}(t) \triangleq \phi(t)^T\hat{\theta}(t+1) = 0$. We also note that the coefficient of $u(t)$ in the equation $\phi(t)^T\hat{\theta}(t) = 0$ is the estimate of $(\beta_0 - \lambda/\beta_0) = [1 + \lambda/\beta_0^2]$. Thus provided that the sign of β_0 is known, we can use a projection facility (see Chapter 3) to ensure that this parameter is bounded away from zero.]

Establish the following preliminary properties of the algorithm:

(a) $\eta(t) = \dfrac{e(t)}{1 + \phi(t-1)^TP(t-2)\phi(t-1)}$

where

$$e(t) \triangleq y(t) - y^*(t) + \frac{\bar{\lambda}}{\beta_0}u(t-1)$$

$$= y(t) - y^*(t) + \frac{\bar{\lambda}}{\beta_0}u(t-1) - \phi(t-1)^T\hat{\theta}(t-1)$$

$$\eta(t) \triangleq y(t) - y^*(t) + \frac{\bar{\lambda}}{\beta_0}u(t-1) - \phi(t-1)^T\hat{\theta}(t)$$

(b) $C(q^{-1})z(t) = b(t)$

where

$$z(t) = \eta(t) - \omega(t)$$
$$b(t) = -\phi(t-1)^T\tilde{\theta}(t)$$

(c) Subject to

$$B'(q^{-1}) + \frac{\lambda}{\beta_0}A(q^{-1}) \text{ is asymptotically stable}$$

Show that:

(i) $\displaystyle\lim_{N\to\infty}\frac{1}{N}\sum_{t=1}^{N}y(t)^2 \le \lim_{N\to\infty}\frac{K_1}{N}\sum_{t=1}^{N}[e(t)-\omega(t)]^2 + K_2$

(ii) $\displaystyle\lim_{N\to\infty}\frac{1}{N}\sum_{t=1}^{N}u(t)^2 \le \lim_{N\to\infty}\frac{K_3}{N}\sum_{t=1}^{N}[e(t)-\omega(t)]^2 + K_4$

(iii) $\displaystyle\lim_{N\to\infty}\frac{1}{N}\sum_{t=1}^{N}\bar{\varepsilon}(t)^2 \le \lim_{N\to\infty}\frac{K_5}{N}\sum_{t=1}^{N}[e(t)-\omega(t)]^2 + K_6$

[*Hint:* Show that

$$\left[\frac{\bar{\lambda}}{\beta_0}A(q^{-1}) + B'(q^{-1})\right]y(t) = B'(q^{-1})e(t) + B'(q^{-1})y^*(t) + \frac{\lambda}{\beta_0}C(q^{-1})\omega(t)$$

$$\left[\frac{\bar{\lambda}}{\beta_0}A(q^{-1}) + B'(q^{-1})\right]u(t-1) = A(q^{-1})e(t) + A(q^{-1})y^*(t) - C(q^{-1})\omega(t)$$

$$\bar{\varepsilon}(t) = y(t) - y^*(t) + \frac{\bar{\lambda}}{\beta_0}u(t-1) - \eta(t)$$

$$= y(t) - y^*(t) + \frac{\bar{\lambda}}{\beta_0}u(t-1) - \frac{e(t)}{1 + \phi(t-1)^T P(t-2)\phi(t-1)}$$

$$\bar{\varepsilon}(t)^2 \le 4y(t)^2 + 4y^*(t)^2 + 4\frac{\bar{\lambda}}{\beta_0^2}u(t-1)^2 + 4e(t)^2]$$

11.10. Consider the setup in Exercises 11.8 and 11.9 and establish the following convergence result.

Theorem. Subject to Assumptions 11.3.A and 11.3.B, parts 1, 2, and 3 and the additional assumption given in Exercise 11.9, part (c), then provided that $[1/C(z) - \frac{1}{2}]$ is very strictly passive, the adaptive controller of Exercise 11.9 ensures that with probability 1:

(i) $\displaystyle\limsup_{N\to\infty}\frac{1}{N}\sum_{t=1}^{N}y(t)^2 < \infty$ a.s.

(ii) $\displaystyle\limsup_{N\to\infty}\frac{1}{N}\sum_{t=1}^{N}u(t)^2 < \infty$ a.s.

(iii) $\displaystyle\lim_{N\to\infty}\frac{1}{N}\sum_{t=1}^{N}[u(t)-u^*(t)]^2 = 0$ a.s.

where $u^*(t)$ is the optimal one-step-ahead input satisfying

$$-\frac{\lambda}{\beta_0}u^*(t) = y^0(t+1|t) - y^*(t+1)$$

and

$$\frac{\lambda}{\beta_0} = \frac{\bar{\lambda}}{\bar{\beta}_0}$$

where $\bar{\lambda}$ and $\bar{\beta}_0$ are used in the algorithm.
(*Hint:* Follow the proof of Theorem 11.3.1.)

11.11. (a) In adaptive pole placement, one must, in theory, be careful to avoid unstable pole–zero cancellations in the estimated model. Does a problem of a similar nature occur in adaptive optimal control? (Why is stabilizability important in optimal control?)

(b) Comment on the use of different canonical forms (e.g., observer or controller forms) in adaptive optimal control (refer to Fig. 11.4.1). Discuss the relationship between the choice of canonical form and the control law and filter synthesis.

11.12. Consider a general stochastic linear model of the form

$$A(q^{-1})y(t) = q^{-d}B'(q^{-1})u(t) + C(q^{-1})\omega(t)$$

and the following multiple recursion algorithm:

$$\hat{\theta}(t) = \hat{\theta}(t - d) + \frac{\bar{a}}{\bar{r}(t - d)}\phi(t - d)[y(t) - \phi(t - d)^T\hat{\theta}(t - d)]$$

$$\bar{r}(t - d) = \bar{r}(t - 2d) + \phi(t - d)^T\phi(t - d)$$

$$\phi(t)^T\hat{\theta}(t) = y^*(t + d)$$

where

$$\phi(t - d)^T = [y(t - d), \ldots, y(t - d - n + 1), u(t - d), \ldots,$$
$$u(t - m - d + 1), -y^*(t - 1), \ldots, -y^*(t - l + 1)]$$

(a) Show that for $C(q^{-1}) = 1$, the algorithm above is globally convergent with $0 < \bar{a} < 2$ and $B'(q^{-1})$ stable.

(b) What condition on $C(q^{-1})$ would ensure convergence for the case $C(q^{-1}) \neq 1$? Why is this made difficult by the multiple recursion form of the algorithm? Discuss and suggest remedies.

11.13. Consider the following line of reasoning. Assume it is true that

$$E\{X \mid A\} + B \leq X + C$$

It then follows that

$$E\{X\} + E\{B\} \leq E\{X\} + E\{C\}$$

Under what circumstances is it not necessarily true that

$$E\{B\} \leq E\{C\}?$$

11.14. Consider the ARMAX model with both deterministic and stochastic disturbances:

$$\tilde{A}(q^{-1})D(q^{-1})y(t) = \tilde{B}(q^{-1})D(q^{-1})u(t) + \tilde{C}(q^{-1})D(q^{-1})\omega(t)$$

Let the set point be described by

$$S(q^{-1})y^*(t) = 0$$

(a) Show that a robust (unity feedback) control law can be obtained from:

$$L(q^{-1})S(q^{-1})D(q^{-1})\tilde{A}(q^{-1}) + P(q^{-1})\tilde{B}(q^{-1}) = C_s(q^{-1})A^*(q^{-1})$$

$$L(q^{-1})S(q^{-1})D(q^{-1})u(t) = P(q^{-1})[y^*(t) - y(t)]$$

(b) Show that the tracking error, $y(t) - y^*(t)$, converges to:

$$y(t) - y^*(t) = \frac{L(q^{-1})S(q^{-1})}{A^*(q^{-1})}\omega(t) + \frac{L(q^{-1})S(q^{-1})}{A^*(q^{-1})}\left[\frac{C_e(q^{-1})}{C_s(q^{-1})}\right]\omega(t)$$

where $\quad \tilde{C}(q^{-1})D(q^{-1}) = C_s(q^{-1}) + C_e(q^{-1}); \ C_s$ stable, C_e small.

(c) Indicate how this controller can be made adaptive.

APPENDIX A

A Brief Review of Some

Results from Systems Theory

A.1 STATE-SPACE MODELS

In this section we briefly review some basic notions regarding state-space models. These models allow a complete description of both the internal and external characteristics of a system. The basic idea of *state* is that, given the state of a system at time t, the future response of the system is determined by the input only.

In the linear finite-dimensional continuous-time case, the state-space model has the following form:

$$\frac{dx'(\tau)}{dt} = Fx'(\tau) + Gu'(\tau); \qquad x'(\tau_0) = x'_0 \qquad (A.1.1)$$

$$y'(\tau) = Hx'(\tau) \qquad (A.1.2)$$

where $x'(\tau)$ denotes an n-dimensional state vector
$u'(\tau)$ denotes an r-dimensional input vector
$y'(\tau)$ denotes an m-dimensional output vector

The dimension of the state vector, n, is called the *order* of the model. As is readily verified by substitution, the solution to (A.1.1) may be expressed as

$$x'(\tau_f) = e^{F(\tau_f - \tau_0)}x'_0 + \int_{\tau_0}^{\tau_f} e^{F(\tau_f - s)}Gu'(s)\,ds \qquad (A.1.3)$$

In the linear discrete-time case, the state-space model takes the following form:

$$x(t + 1) = Ax(t) + Bu(t); \qquad x(0) = x_0 \qquad (A.1.4)$$

$$y(t) = Cx(t) \qquad (A.1.5)$$

where $t \in \{0, 1, 2, \ldots\}$.

472

One way that a discrete-time model can arise is from a linear continuous-time system under discrete control as shown in Fig. A.1.1. To illustrate the relationship between the continuous and discrete models in this case, let the continuous-time system be described by (A.1.1) and (A.1.2) and assume that the input is generated by a zero-order hold as in Fig. A.1.1. Then

$$u'(\tau) = u(t) \qquad \text{for } t\Delta \leq \tau < (t+1)\Delta$$

where $u(t)$ denotes the output of the control algorithm at sampling instant t.

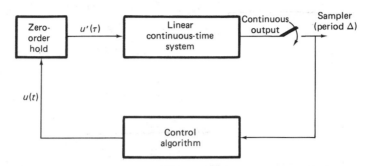

Figure A.1.1 Continuous system with discrete control law.

The output samples of the continuous-time system (A.1.1)–(A.1.2) then satisfy a discrete-time model of the form of (A.1.4)–(A.1.5) with $x(t) = x'(t\Delta)$, $y(t) = y'(t\Delta)$ and where the relationship between A, B, C and F, G, H is

from cont. to discrete

$$A = e^{F\Delta} = \sum_{k=0}^{\infty} \frac{F^k \Delta^k}{k!} \tag{A.1.6}$$

$$B = \sum_{k=0}^{\infty} \frac{F^k \Delta^{k+1} G}{(k+1)!} = F^{-1}(A - I)G \qquad \text{(when } F \text{ is nonsingular)} \tag{A.1.7}$$

$$C = H \tag{A.1.8}$$

The relationships above are readily derived from (A.1.3) by putting $x'_0 = x(t)$, $\tau_0 = t\Delta$, and $\tau_f = (t+1)\Delta$.

There are four important properties of the state-space model (A.1.4)–(A.1.5). These are the properties of controllability, reachability, observability, and reconstructibility.

Definition A.1.A. A particular state $\bar{x}$ of the system (A.1.4)–(A.1.5) is said to be *controllable* if there exists an N and a control function

$$u(t), \, t = 0, \, \ldots, \, N-1$$

which drives the system from the initial state $x(0) = \bar{x}$ to $x(N) = 0$. The system is (completely) controllable if every state is controllable.

Definition A.1.B. A state $\bar{x}$ of the system (A.1.4)–(A.1.5), is said to be *reachable* if there exists an N and a control function

$$u(t), \, t = 0, \, \ldots, \, N-1$$

which drives the system from $x(0) = 0$ to $x(N) = \bar{x}$. The system is (completely) reachable if every state is reachable.

Definition A.1.C.   A particular state $\bar{x}$ of the system (A.1.4)–(A.1.5) is said to be *unobservable* if for any $N > 0$ and $u(t) = 0$, $0 \leq t \leq N$ the initial state $x(0) = \bar{x}$ produces a zero output:

$$y(t) = 0; \qquad 0 \leq t \leq N$$

The system is said to be (completely) observable if no state (except 0) is unobservable.

Definition A.1.D.   An initial state $x(0) = \bar{x}$ of the system (A.1.4)–(A.1.5) is said to be *unreconstructible* if for any $N > 0$ and $u(t) = 0 \; -N \leq t \leq 0$,

$$y(t) = 0; \qquad -N \leq t \leq 0$$

The system is said to be (completely) reconstructible if no state (except 0) is unreconstructible.

We shall also find use for the special matrix functions defined below.

Definition A.1.E.   The *controllability matrix* for the system (A.1.4) is

$$\mathfrak{C} \triangleq [B \quad AB \quad \cdots \quad A^{n-1}B]$$

where $n =$ state dimension.

Definition A.1.F.   The *observability matrix* for the system (A.1.4)–(A.1.5) is

$$\Theta = \begin{bmatrix} C \\ CA \\ \cdot \\ \cdot \\ \cdot \\ CA^{n-1} \end{bmatrix}$$

Definition A.1.G.   The system (A.1.4), (A.1.5) is said to be *stabilizable* if all uncontrollable modes have corresponding eigenvalues strictly inside the unit circle.

Definition A.1.H.   The system (A.1.4), (A.1.5) is said to be *detectable* if all unobservable modes have corresponding eigenvalues strictly inside the unit circle.

We then have the following preliminary result:

Lemma A.1.1.   The system (A.1.4)–(A.1.5) of order n is

(i) Completely controllable:
  if rank $\mathfrak{C} = n$ (the condition is necessary if A is nonsingular)
(ii) Completely reachable
  if and only if rank $\mathfrak{C} = n$
(iii) Completely observable
  if and only if rank $\Theta = n$

(iv) Completely reconstructible
   if rank $\Theta = n$ (the condition is
   necessary if A is nonsingular)

Proof. See Exercise 2.4.

▼▼▼

We now have the following result, which shows how the state-space model can be transformed to isolate the uncontrollable part (if any).

Lemma A.1.2. Consider a state-space model (A.1.4)–(A.1.5) of order n. If the controllability matrix, $\mathcal{C}$, has rank $k(k < n)$, there exists a linear transformation, $\bar{x} = Qx$, which transforms the state-space model into the form

$$\begin{bmatrix} \bar{x}^1(t+1) \\ \bar{x}^2(t+1) \end{bmatrix} = \begin{bmatrix} \bar{A}_{11} & \bar{A}_{12} \\ 0 & \bar{A}_{22} \end{bmatrix} \begin{bmatrix} \bar{x}^1(t) \\ \bar{x}^2(t) \end{bmatrix} + \begin{bmatrix} \bar{B}_{11} \\ 0 \end{bmatrix} u(t) \tag{A.1.9}$$

$$y(t) = [\bar{C}_{11} \quad \bar{C}_{12}] \begin{bmatrix} \bar{x}^1(t) \\ \bar{x}^2(t) \end{bmatrix} \tag{A.1.10}$$

where $\bar{x}^1(t) \in R^k$ and $\bar{x}^2(t) \in R^{n-k}$. In addition, the k-dimensional subsystem formed as follows:

$$\begin{cases} \bar{x}^c(t+1) = \bar{A}_{11}\bar{x}^c(t) + \bar{B}_{11}u(t) & \text{(A.1.11)} \\ \qquad y(t) = \bar{C}_{11}\bar{x}^c(t) & \text{(A.1.12)} \end{cases}$$

is:

(i) Completely reachable
(ii) Zero-state equivalent (has identical input–output properties for zero initial conditions) to the original state-space model

Proof. See Exercise 2.16.

Similarly, for the case of the state-space model with unobservable parts, we have the following result:

Lemma A.1.3. Consider the state-space model (A.1.4)–(A.1.5) of order n. If the observability matrix, Θ, has rank k ($k < n$), there exists a linear transformation, $\bar{x} = Qx$, which transforms the state-space model into the form

$$\begin{bmatrix} \bar{x}^1(t+1) \\ \bar{x}^2(t+1) \end{bmatrix} = \begin{bmatrix} \bar{A}_{11} & \bar{A}_{12} \\ 0 & \bar{A}_{22} \end{bmatrix} \begin{bmatrix} \bar{x}^1(t) \\ \bar{x}^2(t) \end{bmatrix} + \begin{bmatrix} \bar{B}_{11} \\ \bar{B}_{21} \end{bmatrix} u(t) \tag{A.1.13}$$

$$y(t) = [0 \quad \bar{C}_{12}] \begin{bmatrix} \bar{x}^1(t) \\ \bar{x}^2(t) \end{bmatrix} \tag{A.1.14}$$

where $\bar{x}^1(t) \in R^{n-k}$ and $\bar{x}^2(t) \in R^k$.

In addition, the k-dimensional subsystem formed as follows:

$$\begin{cases} \bar{x}^0(t + 1) = \bar{A}_{22}\bar{x}^0(t) + \bar{B}_{21}u(t) & \text{(A.1.15)} \\ \quad y(t) = \bar{C}_{12}\bar{x}^0(t) & \text{(A.1.16)} \end{cases}$$

is:

(i) Completely observable
(ii) Input–output equivalent (has identical input–output properties for all initial states) as the original system (given, of course, that the initial states are transformed consistently)

Proof. See Exercise 2.17.

▾▾▾

✶ Combining the results above for the decomposition of uncontrollable and unobservable systems, we have the following important theorem:

✶⟹ **Theorem A.1.1 (Canonical Structure Theorem).** We can always find an invertible state transformation, $x = T\bar{x}$, such that the system (A.1.4)–(A.1.5), when expressed in the new coordinate basis, becomes

$$\bar{x}(t + 1) = \bar{A}\bar{x}(t) + \bar{B}u(t)$$
$$y(t) = \bar{C}\bar{x}(t)$$
$$\begin{cases} \bar{A} = T^{-1}AT \\ \bar{B} = T^{-1}B \\ \bar{C} = CT \end{cases}$$

such that

(i)
$$\bar{A} = \begin{bmatrix} \bar{A}_{11} & 0 & \bar{A}_{13} & 0 \\ \bar{A}_{21} & \bar{A}_{22} & \bar{A}_{23} & \bar{A}_{24} \\ 0 & 0 & \bar{A}_{33} & 0 \\ 0 & 0 & \bar{A}_{43} & \bar{A}_{44} \end{bmatrix}; \quad \bar{B} = \begin{bmatrix} \bar{B}_1 \\ \bar{B}_2 \\ 0 \\ 0 \end{bmatrix};$$

$$\bar{C} = [\bar{C}_1 \quad 0 \quad \bar{C}_3 \quad 0]$$

(ii) The subsystem $R_1 = [\bar{A}_{11} \quad \bar{B}_1 \quad \bar{C}_1]$ is completely observable and completely reachable.

(iii) The subsystem

$$R_{12} = \left[\begin{bmatrix} \bar{A}_{11} & 0 \\ \bar{A}_{21} & \bar{A}_{22} \end{bmatrix}, \begin{bmatrix} \bar{B}_1 \\ \bar{B}_2 \end{bmatrix}, [\bar{C}_1 \quad 0] \right]$$

is completely reachable.

(iv) The subsystem

$$R_{24} = \left[\begin{bmatrix} \bar{A}_{22} & \bar{A}_{24} \\ 0 & \bar{A}_{44} \end{bmatrix}, \begin{bmatrix} \bar{B}_2 \\ 0 \end{bmatrix}, [0 \quad 0] \right]$$

is completely unobservable.

(v) The subsystem $R_4 = [\bar{A}_{44} \quad 0 \quad 0]$ is completely unreachable and unobservable.

In addition:

(vi) The subsystem

$$R_{13} = \left[\begin{bmatrix} \bar{A}_{11} & \bar{A}_{13} \\ 0 & \bar{A}_{33} \end{bmatrix}, \begin{bmatrix} \bar{B}_1 \\ 0 \end{bmatrix}, [\bar{C}_1 \quad \bar{C}_3] \right]$$

is input–output equivalent (has identical input–output properties for all initial states) as the original systems (provided that the initial states are transformed consistently).

(vii) The subsystem R_1 is zero-state equivalent (has identical input–output properties for zero initial states).

Proof. Immediate from Lemmas A.1.2 and A.1.3.

▼▼▼

A.2 NOTES ON Z-TRANSFORMS

The *two sided z-transform* of a sequence $u(i)$ is defined as follows:

$$\bar{Z}\{u(i)\} \triangleq \bar{U}(z) = \sum_{i=-\infty}^{\infty} u(i)z^{-i} \tag{A.2.1}$$

If $u(i) = 0$ for $i < 0$, then (A.2.1) reduces to the *one-sided z-transform*:

$$Z\{u(i)\} \triangleq U(z) = \sum_{i=0}^{\infty} u(i)z^{-i} \tag{A.2.2}$$

It is important when discussing z-transforms that the region of convergence be specified. We have the following important result from analysis:

Lemma A.2.1 (Power Series). Every power series

$$\sum_{i=0}^{\infty} a_i z^{-i} \tag{A.2.3}$$

has a radius of convergence R^- with $0 \le R^- \le \infty$. The series is absolutely convergent for $|z| > R^-$ and is divergent for $|z| < R^-$.

Proof. See Rudin (1960, p. 60).

It follows from Lemma A.2.1 that every two-sided z-transform has an annulus of convergence $R^- < |z| < R^+$. If $R^- > R^+$ the series diverges everywhere. The relationship between the region of convergence of a z-transform and the corresponding power series is further explored via a simple example in Exercise 2.5.

Elementary properties of the z-transform are described in

Lemma A.2.2. For the two-sided z-transform we have

(1) Linearity
$$\bar{Z}\{\alpha f_1(t) + \beta f_2(t)\} = \alpha \bar{Z}(f_1(t)) + \beta \bar{Z}\{f_2(t)\} \tag{A.2.4}$$

(2) Convolution of time series

$$\bar{Z}\left\{\sum_{j=-\infty}^{\infty} f_1(j)f_2(k-j)\right\} = \bar{Z}\{f_1(j)\}\bar{Z}\{f_2(j)\} \qquad \text{(A.2.5)}$$

(3) Time shift

$$\bar{Z}\{f(k+n)\} = z^n\bar{Z}\{f(k)\} \qquad \text{(A.2.6)}$$

(4) Scaling

$$\bar{Z}\{r^{-k}f(k)\} = \bar{F}(rz), \qquad \text{where } \bar{F}(z) = \bar{Z}\{f(k)\} \qquad \text{(A.2.7)}$$

(5) Final value theorem: If $\bar{F}(z)$ converges for $|z| > 1$ and all poles of $(1-z)\bar{F}(z)$ are inside the unit circle, then

$$\lim_{k\to\infty} f(k) = \lim_{z\to 1} (z-1)\bar{F}(z) \qquad \text{(A.2.8)}$$

Proof. See Exercise 2.6 and Franklin and Powell (1980, p. 39).

The results above apply *mutatis mutandis* to the one-sided z-transform. However, for the one one-sided z-transform, we need to be careful with initial conditions. For example, if $x(i) = 0$ for $i < 0$, then as is readily checked, we have the following result for the one-sided z-transform of $\{x(i+1)\}$:

$$Z\{x(i+1)\} = Z\{q\{x(i)\}\} \qquad \text{(A.2.9)}$$
$$= zZ\{x(i)\} - zx(0) \qquad \text{(A.2.10)}$$

We see that the one-sided z-transform of a shifted sequence is essentially obtained by replacing q by z (but some extra terms arise due to the initial conditions).

Thus if we consider a simple first-order difference equation

$$(q+a)\{y(t)\} = \{u(t)\}; \qquad y(0) \text{ given} \qquad \text{(A.2.11)}$$

then taking one-sided z-transforms,

$$(z+a)Y(z) - zy(0) = U(z); \qquad |z| > R^- \qquad \text{(A.2.12)}$$

For reverse shifts we have

$$Z\{x(i-1)\} = z^{-1}Z\{x(i)\} \qquad \text{(A.2.13)}$$

provided that $x(i) = 0$; $i < 0$. Using the notation of Section 2.3 yields

$$Z\{q^{-1}\{x(i)\}\} = z^{-1}Z\{x(i)\} \qquad \text{(A.2.14)}$$

As an example, we note that $Z\{a^k; k \geq 0\} = 1/(1-az^{-1})$; $|z| > |a|$, thus $1/(z-a)$; $|z| > |a|$ is the one-sided z-transform of $\{y(t)\}$ where

$$y(t) = \begin{cases} a^{t-1}, & t \geq 1 \\ 0, & t < 1 \end{cases}$$

or

$$y(t) = z^{t-1}S(t-1)$$

where $S(j)$ is the unit step function satisfying

$$S(j) = \begin{cases} 1, & j \geq 0 \\ 0, & j < 0 \end{cases}$$

Application of the principles above to a simple example is given in Exercises 2.11 and 2.12.

A.3 PROPERTIES OF POLYNOMIAL MATRICES

For simplicity we shall describe the results only for either left or right operations. The corresponding results for the other case can be obtained by simply substituting right, row, column, premultiply, and so on, by left, column, row, postmultiply, and so on, respectively.

Definition A.3.A. *Elementary row operations* for polynomial matrices may involve:

1. Interchange of any two rows
2. Addition to any row of a polynomial multiple of any other row
3. Scaling any row by any nonzero real or complex number

Note that elementary row operations on a polynomial matrix $P(z)$ can be implemented by performing the same operation on an identity matrix I and then premultiplying $P(z)$ by the resulting matrix.

Definition A.3.B. A *unimodular matrix U(z)* is defined as any square matrix that can be obtained from the identity matrix by a finite number of elementary row and column operations. The determinant of a unimodular matrix is therefore a nonzero real or complex scalar, and conversely any polynomial matrix whose determinant is a nonzero real or complex scalar is a unimodular matrix.

Definition A.3.C. A *greatest common left divisor* (gcld) of two polynomial matrices $\{D_L(z), N_L(z)\}$ with the same number of rows is any matrix $R(z)$ with the following properties:

1. $R(z)$ is a *left divisor* of $D_L(z)$ and $N_L(z)$; that is, there exists polynomial matrices $\bar{D}_L(z), \bar{N}_L(z)$ such that

$$D_L(z) = R(z)\bar{D}_L(z) \qquad (A.3.1)$$

$$N_L(z) = R(z)\bar{N}_L(z) \qquad (A.3.2)$$

2. If $R_1(z)$ is any other left divisor of $D_L(z)$ and $N_L(z)$, then $R_1(z)$ is a left divisor of $R(z)$; that is, there exists a polynomial matrix $W(z)$ such that

$$R(z) = R_1(z)W(z) \qquad (A.3.3)$$

Definition A.3.D. Two matrices $\{D_L(z), N_L(z)\}$ with the same number of rows are said to be *relatively left prime* (or *left coprime*) if and only if their gcld is unimodular.

Lemma A.3.1. Any $m \times l$ polynomial matrix $P(z)$ of rank s can be reduced by elementary column operations to a lower left triangular form in which

1. If $l > s$, the last $l - s$ columns are identically zero.
2. In row j, $1 \leq j \leq s$, the diagonal element is monic and of higher degree than any (nonzero) element to the left of it.
3. In row j, $1 \leq j \leq s$, if the diagonal element is unity, all elements to the left of it are zero.

Proof. If the first row of $P(z)$ is not identically zero, we can choose the polynomial having least degree and, by permuting columns, make it the new $(1, 1)$ entry, $\tilde{P}_{11}(z)$. We then divide every other nonzero element $\tilde{P}_{1i}(z)$ to form $\tilde{g}_{1i}(z)$ and $\tilde{r}_{1i}(z)$ as follows: $\tilde{P}_{1i}(z) = \tilde{P}_{11}(z)\tilde{g}_{1i}(z) + \tilde{r}_{1i}(z)$ where either $\tilde{r}_{1i}(z) = 0$ or degree $r_{1i}(z) <$ degree $\tilde{P}_{11}(z)$. We then subtract from each nonzero ith column the first column multiplied by $\tilde{g}_{1i}(z)$. If not all the $r_{1i}(z)$ are zero, we choose the one of least degree and make it the $(1, 1)$ entry by permutating columns. By repeating the process, the degree of the $(1, 1)$ element is continuously reduced and since its degree is finite, the procedure terminates when the remaining elements in the first row are all zero. We then proceed to the second row of the altered matrix and apply the foregoing procedure in the same manner, beginning with the second column and row. In this way we bring to zero all the elements to the right of the $(2, 2)$ entry. If the $(2, 1)$ element has degree greater than that of the $(2, 2)$ element, the above division procedure can be used to reduce the $(2, 1)$ element to the remainder term resulting from the division of the $(2, 1)$ entry by the $(2, 2)$ entry. Continuing in this way, we can arrive at the desired result.

▼▼▼

Lemma A.3.2. Given $m \times m$ and $m \times r$ polynomial matrices $D_L(z)$ and $N_L(z)$, if we reduce $[D_L(z) \quad N_L(z)]$ to lower left triangular form, $[R(z) \quad 0]$, as in Lemma A.3.1, then $R(z)$ is a gcld of $D_L(z)$ and $N_L(z)$.

Proof. From Lemma A.3.1, there exists a unimodular matrix, $U(z)$, such that

$$[D_L(z) \quad N_L(z)]\begin{bmatrix} U_{11}(z) & U_{12}(z) \\ U_{21}(z) & U_{22}(z) \end{bmatrix} = [R(z) \quad 0] \tag{A.3.4}$$

Now let

$$\begin{bmatrix} U_{11}(z) & U_{12}(z) \\ U_{21}(z) & U_{22}(z) \end{bmatrix}^{-1} = \begin{bmatrix} V_{11}(z) & V_{12}(z) \\ V_{21}(z) & V_{22}(z) \end{bmatrix} \tag{A.3.5}$$

Then it follows that

$$D_L(z) = R(z)V_{11}(z); \qquad N_L(z) = R(z)V_{21}(z) \tag{A.3.6}$$

Thus $R(z)$ is a left divisor of $D_L(z)$ and $N_L(z)$. Also,

$$R(z) = D_L(z)U_{11}(z) + N_L(z)U_{21}(z) \tag{A.3.7}$$

so that if $R'(z)$ is another left divisor, then by definition

$$D_L(z) = R'(z)D'_L(z) \tag{A.3.8}$$

$$N_L(z) = R'(z)N'_L(z) \tag{A.3.9}$$

Then from (A.3.7),

$$R(z) = R'(z)[D'_L(z)U_{11}(z) + N'_L(z)U_{21}(z)] \tag{A.3.10}$$

so that $R'(z)$ is a left divisor of $R(z)$ and consequently $R(z)$ is a gcld.

▼▼▼

Definition A.3.E. Consider a polynomial matrix $P(z)$ and denote by r_i the highest degree in the ith row of $P(z)$. We say that $P(z)$ is *row reduced* if

$$\text{degree det } P(z) = \sum_{i=1}^{m} r_i \tag{A.3.11}$$

[Note that it is always true that degree det $P(z) \le \sum_{i=1}^{m} r_i$.]

We can write a general polynomial matrix, $P(z)$, in the form

$$P(z) = S(z)P_0 + L(z) \tag{A.3.12}$$

where

$$S(z) = \text{diag} \, [z^{r_1} \quad \cdots \quad z^{r_m}]$$

$$P_0 = \textit{leading row coefficient matrix}$$

$$L(z) = \text{remaining terms and is a polynomial matrix with}$$

$$\text{row degrees strictly less than those of } P(z)$$

Hence

$$\det P(z) = (\det P_0)z^{\sum_{i=1}^{m} r_i} + \text{terms of lower degree in } z \tag{A.3.13}$$

Thus we have

Lemma A.3.3. A nonsingular polynomial matrix is row-reduced if and only if its leading row coefficient matrix is nonsingular.

▼▼▼

Proof. Immediate from (A.3.13).

▼▼▼

We also have

Lemma A.3.4. Any nonsingular polynomial matrix, $P(z)$, can be transformed to row-reduced form by elementary row operations. Moreover, the resulting leading row coefficient matrix can be taken to be lower triangular with unity entries on the diagonal.

Proof. A constructive proof is possible whereby elementary row operations are successively used to reduce the individual row degrees until row-reducedness is achieved. See Wolovich (1974, p. 27), Antoulas (1981), and Beghelli and Gruidorzi (1976).

▼▼▼

A.4 RELATIVELY PRIME POLYNOMIALS

We recall from Definition A.3.D that two polynomial matrices $(D_L(z), N_L(z))$ with the same number of rows are relatively left prime if their gcld is unimodular.

An important characterization of relative primeness is described in the following result:

Lemma A.4.1 (The Bezout Identity). Two polynomial matrices $(D_L(z), N_L(z))$ having the same number of rows are left coprime if and only if there exist polynomial matrices $A(z)$, $B(z)$ such that

$$D_L(z)A(z) + N_L(z)B(z) = I \qquad \text{(A.4.1)}$$

Proof. Straightforward; see Exercise 2.8.

▼▼▼

Of course, in the case of scalar polynomials, left and right coprimeness are equivalent and correspond to the fact that the polynomials have no common zeros.

A further important characterization of coprimeness is presented below [we quote the result for scalar polynomials, but a similar result also applies to polynomial matrices; see Wolovich (1974)].

Theorem A.4.1 (Sylvester's Theorem). Two polynomials $(A(q^{-1}), B(q^{-1}))$ of order n are relatively prime if and only if their *eliminant matrix M_e* is nonsingular where M_e is defined to be the following $2n \times 2n$ matrix:

$$
M_e =
\begin{bmatrix}
a_0 & & & & b_0 & & & \\
a_1 & a_0 & & & & \cdot & & \cdot \\
\cdot & & \cdot & & & & \cdot & \\
\cdot & & & a_0 & b_{n-1} & & & b_0 \\
a_n & & & & b_n & & & \cdot \\
& a_n & & & & \cdot & & \cdot \\
& & \cdot & & & & \cdot & \\
& & & \cdot & & & & b_{n-1} \\
& & & a_n & & & & b_n
\end{bmatrix}
\begin{array}{l} \Big\} \, 2n \end{array}
\qquad \text{(A.4.2)}
$$

$$\longleftarrow n \longrightarrow \quad \longleftarrow n \longrightarrow$$

where

$$A(q^{-1}) = a_0 + a_1 q^{-1} + \cdots + a_n q^{-n}$$
$$B(q^{-1}) = b_0 + b_1 q^{-1} + \cdots + b_n q^{-n}$$

Proof. Only if: Suppose that there is a common root γ:

$$A(q^{-1}) = (q^{-1} - \gamma)[a'_0 + \cdots + a'_{n-1} q^{-n+1}]$$
$$B(q^{-1}) = (q^{-1} - \gamma)[b'_0 + \cdots + b'_{n-1} q^{-n+1}]$$

Eliminating $(q^{-1} - \gamma)$ gives

$$A(q^{-1})[b'_0 + \cdots + b'_{n-1} q^{-n+1}] - B(q^{-1})[a'_0 + \cdots + a'_{n-1} q^{-n+1}] = 0$$

Equating coefficients of q^{-i} on both sides gives

$$M_e \theta' = 0$$

where

$$\theta' = [b'_0, \ldots, b'_{n-1}, -a'_0, \ldots, -a'_{n-1}]^T$$

However, the equation above has a nontrivial solution for θ' if and only if $\det M_e = 0$.

If: By reversing the argument above.

▼▼▼

A.5 PROPERTIES OF GENERALIZED EIGENVALUES AND EIGENVECTORS

If λ_j is an eigenvalue of M of multiplicity p_j, the generalized eigenvectors z_{j+r}, $r = 0, 1, \ldots, p_j - 1$ with rank $1, 2, \ldots, p_j$ associated with the eigenvalue λ_j satisfy the set of nontrivial solutions

$$(M - \lambda_j I)z_j = 0$$

$$(M - \lambda_j I)z_{j+r} = z_{j+r-1}; \qquad r = 1, 2, \ldots, p_j - 1$$

A row vector $\omega_i \neq 0$ is called a left eigenvector of M associated with λ_i if

$$\omega_i(M - \lambda_i I) = 0$$

and $\omega_{i+s} \neq 0$ is called a generalized left eigenvector of rank $(s + 1)$ associated with an eigenvalue λ_i of M if

$$\omega_{i+s}(M - \lambda_i I) = \omega_{i+s-1}; \qquad s = 1, 2, \ldots, p_i - 1$$

Without loss of generality, the eigenvectors can be chosen such that

$$\omega_{i+s}z_{j+r} \neq 0, \qquad i = j \quad \text{and} \quad s + r = p_i - 1 = p_j - 1$$

$$= 0, \qquad \text{otherwise}$$

APPENDIX B

A Summary Of Some
Stability Results

B.1 DEFINITIONS

Here we will consider a discrete-time nonlinear system of the form

$$x(t + 1) = f(x(t), u(t), t) \qquad (B.1.1)$$
$$y(t) = h(x(t), t) \qquad (B.1.2)$$

where $x(t)$, $u(t)$, and $y(t)$ denote the state, input, and output, respectively. In the linear time-invariant case we shall express (B.1.1)–(B.1.2) in the form

$$x(t + 1) = Ax(t) + Bu(t) \qquad (B.1.3)$$
$$y(t) = Cx(t) \qquad (B.1.4)$$

Definition B.1.A. A state x_e is said to be an *equilibrium state* of the system (B.1.1)–(B.1.2) if

$$x_e = f(x_e, 0, t) \qquad \text{for all } t \geq t_0 \qquad (B.1.5)$$

Definition B.1.B. The equilibrium state x_e is said to be *stable* (in the sense of Lyapunov) if for arbitrary t_0 and $\epsilon > 0$ there exists a $\delta(\epsilon, t_0)$ such that $\| x(t_0) - x_e \| < \delta$ implies that $\| x(t) - x_e \| < \epsilon$ for all $t \geq t_0$, $(u(t) = 0)$.

Definition B.1.C. The equilibrium state x_e is said to be *uniformly stable* if it is stable and if $\delta(\epsilon, t_0)$ does not depend on t_0.

Definition B.1.D. The equilibrium state x_e is said to be *asymptotically stable* if it is stable and if for arbitrary t_0, there exists a $\delta(t_0)$ such that $\| x(t_0) - x_e \| < \delta$ implies that $\lim_{t \to \infty} \| x(t) - x_e \| = 0$, $(u(t) = 0)$.

Definition B.1.E.   The equilibrium state, x_e, is said to be *uniformly asymptotically stable* if it is asymptotically stable with δ independent of t_0.

Definition B.1.F.   The equilibrium state, x_e, is said to be *globally asymptotically stable* if $\delta(t_0)$ can be arbitrarily large.

Definition B.1.G.   The system (B.1.1)–(B.1.2) is said to be *weakly finite gain stable* if

$$\| y(t) \|_T \leq k \| u(t) \|_T + \beta(x(t_0)) \tag{B.1.6}$$

where $x(t_0) \in \Omega, k \in \mathbb{R}, \beta: \Omega \to \mathbb{R}$, and

$$\| y(t) \|_T = \sum_{t=t_0} y(t)^2 \tag{B.1.7}$$

Definition B.1.H.   The equilibrium state, x_e, is said to be *unstable* if it is neither stable nor asymptotically stable.

B.2 LYAPUNOV THEOREMS

Definition B.2.A. $V(x)$ is *positive definite* if $V(0) = 0$ and $V(x) > 0$ for all $x \neq 0$. Similarly, $V(x)$ is positive semidefinite if $V(0) = 0$, $V(x) \geq 0$, for all $x \neq 0$.

We consider the system (B.1.1)–(B.1.2) in the time-invariant autonomous case; that is, $f(x(t), u(t), t)$ is replaced by $f(x(t))$ with $u(t) \equiv 0$. We also take x_e to be 0 without loss of generality. We note that along the trajectories of (B.1.1)–(B.1.2) we can compute the change of V as

$$\begin{aligned} \Delta V(x(t)) &\triangleq V(x(t + 1)) - V(x(t)) \\ &= V(f(x(t)) - V(x(t)) \end{aligned} \tag{B.2.1}$$

for the system

$$x(t + 1) = f(x(t)) \tag{B.2.2}$$

We then have

Theorem B.2.1.   If there exists some closed bounded region containing the origin and a positive definite function $V(x)$ continuous in x satisfying

(1) $\Delta V \leq 0$ along the trajectories of (B.2.2), then $x_e = 0$ is stable.
(2) If in addition
    (i) $\Delta V < 0$ along the trajectories of (B.2.2) or
    (ii) ΔV is not identically zero along any trajectory of (B.2.2), then $x_e = 0$ is asymptotically stable.
(3) If, in addition, $\Omega = \mathbb{R}^n$ and $\lim (\| x \| \to \infty) V(x) = \infty$ [i.e., $V(x)$ is radially unbounded], then $x_e = 0$ is globally asymptotically stable.

Proof. See Hahn (1963), Willems (1970), and Desoer and Vidyasagar (1975).

▼▼▼

Theorem B.2.2. For the system (B.2.2), if there exists a scalar function $V(x)$ continuous in x such that

$$\Delta V(x) < 0$$

then the system is unstable in the finite region for which V is not positive semidefinite.

Proof. See Hahn (1963) and Willems (1970).

▼▼▼

B.3 LINEAR TIME-INVARIANT SYSTEMS

Here we specialize to the linear time-invariant case (of order n, with r inputs and m outputs.)

$$x(t + 1) = Ax(t) + Bu(t); \qquad x(0) = x_0 \tag{B.3.1}$$

$$y(t) = Cx(t) + Du(t) \tag{B.3.2}$$

We first consider the autonomous system

$$x(t + 1) = Ax(t) \tag{B.3.3}$$

We then have

Lemma B.3.1

(a) The equilibrium state $x_e = 0$ of the system (B.3.3) is globally asymptotically stable if and only if $|\lambda_i(A)| < 1, i = 1, \ldots, n$.

(b) The equilibrium state $x_e = 0$ of the system (B.3.3) is stable if and only if A has no eigenvalues outside the unit cricle and if the eigenvalues on the unit circle correspond to Jordon blocks of order 1.

Proof. Immediate from the solution of (B.3.3). See also Willems (1970, p. 49).

▼▼▼

We also have:

Lemma B.3.2 (The Discrete-Time Lemma of Lyapunov).
Given any $Q \triangleq GG^T \geq 0$ such that (A, G) is completely reachable, the equilibrium state $x_e = 0$ of the system (B.3.3) is globally asymptotically stable if and only if there exists a unique $P > 0$ such that

$$A^T PA - P = -Q \tag{B.3.4}$$

Proof. See Anderson and Moore (1979, p. 64).

▼▼▼

Lemma B.3.3. Consider the system (B.3.1)–(B.3.2). Provided that the following conditions are satisfied:

(i) $|\lambda_i(A)| \leq 1$; $i = 1, \ldots, n$

(ii) All controllable modes of (A, B) are inside the unit circle.

(iii) Any eigenvalues of A on the unit circle have a Jordan block of size 1.

Then

A Summary of Some Stability Results   App. B

(a) There exist constants K_1 and K_2 $(0 \leq K_1 < \infty, 0 \leq K_2 < \infty)$ which are independent of N such that

$$\sum_{t=1}^{N} \| y(t) \|^2 \leq K_1 \sum_{t=1}^{N} \| u(t) \|^2 + K_2 \qquad \text{for all } N \geq 0 \qquad \text{(B.3.5)}$$

that is, the system is finite gain stable.

(b) There exist constants $0 \leq m_3 < \infty, 0 \leq m_4 < \infty$ which are independent of t such that

$$|y_i(t)| \leq m_3 + m_4 \max_{1 \leq \tau \leq N} \| u(\tau) \| \qquad \text{for all } 1 \leq t \leq N;$$
$$i = 1, \ldots, m \qquad \text{(B.3.6)}$$

Proof. We first note from Lemma A.1.2 that the system can be split into controllable and uncontrollable parts. The solution to the uncontrollable part for arbitrary bounded initial state is bounded since the uncontrollable part has roots either inside the unit circle or on the unit circle with a Jordan block of size 1 [Lemma B.3.1(b)]. Thus by superposition, we need only consider the completely controllable part of the system (which has roots inside the unit circle by assumption).

(a) By successive substitution

$$y(t) = CA^t x_0 + Du(t) + \sum_{j=1}^{t} CA^{j-1} Bu(t-j) \qquad \text{(B.3.7)}$$

$$\| y(t) \|^2 \leq 3 \left\{ \| C \|^2 \| A^t \|^2 \| x_0 \|^2 + \| D \|^2 \| u(t) \|^2 \right.$$

$$\left. + \left[\sum_{j=1}^{t} \| C \| \, \| A^{j-1} \| \| B \| \, \| u(t-j) \| \right]^2 \right\}$$

$$\leq K_3 \lambda^{2t} + K_4 \| u(t) \|^2 + K_5 \left[\sum_{j=1}^{t} \lambda^{j-1} \| u(t-j) \| \right]^2$$

where K_3, K_4 and K_5 are finite, and we have used the fact that if A is asymptotically stable, then $\| A^j \| \leq K\lambda^j, 0 \leq \lambda < 1$ and $0 \leq K < \infty$ (Willems, 1970, p. 174). Thus

$$\| y(t) \|^2 \leq K_3 \lambda^{2t} + K_4 \| u(t) \|^2 + K_5 \left[\sum_{j=1}^{t} \lambda^{(j-1)/2} \lambda^{(j-1)/2} \| u(t-j) \| \right]^2$$

$$\leq K_3 \lambda^{2t} + K_4 \| u(t) \|^2 + K_5 \left[\sum_{j=1}^{t} \lambda^{j-1} \right] \left[\sum_{j=1}^{t} \lambda^{j-1} \| u(t-j) \|^2 \right]$$

So

$$\sum_{t=1}^{N} \| y(t) \|^2 \leq K_6 + K_4 \sum_{t=1}^{N} \| u(t) \|^2 + K_7 \sum_{t=1}^{N} \sum_{j=1}^{t} \lambda^{j-1} \| u(t-j) \|^2$$

Introducing $\tau = t - j$ yields

$$\sum_{t=1}^{N} \| y(t) \|^2 \leq K_6 + K_4 \sum_{t=1}^{N} \| u(t) \|^2 + K_7 \sum_{\tau=1}^{N-1} \sum_{t=\tau+1}^{N} \lambda^{t-\tau-1} \| u(\tau) \|^2$$

$$\leq K_6 + K_4 \sum_{t=1}^{N} \| u(t) \|^2 + K_8 \sum_{\tau=0}^{N-1} \| u(\tau) \|^2$$

$$\leq K_1 + K_2 \sum_{t=0}^{N} \| u(t) \|^2$$

(b) Consider again the solution (B.3.7); then

$$|y_i(t)| \leq \|y(t)\|$$

$$\leq \|C\| \|A^t\| \|x_0\| + \|D\| \|u(t)\|$$

$$+ \sum_{j=1}^{t} \|C\| \|A^{j-1}\| \|B\| \|u(t-j)\|$$

$$\leq K_9 \lambda^t + m\|D\| M_1 + \sum_{j=1}^{t} m\|C\| \lambda^{j-1}\|B\| M_1$$

where $M_1 = \max_{1 \leq \tau \leq t} \|u(\tau)\|$, and where we have again used the fact that if A is asymptotically stable, then $\|A^j\| < K\lambda^j$, where $0 \leq K < \infty$ and $0 < \lambda < 1$. Hence

$$|y_i(t)| \leq K_9 \lambda^t + m\|D\| M_1 + \frac{m\|C\| \|B\| M_1}{1 - \lambda}$$

$$= K_9 \lambda^t + m_4 M_1$$

▼▼▼

APPENDIX C

Passive Systems Theory

C.1 INTRODUCTION

A very basic fact in elementary circuit theory is that a linear time-invariant passive circuit (having positive resistance, inductance, and capacitance values) has a positive real impedance function (Desoer and Kuh, 1969). In this case, it is immediate that the circuit possesses an energy storage function which is a quadratic function of the state variables. It is the abstraction of these connections between input–output behavior and properties of energy functions for general linear and nonlinear systems which forms a basis for passive or dissipative systems theory (Willems, 1972; Anderson and Vongpanitlerd, 1973; Hill and Moylan, 1980). In this appendix, a brief summary of various passivity concepts, their interrelationships, and frequency-domain characterizations will be given. The results here are given for discrete-time systems; counterpart results are given for continuous-time systems in the references above.

C.2 PRELIMINARIES

There are two important representations of dynamical systems: the input–output description via an operator on a function space (Desoer and Vidyasagar, 1975) and the state-space description.

First, we introduce the input–output description. The set of time instants which are of interest is $Z_+ = \{0, 1, 2, \ldots\}$. Consider the Hilbert space $l_2^n(Z_+)$ of sequences $v: Z_+ \longrightarrow \mathbb{R}^n$ with inner-product $\langle \cdot, \cdot \rangle$ defined by

$$\langle u, v \rangle \triangleq \sum_{k=0}^{\infty} u^T(k)v(k)$$

Let P_T denote the operator that truncates the signal v at time T. We have

$$P_T v(t) \triangleq \begin{cases} v(t), & t < T \\ 0, & t \geq T \end{cases}$$

The basic signal space $l_{2_e}^n(Z_+)$ is given by an expansion of $l_2^n(Z_+)$ according to

$$l_{2_e}^n(Z_+) \triangleq \{v : Z_+ \longrightarrow \mathbb{R}^n \mid \forall T \in Z_+, P_T v \in l_2^n(Z_+)\}$$

It is convenient to use the notation $v_T \triangleq P_T v$ and $\langle u, v \rangle_T \triangleq \langle u_T, v_T \rangle = \langle u_T, v \rangle = \langle u, v_T \rangle$. A norm on $l_2^n(Z_+)$ is defined by $\|v\| \triangleq \langle v, v \rangle^{1/2}$. A *dynamical system input–output representation* is an operator $G : l_{2_e}^n(Z_+) \longrightarrow l_{2_e}^p(Z_+)$.

An alternative description of a dynamical system is provided by a state-space representation. The state space is taken to be $\mathbb{R}^n$. A time-invariant state-space description is defined by a state transition mapping $\psi : Z_+^2 \times \mathbb{R}^n \times l_{2_e}^m(Z_+) \longrightarrow \mathbb{R}^n$ and a readout mapping $r : \mathbb{R}^{n+m} \longrightarrow \mathbb{R}^p$ satisfying well-known axioms (Willems, 1972). These mappings describe the time evolution of the state and output according to $x(t) = \psi(t, t_0, x(t_0), u)$ and $y(t) = r(x(t), u(t))$. It is clear that an input–output representation involves a mapping that depends on the initial state. That is, we write $y = G(x_0)u$, where $x_0 = x(0)$. For any subset $\Omega \subseteq \mathbb{R}^n$, $G(\Omega)$ denotes the family $\{G(x_0) : x_0 \in \Omega\}$.

The passivity concepts that will be considered are all special cases of the property of dissipativeness. Introduce now the *energy supply function* $E : l_{2_e}^m(Z_+) \times l_{2_e}^p(Z_+) \times Z_+ \longrightarrow \mathbb{R}$. A useful form of this function is quadratic in the sense

$$E(u, y, T) = \langle y, Qy \rangle_T + 2\langle y, Su \rangle_T + \langle u, Ru \rangle_T \tag{C.2.1}$$

where Q, S, and R are appropriately dimensioned matrices with both Q and R symmetric. In general, E should be continuous and satisfy $E(0, 0, T) \equiv 0$.

Definition C.2.A (Hill and Moylan, 1980). Dynamical system $G(\Omega)$ with energy supply E is *dissipative* iff

$$E(u, Gu, T) \geq 0 \tag{C.2.2}$$

$$\text{for all } u \in l_{2_e}^m(Z_+) \quad \text{for all } T \in Z_+ \text{ and for all } x_0 \in \Omega$$

Typically, we take $\Omega = \{0\}$ or some obvious "rest state." For nonlinear systems, there may be many points in $\mathbb{R}^n$ at which inequality (C.2.2) is satisfied.

Useful special cases of dissipativeness obtained by choosing values for Q, S, and R in (C.2.1) are as follows where, ϵ and δ refer to arbitratily small positive scalars and I is the identity matrix:

1. $Q = R = 0$, $S = \frac{1}{2}I$ *passivity (P)*
   Inequality (C.2.2) becomes $\langle y, u \rangle_T \geq 0$.
2. $Q = 0$, $S = \frac{1}{2}I$, $R = -\epsilon I$ *input strict passivity (ISP)*
   Inequality (C.2.2) becomes $\langle y, u \rangle_T \geq \epsilon \|u\|_T^2$. Note that this property is equivalent to requiring $G - \epsilon I$ to be passive.
3. $Q = -\delta I$, $S = \frac{1}{2}I$, $R = 0$ *output strict passivity (OSP)*
   Inequality (C.2.2) becomes $\langle y, u \rangle_T \geq \delta \|y\|_T^2$.

4. $Q = -\delta I,\ S = \frac{1}{2}I,\ R = -\epsilon I$ *very strict passivity* (VSP)
   Inequality (C.2.2) becomes $\langle y, u \rangle_T \geq \epsilon \|u\|_T^2 + \delta \|y\|_T^2$.

Another important input–output property is finite-gain stability for which there exists a constant k such that

$$\|y\|_T \leq k \|u\|_T$$

for all $u \in l_{2e}^m(Z_+)$, all $T \in Z_+$ and for all $x_0 \in \Omega$. This is equivalent to the following dissipativeness property.

5. $Q = -I,\ S = 0,\ R = k^2 I$ *finite-gain stable* (FGS)
   Inequality (C.2.2) becomes $\|y\|_T^2 \leq k^2 \|u\|_T^2$

Clearly, a finite-gain stable system gives bounded signal outputs for small signal inputs where boundedness is with respect to the l_2 norm. A further form of strict passivity that could be considered is the combination of the ISP and stability properties (Caines, 1980). We denote this property by *ISPS*. For convenience, we use the acronyms ISP and so on to denote the class of systems having the corresponding passivity property.

The following result summarizes some useful facts about the input–output properties above.

Lemma C.2.1.   Some direct relationships between the passivity properties are:

(i) $G \in \text{ISP} \cup \text{OSP} \cup \text{VSP} \Rightarrow G \in P$
(ii) $G \in \text{VSP} \Leftrightarrow G \in \text{ISP} \cap \text{OSP}$
(iii) $G \in \text{OSP} \Rightarrow G \in \text{FGS}$
(iv) $G \in \text{ISPS} \Leftrightarrow G \in \text{VSP}$

If the system inverse G^{-1} exists, we have

(v) $G \in P\,(\text{VSP}) \Rightarrow G^{-1} \in P\,(\text{VSP})$
(vi) $G \in \text{ISP}\,(\text{OSP}) \Rightarrow G^{-1} \in \text{OSP}\,(\text{ISP})$
(vii) G and G^{-1} ISP $\Leftrightarrow G$ and G^{-1} OSP $\Leftrightarrow G$ VSP

Proof. Relationships (i), (ii), and (v) to (vii) are trivial consequences of the definitions. Relationship (iii) follows directly after application of the Schwarz inequality. (iv) $\Rightarrow$ inequality (C.2.2) becomes

$$\langle u, y \rangle_T \geq \epsilon \|u\|_T^2$$

$$\geq \frac{\epsilon}{k^2} \|y\|_T^2$$

This shows that an ISPS system is OSP and ISP. $\Leftarrow$ follows from (ii) and (iii).

▼▼▼

Note that the properties of passivity, VSP (and ISPS), are also carried by the system inverse if it exists.

C.3 FREQUENCY-DOMAIN PROPERTIES

An important result for linear systems is the frequency-domain characterization of dissipativeness. Consider linear time-invariant systems with state equations

$$x(k + 1) = Ax(k) + Bu(k); \qquad x(0) = x_0$$
$$y(k) = Cx(k) + Du(k) \tag{C.3.1}$$

and transfer function description

$$G(z) = D + C(zI - A)^{-1}B$$

The hermitian transpose of $G(z)$ is denoted $G^H(z)$. The values of x, u and y lie in euclidean spaces $\mathbb{R}^n$, $\mathbb{R}^m$, and $\mathbb{R}^p$. For any $x(0) \in \mathbb{R}^n$ and $u \in l_{2_e}^m(Z_+)$, we have $y \in l_{2_e}^p(Z_+)$. Assume that the system (C.3.1) is completely controllable.

We will consider dissipativeness with energy supply function of the form (C.2.1). Then

$$E(u, y, T) = \sum_{k=0}^{T-1} \omega(u(k), y(k))$$

where

$$\omega(u, y) = y^T Q y + 2y^T S u + u^T R u$$

Assume throughout the sequel that $\omega(\cdot, \cdot)$ has the property that for all $y \neq 0$ there exists u such that $\omega(u, y) < 0$. This is certainly true for all the properties defined in Section C.2.

Theorem C.3.1. System (C.3.1) is dissipative iff

$$M(z) = R + G^H(z)S + S^T G(z) + G^H(z)QG(z) \geq 0 \tag{C.3.2}$$

for all z in $|z| \geq 1$ [where $G(z)$ exists].

The proof of this result follows the same steps as for the corresponding continuous-time result. It follows either from results in linear optimal control (Hill and Moylan, 1975) or more directly via Parseval's theorem (Gannett and Chua, 1978). Controllability and the above-mentioned restriction on $\omega(\cdot, \cdot)$ are only used in the "if" part of the result.

A useful result for applying Theorem C.3.1 is as follows. It is similar to a result given in (Gannett and Chua, 1978) and helps identify constraints on system poles imposed by dissipativeness.

Lemma C.3.1. Let Λ be a set of isolated points that contains all the poles of $G(\cdot)$. Suppose that $M(z) \geq 0$ for all z in $|z| \geq 1$ except in Λ. If $Q \leq 0$, then $S^T G(z)$ is analytic for all z in $|z| > 1$.

A very important class of transfer functions corresponds to passive systems.

Definition C.3.A (Hitz and Anderson] 1969).   The square matrix $G(z)$ of real rational functions is (discrete) positive real iff:

1. $G(z)$ has elements analytic in $|z| > 1$.
2. $G^H(z) + G(z) > 0$ in $|z| > 1$

To establish that passivity corresponds to positive realness, we use Lemma C.3.1 and note that condition 2 in Definition C.3.A implies, by a limiting operation, that

$$G^T(e^{-j\omega}) + G(e^{j\omega}) \geq 0$$

for all real ω such that no element of $G(z)$ has a pole at $z = e^{j\omega}$.

In the obvious way, we will also refer to input strict positive real matrices (ISPR), and so on. To get a feel for the differences between the various passivity concepts, it is worthwhile to look at the constraints on $G(e^{j\omega})$ in each case for single-input single-output systems.

Lemma C.3.2.   Suppose that system (C.3.1) is (Q, S, R) dissipative; then:

1. If $Q < 0$, the graph of $G(e^{j\omega})$ lies inside the circle on the complex plane with center $S/|Q|$ and radius $(1/|Q|)\sqrt{S^2 + R|Q|}$.
2. If $Q = 0$, the graph of $G(e^{j\omega})$ lies to the right (if $S > 0$) or the left (if $S < 0$) of the vertical line Re $z = -R/2S$.

A more complete set of necessary and sufficient conditions in terms of $G(e^{j\omega})$ for dissipativeness with Q of arbitrary sign can be obtained. The constraints on $G(e^{j\omega})$ for the different versions of passivity are illustrated in Fig. C.3.1.

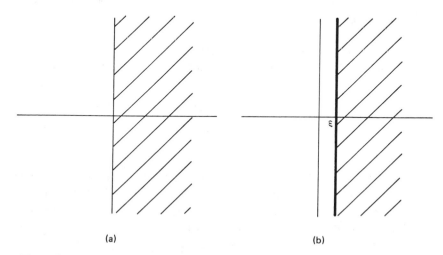

(a)   (b)

Figure C.3.1   Constraints on $G(e^{j\omega})$ for passive systems: (a) passivity; (b) ISP; (c) OSP; (d) VSP. Shaded part is allowable for $G(e^{j\omega})$.

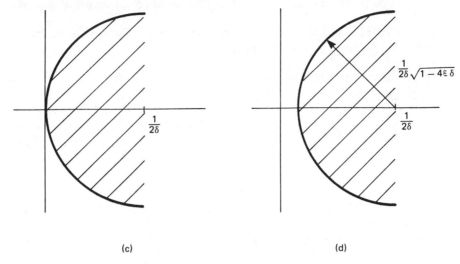

$$\frac{1}{2\delta}\sqrt{1-4\varepsilon\delta}$$

$\frac{1}{2\delta}$

(c)　　　　　　　　　　　　　　(d)

Figure C.3.1 (*Cont.*)

C.4 STORAGE FUNCTIONS

The central result in the passivity theory of linear continuous-time systems is the Kalman–Yakubovich–Popov lemma (or the positive real lemma, as the multivariable system version is often called). (Anderson and Vongpanitlerd, 1973). The discrete-time version has been proved in (Hitz and Anderson, 1969) via the continuous-time result and a bilinear transformation. In (Hill and Moylan, 1980). the positive real lemma is presented as a straightforward corollary of a general result for nonlinear systems. After stating this result as a lemma, a generalization of the discrete-time positive real lemma is given.

Lemma C.4.1 (Hill and Moylan, 1980). Assume that set $X \subseteq \mathbb{R}^n$ is reachable from $\Omega \subseteq X$. Then system $G(\Omega)$ with energy supply E is dissipative iff there exists a function $\phi \colon X \longrightarrow \mathbb{R}$, with $\phi(x) \geq 0$ for all $x \in X$, $\phi(x) = 0$ for all $x \in \Omega$, such that

$$\phi(x_0) + E(u, Gu, T) \geq \phi(x_1) \tag{C.4.1}$$

for all $x_0 \in X$, all $u \in l_{2e}^m(Z_+)$ and all $T \geq t_0$, where $y = G(x_0)u$ and $x_1 = \psi(T, t_0, x_0, u)$.

This result is valid for any system satisfying the general axioms of a dynamical system. Reachability reduces to complete controllability for linear systems of the form (C.3.1). The proof of Lemma C.4.1 follows easily after nominating a suitable candidate for $\phi(\cdot)$. One function that can be used is

$$\phi(x_0) = - \inf_{\substack{u \in l_{2e}^m(Z_+) \\ T \geq 0}} E(u, G(x_0)u, T) \tag{C.4.2}$$

This represents the largest amount of "energy" that can be extracted from the system.

In general, there is a convex set of functions ϕ satisfying (C.4.1). These are called *storage functions*. The function given by (C.4.2) is the minimal element of this set.

The main result of this section comes from applying Lemma C.4.1 to system (C.3.1).

Theorem C.4.1. Suppose that the system (C.3.1) is minimal. Consider an energy supply function of the form (C.2.1); then $M(z) \geq 0$ for all z in $|z| \geq 1$ [where $G(z)$ exists] iff there exists matrix $P > 0$ such that

$$\sum_{k=0}^{T-1} (y^T(k)Qy(k) + 2y^T(k)Su(k) + u^T(k)Ru(k)) + x^T(0)Px(0) \geq 0 \qquad (C.4.3)$$

$$\text{for all } x(0), \text{ for all } u \in l_{2e}^m(Z_+), \text{ and for all } T \geq 0$$

Proof. Theorem C.3.1 has already established the equivalence between positive semidefiniteness of $M(z)$ and dissipativeness.

The function ϕ becomes, in view of the assumed property for $\omega(\cdot, \cdot)$ (see Section C.3),

$$\phi(x_0) = - \inf_{u \in l_{2e}^m(Z_+)} \sum_{k=0}^{\infty} \omega(u(k), y(k))$$

where y is the system output when input u drives the system from the initial state x_0. It is well known (Clements and Anderson, 1978) that such linear optimal control problems have quadratic optimal cost functions. In fact, we can write

$$\phi(x_0) = x_0^T P x_0$$

where $P > 0$. Positive definiteness of P follows from observability of system (C.3.1).

The inequality (C.4.1) can be written

$$x^T(0)Px(0) + \sum_{k=0}^{T-1} (y^T(k)Qy(k) + 2y^T(k)Su(k) + u^T(k)Ru(k)) \geq x^T(T)Px(T) \qquad (C.4.4)$$

for trajectories emanating from $x(0)$. Inequality (C.4.3) is clearly satisfied since $P < 0$.

▼▼▼

Theorem C.4.1 provides a powerful connection between input–output and state-space properties of the system. The positive real lemma gives a set of algebraic equations that can be solved for the matrix P. The generalization to dissipative systems is as follows (the proof is straightforward):

Lemma C.4.2 (Generalized Discrete Positive Real Lemma). Suppose that the system (C.3.1) is minimal. Consider an energy supply function of the form (C.2.1). Then $M(z) \geq 0$ for all z in $|z| \geq 1$ [where $G(z)$ exists] iff there exist matrices P, L, and W with $P > 0$ such that

$$A^T P A - P = \hat{Q} - LL^T$$
$$A^T P B = \hat{S} - LW \qquad (C.4.5)$$
$$B^T P B = \hat{R} - W^T W$$

where

$$\hat{Q} = C^T Q C$$
$$\hat{S} = C^T Q D + C^T S$$
$$\hat{R} = R + D^T S + S^T D + D^T Q D$$

APPENDIX D

Probability Theory and

Stochastic Processes

This appendix gives a brief summary of some ideas from probability and the theory of stochastic processes. See also Burrill (1973), Lamperti (1977), Loeve (1963), Papoulis (1965), Chung (1968), Neveu (1975), Briepohl (1970), and Hall and Heyde (1980).

D.1 PROBABILITY THEORY

An *event* is the occurrence or nonoccurrence of a phenomenon. We denote the *sure event* by Ω and the *impossible event* by ϕ. We let Ω be a space of points, ω, and we denote by $\mathfrak{F}$ a nonempty class of sets in Ω ($\mathfrak{F}$ will be called set of *events*). Let P be a *probability measure* (i.e., a function mapping $\mathfrak{F} \rightarrow \mathbb{R}$). The triplet $(\Omega, \mathfrak{F}, P)$ is called a *probability space*.

We require that $\mathfrak{F}$ be a *sigma field* (or *sigma algebra*) of subsets of Ω, meaning that it is a collection of subsets which contains the empty set ϕ and is closed under complements and countable unions.

We also require that P satisfy:

1. $P(A) \geq 0$
2. $P(\cup A_i) = \sum_i P(A_i)$ provided that $A_i \in \mathfrak{F}$ and $A_i \cap A_j = \phi$ for $i \neq j$
3. $P(\Omega) = 1$

D.2 RANDOM VARIABLES

Let $(\Omega, \mathfrak{F}, P)$ be a probability space and let $X(\omega)$ be a real valued function on Ω. If X is *measurable* with respect to the field $\mathfrak{F}$ [i.e., $\{\omega: X(\omega) \in \mathfrak{S}$ is a member of $\mathfrak{F}$ for every open set, $\mathfrak{S}$, of the reals], then X is a *random variable*.

The *expected value* or *mean* of a random variable is defined as

$$E\{X\} = \int_\Omega X(\omega)\, dP$$

A random variable on $(\Omega, \mathfrak{F}, P)$ induces a probability measure, P_x, on $\mathbb{R}$:

$$P_x(\mathfrak{S}) = P\{\omega: X(\omega) \in \mathfrak{S}\}$$

for $\mathfrak{S} \in \mathfrak{B}$ where $\mathfrak{B}$ is the *Borel field* (i.e., the smallest σ-field formed from open sets of the real line).

We define a function F as follows:

$$F(x) = P\{\omega: X(\omega) \leq x\}$$

We call $F(x)$ the *distribution function* of P. $F(x)$ is clearly nondecreasing, continuous from the right, and satisfies $\lim_{x \to -\infty} F(x) = 0$, $\lim_{x \to \infty} F(x) = 1$.

If $F(x)$ is continuous, we can express it in terms of a *probability density function*, $p(y)$, as follows:

$$F(x) = \int_{-\infty}^{x} p(y)\, dy$$

If $X(\omega)$ is a *vector random variable* (of dimension n), we define the vector distribution function as

$$F(x) = P(\omega: X_i(\omega) \leq x_i, \quad i = 1, \ldots, n)$$

The corresponding probability density function (if it exists) is

$$F(x) = \int_{-\infty}^{x_1} \cdots \int_{-\infty}^{x_n} p(y)\, dy_1, \ldots, dy_n$$

D.3 INDEPENDENCE AND CONDITIONAL PROBABILITIES

The events $A_1, \ldots, A_n \in \mathfrak{F}$ are *independent* if

$$P(A_1 \cap A_2 \cdots \cap A_n) = P(A_1) \cdots P(A_n)$$

We also define the *conditional probability* of A_1 given A_2 as

$$P(A_1 | A_2) = \frac{P(A_1 \cap A_2)}{P(A_2)} \qquad \text{provided that } P(A_2) > 0$$

Note that $P(A_1 | A_2) = P(A_1)$ if and only if A_1 and A_2 are independent. $P(\cdot | A_2)$ is itself a probability measure. Thus if X is a random variable defined on $(\Omega, \mathfrak{C}, P)$, we can define the *conditional mean* of X as

$$E\{X | A\} = \int_\Omega X(\omega)P(d\omega | A)$$

More generally, if $\mathfrak{F}_1$ is a *σ-additive subalgebra* of $\mathfrak{F}$ (i.e., $\mathfrak{F}_1$ contains ϕ and is closed under complements and countable unions), the *conditional expectation* of X *with respect to* $\mathfrak{F}_1$ is written $E\{X|\mathfrak{F}_1\}$ and is a function such that

1. $E\{X|\mathfrak{F}_1\}$ is $\mathfrak{F}_1$ measurable.

2. $\int_A E\{X|\mathfrak{F}_1\}\, dP = \int_A X\, dP$ for all $\mathfrak{a} \in \mathfrak{F}_1$

If X is the characteristic function of a set B [i.e., $X(\omega) = 1$, $\omega \in B$; $X(\omega) = 0$, $\omega \notin B$], then $E\{X|\mathfrak{F}_1\}$ is called the *conditional probability of B*. If $\mathfrak{F}_1$ is the sigma algebra generated by a set of random variable $\{Y_1, \ldots, Y_n\}$, we write the conditional expectation as $E\{X|Y_1 \cdots Y_n\}$.

Properties of conditional expectation are

1. If $A = E\{X|\mathfrak{F}_1\}$, $B = E\{X|\mathfrak{F}_1\}$, then $A = B$ a.s. (where the symbols a.s. mean *almost surely*, i.e., save on a set having probability measure zero).

2. If X is measurable with respect to $\mathfrak{F}_1$, then
$$E\{X|\mathfrak{F}_1\} = X \quad \text{a.s.}$$

3. $E\{XY|\mathfrak{F}_1\} = E\{X\}E\{Y|\mathfrak{F}_1\}$ a.s. if X is independent of every set in $\mathfrak{F}_1$.

4. If $\mathfrak{F}_{n-1}$, $\mathfrak{F}_n$ are sub-sigma algebras of $\mathfrak{F}$ with $\mathfrak{F}_{n-1} \subset \mathfrak{F}_n$, then
$$E\{E\{X|\mathfrak{F}_{n-1}\}|\mathfrak{F}_n\} = E\{X|\mathfrak{F}_{n-1}\} \quad \text{a.s.}$$
$$E\{E\{X|\mathfrak{F}_n\}|\mathfrak{F}_{n-1}\} = E\{X|\mathfrak{F}_{n-1}\} \quad \text{a.s.}$$

We call the results above the *smoothing properties* of conditional expectations. In particular,
$$E\{E\{X|Y_1, \ldots, Y_{n-1}\}|Y_1 \cdots Y_n\} = E\{X|Y_1 \cdots Y_{n-1}\} \quad \text{a.s.}$$
$$E\{E\{X|Y_1, \ldots, Y_n\}|Y_1 \cdots Y_{n-1}\} = E\{X|Y_1 \cdots Y_{n-1}\} \quad \text{a.s.}$$

If $\mathfrak{F}_{n-1} \subset \mathfrak{F}_n \subset \mathfrak{F}$ for $n = 1, 2, \ldots$, we say that $\{\mathfrak{F}_n\}$ is an *increasing sequence of sub-sigma algebras*.

D.4 STOCHASTIC PROCESSES

A *stochastic process* is defined to be a collection of random variables defined on a common probability space $(\Omega, \mathfrak{F}, P)$ and indexed by the integers (for a discrete stochastic process). We use the notation $X(t, \omega)$ [sometimes we simplify this to $X(t)$] for a stochastic process. For fixed t, we obtain a random variable measurable with respect to $\mathfrak{F}$. If we fix ω we obtain a function mapping the integers into the reals. This function is called a *sample path*.

We define the *mean*, $\mu(t_1)$, and *covariance*, $C(t_1, t_2)$, of the process as follows:
$$\mu(t_1) = E\{X(t_1, \omega)\}$$
$$C(t_1, t_2) = E\{[X(t_1, \omega) - \mu(t_1)][X(t_2, \omega) - \mu(t_2)]^T\}$$

If $\mu(t)$ does not depend on t and $C(t_1, t_2)$ is a function only of $|t_1 - t_2|$, we say that the process is *wide-sense stationary*. If the joint probability distribution of $X(t_1, \omega)$ and $X(t_2, \omega)$ depends only on $|t_1 - t_2|$, we say that the process is *stationary*. In the wide-sense-stationary case we write $C(\tau)$ in place of $C(t_1, t_2)$, where $\tau = t_1 - t_2$. The Fourier transform of $C(\tau)$ is known as the *spectral density function* (where we assume that the transform is continuous). If $C(\tau) = \delta(\tau)$ (a dirac delta), the spectral density function is constant. We then say the process is *white* by analogy with white light, which also has a constant spectrum.

Let $(\Omega, \mathfrak{F}, P)$ be a probability space and suppose that for each t there is a sub-sigma algebra $\mathfrak{F}_t$ of $\mathfrak{F}$ such that $\mathfrak{F}_t \subset \mathfrak{F}_s$ if $t \leq s$. Let $X(t)$ be a stochastic process on $(\Omega, \mathfrak{F}, P)$; then $X(t)$ is said to be *adapted* to the sequence of increasing sigma algebras $\{\mathfrak{F}_t\}$ if $X(t)$ is $\mathfrak{F}_t$ measurable for every t. Of course, every stochastic process is adapted to its own past.

Let $(\Omega, \mathfrak{F}, P)$ be a probability space, $\{\mathfrak{F}_t\}$ an increasing sequence of sub-sigma algebras, and $X(t)$ a sequence of real random variables adapted to $\{\mathfrak{F}_t\}$. Then $\{X(t), \mathfrak{F}_t\}$ is a *martingale* provided that:

1. $E\{|X(t)|\} < \infty$ a.s.
2. $E\{X(t+1)|\mathfrak{F}_t\} = X(t)$ a.s. for all t.

If $E\{X(t+1)|\mathfrak{F}_t\} \geq X(t)$ a.s. for all t, we say $\{X(t), \mathfrak{F}_t\}$ is a *submartingale*. Alternatively, if $E\{X(t+1)|\mathfrak{F}_t\} \leq X(t)$ a.s., we say that $\{X(t), \mathfrak{F}_t\}$ is a *supermartingale*.

D.5 CONVERGENCE

If $X(t, \omega)$ is a sequence of random variables (i.e., stochastic process) defined on a probability space $(\Omega, \mathfrak{F}, P)$, we say that:

1. $X(t, \omega)$ converges almost surely if
$$P\{\omega: \lim_{n \to \infty} X(t, \omega) \text{ exists}\} = 1$$
[Equivalently, we say that $X(t, \omega)$ converges along almost all sample paths.] We write $\lim_{n \to \infty} X(t, \omega) = X(\omega)$ a.s.
2. $X(t, \omega)$ converges *in probability to* $X(\omega)$ if
$$\lim_{n \to \infty} P(\{\omega: |X(t, \omega) - X(\omega)| > \epsilon\}) = 0$$
for each $\epsilon > 0$.
3. $X(t, \omega)$ *converges in mean-p to* $X(\omega)$ if
$$\lim_{n \to \infty} E\{|X(t, \omega) - X(\omega)|^p\} = 0 \qquad \text{for some } p > 0$$

Note:

1. Almost sure convergence implies convergence in probability.
2. Convergence in mean-p for any $p > 0$ implies convergence in probability.

In general, the converse of these statements is false, and nothing definite can be said about the relationship between almost sure convergence and mean-p convergence.

We are also sometimes interested in the convergence of the distribution of a sequence of random variables. We say that a sequence of probability measures P_n converges (weakly) to a measure P if

$$\lim_{n \to \infty} \int_R f(x) \, dP_n = \int_R f(x) \, dP$$

for every bounded, continuous, real-valued function $f(x)$.

A key convergence theorem which is used extensively in Part II is the following:

Theorem D.5.1 (Doob, 1953).   Let $\{X(t), \mathcal{F}_t\}$ be an L^1 bounded *submartingale*; then there exists a random variable X such that $X(t)$ converges to X almost surely and $E|X| \leq \lim \inf E|X(t)| < \infty$. If the submartingale is uniformly integrable, then $X(t)$ converges to X in L^1, and if $\{X(t), \mathcal{F}_t\}$ is an L^2 bounded martingale, then $X(t)$ converges to X in L^2.

Related results are:

Lemma D.5.1 (Neveu, 1975).   Let $\{X(t)\}$ be a zero conditional mean sequence of random variables adapted to $\{\mathcal{F}_t\}$. If

$$\sum_{t=0}^{\infty} \frac{1}{t^2} E\{X(t)^2 \mid \mathcal{F}_{t-1}\} < \infty \quad \text{a.s.}$$

then

$$\lim_{N \to \infty} \frac{1}{N} \sum_{t=1}^{N} X(t) = 0 \quad \text{a.s.}$$

Lemma D.5.2.   Let $\{X(t), \mathcal{F}_t\}$ be a Martingale difference sequence, that is, such that $E\{X(t) \mid \mathcal{F}_{t-1}\} = 0$. If

$$E\{X(t)^2 \mid \mathcal{F}_{t-1}\} = \sigma^2$$

and

$$E\{X(t)^4 \mid \mathcal{F}_{t-1}\} < k < \infty$$

then

$$\lim_{N \to \infty} \frac{1}{N} \sum_{t=1}^{N} x(t)^2 = \sigma^2 \quad \text{a.s.}$$

Proof. The result follows from Lemma D.5.1 as follows: Consider $Y(t) = X(t)^2 - E\{X(t)^2 \mid \mathcal{F}_{t-1}\} = X(t)^2 - \sigma^2$. Then $E\{X(t)^2 - \sigma^2 \mid \mathcal{F}_{t-1}\} = 0$ [i.e., $Y(t)$ has zero conditional mean]. Also,

$$\sum_{t=1}^{\infty} \frac{1}{t^2} E\{(X(t)^2 - \sigma^2)^2 \mid \mathcal{F}_{t-1}\} = \sum_{t=1}^{\infty} \frac{1}{t^2} [E\{X(t)^4 \mid \mathcal{F}_{t-1}\} - 2\sigma^4 + \sigma^4]$$

$$< \sum_{t=1}^{\infty} \frac{1}{t^2}(k - \sigma^4)$$

$$< \infty$$

The result follows from Lemma D.5.1.

▼▼▼

Lemma D.5.3. Let $\{X(t)\}$ be a sequence of nonnegative random variables adapted to an increasing sequence of sub-sigma fields $\{\mathcal{F}_t\}$. If

$$E\{X(t+1)|\mathcal{F}_t\} \leq (1 + \gamma(t))X(t) - \alpha(t) + \beta(t) \quad \text{a.s.}$$

where $\alpha(t) \geq 0$, $\beta(t) \geq 0$, and $E\{X(0)\} < \infty$, $\sum_{j=1}^{\infty} |\gamma(t)| < \infty$, $\sum_{j=0}^{\infty} \beta(j) < \infty$ a.s., then $X(t)$ converges almost surely to a finite random variable and

$$\lim_{N \to \infty} \sum_{t=0}^{N} \alpha(t) < \infty \quad \text{a.s.}$$

▼▼▼

This implies the following *positive supermartingale convergence* result:

Corollary D.5.1. If $\{V(t, \omega)\}$ is a sequence of nonnegative random variables adapted to an increasing sequence of sub-sigma algebras $\{\mathcal{F}_t\}$ and if

$$E\{V(t+1)|\mathcal{F}_t\} \leq V(t) - \alpha_j + \beta_j \quad \text{a.s.}$$

where

$$\alpha_j \geq 0, \beta_j \geq 0 \text{ for all} \quad \text{and} \quad E\{V(0)\} < \infty, \quad \sum_{j=0}^{t} \beta_j < \infty \quad \text{a.s.}$$

then $V(t)$ converges almost surely to a finite random variable and $\lim_{N \to \infty} \sum_{t=0}^{N} \alpha_t < \infty$ a.s.

▼▼▼

Rates of convergence in the martingale convergence theorem can be obtained from the analogs of the *central limit theorem* and *law of the iterated logarithm*.

Let $\{Z(n), \mathcal{F}_n\}$ be an L^2 bounded martingale. Then Theorem D.5.1 implies the existence of Z such that $Z(n)$ converges to Z a.s. and in L^2. We can write

$$Z(n) = \sum_{i=1}^{n} X(i) \quad \text{where } X(i) = Z(i) - Z(i-1), Z(0) = 0$$

$\sum X(t)$ converges a.s. and in L^2 to Z and the difference $Z - Z(n)$ can be written

$$Z - Z(n) = \sum_{n+1}^{\infty} X(t)$$

We then have:

Theorem D.5.2 (Brown, 1971). Let

$$V(n)^2 = \sum_{n+1}^{\infty} E\{X(t)^2 | \mathcal{F}_{t-1}\}$$

$$S(n)^2 = E\{V(n)^2\} = E\{(Z - Z(n))^2\}$$

Suppose that:

(i) $\dfrac{V(n)^2}{S(n)^2}$ converges in probability to 1.

(ii) $\lim_{n \to \infty} \dfrac{1}{S(n)^2} \sum_{n+1}^{\infty} E\{X(t)^2 I(|X(t)| \geq \epsilon S(n))\} = 0$

for all $\epsilon > 0$ [where $I(\cdot)$ denotes the indicator function]; then

$$\frac{Z - Z(n)}{S(n)}$$

converges in distribution to an $N(0, 1)$ gaussian random variable.

Theorem D.5.3 (Stout, 1970; Heyde, 1977).   With $V(n)^2$, $S(n)^2$ as in Theorem D.5.2, if:

(i) $\dfrac{V(n)^2}{S(n)^2}$ converges almost surely to 1

(ii) $\displaystyle\lim_{n\to\infty} \sum_1^n \frac{1}{S(t)} E\{|X(t)| I(|X(t)| > \epsilon S(t))\} < \infty$ for all $\epsilon > 0$

(iii) $\displaystyle\lim_{n\to\infty} \sum_1^n \frac{1}{S(t)^4} E\{X(t)^4 I(|X(t)| \leq \delta S(t))\} < \infty$ for some $\delta > 0$

then

$$Z - Z(n) = \xi(n)(2S(n)^2 \log|\log S(n)|)^{1/2}$$

where

$$\limsup \xi(n) = 1 \quad \text{a.s.}$$
$$\liminf \xi(n) = -1 \quad \text{a.s.}$$

▼▼▼

The following propositions on sequences are frequently used in establishing convergence results.

Lemma D.5.4 (Toeplitz Lemma).   Let a_{nk}, $k = 1, \ldots, k_n$ be numbers such that, for fixed k, $a_{nk} \longrightarrow 0$ and, for all n, $\sum_{k=1}^{\infty} |a_{nk}| \leq C < \infty$; let

$$y_n = \sum_{k=1}^{\infty} a_{nk} x_k$$

where $\{x_k\}$ is a sequence of reals. Then:

(i) $\displaystyle\lim_{k\to\infty} x_k = 0$ implies that $\displaystyle\lim_{n\to\infty} y_n = 0$

(ii) (a) $\displaystyle\lim_{n\to\infty} \sum_{k=1}^{\infty} a_{nk} = 1$ $\left.\vphantom{\begin{array}{c}1\\1\end{array}}\right\}$ implies that $\displaystyle\lim_{n\to\infty} y_n = x$
    (b) $\displaystyle\lim_{k\to\infty} x_k = x < \infty$

(iii) In particular, if $b_n = \sum_{k=1}^n a_k$,
    (a) $\displaystyle\lim_{n\to\infty} b_n = \infty$
    (b) $\{a_k\}$ nonnegative $\left.\vphantom{\begin{array}{c}1\\1\\1\end{array}}\right\}$ implies that $\displaystyle\lim_{n\to\infty} \frac{1}{b_n} \sum_{k=1}^n a_k x_k = x$
    (c) $\displaystyle\lim_{k\to\infty} x_k = x < \infty$

Proof. (i) each $\epsilon > 0$, there exists a $k(\epsilon)$ such that for all $k \geq k(\epsilon)$, $|x_k| < \epsilon/C$, so that

$$y_n = \sum_{k=1}^{k(\epsilon)} a_{nk} x_k + \sum_{k=k(\epsilon)+1}^{\infty} a_{nk} x_k$$

$$|y_n| \le \sum_{k=1}^{k(\epsilon)} |a_{nk} x_k| + \epsilon$$

Letting $n \longrightarrow \infty$ and noting that ϵ is arbitrary gives (i).

(ii) $y_n = \sum_{k=1}^{\infty} a_{nk} x + \sum_{k=1}^{\infty} a_{nk}(x_k - x)$

The result follows immediately from (i).

(iii) Putting $a_{nk} = a_k/b_n$, $k \le n$, gives (iii).

▼▼▼

A particular case of the Toeplitz lemma is the Kronecker lemma.

Lemma D.5.5 (Kronecker Lemma)

(a) $\sum_{k=1}^{n} x_k$ converges

(b) $\{b_n\}$ nondecreasing $\left.\rule{0pt}{40pt}\right\}$ implies that $\lim_{n \to \infty} \dfrac{1}{b_n} \sum_{k=1}^{n} b_k x_k = 0$

(c) $\lim_{n \to \infty} b_n = \infty$

Proof. Put $b_0 = 0$

$$a_k = b_k - b_{k-1}$$

$$S_{n+1} = \sum_{k=1}^{n} x_k$$

Then

$$\frac{1}{b_n} \sum_{k=1}^{n} b_k x_k = \frac{1}{b_n} \sum_{k=1}^{n} b_k(S_{k+1} - S_k)$$

$$= \frac{1}{b_n} \sum_{k=1}^{n} (b_k S_{k+1} - b_{k-1} S_k + b_{k-1} S_k - b_k S_k)$$

$$= S_{n+1} - \frac{1}{b_n} \sum_{k=1}^{n} a_k S_k$$

Applying Lemma D.5.4, part (iii) gives

$$\lim_{n \to \infty} \frac{1}{b_n} \sum_{k=1}^{n} b_k x_k = S - S = 0$$

▼▼▼

D.6 GAUSSIAN STOCHASTIC PROCESSES

A random vector X of dimension n is said to have a *gaussian probability density* of mean μ and covariance Σ if

$$p(x) = [(2\pi)^n |\Sigma|]^{-1/2} \exp\left[-\tfrac{1}{2}(x - \mu)\Sigma^{-1}(x - \mu)\right]$$

Elementary properties of a gaussian random variable are described in the following lemmas.

Lemma D.6.1. If X is an N-variate gaussian random variable of mean μ and covariance Σ, then if $Y = BX + v$, Y is a gaussian random variable of mean $B\mu + v$ and covariance $B\Sigma B^T$.

Lemma D.6.2. If X is an N-variate gaussian random variable of mean μ and covariance Σ and if X, μ, and Σ are partitioned as

$$X = \begin{bmatrix} X_1 \\ X_2 \end{bmatrix}, \quad \mu = \begin{bmatrix} \mu_1 \\ \mu_2 \end{bmatrix}, \quad \Sigma = \begin{bmatrix} \Sigma_{11} & \Sigma_{12} \\ \Sigma_{21} & \Sigma_{22} \end{bmatrix}$$

then X_1 is a gaussian random variable of mean μ_1 and covariance Σ_{11}.

Lemma D.6.3. If X is as in Lemma D.6.2, the conditional distribution for X_1 given $X_2 = x_2$ is gaussian with mean $\mu_1 + \Sigma_{12}\Sigma_{22}^{-1}(x_2 - \mu_2)$ and covariance $\Sigma_{11} - \Sigma_{12}\Sigma_{22}^{-1}\Sigma_{21}$.

D.7 MINIMUM VARIANCE ESTIMATORS

Frequently, we are given two random variables X and Y and we want to know what Y tells us about X. An answer to this question is provided by the conditional probability distribution function $p(X \mid Y)$. We have seen in Section D.6 that in the case of gaussian random variables, this conditional distribution can be determined in full. However, in general, it is difficult to evaluate the complete conditional distribution. In this case, the first few moments (e.g., mean and covariance) are often used to describe what Y tells us about X. We shall call a function such as the conditional mean, a *point estimator* since it can be thought of as a single-valued *estimator* of X given the data $Y = y$. If we consider other functions of the data, e.g., $g(y)$, as providing an estimate of x, the conditional mean has the interesting property that it gives the *minimum variance estimator* (MVE). To see this, note that

$$E\{\|g(Y) - X\|^2\} = E\{E\{\|g(Y) - X\|^2 \mid Y\}\}$$
$$= E\{E\{\|g(Y) - E\{X \mid Y\} + E\{X \mid Y\} - X\|^2 \mid Y\}\}$$
$$= E\{E\{\|g(Y) - E\{X \mid Y\}\|^2\} + E\{\|X - E\{X \mid Y\}\|^2 \mid Y\}\}$$

This expression is clearly minimized when

$$g(Y) = E\{X \mid Y\}$$

On some occasions, it is difficult to evaluate the conditional mean. In this case it often suffices to arbitrarily select the form of the function $g(Y)$ and then to estimate parameters in that function. For example, we could fix $g(Y)$ to be a linear function of Y and then choose the parameters in that linear function to minimize the error covariance. This gives a linear minimum variance estimator.

We denote a linear estimator in the form

$$\hat{X} = AY + b$$

The matrix A and the vector b in the linear minimum variance estimator can be expressed in terms of the first two moments of the joint distribution of

$$\begin{bmatrix} X \\ Y \end{bmatrix}$$

This is shown in the following result:

Lemma D.7.1. Let the random vector

$$\begin{bmatrix} X \\ Y \end{bmatrix}$$

have mean

$$\begin{bmatrix} \mu_1 \\ \mu_2 \end{bmatrix}$$

and covariance

$$\begin{bmatrix} \Sigma_{11} & \Sigma_{12} \\ \Sigma_{21} & \Sigma_{22} \end{bmatrix}$$

Then the linear minimum variance estimator of X given Y is (provided that $\Sigma_{22} > 0$)

$$X^* = A_0 Y + b_0; \qquad A_0 = \Sigma_{12}\Sigma_{22}^{-1}, \quad b_0 = \mu_1 - \Sigma_{12}\Sigma_{22}^{-1}\mu_2$$
$$= \mu_1 + \Sigma_{12}\Sigma_{22}^{-1}(Y - \mu_2)$$

▼▼▼

A very important observation is that the expressions for the linear minimum variance estimator coincide with the expression for the minimum variance estimator when the latter is derived under a gaussian assumption (compare Lemma D.7.1 with Lemma D.6.3). Interesting properties of the linear minimum variance estimator are:

1. It also minimizes the error covariance matrix in the vector case.
2. It is unbiased in the sense that

$$E\{X - X^*\} = 0$$

3. The estimation error is orthogonal to Y in the sense that

$$E\{(X - X^*)Y^T\} = 0$$

[*Note:* $E\{(X - X^*)Y^T\} = E\{[(X - \mu_1) - \Sigma_{12}\Sigma_{22}^{-1}(Y - \mu_2)]Y^T\}$
$$= 0$$

The properties above demonstrate the close connection between linear minimum variance estimator and minimum variance estimators.

D.8 MAXIMUM LIKELIHOOD ESTIMATION

Given a parametric family of probability density function, $\mathcal{P} = \{p(\cdot \mid \theta): \theta \in \Theta\}$ for a random variable $Y(\omega)$ on a sample space Ω, we define the *log-likelihood function*, $l(y, \theta)$, as

$$l(y, \theta) = \ln p(y \mid \theta)$$

The maximum likelihood estimator, θ^*, of θ given Y is then

$$\theta^*(Y) = \arg\max_{\theta \in \Theta} l(Y, \theta)$$

We illustrate by several examples.

Example D.8.1

Consider a sequence of independent identically distributed random (n-dimensional) vectors $Y(0), \ldots, Y(N)$, where each $Y(i)$ has a gaussian distribution of mean μ and covariance Σ. Then the log-likelihood function is

$$l(y(N), \ldots, y(1), \theta) = -\frac{nN}{2} \ln(2\pi) - \frac{N}{2} \ln \det |\Sigma| - \frac{1}{2} \sum_{t=1}^{N} (y(t) - \mu)^T \Sigma^{-1} (y(t) - \mu)$$

Differentiating with respect to μ and Σ gives the following equations to be solved for the maximum likelihood estimates $\hat{\mu}, \hat{\Sigma}$ of μ and Σ:

$$-\frac{N}{2} \hat{\Sigma}^{-1} + \frac{1}{2} \sum_{t=1}^{N} \hat{\Sigma}^{-1} (y(t) - \hat{\mu})(y(t) - \hat{\mu})^T \hat{\Sigma}^{-1} = 0$$

$$\sum_{t=1}^{N} \hat{\Sigma}^{-1} (y(t) - \hat{\mu}) = 0$$

These equations give

$$\hat{\Sigma} = \frac{1}{N} \sum_{t=1}^{N} (y(t) - \hat{\mu})(y(t) - \hat{\mu})^T$$

$$\hat{\mu} = \frac{1}{N} \sum_{t=1}^{N} y(t)$$

Example D.8.2

Consider a predictor model of the form

$$y(t) = f(\mathcal{Y}_{t-1}, \mathcal{U}_{t-1}, \theta) + \epsilon(t)$$

where $\{\epsilon(t)\}$ is assumed to be an independent and identically distributed sequence with probability density function $p(\epsilon \,|\, \theta); \theta \in \Theta$.

The log-likelihood function can be evaluated as follows:

$$l(y(N), \ldots, y(1), \theta) = \ln\{p_y[y(N), \ldots, y(1), \theta]\}$$

$$= \ln \left\{ \prod_{t=1}^{N} p_y[y(t) \,|\, \mathcal{Y}_{t-1}, \mathcal{U}_{t-1}, \theta] \right\}$$

$$= \sum_{t=1}^{N} \ln p_y[y(t) \,|\, \mathcal{Y}_{t-1}, \mathcal{U}_{t-1}, \theta]$$

$$= \sum_{t=1}^{N} \ln p_\epsilon \{y(t) - f(\mathcal{Y}_{t-1}, \mathcal{U}_{t-1}, \theta) \,|\, \theta\}$$

Note that the *prediction error* appears in the log-likelihood function.

Example D.8.3

Consider the set up of Example D.8.2, where $\epsilon(t)$ has a gaussian distribution of zero mean and covariance Σ. Then

$$l(y(N), \ldots, y(1), \theta) = \sum_{t=1}^{N} \left\{ -\frac{n}{2} \ln(2\pi) - \frac{1}{2} \ln \det |\Sigma| \right.$$

$$\left. -\frac{1}{2} \sum_{t=1}^{N} [y(t) - f(\mathcal{Y}_{t-1}, \mathcal{U}_{t-1}, \theta)]^T \Sigma^{-1} [y(t) - f(\mathcal{Y}_{t-1}, \mathcal{U}_{t-1}, \theta)] \right\}$$

Differentiating with respect to Σ gives, as in Example D.8.1,

$$\hat{\Sigma} = \frac{1}{N} \sum_{t=1}^{N} [y(t) - f(\mathcal{Y}_{t-1}, \mathcal{U}_{t-1}, \theta)][y(t) - f(\mathcal{Y}_{t-1}, \mathcal{U}_{t-1}, \theta)]^T$$

This equation can be used to eliminate Σ from the log-likelihood function, giving

$$l(y(N),\ \ldots,\ y(1), \theta) = -\frac{nN}{2} \ln(2\pi + 1) - \frac{1}{2} \ln \det |\hat{\Sigma}|$$

In the case of scalar observations, this simplifies to

$$l(y(N),\ \ldots,\ y(1), \theta) = \text{constant} - \frac{N}{2} \ln \left(\frac{1}{N} \sum_{t=1}^{N} (y(t) - f(y_{t-1}, u_{t-1}, \theta)) \right)^2$$

Thus maximization of the log-likelihood function, in this case, is achieved by minimizing the logarithm of the determinant of the prediction error covariance.

Example D.8.4

To show that the output prediction error does not always appear directly in the log-likelihood function, consider the following multiplicative model:

$$y(t) = \theta_1 y(t-1)^{\theta_2} \epsilon(t)$$

where $\epsilon(t)$ has a log-normal distribution with mean 0 and variance σ^2.

The log-likelihood function is

$$l(y(N),\ \ldots,\ y(1), \theta) = \text{constant} - \frac{1}{2\sigma^2} \sum_{t=1}^{N} [\ln y(t) - \ln \theta_1 - \theta_2 \ln y(t-1)]^2$$

Thus the maximum likelihood estimates of θ_1 and θ_2 are obtained by minimizing $\sum_{t=1}^{N} [\ln y(t) - \ln \theta_1 - \theta_2 \ln y(t-1)]^2$.

▼▼▼

APPENDIX E

Matrix Riccati Equations

E.1 INTRODUCTION

This appendix presents results on the solutions of the algebraic Riccati equation (ARE) and on the convergence of the solutions of the Riccati difference equation (RDE) to particular solutions of the ARE.

The results presented here are based principally on Chan, Goodwin, and Sin (1982). [For further reading, consult Kalman (1961, 1964), Bucy (1967), Wonham (1968), Martensson (1971), Willems (1971), Callier and Willems (1981), Caines and Mayne (1970), Hewer (1973), Kucera (1972a, 1972b, 1972c), and Anderson and Moore (1981).]

As in Chapter 7, we shall consider the following signal model:

$$x(t + 1) = Ax(t) + v_1(t) \tag{E.1.1}$$

$$y(t) = Cx(t) + v_2(t) \tag{E.1.2}$$

where $v_1(t)$ and $v_2(t)$ are zero mean, stationary stochastic processes satisfying

$$E\left\{ \begin{bmatrix} v_1(t) \\ v_2(t) \end{bmatrix} [v_1(\tau)^T v_2(\tau)^T] \right\} = \begin{bmatrix} Q & S \\ S^T & R \end{bmatrix} \delta(t - \tau); \qquad R > 0, \quad Q \geq 0 \tag{E.1.3}$$

We can take $S = 0$ without loss of generality; see Remark 7.3.1.

We recall from Theorem 7.3.1 that the optimal filter for the system (E.1.1) to (E.1.3) is given by

$$\hat{x}(t + 1) = \bar{A}(t)\hat{x}(t) + K(t)y(t); \quad \hat{x}(0) = x_0 \tag{E.1.4}$$

where

$$\bar{A}(t) \triangleq A - K(t)C \tag{E.1.5}$$

$$K(t) \triangleq A\Sigma(t)C^T(C\Sigma(t)C^T + R)^{-1} \tag{E.1.6}$$

and $\Sigma(t)$ satisfies the following matrix *Riccati difference equation* (RDE):

$$\Sigma(t + 1) = A\Sigma(t)A^T - A\Sigma(t)C^T(C\Sigma(t)C^T + R)^{-1}C\Sigma(t)A^T + Q \tag{E.1.7}$$

$$\Sigma(0) = P_0 \tag{E.1.8}$$

with x_0 and P_0 the mean and covariance of the initial state, $x(0)$.

If $\Sigma(t)$ converges as $t \longrightarrow \infty$, the limiting solution Σ will satisfy the following algebraic Riccati equation (ARE), obtained from (E.1.7) by putting $\Sigma(t + 1) = \Sigma(t) = \Sigma$:

$$\Sigma - A\Sigma A^T + A\Sigma C^T(C\Sigma C^T + R)^{-1}C\Sigma A^T - Q = 0 \tag{E.1.9}$$

By analogy with (E.1.5)–(E.1.6) we also define the steady-state filter state transition matrix, $\bar{A}$, and steady-state filter gain matrix, K, by

$$\bar{A} \triangleq A - KC \tag{E.1.10}$$

$$K \triangleq A\Sigma C^T(C\Sigma C^T + R)^{-1} \tag{E.1.11}$$

For convenience, we also factorize R and Q as

$$R \triangleq (R^{1/2})(R^{1/2})^T, \qquad Q = DD^T \tag{E.1.12}$$

and define

$$\bar{C} = R^{-1/2}C \tag{E.1.13}$$

Then the ARE equation becomes

$$\Sigma - A\Sigma A^T + A\Sigma \bar{C}^T(\bar{C}\Sigma \bar{C}^T + I)^{-1}\bar{C}\Sigma A^T - DD^T = 0 \tag{E.1.14}$$

and $\bar{A}$ can be expressed as

$$\bar{A} = A - A\Sigma \bar{C}^T(\bar{C}\Sigma \bar{C}^T + I)^{-1}\bar{C} \tag{E.1.15}$$

We shall assume in the sequel that A is nonsingular. (The case of singular A can be treated in a similar fashion see De Souza, Goodwin and Chan, 1983).

Section E.2 presents results on the solution of the ARE (E.1.14) while Section E.3 presents results on the solution of the RDE (E.1.7).

E.2 THE ALGEBRAIC RICCATI EQUATION

In the filtering problem $\Sigma(t)$ has the physical interpretation that it is the state error covariance. $\Sigma(t)$ is therefore real, symmetric, and nonnegative definite. We shall therefore be interested in those solutions of the algebraic Riccati equation (ARE) which retain these properties.

We shall be particularly interested in the question of existence and uniqueness of those solutions which give a steady-state filter having roots on or inside the unit circle. We therefore introduce the following definition.

Definition E.2.A. A real symmetric nonnegative definite solution of the ARE is said to be a *strong solution* if the corresponding filter state transition matrix, $\bar{A}$, has all its roots inside or on the unit circle ($|\lambda_i(\bar{A})| \leq 1$; $i = 1, \ldots, n$). (If there are no roots on the unit circle, the strong solution is termed the *stabilizing solution*.)

▼▼▼

The following lemma summarizes the interrelationships that exist between the solutions of the ARE and the generalized eigenvectors of the *Hamiltonian* matrix defined as follows:

$$M = \begin{bmatrix} A^T + \bar{C}^T \bar{C} A^{-1} Q & -\bar{C}^T \bar{C} A^{-1} \\ -A^{-1} Q & A^{-1} \end{bmatrix} \tag{E.2.1}$$

Lemma E.2.1. (i) Let

$$\begin{bmatrix} x_i \\ y_i \end{bmatrix}; \quad i = 1, 2, \ldots, 2n$$

denote the generalized eigenvectors of M. We construct X and Y from a selection of these vectors as

$$X = [x_{i_1} \quad \cdots \quad x_{i_n}] \tag{E.2.2}$$

$$Y = [y_{i_1} \quad \cdots \quad y_{i_n}] \tag{E.2.3}$$

subject to the following constraint: If λ_i is an eigenvalue of M of multiplicity p_i and if one of its generalized eigenvectors

$$\begin{bmatrix} x_{i+k-1} \\ y_{i+k-1} \end{bmatrix}$$

of rank $k \leq p_i$ constitutes one column in the matrix $\begin{bmatrix} X \\ Y \end{bmatrix}$, then the eigenvectors

$$\begin{bmatrix} x_i \\ y_i \end{bmatrix}, \ldots, \begin{bmatrix} x_{i+k-2} \\ y_{i+k-2} \end{bmatrix}$$

with rank $1, \ldots, k-1$ are also columns in $\begin{bmatrix} X \\ Y \end{bmatrix}$.

Then,

(a) If Σ is a solution to the ARE, there exists a choice of $\begin{bmatrix} X \\ Y \end{bmatrix}$ such that X is nonsingular and such that

$$\Sigma = YX^{-1} \tag{E.2.4}$$

(b) For every possible choice $\begin{bmatrix} X \\ Y \end{bmatrix}$ such that X is nonsingular, then $\Sigma = YX^{-1}$ is a solution of the ARE.

(ii) Let λ_i be an eigenvalue of M of multiplicity p_i and

$$\begin{bmatrix} x_i \\ y_i \end{bmatrix}, \begin{bmatrix} x_{i+1} \\ y_{i+1} \end{bmatrix}, \ldots, \begin{bmatrix} x_{i+p_i-1} \\ y_{i+p_i-1} \end{bmatrix}$$

be the corresponding chain of generalized right eigenvectors. Then $(\lambda_i^*)^{-1}$ is also an eigenvalue of M and

$$\begin{bmatrix} -y_i \\ x_i \end{bmatrix}^*; \quad (-1)^k [\lambda_i^{k+1} P_{k-1}^i]^*; \quad k = 1, 2, \ldots, p_i - 1$$

where P_{k-1}^i is a vector polynomial in λ_i of the form:

$$P_{k-1}^i = \sum_{j=0}^{k-1} \frac{(k-1)!}{j!(k-1-j)!} \lambda_i^{k-j-1} \begin{bmatrix} -y_{k-j+i} \\ x_{k-j+i} \end{bmatrix}$$

is a corresponding chain of generalized left eigenvectors. (See Section A.5 for definition and properties of left and right eigenvectors.)

(iii) If $\begin{bmatrix} X \\ Y \end{bmatrix}$ generates a solution to the ARE, so does $\begin{bmatrix} X \\ Y \end{bmatrix} \Gamma$, where Γ is any nonsingular transformation. (In particular, the order of the eigenvectors is immaterial.)

(iv) Let $\begin{bmatrix} x_i \\ y_i \end{bmatrix}$, $i = 1, 2, \ldots, n$, be the generalized eigenvectors of M corresponding to eigenvalues $\lambda_1, \lambda_2, \ldots, \lambda_n$ (not necessarily distinct). If $\Sigma = [y_1 \ y_2 \ \cdots \ y_n][x_1 \ x_2 \ \cdots \ x_n]^{-1}$ is a solution of the ARE, the matrix $\bar{A}$ as defined in (E.1.15) has eigenvalues $\lambda_1, \lambda_2, \ldots, \lambda_n$ and corresponding generalized eigenvectors $x_1, x_2, \ldots, x_n$. In particular, the eigenvalues of M are the union of the eigenvalues of $\bar{A}$ and $(\bar{A})^{-1}$.

(v) If A is nonsingular, there are no zero eigenvalues of M and hence no zero eigenvalues in any $\bar{A}$. Thus A nonsingular implies that both M and $\bar{A}$ are nonsingular.

(vi) The solution $\Sigma = YX^{-1}$ of the ARE is *real* if and only if one of the following conditions is satisfied:

(a) All generalized eigenvectors used to construct Σ are real.
(b) If the complex generalized eigenvector, a_i, of rank k corresponding to the eigenvalue λ_i is used to construct Σ, then a_i^* of rank k corresponding to λ_i^* is also used.

(vii) If λ is an uncontrollable mode of (A, D) having multiplicity p such that

$$(A^T - \lambda I)x_i = x_{i-1}; \qquad i = 0, 1, \ldots, p-1, \quad x_{-1} = 0$$
$$D^T x_i = 0$$

then λ is an eigenvalue of M having multiplicity p such that

$$(M - \lambda I)\begin{bmatrix} x_i \\ 0 \end{bmatrix} = \begin{bmatrix} x_{i\,1} \\ 0 \end{bmatrix}; \qquad i = 0, 1, \ldots, p-1$$

(and conversely).

(viii) If λ is an unobservable mode of (C, A) having multiplicity p such that

$$(A - \lambda I)y_i = y_{i-1}; \qquad i = 0, 1, \ldots, p-1, \quad y_{-1} = 0$$
$$Cy_i = 0$$

then λ^{-1} is an eigenvalue of M of multiplicity p, and

$$(M - \lambda^{-1}I)\begin{bmatrix} 0 \\ z_i \end{bmatrix} = \begin{bmatrix} 0 \\ z_{i-1} \end{bmatrix}; \qquad i = 0, 1, \ldots, p-1$$

where $z_i = (-1)^i \lambda^{i+1} A^{i-1} y_i$
(and conversely).

(ix) M will have an eigenvalue λ of multiplicity $2p$ on the unit circle if and only if:

(a) There is an uncontrollable mode λ of multiplicity p of (A, D) on the unit circle such that

$$(A^T - \lambda I)x_i = x_{i-1}, \qquad i = 0, 1, \ldots, p-1, \quad |\lambda| = 1$$
$$D^T x_i = 0, \qquad x_{-1} = 0$$

or

(b) There is an unobservable mode λ of multiplicity p of (C, A) on the unit circle such that

$$(A - \lambda I)y_i = y_{i-1}, \qquad i = 0, \ldots, p-1, \quad |\lambda| = 1$$
$$Cy_i = 0, \qquad y_{-1} = 0$$

(x) The matrix (X^*Y) is hermitian if and only if:

(a) For each pair of eigenvalues α_i, β_i of multiplicity p_i where $|\alpha_i| \neq 1$ and $\alpha_i = (\beta^*)^{-1}$, then exactly p_i of the total of $2p_i$ eigenvectors are used in the construction of $\begin{bmatrix} X \\ Y \end{bmatrix}$.

(b) For each eigenvalue γ_j such that $|\gamma_j| = 1$, the multiplicity is even, and exactly p_j of the total of $2p_j$ eigenvectors are used in the construction of $\begin{bmatrix} X \\ Y \end{bmatrix}$.

In particular, if X is nonsingular, then conditions (a) and (b) give necessary and sufficient conditions for the solution $\Sigma = YX^{-1}$ to be hermitian.

Proof. See Chan, Goodwin and Sin (1983).

▼▼▼

The lemma above deals in general with the solutions of the ARE. Properties of the strong solution are discussed in the following theorem.

Theorem E.2.1. Provided that (C, A) is detectable:

(i) The strong solution of the ARE exists and is unique.
(ii) If (A, D) is stabilizable, the strong solution is the only positive semi definite solution of the ARE.
(iii) If (A, D) has no uncontrollable modes on the unit circle, the strong solution coincides with the stabilizing solution.
(iv) If (A, D) has an uncontrollable mode on the unit circle, then although the strong solution exists, there is no stabilizing solution.
(v) If (A, D) has an uncontrollable mode inside, or on, the unit circle, the strong solution is not positive definite.
(vi) If (A, D) has an uncontrollable mode outside the unit circle, then as well as the strong solution, there is at least one other positive semidefinite solution of the ARE.

Proof. See Martensson (1971), Kucera (1972a, b, c), and Chan, Goodwin, and Sin (1983).

▼▼▼

E.3 THE RICCATI DIFFERENCE EQUATION

Here we will consider the convergence of the solutions of the Riccati difference equation (RDE) to the strong solution of the algebraic Riccati equation. We present three theorems; the first is well known and assumes that the system is stabilizable, whereas the second and third are relatively new.

Theorem E.3.1. Subject to

(i) (A, D) is stabilizable.

(ii) (C, A) is detectable.

(iii) $\Sigma_0 \geq 0$

then

$$\lim_{t \to \infty} \Sigma(t) = \Sigma_s \qquad \text{(exponentially fast)}$$

where $\Sigma(t)$ is the solution of the Riccati difference equation with initial condition Σ_0 and Σ_s is the (unique) stabilizing solution of the ARE.

Proof. See Caines and Mayne (1970, 1971).

▼▼▼

Theorem E.3.2. Subject to:

(i) There are no uncontrollable modes of (A, D) on the unit circle,

(ii) (C, A) is detectable,

(iii) $\Sigma_0 > 0$,

then

$$\lim_{t \to \infty} \Sigma(t) = \Sigma_s \qquad \text{(exponentially fast)}$$

where $\Sigma(t)$ is the solution of the Riccati difference equation with initial condition Σ_0 and Σ_s is the (unique) stabilizing solution of the ARE.

Proof

Step 1: Assumption (ii) and Theorem E.2.1 (i) imply that the strong solution, Σ_s, of the ARE exists and is unique. Assumption (i) and Theorem E.2.1(iii) imply that the strong solution coincides with the stabilizing solution and hence

$$|\lambda_i(\bar{A}_s)| < 1 \qquad \text{for } i = 1, 2, \ldots, n \qquad (E.3.1)$$

where $\bar{A}_s$ is the steady-state filter gain corresponding to Σ_s.

Step 2: Some lengthy manipulations show that

$$\Sigma(t) - \Sigma_s = (\bar{A}_s)^t[\Sigma_0 - \Sigma_s]\psi(t - 1) \qquad (E.3.2)$$

where

$$\psi(t - 1) = \bar{A}(0)^T \bar{A}(1)^T \cdots \bar{A}(t - 2)^T \bar{A}(t - 1)^T$$

We now show that $\psi(t)$ is bounded by noting that

$$\Sigma(t) = \bar{A}(t - 1)\Sigma(t - 1)\bar{A}(t - 1)^T$$
$$+ A\Sigma(t - 1)\bar{C}^T[\bar{C}\Sigma(t - 1)\bar{C}^T + I]^{-1}[\bar{C}\Sigma(t - 1)\bar{C}^T + I]^{-1}\bar{C}\Sigma(t - 1)A^T + Q$$

Thus
$$\Sigma(t) = \psi(t-1)^T \Sigma_0 \psi(t-1) + \text{nonnegative definite terms}$$

Now because (C, A) is detectable and R is nonsingular, we have that $(\bar{C}, A)$ is detectable and thus there exists a $\bar{K}$ such that $(A - \bar{K}\bar{C})$ is exponentially stable. Hence the corresponding suboptimal filter based on $\bar{K}$ gives estimates with bounded error covariance. Since it is a suboptimal filter, its error covariance overbounds the error covariance of the optimal filter. Hence $\Sigma(t)$ is bounded for all t and for any fixed initial value Σ_0. With $\Sigma(t)$ bounded and $\Sigma_0 > 0$, we see that $\psi(t)$ is bounded for all t.

Step 3: Substituting (E.3.1) into (E.3.2) immediately gives the result.

▼▼▼

Theorem E.3.3. Subject to:

(i) (C, A) is observable,
(ii) $(\Sigma_0 - \Sigma_s) > 0$ or $\Sigma_0 = \Sigma_s$

then
$$\lim_{t \to \infty} \Sigma(t) = \Sigma_s$$

where $\Sigma(t)$ is the solution of the Riccati difference equation with initial condition Σ_0 and Σ_s is the (unique) strong solution of the ARE.

Proof. We first treat the case $\Sigma_0 > \Sigma_s > 0$.
We define
$$Q(t) \triangleq \Sigma(t) - \Sigma_s \tag{E.3.3}$$
$$H_s \triangleq (I + \bar{C}\Sigma_s\bar{C}^T) \tag{E.3.4}$$

Then using the ARE and RDE we can show (after some lengthy algebra) that
$$Q(t+1) = \bar{A}_s[Q(t) - Q(t)\bar{C}^T(H_s - \bar{C}Q(t)\bar{C}^T)^{-1}\bar{C}Q(t)]A_s^T \tag{E.3.5}$$
where
$$\bar{A}_s = A - A\Sigma_s\bar{C}^T(\bar{C}\Sigma_s\bar{C}^T + I)^{-1}\bar{C} \tag{E.3.6}$$
Using the matrix inversion lemma, we obtain
$$Q(t+1) = \bar{A}_s[Q(t)^{-1} + \bar{C}^T H_s^{-1}\bar{C}]^{-1}\bar{A}_s^T \tag{E.3.7}$$
and
$$Q(t+1)^{-1} = (\bar{A}_s^T)^{-1}Q(t)^{-1}(\bar{A}_s)^{-1} + (\bar{A}_s^T)^{-1}\bar{C}^T H_s^{-1}\bar{C}(\bar{A}_s)^{-1} \tag{E.3.8}$$

$[(\bar{A}_s)^{-1}$ exists since our standing assumption is that $\det A \neq 0$ and hence by Lemma E.2.1(v), $\det \bar{A}_s \neq 0$. Also, $Q(0)$ is nonsingular in view of (ii) in the theorem statement and we can argue by induction that $Q(t)$ nonsingular implies $Q(t+1)$ nonsingular for all t.]
Solving (E.3.6) gives
$$Q(t+1)^{-1} = \{(\bar{A}_s^T)^{-1}\}^{t+1}Q(0)^{-1}\{(\bar{A}_s)^{-1}\}^{t+1} + \sum_{i=1}^{t+1}\{(\bar{A}_s^T)^{-1}\}^i\bar{C}^T H_s^{-1}\bar{C}\{(\bar{A}_s)^{-1}\}^i \tag{E.3.9}$$
Let
$$S(t) \triangleq \sum_{i=1}^{t}\{(\bar{A}_s^T)^{-1}\}^i\bar{C}^T H_s^{-1}\bar{C}\{(\bar{A}_s)^{-1}\}^i \tag{E.3.10}$$

Now by assumption (i) and the fact that R is nonsingular we have that $(\bar{C}, (\bar{A}_s)^{-1})$ is also completely observable. Thus

$$\text{rank}\,[\bar{C}^T, (\bar{A}_s^T)^{-1}\bar{C}^T, \ldots, (\bar{A}_s^T)^{-(n-1)}\bar{C}^T]^T = n \qquad \text{(E.3.11)}$$

Hence

$$\bar{C}(\bar{A}_s)^{-j}x \text{ is nonzero for at least one } j \in (0, 1, \ldots, n-1) \quad \text{for any } x \neq 0 \qquad \text{(E.3.12)}$$

Since

$$|\lambda_i(\bar{A}_s)^{-1}| \geq 1$$

then

$$x_n = (\bar{A}_s)^{-n}x \neq 0 \qquad \text{for any } x \neq 0$$

Now

$$x^T S(Kn)x = \sum_{k=0}^{K} \sum_{j=1}^{n} x^T(\bar{A}_s^T)^{-kn}(\bar{A}_s^T)^{-j}\bar{C}^T H_s^{-1}\bar{C}(\bar{A}_s)^{-j}(\bar{A}_s)^{-kn}x$$

$$= \sum_{k=0}^{K} \left[\sum_{j=1}^{n} x_{nk}^T(\bar{A}_s^T)^{-j}\bar{C}^T H_s^{-1}\bar{C}(\bar{A}_s)^{-j}x_{nk} \right]$$

The term within the square brackets is greater than zero for any $x \neq 0$ using (E.3.12). Hence

$$\lim_{t\to\infty} x^T S(t)x = \infty \qquad \text{for all } x \neq 0 \qquad \text{(E.3.13)}$$

Thus the minimum eigenvalue of $S(t)$ diverges. Hence from (E.3.9), on noting that the first term is nonnegative definite, we can conclude

$$\lim_{t\to\infty} \lambda_{\min} Q(t+1)^{-1} = \infty \qquad \text{(E.3.14)}$$

Thus

$$\lim_{t\to\infty} \lambda_{\max} Q(t+1) = 0 \qquad \text{(E.3.15)}$$

The result then follows from (E.3.3).

Finally, the case $\Sigma_0 = \Sigma_s$ is trivial.

▼▼▼

References

AHMED, N., and D. H. YOUN (1980). "On a realization and related algorithm for adaptive prediction," *IEEE Trans. Acoust., Speech Signal Process.*, Vol. ASSP-28, No. 5, pp. 493–497.

AKAIKE, H. (1974). "A new look at the statistical model identification," *IEEE Trans. Autom. Control*, Vol. AC-19, pp. 716–723.

AKAIKE, H. (1976). "Canonical correlation analysis of time series and the use of an information criterion," in *System Identification: Advances and Case Studies*, Academic Press, New York.

AKAIKE, H. (1980). "Seasonal adjustment by a Bayesian modelling," *J. Time Ser. Anal.*, Vol. 1, No. 1, pp. 1–13.

ALBERT, A. E., and L. A. GARDNER (1967). *Stochastic Approximation and Nonlinear Regression*, MIT Press, Cambridge, Mass.

ALLIDINA, A. Y., and F. M. HUGHES (1980). "Generalized self-tuning controller with pole assignment," *Proc. IEE*, Vol. 127, No. 1, pp. 13–18.

ANDERSON, B. D. O. (1982). "Exponential convergence and persistent excitation," *21st Conf. Decis. Control*, Orlando, Fla.

ANDERSON B. D. O., and C. R. JOHNSON, JR. (1980). "Exponential convergence of adaptive identification and control algorithms," Tech. Rep., University of Newcastle, Newcastle, New South Wales, Australia.

ANDERSON, B. D. O., and J. B. MOORE (1971). *Linear Optimal Control*, Prentice-Hall, Englewood Cliffs, N.J.

ANDERSON, B. D. O., and J. B. MOORE (1979). *Optimal Filtering*, Prentice-Hall, Englewood Cliffs, N.J.

ANDERSON, B. D. O., and J. B. MOORE (1981). "Detectability and stabilizability of time varying discrete time linear systems," *SIAM J. Control Optim.*, Vol. 19, No. 1, pp. 20–32.

ANDERSON, B. D. O., and S. VONGPANITLERD (1973). *Network Analysis*, Prentice-Hall, Englewood Cliffs, N.J.

ANTOULAS, A. C. (1981). "On canonical forms for linear constant systems," *Int. J. Control*, Vol. 33, No. 1, pp. 95–122.

ASHER, R. B., I. I. ANDRISANI, and P. DORATO (1976). "Biography on adaptive control systems," *Proc. IEEE*, Vol. 64, No. 8, pp. 1226–1240.

ÅSTRÖM, K. J. (1970). *Introduction to Stochastic Control Theory*, Academic Press, New York.

ÅSTRÖM, K. J. (1980). "Self-tuning regulators—design principles and applications," in *Applications of Adaptive Control*, ed. K. S. Narendra and R. V. Monopoli, Academic Press, New York.

ÅSTRÖM, K. J. (1981). "Theory and applications of adaptive control," *IFAC Congr.*, Kyoto, Japan.

ÅSTRÖM, K. J., and P. EYKHOFF (1971). "System identification—a survey," *Automatica*, Vol. 7, pp. 123–162.

ÅSTRÖM, K. J., and B. WITTENMARK (1973). "On self-tuning regulators," *Automatica*, Vol. 9, pp. 195–199.

ÅSTRÖM, K. J., and B. WITTENMARK (1974). "Analysis of a self-tuning regulator for non-minimum phase systems," *IFAC Symp. Stoch. Control*, Budapest.

ÅSTRÖM, K. J., and B. WITTENMARK (1980). "Self-tuning controllers based on pole-zero placement," *IEE Proc.*, Vol. 127, Pt. D, No. 3, pp. 120–130.

ÅSTRÖM, K. J., P. HAGANDER, and J. STERNBY (1980). "Zeros of sampled systems," *Proc. 19th IEEE Conf. Decis. Control*, Albuquerque, N. Mex.

ÅSTRÖM, K. J., B. WESTERBURG, and B. WITTENMARK (1978). "Self-tuning controllers based on pole-placement design," Lund Rep. LUTFD2/(TFRT-3/48)/1-052, Lund Institute of Technology, Lund, Sweden.

ÅSTRÖM, K. J., U. BORISSON, L. LJUNG, and B. WITTENMARK (1977). "Theory and application of self tuning regulators," *Automatica*, Vol. 13, No. 5, pp. 457–476.

BALAS, M. (1982). "The spatial order reduction problem and its effect on adaptive control of distributed parameter systems," *21st Conf. Decis. Control*, Orlando, Fla.

BAR-SHALOM, Y., and E. TSE (1974). "Dual effort, certainty equivalence and separation in stochastic control," *IEEE Trans. Autom. Control*, Vol. AC-19, pp. 494–500.

BAR-SHALOM, Y., and E. TSE (1976). "Concepts and methods in stochastic control," in *Control and Dynamic Systems: Advances in Theory and Applications*, Vol. 12, ed. C. T. Leondes, Academic Press, New York.

BAR-SHALOM, Y., and K. D. WALL (1980). "Dual adaptive control and uncertainty effects in macroeconomic modelling," *Automatica*, Vol. 16, No. 2, pp. 147–156.

BEGHELLI, S. and R. GUIDORZI (1976). "A new input–output canonical form for multivariable systems," *IEEE Transactions on Automatic Control*, Vol. 21, October, pp. 692–696.

BELLMAN, R. E. (1957). *Dynamic Programming*, Princeton University Press, Princeton, N.J.

BERGMANN, S., and K. H. LACHMANN (1980). "Digital parameter adaptive control of a pH process," *Joint Autom. Control Conf.*, San Francisco.

BIERMAN, G. J., and M. W. NEAD (1977). "A parameter estimation subroutine package," JPL 77–26, NASA JPL Report.

BITMEAD, R. R. (1979a). "Convergence properties of discrete-time stochastic adaptive estimation algorithms," Ph.D. thesis, Department of Electrical Engineering, University of Newcastle, Newcastle, New South Wales, Australia.

BITMEAD, R. R. (1979b). "Convergence in distribution of LMS-type adaptive parameter estimates," Tech. Rep., Department of Electrical Engineering, University of Newcastle, Newcastle, New South Wales, Australia.

BITMEAD, R. R., and B. D. O. ANDERSON (1980a). "Performance of adaptive estimation algorithms in dependent random environments," *IEEE Trans. Autom. Control*, Vol. AC-25, No. 4, pp. 788–794.

BITMEAD, R. R., and B. D. O. ANDERSON (1980b). "Lyapunov techniques for the exponential stability of linear difference equations with random coefficients," *IEEE Trans. Autom. Control*, Vol. AC-25, No. 4, pp. 782–788.

BLUM, J. R. (1954). "Multidimensional stochastic approximation methods," *Ann. Math. Stat.*, Vol. 25, pp. 737–744.

BOGNER, R. E., and A. C. CONSTANTINIDES (1975). *Introduction to Digital Filtering*, Wiley, New York.

BORISSON, U. (1979). "Self-tuning regulators for a class of multivariable systems," *Automatica*, Vol. 15, No. 2, pp. 209–217.

BORISSON, U., and R. SYDING (1976). "Self-tuning control of an ore crusher," *Automatica*, Vol. 12, pp. 1–7.

BORISSON, U., and B. WITTENMARK (1974). "An industrial application of a self-tuning regulator," *IFAC Symp. Digital Comput. Appl. Process Control*, Zurich.

BOX, G. E. P., and G. M. JENKINS (1970). *Time Series Analysis: Forecasting and Control*, Holden-Day, San Francisco.

BRIEPOHL, A. M. (1970). *Probabilistic Systems Analysis*, Wiley, New York.

BROCKETT, R. E. (1970). *Finite Dimensional Linear Systems*, Wiley, New York.

BROWN, B. M. (1971). "Martingale central limit theorems," *Ann. Math. Stat.*, Vol. 42, pp. 59–66.

BROWN, J. E., III (1970). "Adaptive estimation in nonstationary environments," Tech. Rep. SU-SEL-71-004, Stanford University Center for Systems Research, Stanford, Calif.

BROWN, R. G. (1962). *Smoothing, Forecasting and Prediction of Discrete Time Series*, Prentice-Hall, Englewood Cliffs, N.J.

BUCY, R. S. (1967). "Global theory of the Riccati equation," *J. Comput. Syst. Sci.*, Vol. 1, pp. 349–361.

BURG, J. (1975). "Maximum entropy spectral analysis," Ph.D. dissertation, Stanford University, Stanford, Calif.

BURHOLT, F., and M. KUMMEL (1979). "Welf-tuning control of a pH-neutralization process," *Automatica*, Vol. 15, pp. 665–671.

BURRILL, C. W. (1973). *Measure, Integration and Probability*, McGraw-Hill, New York.

BURTON, D. M. (1967). *Introduction to Modern Abstract Algebra*, Addison-Wesley, Reading, Mass.

CAINES, P. E. (1972). "Relationship between Box–Jenkins–Astrom control and Kalman linear regulator," *Proc. IEE*, Vol. 119, No. 5, pp. 615–620.

CAINES, P. E. (1980). "Passivity, hyperstability and positive reality," *Conf. Inf. Sci. Syst.*, Princeton, N.J.

CAINES, P. E. (1981). "Stochastic adaptive control: randomly varying parameters and continually disturbed controls," *IFAC Congr.*, Kyoto, Japan.

CAINES, P. E., and D. DORER (1980). "Adaptive control of systems subject to a class of random parameter variations and disturbances," Tech. Rep., Department of Electrical Engineering, McGill University, Montreal.

CAINES, P. E., and S. LAFORTUNE (1981). "Adaptive optimization with recursive identification for stochastic linear systems," Tech. Rep., Department of Electrical Engineering, McGill University, Montreal.

CAINES, P. E., and D. Q. MAYNE (1970). "On the discrete time matrix Riccati equation of optimal control," *Int. J. Control*, Vol. 12, No. 5, pp. 785–794.

CAINES, P. E., and D. Q. MAYNE (1971). "On the discrete time matrix Riccati equation of optimal control—a correction," *Int. J. Control*, Vol. 14, No. 1, pp. 205–207.

CALLIER, F. M., and J. L. WILLEMS (1981). "Criterion for the convergence of the solution of the Riccati differential equation," *IEEE Trans. Autom. Control*, Vol. AC-26, No. 6, pp. 1232–1242.

CANTONI, A. (1980). "Fast algorithms for time domain adaptive array processing," NATO Advanced Study Institute on Underwater Acoustics and Signal Processing, pp. 31-1 to 31-12. (Also Tech. Rep. EE8003, Department of Electrical and Computer Engineeing, University of Newcastle, Newcastle, New South Wales, Australia.)

CARTER, T. E. (1978). "Study of an adaptive lattice structure for linear prediction analysis of speech." *Proc. IEEE ICASSP*, pp. 27–30, Tulsa, Oklahoma.

CARVALHAL, F. T., P. E. WELLSTEAD, and J. J. PEREIRA (1976). "Identification and adaptive control in earthquake and vibration testing of large structures," *IFAC Symp. Identif. Parameter Estim.*, Tbilisi, USSR.

CEGRELL, T., and T. HEDQVIST (1975). "Successful adaptive control of paper machines," *Automatica*, Vol. 11, pp. 53–59

CHAN, S W., G. C. GOODWIN, and K. S. SIN (1982). "Convergence properties of the Riccati difference equation in optimal filtering of nonstabilizable systems," Tech. Rep. EE8201, Department of Electrical and Computer Engineering, University of Newcastle, Newcastle, New South Wales, Australia, also *IEEE Trans. Auto. Control*, Vol. 28, No. 12, December.

CHANG, A., and J. RISSANEN (1968). "Regulation of incompletely identified linear systems," *SIAM J. Control*, Vol. 6, No. 3, pp. 327 348.

CHANG, J. H., and F. B. TUTEUR (1971). "A new class of adaptive array processors," *J. Acoust. Soc. Am.*, Vol. 49, No. 3 (Part I), pp. 639–649.

CHEN, C. T. (1970). *Introduction to Linear System Theory*, Holt, Rinehart and Winston, New York.

CHEN, C. T. (1979). *One-Dimensional Digital Signal Processing*, Marcel Dekker, New York.

CHEN, H. F. (1982). "Recursive system identification and adaptive control by use of the modified least squares algorithm," Tech. Rep., McGill University, Montreal.

CHUNG, K. L. (1968). *A Course in Probability Theory*, Harcourt Brace & World, New York.

CLARKE, D. W., and J. P. GAWTHROP (1975). "Self tuning controller," Proc. *IEE*, Vol. 122(a), pp. 929–934.

CLARKE, D. W., S. N. COPE, and J. P. GAWTHROP (1975). "Feasibility study of the application of microprocessors to self tuning controllers," Tech. Rep. 1137/75, Department of Engineering Science, Oxford University, Oxford.

CLARKE, D. W., D. A. J. DYER, R. HASTINGS-JAMES, R. P. ASHTON, and J. B. EMERY (1973). "Identification and control of a pilot scale boiling rig," *IFAC Symp. Identif.*, The Hague, pp. 355–366.

CLEMENTS, D. J., and B. D. O. ANDERSON (1978). *Singular Optimal Control: The Linear-Quadratic Problem*, Springer-Verlag, New York.

COHN, A. (1922). "Über die Anzahl der Wurzeln einer algebraischen Gleichung in einem Kreise," *Math 2*, pp. 110–148.

CORDERO, A. O., and D. Q. MAYNE (1981). "Deterministic convergence of a self-tuning regulator with variable forgetting factor," *Proc. IEE*, Vol. 128, Pt. D, No. 1, pp. 19–23.

COX, H. (1964). "On the estimation of state variables and parameters for noisy dynamic systems," *IEEE Trans. Autom. Control*, Vol. AC-9, pp. 5–12.

COX H., and J. MILLER (1965). *Stochastic Processes*, Chapman & Hall, London.

DANIELL, T. P. (1970). "Adaptive estimation with mutually correlated training sequences," *IEEE Trans. Syst. Sci. Cybern.*, Vol. SSC-6, No. 1, pp. 12–19.

DAVISSON, L. D. (1970). "Steady-state error in adaptive mean-square minimization," *IEEE Trans. Inf. Theory*, Vol. IT-16, No. 4, pp. 382–385.

DE KEYZER, R. M. C., and A. R. VAN CAUWENBERGHE (1979). "A self-tuning predictor as operator guide," *IFAC Symp. Identif. Syst. Parameter Estim.*, Darmstadt, West Germany.

DE SOUZA, C. and G. C. GOODWIN (1983). "Robustness effects of sampling in mode reference and minimum variance control," *22nd CDC Conference*, San Antorio, Texas, December.

DE SOUZA, C., G. C. GOODWIN and S. W. CHAN (1983). "Optimal filtering of nonstabilizable systems having singular state transition matrices," Tech. Rep., Department of Electrical and Computer Engineering, University of Newcastle, Newcastle, New South Wales, Australia, June.

DERMAN, C., and J. SACKS (1959). "On Dvoretsky's stochastic approximation theorem," *Ann. Math. Stat.*, Vol. 30, pp. 601–606.

DESOER, C. A. (1970). *Notes for a Second Course on Linear Systems*, Van Nostrand Reinhold, New York.

DESOER, C. A., and E. S. KUH (1969). *Basic Circuit Theory*, McGraw-Hill, New York.

DESOER, C. A., and M. VIDYASAGAR (1975). *Feedback Systems: Input–Output Properties*, Academic Press, New York.

DOOB, (1953). *Stochastic Processes*, John Wiley, New York.

DUGARD, L., G. C. GOODWIN and C. DE SOUZA (1983). "Prior knowledge in model reference adaptive control of multi-input multi-output systems," *22nd IEEE Decision and Control Conference*, San Antorio, Texas.

DUMONT, G. A. (1982). "Self-tuning control of a chip refiner motor load," *Automatica*, Vol. 18, No. 13, pp. 307–314.

DUMONT, G. A., and P. R. BÉLANGER (1978a). "Control of titanium dioxide kiln," *IEEE Trans. Autom. Control*, Vol. AC-23, No. 4, pp. 521–531.

DUMONT, G. A., and P. R. BÉLANGER (1978b). "Self tuning control of a titanium dioxide kiln," *IEEE Trans. Autom. Control*, Vol. AC-23, No. 4, pp. 532–538.

DUPAC, V. (1965). "A dynamic stochastic approximation method," *Ann. Math. Stat.*, Vol. 36, pp. 1695–1702.

DURBIN, J. (1960). "The fitting of time-series models," *Rev. Inst. Int. Stat.*, Vol. 28, No. 3, pp. 233–243.

DUTTWEILER, D. L. (1980). "Bell's echo-killer chip," *IEEE Spectrum*, Vol. 17, No. 10, pp. 34–37.

DVORETZKY, A. (1956). "On stochastic approximation," *Proc. 3rd Berkeley Symp. Math. Stat. Probab. 1*, University of California Press, Berkeley, Calif., pp. 39–55.

EGARDT, B. (1978). "A unified approach to model reference adaptive systems and self tuning regulators," Tech. Rep., Department of Automatic Control, Lund Institute of Technology, Lund, Sweden.

EGARDT, B. (1980a). "Unification of some discrete-time adaptive control systems," *IEEE Trans. Autom. Control*, Vol. AC-25, No. 4, pp. 693–697.

EGARDT, B. (1980b). "Stability analysis of discrete-time adaptive control schemes," *IEEE Trans. Autom. Control*, Vol. AC-25, No. 4, pp. 710–717.

ELLIOTT, H. (1980). "Hybrid adaptive control of continuous time systems," Tech. Rep. JL80-DELENG-1, Colorado State University, Fort Collins, Colo. Also *IEEE Trans. Autom. Control*, Vol. AC-27, No. 2, April.

ELLIOTT, H., R. CHRISTI and M. DAS (1982). "Global stability of a direct hybrid adaptive pole placement algorithm," Tech. Rep. #UMASS-ECE—No. 81-1, University of Massachusetts, Amherst.

ELLIOTT, H., and W. A. WOLOVICH (1979). "Parameter adaptive identification and control," *IEEE Trans. Autom. Control*, Vol. AC-24, No. 4, pp. 592–599.

EVERLEIGH, W. (1967). *Adaptive Control and Optimization Techniques*, McGraw-Hill, New York.

EYKHOFF, P. (1974). *System Identification: Parameter and State Estimation*, Wiley, Chichester, England.

FARDEN, D. C., J. C. GODING, and K. SAYWOOD (1979). "On the 'desired behaviour' of adaptive signal processing algorithms," *Proc. Int. Conf. on Acoustics, Speech and Signal Processing*, Washington D.C., April.

FARISON, J. B., R. E. GRAHAM, and R. C. SHELTON (1967). "Identification and control of linear discrete systems," *IEEE Trans. Autom. Control*, Vol. AC-12, pp. 438–442.

FEINTUCH, P. L. (1976). "An adaptive recursive LMS filter," *Proc. IEEE*, Vol. 64, pp. 1622–1624.

FEL'DBAUM, A. A. (1965). *Optimal Control Systems*, Academic Press, New York.

FEUER, A., and S. MORSE (1978). "Adaptive control of single-input single-output linear systems," *IEEE Trans. Autom. Control*, Vol. AC-23, No. 4, pp. 557–570.

FORNEY, D. D., JR. (1972). "Maximum-likelihood sequence estimation of digital sequences in the presence of intersymbol interference," *IEEE Trans. Inf. Theory*, Vol. IT-18, pp. 363.

FORTESCUE, T. R., L. S. KERSHENBAUM, and B. E. YDSTIE (1981). "Implementation of self tuning regulators with variable forgetting factors," *Automatica*, Vol. 17, pp. 831–835.

FORTMANN, T., and K. HITZ (1977). *An Introduction to Linear Control Systems*, Marcel Dekker, New York.

FRANCIS, B. A., and W. M. WONHAM (1976). "The internal model principle of control theory," *Automatica*, Vol. 12, pp. 457–465.

FRANKLIN, G. F., and J. D. POWELL (1980). *Digital Control of Dynamic Systems*, Addison-Wesley, Reading, Mass.

FROST, O. L., III (1972). "An algorithm for linearly constrained adaptive array processing," *Proc. IEEE*, Vol. 60, pp. 926–935.

FUCHS, J. J. J. (1980a). "Discrete adaptive control: a sufficient condition for stability and applications," *IEEE Trans. Autom. Control*, Vol. AC-25, No. 5, pp. 940–946.

FUCHS, J. J. J. (1980b). "The recursive least-squares algorithm revisited," Rep. IRISA Laboratoire Automatique, Rennes, France.

FUNAHASHI, Y. (1979). "An observable canonical form of discrete-time bilinear systems," *IEEE Trans. Autom. Control*, Vol. AC-24, pp. 802–803.

FURHT, B. P. (1973). "New estimator for the identification of dynamic processes," IBK Rep., Institut Boris Kidric Vinca, Belgrade, Yugoslavia.

GABRIEL, W. F. (1976). "Adaptive arrays—an introduction," *Proc. IEEE*, Vol. 64, No. 2, pp. 239–272.

GANNETT, J. W., and L. O. CHUA (1978). "Frequency domain passivity conditions for linear time-invariant lumped networks," Memo. UCB/ERL M78/21, Electronics Research Laboratory, University of California, Berkeley, Calif.

GAWTHROP, P. J. (1980). "On the stability and convergence of a self-tuning controller," *Int. J. Control*, Vol. 31, No. 5, pp. 973–998.

GEORGE, D. A., R. R. BOWEN, and J. R. STOREY (1971). "Adaptive decision feedback equalization for digital communications over dispersive channels," *IEEE Trans. Commun.*, Vol. COMM-19.

GERSHO, A. (1969). "Adaptive equalization of highly dispersive channels for data transmission," *Bell Syst. Tech., J.*, Vol. 48, No. 1, pp. 55–70.

GERSOVITZ, M., and J. G. MACKINNON (1978). "Seasonality in regression: an application of smoothness priors," *J. Am. Stat. Assoc.*, Vol. 73, pp. 264–273.

GERTLER, J., and C. BANYASZ (1974). "A recursive (on-line) maximum likelihood identification method," *IEEE Trans. Autom. Control*, Vol. AC-19, pp. 816–820.

GEVERS, M. R., and V. J. WERTZ (1980a). "A d-step predictor in lattice and ladder form," Tech. Rep., Department of Electrical and Computer Engineering, University of Newcastle, Newcastle, New South Wales, Australia.

GEVERS, M. R., and V. J. WERTZ (1980b). "A recursive least squares d-step ahead predictor in lattice and ladder form," Tech. Rep., Department of Electrical and Computer Engineering, University of Newcastle, Newcastle, New South Wales, Australia.

GLOVER, J. R., JR. (1977). "Adaptive noise cancelling applied to sinusoidal interference," *IEEE Trans. Acoust. Speech Signal Process.*, Vol. ASSP-25, No. 6, pp. 484–491.

GOODWIN, G. C., and S. W. CHAN (1982). "Restricted complexity predictors for time series with deterministic and non-deterministic components," *Utilitas Math.*, May.

GOODWIN, G. C. and L. DUGARD. "Stochastic adaptive control with known and unknown interactor matrices," *IFAC Workshop on Adaptive Control*, San Francisco, June.

GOODWIN, G. C., and R. S. LONG (1980). "Generalization of results on multivariable adaptive controls," *IEEE Trans. Autom. Control*, Vol. AC-25, No. 6.

GOODWIN, G. C., and R. L. PAYNE (1977). *Dynamic System Identification: Experiment Design and Data Analysis*, Academic Press, New York.

GOODWIN, G. C., and K. S. SIN (1981). "Adaptive control of non-minimum phase systems," *IEEE Trans. Autom. Control*, Vol. 26, No. 2.

GOODWIN, G. C., R. JOHNSON, and K. S. SIN (1981). "Global convergence for adaptive one step ahead optimal controllers based on input matching," *IEEE Trans. Autom. Control*, Vol. AC-26, No. 6, pp. 1269–1273.

GOODWIN, G. C., R. S. LONG, and B. MCINNIS (1980). "Adaptive control of bilinear systems," Tech. Rep. 8017, Department of Electrical and Computer Engineering, University of Newcastle, Newcastle, New South Wales, Australia.

GOODWIN, G. C., B. MCINNIS, and R. S. LONG (1981). "Adaptive control algorithms for waste water treatment and pH neutralization," Tech. Rep. EE8112, Department of Electrical and Computer Engineering, University of Newcastle, Newcastle, New South Wales, Australia.

GOODWIN, G. C., P. J. RAMADGE, and P. E. CAINES (1978a). "Discrete time multivariable adaptive control," Tech. Rep., Harvard University, Cambridge, Mass. (See also *IEEE Trans. Autom. Control*, Vol. AC-25, pp. 449–456.)

GOODWIN, G. C., P. J. RAMADGE, and P. E. CAINES (1978b). "Stochastic adaptive control," Tech. Rep., Harvard University, Cambridge, Mass. (Also *SIAM J. Control Optim.*)

GOODWIN, G. C., K. K. SALUJA, and K. S. SIN (1979). "An adaptive infinite impulse response digital filter," Tech. Rep. EE7914, Department of Electrical and Computer Engineering, University of Newcastle, Newcastle, New South Wales, Australia.

GOODWIN, G. C., K. K. SALUJA, and K. S. SIN (1980). "A self tuning fixed lag smoother," Tech. Rep. EE8013, Department of Electrical and Computer Engineering, University of Newcastle, Newcastle, New South Wales, Australia.

GOODWIN, G. C., K. S. SIN, and K. K. SALUJA (1980). "Stochastic adaptive control and prediction: the general delay-coloured noise case," *IEEE Trans. Autom. Cont.*, Vol. AC-25, No. 5, pp. 946–950.

GOODWIN, G. C., and E. K. TEOH, (1983). "Adaptive control of a class of linear time varying systems," *IFAC Workshop on Adaptive Systems in Control and Signal Processing*, San Francisco, June 20–22.

GRAY, A. H., and J. D. MARKEL (1973). "Digital lattice and ladder filter synthesis," *IEEE Trans. Audio Electroacoust.*, Vol. AU-21, No. 6, pp. 491–500.

GREGORY, P. C., Ed. (1959). "Proceedings of the Self Adaptive Flight Control Systems Symposium," WADC Tech. Rep. 59-49, Wright-Patterson Air Force Base, Ohio.

GRENANDER, N., and G. SZEGO (1958). *Toeplitz Forms and Their Application*, University of California Press, Berkeley, Calif.

GRIFFITHS, L. J. (1969). "A simple adaptive algorithm for real-time processing in antenna arrays," *Proc. IEEE*, Vol. 57, pp. 1696–1704.

GRIFFITHS, L. J. (1975). "Rapid measurement of instantaneous frequency," *IEEE Trans. Acoust., Speech Signal Process.*, Vol. ASSP-23, pp. 209–222.

GRIFFITHS, L. J. (1977). "A continuously adaptive filter implemented as a lattice structure," *Int. Conf. Acoust. Speech Signal Process.*, Hartford, Connecticut.

GRIFFITHS, L. J., and R. S. MEDAUGH (1978). "Convergence properties of an adaptive noise cancelling lattice structure," *Proc. 1978 IEEE Conf. Decis. Control*, San Diego, Calif., pp. 1357–1361.

GUIDORZI, R. P. (1981). "Invariants and canonical forms for systems structural and parametric identification," *Automatica*, Vol. 17, No. 1, pp. 117–133.

GUPTA, N. K., and R. K. MEHRA (1974). "Computational aspects of maximum likelihood estimation and reduction in sensitivity function calculations," *IEEE Trans. Autom. Control*, Vol. AC-19, pp. 774–784.

HAGANDER, P., and B. WITTENMARK (1977). "A self tuning filter for fixed-lag smoothing," *IEEE Trans. Inf. Theory*, Vol. IT-23. No. 3, pp. 377–384.

HAHN, W. (1963). *Theory and Application of Lyapunov's Direct Method*, Prentice-Hall, Englewood Cliffs, N.J.

HALL, P., and C. C. HEYDE (1980). *Martingale Limit Theory and Its Applications*, Academic Press, New York.

HANNAN, E. J. (1970). *Multiple Time Series*, Wiley, New York.

HANNAN, E. J. (1978). "Recursive estimation based on ARMA models," Tech. Rep., Australian National University, Canberra, Australia.

HARRIS, C. J., and S. A. BILLINGS (1981). *Self-Tuning and Adaptive Control: Theory and Applications*, Peregrinus, Stevenage.

HERSH, M. A., and M. B. ZARROP (1981). "Stochastic adaptive control of nonminimum phase systems," Tech. Rep., Control Systems Centre, University of Manchester Institute of Science and Technology, Manchester, England.

HEWER, G. A. (1973). "Analysis of a discrete matrix Riccati equation of linear control and Kalman filtering," *J. Math. Anal. Appl.*, Vol. 42, pp. 226–236.

HEYDE, C. C. (1977). "On central limit and iterated logarithm supplements to the martingale convergence theorem," *J. Appl. Probab.*, Vol. 14, pp. 758–775.

HILL, D. J., and P. J. MOYLAN (1975). "Cyclo-dissipativeness, dissipativeness and losslessness for nonlinear dynamical systems," Tech. Rep. EE7526, Department of Electrical and Computer Engineering, University of Newcastle, Newcastle, New South Wales, Australia.

HILL, D. J., and P. J. MOYLAN (1980). "Dissipative dynamical systems: basic input-output and state properties," *J. Franklin Inst.*, Vol. 309, No. 5.

HITZ, K. L., and B. D. O. ANDERSON (1969). "Discrete positive real functions and their applications in system stability," *Proc. IEE*, Vol. 116, pp. 153–155.

HOLST, J. (1977). "Adaptive prediction and recursive parameter estimation," Rep. TFRT-1013-1-206. Department of Automatic Control, Lund Institute of Technology, Lund, Sweden.

IOANNOU, P. A., and P. V. KOKOTOVIC (1982). "Singular perturbations and robust redesign of adaptive control," *21st Conf. Decis. Control*, Orlando, Fla.

IONESCU, T., and R. V. MONOPOLI (1977). "Discrete model reference adaptive control with an augmented error singnal," *Automatica*, Vol. 13, No. 5, pp. 507–518.

ITAKURA, F., and S. SAITO (1971). "Digital filtering for speech analysis and synthesis," *Proc. 7th Int. Conf. Acoust.*, Budapest, Vol. 25-C-1, pp. 261–264.

JACOBS, O. L. R., and J. W. PATCHELL (1972). "Caution and probing in stochastic control," *Int. J. Control*, Vol. 16, pp. 189–199.

JACOBS, O. L. R., and P. SARATCHANDRAN (1980). "Comparison of adaptive controllers," *Automatica*, Vol. 16, pp. 89–97.

JEANNEAU, J. L., and P. DE LARMINANT (1975). "A method for adaptive regulation of nonminimum phase systems," *Annales, École Nationale Supérieure de Mécanique* (in French).

JOHANSSON, R. (1982). "Parametric models of linear multivariable systems for adaptive control," *Proc. 1982 Conference on Decision and Control*, Orlando, Fla.

JOHNSON C. R., JR. (1979). "A convergence proof for a hyperstable adaptive recursive filter," *IEEE Trans. Inf. Theory*, Vol. IT-25, No. 6, pp. 745–749.

JOHNSON C. R., JR. (1980). "A stable family of adaptive IIR filters," *Proc. 1980 IEEE Int. Conf. Acoust., Speech Signal Process.*, Denver, Colo., pp. 1001–1004.

JOHNSON, C. R., JR., and G. C. GOODWIN (1982). "Robustness issues in adaptive control," *21st Conf. Decis. Control*, Orlando, Fla.

JOHNSON, C. R., JR., and T. TAYLOR (1979). "Failure of parallel adaptive identifier with adaptive error filtering," Tech. Rep. EE7912, Department of Electrical Engineering, Virginia Polytechnic Institute and State University, Blacksburg, Va.

JOHNSON, C. R., JR., and E. TSE (1978). "Adaptive implementation of one-step-ahead, optimal control via input matching," *IEEE Trans. Autom. Control*, Vol. AC-23, No. 5) pp. 865–872.

JOHNSON, C. R., JR., M. G. LARIMORE, J. R. TREICHLER, and B. D. O. ANDERSON (1981). "SHARF convergence properties," *IEEE Trans. Acoust. Speech Signal Process.*, Vol. ASSP-29, No. 3, pp. 659–670.

JONES, S. K. (1973). "Adaptive filtering with correlated training samples: a mean square error analysis," Ph.D. thesis, Electrical Engineering Department, Southern Methodist University, Dallas, Tex.

JOSEPH, P. D., and J. T. TOU (1961). "On linear control theory," *Trans. Am. Inst. Electr. Eng.*, Pt. 2, Vol. 80, pp. 193–196.

KAILATH, T. (1980). *Linear Systems*, Prentice-Hall, Englewood Cliffs, N.J.

KÄLLSTRÖM, C. G., K. J. ÅSTRÖM, N. E. THORELL, J. ERIKSSON, and L. STEN (1979). "Adaptive autopilots for tankers," *Automatica*, Vol. 15, No. 3, pp. 241–254.

KALMAN, R. E. (1958). "Design of a self-optimizing control system," *Trans. ASME*, Vol. 80, pp. 468–478.

KALMAN, R. E. (1960). "A new approach to linear filtering and prediction problems," *J. Basic Eng., Trans. ASME*, Ser. D, Vol. 82, No. 1, pp. 35–45.

KALMAN, R. E. (1961). "Contributions to the theory of optimal control," *Bol. Soc. Mat. Mex.*, Vol. 5, pp. 102–119.

KALMAN, R. E. (1963). "New methods in Wiener filtering theory," *Proc. Symp. Eng. Appl. Random Funct. Theory Probab.* ed. J. L. Bogdanoff and F. Kozin, Wiley, New York.

KALMAN, R. E. (1964). "When is a linear control system optimal?" *J. Basic Eng., Trans. ASME*, Vol. 86, March. pp. 51–60.

KALMAN, R. E., and R. S. BUCY (1961). "New results in linear filtering and prediction theory," *J. Basic Eng., Trans. ASME*, Ser. D, Vol. 83, No. 3, pp. 95–108.

KEVICZKY, L., and K. S. P. KUMAR (1979). "On the choice of sampling interval applying certain optimal regulators," Tech. Rep., Center for Control Sciences, University of Minnesota, Minneapolis, Minn.

KEVICZKY, L., J. HETTHESSY, M. HILGER, and J. KOLOSTORI (1978). "Self-tuning adaptive control of cement raw material blending," *Automatica*, Vol. 14, No. 6, pp. 525–532.

KIEFER, J., and J. WOLFOWITZ (1952). "Stochastic estimation of the maximum of a regression function," *Ann. Math. Stat.*, Vol. 23, pp. 462–466.

KIKUCHI, A., S. OMATI, and T. SOEDA (1979). "Application of adaptive digital filtering to the data processing for the environmental system," *IEEE Trans. Acoust. Speech Signal Process.*, Vol. ASSP-27, No. 6, pp. 790–803.

KIM, J.-K., and L. D. DAVISSON (1975). "Adaptive linear estimation for stationary M-dependent processes," *IEEE Trans. Inf. Theory*, Vol. IT-21, No. 1, pp. 23–31.

KNOPP, K. (1956). *Infinite Sequences and Series*, trans. F. Bagemihl, Dover, New York.

KO, K. (1980). M.Sc. thesis, Electrical Engineering Department, University of Houston, Houston, Tex.

KOLMOGOROV, A. N. (1941). "Interpolation and extrapolation of stationary random sequences," *Bull. Acad. Sci USSR Ser. Math.*, Vol. 5. Transl: Rand Corp., Santa Monica, Calif., Memo RM-3090 PR.

KOPP, R. E., and R. J. ORFORD (1963). "Linear regression applied to system identification and adaptive control systems," *AIAA J.*, Vol. 1, No. 10, pp. 2300–2306.

KOSUT, R. L., and B. FRIEDLANDER (1982). "Performance robustness properties of adaptive control systems," *21st Conf. Decis. Control*, Orlando, Fla.

KREISSELMEIER, G. (1980). "Adaptive control via adaptive observation and asymptotic feedback matrix synthesis," *IEEE Trans. Autom. Control*, Vol. AC-25, No. 4.

KUCERA, V. (1972a). "On nonnegative definite solutions to matrix quadratic equations," *Automatica*, Vol. 8, pp. 413–423.

KUCERA, V. (1972b). "The discrete Riccati equation of optimal control," *Kobern. Dok.*, Vol. 8, pp. 430–447.

KUCERA, V. (1972c). "A contribution to matrix quadratic equations," *IEEE Trans. Autom. Control*, Vol. AC-17, pp. 344–347.

KUMAR, R. (1981). "Almost sure convergence of adaptive identification, prediction and control algorithms," Tech. Rep. LCDS 81-8. Division of Applied Mathematics, Brown University, Providence, R.J.

KUMAR, R., and J. B. MOORE (1979). "Convergence of adaptive minimum variance algorithms via weighting coefficient selection," Tech. Rep. 7917, Department of Electrical and Computer Engineering, University of Newcastle, Newcastle, New South Wales, Australia.

Kumar, R., and J. B. Moore (1980). "Adaptive equalization via fast quantized state methods," Tech. Rep. EE8008, Department of Electrical and Computer Engineering, University of Newcastle, Newcastle, New South Wales, Australia.

KUO, B. C. (1980). *Digital Control Systems*, Holt, Rinehart and Winston, New York.

KUSHNER, H. J. (1977). "General convergence results for stochastic approximations via weak convergence theory," *J. Math. Anal. Appl.*, Vol. 61, pp. 490–503.

KUSHNER, H. J. (1978a). "Rates of convergence for sequential Monte Carlo optimization methods," *SIAM J. Control Optim.*, Vol. 10, pp. 150–168.

KUSHNER, H. J. (1978b). "Convergence of recursive adaptive and identification procedures via weak convergence theory," *IEEE Trans. Autom. Control*, Vol. 22, pp. 921–930.

KUSHNER, H. J., and KUMAR, R. (1981). "Convergence and rate of convergence of a recursive identification and adaptive control method which uses truncated estimators," Tech. Rep. LCDS 81-4, Division of Applied Mathematics, Brown University, Providence, R.J.

KUTZ, J., R. ISERMANN, and R. SCHUMANN (1980). "Experimental comparison and application of various parameter adaptive control algorithms," *Automatica*, Vol. 16, No. 2, pp. 117–133.

KWAKERNAAK, H., and R. SIVAN (1972). *Linear Optimal Control Systems*, Wiley, New York.

LAFORTUNE, S. (1982). "Adaptive control with recursive identification for stochastic linear systems," M.Sc. thesis, McGill University, Montreal.

LAINIOTIS, D. G. (1976a). "Partitioning: a unifying framework for adaptive systems I: estimation," *Proc. IEEE*, Vol. 64, No. 8, pp. 1126–1142.

LAINIOTIS, D. G. (1976b). "Partitioning: a unifying framework of adaptive systems II: control," *Proc. IEEE*, Vol. 64, No. 8, pp. 1182–1197.

LAMPERTI, J. (1977). *Stochastic Processes; A Survey of the Mathematical Theory*, Springer-Verlag, New York.

LANDAU, I. D. (1974). "A survey of model reference adaptive techniques—theory and applications," *Automatica*, Vol. 10, pp. 353–379.

LANDAU, I. D. (1976). "Unbiased recursive identification using model reference adaptive techniques," *IEEE Trans. Autom. Control*, Vol. AC-21, pp. 194–202.

LANDAU, I. D. (1978a). "An addendum to 'Unbiased recursive identification using model reference adaptive techniques'," *IEEE Trans. Autom. Control*, Vol. AC-23, No. 1, pp. 97–99.

LANDAU, I. D. (1978b). "Elimination of the real positivity condition in the design of parallel MRAS," *IEEE Trans. Autom. Control*, Vol. AC-23, No. 6, pp. 1015–1020.

LANDAU, I. D. (1979). *Adaptive Control—The Model Reference Approach*, Marcel Dekker, New York.

LANDAU, I. D. (1981). "Combining model reference adaptive controllers and stochastic self-tuning regulators'" *IFAC Congr.*, Kyoto, Japan.

LARIMORE, M. G., J. R. TREICHLER, and C. R. JOHNSON, JR. (1980). "SHARF: an algorithm for adapting IIR digital filters," *IEEE Trans. Acoust. Speech Signal Process.*, Vol. ASSP-28, No. 4.

LEE, D. T. L., and M. MORF (1980). "Recursive square-root ladder estimation algorithm," *Proc. 1980 IEEE Int. Conf. Acoust. Speech Signal Process.*, Denver, Colo.

LEE, D. T. L., M. MORF, and B. FRIEDLANDER (1981). "Recursive least squares ladder estimation algorithms," *IEEE Trans. Circuits Syst.*, Vol. CAS-28, No. 6, pp. 467–481.

LEUNG, V., and L. PADMANABHAN (1973). "Improved estimation algorithms using smoothing and relinearization," *Chem. Eng. J.*, Vol. 5, pp. 197–208.

LEVINSON, N. (1947). "The Wiener rms (root mean square) error criterion—filter design and prediction," *J. Math. Phys.*, Vol. 25, pp. 261–278.

LJUNG, L. (1974). "Convergence of recursive stochastic algorithms," Rep. 7403, Department of Automatic Control, Lund Institute of Technology, Lund, Sweden.

LJUNG, L. (1975). "Theorems for the asymptotic analysis of recursive stochastic algorithms," Rep. 7522, Department of Automatic Control, Lund Institute of Technology, Lund, Sweden.

LJUNG, L. (1976). "Consistency of the least squares identification method," *IEEE Trans. Autom. Control*, Vol. AC-21, No. 5, pp. 779–781.

LJUNG, L. (1977a). "Analysis of recursive stochastic algorithms," *IEEE Trans. Autom. Control*, Vol. AC-22, No. 4, pp. 551–575.

LJUNG, L. (1977b). "On positive real functions and the convergence of some recursive schemes," *IEEE Trans. Autom. Control*, Vol. AC-22, pp. 539–551.

LJUNG, L. (1978a). "On recursive prediction error identification algorithms," Rep. LiTH-ISY-I-0226, Linköping University, Linköping, Sweden.

LJUNG, L. (1978b). "Strong convergence of a stochastic approximation algorithm," *Ann. Stat.*, Vol. 6, No. 3, pp. 680–696.

LJUNG, L. (1979a). "Convergence of recursive estimators," *IFAC Symp.*, Darmstadt, West Germany.

LJUNG, L. (1979b). "Asymptotic behaviour of the extended Kalman filter as parameter estimator for linear systems," *IEEE Trans. Autom. Control*, Vol. AC-24, No. 1, pp. 36–51.

LJUNG, L. (1981). "Analysis of a general recursive prediction error identification algorithm," *Automatica*, Vol. 17, No. 1, pp. 89–99.

LJUNG, L., and P. CAINES (1979). "Asymptotic normality of prediction error estimators for approximate system models," *Automatica*, Vol. 3, pp. 29–46.

LJUNG, L., and J. RISSANEN (1976). "On canonical forms, parameter identifiability and the concept of complexity," *IFAC Symp. Identif.*, Tbilisi, USSR.

LJUNG, L., and T. SÖDERSTRÖM (1982). *Theory and Practice of Recursive Identification*, MIT Press, Cambridge, Mass.

LJUNG, L., and B. WITTENMARK (1974). "Asymptotic properties of self tuning regulators," Rep. 7404, Division of Automatic Control, Lund Institute of Technology, Lund, Sweden.

LJUNG, L., and B. WITTENMARK (1976). "On a stabilizing property of adaptive regulators," Preprint, *IFAC Symp. Identif.*, Tbilisi, USSR.

LOEVE, M. (1963). *Probability Theory*, Springer-Verlag, New York.

LONG, R. S., G. C. GOODWIN, and E. K. TEOH (1982). "Modelling of discrete time bilinear systems in state space and input–output formats," Tech. Rep., Department of Electrical and Computer Engineering, University of Newcastle, Newcastle, New South Wales, Australia.

LOZANO, L. R. (1981). "Adaptive control with forgetting factor," *IFAC World Congr.*, Kyoto, Japan.

LUCKY, R. W. (1966). "Techniques for adaptive equalization of digital communication systems," *Bell Syst. Tech. J.*, Vol. 45, pp. 255–256.

LUENBERGER, D. G. (1973). *Introduction to Linear and Nonlinear Programming*, Addison-Wesley, Reading, Mass.

MAKHOUL, J. (1977). "Stable and efficient lattice methods for linear prediction," *IEEE Trans. Acoust., Speech Signal Process.*, Vol. ASSP-25, No. 5, pp. 423–428.

MAKHOUL, J. (1978). "A class of all-zero lattice digital filters: properties and applications," *IEEE Trans. Acoust. Speech Signal Process.*, Vol. ASSP-26, No. 4, pp. 304–314.

MAKHOUL, J. I., and L. K. COSELL (1981). "Adaptive analysis of speech," *IEEE Trans. Acoust. Speech, Signal Process.*, Vol. ASSP-29, No. 3, pp. 654–659.

MARTENSSON, K. (1971). "On the matrix Riccati equation," *Inf. Sci.*, Vol. 3, pp. 17–49.

MARTIN-SANCHEZ, J. M. (1976). "A new solution to adaptive control," *Proc. IEEE*, Vol. 64, No. 8, pp. 1209–1218.

McAVOY, T. J. (1972). "Time optimal and Ziegler-Nichols control," *Ind. Eng. Chem. Process Des. Dev.*, Vol. 11, No. 1.

McAVOY, T. J., E. HSU, and S. LOWENTHAL (1972). "Dynamics of pH in a controlled stirred tank reactor," *Ind. Eng. Chem. Process Des. Dev.*, Vol. 11, No. 1.

McINNIS, B. C., C. LIN, and P. BUTLER (1979). "Adaptive microcomputer dissolved oxygen controller for wastewater treatment," Tech. Rep., Department of Electrical Engineering, University of Houston, Houston, Tex.

McINNIS, B. C., J. C. WANG, and T. AKUTSU (1981). "Adaptive controls for the artificial heart," *Proc. IEEE Front. Eng. Health Care*, Houston, Tex.

McINNIS, B. C., R. L. EVERETT, J. C. WANG, and B. VAJAPEYAM (1981). "A microcomputer based adaptive control system for the artificial heart," *IFAC Congr.*, Kyoto, Japan.

MENDEL, J. M. (1973). *Discrete Techniques of Parameter Estimation*, Marcel Dekker, New York.

MESSERSCHMITT, D. G. (1980). "A class of generalized lattice filters," *IEEE Trans. Acoust. Speech Signal Process.*, Vol. ASSP-28, No. 2.

MONOPOLI, R. V. (1974). "Model reference adaptive control with an augmented error signal," *IEEE Trans. Autom. Control*, Vol. AC-19, pp. 474–482.

• Monzingo, R. A., and T. W. Miller (1980). *Introduction to Adaptive Arrays*, Wiley-Interscience, New York.

Moore, J. B. (1980). "Persistence of excitation in extended least squares," Tech. Rep., Department of Electrical and Computer Engineering, University of Newcastle, Newcastle, New South Wales, Australia.

Moore, J. B., and R. Kumar (1980). "Convergence of weighted minimum variance N-step ahead prediction error schemes," Tech. Rep. 8009, Department of Electrical Engineering, University of Newcastle, Newcastle, New South Wales, Australia.

Moore, J. B., and G. Ledwich (1977). "Multivariable adaptive parameter and state estimation with convergence analysis," Tech. Rep., Department of Electrical Engineering, University of Newcastle, Newcastle, New South Wales, Australia.

Morf, M., and D. T. L. Lee (1979). "Recursive least squares ladder forms for fast parameter tracking," *IEEE Conf. Decis. Control*, Fort Landerdale, Fla.

Morf, M., A. Vieira, and D. T. L. Lee (1977). "Ladder forms for identification and speech processing," *IEEE Conf. Decis. Control*, New Orleans.

Morf, M., D. T. L. Lee, Nicholls, and A. Vieira (1977). "A classification of algorithms for ARMA models and ladder realization," *Int. Conf. Acoust. Speech Signal Process.*, Hartford, Connecticut.

Morse, A. S. (1980). "Global stability of parameter-adaptive control systems," *IEEE Trans. Autom. Control*, Vol. AC-25, No. 3, pp. 433–439.

Muschner, J. L. (1970). "Adaptive filter with clipped input data," Tech. Rep. 6796–1, Stanford University Center for System Research, Stanford, Calif.

Nagumo, J.-I., and A. Noda (1967). "A learning method for system identification," *IEEE Trans. Autom. Control*, Vol. AC-12, No. 3, pp. 282–287.

Narendra, K. S., and Y. H. Lin (1980). "Stable discrete adaptive control," *IEEE Trans. Autom. Control*, Vol. AC-25, No. 3, pp. 456–461.

Narendra, K. S., and R. V. Monopoli, eds. (1980). *Applications of Adaptive Control*, Academic Press, New York.

Narendra, K. S., and B. B. Peterson (1980). "Bounded error adaptive control," *Proc. 19th IEEE Conf. Decis. Control*, Albuquerque, N. Mex., pp. 605–610.

Narendra, K. S., and L. S. Valavani (1978). "Stable adaptive controller design-direct control," *IEEE Trans. Autom. Control*, Vol. AC-23, No. 4, pp. 570–583.

Narendra, K. S., and L. S. Valavani (1980). "A comparison of Lyapunov and hyperstability approaches to adaptive control of continuous systems," *IEEE Trans. Autom. Control*, Vol. AC-25, pp. 243–247.

Narendra, K. S., Y. H. Lin, and L. S. Valavani (1980). "Stable adaptive controller design, Part II: Proof of stability," *IEEE Trans. Autom. Control*, Vol. AC-25, No. 3, pp. 440–449.

Nelson, L. W., and E. Stear (1976). "The simultaneous on-line estimation of parameters and states in linear systems," *IEEE Trans. Autom. Control*, Vol. AC-21, pp. 94–98.

Nevel'son, M. B., and R. E. Z. Khasminskii (1973). "Stochastic approximation and rescursive estimation," in *Translations of Mathematical Monographs*, Vol. 47, American Mathematical Society, Providence, R.I., Chap. 7.

Neveu, J. (1975). *Discrete Parameter Martingales*, North-Holland, Amsterdam.

Olsson, G. (1977). "State of the art in sewerage treatment plant control," *AIChE Symp. Ser.*, Vol. 172, No. 159, pp. 52–76.

Olsson, G. (1980). "Some new results on activated sludge control based on dissolved oxygen profiles," *Joint Autom. Control Conf.*, San Francisco.

OMURA, J. K. (1971). "Optimal receiver design for convolutional codes and channels with memory via a control theoretic approach," *Inf. Sci.*, Vol. 3.

OPPENHEIM, A. V. (1978). *Applications of Digital Signal Processing*, Prentice-Hall, Englewood Cliffs, N.J.

OPPENHEIM, A. V., and R. W. SCHAFER (1975). *Digital Signal Processing*, Prentice-Hall, Englewood Cliffs, N.J.

OWSLEY, N. (1980). "An overview of optimum adaptive control in sonar array processing," in *Applications of Adaptive Control*, ed. K. S. Narendra and R. V. Monopolic, Academic Press, New York.

PALMER, Z. J., and R. Shinnar (1979). "Design of sampled data controllers," *Ind. Eng. Chem. Process Des. Dev.*, Vol. 18, No. 1.

PANUSAKA, V. (1968). "A stochastic approximation method for identification of linear systems using adaptive filtering," *Joint Autom. Control Conf.*, University of Michigan.

PANUSKA, V. (1969). "An adaptive recursive least squares identification algorithm," *Proc. IEEE Symp. Adapt. Process. Decis. Control.*

PAPOULIS, A. (1965). *Probability, Random Variables, and Stochastic Processes*, McGraw-Hill, New York.

PARIKH, D., and N. AHMED (1978). "On an adaptive algorithm for IIR filters," *Proc. IEEE*, Vol. 65, No. 5, pp. 585–587.

PARKS, P. C. (1966). "Liapunov redesign of model reference adaptive control systems," *IEEE Trans. Autom. Control*, Vol. AC-11, pp. 362–367.

POLYAK, B. T. (1976). "Convergence and convergence rate of iterative stochastic algorithms, I: General case," *Autom. Remote Control*, No. 12, pp. 1858–1868.

POLYAK, B. T. (1977). "Convergence and convergence rate of iterative stochastic algorithms, II: The linear case," *Autom. Remote Control*, No. 4, pp. 537–542.

POPOV, V. M. (1972). "Invariant description of linear time invariant controllable systems," *SIAM J. Control*, Vol. 10, pp. 252–264.

POPOV, V. M. (1973). *Hyperstability of Automatic Control Systems*, Springer-Verlag, New York.

PRIESTLEY, M. B. (1978). "Non-linear model in time series analysis," *Statistician*, Vol. 27, No. 3, p. 159.

PRIESTLEY, M. B. (1980). "State dependent models: a general approach to nonlinear time series analysis," *Time Ser. Anal.*, Vol. 1, No. 1, pp. 47–73.

QUAGLIANO, J. V. (1958). *Chemistry*, Prentice-Hall, Englewood Cliffs, N.J.

RABINER, L. R., and B. GOLD (1975). *Theory and Application of Digital Signal Processing*, Prentice-Hall, Englewood Cliffs, N.J.

REDDY, V. U., B. EGARDT, and T. KAILATH (1981). "Optimized lattice-form adaptive line enhancer for a sinusoidal signal in broad-band noise," *IEEE Trans. Acoust. Speech Signal Process.*, Vol. ASSP-29, No. 3, pp. 702–709.

REDMAN, G. V. (1980). "Simulation studies of self-tuning regulators," Honours thesis, Department of Electrical and Computer Engineering, University of Newcastle, Newcastle, New South Wales, Australia.

RISSANEN, J. (1981). "Estimation of structure by minimum description length," *Workshop Ration. Approx. Syst.*, Catholic University, Louvain, France.

ROBBINS, H., and S. MONRO (1951). "A stochastic approximation method," *Ann. Math. Stat.*, Vol. 22, pp. 400–407.

ROBINSON, E. A. (1967). *Multi-channel Time-Series Analysis with Digital Computer Programs*, Holden-Day, San Francisco.

ROHRS, C., L. VALAVANI, M. ATHANS, and G. STEIN (1981). "Analytical verification of undesirable properties of direct model reference adaptive control algorithms," *20th IEEE Conf. Decis. Control*, San Diego, Calif.

ROHRS, C., L. VALAVANI, M. ATHANS, and G. STEIN (1982). "Robustness of adaptive control algorithms in the presence of unmodelled dynamics," *21st Conf. Decis. Control*, Orlando, Fla.

ROSENBROCK, H. H. (1970). *State Space and Multivariable Theory*, Wiley-Interscience, New York.

ROZANOV, YU. A. (1967). *Stationary Random Processes*, Holden-Day, San Francisco.

RUDIN, W. (1960). *Principles of Mathematical Analysis*, McGraw-Hill, New York.

SACKS, J. (1958). "Asymptotic distribution of stochastic approximation procedures," *Ann. Math. Stat.*, Vol. 29, pp. 373–405.

SAGE, A. P., and J. L. MELSA (1971). *System Identification*, Academic Press, New York.

SAGE, A. P., and C. D. WAKEFIELD (1972). "Maximum likelihood identification of time varying random parameters," *Int. J. Control*, Vol. 16. No. 1, pp. 81–100.

SARIDIS, G. N. (1977). *Self-Organizing Control of Stochastic Systems*, Marcel Dekker, New York.

SASTRY, V. A., D. E. SEBORG, and R. K. WOOD (1977). "Self-tuning regulator applied to a binary distillation column," *Automatica*, Vol. 13, pp. 417–424.

SENNE, K. D. (1970). "An exact solution to an adaptive linear estimation problem," Tech. Rep. SRL-TR-70-0014, U.S. Air Force Systems Command, Frank J. Seiler Research Labotatory.

SHINSKY, F. G. (1978). "A self-adjusting system for effluent pH control," *Spring Joint Conf. ISA*, St. Louis, Mo.

SIN, K. S., and G. C. GOODWIN (1982). "Stochastic adaptive control using modified least squares algorithms," *Automatica*, Vol. 18, No. 3, p. 315–321.

SMITH, O. J. M. (1959). ISA J., Vol. 6(2), p. 28.

SÖDERSTRÖM, T. (1973). "An on-line algorithm for approximate maximum likelihood identification of linear dynamic systems," Rep. 7208, Department of Automatic Control, Lund Institute of Technology, Lund, Sweden.

SÖDERSTRÖM, T., L. LJUNG, and I. GUSTAVSSON (1974). "A comparative study of recursive identification methods," Rep. 7427, Department of Automatic Control, Lund Institute of Technology, Lund, Sweden.

SÖDERSTRÖM, T., L. LJUNG, and I. GUSTAVSSON (1978). "A theoretical analysis of recursive identification methods," *Automatica*, Vol. 14, No. 3.

SOLO, V. (1978). "Time series recursions and stochastic approximation," Ph.D. thesis, Australian National University, Canberra, Australia.

SOLO, V. (1979). "The convergence of AML," *IEEE Trans. Autom. Control*, Vol. AC-24, No. 6, pp. 958–962.

SOLO, V. (1981). "The second order properties of a time series recursion," *Ann. Stat.*, Vol. 9, pp. 307–317.

SONDHI, M. M., and D. A. BERKLEY (1980). "Silencing echoes on the telephone network," *Proc. IEEE*, Vol. 68, No. 8, pp. 948–963.

SONDHI, M. M., and D. MITRA (1976). "New results on the performance of a well-known class of adaptive filters," *Proc. IEEE*, Vol. 64, No. 11, pp. 1583–1597.

SPEEDY, C. B., R. F. BROWN, and G. C. GOODWIN (1970). *Control Theory*, Oliver & Boyd, Edinburgh.

STANLEY, W. D. (1975). *Digital Signal Processing*, Reston, Reston, Va.

STERNBY, J. (1977). "On consistency of the method of least squares using martingale theory," *IEEE Trans. Autom. Control*, Vol. AC-22, pp. 346–352.

STOUT, W. F. (1970). "The Hartman–Wintner law of the iterated logarithm for martingales," *Ann Math. Stat.*, Vol. 41, pp. 2158–2160.

STRIEBEL, C. (1975). *Optimal Control of Discrete Time Stochastic Systems*, Lecture Notes in Economics and Mathematical Systems, Vol. 110, Springer-Verlag, New York.

SVORONOS, S., G. STEPHANOPOULOS, and R. ARIS (1981). "On bilinear estimation and control," *Int. J. Control*, Vol. 34, No. 4, pp. 651–684.

TALMON, J. L., and A. J. W. BAN DEN BOOM (1973). "On the estimation of transfer function parameters of process and noise dynamics using a single stage estimator," *IFAC Symp. Identif.*, The Hague.

TREICHLER, J. R. (1980). "Response of the adaptive line enhancer to chirped and doppler-shifted sinusoids," *IEEE Trans. Acoust. Speech Signal Process.*, Vol. ASSP-28, No. 3, pp. 343–348.

TSE, E., Y. BAR-SHALOM, and L. MEIER (1973). "Wide sense adaptive dual control of stochastic nonlinear systems," *IEEE Trans. Autom. Control*, Vol. AC-18, p. 98.

TSUCHIYA, T. (1982). "Improved direct digital control algorithm for microprocessor implementation," *IEEE Trans. Autom. Cont.*, Vol. AC-27, No. 2, pp. 295–305.

TSYPKIN, YA. Z. (1971). *Adaptation and Learning in Automatic Systems*, Academic Press, New York.

UNBEHAUEN, H., ed. (1980). *Methods and Applications in Adaptive Control*, Springer-Verlag, Berlin.

UNBEHAUEN, H., C. SCHMID, and F. KLEIN (1978). "Design and application of an adaptive multivariable controller using a process computer," *Proc. 12th Ann. Asilomar Conf. Circuits Syst. Comput.*, Pacific Grove, California.

UNGERBOECK, G. (1971). "Nonlinear equalization of binary signals in Gaussian noise," *IEEE Trans. Commun. Technol.*, Vol. COM-19, No. 6,

VAN AMERONGEN, J., and A. J. UDINK TEN CATE (1975). "Model reference adaptive autopilots for ships," *Automatica*, Vol. 11, pp. 441–449.

VITERBI, A. J. (1966). *Principles of Communication*, McGraw-Hill, New York.

VITERBI, A. J. (1967). "Error bounds for convolution codes and an asymptotical optimum decoding algorithm," *IEEE Trans. Inf. Theory*, Vol. IT-13, pp. 260–269.

VIZWANATHAN, M. N. (1981). "Effect of restrictions on water consumption levels in Newcastle," M.E. thesis, University of Newcastle, Newcastle, New South Wales, Australia.

VIZWANATHEN, R., and J. MAKHOUL (1976). "Sequential lattice methods for stable linear prediction," *Proc. EASCON*, pp. 155A–155H.

VOGEL, E. G., and T. F. EDGAR (1982). "Application of an adaptive pole-zero placement controller to chemical processes with variable dead time," *Am. Control Conf.*, Washington, D.C.

WEBSTER, I. (1981). "Control strategies for an inverter," Honours thesis, Department of

Electrical and Computer Engineering, University of Newcastle, Newcastle, New South Wales, Australia.

WEISS, A., and D. Mitra (1979). "Digital adaptive filters: conditions for convergence rates of convergence effects of noise and errors arising from the implementation," *IEEE Trans. Inf. Theory*, Vol. IT-25, No. 6, pp. 637–653.

WELLSTEAD, P. E., and D. PRAGER (1981). "Multivariable pole assignment self tuning requlators," *Proc. IEE*, Vol. 128.

WELLSTEAD, P. E., and P. ZANKER (1978). "The techniques of self-tuning," Rep. 432, Control Systems Centre, University of Manchester Institute of Science and Technology, Manchester, England.

WELLSTEAD, P. E., D. PRAGER, and P. ZANKER (1979a). "Pole assignment self tuning regulator," *Proc. IEE*. Vol. 126, No. 8, pp. 781–787.

WELLSTEAD, P. E., D. PRAGER, and P. ZANKER (1979b). "Pole/zero assignment self tuning regulators," *Int. J. Control*, Vol. 30, pp. 1–26.

WENK, C. J., and Y. BAR-SHALOM (1980). "A multiple model adaptive dual control algorithm for stochastic systems with unknown parameters," *IEEE Trans. Autom. Control*, Vol. AC-25, No. 4, pp. 703–710.

WHITAKER, H. P., J. YAMRON, and A. KEZER (1958). "Design of model reference adaptive control systems for aircraft," Rep. R-164, Instrumentation Laboratory, Massachusetts, Institute of Technology, Cambridge, Mass.

WHITE, S. A. (1975). "An adaptive recursive digital filter," *Proc. 9th Annu. Asllomar Conf. Circuits Syst. Comput.*, pp. 21–25.

WHITTLE, P. (1963). "On the fitting of multivariate autoregressions and the approximate canonical factorization of a spectral density matrix," *Biometrika*, Vol. 50, pp. 129–134.

WIDROW, B. (1970). "Adaptive filters," in *Aspects of Network and System Theory*, ed. R. E. Kalman and N. De Claris, Holt, Rinehart and Winston, New York.

WIDROW, B., and M. E. HOFF, Jr. (1960). "Adaptive switching circuits," *IRE Wescon Conv. Rec.*, Pt. 4, pp. 96–104.

WIDROW, B., J. GLOVER, J. McCOOL, J. KAUNITZ, C. WILLIAMS, R. HEARN, J. ZEIDLER, E. DONG, and R. GOODLIN (1975). "Adaptive noise cancelling: principles and applications," *Proc. IEEE*, Vol. 63, No. 12, pp. 1692–1716.

WIDROW, B., J. M. McCOOL, M. G. LARIMORE, and C. R. JOHNSON, Jr. (1976). "Stationary and nonstationary learning characteristics of the LMS adaptive filter," *Proc. IEEE*, Vol. 64, No. 8, pp. 1151–1162.

WIENER, N. (1949). *Extrapolation, Interpolation and Smoothing of Stationary Time Series*, MIT Press, Cambridge, Mass. (Originally issued 1952 as a classified report.)

WIGGINS, R. A., and E. A. ROBINSON (1965). "Recursive solution to the multichannel filtering problem," *J. Geophys. Res.*, Vol. 70, pp. 1885–1891.

WILLEMS, J. C. (1971). "Least-squares stationary optimal control and algebraic Riccati equation," *IEEE Trans. Autom. Control*, Vol. AC-16, pp. 621–634.

WILLEMS, J. C. (1972). "Dissipative dynamical systems, Part I: General theory; Part II: Linear systems with quadratic supply rates," *Arch. Ration. Mech. Anal.*, Vol. 45, No. 5, pp. 321–393.

WILLEMS, J. L. (1970). *Stability Theory of Dynamical Systems*, Thomas Nelson, Walton-on-Thames, Surrey, England.

WILLIAMSON, D. (1977). "Observation of bilinear systems with application to biological control," *Automatica*, Vol. 13, pp. 243–254.

WITTENMARK, B. (1974). "A self-tuning predictor," *IEEE Trans. Autom. Control*, Vol. 19, pp. 848–851.

WITTENMARK, B. (1979). "Self-tuning PID-controllers based on pole placement," Rep. LUTFD2, Department of Automatic Control, Lund Institute of Technology, Lund, Sweden.

WITTENMARK, B., and K. J. ÅSTRÖM (1982). "Implementation aspects of adaptive controllers and their influence on robustness," *21st Conf. Decis. Control*, Orlando, Fla.

WOLD, H. (1938). *A Study in the Analysis of Stationary Time Series*, Almqvist and Wicksell, Stockholm.

WOLOVICH, W. A. (1974). *Linear Multivariable Systems*, Series in Applied Mathematical Sciences, Vol. 11, Springer-Verlag, New York.

WOLOVICH, W. A., and P. L. FALB (1976). "Invariants and canonical forms under dynamic compensation," *SIAM J. Control Optim.*, Vol. 14, No. 6, pp. 996–1008.

WONHAM, W. M. (1968). "On a matrix Riccati equation of stochastic control," *SIAM J. Control*, Vol. 6, No. 4, pp. 681–692.

WOUTERS, W. R. E. (1977). "Adaptive pole placement for linear stochastic systems with unknown parameters," *IEEE Conf. Decis. Control*. New Orleans, pp. 159–166.

YOUNG, P. C. (1965a). "The determination of the parameters of a dynamic process," *Radio Electron. Eng.*, Vol. 29, No. 6, pp. 345–361.

YOUNG, P. C. (1965b). "Process Parameter Estimation and Self Adaptive Control," in *Theory of Self-Adapting Control System*, ed. P. H. Hammond, Plenum Press, New York, 1966. (*Proc. IFAC Symp.*, Teddington, England.)

YOUNG, P. C. (1968). "The use of linear regression and related procedures for the identification of dynamic processes, *Proc. 7th IEEE Symp. Adapt. Process.*, University of California at Los Angeles.

YOUNG, P. C. (1969). "Adaptive pitch autostabilization of an air-to-surface missile," *South Western IEEE Conf.*, San Antonio, Tex.

YOUNG, P. C. (1974). "Recursive approaches to time series analysis," *Bull. Inst. Math. Appl.*, Vol. 10, No. 5/6, pp. 209–224.

YOUNG, P. C., and A. JAKEMAN (1979). "Refined instrumental variable methods for recursive time series analysis," *Int. J. Control*, Vol. 29, No. 1, pp. 1–30.

YOUNG, P. C., and J. C. WILLEMS (1972). "An approach to the linear multivariable servomechanism problem," *Int. J. Control*, Vol. 15, No. 5, pp. 961–979.

ZANKER P. M., and P. E. WELLSTEAD (1979). "Practical features of self-tuning," Rep. 461, Control Systems Centre, University of Manchester Institute of Science and Technology, Manchester, England.

Index

Least squares (*cont.*)
   convergence, 69, 339
   covariance modification, 67, 224
   covariance resetting, 65
   exponential data weighting, 64, 226
   finite data window, 225
   multi-output systems, 95
   persistent excitation, 72
   selective data weighting, 62
Left difference operator representation, 26
Levinson prediction, 280
Likelihood function, 506
Line enhancer, 386
Linear control form, 192, 201
Linear innovations, 278
Linear minimum variance estimate, 248
Linear quadratic gaussion optimal control, 426
Linear regression, 320
LMS, 348, 387, 404
LQG, 426
Lyapunov theorems, 485
Lyapunov's lemma, 486

M

Markov model, 247
Martingale, 499
Martingale convergence theorem, 327, 500
Matrix inversion lemma, 58, 100
Maximum likelihood estimation, 505
Mean, 498
Minimal realization, 19
Minimal state-space model, 18
Minimum prediction error control, 120, 154, 410
   adaptive version, 182
   stochastic adaptive version, 442
Minimum variance control, 411, 431, 433
   adaptive, 442
   MIMO adaptive, 457

   zeros on unit circle, 449
Minimum variance estimator, 504
Model reference adaptive control, 199
   filtered errors, 238
   stochastic, 456
Model reference control, 129
   adaptive version, 199
   linear control form, 201
   stochastic case, 420
Moving average model, 73

N

NLMS, 52, 348
Noise cancelling, 382
Nonsquare systems, 140
Normal equations, 279, 394
Normalized least mean square, 52, 348

O

Observability form, 15
Observability grammiah, 71
Observability index, 16
Observability matrix, 474
Observable, 474
Observer, 150
Observer form, 16, 27
ODE, 335
Off-line estimation, 303
One-step-ahead control, 120
   adaptive version, 182
   linear control form, 192
   multi-input/multi-output, 132
   multi-input/Multi-output adaptive, 202
Open-loop predictor, 109
Optimal prediction, 261
   concatination property, 269
Orthogonalized projection algorithm, 54
   convergence, 68
   persistent excitation, 72
Output error methods, 82
Output function controllable, 132

⑥ state model

$$\bar{x}(t+1) = \begin{bmatrix} .4 & 0 \\ -.6 & .2 \end{bmatrix} \bar{x}(t) + \bar{v}(t)$$

variance $\Rightarrow R_1 = \begin{bmatrix} 1 & 0 \\ 0 & 2 \end{bmatrix}$

Find stationary covariance for $\bar{x}(t)$

$$P(t) = A P(t) A^T + R_1$$
solve for $P(t)$

---

*Cauley Hamilton: $|ZI - A| = Z^n + a_n Z^{n-1} + \cdots + a_1$

$$= A^n + a_n A^{n-1} + \cdots + a_1 I = 0$$

$$\therefore A^n = -a_1 I - a_2 A - \cdots - a_n A^{n-1}$$

$$\therefore A^n B = -a_1 B - a_2 AB - \cdots - a_n A^{n-1} B$$

---

* $P^{-1} A P = \begin{bmatrix} 0 & 0 & & & & -a_1 \\ 1 & 0 & & & & -a_2 \\ 0 & 1 & & & & -a_3 \\ \vdots & \vdots & & & & \vdots \\ & & & & & -a_n \end{bmatrix} = A^*$ $\Leftarrow$ controllability canonical form

$P^{-1} B = \begin{bmatrix} 1 \\ 0 \\ 0 \\ \vdots \end{bmatrix}$

$\bar{P}^{-1} A \bar{P} = \begin{bmatrix} -a_1 & -a_2 & \cdots & -a_n \\ 1 & 0 & & 0 \\ & 1 & & \\ & & & 0 \end{bmatrix} = A'$ controller canonical form

$\bar{P}^{-1} B = \begin{bmatrix} 1 \\ 0 \\ 0 \\ 0 \\ 0 \end{bmatrix}$ $\underset{A'}{\uparrow}$

$A^* = P'^{-1} A' P' \Rightarrow A' = P' A^* P'^{-1} = P P' A \overset{\bar{P}}{\overbrace{P P'^{-1}}}$

---

*for Markov

$\mu x(t+1) = A \mu x(t) + B u(t)$

variance $P(t+1) = A P(t) A^T + Q$

autco $P(t,s) = A^{t-s} P(s)$

$E \begin{bmatrix} v_1 \\ v_2 \end{bmatrix} [v_1 \ v_2] = \begin{bmatrix} Q & S \\ S^T & R \end{bmatrix}$ $\overset{\text{cov.}}{\underset{\text{var.}}{}}$

* Controllability : $P = [B \vdots AB \vdots \cdots \vdots A^{n-1}B]$

* OBSERVABILITY : $Q = \begin{bmatrix} C \\ CA \\ \vdots \\ CA^{n-1} \end{bmatrix}$

$$\underbrace{C[zI-A]^{-1}B}_{TF}$$

$$\underbrace{[zI-A] = 0}_{poles}$$

---

* Canonical Structure Theorem

$$\begin{cases} \bar{x}(t+1) = A \cdot \bar{x}(t) + B \bar{u}(t) \\ \bar{y}(t) = C \bar{x}(t) \end{cases}$$

$$\exists \text{ a transformation } T \Rightarrow \begin{cases} \bar{q}(t+1) = A' q(t) + B' u(t) \\ \bar{y}(t) = C' q(t) \end{cases}$$

where : $A' = T^{-1}AT = \begin{bmatrix} A_{11} & 0 & A_{13} & 0 \\ A'_{21} & A'_{22} & A'_{23} & A_{24} \\ 0 & 0 & \cdot A_{33} & 0 \\ 0 & 0 & A_{43} & A_{44} \end{bmatrix}$ $B' = T^{-1}B = \begin{bmatrix} B'_1 \\ B'_2 \\ 0 \\ 0 \end{bmatrix}$

$$C' = CT = [C'_1 \quad 0 \quad C'_3 \quad 0]$$

N.B. : $[A'_{11}, B'_1, C'_1]$ con & obs

$\begin{bmatrix} A_{11} & 0 \\ A_{21} & A'_{22} \end{bmatrix}, \begin{bmatrix} B'_1 \\ B'_2 \end{bmatrix}, [C'_1, 0]$ con & ~~obs~~

$\begin{bmatrix} A'_{22} & A'_{24} \\ 0 & A'_{44} \end{bmatrix}, \begin{bmatrix} B'_2 \\ 0 \end{bmatrix}, [0,0]$ ~~con~~ & ~~obs~~

$A'_{44}, 0, 0$ ~~con~~ & ~~obs~~

✓ $\begin{bmatrix} A_{11} & A'_{13} \\ 0 & A'_{33} \end{bmatrix}, \begin{bmatrix} B'_1 \\ 0 \end{bmatrix}, [C'_1, C'_3]$

Random Processes

* $F_{X(t_1), X(t_2)}(x_1, t_1, x_2, t_2) = P[X(t_1) \leq x_1, \ \& \ X(t_2) \leq x_2]$

* Autocorrelation: $R_X(t_1, t_2) = E[(X(t_1) - \mu_x(t_1))(X(t_2) - \mu_x(t_2))]$

* Wide Sense Stationary: - Mean stationary: $E[X(t)] = \mu_x$ (constant)
  - Covariance stationary: $E[X(t_1), X(t_2)] = R_x(t_2 - t_1)$

* Ergodicity: Time ave. yield the same as the ensemble average

$$\underbrace{E[X(t)]}_{\substack{\text{ensemble} \\ \text{ave.}}} = \lim_{T \to \infty} \frac{1}{T} \int_0^T \underbrace{x(t)}_{\text{any member of ensemble}} dt$$

  <u>N.B.</u>: if nonstationary $\Rightarrow$ not ergodic

* white noise (a_n) process: - $\mu_a = E[a_t] = 0$

  $\overset{cov}{=} \gamma_a(k) = E[a_t a_{t-k}] = \sigma_a^2 \quad k=0$, $= 0 \quad k \neq 0$

* Autocorrelation Function: - $\gamma_y(k) = E[(y_t - \mu_y)(y_{t-k} - \mu_y)]$ $\leftarrow$ for stationary (else funct. of tu
  - $\gamma_y(0) = \sigma_y^2$
  - $\gamma_y(k) = \gamma_y(-k)$
  - $|\gamma_y(0)| \geq \gamma_y(k)$

* Time average: - $\hat{\gamma}_y(k) = \frac{1}{N-k} \sum_{t=k+1}^{N} y_t y_{t-k}$
  - $\gamma_y(k) = E[y_t y_{t-k}]$ - $\mu_y = E[y_t]$
  - $\hat{\mu}_y = \frac{1}{N} \sum_{t=1}^{N} y_t$

* Autoregressive Process (order P): $\boxed{y_t = \phi_1 y_{t-1} + \phi_2 y_{t-2} + \ldots + \phi_p y_{t-p} + d_t}$ (AR)

  - Autocorrelation: $E[y_{t-k} y_t] = \phi_1 E[y_{t-k} y_{t-1}] + \ldots + \phi_p E[y_{t-k} y_{t-p}] + E[y_{t-k} d_t]$
    $\gamma_y(k) = \phi_1 \gamma_y(k-1) + \ldots + \phi_p \gamma_y(k-p)$ .... for $k \neq 0$

  - example: $y_t = \phi y_{t-1} + d_t \Rightarrow E[y_{t-k} y_t] = \phi E[y_{t-k} y_{t-1}] + E[y_{t-k} a_t]$
    [N.B. if $k=0$ then: $a_t y_t = d_t d_t + \phi a_{t-1} a_t + \ldots \Rightarrow E[a_t y_t] = \sigma_a^2$]
    $\therefore \gamma_y(k) = \phi \gamma_y(k-1)$

  - Z transform: $* Z[y_{t+1}] = zY(z) - y(0)z$ $* Y(z) = Z[y_t] = \sum_{k=0}^{\infty} y_k z^{-k}$
    $* \frac{Y(z)}{A(z)} = \frac{1}{1 - \phi_1 z^{-1} \ldots \phi_p z^{-p}} = \pi_0 + \pi_1 z^{-1} + \pi_2 z^{-2}$
    $\therefore y_t = \pi_0 d_t + \pi_1 a_{t-1} + \pi_2 a_{t-2}$ (N.B. π_s)